HERRMANN-ALKEMADE
FLAME PHOTOMETRY

CONTENTS

INTRODUCTION

THE BASIS OF FLAME ANALYSIS

APPARATUS AND METHODS OF MEASUREMENT

Control of the Flame and Introduction of the Sample
> The Gases. Gas Regulation. Atomizers and Mixing Chambers. Burners.

Optics and Electronics
> Flame Photometry with Filters. Flame Spectrophotometry. Flame Spectrography.

Analytical Technique
> Use of Equipment. The Flame Emission. Standardization and Measurement.

The Errors of Flame Analysis
> Blank Interferences. Interferences from Concomitants. Precision and Accuracy.

APPLICATIONS

Agricultural Applications

Medicine and Biology

Geology, Industry and Other Fields

Appendices, Introduction to the Bibliography, Bibliography

Index

CHEMICAL ANALYSIS

Vol. I: **The Analytical Chemistry of Industrial Poisons, Hazards, and Solvents**
Second Edition. By Morris B. Jacobs

Vol. II: **Chromatographic Adsorption Analysis**
By Harold H. Strain

Vol. III: **Colorimetric Determination of Traces of Metals**
Third Edition. By E. B. Sandell

Vol. IV: **Organic Reagents Used in Gravimetric and Volumetric Analysis**
By John F. Flagg

Vol. V: **Aquametry. Application of the Karl Fischer Reagent to Quantitative Analyses Involving Water**
By John Mitchell, Jr., and Donald Milton Smith

Vol. VI: **Analysis of Insecticides and Acaricides**
By Francis A. Gunther and Roger C. Blinn

Vol. VII: **Chemical Analysis of Industrial Solvents**
By Morris B. Jacobs and Leopold Scheflan

Vol. VIII: **Colorimetric Determination of Nonmetals**
Edited by David F. Boltz

Vol. IX: **Analytical Chemistry of Titanium Metals and Compounds**
By Maurice Codell

Vol. X: **Chemical Analysis of Air Pollutants**
By Morris B. Jacobs

Vol. XI: **X-Ray Spectrochemical Analysis**
By L. S. Birks

Vol. XII: **Systematic Analysis of Surface-Active Agents**
By Milton J. Rosen and Henry A. Goldsmith

Vol XIII: **Alternating Current Polarography**
By B. Breyer and H. H. Bauer

Vol. XIV: **Chemical Analysis by Flame Photometry**
Second Edition. By R. Herrmann and C. T. J. Alkemade. Translated by Paul T. Gilbert, Jr.

Other volumes in preparation

CHEMICAL ANALYSIS

A SERIES OF MONOGRAPHS ON ANALYTICAL CHEMISTRY AND ITS APPLICATIONS

Editors

P. J. ELVING • I. M. KOLTHOFF

Advisory Board

S. E. Q. Ashley	J. J. Lingane	E. B. Sandell
John F. Flagg	J. Mitchell, Jr.	E. H. Swift
L. T. Hallett	R. H. Müller	Hobart H. Willard

Volume XIV

Chemical Analysis by Flame Photometry

by Roland Herrmann and C. T. J. Alkemade
translated by Paul T. Gilbert, Jr.

INTERSCIENCE PUBLISHERS
A division of John Wiley & Sons
NEW YORK   LONDON

CHEMICAL ANALYSIS BY FLAME PHOTOMETRY

SECOND REVISED EDITION

ROLAND HERRMANN
*Privatdozent in Medical Physics and Lecturer in Spectrochemistry
University of Giessen, Giessen, Germany*

and

C. T. J. ALKEMADE
*Professor of Experimental Physics, Physical Institute
of the University of Utrecht, Holland*

Translated by
PAUL T. GILBERT, JR.
Beckman Instruments, Inc., Fullerton, California

1963

INTERSCIENCE PUBLISHERS
A division of John Wiley & Sons
NEW YORK   LONDON

First published under: Herrmann und Alkemade 'Flammenphotometrie', 2. Auflage,
by Springer-Verlag, Berlin–Göttingen–Heidelberg

Library of Congress Catalog Card Number 63–17446

© by Springer-Verlag Ohg.
Berlin–Göttingen–Heidelberg 1956 and 1960

Made and printed in Great Britain by
William Clowes and Sons, Limited, London and Beccles

FROM THE PREFACE TO THE FIRST GERMAN EDITION

Spectrochemical analysis with flame excitation has advanced with exceptional vigor during the past decade owing to the systematic development and the increasing efficiency of the apparatus. But the literature on this subject is scattered in the technical journals of many branches of science; it is therefore hard to get a general view of the entire field. In the present monograph we undertake to provide such a view, correlating anew the several aspects of the subject.

We emphasize topics of interest to the practical analyst, such as apparatus, technique of measurement, and advantages, drawbacks, and limitations of individual methods. The theory of excitation of atoms and molecules in the flame, the historical development, etc., are treated only briefly. Considerable space is given to the sources and circumvention of error, for control of error is certainly necessary for successful work.

In presenting apparatus and technique we use several commercial instruments as examples. No judgment of the merit of these instruments is thereby implied. Further, certain special applications are described even though the necessary apparatus is not yet commercially available. Perhaps this will stimulate development of more efficient methods and instruments for such applications. With this thought in mind, too, we often detail the limitations within which a method must operate, at least with the means at present available.

To keep this book within reasonable bounds, we could not list among the applications all the analytical procedures worked out to date. More than 700 materials of the most diverse types and origins have already been investigated by flame analysis. For many of these a number of other analytical methods also exist. We therefore review the various applications summarily, giving a few examples and references to the extensive bibliography at the end of the book.

Giessen R. HERRMANN
January 1956

PREFACE TO THE SECOND GERMAN EDITION

The favorable reception accorded in Germany and other countries to the first edition, which was rather quickly sold out, has induced us to leave the character of the book unaltered in this second edition. But many new publications on the theory and applications of flame photometry have meanwhile appeared, considerably extending our knowledge of this field. The book has consequently had to be expanded and completely revised. It is noteworthy that flame photometry, once a rather empirical art, has become a well-established science through many theoretical physicochemical investigations on such matters as the mechanism of radiation interference from concomitant materials in the flame. In this new edition we try to give due heed to these theoretical studies with their intimate bearing upon practical flame photometry. Besides, there are the many advances in apparatus and methods recorded in over 2500 publications in a great variety of journals throughout the world. To survey all of this would have exceeded one man's capacity. The author of the first edition (Herrmann) is therefore grateful to have secured Professor Alkemade of Utrecht as collaborator.

Several sections have been sharply cut. Despite this, an overall expansion of the book could not be avoided. To keep its size within tolerable limits, we again confined ourselves, in the chapter on applications, to matters of chief importance, as in the first edition. The introduction to this chapter presents the viewpoints that have been adhered to. The number of text figures was reduced, to make room for the 74 new recordings of flame spectra in the appendix. Many of the text figures were replaced by new ones. It was not possible to include in the bibliography of the second edition all of the references (over 2500) that we have compiled on the subject of flame photometry; we limited the list to about one third of the total. Our selection of titles is not intended as a judgment of their value. We have emphasized later papers (up to the fall of 1959, last-minute additions up to the spring of 1960) in the most widely distributed journals in English, German, French, and Russian. Works containing many references are marked with an asterisk and will supply the reader with additional references, especially the older ones.

Since recording methods are steadily gaining importance, the photographic flame spectrograms of the first edition have been replaced by the 74 recorded spectrograms.

PREFACE TO THE SECOND GERMAN EDITION

Both authors wish to thank many colleagues and firms in Germany and other countries for numerous corrections and suggestions for improvement, and for supplying reprints, figures and other material. Among these colleagues we mention particularly Mr. Paul T. Gilbert, Jr., of Beckman Instruments, Inc., Fullerton, California, who in translating the first edition into English noted various ambiguities and errors, which have been corrected in the present edition. We acknowledge also his services in expanding Section 81, the table in Section 90, and the wavelength tables in the appendix, and in providing the atlas of recorded spectrograms. Dr. Franziska Pruckner of Munich and Mr. Kamil Marcinka of Bratislava have been of great help in searching out the literature of flame photometry, while Mrs. Margit Herrmann and Miss Ruth Eussner of Giessen have rendered invaluable aid with the bibliography and the tables of flame spectra. We are grateful to Dr. Luise Simon and Mr. Walter Lang of Giessen for many corrections and valuable suggestions.

The Springer-Verlag has cooperated with several printers in overcoming the considerable difficulties in printing and bookbinding entailed in the reproduction of the 26 fold-out plates of the 74 flame spectrograms. We are especially grateful to the publisher and to the associated printing firms.

Giessen and Utrecht R. HERRMANN
Spring, 1960 C. T. J. ALKEMADE

TRANSLATOR'S NOTE

Both Dr. Roland Herrmann and Dr. C. T. J. Alkemade are men of the widest experience in flame photometry. They have worked on a great variety of topics in instrumentation and analysis and have many inventions and innovations to their credit. Dr. Herrmann's interests lie more particularly in the field of physicochemical methods in medical research, while Dr. Alkemade's incline toward the theoretical interpretation of spectrochemical phenomena. Their joint authorship has therefore yielded an exceptionally well-balanced work.

A translation was prepared of the first German edition, but owing to continuing efforts to bring it up to date, it was not completed ahead of the second German edition. It has accordingly seemed much more sensible to present this translation of the second edition with as little further delay as possible. Because the second German edition took cognizance of last-minute developments up until the time it went to press, this translation represents the status of the subject in the spring of 1960, except for Tables 10, 13, 14, and 15, which have been corrected to late 1961.

The two authors have been extraordinarily helpful in clarifying and discussing various questions that arose in the course of the translation; Dr. Herrmann's correspondence concerning the first edition was nearly as bulky as the book itself. I am grateful for their exceptionally fine and friendly cooperation.

I wish to acknowledge the helpful attitude of Beckman Instruments, Inc., whose facilities have enabled me to prepare the atlas of spectrograms and the revised tables. In this work I have been in the debt of many colleagues both at Beckman and at other institutions, who have supplied unpublished information.

To speed the final typing, Mrs. Violet Johnson of Beckman Instruments and my wife, Hazel, contributed many hours of their time. I wish to thank also the Springer-Verlag and Interscience Publishers for their initiative and their patience, which have made this translation possible.

Fullerton, California   PAUL T. GILBERT, JR.
May, 1962

CONTENTS

1. **Introduction** 1
   1. Basis of the Method and Terminology 1
   2. Comparison with Chemical Methods of Analysis 2
   3. Comparison with Other Spectrochemical Emission Methods . . 4
   4. History 6

2. **The Basis of Flame Analysis** 12
   5. The Flame 13
   6. The Introduction of Sample into the Flame 20
   7. Vaporization, Dissociation, and Ionization of the Analyte in the Flame 27
   8. Atomic and Molecular Radiation in the Flame 36
   9. The Background Radiation of the Flame 52
   10. Working Curves 55
   11. Interferences from Concomitants 59

3. **Apparatus and Methods of Measurement** 68
   12. Survey 68
   Control of the Flame and Introduction of the Sample 69
   The Gases 69
   13. The Choice of Gases 69
   14. The Flame Temperature 71
   15. Acetylene and Propane 77
   16. Oxygen and Compressed Air 81
   17. The Use of Compressed Gases 83
   Gas Regulation 86
   18. Control and Monitoring of the Oxidant and Fuel Flow to the Flame 86
   19. Determination of Optimal Flow Rates of Fuel and Oxidant . . 92
   20. The Consumption of Oxygen and Gas 96
   Atomizers and Mixing Chambers 97
   21. Introduction of Sample into the Flame 97
   22. Requirements of the Atomizer 98
   23. Material of Construction of the Atomizer 99
   24. Classification of Atomizers 100
   25. Nonpneumatic Atomizers 101
   26. Pneumatic Atomizers and Spray Chambers 103
   27. Introduction of Sample into the Atomizer 109
   28. Drainage of the Sample 111
   29. Atomizers for Microanalysis 113
   30. The Heated Spray Chamber 115
   31. Injection of the Gas 115

CONTENTS

Burners 117
32. The Burner (with Indirect Atomizers) 117
33. Atomizer-burners 121
34. Size and Shape of the Flame 125
35. The Chimney 128

Optics and Electronics 129
36. General 129
37. The Optical Alignment of the Flame 130
38. The Mirror 133

Flame Photometry with Filters 134
39. Condensers 134
40. Filters 136
41. Diaphragms 143
42. Radiation Receivers 143
43. Adjustment of the Radiation Receiver 149
44. The Amplifier 149
45. Power Supplies 151
46. The Indicating Meter 152
47. Integration Methods 155
48. Control of Sensitivity 156
49. Potentiometric Circuits 157
50. Direct Multiple-Beam Methods 159
51. Internal Standardization 160
52. Auxiliary Method of Internal Standardization 161
53. The Optical–Electric System for Internal Standardization . . 163
54. Multiple-Beam Internal Standardization 164
55. Advantages and Drawbacks of Internal Standardization . . 165
56. Recording Methods with Filter Photometers 168
57. Multiple-Use Flame Photometers 169
58. Limitations of Filter Photometers 170

Flame Spectrophotometry 170
59. General 170
60. Spectral Dispersion 172
61. Requirements of the Monochromator 176
62. The Slit Width 177
63. Flame Accessories for Spectrophotometers 179
64. Alignment of the Flame with Respect to the Monochromator . 180
65. The Photoreceiver in a Monochromator 181
66. Amplifiers, Meters, and Power Supplies 187
67. Multiple-beam Methods in Flame Spectrophotometry . . 187
68. Recording Methods in Flame Spectrophotometry . . . 188
69. Internal Standardization with a Spectrophotometer . . 190

Flame Spectrography 192
70. General 192
71. The Spectrograph 195
72. The Photographic Plate 196
73. The Photographic Process 197

74.	The Exposure	197
75.	Evaluation	197

4. Analytical Technique 199
Use of Equipment 199
76.	The Laboratory	199
77.	Operating Procedure	200
78.	Development of a Procedure	203
79.	The Solvent: Adjustment of the Concentration	205
80.	Containers and their Maintenance	212

The Flame Emission 215
81.	Flame Spectra of the Elements: the Principal Analytical Emissions	215
82.	Choice of a Line or Band	226
83.	Establishment of Working Curves	230

Standardization and Measurement 236
84.	The Preparation of Standards	236
85.	The Technique of Measurement and Standardization	238
86.	Measurements with Overlapping Bands or Lines	241
87.	The Determination of Nonmetals: Indirect Methods	243
88.	Combined Flame-Photometric and Chemical or Physiocochemical Methods	247
89.	Micro and Ultramicro Methods	253
90.	The Detection Limits	255
91.	The Speed of Flame Analysis	276

The Errors of Flame Analysis 278
92.	Introduction	278
93.	Impurity of the Gases and the Ambient Air	279
A.	Blank Interferences	280
94.	Cross-Sensitivity	280
95.	Elimination of the Background	285
B.	Interferences from Concomitants	292
96.	Introduction	292
97.	Specific Radiation Interferences with the Alkali Metals	294
98.	Specific Radiation Interferences with the Alkaline-Earth Metals	300
99.	Catalog of Specific Interferences	310
100.	Detection and Elimination of Specific Interferences	319
101.	Nonspecific Interferences	325
C.	Precision and Accuracy	327
102.	The Random Error	327
103.	Systematic Errors	331
104.	The Aggregate Error of Measurement	332

5. Applications 333
Agricultural Applications 334
105.	Agriculture and Botany	334
106.	Foods and Nutrition	337

Medicine and Biology 339
107. General 339
108. Characteristics of Biological and Medical Sample Material . . 340
109. Ashing and Extraction 341
110. The Removal of Protein 343
111. The Determination of Sodium in Body Fluids 344
112. The Determination of Potassium in Body Fluids 345
113. Blood Sampling for Determination of Serum Electrolytes . . 347
114. The Determination of Calcium in Body Fluids 349
115. The Determination of Other Elements in Biological Materials . 353
Geology, Industry, and Other Fields 354
116. Water Analysis 354
117. Physics, Nuclear Physics, and Electronics 356
118. Chemical Industry and Pharmacy 358
119. Geology and Mineralogy 359
120. The Glass Industry 360
121. The Ceramic Industry 361
122. The Cement Industry 362
123. Petroleum and Fuels 363
124. The Metallurgical Industries: Analysis of Metals and Slags . . 367
125. Criminology 368

Appendix 370

Bibliography 432

Atlas of Flame Spectrograms 471

Index 581

CHAPTER 1

INTRODUCTION

1. *Basis of the Method and Terminology*

If the salts of certain metals are introduced into a bunsen flame, they give rise to a distinct luminous emission as they vaporize. Thus, sodium chloride yields a bright yellow light. Salts of other metals give different flame colors: salts of strontium and lithium color the flame carmine-red; calcium, brick-red; barium, yellow-green; copper, emerald-green; potassium, blue-violet; arsenic, antimony, and lead, pale blue to pale green; and so forth. The flame colors are characteristic of certain cations, sometimes of certain compounds such as CuOH. Consequently, these luminous phenomena are often used as an aid in qualitative chemical analysis. But if salts of several metals are simultaneously introduced, the simple method fails, for the eye cannot separate the mixture of colors; and sometimes the characteristic "color" lies in a region of the spectrum invisible to the eye.

If the mixed light emitted by a multiple colored flame is passed through a spectroscope, it reveals definite spectral lines characteristic of the individual elements or bands (assemblages of lines) characteristic of certain compounds. With the help of tables or charts of spectral lines, one can identify certain constituents of the material introduced into the flame. This well-known qualitative procedure will not be further considered here.

Now, if by suitable means the flame can be kept burning uniformly for an extended period and the rate of introduction of material into the flame can be held constant, the intensity of the observed spectral line or band will be a measure of the concentration of the emitting substance. A calibration can be made with material of the same composition and known concentration. From the qualitative method a quantitative comparative method has now evolved.

Depending upon how the spectral lines or bands are separated from each other and how their intensities are measured, the method is called *flame photometry* (Sections 39–58), *flame spectrophotometry* (Sections 59–69), or *flame spectrography* (Sections 70–75). These methods are grouped under the general heading of *flame analysis*. Thus we exclude the study of pure flames when directed to elucidation of the complex processes in the flame itself. Flame analysis, then, will be defined as comprising quantitative measurements of flame emission for the determination of the concentration of substances introduced into the flame. (Often the term *flame photometry* is used more loosely as equivalent to *flame analysis*.)

The apparatus needed for measuring the intensities of these spectral radiations—filter photometers, spectrophotometers, and spectrographs—inclusive of the flame source, we group under the designation *flame photometer*.

Before entering upon the history of flame analysis in Section 4, we discuss next the distinctions between flame analysis and the related fields of analytical chemistry and spectral analysis with arc or spark.

2. *Comparison with Chemical Methods of Analysis*

Chemical methods of analysis will always keep their status as standardizing and reference methods. Also, they can be applied in every case, whereas flame analysis is possible only for such elements or compounds as can be directly or indirectly detected in the flame. As compared with chemical methods, flame analysis offers certain distinct advantages, but there are disadvantages as well.

1. Most metals can be determined in a single solution without chemical separations. The state of combination of the metal with anions, etc., plays a subordinate role. The observed lines and also many bands (see below) are specific for these elements or for such of their compounds as form in the flame.

2. A sequence of similar flame analyses takes much less time than the corresponding chemical analyses. Flame analysis is therefore well suited for routine determinations on series of similar samples. The saving of time is the greater, the more tedious and cumbersome the corresponding chemical determination is. This advantage is particularly marked with the alkali metals, which are easily determined by flame photometry; their quantitative chemical determination (for example, of potassium as perchlorate) is very troublesome.

3. The amount of material required is often less than for chemical micro- or ultramicroanalysis. This advantage is utilized in special applications such as biology, medicine, and criminology.

4. The accuracy is generally good, sometimes in fact better than that of chemical methods, especially at low concentrations. Accuracy is retained even at relatively low concentrations, except close to the

detection limit (Section 90). Flame analysis is therefore exceptionally well suited for investigating impurities, trace elements, etc., in various materials.

5. The analysis can be done by a technician, except for the original development of the method.

6. Preparation of the samples is usually easier.

7. There are special advantages in certain determinations; for example, ammonia formed from unstable organic substances may interfere in the chemical determination of potassium.

Disadvantages of flame analysis, as compared with chemical methods, are:

1. As in all spectroscopic emission methods, the measurements are relative. Standards must be used for comparison.

2. Costlier apparatus is needed than for chemical methods.

3. Initial establishment of a method for a given application takes more work (preparation of standard solutions, working curves, etc.). It is justified usually only when many similar determinations are to be carried out or when the corresponding chemical method is very tedious.

4. Only about 60 elements can be determined by flame analysis, provided they are present in sufficient concentration (Section 90).* The flame gives very little information concerning the acid radicals or state of combination of the element determined, and usually none concerning elements and radicals not excited in the flame (for exceptions see Section 87).

5. The influences of the elements upon each other in the flame (interferences) introduce difficulties (Sections 11 and 96–100). These effects also depend upon instrumental characteristics such as flame temperature and method of atomization. In difficult cases flame photometry can lead to considerable error, especially when unsuitable methods and apparatus are used. In any case these interferences can be alleviated more easily and quickly than the interferences that may occur in chemical methods of analysis. Each new type of sample generally demands the evaluation of a new flame-photometric procedure. Methods described in the literature can be taken over directly only under the right conditions.

Flame analysis can thus replace chemical methods only in part. But in some cases it offers great advantages, and it is welcomed as a supplement to the chemical methods.

Before undertaking a flame analysis one should consider whether

* The number now exceeds 70 elements. See footnote to Section 81.—*Translator.*

some other analytical method might not be better or quicker. Procurement of the equipment needed for flame analysis will be justified only when its advantages can be put to full use. But if the equipment is already at hand and if one is familiar with the method, flame analysis may often be used in other cases where its advantages are less apparent.

3. *Comparison with Other Spectrochemical Emission Methods*

According to the kind of excitation source, spectral emission analysis is subdivided into emission analysis with arc, with spark, or with flame, among others. Many considerations pertinent to spectral analysis with spark and arc apply also to flame analysis.

It is difficult to compare the flame methods discussed in this book with the other spectrochemical methods (chiefly arc or spark). These different methods of excitation may be suited to entirely different analytical problems or samples. Moreover, the methods of measurement (photoelectric integration, etc.), ancillary to but not a part of the excitation proper, are often basically different for the different methods of excitation.

Nevertheless, to illustrate the differences we offer a comparison below. We make the assumption (not always valid in practice) that the measurement is made directly in each case, that is, without use of an internal standard (Section 51) or integration (Section 47). We further assume that a liquid sample is atomized into the flame, as usual, while with the arc or spark a solid or powdered material is vaporized into the excitation region by the heat of the source. The following advantages of flame analysis will then appear:

1. It is usually more accurate, since atomization into a flame is a more reproducible method of introducing sample into the excitation region. The flame itself can readily be kept constant for hours. It avoids the errors due to differences in the melting and boiling points of the electrode material, consumption of the electrodes, etc.

2. Flame analysis is often faster, since presparking time and the like are unnecessary.

3. There is usually better detection sensitivity for the alkali metals and sometimes also for the alkaline-earth metals, because, in contrast with the situation in the arc or spark, the alkali-metal atoms in the flame are ionized to only a limited extent. Hence their resonance lines are stronger in the flame. The singly ionized alkali atoms and the doubly ionized alkaline-earth atoms emit no spectra in the flame, although the singly ionized alkaline-earth atoms emit fairly intense lines in the hotter flames.

4. Flame spectra have comparatively few lines. Consequently it is easier to resolve and identify the individual elements even in the presence of many other elements.

5. The equipment is appreciably less costly and needs less space than that required for a spectrochemical laboratory using arc or spark sources. Because of the simplicity of flame spectra it is possible to make out with small instruments, often indeed with filter instruments.

6. In flame excitation the emission is often less dependent on concomitant elements in the sample than in the other sources with their higher temperatures.

7. The analytical procedures require less experience to set up and carry out.

8. The preparation of standard solutions for flame analysis is easier than, for instance, the preparation of standard alloys for arc or spark spectroscopy, since the samples are usually liquid for the former but not for the latter.

This comparison is biased in that arc and spark excitation are used almost exclusively with internal standards and integrative methods or photographic plates (spectrography). These methods are also possible with the flame (Sections 47, 51, and 70–75) but scarcely necessary, as the reproducibility of flame emission intensity is sufficient for many applications without use of these artifices.

The following are disadvantages of flame analysis compared with other spectrochemical emission methods:

1. Fewer elements can be directly determined in the conventional flames. But with the oxygen–cyanogen flame conditions approach more nearly those in the arc.

2. Elements not easily excited in conventional flames must be present in higher concentration.

3. Solid substances must be first dissolved by suitable means (exceptions: Section 21).*

4. Instead of sharp lines, bands of compounds such as CaO, CaOH,† and CuOH often appear (Sections 8 and 81). Being diffuse, they can interfere with spectral separation, for instance, the isolation of the yellow sodium line in the presence of the calcium bands. This effect, a

---

* Undissolved, powdered solid samples can be sprayed into an atomizer-burner flame if they are first suspended in a moderately viscous, preferably combustible liquid. The method is very reproducible and can handle high concentrations. See *Anal. Chem.*, **34**, 1025–6 (1962).—*Translator.*

† Besides Ca and CaO, the molecule CaOH [not $Ca(OH)_2$] is stable at the high temperature of the flame. We return to this in Section 8.

type of background interference, spoils the accuracy and sensitivity of the analysis (Sections 11 and 94–95).

Thus, one or another mode of excitation will be chosen depending on the type of sample (solid or liquid, metallic or nonmetallic, etc.), on the element sought and its concentration, and on the concomitant elements and the likelihood of interference. If the sample is in solid, metallic, electrically conductive form, the spark (or arc) will generally be preferred, since the material can then be used directly as electrode, provided the sample is large enough. With nonmetallic solid samples, on the other hand, the spark (or arc) gives less reproducible results, for reasons mentioned above. In this case flame photometry will be preferred if it is at all applicable. With liquid samples flame photometry is generally the method of choice.

A comparison, like the one above, of the flame with the other spectrochemical emission methods can be developed also for the flame method with other physicochemical methods such as spectrophotometry (colorimetry) and polarography. Such a comparison will show that the flame is entirely competitive with these methods in terms of sensitivity and accuracy.[175]

4. *History*

The first beginnings of flame photometry reach back to the middle of the 18th century (e.g., Geoffroy, 1732; Melvill, 1752; Marggraf, 1758). These chemists, usually apropos of other observations, described the colors imparted to alcohol and candle flames by metallic salts and even used this coloration to some extent for distinguishing different materials (Marggraf, 1758). But these forerunners of qualitative flame analysis were not aware of the potentialities of their observations, and these incidental notations were largely ignored. The development of better apparatus (prism, spectroscope) opened new opportunities for research. Thus, Wollaston as early as 1802 described the yellow sodium line but thought it a peculiarity of the flame itself. Fraunhofer not only described the solar spectrum but found its D line to be identical with the yellow flame line, without, however, drawing any further conclusions.

A series of further experiments, incidental, as before, to other matters (e.g., Herschel, 1823), still shed no clear light upon these various individual observations. We must realize, though, that most of the flames used in those days were incandescent candle flames of low temperature and high luminosity, far from suitable for detecting weak spectral lines. Besides, the purity of the available chemicals left much to be desired, a circumstance which made it difficult to ascribe a given line to a particular substance.

Talbot (1826 and 1836) showed insight remarkable for his time; the method of qualitative flame analysis revealed in his publications was basically sound if very briefly described. But his work, too, was without lasting impact.

The comprehensive researches of Kirchhoff and Bunsen, leading to the discovery of new elements, brought the decisive turning-point. Thenceforth the flame was universally recognized as a means of qualitative analysis. In 1855 Bunsen had developed the bunsen burner with its colorless, hot flame, which proved very useful in his later studies of the emission of metal salts. With Kirchhoff he developed a spectroscope of exceptional efficiency for the times. The two men succeeded in demonstrating that the visible spectral lines are due not to compounds but to the elements themselves. This conclusion was widely acknowledged. It laid the cornerstone of the science of spectrochemistry. From Fraunhofer's observations they were able to deduce, for example, that sodium and potassium, but not lithium, occur in the sun.

Kirchhoff and Bunsen in the same year discovered by its spectrum the new element cesium in mineral waters, and rubidium a year later, while thallium was discovered by Crookes. A series of further discoveries soon followed, including indium and gallium. Today all the gaps in the periodic table are filled. Nevertheless, the procedure employed by the pioneers of qualitative flame analysis still retains its utility in the laboratory for quick qualitative investigations, especially when traces of easily excited elements are to be detected as impurities in other substances.

The first experiment in quantitative flame analysis goes back to Champion, Pellet, and Grenier (1873). They described a "spectronatromètre" for quantitative sodium determination. It consisted basically of two flames, of which one was saturated with sodium salts while the sample to be analyzed was introduced into the other. The intensities of yellow light from the two flames were compared. For this purpose a blue wedge, placed before the saturated flame, was displaced until the two sodium flames appeared equally bright. The position of the wedge was a measure of the desired concentration. It was calibrated with materials of known sodium content. The introduction of the substance into the flame appears to have been quite tedious with the older procedures. Clean platinum wires were drawn through the samples, dried, and then moved through the flame slowly and uniformly with a clockwork.

Considerably more modern is the atomizer method described as early as 1879 by Gouy for introducing substances into the flame. However, he was less interested in spectrochemical analysis than in quantitative investigation of the emission processes in the flame. There followed a series of further experiments in quantitative analysis. We mention, among others, Beckmann and Waentig[125] and Klemperer (1910). But this earlier work remained without lasting influence.

Emphasis in spectroscopic research from the turn of the century until 1930 lay in the fields of arc and spark analysis (Lockyer, 1873; Hartley, 1882; Pollack and Leonard, 1907; and others). This was owing to progress in electric technology and to the fact that the search was still on for new elements that were not easy to excite; also, the complex arc and spark spectra were yielding insight into atomic structure. These developments were further speeded by advances in the optical industry, which was providing spectroscopes and spectrographs of ever greater dispersion. But these advances in spectroscopy (utilizing arc and spark excitation), such as the introduction of the internal-standard method by Gerlach (1925), ultimately redounded to the advantage of flame photometry.

Lundegårdh, the Swedish plant physiologist and agricultural chemist, and his associates, through many publications beginning in 1928, first brought to flame spectrophotometry the recognition it merits. Flame analysis was not an end in itself for him, but rather a welcome means of investigating plant metabolism and environmental effects upon it (Section 105). Lundegårdh had to use photographic methods at the start; and in his later work he largely confined himself to spectrography. His observations and results as recorded in his two-volume work[33] are still valid today. Lundegårdh's methods were adopted and refined by Mitchell[39,493] in 1936 in England, by Ells and Marshall in 1939 in the United States, by Hasler in 1940 in Switzerland, in the same year by Schuffelen in Holland, and subsequently by others.

Jansen, Heyes, and Richter in 1935 considerably improved the method of direct photometry of the flame spectrum, already tried by Lundegårdh. Their instrument consisted of a monochromator, photocell, and electrometer. In 1935 they made a preliminary experiment in separating the yellow sodium line and the red potassium line with filters, without recognizing the practical advantage of this. Their later work was done with monochromators. Schuhknecht[631-633] first succeeded, through his investigations and publications, in obtaining proper recognition for flame photometry with filters. He proposed separating the red potassium line with an improved filter combination (Section 40), which permitted excellent potassium determinations even with unfavorable Na:K ratios. On the basis of Schuhknecht and Waibel's later work, commercial instruments were developed (by Siemens and by Zeiss). Thus the way was opened to a wider use of flame photometry proper (that is, with filters). The number of publications increased abruptly, especially in the field of agriculture. These filter photometers had enough optical efficiency to permit use of the simpler selenium barrier-layer photocell of Lange[30] instead of the phototube.

A parallel development began later in the United States. The first practical American filter photometer of Barnes, Richardson, Berry, and Hood[90] in 1945 corresponds roughly to the instrument of Schuhknecht and Waibel. A year later Berry, Chappell, and Barnes[140] described their internal-standard filter photometer. The internal-standard principle originated, as a matter of fact, with arc and spark spectroscopy. It was first described in 1925 by Gerlach, who developed it further with Schweitzer and others. In flame spectrography Lundegårdh appears to have been the first to work with internal standards. Jansen, Heyes, and Richter also used it in flame spectrography. In flame photometry with filters, internal standardization was adopted by Herrmann in Germany and developed into an internal-standard, multiple-beam method.[339,340,342,344]

The range of application of flame photometry was widened considerably in 1949 with the appearance of serviceable flame spectrophotometers (monochromators with flame attachment, Sections 60 and 63) which, owing to their low luminosity,* are best equipped with atomizer-burners yielding intense, hot flames and with sensitive radiation detectors, especially photomultipliers. This made possible the determination of elements having spectra that are relatively weak, rich in lines, or located in the ultraviolet, and the number of elements amenable to routine flame analysis rose markedly. Besides, the atomizer-burners (Section 33) developed for these instruments opened new possibilities: combustible substances such as gasoline and oil could be safely sprayed directly into the flame, and many new applications in the petroleum and fuel industries appeared.

At the same time, the simpler filter flame photometers benefited from advances in interference filters, radiation detectors, and amplifiers. Among other things, it appeared that certain interferences (for example, those of phosphorus and aluminum with the alkaline earths) could be lessened by efficient sorting of the droplets in the spray chamber. These improvements, along with their greater ease of operation, have ensconced these inexpensive instruments securely in a great variety of laboratories for many routine analyses, where monochromator instruments can no longer compete with them.

* See Section 36 for a definition and Section 60 for further discussion.

With spectrophotometers, advances in potentiometric recording have opened new possibilities. There are simple, relatively inexpensive recording attachments for spectrophotometers, which reduce the scanning and recording of a flame spectrum to a matter of routine. This simplifies the simultaneous determination of several elements, the recognition and elimination of errors, the detection of impurities, etc.

Recent years have seen great progress in the adaptation of flame photometry to micro- and ultramicroanalysis, through the introduction of multiple-beam instruments, integrating methods, and similar procedural and electronic innovations.

At present about 60 elements are known to be directly determinable in the conventional flames.* Of these, about 40 have line spectra or very sharp bands, while another 15 have sufficiently characteristic band spectra to permit quantitative determination in the presence of other elements. A few (Ce, Mo, Nb, U, and Zr) show chiefly a continuum with weak bands or lines at best, and for the most part can be determined quantitatively only in the absence of other elements.[307] Of the other 40 elements of the periodic system, many lack flame spectra, although some of these can be determined by indirect methods of flame analysis (Section 87). A few of those not yet investigated may prove to be determinable, while the rest are too rare and radioactive.

The oxygen–cyanogen flame, recently recommended, has shown the following results (Vallee, Gilbert, and others).

Only 19 elements (H, B, C, Si, N, P, O, S, F, Cl, Br, I, At, He, Ne, Ar, Kr, Xe, Em) yield no useful atomic flame spectra. Another 21 (Fr, Th, Pa, U, Np, Cm, Bk, Cf, Es, Fm, Md, No, Zr, Hf, Nb, Ta, W, Ge, As, Sb, Se) either have not been investigated, or are too rare to be amenable to flame analysis, or are known to emit weakly. Of the remaining 62 elements, 26 (Li, Na, K, Rb, Cs, Be, Mg, Ca, Sr, Ba, V, Cr, Mo, Mn, Fe, Co, Ni, Cu, Ag, Zn, Cd, Al, Tl, Sn, Pb, Bi) have been shown to emit strong atomic spectra in the oxygen–cyanogen flame, while the other 36 very probably emit equally characteristic spectra.†[304]

The significance of flame analysis today can be summed up in the statement that it permits rapid and accurate determination of many metals. Many types of investigations not hitherto practicable now take little time. In addition to this extension of the method to many areas of application, we now have a better understanding of the principles. The interferences in the flame (Sections 96–100), often very troublesome

* This number has now risen to 77, of which over 60 have lines or sharp bands.

† J. W. Robinson (*Anal. Chem.* 33, 1226 (1961)) reports lines of boron, cerium, gallium, osmium, palladium, tantalum, titanium, tungsten, and zirconium, as well as other elements mentioned above, in oxygen–cyanogen.—*Translator*.

to the practical analyst, have stimulated physicochemical investigations, to be discussed in Sections 5–11. In these theoretical and experimental studies, improvements in instrumental technique, especially the notable enhancement of accuracy, have greatly benefited pure research.

The present rate of progress makes it seem not unlikely that within a few years flame photometry in the original sense, at least in certain applications, will be obsolete. Thus, the flame with its very complex processes could be replaced by another source of energy offering reproducible excitation. Another possibility is absorption flame photometry (atomic-absorption spectroscopy); since the first edition of this book, it has been applied repeatedly in practical analysis, and now, at the close of this historical survey, we note that commercial instruments have actually become available for the expanded application of absorption flame photometry.

For a more detailed study of the earlier history of flame photometry we refer to the literature (e.g., refs. 26, 33, 37, 39, 152, 469).

CHAPTER 2

THE BASIS OF FLAME ANALYSIS

Although this book is intended mainly as an introduction to the technique of analytical flame photometry, and although at first glance the flame-photometric method looks attractively simple, we must start with a discussion of the fundamentals. A closer practical acquaintance with flame-photometric analysis reveals many complications, which are difficult or impossible to allay without knowledge of the theory. Besides, for the novice unfamiliar with the theory, these interferences, generally unsuspected, can cause large analytical errors in careless work. To cure these difficulties by purely empirical means, a tedious preliminary study of possible error must be carried out in every case. The resulting procedure will be unduly detailed and of very limited scope, for the best method of eliminating the errors cannot be prescribed without knowledge of the causes. An analyst working out a method in ignorance of the fundamentals will be troubled by uncertainty, never knowing whether he has perhaps overlooked some essential point. If an empirical procedure is to be adapted to another type of sample or another instrument, the errors arising under the new circumstances must be reanalyzed without the help of rules governing the occurrence or removal of such errors.

But an analyst fairly well acquainted with the fundamentals of the method will know at once how to cope with the situation. He will quickly understand the otherwise disturbing multiplicity of sources of error, which can be better classified and examined in the light of theoretical information. He can then more easily judge whether a procedure set up in another connection may be transferred to the case at hand, which features can be carried over, which additional points must be heeded, and which can be ignored. And he is less likely to communicate ill-digested observations, describing, for instance, the errors of his own instrument, and perhaps publishing these as having general validity.

Despite the considerable advances of the last few years in the study of the fundamentals of flame photometry, it must be admitted that many an unsolved problem remains. This should spur the physico-chemically inclined scientist to attack some of these problems, which are certainly of great importance for practical flame photometry.

5. *The Flame*

To excite atoms to emission of light, a medium of high temperature is required, of sufficient heat content to avoid its being too greatly cooled by the evaporation and dissociation of the sample material introduced into it. The temperature must be high enough to evaporate the liquid components of the sample entering the flame in the form of mist, vaporize the residual solid particles, and, as completely as possible, dissociate the molecules (which usually do not emit) to atoms. Moreover, the medium itself should emit as little light as possible, so that the emission of the sample introduced into it may be observed readily. In the simplest case, the nonluminous bunsen flame constitutes such a medium, in which a fuel gas is burned with air or oxygen. A flame photometer includes such a flame, or a similar one (see below), as well as means of introducing the sample into the flame, of measuring the emission from the flame, etc. This section will be concerned with the flame.

If thermodynamic equilibrium prevails in the observed part of the flame, then specification of its temperature alone suffices to account for the emission phenomena and there is no need to consider the manner in which the temperature is generated in the combustion process. But since in some applications the inner cone (the primary combustion zone; Fig. 1) is of interest, and since certain secondary phenomena seen in analyses utilizing the region above the combustion zone originate directly in it, we shall have to take the combustion zone into consideration too. In this connection we have to admit that despite the vast number of publications on technically important combustion processes in the fields of heating, motor propulsion, jet propulsion, etc., a great many things are still obscure, and the remark by Gaydon and Wolfhard[11] remains true, that we know more about the processes in stellar atmospheres than about those in a bunsen flame. We shall not here be entering into these detailed problems, but we shall outline such knowledge of the flame as seems pertinent to an understanding of subsequent sections of the book. For further information see, for example, refs. 11, 12, 25, 31 and 37.

The flames most commonly used in flame photometry may be divided into two groups:

1. Flames in which the inflowing gas and air or oxygen are well

mixed before reaching the combustion zone are called *premixed flames*. Since premixed flames used in flame photometry generally exhibit laminar flow* (see below), they may also be called *laminar flames*.

2. Flames in which the gas and oxygen or air are first mixed in the flame itself are called *un-premixed flames*.† Since the un-premixed flames used in flame photometry generally exhibit strong turbulence, they are called also *turbulent flames*.

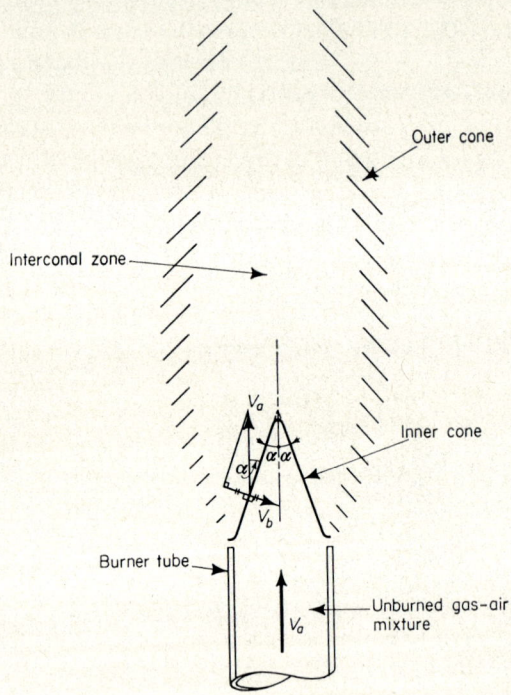

Fig. 1. Schematic vertical section of a premixed laminar bunsen flame. The vectors marked with two strokes (//) are equal. See text for explanation.

With premixed flames (type 1), the flow of gas mixture from the burner is generally kept as laminar as possible, to yield a stable, noiseless flame. The sample is mixed with the air stream by means of an

---

* In laminar flow the streamlines run parallel. In turbulent flow, on the other hand, the elements of volume pursue a crooked, irregular course with respect to each other; they may thus move at right angles to the general forward flow of the gas stream.

† Flames of intermediate character are also possible, e.g., partially premixed flames;[513,728] but they are not common.

"indirect" atomizer* (Section 26), which forms a separate part of the instrument. This air–mist mixture or aerosol is next well mixed with the fuel gas and passed upward through the burner tube in laminar flow (cf. Fig. 9, page 72).

Atomizer-burners operate in a different way (cf. Section 33 and Fig. 21). In these the gases are not premixed. The un-premixed, turbulent flame burns immediately above the concentric gas and oxygen ports, while the sample is atomized directly into the flame from a third orifice at the center. This type of atomizer may be called a direct atomizer, but the entire device is oftener called an atomizer-burner. It has grown increasingly popular in recent years, since it can safely burn highly explosive gas mixtures (oxyhydrogen, oxyacetylene, etc.) with their hotter flames, and can also directly spray combustible samples such as gasoline. It has a few other advantages besides, to which we return in Section 33. The flame burning above an atomizer-burner is strongly turbulent, owing to the high speed with which the oxygen emerges from the very narrow annular oxygen port, simultaneously atomizing the liquid. This turbulence incidentally aids the mixing of gas and oxygen in the flame, promoting complete combustion.

We consider next the simpler *laminar premixed flame*, which reveals a distinct structure. The simplest example is the nonluminous bunsen flame. Three zones can be distinguished, as shown in Fig. 1:

1. The *primary combustion zone*. This is called also the reaction zone or inner cone, since with a round burner port, as in the bunsen burner, it has a conical form. The combustion zone is about 0.1 mm. thick at atmospheric pressure with laminar flow. In flames burning gases containing carbon and hydrogen, such as illuminating gas, this conical zone is visible by virtue of its strong blue-green light, ascribed to the radicals C_2 and CH (Section 9). These and similar radicals play an important but as yet imperfectly understood intermediary role in the combustion process. Thermodynamic equilibrium does not prevail in the inner cone. Light emission, ionization, and the concentration of radicals are extraordinarily high in this zone. It is used for flame-photometric analysis only in exceptional cases.

The surface of the combustion zone propagates itself toward the unburned gas mixture with a characteristic velocity, the *burning velocity*, which is determined largely by the thermal conductivity of the gas mixture. Through thermal conduction the inflowing cold gas mixture is preheated to the ignition temperature. In some cases the burning velocity is governed also by the diffusion of radicals such as H out of the

* So called because it does not inject sample directly into the flame.

combustion zone into the incoming gas. In practical burners, a stable state is established for the conical flame front, in which at each point the component v_a of the gas flow velocity normal to the combustion front is balanced against the burning velocity v_b, which is always normal to the combustion front. In Fig. 1 the components of the flow velocity and the balancing of the velocities (marked with two strokes) are shown for one point on the flame front.

Quantitatively, in Fig. 1, we assume that the inflowing air–gas mixture has the same velocity v_a across the entire diameter of the burner, and that the burning velocity has the same value v_b along the entire combustion front. Then

$$\sin \alpha = v_b/v_a$$

Hence, with a round burner port the primary combustion zone must take the form of a cone with apical angle 2α. Departures from the above assumptions result in a rounding of the inner cone. The equation shows that increasing efflux velocity v_a must make the cone sharper, until with excessive velocity the flame is blown off the burner. But if the efflux velocity is less than the combustion velocity ($v_b/v_a > 1$), there is again no stable state and the flame flashes back into the burner tube. With large burner ports, about 1 cm. or more, the flashback may take the form of a noisy but generally harmless explosion. Flashback is impeded by covering the burner port with a metallic burner cap having many small orifices, as in the Méker burner (Section 32). The heat loss in this metallic cap, when the orifices are small enough and the thermal conductivity of the metal high enough, is so large that the flame cannot burn in the orifices. The greatest permissible orifice diameter for a given gas mixture, wall material, and orifice length, that will avoid flashback, may be found in certain compendia.[31]

The stability of the flame front and hence of the flame itself is governed, incidentally, not alone by the ratio v_b/v_a as indicated by the simplified argument above, but also by other secondary effects, such as may arise at the upper edge of the burner port. The practical limits of stability have been measured for various burner forms and various gas mixtures.[31]

2. The *interconal zone*. Immediately adjacent to the primary combustion zone is the interconal zone (or reaction-free zone) (Fig. 1), which can extend to a considerable height. It is the part of the flame usually employed for flame-photometric analysis. Since this zone enjoys nearly complete thermodynamic equilibrium, its state from the viewpoint of flame photometry can be characterized in general by specifying its temperature. This will determine, for a given instrument design and

mode of atomization, which elements can be detected and which spectral lines are intense enough to be used for analysis. The role of flame temperature in flame-photometric practice will be discussed further in Section 14. However, a few additional remarks are in order here. The temperature is determined on the one hand by the chemical energy liberated in the reaction and on the other by the amount of heat absorbed by the burned gas. Appreciable energy is lost also as work of expansion against the atmospheric pressure. A consequence of the interplay of these three effects is that on continuous variation of the gas/air mixture ratio, the temperature passes through a maximum (Fig. 10, page 74). At temperatures much above 2000°K. one has to reckon also with the energy required for dissociation (of carbon dioxide into carbon monoxide and oxygen, for example), which reduces the expected temperature. This dissociation may cause the mixing ratio for maximal temperature to depart from the stoichiometric.

The cold inflowing gas mixture expands appreciably after passing the inner cone, for two reasons: (1) thermal expansion on heating, and (2) in the case of hydrocarbons, an increase in the number of moles of gas on combustion. This expansion can broaden the flame and increase the vertical rise velocity. The latter is further increased by the buoyancy of the hot gases in air, but is diminished by friction in the outer layers.

We have stated that the interconal zone enjoys nearly complete thermal equilibrium. But the same cannot be said for radiational equilibrium. With complete equilibrium the flame would radiate like a blackbody. If this were the case, no element introduced into the flame could increase the radiation and analytical flame photometry would be impossible. But the emission of the interconal zone, especially in hydrogen flames (Section 9), is extraordinarily low in the visible region. Even in the so-called luminous (sooty) flames the emission is still relatively low compared with that of a blackbody of the same temperature.

There are also other departures from general equilibrium in the interconal zone. Up to about 1 cm. above the inner cone, the concentrations of radicals such as H, OH, O, and NO, and, in hydrocarbon flames, of flame electrons* as well, can considerably exceed the expected equilibrium values. This is due to the fact that these radicals and electrons appear in considerably higher concentrations in the primary combustion zone lying immediately below, and to the sluggishness of

* By flame electrons we mean those generated from the flame gases themselves and not from added metal vapors. In the primary combustion zone there are about 10^{12} electrons per cm.^{3} [11] The partial pressure of atomic hydrogen above this zone in hydrogen flames is about 0.01 atm.[169]

such as oxyhydrogen and oxyacetylene, can be very loud and can grow still more intense not only when a solution is atomized but also when the gas–oxygen mixture is incorrectly adjusted (as when the flame lifts from the burner top). The instrument consequently needs special means for soundproofing (Sections 35 and 63). These pressure impulses may enhance atomization of the liquid droplets already sprayed into the flame (Section 6), raising the concentration of metal atoms and brightening the analytical lines.

An un-premixed turbulent flame changes noticeably in form, size, and background radiation when liquid is atomized (Fig. 25, page 127). The causes of this are not fully clear. It is at any rate certain that spraying an aqueous solution considerably cools the flame.[85,228] Hence, in analytical flame photometry, particularly with this type of flame, the flame background cannot be measured on the empty flame (Section 95); the background intensity must be found by spraying a blank, i.e., the solvent together with any concomitant materials, excluding the element sought.

More detailed investigation of the structure and properties of these turbulent atomizer-burner flames would, we believe, be very useful; in fact, there are a few starts in this direction.[152,228,233]

6. *The Introduction of Sample into the Flame*

The substance to be analyzed must appear in the flame as atomic or molecular vapor. To this end the sample solution must pass through three different processes: (1) *Transport*: the sample must be atomized and carried into the flame as aerosol. (2) *Vaporization*: it must be converted to atomic vapor, in which dissociation and ionization play a part. (3) *Excitation*: the vapor must be excited to emission. These three processes will be discussed in sequence in Sections 6, 7, and 8.

The introduction of analytical substance (process 1) must of course be reproducible and constant while consuming as little liquid as possible. These requirements will be discussed in detail in Sections 21–31 and 33. We are generally concerned here with liquid atomizers, in which a jet of compressed air or oxygen atomizes the sample into a fine, salt-containing fog or liquid aerosol. Nearly all flame photometers today employ these pneumatic atomizers. In this section we deal with only the basic problems of pneumatic atomization.

On atomization, the sample solution is reduced to small droplets, which are carried into the flame either directly (in direct atomizers, Section 33), or after passing through a spray chamber, where the larger droplets separate (indirect atomizers, Sections 24–26). The solvent of the smaller droplets evaporates partly first in the spray chamber and the conduits to the burner, and finally in the flame. The remaining solute crystals then form a solid aerosol, usually only briefly, which on heating in the flame passes over into atomic or molecular vapor. The final evaporation of liquid and solid aerosol in the flame we classify as vaporization, the topic of Section 7.

The transport of sample into the flame, to be considered here, and the subsequent evaporation would be without complications if all of the aspirated sample solution appeared in the flame as vapor. Unfortunately there are various losses. On the one hand, with indirect atomizers much is lost in the spray chamber and the conduits to the burner through separation of the larger drops, and on the other hand, especially in direct atomizers, the larger aerosol drops evaporate incompletely in the flame. Thus a larger or smaller fraction of the aspirated sample solution fails to attain excitation. This diminishes the light intensity and worsens the detection limits. But, further, the differences in physical properties (surface tension, etc.) among the individual samples, never wholly avoidable, sometimes enhance, sometimes reduce these losses. This leads to errors in the analytical result. We speak then of transport interference or evaporation interference, to which we return in Section 11.

The magnitude of these losses and their variation are governed largely by the drop size distribution of the mist. The factors affecting it will therefore be discussed here. Since the influence of these factors upon the atomization and the subsequent vaporization is very complex, and since many instrumental factors such as form and size of the atomizer and the spray chamber enter in too, our discussion of these effects upon the drop size distribution can be only qualitative.

That sample can be lost even with a direct atomizer prior to excitation because of incomplete evaporation is shown by various observations. For instance, a depression of surface tension can brighten the emission in an atomizer-burner.[233,575] Measurements of line and background intensity in a flame with controlled sample flow[347] also point to such losses. Since the losses depend on drop size distribution, the following discussion of the effects of drop size (in indirect atomizers) is pertinent also for direct atomizers.

In a pneumatic atomizer the sample liquid is sucked up from the sample container (Petri dish, watch glass, etc.) through the atomizer capillary generally by the suction caused by the compressed-air jet. The suction (about 100 cm. of water) may greatly exceed the opposing hydrostatic head of the lifted water column in the capillary as well as the force of capillarity.[65,139,182] The suction rate (sample consumption rate, in ml./min.) is then little affected by the level difference between the subjacent sample and the atomizer tip—a desirable situation—or by differences of density or surface tension among samples, but it is affected by viscosity differences (see below). Otherwise, as the liquid level falls in the sample container, the meter reading also falls, and the effects of density and surface tension create larger errors (cf. Section 101).

The flow in the suction capillary is nearly always laminar because of the small Reynolds number (Section 5).[228,233] Hence the Hagen–Poiseuille law applies, according to which the flow rate is proportional to the pressure drop along the capillary and to the fourth power of the capillary diameter d, and inversely proportional to the viscosity. The dependence on d^4 has been confirmed experimentally for an atomizer.[139] But the dependence on viscosity seems in most cases to be less than the theoretical.[139,182,228,718,760] It is sometimes assumed that a change of suction rate causes a proportional change of emission, i.e., a proportional change of atomic concentration in the flame. This is true in a few cases,[133] but it is contradicted by experiment in many others.[65,140,270,513,545] Evidently, in these commoner cases the suction rate has a (usually opposing) effect on the flame temperature (Section 14) and/or on the drop size distribution, which we next consider.

What fraction of the aspirated sample solution ultimately becomes available as metal vapor for excitation in the flame, depends, then, to a great extent on the *drop size distribution* in the atomization process, or, more exactly, on the fraction of the spray consisting of droplets below a certain diameter. The larger drops, comprising a disproportionate fraction of the volume and mass, are thus lost to excitation. Concerning the atomization process itself, little is yet known quantitatively, despite its great importance in technology and industry. The process can be pictured qualitatively, however. The liquid emerging from the suction capillary is shredded into liquid tatters, which then contract into droplets owing to surface tension. Through the action of the strong air stream, or by striking a barrier surface or the like, these drops can shatter into still smaller droplets.[32,37] Both tangential and radial forces act upon a drop in an air stream. The tangential forces arise from friction with the flowing air, and the radial forces from central impingement of air upon the drop. This dynamic pressure is wv^2, where w is the density of the air in g./cm.^{3} and v the relative velocity of the air with respect to the drop. These two forces oppose the forces that hold the drop together, viz., viscosity and surface tension. At sufficiently high relative speeds, the viscous forces can be neglected, and the drop begins to shatter when the dynamic pressure exceeds the pressure of surface tension $2s/r$ (s = surface energy; r = radius of the drop) by a factor of at least 3.[358,442]

Under these assumptions, the critical radius r' is given by

$$wv^2 = 6s/r'$$

The maximal droplet diameter $2r'$ possible during atomization can thus be calculated; with pneumatic atomizers in flame photometry it is of the order of 10 μ. But more important than this result is the relation between the critical diameter and the impact pressure and surface tension revealed by the equation.

However, the conditions prevailing in a pneumatic atomizer depart considerably from those for which the above simple equation for r' holds exactly (the compressed air jet is inhomogeneous, for instance). Hence, it is not easy to calculate in advance the maximal drop diameter, much less the drop size distribution. One has to resort to empirical formulas, which, however, apply to a given atomizer only under specific conditions.[32]

The following is such a formula for the mean drop diameter D_0 (from ref. 520), valid only for subsonic velocities and aqueous solutions with concentric atomizers (Section 26):

$$D_0 = 586\sqrt{s}/v\sqrt{d} + 597(f/\sqrt{sd})^{0.45}(1000Q_1/Q_2)^{1.5}$$

Here D_0 is the diameter of an average drop in μ, the average being so taken that this drop has the same volume/surface ratio as the entire liquid mist (it is then also true that the mass of all drops of diameter less than D_0 is about half the total mass); s is the surface tension in dy./cm.; v is the velocity of the compressed air at the orifice in m./sec.; d is the density of the liquid in g./cm.^{3}; f is the viscosity in dy. sec./cm.^{2}; Q_1 is the liquid flow in cm.^{3}/sec., and Q_2 the air flow in cm.^{3}/sec. This empirical formula is valid for the limits $0.8 < d < 1.2$; $30 < s < 73$; and $0.01 < f < 0.3$. For conditions outside these ranges, other formulas can be obtained.[520] It is interesting that the validity of the formula turns out to be fairly independent of the dimensions of the annular jet. With economical atomization Q_2 exceeds $3000Q_1$.[341, 347] In this case the last term of the formula can be ignored. The size of the spray drops then depends almost entirely on s, d, and v, while the effect of viscosity f vanishes. This conclusion is in line with the earlier equation for the critical radius, in which likewise the viscosity plays no part. Both formulas show that the mist becomes finer with increasing air velocity and decreasing surface tension s. Surface tension especially proves to be very important in flame-photometric work.

For direct atomizers, the drop size distribution at the atomizer is the same as the distribution entering the flame, aside from possible post-atomization by shocks in a turbulent flame. What fraction of the sample material becomes available for excitation then depends only upon the evaporation of the solvent and the solid particles.

The situation is different with indirect atomizers. Here the deposition of the larger drops in the spray chamber and burner connections, or the separation of drops by other means (centrifugation, etc.) influences the size distribution of the drops and the quantity of solvent entering the flame. We have to consider also the effects of post-atomization at a barrier surface (Section 26) and recombination processes (smaller droplets agglomerating to form larger ones), and finally the considerable effect of solvent evaporation in the spray chamber.

Some comments on these opposed recombination and evaporation processes are in order. Recombination occurs only when the probability of encounter of two drops during the residence time (between atomization and entry into the flame) is not too small. With a concentric atomizer, recombination increases the mean drop size appreciably only when Q_2/Q_1 (see above) is less than about 5000.[32,520] This may well be the case with the usual compressed-air atomizer, so that recombination cannot be ignored. Unfortunately this effect is hard to estimate exactly.

It can be shown theoretically and experimentally that evaporation in the spray chamber has an important effect on the size distribution of the drops reaching the flame.[65,139,182,228,241,563,680] Experiments with and without a spray chamber (Section 30), atomization with preheated, dry air or with saturated air, etc., reveal this. Like recombination, this effect of evaporation on the size distribution and hence on the flame emission is also hard to predict.

The effect can, however, be gaged as follows. Consider the evaporative loss of mass per unit time from a stationary, spherical droplet in still, dry air. The Langmuir equation,[444] derived on the assumption that the free path of air molecules is small compared to the drop diameter, applies:

$$-dm/dt = 4\pi M D p_s r/RT$$

Here m is the mass of the drop, t the time, M the molecular weight of the vapor, D the diffusion coefficient of the vapor in air, R the gas constant, T the absolute temperature, p_s the vapor pressure at saturation, and r the radius of the drop. Hence the rate of loss of surface F is

$$-dF/dt = 8\pi M D p_s/wRT$$

where w is the density of the liquid. This equation states that at constant temperature the rate of surface contraction, $-dF/dt$, is constant. For water droplets at room temperature it amounts to about

§ 6] THE INTRODUCTION OF SAMPLE INTO THE FLAME 25

10^{-4} cm.^{2}/sec., so that a drop with a diameter of 50 μ evaporates completely in about 1 second (cf. refs. 55, 685). This is roughly the residence time of the droplet in the spray chamber, so that all droplets below 50 μ at the start, in an indirect atomizer, might be supposed to have lost their solvent entirely before entering the flame. But several factors upset this conclusion: (1) The droplets cool on evaporation—sometimes as much as 14°, reducing the evaporation rate 2.5-fold.[65] (2) The air in the spray chamber gets moister from the evaporation; hence p_s should be replaced by $p_s - p_0$, where p_0 is the aqueous vapor pressure averaged over the spray chamber. (3) As the drop shrinks, the changing radius and salt concentration affect the pressure p_s. (4) The evaporating drops are not quiescent, but are moving with respect to air that is also in motion; therefore they evaporate faster. (5) If the solvent consists of two or more liquids, such as glycol and water, their concentrations change during evaporation, and so $-dF/dt$ changes.[290]

The approximate constancy of $-dF/dt$ implies that with finer atomization at a given rate (smaller drops of given total mass), the quantity of solvent evaporated per unit time increases. An upper limit is imposed by saturation of the air. With a consumption of 10 l. of dry air per minute, no more than about 0.1 g. of water can evaporate per minute at the temperature which is attained. This is of the same order of magnitude as the quantity of more or less evaporated mist that enters the flame.

The merit of an indirect atomizer can be specified by its *efficiency*.[341] Expressed in percent, it is the fraction of the originally aspirated analyte* which actually reaches the flame. The efficiency is usually only about 1–6%,[139,269,341,513,741] but it may reach 10–15%.[65,341] Lowering the suction rate, as by means of a longer or finer capillary tube, appears to improve the efficiency appreciably.[540a]

The efficiency can be determined with accuracy adequate for most purposes by measuring the suction rate at the atomizer and the rate at which liquid condenses on the spray chamber walls (with a graduate). The difference is approximately the fraction reaching the flame. But since the efficiency is defined in terms of the influx of analyte and not of liquid, the value thus obtained must be corrected for the evaporation of the condensed liquid that occurred between atomization and condensation. The extent of this evaporation is easily found with the same flame photometer, by analyzing the condensate. In one instance the concentration was found to have risen about 10%.[65] Some solute is lost also in the tubing and in the burner. This is seen on dismantling a

* The word *analyte* will denote the analytical element (element sought, element to be determined).

burner or connecting tube that has had long service and is incrusted with salt. Thus, the product of suction (sample flow) rate, analyte concentration, and atomizer efficiency as thus determined, is not necessarily identical with the rate of introduction of analyte into the flame.[65]

As the aerosol enters the flame, more or less extensively evaporated, the analyte is vaporized (Section 7) and excited (Section 8), in each drop separately. Consequently the emission of light from the flame is not uniform but exhibits statistical fluctuations. This effect can be compared in a sense with the shot effect (Section 65), which results from the non-zero charge of individual electrons. The relative magnitude of the statistical fluctuation of the intensity of the measured beam, due to the dropwise vaporization and excitation of the analyte in the flame, within the part of the flame included in the observed solid angle, is inversely proportional to $\sqrt{N}$, where N is the effective mean number of aerosol particles entering the observed region per unit time. This fluctuation of intensity, which limits the precision of measurement, therefore depends on the fineness of atomization. In an indirect atomizer with high atomizing pressure, the relative fluctuation of emission (with an instrumental response time of 1 sec.) was found to be about 0.1%.[65] With coarser droplets, such as are more likely with direct atomizers, this value can be considerably higher.

SUMMARY: An atomizer is generally used for introducing sample into the flame. The atomizer is characterized primarily by the drop size distribution generated. With indirect atomizers this distribution is secondarily influenced by more or less extensive evaporation and recombination of the drops in the spray chamber, the connections to the burner, and the burner itself. The primary drop size distribution together with the resulting secondary distribution, which depends on the primary, determine, for indirect atomizers, what fraction of the aspirated analyte enters the flame. Losses occur in the spray chamber, the burner connections, and the burner. These losses are expressed by the atomization efficiency of the overall instrument. With direct atomizers all aspirated liquid enters the flame, and the efficiency is 100%; but appreciable losses can result here from incomplete evaporation of the mist and solute particles in the flame (Section 7). These losses, too, depend on the primary drop size distribution of the mist. This primary distribution depends in every case largely on the pressure used for atomization, the suction rate, and the properties of the solution such as surface tension and viscosity. With indirect atomizers the secondary drop size distribution obtaining at the flame is additionally affected by prior (partial) evaporation of the solvent and by deposition of the larger

7. Vaporization, Dissociation, and Ionization of the Analyte in the Flame

The preceding section has dealt with the introduction of analyte into the flame. We next examine the conversion of the aerosol particles entering the flame into atomic or molecular vapor capable of radiating. The mechanism of radiation will be taken up in Section 8.

If a sodium chloride solution is sprayed into a flame, the solvent in the mist droplets evaporates first. The resulting salt particles next grow hot and evaporate, and the liberated NaCl molecules dissociate, yielding atomic sodium vapor, capable of radiating. But non-radiating sodium compounds (such as NaOH) and sodium ions also form. This combined or ionized sodium is undetectable by flame photometry, a fact that must sometimes be taken into account in quantitative analysis, as when concomitant substances appreciably affect the degree of ionization or formation of compounds. We shall discuss here in sequence the evaporation of the solvent, the vaporization of the solute particles, the dissociation of the molecules, and the ionization.

Vaporization. When the droplets generated by the atomizer reach the flame, they are quickly heated to boiling; the heat delivered by the flame is consumed first in evaporating the solvent from the drops. The time needed for complete evaporation of the solvent depends largely on the initial diameter of the droplets. It has been shown that only very small droplets can lose all solvent while passing through the preheating zone of a laminar, premixed flame. More commonly, in such a flame, evaporation is not complete until the drops have passed through the reaction zone and enter the actual flame. The time t required for complete evaporation is thus determined by the initial diameter d_0, according to van der Held's relation $d_0^2 = Ct$, where C is a constant depending on the flame temperature, the boiling point of the solvent, and the thermal conductivity of the solvent vapor.[65] The same relation $d^2 = C't$ was derived in Section 6 in differential form from the Langmuir formula for the evaporation of droplets in the spray chamber. Van der Held's formula, entailing other physical properties of the solvent, may be more accurate in the present case, since diffusion is unimportant at the high temperature of the flame.

Since the life of the droplets or the solute particles depends on the speed of ascent of the hot gases in the laminar flame, it is possible to calculate at what height in the flame a droplet of given size will finish drying. For water droplets with an initial diameter of 1 μ in an air–acetylene flame ascending at about 10 m./sec., this height is 0.03 mm. But at a height of 3 mm., even drops with an initial diameter of 10 μ will start contributing to the radiation; hence, in these laminar flames with indirect atomizers practically all drops have dried at the customary height of observation, 1 to 3 cm. above the base of the flame. For still larger drops, such as may be expected with atomizer-burners, the short time available in the turbulent flame with its high speed may not suffice to evaporate all the solvent below the usual observation height of 1 to 3 cm. Therefore, analyte may in effect be lost owing to incomplete or slow evaporation of the solvent (cf. Section 6). However, the evaporation rate in a turbulent flame may be greater than in a laminar flame, owing to the turbulent motion of the gases. It would be hard to evaluate the relations in a turbulent flame quantitatively.

The energy required for evaporating the solvent and heating the vapor generally amounts, in a premixed flame with indirect atomizer, to only a small fraction of the heat content of the flame gases.[37, 65] In this case the solvent is, moreover, already partly evaporated before the drops enter the flame (Section 6), so that the requisite energy has been partially supplied by influx of heat from the surroundings in the spray chamber and in the connections to the burner. With an atomizer-burner, in which there is no preliminary evaporation and the liquid flow rate is materially greater, the energy needed for evaporation can be relatively much greater and can markedly cool the flame,[85, 228] especially when the solvent evaporates completely—which, as mentioned, it may not do.

In the case of a sodium chloride solution, when the solvent has dried completely, a sodium chloride particle remains, which is heated to the melting point (ca. 1100°K.) and then to the boiling point (ca. 1750°K.). Since most flames are hotter than the boiling point, the particles or molten drops of salt are not stable. These drops, even when the sodium concentration of the solution was high, should vaporize quickly and completely. The heat required for this process is a wholly negligible fraction of the heat content of the flame.[37]

With more complex salts such as nitrates, carbonates, and phosphates, the flame heat can cause chemical changes prior to vaporization, yielding oxides, metaphosphates, etc. It is true that the flame temperature may be below the melting point of such a compound, as in the case of calcium oxide (2850°K.), but even if the solid calcium oxide vaporizes completely, its partial pressure will usually prove to be still far below its saturation vapor pressure at the temperature of the flame. When the solid phase evaporates rapidly enough, the analyte usually passes completely and promptly into the vapor phase even if the substance has a high boiling point. But for some compounds, such as calcium aluminates and (pyro-)phosphates, which form in the flame when calcium is present together with aluminum or phosphorus in the sample, the residence time in the flame is inadequate for complete vaporization (Section 98).

Molecular dissociation. When a sodium chloride solution is sprayed into a flame, only a part of the sodium appears as free atoms, owing to the limited dissociation of the NaCl molecules. The fraction combined as chloride (which may be very small) cannot yield sodium radiation. We shall examine here the factors affecting the degree of dissociation

$$a = p_{Na}/(p_{Na} + p_{NaCl})$$

(p_{Na} is the partial pressure of Na, etc.), and show its significance for flame photometry.

When a chemical equilibrium exists in the flame,* the kinetics of dissociation and recombination are unimportant for evaluating the degree of dissociation. We merely write the mass action law (which is derivable from statistical thermodynamics) for the dissociation constant:

$$K_p(T) = p_{Na} \cdot p_{Cl}/p_{NaCl}$$

and

$$\log K_p(T) = -1.588 + 1.5 \log (G_1 G_2/G_{12}) + \log (U_1 U_2/U_{12}) + \tfrac{5}{2} \log T - 5040\, V_D/T$$

Here p_{Na} is the partial pressure of Na in atmospheres, etc., V_D is the dissociation energy in e.v., and T is the absolute temperature. G_1, G_2, and G_{12} are in the present instance the atomic or molecular weights of Na, Cl, and NaCl, and U_1, U_2, and U_{12} are the partition functions, characteristic for the reactants and the temperature. Assuming that Na and Cl undergo no other reactions in the flame and that only NaCl is present in the starting solution, we obtain from the above formula for the degree of dissociation

$$a^2/(1-a) = K_p(T)/\bar{p}_{Na}$$

where $\bar{p}_{Na} = p_{Na} + p_{NaCl}$. This equation leads to the following conclusions.

1. The dissociation increases as the concentration of the solution and the total partial pressure $\bar{p}_{Na}$ in the flame (which is proportional to the concentration) diminish. The last formula shows that a diminution of

---

* In the case of NaCl, equilibrium is to be expected because the dissociation reaction of NaCl is extremely rapid at high temperature.[194]

$\bar{p}_{Na}$ has the same effect as an increase of K_p, increasing the dissociation, i.e., raising the degree of dissociation a. It follows further that the working curve should be convex, since the atomic radiation is proportional to $p_{Na} = a\bar{p}_{Na}$, and $\bar{p}_{Na}$ is itself proportional to the salt concentration, while a decreases with increasing $\bar{p}_{Na}$.

2. The dissociation increases with temperature, since K_p increases rapidly with temperature.

3. For equal sodium concentrations in the solution, the radiation can be different for different sodium salts, owing to the effect of $V_D : V_D$ is 4.24 e.v. for NaCl, 3.84 e.v. for NaBr, and 3.11 e.v. for NaI.[29]

If there are other chlorine compounds in the solution, they raise p_{Cl}, the partial pressure of Cl in the flame. By the mass action law, this should lower the degree of dissociation of the NaCl and hence the Na radiation. This effect has been observed.[455] The intensity of the rubidium lines 780 and 795 mμ at a concentration of 0.001% Rb was considerably depressed by the addition of a high concentration (1%) of sodium chloride. In a flame at 2000°K. the dissociation constant K_p should be 6.3×10^{-6} atm. For RbCl at an estimated partial Rb pressure of 10^{-6} atm., the degree of dissociation a should then be 88%. Only this fraction of the Rb can radiate. If p_{Cl} is increased 10-fold (as by addition of NaCl), a will amount to hardly 40%, so that the Rb will radiate less than half as intensely as before.

If excess hydrogen is present in the flame, allowance must be made for the fact that a fairly large part of the chlorine may be combined as HCl in the flame. This reduces the concentration of free Cl and enhances the dissociation of the NaCl. In general, such reactions of the hydrogen in the flame with anions of the alkali salts enhance their dissociation. Hence the above effect (depression by excess Cl, etc.) should not become evident in the usual ranges of concentration. Gouy had found as early as 1877 that the Na radiation in the flame was practically the same for all sodium halides and other sodium salts.

The theoretical calculation of the degree of dissociation with allowance for the formation of HCl in the flame is further embarrassed by the fact that the concentration of atomic hydrogen in the flame can considerably exceed the equilibrium value (see below). The mass action law alone will then not suffice, but all reactions must be heeded that lead to the formation of HCl and NaCl in the flame (cf. ref. 169). Such departures from equilibrium are surely also responsible for the high values found experimentally for the degree of dissociation of alkali chlorides.[65]

For practical flame photometry, we judge that the radiation of an alkali metal will be appreciably lowered by halide formation only when the solution contains high concentrations of other chlorine-containing substances. But the addition of HCl, KCl, etc., can affect the radiation by other mechanisms also, as will be seen in Section 11. Radiation effects due to displacement of the dissociation equilibrium by high concentrations of HCl (ca. 1 mol./l.) have been observed in air–propane, air–acetylene, oxyhydrogen, and oxyacetylene flames.[65,157,233,263,640] The effect on potassium is greater than that on sodium on account of the different dissociation energies [4.40 e.v. for KCl and 4.24 e.v. for NaCl].[29] This effect is weaker in hotter flames.

The effect of the concentration C of HCl on the radiation of an alkali metal, ignoring secondary effects such as formation of Cl_2, is readily expressed as follows:[65,228]

$$J_0/J = 1 + kC$$

where J_0 and J are the radiation without and with HCl, and k is a constant depending only on the analyte and the instrumental conditions (flame, atomizer). Considering now the loss of radiation for two alkali metals, K and Na, and writing for K

$$J_0/J = 1/Z_K$$

and for Na

$$J_0/J = 1/Z_{Na}$$

we obtain by subtraction and use of the above equation

$$1/Z_K - 1/Z_{Na} = C(k_K - k_{Na})$$

where $k_K - k_{Na}$ is positive, since the dissociation energy of KCl exceeds that of NaCl. It can be shown under a fully valid assumption that this last expression is independent of any effect of HCl on the atomization (Section 11), although this is not true of the first formula above.[65] The last equation has been confirmed experimentally in several cases.[65,67]

It is not yet certain whether the reduction of alkali-metal radiation by high concentrations of sulfate is also due to a dissociation mechanism.[233] It is at any rate clear that the complex SO_4 ion is decomposed.

When the halides that are formed themselves emit light, as in the case of CaF and CuCl (but not the alkali halides), the radiation of the compound can be used for the indirect flame-photometric determination of the halogen[277] (Section 87).

Hitherto we have considered compounds, such as NaCl, all of whose constituents (Na and Cl) are brought into the flame in the sample solution. But molecular compounds of the analyte with constituents of the flame gases can also be formed. Examples are LiOH, CaO, CaOH, CuH, etc. The formation of these molecules is important in deciding whether an atomic line is to be used for analysis—in which case any reaction yielding a molecule weakens the line—or whether the radiation of the molecule itself is to be used.

Oxide formation is apparently unimportant with the alkali metals. But they readily form non-radiating hydroxides in the flame. The tendency to form hydroxides increases in the order Na, K, Rb, Cs, Li.[391, 663] In air–propane, air–hydrogen, and of course the hotter flames, the formation of NaOH can probably be ignored;[65, 663] but with lithium and the heavier alkali metals hydroxide formation is appreciable. In an air–acetylene flame with indirect atomizer, a loss of 90% of the atomic Li concentration has been measured.* Since the concentration of OH in flame gases considerably exceeds that of added alkali metal, and since the hydroxide contains only one metal atom, the degree of dissociation should be independent of the alkali concentration. Hydroxide formation should therefore cause no bending of the working curve (compare the earlier remarks on NaCl); the percentage loss of radiating atoms is constant.

In theoretical calculations of hydroxide formation, there is the difficulty that the concentrations of OH, H, and O radicals immediately above the reaction zone considerably exceed the equilibrium values. This is because the recombination of these radicals, which are formed in very high concentration in the reaction zone, proceeds slowly, as it requires three-body collisions.[171] Since equilibrium is lacking, the actual individual reactions leading to hydroxide formation must be considered. For lithium the situation is as follows.[169] At every point in the flame the LiOH is in (partial) equilibrium with the excess atomic hydrogen present:

$$Li + H_2O \rightleftharpoons LiOH + H$$

The high H concentration immediately above the reaction zone therefore keeps the hydroxide rather below the concentration corresponding to true equilibrium. But since the H concentration diminishes with increasing height in the flame owing to recombination, it approaches an equilibrium value,[169, 171] as does also the LiOH concentration; hence, the latter increases with height in the flame. The atomic Li radiation therefore decreases with increasing height. Incidentally, the H_2O content, which greatly exceeds the H content, remains practically constant.

As for the alkaline-earth metals, although at least some of their visible bands are ascribable to the hydroxides (Section 8), still most of

* Unpublished measurements of G. R. D. Zyderveld, Utrecht.

the alkaline earth exists as oxide in the flame.[393] The dissociation energies of the alkaline-earth oxides and hydroxides are not yet unequivocally certain. But it is clear that, for instance in air–acetylene, an appreciable fraction of the calcium exists also as atoms, so that the atomic Ca radiation can be utilized for analysis. However, some metals, such as uranium, vaporize incompletely and yield no dissociated atoms in the flame. The spectrum then shows only a continuum, possibly due to the unvaporized particles, or a weak band spectrum superposed on the continuum. But in the very hot oxycyanogen flame most metals occur only as atoms or ions.[304]

Ionization. Since the time of Arrhenius (1891) it has been known that ionization occurs in flames, and that it is enhanced by metal salts added to the flame. Of all metals, the alkalies show the strongest ionization. A positive alkali ion and an electron are formed. Since the electron shell of an alkali ion resembles that of a noble gas, the ion emits no light in the flame. Because it reduces the concentration of radiating atoms, ionization of the alkali metals is always disadvantageous for practical flame photometry. With the alkaline-earth metals, which have two valence electrons, the singly ionized atom can also radiate, but in practical flame photometry this is seldom important, as the alkaline earths are less strongly ionized than the alkali metals.

The ionization of a metal and the loss of atomic radiation caused by it are measured by the degree of ionization

$$b = p_{K^+}/(p_{K^+} + p_K)$$

where p_{K^+} and p_K are the partial pressures of the K ion and atom. Analogous equations apply to the other alkali metals. The sum of the partial pressures is again proportional to the concentration of metal in the sample solution. If thermal equilibrium is assumed, the degree of ionization can be calculated from the Saha equation

$$J(T) = p_{K^+} \cdot p_e/p_K$$

where p_e is the partial pressure (in atm.) of free electrons and J is the ionization constant, which is a function of the absolute temperature. The temperature dependence is given by the equation

$$\log J(T) = -6.18 + \log(U_i/U_a) + \tfrac{5}{2} \log T - 5040 V_i/T$$

Here the partition functions U_i and U_a, for the ion and atom, equal 1 and 2 respectively. V_i is the ionization potential in e.v. The resemblance of this formula to that for molecular dissociation is noteworthy.

If, for example, only the reaction

$$K^+ + e \rightleftharpoons K$$

is occurring in the flame, then b is given by

$$b^2/(1-b) = J(T)/\bar{p}_K$$

where $\bar{p}_K = p_K + p_{K^+}$ is the total content of potassium in atm. Thus, if $\bar{p}_K = 10^{-6}$ atm. and $T = 2500°K.$, about half the potassium is ionized, whereas in the case of sodium under the same conditions only 7% is present as ions in the flame, and for calcium only about 1%.[11] These wide differences in the degree of ionization result from the differences of ionization potential (K, 4.34; Na, 5.14; Ca, 6.11 e.v.). The ionization potentials of some of the elements of interest in flame photometry are listed in Table 1.

We have already seen that the degree of dissociation a depends on temperature and the total concentration of metal. The same is true of the degree of ionization b. However, these temperature and concentration dependences have different effects in the two cases (dissociation and ionization), for a rise of b means a fall of atomic radiation. Rising temperature increases the loss of atomic concentration due to ionization in the flame; this increase can even balance or outweigh the increase of thermal excitation (see Sections 8 and 14). The percentage loss of atomic

TABLE 1. The ionization and excitation energies and the wavelengths of the resonance lines of some of the more important elements (from refs. 29 and 37a)

Element	Wavelength of the resonance line, mμ	Ionization energy, e.v.	Excitation energy of the resonance line, e.v.
Lithium	670.8	5.39	1.84
Sodium	589.0, 589.6	5.14	2.10
Potassium	766.5, 769.9	4.34	1.61
Rubidium	780.0, 794.8	4.18	1.56
Cesium	852.1, 894.3	3.89	1.39
Copper	324.8, 327.4	7.72	3.80
Silver	328.1, 338.3	7.57	3.67
Magnesium	285.2	7.64	4.35
Calcium	422.7	6.11	2.93
Strontium	460.7	5.69	2.69
Barium	553.6	5.21	2.24
Zinc	(307.6)	9.39	(4.04)
Cadmium	(326.1)	8.99	(3.80)

concentration also rises with dilution. This effect makes the working curve concave (Section 10). Lastly, the general Saha equation shows that addition of another metal (e.g., cesium) which ionizes readily in the flame, raising p_e, will lower the degree of ionization of the element in question (potassium). Its atomic radiation then rises, and we have a case of radiation interference, which causes error in practical flame photometry (Sections 11 and 97). As all of these effects are common in flame photometry, there have been frequent attempts at quantitative interpretation.[65,179,233,371,603,612,660,661,675] Flame ionization has been studied also by high-frequency electrical measurements and mass spectrometry.[127,423,425,694]

The quantitative elucidation of ionization is difficult in many respects. For one thing, it is not yet clear to what extent, in any given case, equilibrium may prevail in ionization. The kinetics of ionization and recombination also require attention.[180,423,536] Further, one has to allow for concomitant substances such as negative ions (OH^-, O^-, and Cl^-), which affect the ionization; their effect is greater in cooler flames.[11,127,535] At equilibrium these ions should in fact obey the Saha equation, but the electron affinity is not always known accurately enough. Besides, as shown above, the concentrations of the corresponding radicals (OH, O, Cl) are not in equilibrium, especially just above the reaction zone. Moreover, mass spectrometry has revealed complex ionized molecules such as $(BaOH)^+$ in flames,[91,424,695] and these make the situation even more complex.

There is another special difficulty. In air–acetylene flames (perhaps also in other flames—see below) free electrons, the "flame electrons", can arise by ionization of the flame gases themselves, without added metal salts. The concentration of these electrons can considerably exceed the thermal-equilibrium value.[63,65] The latter should be very low in a pure flame (without metallic contamination or soot), in view of the high ionization potentials of the constituents of the flame gases.[11,371] The large excess of the observed electron concentration above the expected may be due to slow recombination of certain ions with electrons.[11,65] The concentration diminishes with increasing height in the flame (increasing distance from the reaction zone), approaching equilibrium. It is interesting that flame ions have been found by mass spectrometry above the reaction zone.[424,425] It is not yet fully known to what extent flame electrons occur in other hydrocarbon flames and in turbulent flames, and to what extent they affect analytical flame photometry[65,233] (see below).

When free flame electrons occur, they affect practical analysis in several ways, always by lowering the ionization of the analyte.[65] At low analyte concentrations, where the concentration of metal ions is very small compared with that of the flame electrons, a fixed value of p_e can be used in the Saha equation above, and the metal concentration plays no part in this range. The degree of ionization of the metal is then practically independent of its concentration, and the working curve is therefore straight in this range; the characteristic concavity caused by ionization is absent. This suppression of the ionization in the lowest concentration range improves the actual detection limits and diminishes the mutual ionization interference (as of potassium and sodium; see Section 97). Since the concentration of flame electrons falls with rising height in the flame, their beneficial effect diminishes and, for constant temperature, the ionization of the metal increases. Correspondingly, the atomic radiation of the metal decreases with increasing height of observation, and the shape of the working curve changes.

In summary, dissociation and ionization can occur jointly; their effects are hard to separate, as in the case of KCl. The optimal flame temperature, height of observation, concentration, etc., derivable from consideration of dissociation on the one hand and ionization on the other, are in partial conflict and require compromise.

8. *Atomic and Molecular Radiation in the Flame*

We have examined the transport of sample solution into the flame and its conversion there into atomic or molecular vapor, amenable to excitation. Next we examine the origin and properties of the radiation emitted by these atoms and molecules, taking up first its spectral composition and later its intensity and the mechanism of excitation. Plainly, we can offer only a short outline of the main concepts from atomic and molecular physics; for a fuller treatment see the special literature.[5,12,17,19,37,47,54]

The spectral composition. The basic essentials will be presented using as an example the sodium atom, which consists of a positive nucleus and 11 electrons (11 is the atomic number of sodium), which move about the nucleus in definite paths or orbits, to be thought of as arranged in separate shells. Ten of these electrons, at the temperatures encountered in flames, remain unaltered in their closed inner shells. The eleventh outer electron, the valence electron, is alone responsible for the radiative process to be discussed here. To visualize this process, we can use the Rutherford–Bohr atomic model. In terms of this model, the process of excitation consists in the transition of the valence electron from its lowest orbit, of energy E_0, to a more distant orbit, of higher energy E_i. The atom as a whole is then said to be in an excited state.

Any given kind of atom can have only one or another of a set of perfectly discrete states of excitation or energy levels with their associated discrete energies $E_1, E_2, E_3, \ldots, E_i, \ldots$, which are characteristic for this atom. The law of energy conservation demands that in such a transition a quantity of energy $E_i - E_0$ be supplied in some way from outside. If the valence electron now falls back from an outer, more energetic orbit of energy E_i to an inner orbit of energy E_j, the energy $E_i - E_j$ is released. This liberated energy can be radiated in the form of a photon or quantum of light. Its energy, $h\nu$, obeys the Bohr frequency condition $h\nu = E_i - E_j$, where h is Planck's constant or quantum of action, equal to 6.60×10^{-27} erg sec., and ν is the frequency of the radiated light. Since the energy values E_i and E_j are specifically characteristic for each atomic species, the various energy differences $E_i - E_j$ and therefore the frequencies and the wavelengths $\lambda = c/\nu$ (c being the velocity of light) of the radiated light are also characteristic. This fact is the basis of qualitative flame analysis, in which the type of atom present in the flame is identified by the wavelengths of the emitted radiation.

Thus the arrangement of the energy levels of the atom, together with rules that forbid certain transitions, determines which wavelengths can appear in the spectrum. The reason why only certain electron orbits of the atom are possible and why certain transitions are forbidden is elucidated by wave mechanics and will not be discussed here. We show here as an example the energy level diagram (term diagram, term scheme) of the sodium atom in Fig. 2. The excitation energies in e.v. of the resonance lines of some of the principal elements were listed in Table 1.

In Fig. 2, each of the terms representing a possible energy state of the atom (a possible electron orbit) is shown by a horizontal line, the height of which corresponds to its energy. This energy, referred to that of the ground (lowest) state, is measured in e.v. (electron volts), 1 e.v. being equal to 1.60×10^{-12} erg. The arrows between the various terms show which transitions are allowed. Thus, to each transition or arrow corresponds a definite spectral line, whose wavelength is given in A.* beside the arrow. The well-known yellow double line 5890–5896 A. of sodium is seen in Fig. 2, representing a transition from the first excited state to the ground state. This line appears as a doublet (two separate lines close together), for the first excited level is split into two very close but separate terms. The ultraviolet doublet 3302–3303 A. is another such transition to the ground state. Unlike these, the infrared

---

* Hereafter we shall be giving wavelengths in the customary engineering unit mμ or nm. (nanometer), equal to 10^{-7} cm. or 10 A.

doublet 8183–8195 A. represents a transition from a higher level to a lower excited level, not to the ground state. Even this line is detectable in flames (see the wavelength table in the Appendix). The line or lines of longest wavelength representing a transition to the ground state are called resonance lines. They are active in the absorption processes discussed later.

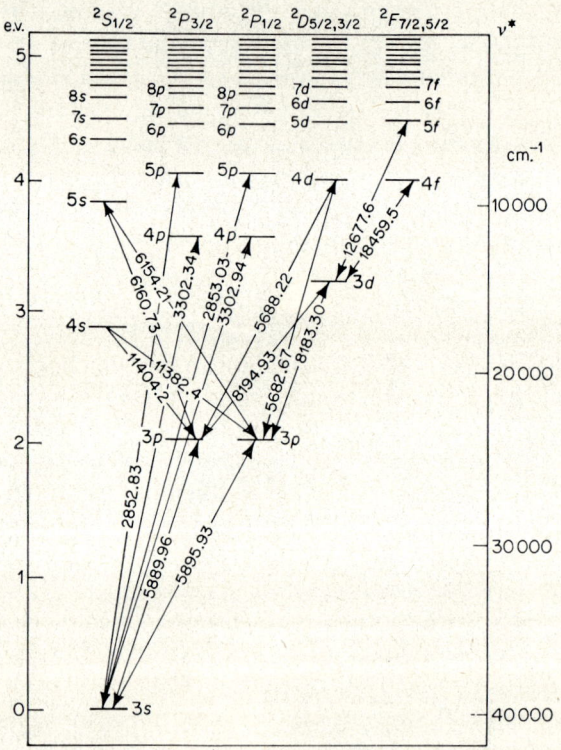

Fig. 2. Term diagram of the sodium atom (from ref. 15).

Each energy level in the term scheme of Fig. 2 is characterized by three quantum numbers: the principal quantum number n, the azimuthal quantum number l, and the inner quantum number j. These quantum numbers define the quantization of the energy, the angular momentum, and the spin (axial rotation), respectively, of the electron. The principal quantum number denotes the shell in which the valence electron lies; it is written ahead of the term symbol. In our example,

$n = 3, 4, 5, \ldots$, since the two inner shells are closed (fully occupied by electrons) and do not enter into consideration for our purposes, as mentioned earlier. The azimuthal quantum number denotes the eccentricity of the electron orbit. It can have the values $l = 0, 1, 2, 3, \ldots$, $(n-1)$. These values are indicated by the letters s, p, d, f, etc., in the term symbol or by S, P, D, F, etc., in the symbol for the state of the atom. The inner quantum number j indicates the manner of coupling of the angular momentum and the spin of the valence electron through their associated magnetic fields. If the directions of the orbital motion and the spin are in the same sense, $j = l + \frac{1}{2}$, but if they are opposed, $j = l - \frac{1}{2}$. In the case of the sodium atom there are no other possibilities. The value of j is written as subscript of the term symbol. The closely adjacent terms of the lower excited levels of sodium, $3p_{3/2}$ and $3p_{1/2}$ (Fig. 2), thus differ only in the j value. A slight difference in excitation energy corresponds to these different j values, and hence the associated lines differ slightly in wavelength. The yellow sodium line consequently appears double. In fact, all the alkali-metal lines representing transitions to the ground state are doublets.

The spectra of the alkali metals are very similar, since each has only one valence electron. The spectra of the alkaline-earth elements, which have two electrons in the outer, unfilled shell, also resemble each other but differ from those of the alkali metals. Unlike the ionized alkali atom, the singly ionized alkaline-earth atom emits a line spectrum in the flame, which resembles the spectrum of a neutral alkali atom. The term diagram of an alkaline-earth atom differs basically from that of an alkali atom (Fig. 2) in that the resonance line (e.g., the blue calcium line at 4227 A.) is single instead of double.

An alkaline-earth metal, moreover, emits a molecular band spectrum in the flame. These bands are generally more important for analytical flame photometry than the lines. We therefore describe here briefly the appearance and origin of molecular spectra. They occur, incidentally, also in the radiation of the flame itself; we return to this in Section 9. To supplement the following outline, see refs. 5, 12, 19, 37, 41, 54.

As with atoms, the frequencies radiated by *molecules* or *radicals* (such as OH) are also determined by $h\nu = E_i - E_j$, where E_i is the (internal) energy of the excited molecule and E_j is its energy after radiation of a photon. Here, too, there are selection rules which forbid certain energy transitions. Again, molecules have discrete states of excitation and hence discrete spectral frequencies, which are specific for the molecule. But the energy levels in the term diagram are generally much more crowded than those of atoms, so that the molecular spectrum is very rich in lines. If the spectroscope has insufficient resolving power or if the lines are strongly broadened, these closely grouped lines cannot be separated, and only a diffuse peak of intensity can be observed; such a group is called a (molecular) band, and the assemblage of bands is known as a band spectrum, in contrast with the line spectrum of an atom. Recordings of molecular band spectra are seen in Fig. 6 and in the spectrograms at the end of the book (Spectrograms 12, 52, 64, 74, 80 83, 94 and 95).

The complexity of a molecular spectrum and the profusion of energy levels are explained as follows. The energy of a molecule comprises three independent parts:

1. The electronic excitation energy, corresponding to that of an atom. As in atoms, the electrons can make certain quantum jumps between different orbits (terms, levels).
2. The vibrational energy, due to vibration of the atomic nuclei composing the molecule. The atoms can vibrate in opposition to each other, about axes of symmetry, or in other ways. As with electronic excitation, only discrete vibrational states are possible, each with a corresponding vibrational energy, determined by the vibrational amplitude. Complex molecules exhibit an immense variety of vibrational possibilities.
3. The rotational energy. A dumbbell-shaped molecule, for example, can rotate about an axis perpendicular to the molecular axis at the center of gravity. The rotational states are also quantized, in accordance with wave mechanics; that is, there are only discrete values of the rotational energy, each corresponding to a distinct rotational frequency. However, the energy differences between two consecutive rotational states are generally much smaller than those between vibrational states, which in turn are generally much smaller than those between two consecutive excitation states of the radiating electron.

The possible discrete values of the total energy of a simple diatomic (dumbbell) molecule are shown schematically in the term diagram of Fig. 3. The orbit of the radiation electron is denoted here by $^1\Sigma$ (ground state) or $^1\Pi$ (for interpretation of these symbols see the literature cited above). These electronic states can appear in combination with any vibrational state, designated by $v = 0, 1, 2, \ldots$. Finally, each vibrational state can be associated with any rotational state, denoted by the quantum number $J = 0, 1, 2, 3, \ldots$. When the electronic and vibrational states remain unchanged, transitions can occur between rotational states ($J' \to J''$); these are observable usually only in the far infrared, since the energy differences are very small. Combined vibrational and rotational transitions ($v' \to v''$; $J' \to J''$) for a fixed electronic state yield radiation chiefly in the near infrared. However, for the molecular bands in the visible and ultraviolet, important in flame photometry, electronic transitions are always involved, associated with superposed rotational and vibrational transitions.

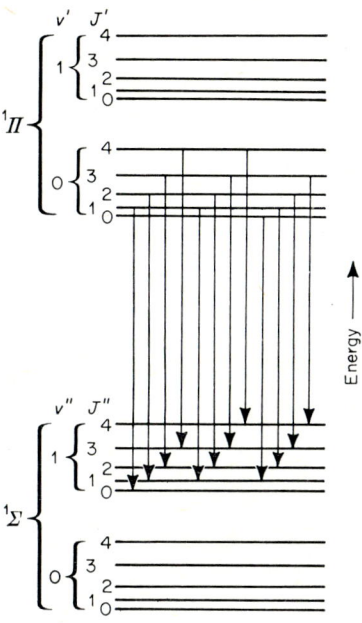

Fig. 3. Schematic term diagram of a diatomic molecule. A band is formed from the permitted rotational transitions (some of which are drawn in) for a given electronic transition $^1\Pi \to {}^1\Sigma$ and a given vibrational transition $v'=0 \to v''=1$. The intervals between vibrational and rotational levels are exaggerated for clarity.

In the example of Fig. 3, such a band (marked with arrows) is composed of all the molecular lines due to a transition ($^1\Pi \to {}^1\Sigma$; $v' \to v''$; $J' \to J''$) having specified values of v' and v'' ($v'=0$, $v''=1$) but variable J'' and J' values. The J'', J' combinations are subject to certain selection rules. Since the different J values correspond to energies that differ only slightly, the lines of a band lie very close together. Their wavelengths generally converge to a particular point in the spectrum, where the piled-up lines form what is called a band head; consequently the band, viewed as a whole, shows an edge at the band head and a shading in the direction of either longer or shorter wavelengths—it is said to be degraded to the red or to the violet. Each permissible $v' \to v''$ combination yields a distinct band.

The band spectra of the flame gases will be discussed in Section 9. The origin of the alkaline-earth bands in the visible (see also Section 81) is not yet fully agreed upon in the literature. Although the alkaline-earth elements are present largely as oxides in the usual flames,[393] and certain bands are doubtless due to oxides (e.g., the blue, ultraviolet, and infrared bands of CaO and SrO[41,393]), at least some of the bands must be ascribed to CaOH, SrOH, and BaOH. These compounds, which do not exist at room temperature, can be stable in the flame. That the calcium bands at 554 and 623 mμ, the strontium bands at 605 and 660 mμ, and the barium bands around 500 mμ must be due to hydroxides, is shown by intensity measurements at constant temperature but with varied OH content,[170,393] and by the isotope shift in these spectra when deuterium replaces hydrogen.[186,285,286,376,441] However, some observers[376] disagree with these assignments. We wish to take no stand, and will tentatively designate the emitters as CaO(H), SrO(H), etc. It seems certain, at any rate, that most of the alkaline-earth bands cannot be ascribed to polymer molecules such as Ca_2 or Ca_2O.[375,393] The bands of the alkaline-earth chlorides do not seem to occur in analytical flame photometry, at least at the usual chlorine concentrations.[361]

Various elements, such as molybdenum and nickel, radiate a weak *continuum** in the flame, in addition to a line or band spectrum. The continuum is noticed in flame photometry as an increase of the background radiation. A continuum of this kind can be due to the thermal (blackbody) radiation of incompletely vaporized particles of an oxide of the element. It can be due also to recombination processes in the flame, as between K^+ ions and free electrons[6] or between Na atoms and OH radicals.[71,395,534] The formation of a neutral K atom or NaOH molecule liberates energy, comprising the discrete ionization or dissociation energy plus the continuously distributed initial kinetic energy of the reactants. This energy can then be radiated as photons, whose frequencies or wavelengths are continuously distributed (within limits) owing to the continuous distribution of the kinetic energy. These continua are of course to be distinguished from those apparent continua that result from instrumental inadequacies such as stray light (Section 94). The true continuum of the flame gases will be considered in Section 9.

Intensity of the radiation. We have discussed the relation between the

---

* A continuum is a radiation which is continuous with respect to wavelength, like that of a thermal radiator such as incandescent tungsten, and is not composed of discrete wavelengths (spectral lines).

structure of atoms or molecules and the wavelengths of their line or band spectra. Since quantitative flame photometry rests upon measurement of the intensity of lines and bands, we take up next the factors governing the intensity.

We consider again the case of a sodium salt introduced into the flame. Let the number of free Na atoms per cm.^{3} in the flame be N, the flame thickness L, and the area of observation S. Then NLS Na atoms are contained in the volume LS of the flame. These atoms collide with the molecules of the flame gases and derive energy from them. The Na atoms are thereby frequently raised from the ground state to one of the excited states shown in Fig. 2, while excited Na atoms keep falling back to the ground state. If we make the assumption, valid in many cases, that thermodynamic equilibrium prevails in the flame, we can derive from statistical thermodynamics the following equation (the Boltzmann equation) for the number of excited atoms N^* per cm.^{3}:

$$N^* = N(g_a/g_0) \exp(-E_a/kT)$$

in which E_a is the excitation energy (see Table 1), k is the Boltzmann constant, T is the absolute temperature, and g_a and g_0 are the statistical weights of the excited and ground states. The energies are reckoned, as usual, from the ground state as zero. This somewhat simplified formula is approximately true under the assumption (certainly justified) that the number N^* of excited atoms is small compared with the total N.

The statistical weights depend on the number of possible quantized states of the atom in a magnetic field. For an atomic state with inner quantum number J, the statistical weight $g = 2J + 1$; hence g is usually in the range 1–10 (compare the term diagram in Fig. 2). We cannot go into further details here.

For each transition from a higher to a lower term there exists a characteristic temperature-independent quantity, the *Einstein transition probability* A. It is the probability that the excited atom will spontaneously make the transition in unit time (1 sec.), with emission of a photon. If on the average N^* atoms per cm.^{3} exist in the excited state, then N^*A of them emit a photon of energy $h\nu$ per second. Hence, a flame volume SL will radiate a spectral line of frequency ν with the intensity

$$I = LSAh\nu N(g_a/g_0) \exp(-E_a/kT) \text{ erg/sec.}$$

into the entire surrounding space; in the special case of a resonance line, $E_a = h\nu$. Of this radiation, the photodetector collects only a part, determined by the solid angle within which the flame "sees" the detector and by the fraction lost in the optical elements (filters, lenses, etc.) through absorption, scattering, etc.

If the excitation energy is expressed in e.v., the exponential factor in the above equations becomes $10^{-5040\,V_a/T}$. For the yellow sodium line with $V_a = 2.10$ e.v., this factor is 5.8×10^{-5} at 2500°K., while for the sodium line at 330 mμ with $V_a = 3.7$ e.v., the factor is 4×10^{-8}. It is evident that under constant conditions (of temperature, etc.), the intensities of the ground-state lines fall rapidly with decreasing wavelength. Another reason why the Na 330 mμ line is much weaker than the yellow line is that its A value is about 100 times lower. Further, the exponential factor decreases rapidly with decreasing temperature; for the yellow sodium line at 2000°K. it is only 5×10^{-6} (Section 14). It follows also that the resonance lines (representing the transition from the lowest excited states to the ground state) are generally the strongest lines in the spectrum of an element. The yellow sodium doublet is a prime example. These lines are known as *raies ultimes*, since they are the last to disappear on continuing dilution of the analyte. The resonance lines of some of the chief elements and their excitation energies are listed in Table 1. Wavelengths of the resonance lines of many of the other elements can be found in the table in the Appendix.

To be sure, the excitation energy E_a and the other quantities (T, A, etc.) in the above formula for I are not the only factors governing the intensity of the radiation for a given concentration of analyte in the sample solution. The intensity depends also on the vaporization, dissociation, and ionization processes discussed in the preceding section. Thus, the resonance line of uranium, despite its low excitation potential of 1.44 e.v., does not appear, because the uranium is combined as molecules that hardly dissociate.

The intensity of a molecular line, or of a band (usually measured integrally), is given by an equation like that above for atomic lines. But the relationships are considerably more complicated on account of the complexity of the molecular spectrum. We cannot discuss this here, except to mention that the intensity distribution within a band and, in particular, the wavelength of greatest intensity depend directly on the flame temperature.[11]

Till now we have tacitly assumed that each photon emitted by an atom or molecule in the flame leaves the flame unimpeded. But it is universally true that every medium that emits radiation of a given frequency can absorb radiation of the same frequency. Absorption of a photon entails the same process as emission, but in reverse. The atom in a state of lower energy, e.g., the ground state, passes on absorption to an excited state. Hence, some of the photons generated in the flame will be reabsorbed, the absorbing atom passing into the higher state. Since an excited atom in a flame loses its energy by inelastic collision with a flame molecule oftener than by radiation of a photon, absorption generally means loss of light. The above equation for intensity should be modified accordingly. The probability of absorption of a photon increases with the concentration of the absorbing species of atom in its lower energy state (usually the ground state). Hence, the probability of absorption increases with the concentration of analyte in the solution, and makes the working curve convex (Section 10). It increases also with flame thickness, which determines the average path length that a photon traverses before it leaves the flame; and it depends on the Einstein transition probability A (see above). Since most atoms are in the ground state at the relatively low temperatures of flames (note the very small Boltzmann factors given above for such temperatures), only ground-state lines show appreciable absorption. The type of absorption we are discussing here is called *self-absorption*, since it is caused by atoms of the same kind as those emitting the light absorbed (cf. Section 10).

Self-absorption affects not only the total intensity of a spectral line, but also the spectral distribution of intensity within the line, especially its spectral width. Even in so-called monochromatic line emission we are dealing not with a single wavelength, but with a wavelength range, which, to be sure, is quite narrow (say, of $\Delta\lambda < 0.2$ mμ). The causes reside in collision broadening and Doppler broadening.* Fig. 4 shows the intensity distribution of a broadened line. With its help we can obtain a qualitative view of self-absorption in relation to line broadening; for details see the literature.[38,51,538] Take one line of the yellow resonance doublet of sodium as example. At low sodium concentrations, where self-absorption can be neglected, the line has the bell-shaped spectral distribution given by collision and Doppler broadening. In this range the line has the same half-width† at all concentrations. If the concentration is doubled, all ordinates on the intensity curve and hence

---

* Collision (or Lorentz) broadening is line broadening due to a disturbance of the radiation process by a collision with a molecule (e.g., a flame molecule) occurring during the radiation. The Doppler effect is a shifting of the observed wavelength due to motion of the emitting atom relative to the (stationary) observer. The disordered thermal motion of the atoms in the flame carries them in all directions at various velocities, so that the line appears spread over a (narrow) wavelength range.

† The half-width (line width, half-intensity width) is the width of the spectral line in mμ measured at that point on the intensity curve (Fig. 4) where the intensity is half the maximum.

the total intensity of the line are doubled. The total intensity is proportional to the area under the bell-shaped curve, like the hatched area a in Fig. 4. This total intensity at low concentration is given correctly by the intensity formula derived above; the line width here plays no part.

At higher sodium concentrations, however, self-absorption becomes noticeable. It is most marked at the wavelength of peak intensity in the line, since by a general law the emitting and absorbing power are closely related; see curves c and d in Fig. 4. The spectral distribution of intensity then shows a certain flattening. The emission at the peak still increases with concentration, but less rapidly than in the absence of

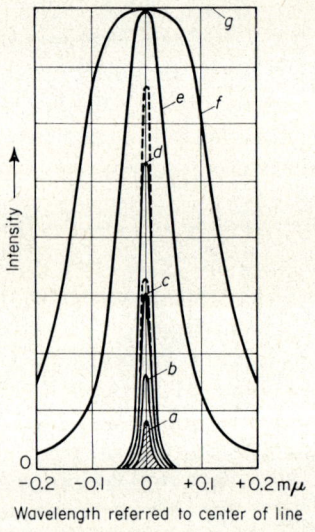

Fig. 4. Schematic profiles of a resonance line with atomic concentration in the flame increasing in the sequence a, b, c, d, e, f. The line g at the top of the figure represents a narrow segment of the Planck radiation distribution. Curves a–d are drawn to the same intensity scale, while curves e–g are drawn to a much smaller intensity scale. The dashed curves above c and d show the profiles in the absence of self-absorption. The hatched area shows the total intensity for curve a as measured in flame photometry.

self-absorption (see the dashed curves above c and d). At still higher concentrations the emission at the center of the broadened line attains saturation (curves e and f), corresponding to the intensity of blackbody radiation at this temperature and wavelength (curve g), normally incapable of being exceeded. The radiant intensity at the line maximum is then independent of concentration and is given by the Planck radiation law. With further increase of concentration the line can increase no longer in central intensity, but only in width. True, the total intensity (the area under the curve) continues to increase with concentration, but less than proportionally. We return to this in the discussion of working curves in Section 10. We may note, finally, that even at the highest practically attainable sodium concentrations the width of the yellow line hardly exceeds 0.1 mμ. The line broadening is thus too small to be detected with the monochromators usually used in flame photometry.

Hitherto we have assumed that the part of the flame colored by metal vapor is of uniform temperature. But the flames customarily used in flame photometry have a cooler peripheral zone which also contains metal vapor. At high sodium concentrations the self-absorption of the sodium light can become so strong that at the peak of the line nearly all

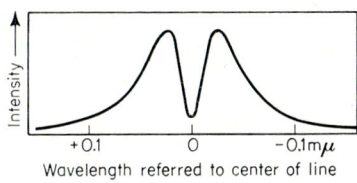

Fig. 5. Recorded profile of *one* of the two lines of the yellow sodium doublet at high concentration (0.1 molar), showing marked self-reversal (ref. 372).

of the light radiated from the hotter, inner part of the flame is absorbed in the outer zone and is thus lost. The absorbed light is, in fact, replaced by the light emitted by the outer zone itself, but since this zone is cooler, the replacement is insufficient to compensate the loss of the radiation from the inner part of the flame. Hence, at the location of the former maximum of the spectral intensity distribution of the line, a depression now appears (Fig. 5). This phenomenon, in which the maximum of the original line appears reversed, is called *self-reversal*. Like self-absorption, it is expected only with ground-state lines. Self-reversal reduces the total intensity of a line. This weakening increases with concentration, so that the convexity of the working curve is even greater than in the case of a flame of uniform temperature, in which only self-absorption without self-reversal occurs (Section 10).

The dependence of absorption on the concentration of the absorbing atoms is the basis of the newest branch of flame analysis, *absorption flame photometry* (oftener called atomic-absorption spectroscopy).[66,72,]

353,608,722* In this method, the monochromatic radiation of a discharge lamp containing vapor of sodium, cadmium, magnesium, or the like, passes through a flame into which the sample (to be analyzed for the same metal) is sprayed. The absorption of the radiation is measured by the difference between the readings of a photometer behind the flame in the presence and absence of the sample, or else by means of a double-beam method of the sort commonly used in other types of absorptiometry, in which one beam is empty or occupied by an empty flame for compensation, while the other beam contains the absorbing flame. To prevent the photometer from responding to the emission of the metal vapor in the absorbing flame, use is made of a discriminative method of light modulation, for example, with a phase- and frequency-sensitive alternating-current meter; see Section 65. The concentration of analyte is found from the absorbance by means of a working curve. All working curves are necessarily curved and at higher concentrations show saturation as 100% absorbance is approached. Absorption flame photometry, in which only ground-state lines can be used, has the advantage that the sensitivity and detection limits are independent of the excitation energy, since the absorption represents a transition from the ground state, and nearly all the atoms in a flame exist in this state. For the same reason, the absorption is far less affected by variations of the flame temperature than is the radiant intensity in emission flame photometry. Consequently, certain interferences are absent, but others such as that due to ionization are still present (Section 11). Absorption flame photometry is especially useful for elements such as magnesium, zinc, and cadmium, which are hard to excite and whose resonance lines lie well down in the ultraviolet. The detection limits for magnesium, zinc, and cadmium are about 0.1 p.p.m.† in absorption; compare the corresponding limits for emission flame photometry in Section 90. Another advantage is the fact that a monochromator is not indispensable in absorption flame photometry.[66]

* For recent reviews of atomic-absorption spectroscopy including absorption flame photometry, see A. Walsh, pp. 1–22 of *Advances in Spectroscopy*, Vol. II, edited by W. H. Thompson, Interscience, New York, 1961; J. E. Allan, *Spectrochim. Acta*, **18**, 605–14 (1962); and P. T. Gilbert, Jr., *Anal. Chem.*, **34**, 210R–220R (1962) and the review in *Proceedings of the Sixth Conference on Analytical Chemistry in Nuclear Reactor Technology* (Gatlinburg, Tenn., Oct., 1962), available from the Office of Technical Services, early 1963. These last two reviews contain, between them, a complete bibliography and index to the literature of atomic-absorption spectroscopy.—*Translator*.

† Recent work has brought these limits down to about 0.003 p.p.m. for Mg, 0.0006 p.p.m. for Zn, and 0.0004 p.p.m. for Cd (K. Fuwa and B. L. Vallee, International Conference on Spectroscopy, University of Maryland, June 18, 1962).—*Translator*.

By virtue of Kirchhoff's law, the emissivity (the radiating power relative to that of a blackbody) always equals the absorptivity (the fraction of light absorbed). On this basis it would seem superficially inconsequential whether a measurement is made by emission or by absorption. But closer consideration shows the following advantages for absorption flame photometry:

1. In contrast with emission, temperature fluctuations of the flame scarcely affect the absorbance and hence the analytical result, aside from the effects of droplet evaporation, dissociation, etc. It is easier to keep the intensity of the discharge lamp behind the flame constant than to keep the flame temperature constant in the emission method.

2. Since the effective radiation temperature of the discharge lamp can be kept quite high, the intensity to be measured (the lamp intensity times the transmittance of the flame) can be much greater than the intensities encountered in flame emission. Absorption flame photometry hence requires a less sensitive photometer, and it can utilize a cooler flame, which offers certain advantages such as smaller ionization interference.

These two features can be summarized by remarking that the relatively low flame temperatures of conventional emission flame photometry are replaced by the constant and effectively higher temperature of the discharge lamps used in absorption flame photometry. The high effective radiant temperature is due to excitation by accelerated electrons in the lamp.

The excitation mechanism. We have assumed in the foregoing that thermodynamic equilibrium prevails and that the population of an excited state is given by the Boltzmann equation. This population governs the intensity of the corresponding spectral line. The Boltzmann equation is valid under equilibrium conditions, regardless of the nature of the excitation mechanism accounting for the transition of the atom to the excited state. Excitation mechanisms, which are as yet rather ill understood, would not be worth discussing here if equilibrium always prevailed. But departures from equilibrium can occur in the flame, affecting the excitation of metal atoms. We therefore devote a little space here to these effects and excitation mechanisms in general.

The following excitation mechanisms can be distinguished:

1. Absorption of radiation.
2. Collision (conversion of translational to excitation energy).
3. Conversion of internal energy (vibrational energy of a molecule) to excitation energy.
4. Chemiluminescence.

We discuss these next in order, pointing out their significance for flame photometry.

The largest departure from thermal equilibrium is the departure of the observed radiation density in the flame at almost all wavelengths from the theoretical equilibrium value. Complete equilibrium would exist only if the flame were to burn in a *hohlraum** having the same temperature as the flame, instead of in surroundings at room temperature. Only at the wavelength of the intensity maximum of a ground-state line at high metal concentrations does the radiation density within the flame approach that of a *hohlraum* radiator of the same temperature (see the preceding subsection).

Now, to what extent does this nonequilibrium of the radiation density affect the excitation leading to line emission? If the chief mechanism of excitation of the atom from the ground state were *absorption* of radiation, the population of the excited state would be much smaller than that given by the Boltzmann equation, especially at low atomic concentrations, where the radiation density and hence the probability of exciting an atom by absorption are very small. In the usual flames at low concentrations, departures from the Boltzmann population are found to be slight. Hence, there must be another mechanism more effective than absorption, which raises the population of the excited state nearly to the equilibrium value. The unimportance of absorption for the excitation of the yellow sodium doublet in an air–acetylene flame has been shown indirectly by measurement of fluorescence.[147]

Owing to the relatively low concentration of free electrons, excitation by *electron impact*, which plays an essential role in the arc, is unimportant in the flame, except in the inner cone. Similarly, it has been shown by fluorescence that excitation by transfer of *translational energy* on collision with atoms, molecules, or radicals, is much too infrequent a process.† Fluorescence experiments have also shown, however, that excitation of a metal atom by conversion of internal molecular *vibrational energy* of a flame molecule into energy of the radiating electron of the atom on collision is of greater importance.[11,17,284] This energy transfer is facilitated by the fact that there is nearly always an energy level of the excited molecule that closely matches the excited level of the metal atom, owing to the large number of possible energy levels in a molecule (see the discussion of band spectra). There is thus a sort of

---

* That is, an enclosed space at thermal equilibrium, in which blackbody radiation prevails.

† Even classical theory shows that collisional energy transfer between an atom and a radiating electron is a process of low efficiency because of the great difference in mass.

resonance, in which good efficiency may be expected. On the other hand, the conversion of vibrational to translational energy on collision between flame molecules is facilitated probably by the fact that the energy is transferred stepwise, the molecule passing each time to the next lower vibrational state and yielding only a small part of its vibrational energy. By this indirect route the electronic excitation of the metal atom depends ultimately upon the distribution of translational energy, i.e., upon the kinetic temperature of the flame. Since the processes are all reversible, it must be assumed that excited metal atoms can be deactivated by conversion of their excitation energy into molecular vibrational energy on collision. It is this interplay between activation and deactivation that permits the resulting equilibrium to be described by the Boltzmann formula for the population of the excited states.

The interconversion of vibrational and translational energy proceeds with finite speed. The rising gases on passing the inner cone experience an abrupt increase of mean translational energy with the sudden rise of temperature, but it takes a few milliseconds before the mean vibrational energy of the molecules has fully equilibrated. This delay entails a simultaneous delay in the metal emission just above the inner cone, since excitation of the metal atoms depends on the molecular vibrational energy.[65,194,447]

Chemical processes can also contribute to the excitation of metal atoms in a flame.[11] The chemical energy of combustion is converted to atomic excitation energy. An example is

$$Na + H + H \rightarrow Na^* + H_2$$

where Na^* is an excited Na atom.[534] The converse quenching reaction (deactivation) occurs as well. But since the concentration of H atoms just above the inner cone considerably exceeds the equilibrium value (Section 7), excitation is much more frequent than quenching here. The gradual reduction of the excess H concentration with rising height entails a corresponding drop in the sodium emission; this fading with height has been observed also with other elements (thallium, silver, manganese).[534] These *chemiluminescent* effects are more readily observed in cooler flames (below 2000°K.), since they are masked in hotter flames by the stronger thermal excitation.* We encountered a type of chemiluminescence earlier in connection with the continuous emission of metals.

---

* For practical applications of chemiluminescence see the footnote to Section 81 and the paper by P. T. Gilbert, Jr., in *Proceedings of the Xth Colloquium Spectroscopicum Internationale*, Spartan Books, Washington, D.C.—*Translator*.

52   THE BASIS OF FLAME ANALYSIS

The emission of the flame itself (Section 9) will yield other examples.

That the flame processes are complex and incompletely understood does not, of course, detract from the utility of the flame as an excitation source for chemical analysis. The excitation conditions in a flame can remain highly reproducible even when there are departures from equilibrium.

9. *The Background Radiation of the Flame*

The flame radiation collected by the optical–electric system of a flame photometer consists essentially of two parts, the radiation of the flame gases themselves, and the superposed radiation of the sample material.

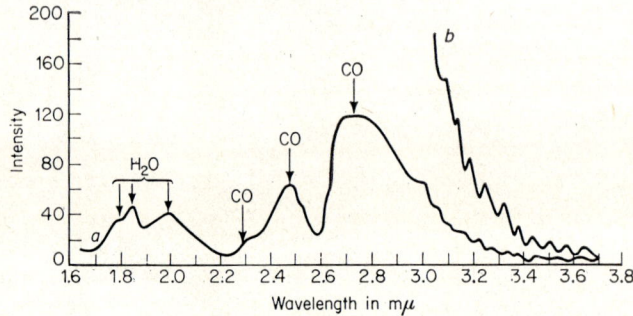

Fig. 6. Emission spectrum of an oxyacetylene flame: a, above the inner cone; b, also above the inner cone but with much higher oxygen content. In b many rotational lines of OH and other molecules appear. From ref. 555.

The former—the flame background radiation—will be dealt with here. The basic considerations of Section 8 apply equally to the origin of this radiation. Unlike the radiation of the added metal atoms, the background radiation is usually unwanted and is almost entirely due to molecules. Recorded spectra of the flame background are found in Fig. 6 and the Appendix. We shall cover here only such aspects of the flame background radiation as pertain to analytical flame photometry; for further information see the literature.[11,12,31,37,41,227]

The unwanted flame background is of concern in practical analysis for several reasons. It affects the choice of analytical lines and bands (Section 82). A background emission must not mask a line or band too greatly; in other words, the ratio of the net intensity of the analytical emission to the interfering intensity of the flame background (within the spectral range transmitted by the instrument) must be made as high as possible by proper choice of flame and wavelength. There is also the fact that even under the most uniform conditions, the flame background fluctuates (flickers) more or less rapidly. The flicker limits the precision of a measurement and fixes the lower limit of concentration that can be detected (detection limit; see Section 90). Various other errors can enter the analysis through inconstancy of the flame background (Section 11). The background intensity depends upon the instrumental parameters, some of which can be varied; and it can be affected by concomitant materials in the analytical solution. For this reason too, a high ratio of analyte intensity to background intensity is desirable.

Thus, in practical flame analysis we have the desiderata of maximal excitation of the analytical line or band in the flame to be chosen (demanding high temperature), plus minimal flame background emission within the spectral range transmitted by the optics (filter or monochromator) at the wavelength selected for measurement. Compromise is often necessary, the ratio of net analyte to background emission serving as a criterion of the utility of a flame (or any other) source. Thus, the oxyacetylene flame is very hot and excites the entering atoms well (Section 14); but the somewhat cooler oxyhydrogen flame, with its weaker excitation, is often preferred because it offers much lower background emission and a higher ratio of analyte to background emission.

That a (nonluminous or non-sooty) flame makes a very favorable excitation source with respect to its background emission is obvious from its emissivity—the ratio of the flame radiation to that of a blackbody of the same temperature. In an air–acetylene flame the background emissivity is only 0.0001% at 589 mμ and 0.1% at 345 and 285 mμ.[65] This low background emissivity is related to the high transparency of the flame. According to the Kirchhoff law the absorptivity of any substance equals its emissivity. Only at the wavelengths of the strong OH bands near 306 mμ and in a few regions of the infrared is the flame emission roughly comparable with that of a blackbody.[12, 373]

It is hard to give a general picture of the flame background radiation, since its spectral distribution depends on the nature of the gas mixture and many other factors. The mixing ratio (gas/air or gas/oxygen), the purity of the gases (Section 15), the type of burner, the mode of gas flow (laminar or turbulent), the solvent sprayed into the flame (Section 95), and the height of observation in the flame are all involved. Further, the appearance of a photographic or recorded spectrogram of the empty flame depends on the resolving power of the spectroscope. With wider slits the fine structure is obscured and molecular bands look more like continua.

The inner cone is seldom used for flame analysis (Sections 32, 87). In hydrocarbon and carbon monoxide flames it emits a bright blue-green light, due to the C_2 molecule, which emits the familiar Swan bands between 436 and 686 mμ (see Spectrograms 3, 84 and 95), and the CH radical with its bands in the regions 310–320 mμ and 387–438 mμ; the ultraviolet bands of the OH radical are also present (see below). The exceptional intensity of this inner-cone radiation results from chemiluminescence and represents nonequilibrium conditions.

The flame background radiation above the inner cone, which is of greater interest for our purposes, comes largely from the second combustion zone (mantle or outer cone; Section 5). In the usual flames the radiation of the interior region (the reaction-free or interconal region) cannot be separated from the radiation of the outer combustion zone. We shall consider them both together. They cannot in any case be separated in a turbulent flame, for the outer combustion zone penetrates deeply into the flame, obliterating the distinction.

The radiation of the interconal region can be observed by means of an arrangement due to Smithells (Section 32) or by means of a flame burning in a nitrogen atmosphere. This radiation is then seen to be very weak, consisting chiefly of OH bands. This shows that most of the emissions described below arise from chemiluminescence in the outer reaction zone.

That part of a hydrogen flame lying above the inner cone emits the OH bands with heads at 281, 306 (strongest), and 343 mμ (see Spectrograms 12, 52, 57 and 64), and also O_2 bands between 250 and 400 mμ. It may show also a weak continuum, arising from the luminescent recombination reactions $NO + O$ and $H + OH$ (Section 8). A slight excess of hydrogen appreciably weakens the O_2 band emission and the NO continuum. The strong H_2O emission at 0.9–1.1 μ and 1.8–3.0 μ makes this spectral region unsuitable for flame analysis.

Hydrocarbon flames such as oxyacetylene show the bands and continua of both the hydrogen and the carbon monoxide flames. Besides the OH bands, they emit a marked continuum between 300 and 500 mμ (see Spectrogram 17), which is overlaid by diffuse bands. This continuum, occurring also in carbon monoxide flames, arises from the chemiluminescence of the reaction $CO + O \rightarrow CO_2 + h\nu$, which is responsible for the familiar blue color of the flame. Its intensity depends in part upon the completeness of the primary combustion, which in turn depends on the flow rate of primary oxygen. In the infrared, besides the H_2O bands (Spectrograms 5 and 6), there are CO bands at 2.3–2.8 μ and CO_2 bands around 4.5 μ, which further limit the possible utility of the flame as an excitation source in this region (Fig. 6).

When organic solvents are sprayed, bands may appear which occur otherwise only in the inner cone; the Swan bands are an example.[307] The oxyacetylene flame can also emit the carbon line at 248 mμ. The background emission of the very hot oxygen–cyanogen flame is not as intense as might be expected; in particular, the OH bands, absent from the pure flame, are relatively weak even with a hydrogen-containing solvent.[304]

10. *Working Curves*

As the concentration of an element sprayed into a flame photometer increases, so does the meter reading. Ideally, this reading (after subtraction of the reading on the flame background: Section 9) is propor-

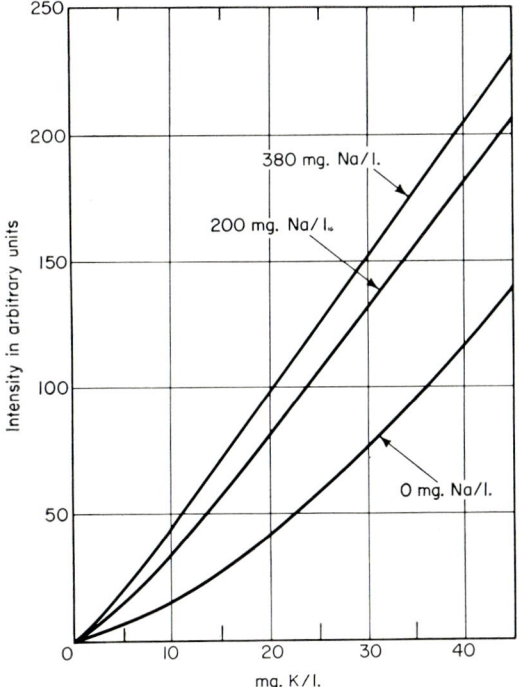

Fig. 7. Working curves of potassium for various concentrations of sodium (from ref. 661).

tional to the concentration. The working curve (sometimes called calibration curve), which graphs the reading against the concentration, is then a straight line passing through the origin of the coordinate system. To establish it, only one point of calibration is needed: a single standard solution serves. But oftener the working curve is convex or concave (Figs. 7 and 50, page 231). Both types of curvature may combine to give a sigmoid (S-shaped) working curve. We shall discuss here the various causes of this curvature, describing first the causes of nonproportionality between analyte concentration and radiation intensity, and later the causes of nonproportionality between intensity and meter reading.

In Sections 6, 7, and 8 we showed how the radiation from a sample sprayed into a flame is affected by various factors and processes. Recall the factors bearing upon transport of the analyte to and its vaporization in the flame: evaporation of the solvent and the solid particle, molecular dissociation, ionization, and self-absorption. If these depend upon the concentration of the analyte, we may expect deviations from proportionality between the concentration and the emission intensity.

(a) *Transport*. High solute concentrations, above 10 000 p.p.m., can alter the physical properties of the solution, such as viscosity and vapor pressure, sufficiently to change the suction rate and the atomization efficiency, usually downward (cf. Sections 6 and 11). This dependence of transport of the analyte into the flame upon the solute concentration makes the working curve convex in the region of very high concentration. Such curvature has been observed with both direct and indirect atomization.[233]

Curvature of the working curve due to concentration dependence of the degree of completeness of vaporization in the flame appears to be rare in practice. This effect might be expected, for example, when an involatile complex compound, as of Ca–Al–O, is formed (Section 98).

The concentration dependence of the degree of dissociation of a molecule such as KCl in the flame (Section 7) is probably less significant as a cause of convexity of the working curve. As shown earlier, the concentration of KCl in the flame and its effect on the curvature are extensively suppressed by the formation of HCl with the excess of atomic H present.

(b) *Ionization*. More important in practice is the frequently observed concavity of the working curve due to a concentration-dependent ionization of the element (Fig. 7). As shown in Section 7, this accelerated ascent of the working curve can be accounted for as follows. At low concentrations the element (potassium, commonly) is extensively ionized in the flame, so that a large part of the potassium becomes inaccessible to determination by any of its atomic lines (usually a resonance line). The Saha equation shows that with rising potassium concentration the ionized fraction should diminish.[67] This means that at low concentrations the atomic potassium concentration in the flame and hence its atomic emission must rise faster than proportionally with concentration of the solution, yielding the observed concave working curve. At high concentrations ionization is practically negligible. Hence, this type of curvature is noticeable at low concentrations with elements of low ionization potential in hot flames.[64, 513, 575]

As seen in Fig. 7, the concavity of the potassium working curve depends also on the concentration of sodium simultaneously present. Addition of sodium, lithium, etc., both raises the potassium emission (Section 11) and straightens the working curve.[632] This can be ascribed to the increase in concentration of free electrons in the flame resulting from the presence of sodium or lithium, which are themselves partly ionized. This increase of electron concentration represses the ionization of potassium and mitigates or entirely removes the concavity of the potassium working curve. Furthermore, in the absence of a concomitant element, the working curve can become straight at very low potassium concentrations when the flame gases themselves contain enough free electrons (Section 7).

When the temperature and/or the content of free flame electrons changes with height in the flame, the degree of concavity of the working curve should change correspondingly with height of observation.[65] Lastly, we note that an ionization concavity of the working curve should be observable both on the atomic lines and on the oxide or hydroxide bands of the analyte,[361, 513] depending on the chemical equilibrium between the metal atom and the oxide or hydroxide; it is necessary to keep in mind that free oxygen and OH are always present in the flame in great excess.

(c) *Self-absorption*. Unlike ionization, which causes concavity at low concentrations, self-absorption makes the working curve convex at higher concentrations (Fig. 50, page 231).

As shown in Section 8, self-absorption causes loss of a fraction of the radiation which increases with the concentration of the atoms. Consequently, the atomic emission in the higher concentration range increases less than proportionally with concentration. This can be seen in the logarithmic graph of Fig. 50 where the slope of the working curve is less than 45°. Characteristically, when self-absorption is strong, the emission increases as the square root of the concentration.[38, 667] The working curve is then a parabola with horizontal axis. Correspondingly, in the logarithmic presentation of Fig. 50 the working curve above 10 p.p.m. sodium is linear but with a slope of $\frac{1}{2}$. Self-absorption and convexity of the working curve are to be expected only with ground-state lines (lines representing a transition to the ground state), especially resonance lines. Convexity therefore occurs with the lines Na 589, K 770, Cs 852, Li 671, Mg 285, Ca 423, Sr 461, Cu 327 mµ, etc., but not with Na 819 mµ (cf. Fig. 2) and similar lines.[152,274,361,513,575,667,734] Bands show no self-absorption, and their working curves show no curvature from this source.

The amount of self-absorption for a given ground-state line depends on the product NL (N being the number of atoms per cm.^{3} in the flame and L the thickness of the flame in cm.; cf. Section 8). Thicker flames hence normally show more self-absorption. For a given value of NL the intensity of self-absorption depends further on a parameter a (related to line broadening) and a quantity f, the oscillator strength, which is proportional to the Einstein transition probability (Section 8).[38] These parameters can vary considerably among ground-state lines. Hence the self-absorption curvature may be much less marked for one line, such as Na 330 mµ,[349] than for another; the oscillator strength of Na 330 mµ is much smaller than that of Na 589 mµ.[29]

The onset of self-absorption is revealed in logarithmic presentation (Fig. 50) by the more or less marked knee in the working curve. (On closer inspection it is evident that the self-absorption does not, of course, commence abruptly, but increases gradually with concentration.) For a given flame, line, and sample concentration, the knee is lower (occurs at lower concentration), the more metal atoms are supplied to the flame by the atomizer per unit time. Hence, from the position of the working curve and especially from the location of the knee in the logarithmic graph, some estimate can be made of the absolute concentration of atoms in the flame and hence of the efficiency of the atomizer.[356,391,394,667]

Our earlier statement regarding the parabolic curvature of the working curve at high concentrations in the presence of strong self-absorption is valid only when the flame temperature is uniform throughout the entire thickness containing metal atoms. But if the temperature is lower in the outer zone, self-reversal can occur (Section 8). The emission then increases more slowly than the square root of concentration.[667] This can be visualized as follows. With rising concentration the effective center of the radiating region moves outward toward the cooler border zone. Of this zone only the part encompassed by the photodetector is observed. Hence the effective radiation temperature falls with rising concentration, imparting an additional convexity to the working curve. This effect of self-reversal is, however, not usually severe in practice.*

* It may be observed in oxygen flames for Mg 285, Cu 325, or Zn 214 mµ; the curvature passes from parabolic to cubic (more or less) at rather low concentrations.—*Translator*.

It can easily be confused with the transport effect discussed above.

(d) *The photodetector*. Departures from proportionality between solution concentration and meter reading can be caused also by lack of proportionality between the radiant flux striking the photosensitive surface of the detector and the photocurrent. This is due to saturation and ordinarily produces a convex working curve. We return to this in Section 42. These effects appear only at very high light levels.

(e) *The meter*. Finally, the meter reading may not be strictly proportional to the applied current; the working curve is then distorted. Similar nonlinearities can occur in amplifiers or, in alternating-current instruments, in the rectifier (Section 65).

In Section 46 we give a method for controlling simultaneously all of the nonlinearities mentioned under (d) and (e). But in a good instrument these departures should be smaller than the reading precision of the meter.

11. *Interferences from Concomitants*

In flame-photometric analysis, the relation between spectral intensity (meter reading) and the sought concentration of the analyte in the atomized solution is often equivocal. This relation is affected on the one hand by various instrumental factors (gas and air flow, height of observation, etc.), and on the other hand by any concomitants* that may be present. When these factors are constant, they can be absorbed in the working curve. But if the concomitants vary in type or concentration, they may *interfere* with the reading on the analyte; that is, their effect upon the reading is not constant, but varies in magnitude or sign. It is then difficult to establish the concentration of the analyte within the permissible limits of error. In this section we discuss the fundamentals of interference and classify the known kinds of interference. In later sections, especially 94–101, we present proven methods of eliminating or correcting for these interferences.

In the following discussion, for simplicity, we assume constancy of the instrumental conditions, sample temperature, and suction height (the difference of level between the sample surface and the atomizer nozzle); these assumptions are not usually entirely valid in practice (see Sections 26 and 27). The flame background emission will be taken into account only when its intensity and/or spectral distribution depends on

---

* We use the term *analyte* to denote the element sought, and *concomitant* to denote any other element, radical, or solute accompanying the analyte in an analytical solution (sample, standard, or blank). When the concomitant interferes, we call it an *interferent*.

3*

the composition of the atomized solution. We assume further a fixed height of observation in the flame. However, the *effective* height of observation, which depends on the time spent by the particles in the flame before they reach the point of observation, can depend on the composition of the solution.[37] Since the analyte emission depends on the effective height of observation (because the evaporation of solvent and solute, hydroxide formation, flame temperature, ionization, excitation, and chemiluminescence all depend on height), concomitants can interfere by affecting the height dependence. We shall ignore these indirect, circuitous influences and confine ourselves for simplicity to direct interferences.

It is very hard to arrive at a suitable and unambiguous classification and nomenclature of the various interferences. Earlier investigators not only built their own flame photometers but coined their own terms for the interferences that they observed. These various classifications and terminologies overlap considerably, so that a term such as "excitation interference" can at present have various distinct meanings. There is another difficulty in that interferences can be classified from entirely different viewpoints. For example, they can be classified, according to type of interfering concomitant, into cation and anion interferences; or, according to the part of the flame photometer in which the interference occurs, into atomizer effects, filter effects, etc.; or, according to the mechanism of the interference, into drop-size interference, ionization interference, etc. Again, the nomenclature may express an evaluation, in designating them interferences of the first and second kind, or genuine and apparent interferences. The terminology can also be based on the manner of correcting the interference, by referring to it as multiplicative or additive. Our classification below, like any other scheme, is not free from shortcomings, but we have tried to construct an unambiguous classification based on the causes of interference, and to select the simplest and clearest terms. There may be some inevitable overlap with earlier terms used in other senses.

Table 2 presents a survey of the various types of interference due to concomitants. We shall briefly explain this table here, referring to other sections that deal more fully with the basic causes of the interferences and with practical methods of eliminating them. Divisions I and II of the table distinguish the two components of the measured radiation, viz., the radiation of the analytical line or band and the unwanted radiation transmitted by the device used for spectral separation (filter or monochromator). The latter radiation will be called the *blank*, in conformance with the customary terminology of absorptiometry. The meter reading thus basically always comprises parts I and II, and an

§ 11] INTERFERENCES FROM CONCOMITANTS

interference can supervene through an effect of a concomitant either upon the blank (I) or upon the intensity of the analytical line or band (II).

TABLE 2. Interferences from concomitants[a]

I. Blank interferences
   A. Cross-sensitivity
   B. Background interferences
      1. Flame background interference
      2. Spectral interference (spectral overlap)
II. Radiation interferences with the analytical line or band
   A. Nonspecific radiation interferences
      1. Sample transport interference
      2. Solvent evaporation interference
      3. Flame shape interference
   B. Specific radiation interferences
      1. Solute vaporization interference
      2. Dissociation interference
      3. Ionization interference
      4. Excitation interference

---

[a] The terminology adopted here is compact and must be interpreted circumspectly. "Interference" can mean the observed phenomenon, the mechanism, or the associated analytical error, depending on viewpoint and context. "Flame background interference" can thus mean the effect of a concomitant upon an analysis through the intermediation of the flame background; or the action of a concomitant upon the flame background, which incidentally affects the analytical result; or the analytical error resulting from the action of a concomitant upon the flame background.—*Translator*.

Next, as to the subdivisions of I: the blank interference can arise from two different causes:

I.A. The *cross-sensitivity*, due to instrumental inadequacy; interfering radiation consisting of unwanted transmission by the filter (filter leakage), stray light in the monochromator, or the like, is superposed upon the reading. We return to this in Sections 94 and 95.

I.B. *Background interferences* which, unlike I.A., cannot in principle be entirely eliminated by the instrument, however efficient; for background is always present at the wavelength of the analytical line or band. This interference can be subdivided further according to the cause:

I.B.1. *Flame background interference*, due to an effect of concomitants upon the continuous flame background (Section 9) and/or upon the band emission of the flame gases.

I.B.2. *Spectral interference*, usually consisting of overlapping continuous radiation from the concomitants. Many elements, such as molybdenum, display a nearly continuous emission over a wide range of the spectrum, apart from their specific line emission. Such elements can more or less raise the background at the analytical wavelength.

It must be emphasized that the two interferences, I.B.1 and I.B.2, cannot be entirely relieved even with the best possible monochromators (Section 95). These interferences are to be distinguished from those encountered when an instrument has to be used that has poor means of spectral separation such as a color filter, which may be unable to separate spectral emissions of analyte and interferent. Such interferences are classed under I.A.; they can be completely eliminated instrumentally, by improving the spectral isolation. The same is true of filter leakage, residual transmission by a filter of a part of the spectrum where a strong interfering line of another element occurs.

Interferences of class II, radiation interferences, are distinguished from those of class I in that they act not upon the blank but upon the analytical line or band. We distinguish, further, specific and nonspecific radiation interferences. A *nonspecific interference* is one which exercises the same relative (percentage) effect upon all the emissions of various elements and upon any given emission at different concentrations of the element.

Here again we exclude secondary effects. Thus, an interference affecting suction rate will react on sodium and lithium emission to about the same extent and hence belongs to group II.A (specifically, II.A.1). This fact is, incidentally, utilized in internal standardization (Sections 51 and 101), which compensates such errors to a large extent. But a closer examination, by way of a second approximation, shows that this nonspecific interference does not have exactly the same (relative) effect upon different elements or even upon the same element at different concentrations. This can happen when the two lines under comparison have working curves of different curvature owing to differences in self-absorption or ionization (Section 7). In internal-standard operation, such secondary effects can be eliminated by converting the readings to apparent concentrations by means of the (curved) working curve. We return to this in Sections 51 and 106. But for simplicity we shall here ignore these secondary effects, and, despite the fact that they are somewhat specific for element and concentration, classify these interferences as nonspecific, under class II.A.

Accordingly, all factors influencing the transport of analyte to the flame (Section 6) will be considered nonspecific and grouped under II.A.1. Other nonspecific interferences are those in which the rate of evaporation of the solvent (II.A.2) and the form of the flame (II.A.3) are affected, occurring especially in atomizer-burners (Section 33). Recall that coarse drops cannot always evaporate quickly enough, especially in the turbulent flame of an atomizer-burner. Concomitants that affect the solvent evaporation rate thus indirectly affect the quantity of solute that becomes promptly available to the flame (Section 7).

Referring to Section 6, one would expect concomitants that affect the surface tension, density, viscosity, and/or vapor pressure to cause nonspecific interference (II.A.1) in both indirect and, to some extent, in direct atomizers. These factors were seen to affect the suction rate and the drop size distribution of the mist and hence the rate of entrance of analyte into the flame, where it is vaporized and excited. Published statements denying the effect of surface tension in experiments in which wetting agents that greatly depress surface tension were used, can be explained as follows. The molecules of wetting agent collect at the surface of the liquid, a process requiring a certain amount of time. On abrupt atomization most of the new surface molecules come from the interior of the liquid stream, where wetting-agent molecules are scarcer. Such a wetting agent, then, does not have time to exert its effect.[65,442] But with additives such as alcohol which are distributed uniformly throughout the liquid, being present in true solution, there is a conspicuous effect of surface tension on atomization.

In contrast with class II.A, the *specific interferences* of class II.B act differently upon different elements and upon a given element at different concentrations; the mechanism of interference involves factors which are specific for the element suffering interference and which may depend on its concentration. Among these interferences we distinguish those that affect the number of atoms made available for excitation in the flame (II.B.1–II.B.3; Section 7), and those that affect the excitation itself (II.B.4; Section 8). Hydrochloric acid, for instance, affects the degree of dissociation of potassium chloride and sodium chloride and thereby influences the atomic emission of sodium and potassium (II.B.2). But since the chlorides have different dissociation energies, this interference should manifest itself to different degrees in the two cases, as is in fact observed. Similarly, cesium affects the degrees of ionization of sodium and potassium, since it supplies electrons to the flame, which depress the ionization; but, again, this effect of the cesium differs with sodium and potassium, because of their different ionization potentials (Section 97).

Since the excitation of atomic lines and molecular bands usually occurs nearly at thermal equilibrium (Section 8), it is affected only by a change of flame temperature. Excitation interference due to a shift of temperature is likely when the atomized solution contains a combustible substance such as alcohol (Section 79). Incombustible concomitants can also change the temperature, especially in atomizer-burners, if only because they affect the flow rate or the evaporation of the solvent in the flame. They thus alter the extent to which the flame is cooled (Section 14). But we may consider these effects specific, since the temperature dependence of line and band emission varies with the excitation potential. A change of flame temperature naturally changes also the degrees of dissociation and ionization, since chemical equilibria are temperature-dependent. The flame background emission also varies with temperature. Ordinarily, though, the effect on excitation is the most pronounced, so that excitation interference (II.B.4) may be regarded as a thermal effect.

Another kind of excitation interference is sometimes mentioned in the literature. When the excitation energy of an emission of the interfering element is close to that of the emission interfered with, the latter can be influenced by a resonance exchange of energy. But if thermodynamic equilibrium prevails, as it usually does, the population of the energy levels will under any circumstances be given by the Boltzmann equation (with T as the flame temperature), and this kind of interference is theoretically excluded.

A common type of specific radiation interference occurs when the analyte (e.g., calcium) forms a compound of low volatility with an interferent (e.g., aluminum) in the aerosol particle (Section 98). The concentration of calcium atoms or molecules capable of emitting in the flame is thereby reduced, and the calcium emission drops. Since elements differ greatly in their tendency to form nonvolatile complexes with aluminum, phosphorus, etc., and the complexes differ in volatility, we must regard this effect, too, as a specific interference (II.B.1).

The magnitude of the various interferences depends obviously not only on the kind and concentration of analyte and interferent, but also on the presence of other concomitants and the instrumental conditions. Thus, transport interference depends on air pressure or flow rate, the form and arrangement of the orifices, the shape of the spray chamber, the temperature of the air and the sample, etc. Ionization interference depends also on flame temperature, the rate of sample flow into the flame, the concentration of free flame electrons, etc. These factors depend in turn on the quantitative and qualitative composition of the gas–oxidant mixture, the height of observation in the flame, the form of the burner, etc. Quantitative data on interferences, therefore, cannot be transferred directly from one instrument to another. From this it is clear that the most uniform and reproducible types of equipment are requisite if the results of one author are to be taken over by others without repeated preliminary work (cf. Sections 19 and 83). As yet, unfortunately, there has been no evidence of standardization of the components of flame photometers.

Up to this point we have classified and discussed the interferences in terms of their mechanisms. We shall now offer a classification of interferences according to the kind of interferent, reviewing in succession the typical interferences due to common salts, acids, cations, anions, and organic materials.

In mutual *cation interferences* both partners (interferent and analyte) can be alkali metals. In hot flames a positive ionization interference (II.B.3) of this kind is likely (Section 97). This interference can properly be called mutual (the term "mutual interference" has been commonly misused for unidirectional interferences). Cation interferences of all kinds, not alone of the alkali metals, very often take the form of cross-sensitivity (I.A) or background interferences (I.B.2) (Sections 94 and 95); many metals, such as nickel, form compounds in the flame which emit bands over an extended spectral range. Other metals, such as aluminum, which themselves emit little light, can cause a negative cation interference (a depression of emission) by forming complexes of low volatility with the analyte (II.B.1). The alkaline earths are especially sensitive to this interference (Section 98), which characteristically becomes noticeable at relatively low concentrations of the interferent.

Salts, acids, or *organic substances* can alter the physical properties of the solution (viscosity, surface tension, density, vapor pressure) and thereby cause a transport interference (II.A.1) (Section 101), which can be accompanied by changes in the flame temperature, the rate of evaporation of solvent and solute, and the form of the flame. The change of flame temperature, which can be particularly marked with atomizer-burners, in turn alters the flame background (I.B.1) and also affects the emission of the analyte (excitation interference, II.B.4). With indirect atomizers, in which the flow rate of solution ordinarily has little effect on the flame, transport interferences are dominant; these are noticeable usually only with fairly high concentrations of interferent, and it is commonly found that the amount of interference per mole of interferent is greater at low concentrations of interferent than at higher concentrations. The interference curve showing emission vs. interferent concentration then has steeper slope at the low end. Interferences due to differences of surface tension, vapor pressure, etc., have been found also with direct atomizers, although to a lesser extent.[182,232,241,563] These observations contradict the statement, occasionally met, that transport interference cannot occur in a direct atomizer.

Interference due to changes in the physical properties of the solution can be positive or negative. Alcohol, for instance, raises the vapor pressure and reduces the surface tension, and thus raises the emission (Section 79). Acids and salts, on the other hand, which lower the vapor pressure or raise the surface tension or both, diminish the emission.

Organic substances, aside from their above-mentioned effects on the properties of the solution, can interfere in another way. Their combustion affects the flame temperature and the concentrations of carbon monoxide, carbon dioxide, and water in the flame (Section 79). The flame background is thus altered (I.B.1), and specific interferences, especially excitation interference (II.B.4), can also arise. The processes in the flame become still more complicated when alcohol, for example, enters not as vapor alone but also as liquid droplets. Around each drop a little diffusion flame forms through evaporation and combustion of the alcohol–water mixture, and the resulting conditions (temperature, emission, etc.) can no longer be regarded as in equilibrium.

Anions can interfere not only as components of salts or acids (see above), but sometimes also in their own right. Thus, for instance, the chloride ion causes a dissociation interference (II.B.2) when the analyte, especially an alkali metal, forms chloride to an extent that depends on the atomic chlorine concentration in the flame. Part of the analyte is then withdrawn from emission. Chlorine can also affect the ionization through the formation of negative chloride ions in the flame. Their concentration affects the concentration of free electrons and hence the degree of ionization of the analyte.[233]

Other anions such as phosphate and sulfate can cause vaporization interference (II.B.1) with alkaline-earth elements through the formation of nonvolatile alkaline-earth (pyro)phosphates or sulfates in the aerosol particles (Section 98). When phosphate is present in large excess, it generates also a continuous background emission due to unvaporized incandescent particles of P_2O_5 or other phosphorus compound, causing background interference (I.B.2). Nitrate, on the other hand, causes little interference. For this reason many analysts prefer to convert the sample to nitrate before analysis.[233,289] Incidentally, not all anion interferences have yet been explained. Thus, nothing is yet known of the cause of the perchloric acid effect on calcium[83,228] or of the enhancement of potassium by ammonium chloride and carbonate.[289]

Absorption flame photometry (Section 8) has the advantage over emission flame photometry of being immune to blank interferences (I) and excitation interferences (II.B.4). But the other interferences discussed above affect both emission and absorption flame photometry.

It is not always easy in practice to decide which of the interferences classified above is active, especially in working with a simple filter photometer. A few suggestions for distinguishing them may be useful. Blank interferences (I) are most easily recognized by means of blank solutions containing the interferents in various concentrations but not the analyte (Sections 94 and 95). The ranges of interferent concentration must correspond to those found in the samples.

If a radiation interference (II) is present, the following method will enable one to decide whether the effect is one of altered temperature or altered concentration of available analyte. The interference is measured for lines of different excitation potential. If the effect, expressed as a percentage change in apparent concentration, is the same for all lines, then a nonspecific interference (II.A) is present; otherwise a specific interference (II.B) is at work. The apparent concentration is the concentration that corresponds to the emission in the presence of the interference, as found from a working curve representing emission free from interference, constructed from standards containing no interferent. Nonspecific interference can be tested for also by measuring the interference at two concentrations of the analyte, at which self-absorption is negligible and strong, respectively. Further, the interference should be observed with different metals; if it is nonspecific, the percentage change in apparent concentration will be the same in every case. In this way, for example, the effect of hydrochloric acid on the transport of sample can be distinguished from its effect on dissociation. With indirect atomization, interference of type II.B.1 can be detected by means of two parallel atomizers as described in Section 98, interferent and analyte being sprayed simultaneously but separately into the flame by the two atomizers.

Other such tests for differentiating interferences can be given; they have, as emphasized at the start of Section 5, more than academic interest. The identification of interference mechanisms suggests methods for their most expeditious elimination.

CHAPTER 3

APPARATUS AND METHODS OF MEASURMENT

12. *Survey*

The success of flame analysis depends upon proper application of the basic principles and upon the sensitivity and reliability of the apparatus. Apparatus and method cannot be considered separately since they are intimately related. In the following sections we therefore discuss aspects of the method and apparatus together and present the various types of instrument. Equipment for absorption flame photometry will not be dealt with here; see Section 8.

There are a great many commercial makes of flame photometer. All those known to us, together with their chief characteristics, are listed in the Appendix. In this chapter we have selected as examples a few representative makes from among them. These commercial flame photometers are intended for a wide range of application in many fields. But even the best commercial instruments are not always adequate in special applications, and it is not surprising that many methods and special instruments have been devised for particular tasks. Some of these special methods and instruments will also be described, though more briefly.

Among the divisions of this subject, we take up first the controlling and monitoring devices needed for maintaining the flame, and the devices for introducing the material to be analyzed into the flame, since basically any of these can be used in conjunction with any optical–electrical system. The dispersion of the light into its spectral components and their measurement can be carried out in various ways. According to the method of dispersion and measurement we distinguish *flame photometry*, *flame spectrophotometry*, and *flame spectrography*.

We survey first the over-all construction of a flame photometer, before proceeding to details of the separate functions. The basic operation of the several parts is shown in Fig. 8. The parts are ordered in the sequence followed by the gas or air stream. Thus, beginning with the pressure tanks, we proceed through the regulating and metering system to the flame. Our later detailed discussion will follow the same sequence (Sections 16–35). In the optical–electrical system the flame comes first and the indicating meter last. Certain departures from this sequence are possible, as will be indicated in due course.

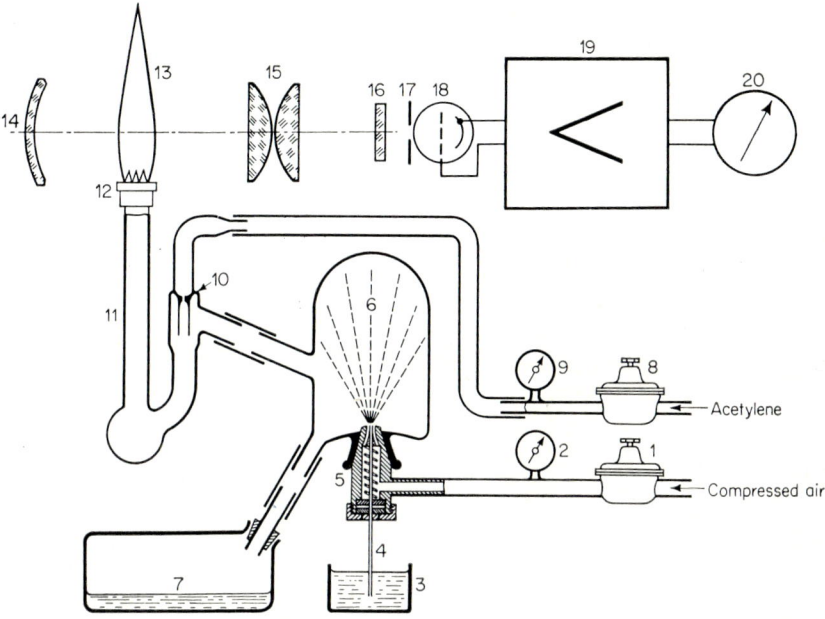

Fig. 8. Example of the design of a filter flame photometer (Model PF 5 of C. Zeiss, Oberkochen). 1, Second low-pressure regulator for the compressed air; 2, air pressure gage (Bourdon tube); 3, sample in its container; 4, suction capillary; 5, atomizer; 6, spray chamber for separating drops; 7, waste container; 8, second low-pressure regulator for the acetylene; 9, acetylene pressure gage (Bourdon tube); 10, gas mixing jet; 11, burner tube; 12, burner port; 13, flame; 14, backing mirror; 15, condenser; 16, filter, exchangeable for various spectral lines or bands or elements; 17, diaphragm; 18, photomultiplier, shown as a phototube; 19, amplifier; 20, indicating meter.

Control of the Flame and Introduction of the Sample

THE GASES

13. *The Choice of Gases*

Various flames can be used for flame photometry—air–methane, air–illuminating gas, air–propane, oxygen–propane, air–gasoline, air–acetylene, oxyacetylene, oxyhydrogen oxygen–cyanogen, etc. The choice will depend on the task at hand and on local circumstances. Commercial instruments, however, are designed usually for only one gas mixture, less often for two or more mixtures. The choice of mixture is then quite restricted and it will seldom be possible to work under really optimal conditions. But if one is free to choose one of the many commercial instruments with regard for the type of gas mixture or flame and its suitability for a particular application, or if one is to build or modify one's own instrument, then the following discussion is to the point.

Various matters govern the choice:

1. The flame temperature or conditions of excitation (Sections 8 and 14).
2. Emission of the flame itself in the spectral region of interest (Section 9).
3. Constancy of composition.
4. Controllability of pressure and flow rate.
5. Hazardous nature of the mixture.
6. The availability of the gas.
7. The cost per analysis or per hour.

We consider these now further:

1. The chief points regarding flame temperature are discussed in the next section. If a specific gas mixture has been decided upon, the temperature can still be varied to some extent by varying the gas/air or gas/oxygen ratio (Section 19).
2. Especially in working close to the detection limit one will try to maximize the ratio of net intensity of the analytical line or band to intensity of the flame background. For example, this ratio is usually about 10 times as high for oxyhydrogen as for oxyacetylene, despite the higher temperature and consequently better excitation of lines in oxyacetylene; this is due to the fact that the flame background of oxyacetylene is usually more than 10 times as intense as that of the much paler oxyhydrogen flame. But in working at higher concentrations, this matter is less important. If the available instrument has only moderate photometric sensitivity, oxyacetylene will be preferred for its greater line intensity.
3. Constancy of composition cannot be guaranteed with city gas. Even with bottled gas the composition can vary among tanks and also with the tank pressure (Section 15).
4. Good control demands a sufficiently high pressure ahead of the regulator. City gas is poorer in this respect than compressed gas. At all events, pressure settings can be found for which small fluctuations of

pressure hardly affect the readings (Section 19). Residual fluctuations can be reduced further by internal standardization (Section 51).

5. The hazard of a gaseous mixture will enter into the choice especially when routine analyses are to be conducted by technicians. Acetylene, happily, is promptly detectable even at low concentrations by its characteristic odor (due, however, to impurities) in case it escapes from leaky tubing or valves. Hydrogen, on the contrary, is odorless. However, it has the advantage of rising owing to its low density, so that an open skylight affords some protection against explosions in the room.

Other gases, such as butane or propane, have the disadvantage of forming layers on the ground owing to their greater-than-air density (see Table 3 in Section 14). Such gases should therefore not be used in basement laboratories lacking vents for them. For other matters reference should be made to published instructions for working with compressed gases and to Section 17.

6. If one of the gases is obtainable only with great difficulty, this limits the choice. For instance, cyanogen is not at present available commercially.* Sufficiently pure acetylene may not always be easily obtainable. We return to this in Section 15.

7. The cost of the fuel gas, reckoned per analysis, is generally slight. Sometimes the cost of oxygen is more critical. In any case, it is abundantly compensated by the saving in time achieved with flame analysis, and the reagents needed for chemical analysis are also saved.

It would be advantageous if all flame photometers could be quickly converted for different gas mixtures by interchange of various parts, such as reducing valves (Section 18), gas mixing valves (Section 30), or burner or burner top (Section 32), as this would allow optimization of conditions for each application. In principle this is possible. Unfortunately, few instrument manufacturers incorporate such flexibility.

14. *The Flame Temperature*

We discussed flame temperature briefly in Section 5; we shall now consider the optimization of temperature for practical analysis. We take up first the factors that govern the temperature of the flame.

First of all, the temperature depends on the type of gas mixture and on the mixing ratio of the components. As a function of this ratio, the temperature would be expected to have a maximum at the stoichiometric composition, falling off on either side. Measurements of flame

---

* See footnote, p. 77.

temperature confirm this qualitatively; Fig. 9 shows an example. But the flame temperature calculated from the composition and initial temperature of the gases usually departs significantly from the measured value, even with allowance for uncertainties in the physicochemical

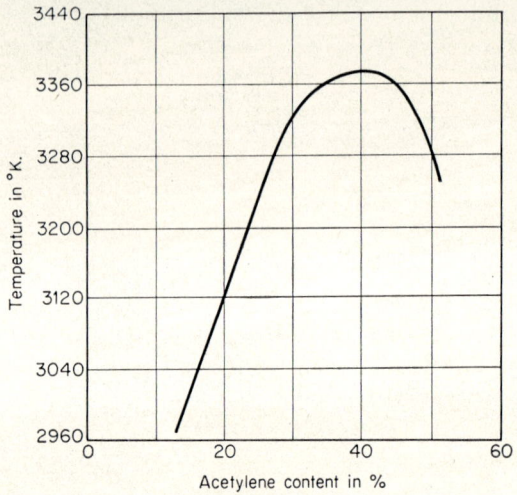

Fig. 9. Temperature of an oxyacetylene flame as a function of the acetylene content of the mixture (from ref. 31).

constants employed (heat of combustion, etc.). The flame temperature depends, in fact, also upon the following secondary factors:

1. Immediately above the reaction zone chemical equilibrium has probably not been attained. This is especially likely with the H and OH radicals, which recombine very sluggishly (Section 7). These relatively slow reactions liberate additional heat higher in the flame. On the other hand, a delay in O_2 dissociation can raise the temperature just above the reaction zone.[194]

2. Since the flames used in flame photometry burn in air at room temperature, heat is lost by convection and conduction as well as radiation. Heat is lost also by conduction through the burner material; this loss depends of course on the design of the burner, but it is partly recovered by the flame, for the entering gas mixture is prewarmed in the hot burner tube.*

* In a Méker burner the heat transfer to the burner has been measured at 10 kcal. per mole of acetylene. The temperature of the gas mixture thus warmed was 60° just above the burner port.[65] About half the lost heat is thus restored to the flame.

§ 14]   THE FLAME TEMPERATURE   73

3. The above losses are opposed by the heat of reaction liberated in the secondary combustion zone (Section 5).

4. The temperature actually attained depends further upon the efficiency of the prior or subsequent mixing of the air or oxygen with the fuel, the moisture content of the gases, and the quantity of air entrained from the surroundings for secondary combustion.

5. Especially in atomizer-burners, the evaporation of the entering liquid and the heating of the vapor can cool the flame considerably.[228] This effect is much smaller with indirect atomizers.[65] If the atomized liquid contains combustible substances such as acetone or alcohol, the flame temperature can be raised further, provided an excess of air is present, so that the mixture lies on the left side of the curve of Fig. 9 (see also Section 79).

6. The partial pressure of acetone in acetylene, which rises as the tank is depleted (Section 15), affects the flame temperature under otherwise constant conditions of pressure, etc. In one instance a reduction of 55° in the flame temperature was measured.[65]

It is obvious from these remarks on the processes in the flame that the temperature depends on the location in the flame at which it is measured, especially the height above the burner and the lateral distance from the axis of the flame. In the laminar, stably burning flame above a Méker burner this dependence is less pronounced than in the welding-torch flame (Fig. 10), which is more like the flame of an atomizer-burner.

These considerations all point up the difficulty of theoretically calculating the temperature of a given flame. Flame temperatures not only have academic interest, but can be of great help in interpreting many of the interferences (Sections 92 ff.) that are common in analytical flame photometry. Experimental methods are accordingly better suited for providing temperature data. We refer to the line-reversal method,[11,25,31,37,338,371] generally considered valid; we lack space for discussing it here.

Of greater practical interest is the question of optimizing the flame temperature for a given analytical task. We first recapitulate the temperature-dependent processes that are important for flame photometry:

1. The excitation of a spectral line depends on the temperature through the Boltzmann factor, $\exp(-E_a/kT)$ (Section 8). For example, this factor raises the sodium emission about 10% if the temperature rises 1%. But the intensity of a line of higher excitation energy E_a rises with temperature faster than this. In general, the intensity increases (E_a/kT) percent for a 1% temperature rise.

2. A hotter flame better vaporizes the sample, especially when involatile compounds such as calcium aluminates or phosphates form (Section 98).

3. Higher temperature promotes dissociation of non-emitting compounds such as alkali hydroxides or alkali halides (Section 7), thus raising the proportion of atoms excitable in the flame.

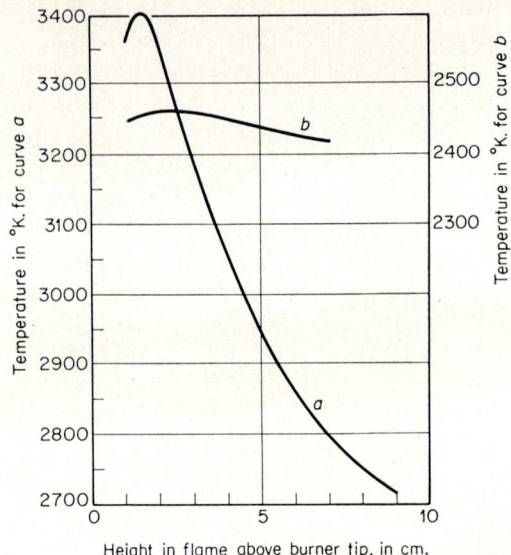

Fig. 10. Dependence of temperature on height in an oxyacetylene welding-torch flame (curve *a*, from ref. 338) and an air–acetylene Méker burner (curve *b*, from unpublished measurements by G. R. D. Zyderveld, Utrecht).

These factors point to the use of the hottest possible flame. But the following factors contra-indicate this, suggesting use of the coolest possible flame.

4. From 1 it follows that with rising temperature the weak, less easily excited lines will intensify relatively faster than the stronger lines. This makes the spectrum richer in lines and the relative intensity differences among the lines diminish. Hence, with multicomponent samples a high flame temperature makes it harder, perhaps impossible, to separate the analytical line from adjacent interfering lines, especially when the available spectral equipment is simple, as in the case of color filters.

5. Ionization increases with temperature, other things being equal. This weakens the line emission of an ionizable element and makes the

working curve undesirably concave (Section 82). Besides, the troublesome mutual interference between ionizable elements is enhanced (Sections 7, 10, 97).

6. Owing to thermal expansion, the burned gases are less dense at higher temperature. Other things being equal, this entails a diminution of radiation density and, for constant flame thickness, a loss of emission. However, the expansion is effective not only vertically, increasing the rise rate, but also laterally, thickening the flame somewhat. The loss of emission is thereby compensated partly but not completely, since the photodetector is benefited only by the lateral expansion. A loss of intensity is therefore inevitable. But this effect and the effect of temperature on self-absorption,[65] not further considered here, are much less important for flame photometry than effect No. 1 above.

7. Gas mixtures yielding hotter flames are more explosive than those that burn cooler, and therefore demand more care in handling.

Thus, there exists an optimal temperature or temperature range, depending, of course, on the nature of the application.[263,513,575] With low line/background ratios, the temperature dependence of the background also plays a part; the background depends not merely on the temperature but on the composition of the flame gases as well.

If ionization alone is considered, it is easily seen that the emission–temperature curve must have a maximum. At lower temperatures where ionization can be neglected, the emission rises as $\exp(-E_a/kT)$ with rising temperature (Section 7). But ionization supervenes at higher temperatures and weakens the emission. At very high temperatures where ionization is nearly complete, the concentration of neutral atoms of an element varies with temperature, by the Saha equation, approximately as $\exp(E_i/kT)$, where E_i is the ionization potential. The atomic emission then falls with rising temperature proportionally to $\exp[(E_i-E_a)/kT]$, for E_i exceeds the excitation energy E_a. Thus the emission–temperature curve rises at first, attains a maximum, and finally falls.

The optimal temperature should be established for any given case by trying different fuel gases and/or mixing ratios (cf. Section 19). The flame temperature can also be controlled by preheating the entering gases.[404] As a general rule we can state that potassium and sodium are best determined at a temperature of about 2100°K., most easily provided by a flame of city gas, propane, or butane with air[65,594] (Section 97).

A well-equipped laboratory should, accordingly, have several instruments offering different flames, or the equipment should be adaptable to flames of different temperatures. For practical reasons manufacturers usually do not incorporate means of conversion. Hence, the hotter flames with good spectroscopic instruments are naturally more versatile. But with these, one must often tolerate stronger interelement effects, less linear working curves, and greater difficulty of separating lines.

In Table 3 below, we list the flame temperatures and burning velocities (Section 5) of a few of the gas mixtures more commonly used in flame photometry; but the reservations made at the start of this section must be kept in mind.

In the determination of the less easily excited elements, the usual flames have the disadvantage of being limited substantially to 3000°C. These elements are, accordingly, detectable only at higher concentrations (Section 90). To circumvent this limitation, the following special flames or excitation sources might be employed, none of which, however, has yet found general practical application.

1. In the Langmuir atomic-hydrogen torch (3200–4100°C.), hydrogen is passed through an electric arc which dissociates the H_2 molecules to H atoms. These H atoms recombine above the arc, producing a flamelike phenomenon and liberating considerable heat.[37] Closely related to the Langmuir torch is the electronic torch, in which gas molecules are dissociated at a single electrode by the energy of a high-frequency or microwave discharge.[37,195]*

2. In the spark-in-flame method, a spark discharge through the flame supplies additional energy.[377,384,682]

3. The plasma jet is produced by a direct-current arc constricted by thermal (and magnetohydrodynamic) pinch effects, forming a flamelike stream with temperatures up to 15 000°C.[458a]†

4. A true chemical flame reaching temperatures of 4600°C. is the oxygen–cyanogen ("oxycyanogen") flame. The liquid sample, to be sure, cools the flame appreciably. Its suitability for flame photometry has been investigated in some detail.[84,85,85a,304,711] This flame might be expected to rival the arc, in yielding a very good line/background ratio.

---

* R. Mavrodineau and R. C. Hughes (International Conference on Spectroscopy, University of Maryland, June 19, 1962) have used a 2450-mc. discharge for exciting many elements introduced as spray, including Au, Be, Zn, Cd, Hg, Si, Ge, Sn, Ti, Zr, Th, P, As, Sb, Bi, V, Nb, Ta, S, Se, Te, Mo, W, I, Os, Ir, Pt, the rare earths, and other elements more easily excited in flames.—*Translator.*

† Plasma jet sources for analyzing solutions are now available from Spex Industries, Inc., Scotch Plains, N.J., and from National Spectrographic Laboratories, Cleveland, Ohio.—*Translator.*

It can find wide use only if cyanogen is made available commercially.*
Owing to the high toxicity of cyanogen, carefully tested safety precautions would be needed, such as valves that close automatically in case a limiting concentration is exceeded in the room, first-aid supplies in case of poisoning, etc. Chemical flames of even higher temperature can be obtained by substituting ozone for oxygen (with cyanogen) or by raising the pressure. An increase of pressure from 1 to 40 atm. raises the flame temperature by about 500°C. For further information on these very hot chemical flames see the literature.[307a, 470]

TABLE 3

Gas mixture	Formula of the fuel	Density of fuel,[a] kg./m.^{3}	Density relative to air	Temperature,[b] °C., from ref. 31	Burning velocity,[c] cm./sec., from ref. 37
Methane–air	CH_4	0.72	0.55	1875	70
n-Butane–air	C_4H_{10}	2.70	2.09	1895	83
City gas–air	—[d]			1918	ca. 55
Propane–air	C_3H_8	2.00	1.55	1925	82
Acetylene–air	C_2H_2	1.17	0.91	2325	266
Acetylene–oxygen	C_2H_2	1.17	0.91	3137	2480
Hydrogen–oxygen	H_2	0.09	0.07	2660	3680

[a] From ref. 3 at 0°C. and 760 mm.
[b] The maximal flame temperature is given, as measured by line reversal with sodium vapor. The fuel gas is assumed to be dry.
[c] The maximal burning velocity for laminar flow is given.
[d] Composition 49.8% H_2, 11.8% CO, 25.8% CH_4, 1.5% C_2H_6 and incombustibles.

15. Acetylene and Propane

Acetylene. The gas most commonly used in flame photometry is acetylene. It cannot be handled commercially in highly compressed form as are hydrogen, oxygen, and nitrogen (at 2200 p.s.i.), for it tends to decompose and then explode. Instead, acetylene tanks are filled with a porous mixture of kieselguhr, wood charcoal, and a binder

* Cyanogen is now available from The Matheson Co., Inc., East Rutherford, New Jersey.—*Translator.*

(Fig. 11) or with a similar filler, which is soaked with acetone. The filler keeps the acetone in place for any position of the tank and inhibits explosion in case the flame flashes back. It occupies about 25% of the volume of the tank, the acetone 38%; 29% is reserved for expansion when the acetylene dissolves in the acetone, and 8% is reserve space for

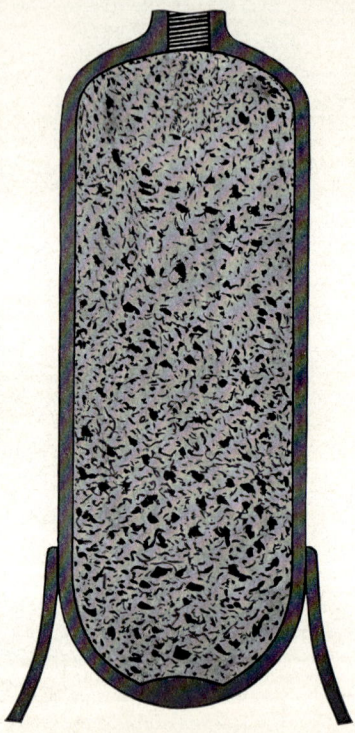

Fig. 11. Cross-section of an acetylene tank, showing the porous filler for holding the acetone.

thermal expansion of the liquid. Acetone can dissolve large quantities of acetylene; one liter of acetone at the normal filling pressure of 230 p.s.i. and 15°C. dissolves about 400 l. of acetylene. A commercial steel tank with filler and acetone content of about $1\frac{1}{4}$ ft.³ will hold about 300 ft.³ of acetylene under these conditions. The tank pressure is strongly dependent upon the temperature. In warm weather (35°C.) a full tank can rise to 360 p.s.i., and the pressure drops in the cold. The pressure is therefore no unique indication of the acetylene content, which is instead determined by weighing the tank with filler and acetone before and after filling. Acetylene is accordingly sold by weight. A tank of 300 ft.^{3} nominal capacity contains about 20 lb. of acetylene.

Unlike acetylene prepared in gas generators from calcium carbide and water, bottled acetylene is free from harmful impurities such as phosphine, hydrogen sulfide, ammonia, and moisture. This is important because the attainable temperature is very dependent upon the content of moisture and other impurities. Further, the flame background radiation, especially in oxyacetylene, rises undesirably with the phosphine content. But even in bottled gas phosphine may not be sufficiently low. This can happen if the tank is filled with acetylene taken just before renewal of the depleted scavenger material in the scrubbing tower of the acetylene plant. Such contamination can be recognized by the milky appearance of the flame. It can be checked also by measuring the acetylene flame background with each change of tank, by taking a reading on a suitable blank solution (Section 95) with the same air or oxygen and acetylene pressures, the same filter (or wavelength and slit width in a monochromator), and the same electronic sensitivity. This will easily distinguish good from poor tank charges. Excessive phosphine in the acetylene can preclude sufficiently accurate measurement on dilute solutions near the detection limit. The following countermeasures can be taken:

1. If several tanks of acetylene are on hand, the purest is chosen.

2. A commercial charcoal filter can be installed in the acetylene connection. Some manufacturers (e.g., ref. 513) provide a filter as standard equipment. But it loses its adsorptive power rather quickly, since it traps also the acetone that distills over. To be sure, it can be regenerated by blowing out with compresssed air for a time, but this is tedious, and an inept procedure can lead to poisoning.

3. The tank can be ordered from the nearest acetylene plant with the request that it be filled shortly after the scrubbing tower is recharged; or else an analytical grade of acetylene can be procured.

4. One can purify the acetylene oneself by means of suitable chemicals in a tower.[37]

It is advisable to acquire outright at least one acetylene tank and preferably a spare. This will avoid tank rental (if any); further, old tanks supplied on loan, even when filled with a clean charge, may yield a more or less impure acetylene owing to residual impurities from previous filling retained by the acetone and released only slowly. It is possible to ensure that one's personal tanks are filled only with pure acetylene.

Acetylene tanks used for flame photometry must never be drained to exhaustion, since the partial pressure of acetone rises considerably as the tank is depleted and affects the flame temperature and excitation. The temperature can drop by as much as 50°.[65]

Propane. Propane[2] is also often used for flame photometry. It offers advantages over acetylene when a low-temperature flame is called for (Section 14). Propane is supplied commercially in steel bottles, usually in the form of liquefied petroleum gas (LPG). It is derived as a by-product both in petroleum refining and in the processing of coal products. Commercial propane consists of the more volatile fractions of casing-head gas, chiefly, indeed, of propane itself (70–80%), together with butane and minor amounts of a few other low-boiling hydrocarbons. These gases liquefy on compression and evaporate on discharge from the tank. Since the pressure within the tank is the vapor pressure of the liquid phase, it remains constant (aside from the effect of temperature) on depletion of the tank, so long as liquid remains. Hence the tank pressure is not a measure of the contents. The propane, having a somewhat lower boiling point, evaporates first, the butane later. This causes a slight shift of tank pressure, due to the difference in vapor pressures, as the propane evaporates, leaving the butane behind. Consequently, the gas pressure at the flame photometer will drift, but this effect shows up only with a nearly empty tank, which should be weighed to check the contents. The drift can be avoided by procuring technically pure propane (at least 95%), or (if the more easily available commercial propane is bought) by not using up the entire contents of the tank.

In common use are 35- or 100-lb. tanks of commercial propane. These tanks are not filled completely, to leave room for thermal expansion of the liquid. They are equipped also with a safety relief valve (older tanks have a safety diaphragm). Since propane is heavier than air, it readily forms layers on the floor and therefore requires certain precautions (Section 17). Such pools of gas can persist for some time, since spontaneous diffusion into the air is rather slow.

Propane has the following advantages over city gas:

1. Access to the gas mains is unnecessary; propane can be used in rural areas without a gas supply.
2. Its heating value per cubic meter is about 6 times (for butane, 8 times) as great as that of city gas. Hence, smaller volumes of gas are needed.
3. The condensibility of propane, along with item 2, makes for easy transportation and small space requirement.
4. Owing to the small explosive range and low burning velocity of air–propane mixtures, flashback into the burner is less likely than with city gas.

5. The tank pressure is relatively high (about 7 atm.), permitting use of regulators that provide a nearly constant flow of gas to the flame. But the pressure of city gas is very low and variable, and regulation is more difficult (Section 18).

But propane has some disadvantages:

1. Accidents can result from improper transport, storage, or use. But with proper handling, propane is not more dangerous than city gas, although a few special precautions are needed (Section 17); in particular, propane must not be used in basement laboratories.

2. There is the trouble of changing the tank after about every 200 hours of operation, and procuring a fresh tank.

16. *Oxygen and Compressed Air*

The oxygen or compressed air needed for the flame is often purchased in tanks. The commercial gases are generally pure enough for flame photometry. For reducing the high tank pressure to the constant working pressure at the atomizer, a regulating valve is best (Section 18). Oxygen tanks have to be changed fairly often, since oxygen consumption usually exceeds that of fuel. For premixed, laminar flames, compressed-air tanks have to be replaced about 5 times as often because of the low oxygen content of air. This inconvenience can be avoided with a compressor. But if the flame is to be operated in a protective atmosphere of purified air, to avoid contamination by impurities from the surroundings (dust particles cause flashes of light), the compressed-air requirement is about 3 times normal and about 15 times that of oxygen. Use of bottled air is then impractical, and a compressor will almost invariably be used.

The following requirements are imposed upon compressors used for operating flame photometers.

(*a*) *Safety.* The compressor must run as steadily as possible, without disturbances. This requirement is hardly attainable with a line-operated electric motor. If the power fails briefly, then in the absence of a ballast tank the flame receives no air, and in this event an acetylene flame produces an unpleasant soot. If someone is present to start the compressor immediately after such an interruption, or if it starts again by itself, no harm will have been done provided the burner cap is so designed that the flame cannot flash back on interruption of the compressed air. But failure of a compressor that cannot start itself again can have very unpleasant consequences if the flame photometer is untended, as during warm-up. The laboratory can be blackened by soot in a few minutes, or if the flame flashes back and keeps burning, the apparatus can be badly damaged. Hence, with a non-self-starting compressor, the acetylene or gas line should have a solenoid valve, which will shut off the fuel automatically in the event of power failure but will not open again on restoration of power. In the absence of such a safety feature, the instrument should not be left untended when the flame is burning. We should mention also the matter of testing the compressor, tank, etc., for tightness.

(b) *Pressure stability*. In some instruments the compressor generates just the pressure needed at the atomizer of the flame photometer. A reducing valve is then neither required nor permissible. But the delivered air pressure must be kept independent of line voltage fluctuations and the condition of the compressor; note that falling oil viscosity affects the running speed during warm-up. These requirements cannot be satisfied completely.

With larger compressors a higher pressure, 5–10 atm., is usually maintained automatically in a storage ballast tank by means of a pressure-activated switch, which turns the compressor motor on or off when the pressure falls below or exceeds previously set limits. A regulating valve then provides constant pressure at the atomizer jet. The ballast tank between compressor and flame photometer precludes the accidents mentioned in paragraph (a).

(c) *Temperature stability*. The temperature of the compressed air used for atomizing has a large effect on the evaporation of droplets in the spray chamber and hence on the efficiency of atomization, as described in Sections 6 and 26. Hence, the air delivered by the compressor must be of constant temperature during an analysis. This requires the air consumed by the compressor to be of uniform temperature. Thus, for instance, a window next to the compressor, especially in winter, should not be opened during an analysis.

(d) *Air purity*. The compressed air supplied to the flame photometer must be sufficiently clean. The chief impurities are likely to be oil from the compressor and water from the air, which is partly condensed on compression of the air, as well as dust in the air consumed by the compressor. Diaphragm compressors deliver an oil-free air, but not most other types. Compressors of every type, moreover, discharge more or less water into the compressed-air line, and the air contains the dust drawn into the compressor. If the air line is improperly arranged, it can be blocked completely by water. Means for removing the oil, water and dust are therefore essential components of a satisfactory compressor unit for a flame photometer. The various types of separators and filters will not be discussed here. Oil and water are generally separated by an installation in the air line offering a tortuous passage; at each bend the entrained drops are separated by centrifugal and other forces. The removal of dust is more difficult; filters of porous clay, fritted glass, etc., are often used. Such filters can usually be regenerated by washing with a solvent such as gasoline or benzene. The filter must not shed interfering material, as may happen with cotton or glass wool. Electrical purification of the air has been recommended.[152, 527]

(e) *Compressor noise.* Some kinds of compressor are very noisy, others much less so. The noise can be abated by a suitable muffler and proper mounting. We return to this in Section 76.

17. The Use of Compressed Gases

For the prevention of accidents in working with compressed gases, safety regulations have been issued in Germany.[42, 60, 432, 705] The most important aspects of these are summarized below.

(a) *Transport.* For the distribution of compressed gases, steel tanks must be used which comply with certain specifications. These containers must be submitted by the proprietor at regular intervals for inspection by the board of control. The tanks must never be thrown, dropped, or rolled during transportation, whether empty or full (since they are reused). They are transported in special carts or carrying devices, in which the tanks are strapped.

(b) *Storage.* Juxtaposition of many tanks in a battery or storage close to inflammable materials is inadvisable. Large propane tanks (over 25 lb.) and compressed gases denser than air may not be kept in a closed room, such as a basement, without adequate floor ventilation. Ventilation for propane demands a hatch at least 10×10 in., always open and above the ground. Tanks are damaged by hard frost below $-10°$C. and by heating above $40°$C.; they should not be stored close to heaters or in direct sunlight. Acetylene tanks should stand upright, so that the acetone in them cannot reach the valve. Tanks should be fastened by clamps or chains so that they cannot fall over. They should be kept from unauthorized persons and children, and should therefore not be kept in stairwells or halls.

(c) *Installation.* The above remarks apply also to installation. In addition, the tank should not be placed directly next to the flame photometer; it should be at a safe distance from the flame of the flame photometer and from any other flame, such as a bunsen burner in the laboratory. However, the tank should be in the same room as the flame photometer so that the pressure gage can be watched (see the next two paragraphs) and the main valve can be quickly closed in case of trouble.

(*d*) *Connection.* The protective cap should not be removed until the tank is put into service. The gasket in the connecting fitting should be checked to see that it is of suitable material, undamaged, and correctly positioned. The gasket should be of fat-free fiber only, never of rubber, leather, or the like, since these explode on contact with compressed oxygen. For the same reason screw-valves on tanks, reducing or regulating valves, and the ends of tubing must never be oiled, greased, or wetted with glycerine. Next, the tank valve should be opened slowly for a moment (about 1 sec.), to blow any dirt particles out of the vent. Such particles might otherwise get into the regulating valves and cause trouble or, with compressed oxygen, an explosion. For the same reason the tubes should be blown out when the tank is first connected. The tank valve is then closed again.

(*e*) *Connection of the high-pressure regulator.* One should be sure that the correct pressure regulator is at hand. For instance, an acetylene regulator must never be used on an oxygen tank. To prevent interchange, the connections of the various types of regulator are provided with different screw threads. They are distinguished also by color (oxygen, blue; acetylene, yellow; fuel gas, red; compressed air, gray). A regulator that does not fit must never be connected via an adapter, and high-pressure valves or fittings must never be interchanged with low-pressure ones. This can happen easily with low-pressure valves that have hose nozzles on both sides. Regarding the gaskets, see the preceding paragraph. Before connecting the regulator, fully unscrew the knob RS (Fig. 13, page 91) to unload the diaphragm, and close the low-pressure shut-off valve. Then slowly open the tank valve and note the pressure reading on the gage (to check the fullness of the tank). The tank valve should always be opened slowly, to prevent ignition of the regulator diaphragm by oxygen, or expulsion of acetone with acetylene or of liquid fuel with propane. Next close the tank valve again and watch the high-pressure gage. If the reading drops, the regulator or its connection is leaky.

(*f*) *Detection and elimination of leaks.* Leaks can be quickly located by painting the suspected area with soapy water. They should never be hunted with an open flame. They can usually be repaired by tightening a nut or a hose clamp (see below). If the tank valve itself leaks at the packing gland, it usually suffices to turn the handle all the way out; the valve spindle then compresses the packing. If this does not work, the tank should be replaced by one with a tight valve packing. The leaky tank should be marked. It is dangerous to work on the tank valve with tools; too much force can break the valve.

(*g*) *Hose connections.* The high-pressure regulator reduces the tank pressure to about 3 to 5 atm. gage. High-pressure hose of suitable material with cloth insert may then be connected to the regulator. Only with pressures below about 0.5 atm., as usually obtained with a second low-pressure regulator, can ordinary gas tubing of rubber or plastic be used. To prevent interchange, the high-pressure tubing is coded by color (oxygen, green; fuel, red). It can be connected only with the proper fittings and nuts. Their thread must fit that of the regulator; different threads are used for different gases. Left-handed threads are marked with a notch on the nut. The ends of the high-pressure tubing must be secured to the fittings with hose clamps. Pure copper cannot be used in tubing for acetylene, but steel tubing and low-copper alloys such as brass with less than 17% copper are suitable. The tubing is tested for leaks as above. If the tubing is damaged, it should be replaced, never patched with electrical tape, rubber cement, etc. If needle valves or other parts of the low-pressure connections require lubrication, only explosion-proof (silicone) greases should be employed.

(*h*) *Connection of the low-pressure regulator.* This regulator is generally placed in the flame photometer itself next to the pressure gage, as in Fig. 29 (page 136). The same remarks as above, where pertinent, apply to the connection and leak-testing of the low-pressure regulator.

(*i*) *Adjustment of the gas flow.* After all parts are duly connected and checked for leaks, the oxygen and gas are adjusted as follows (note the sequence: oxygen or compressed air first, then gas). (1) Open the tank valve. (2) Check tank pressure, change the tank if necessary. (3) Adjust the intermediate pressure between high- and low-pressure regulators by means of the high-pressure regulator. This adjustment is often prescribed in the instructions supplied by the instrument manufacturer. (4) Set the low-pressure regulator to the working pressure at the atomizer or gas jet. This adjustment will be discussed in Section 19. Leaks developing subsequently can be recognized by increased gas consumption, contraction of the flame, or sometimes smell. A hydrogen leak cannot be smelled; frequent tests with soapy water are advisable. An open transom offers some protection against explosions from gradual escape of hydrogen.

(*j*) *Shutting down.* Close the tank valves; be sure to turn off the fuel gas first, and then the oxygen or air. Close the regulator valves and then any low-pressure shut-off valves. The tank valve alone should not be relied upon, since it may leak when apparently closed.

(*k*) *Acetylene fires.* If escaping acetylene should catch fire at a leaky tank valve, regulator, or tube, the tank valve should be closed at once if possible. An acetylene fire away from the tank itself should be combated with sand or, better, a carbon dioxide fire extinguisher or dry powder extinguisher; foam extinguishers are unsuitable. The chief difficulty in extinguishing such a fire occurs when the quenching material or foam is blown away by the escaping acetylene. In this case it is better to use a denser material like sand, which smothers the fire, or a quenching agent like carbon dioxide, a sharp jet of which will blow the flame out.

It may rarely happen that in an acetylene fire away from the tank, autoignition sets in and propagates to the interior of the tank. The tank does not then explode at once, but there may be a gradual increase of temperature and pressure, lasting sometimes for hours, which finally bursts the tank. The hazardous tank should be removed promptly and chilled strongly, as by throwing it into a pond or river.

GAS REGULATION

18. *Control and Monitoring of the Oxidant and Fuel Flow to the Flame*

A reproducible and steady flame is a prerequisite for quantitative flame photometry. To this end the flow of fuel gas and of oxygen or air to the flame must be kept as uniform as possible. The needed precision of control, i.e., the greatest permissible fluctuation of flow, depends essentially on the analyte, its concentration, the chosen spectral line (Section 82), and a few other factors such as the type of flame, the burner design, height and width of the observed part of the flame. Usually the flow will have to be controlled to better than 1%. If unfavorable circumstances require operation on the side of an emission–gas pressure curve (Figs. 14, 15) and an accurate analysis is wanted, the precision of control will have to be even better. To achieve this control, automatic devices are necessary, which keep the gas or oxidant flow to the flame as constant as possible at a preset value, even while the pressure in the storage tank keeps declining as the gas is consumed or the pressure in the ballast tank of a compressor keeps changing. Gages are needed also, which indicate the gas and air flow or enable the optimal flow rates to be found. The method of finding these optima for fuel and oxidant will be taken up in Section 19. In the present section we discuss first the measurement and monitoring of flow and then its automatic regulation.

In principle any method of flow measurement is applicable; but the

§ 18]   CONTROL OF THE OXIDANT AND FUEL FLOW TO THE FLAME   87

device cannot be installed in the gas or air stream beyond the atomizer jet, for then all of the spray would have to pass through the flowmeter. A permanently installed flowmeter must therefore be placed ahead of the atomizer or jet, where an appreciably higher pressure prevails. But since the meter reading depends on the pressure, temperature, and type of gas, a special (and expensive) flowmeter is required, which is calibrated for the gas pressure employed. In practice such flowmeters are seldom used; instead, a measurement of pressure ahead of the atomizer or the gas-mixing jet usually suffices. The rate of flow of gas through a nozzle into the open atmosphere depends on the pressure in the nozzle (more exactly, the pressure difference across the nozzle as well as the barometric pressure). Hence, the measurement of pressure ahead of a nozzle provides an indirect flow measurement. But this method, although used almost exclusively in practice, has the following disadvantages:

1. Since nozzle diameters and lengths vary a great deal among different types of flame photometer and sometimes even among instruments of the same make, especially with glass nozzles, the optimal pressures for a given application found by different authors cannot be intercompared. Yet uniformity in this regard would be very helpful for arriving at simple and uniform procedures and for exchange of experience among analysts. The data on pressure settings published by most authors are useless; the equivalent flow rates, determined by an auxiliary calibration, should be given (see below).

2. The flow of gas or air to the flame remains constant under constant pressure at the point of constriction only when the properties of the constriction or nozzle, the downstream resistance to flow, and the properties of the flowing gas (composition, moisture, temperature) remain constant. But these conditions cannot always be maintained in practice. Nozzles can become partially blocked or corroded; the flow resistance beyond the nozzle may be changed by deposits of sample material in the orifices of a Méker burner top, or by a kink in the tube between spray chamber and burner acquired during cleaning; and the composition of acetylene and propane (usually propane–butane mixtures) can change. The flow rate of the gas leaving the fuel-mixing jet (Fig. 8, page 69) depends also on the pressure in the burner tube where the gas enters, and this pressure in turn depends markedly on the air pressure in use and its fluctuations. The air flow thus indirectly influences the gas flow.

Some of these difficulties in the use of pressure to measure flow rate can be alleviated as follows.

1. Standard nozzles, well polished inside and with close dimensional

tolerances, might be used. (Unfortunately, standard specifications do not yet exist.) The pressures determined as optimal for such a nozzle will have general validity. But some of the above difficulties, such as the possibility of clogging, still remain.

2. The gage is calibrated by preparing a curve of nozzle pressure vs. gas or air flow. Such curves are usually convex toward the flow axis.

Ordinary gas meters or other flowmeters can be used for calibration; they must be standardized for the gas in question as well as for the pressure range and temperature prevailing. The meter can be connected with a rubber tube to the open burner barrel (the cap of a Méker burner being removed). During calibration, of course, nothing is sprayed through the atomizer.

Instead of using a commercial gas meter or flowmeter, one can improvise a gasometer from laboratory equipment. The gas to be measured is passed through a rubber tube into an inverted graduated cylinder filled with water, with its top in a dish of water, so that as the gas bubbles into the cylinder it displaces the water. It should be kept in mind that some gases dissolve appreciably in water; in this case the water should be presaturated with the gas.

These expedients can be dispensed with by installing commercial flowmeters permanently in the connections ahead of the nozzles. But an adequate flowmeter can be expensive. However, a flowmeter is easily made from laboratory equipment as follows.

A capillary constriction is installed in the air line and the pressure drop across it is measured with a closed differential U-tube manometer. Proper choice of the diameter and length of the capillary yields a convenient reading on the manometer at the usual operating pressures and flow rates. This meter needs to be calibrated as above, the calibrating device being attached to the (atomizer) nozzle. Even these calibration curves are not linear. The closed U-tube manometer has the advantage over an open-end manometer used as a pressure gage, that a change in the flow resistance beyond the manometer has less effect on the reading. But in other respects this method has the same drawbacks as the measurement of pressure at a constriction. Besides, the differential manometer is very sensitive to sudden increases of air or gas flow, such as may happen during cleaning under pressure or changing atomizers, on development of a hose leak, etc. To test for clogging or corrosion, the meter should be occasionally recalibrated.

As for the pressure gages employed in the usual method of measuring flow rate discussed above, only two types are used:

1. At low pressures, open U-tube manometers are used, filled with a

§ 18] CONTROL OF THE OXIDANT AND FUEL FLOW TO THE FLAME   89

barrier liquid such as colored water with a wetting agent, or mercury covered with a little alkaline water. These additives prevent the meniscus from sticking. The range and sensitivity can be varied by choice of the arm length of the U-tube or of the density of the liquid (oil, water, dibromoethane, mercury) or by tilting the tube.

2. For either low or high pressures commercial Bourdon-tube or diaphragm gages are often used. Their readings depend on temperature, however, unless they are specially compensated. The simplest such gages, moreover, show hysteresis; the reading at constant pressure depends on whether a higher or lower pressure prevailed earlier. These effects may cause errors.

Neither type of gage can tolerate an appreciably higher pressure than that corresponding to the top of its range. With the first type the barrier liquid is ejected from the open end, and with the second the bellows can be damaged.

Next, as to *control* of the gas and air flow to the flame. The situation is basically the same as for flow measurement; that is, for practical reasons it is the pressure at a constriction, not the flow rate itself, that is controlled. This method of flow control has the drawbacks listed above, which, however, are relatively less serious for control than for measurement when a suitable technique is employed.

Fluctuations of pressure at the constriction include fairly rapid changes—with a period of 5 sec. or less—and slower changes such as a gradual drift on depletion of the gas tank. The quicker fluctuations are more easily corrected and less important. They will be dealt with first.

Causes of rapid fluctuation of gas or air pressure include compression strokes of the compressor, oscillation of a regulator diaphragm (see below), and periodic pressure surges at nozzles, corners, etc., in the gas or air connections to the burner. This short-term unsteadiness is easily cured by installing a sufficiently large buffer flask in the gas or air line. The unsteadiness can be smoothed also at the meter of the instrument or in the amplifier by means of a higher time constant (resistance times capacitance; Sections 44, 46, and 65).

If a damping flask is to be installed in the gas or air line, the volume V required for stabilization can be found from the following relation:[65]

$$S_1 = (1+C_1^2\omega^2)^{1/2} \quad \text{where } C_1 = V(P_0-B)/BQ\alpha$$

Here S_1 is the stabilization factor; $1/S_1$ equals the fraction to which the amplitude of a periodic disturbance is reduced after passing the flask; $\omega = 2\pi/\tau = 2\pi\nu$ is the angular frequency (radians/sec.) of the disturbance of period τ and frequency ν; Q is the mean flow rate of gas or air to the flame, measured at the barometric pressure B; α characterizes the flow of air or gas in the connections between the buffer flask and open air in accordance with the relation $QR = (P_0 - B)^\alpha$, R being the flow resistance; P_0 is the average pressure in the flask. At low pressures, α is about 0.57, and at high pressures about 1.6. The following practical example will give an idea of the magnitudes involved. If α is taken as 1 on the average, and P_0 as 1 atm. gage, then $(P_0 - B)/B = 1$, and $C_1 = V/Q$. If the fluctuation has a frequency of 2 cycles per sec., $\omega = 12.6$. If the amplitude is to be reduced to $\frac{1}{10}$, $S_1 = 10$. For an air flow of $\frac{1}{6}$ l./sec. (10 l./min.), the buffer flask must have a volume of about $\frac{1}{8}$ l.

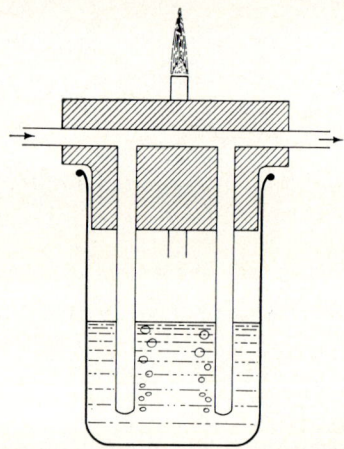

Fig. 12. Overflow regulator for maintaining constancy of gas pressure (from ref. 443).

Automatic maintenance of constancy of the gas or air flow (at the desired optimum) for an extended time is less easy. One method, hardly used any more, is to regulate the flow at the constriction by hand, as with a needle valve. Continual attention is necessary when the pressure is falling. Besides, the needle valve is easily blocked.

Another infrequently used method employs a relief valve that opens when the pressure exceeds a preset value and thus maintains automatic constancy at the constriction. This wastes air, of course, and makes noise. If this method is to be used for gas, it generally takes the form shown in Fig. 12. The gas after passing a throttle flows from the left through a tube, being taken off at the right for use. The two branch tubes dip into a beaker of water, closed at the top; in the cover is a vent.

The pressure of the incoming gas must exceed the working pressure. The excess gas bubbles through the water, and can be vented through a hose from the exit tube to an outlet; in the illustration the excess gas is burned above the vent. By varying the head of water at the lower ends of the immersed tubes, as by adding water, the pressure can be adjusted at the point of use (the burner of the flame photometer). A needle valve or the like between this regulator and the flame will provide continuous adjustability of the gas flow without the need of changing the water depth. It is then merely necessary to use higher pressure in the regulator ahead of the valve. This simple device, however, has disadvantages:

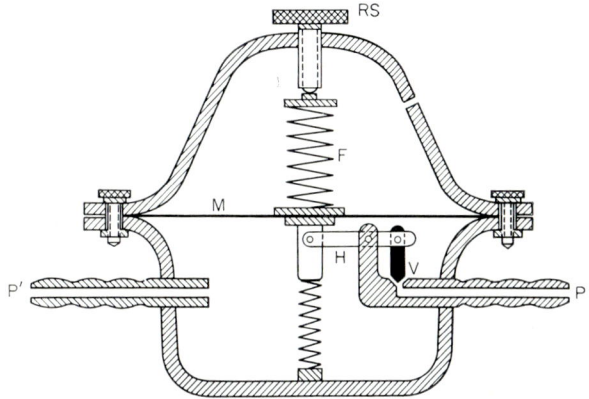

Fig. 13. Construction of a pressure regulator (schematic). M, Membrane; H, lever; V, valve; F, spring; RS, regulating screw; P, high entrance pressure; P′, low exit pressure.

(1) Gas is wasted in regulation. (2) If the pressure temporarily drops too low, the flame goes out, and gas escapes into the room. (3) The release of gas bubbles at the lower ends of the tubes causes small fluctuations of pressure, and hence flame flicker and an unsteady meter reading.

For automatic maintenance of constant pressure at the constriction, regulating valves have become increasingly popular. These are valves which automatically adjust their opening according to the difference between the ambient air pressure, P, and the working pressure, P', between the regulator and the throttle. This pressure difference, $P-P'$ (Fig. 13), is balanced by a membrane M. If $P-P'$ changes, the membrane bulges more or less to one side or the other and opens or closes the valve V via the lever H. The pressure difference is adjustable by loading the membrane with the spring F through the screw RS.

Regulating valves should be chosen with the largest permissible membrane diameter, so that the forces which displace the membrane and effect the adjustment may be large. If the working pressure is relatively low (below 15 p.s.i.), a small high-pressure membrane provides only small displacement forces for opening and closing the valve. A larger membrane is inadvisable for mechanical reasons with high fore-pressure in the tanks (for example, 2000 p.s.i. for bottled compressed air). In this case, two regulators in series are more effective, of which the first, at the tank, has a smaller and very stiff membrane (diameter about 10 cm.) and reduces the pressure from 2000 p.s.i. to 75 p.s.i., for example, while the second regulator, on the low-pressure side, with a larger and thinner membrane, takes the pressure further down to perhaps 3 p.s.i. Three regulators have sometimes been used in series. With high working pressures of about 75 p.s.i., we have found a single regulator sufficient, provided the membrane diameter is not too small.

For best regulation, the pressure difference should not be too small between the working pressure above the jet and the pressure in the tank or above the regulator. City gas has a very low and irregular pressure, and is therefore hard to regulate. Moreover, it sometimes changes composition.

A stopcock should be provided at a point just before entry of the compressed air or oxygen into the atomizer or of the gas into the mixing chamber or burner, to permit shutting down quickly in case of danger. The tank valves are ill suited for this, for the tanks should stand at least 10 ft. from the flame for reasons of safety, and cannot be reached as quickly. Moreover, closing these valves does not extinguish the flame at once, owing to hold-up of gas in the connections.

19. *Determination of Optimal Flow Rates of Fuel and Oxidant*

For a given instrument and sample, both the intensity and the steadiness of the radiation received by the photodetector depend upon the flow rates of fuel and oxidant. In most instruments these are freely adjustable within limits. The question thus arises of the best settings for these flow rates. The optimal flow rates depend, aside from factors to be considered later, upon the requirements of the application at hand. For instance, to attain the best detection sensitivity in trace analysis, the fuel and oxidant flow will be chosen to maximize the ratio of net line or band to background intensity while keeping the unsteadiness of the reading small. Again, in the analysis of complex solutions the

§ 19]   OPTIMAL FLOW RATES OF FUEL AND OXIDANT

flow rates may be chosen to minimize certain interferences. Or perhaps a solution of simple composition and ordinary concentration is to be analyzed with the highest possible precision; the flow rates must then be chosen to minimize unreproducibility.

But the flame photometer is generally supplied not with flowmeters but with pressure gages, so that it is less a question of finding the best fuel and oxidant flow rates as such (in l./min.) than the practical one of finding the best fuel and oxidant pressures. Published optimal pressures for a particular application cannot be safely adopted even for the same type of instrument. The flow rates depend not only on the pressures but also on the dimensions of the constrictions (nozzles) forming part of the equipment, as discussed in Section 18. Different instruments of the same make can differ greatly in nozzle properties. Indeed, the optimal pressures can change with time in one and the same instrument, if the constrictions become obstructed or corroded. Hence, the optimal values for a given instrument and application must inevitably be ascertained by experiment and, owing to the likelihood of corrosion, etc., checked occasionally. Some index of the optimal values is afforded by the shape, color, and stability of the flame (Section 34) and by the manufacturer's data.

Eventual standardization of flame photometers would be very beneficial. In particular, specification of the shape and dimensions of nozzles to the closest practicable tolerances would make universal operating instructions possible. Since the relation between pressure and flow rate would then be fixed for each gas, the relations discussed below could be embodied in statements having universal validity.

When the flow of fuel gas or air (oxygen) is altered, various other factors change, usually in a complex manner, affecting the analytical emission and flame background in various ways. If the gas flow is changed, leaving the air flow constant, the mixture ratio and therefore the temperature change, greatly affecting the flame emission (Section 14). The shape and size of the flame, especially its width (Fig. 14), and the composition of the flame gases, including the concentrations of the flame ions and electrons, all change. These changes also affect the emission, for instance, when the analyte is appreciably bound as oxide in the flame. An increase of fuel gas flow lowers the concentration of free oxygen and hence the tendency toward oxide formation in the flame. A change of air flow, further, changes the suction rate and drop size and hence the rate of transport of analyte and solvent to the flame. This change of transport, especially with direct atomizers and with organic solvents, alters the shape and temperature of the flame (Section 5). An increase of air flow also raises the velocity of the flame gases, altering the effective height of observation in the flame. This in turn affects the emission, interference from concomitants, etc., to the extent that these are sensitive to height (Sections 11, 97, and 98).

In the simplest case, which is common, the temperature effect is the dominant factor in the dependence of emission on gas flow. Fig. 14 shows an example, in which both temperature and sodium emission at first rise with rising gas flow, but then pass through maxima (at nearly the same gas flow), falling at still higher flow. The thickness of the flame increases too, but monotonically; this effect displaces the emission maximum a bit to the right of the temperature maximum in this

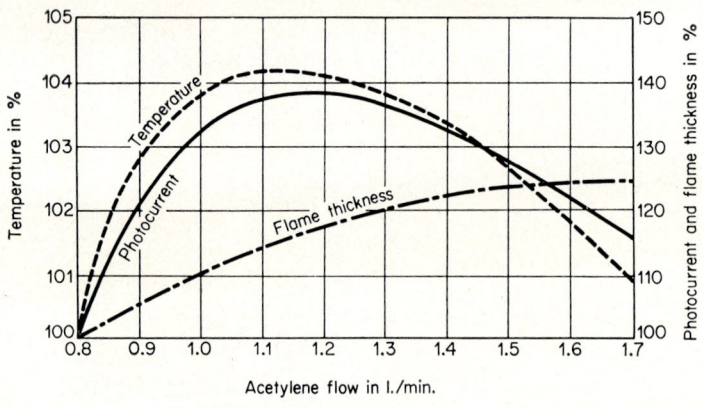

Fig. 14. Photocurrent, flame temperature and flame thickness as functions of acetylene flow for constant air flow (11.2 l./min.) and constant sodium concentration in the solution (250 p.p.m.) for a Méker burner with indirect atomizer. The maximal temperature was 2410°K. (From unpublished measurements of H. J. v. Ark, Utrecht.)

particular instrument. The location of the emission maximum, which varies among elements and spectral lines, depends generally also on the analyte concentration, solvent, physical properties of the solution, etc., and of course on the air flow rate. The maximum can be lacking entirely, when other effects besides temperature, such as dissociation, are important. The emission then rises steadily with gas flow until the flame at length becomes yellow, the maximum being unattainable. Whether and at what gas flow a maximum occurs can be found only by systematic test in each case.

In a case like that of Fig. 14, the gas pressure should be set at the emission maximum, unless other considerations such as flame background call for attention (see below). The peak of the intensity–gas flow curve offers (besides maximal intensity) the advantage that fluctuations of gas pressure (which can never be wholly avoided) have hardly any effect on the emission, since the curve is horizontal there.

These considerations apply to optimization of the gas pressure for a given air pressure. We next inquire how the optimal gas pressure varies with the air pressure, i.e., how the optimal gas and air pressures can be found jointly. Fig. 15 shows an experiment like that of Fig. 14, in which, however, the gas flow is held constant for each curve while the air flow is varied. The optimal air flow is seen to move toward higher flow as the

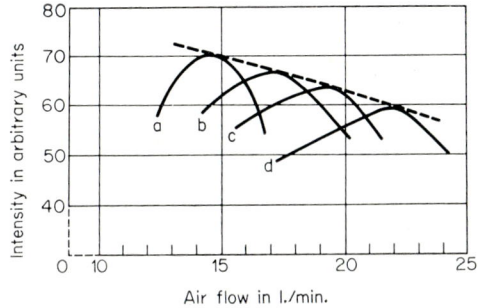

Fig. 15. Photocurrent (intensity) as a function of air flow for various propane flow rates, in l./min.: a, 0.57; b, 0.67; c, 0.74; d, 0.84. (From ref. 65.)

gas flow rises. This is easily accounted for by the dependence of the flame temperature on the gas/air ratio. From the dashed envelope of these curves it appears, however, that the best gas and air flows lie not toward the right but toward the left; that is, when gas and air pressures are varied simultaneously, the best conditions are found at relatively low flow rates.

But this is not always the situation; in the case of Fig. 15 it is due to the fact that all the curves represent very high atomizer pressures and supersonic velocities, where dilution of the aerosol by the increasing air flow outweighs the gain in analyte transport (suction rate times atomizer efficiency) resulting from the higher pressure. The effective height of observation can also play a part. With subsonic velocities (pressure below 0.9 atm. gage) the situation is usually reversed; that is, the envelope of the curves rises toward the right and the best gas and air pressures are fairly high. See ref. 298 for an example. In a case like that of Fig. 15, however, the air pressure cannot be reduced indefinitely, since sooner or later the atomizer will quit or the flame flash back. And in the contrary case the pressure cannot be raised indefinitely, for the flame will blow off the burner. In summary, we can experimentally find the gas and air pressures yielding the highest over-all intensity for a given element, while ignoring other variables such as flame background and flicker.

But in practice the situation is not always as simple as this. Compromises must then be sought among incompatible requirements. If several elements are to be determined in a sample, and if the best gas and air (or oxygen) pressures differ among them, further compromise is necessary if the pressures are not to be readjusted for each element. In a multiple-beam instrument, which determines several elements simultaneously (Section 54), or in a recording instrument (Section 68), which records a spectrum comprising many lines, optimal adjustment is basically impossible and a compromise is always required. This compromise will generally favor the element hardest to detect.

In working close to a detection limit, the blank reading (flame background plus response due to cross-sensitivity, Sections 94 and 95) must be allowed for. The ratio of net line or band intensity to blank should then be measured, instead of the total intensities as in Figs. 14 and 15. With a monochromator these ratios (and their dependence on flow) depend on the slit width (Section 62). Several such sets of data for various slit widths can then be obtained. There is a further complication in working close to a detection limit; the maximal line/blank ratio as governed by gas and air flow and slit width is not the sole criterion of optimization, but the fluctuation of the reading (flicker) as governed by these factors must also be taken into account. We return to this in Section 90.

20. *The Consumption of Oxygen and Gas*

The consumption of oxygen or compressed air and fuel gas depends on the shape and size of the burner (Section 32), the type of gas, and the flow rates. The oxygen or air consumption depends also on the type of atomizer (pneumatic only). Table 4 gives orders of magnitude.

From these data one can judge how often the tanks must be changed, knowing the content of a tank (the commonest size holds 240 ft.^{3} at 2200 p.s.i.) and the number of working hours. Thus, if air is consumed at 20 ft.^{3}/hr., a cylinder of compressed air will last 12 hr.

If the air or oxygen requirement is reckoned on the basis of the stoichiometric relation, it will be noted that the flame apparently uses too little air or oxygen. The reason is that flames derive a part of their oxygen from the surrounding air, by drawing it laterally into the diffusion flame zone and using it for combustion.

TABLE 4

Gas mixture	Consumption, ft.^{3}/hr.	
	Oxygen or air	Gas
Air–propane	4–20	0.5–3
Air–acetylene	5–20	1–4
Oxygen–acetylene	10–20	6–12
Oxygen–hydrogen	4–20	6–30

ATOMIZERS AND MIXING CHAMBERS

21. *Introduction of Sample into the Flame*

The uniform feeding of analytical sample or standard into the region of excitation is paramount for the accuracy of every spectrochemical procedure. For flame analysis the substance is generally dissolved in acid, if it is not already available in liquid form, and sprayed with an atomizer. The aerosol of liquid droplets in air or oxygen is then mixed with the gas.

Atomization is most commonly used today. But before describing the methods of atomization, we shall mention some other techniques.

1. The platinum wire was used long ago in the experiments of Kirchhoff and Bunsen. The powdered or liquid substance is taken up with the wire and placed in the flame. Champion, Pellet, and Grenier showed that this method can be refined so as to be useful for quantitative work. More recently it has again been used for ultramicro determinations[585] with integrating circuitry (Section 47).

2. In the capillary action method, a bundle of platinum wires is placed with one end in the flame and the other end in contact with the solution. The liquid passes into the flame by capillarity (Mitscherlich, 1862).

3. A disk with platinum wires at the periphery or an asbestos disk rotates slowly on an axis inclined at 45°. The lower edge of the disk runs through a container with the liquid, while the upper edge passes tangentially through the flame. The rotation continually carries fresh material from the container into the flame (Eder and Valenta, 1893). Lundegårdh[33] later used a similar system with a horizontal rotating

disk of asbestos to bring ore powder slowly and uniformly into the flame.

4. In the filter paper method, a small roll of ash-free filter paper, about 5 ×100 mm., is soaked as uniformly as possible with the sample solution and dried, or else it is sprinkled with the powdered substance. A mechanical arrangement moves this roll uniformly into the flame (Stewart and Harrison, 1939).

5. In another scheme, the sample is pulverized and spread along a channel or groove. Immediately above the channel is a powder atomizer, which sucks up the powder with the air stream in the manner of a liquid atomizer. The air stream carries the powder into the flame. A motor slowly moves the channel forward underneath the atomizer.[178]*

6. Evaporation has been employed. To avoid the trouble of pulverizing or dissolving a solid sample in preparation for a flame analysis, the sample may be formed into two electrodes, between which a spark or arc is passed. The luminous phenomena at the discharge are not observed as in spectral analysis with spark or arc; instead, the arc or spark is used only as a means of evaporating the substance and mixing it with the air stream passing by. The vapor-laden air after mixing with gas goes to the burner and the analysis is conducted by observing the flame in the usual way.[125]

In summary, the above devices have not been used nearly so widely as the pneumatic atomizer (Section 26). Their poorer reproducibility accounts for this fact. They may, however, find justification in special applications. The next few sections will describe atomizers and their use for introducing sample into the flame.

22. *Requirements of the Atomizer*

The great number of publications concerning new kinds of atomizers that continue to appear indicates that much here remains to be done. As compared with the means commonly employed in spectral analysis with arc or spark, atomizers offer the great advantage of a more constant rate of introduction of sample into the flame, and permit quicker changing of samples. An efficient atomizer must meet the following general requirements.

1. A steady rate of spraying, independent of the height of suction (hydraulic head), that is, unaffected by the emptying of the container during operation, and, as nearly as possible, independent of density, surface tension, and viscosity of the sample solutions.

* Powdered samples, suspended in a liquid, can be introduced into a flame by means of a pneumatic atomizer. See the footnote in Section 3.—*Translator*.

2. Constant and minimal mean drop size. Drop size considerably affects the intensity of the spectral lines and the detection limits (Sections 6 and 90).

3. Constant flow of the mist from the atomizer into the flame. Condensation on the walls of the spray chamber and the mixing chamber (Section 26) between the atomizer and flame must not be irregular.

4. The highest possible salt concentration in the gases at the burner port or bottom of the flame, combined with the lowest possible suction rate (economy of sample). In other words, the ratio of atomizer efficiency to air consumption should be maximal.

5. Short delay time between the start of aspiration of the liquid and equilibration of the aerosol or attainment of constant analyte concentration in the flame, and quick clearance of residual mist from the preceding sample out of the spray chamber, mixing chamber, etc.

6. Freedom from tendency to plug up, become incrusted, etc.

7. Ease of cleaning, and corrosion resistance of the atomizer material to acids, alkalies, organic solvents, etc.

8. Close tolerances in manufacture, assuring close similarity among atomizers of the same type.

9. Freedom from carry-over of the residue from one sample to the next.

How well these requirements are met by several of the important types of atomizers will be considered in Section 24.

23. *Material of Construction of the Atomizer*

Since the question of material is common to all types of atomizers, it will be taken up first. Glass, quartz, and a few plastics best meet the requirement of corrosion resistance against the commonest solutions, including acids, alkalies, and organic solvents. The entry of the sample and the formation of mist can be watched in these materials. They are easily cleaned, as with chromic–sulfuric acid cleaning solution. Glass blowing is generally used for forming them. But this method of fabrication can hardly avoid appreciable variations, from one atomizer to the next, in the nozzle diameter and relative position, and hence in the atomizing characteristics. Thus, published results obtained with a given instrument (with glass atomizer) are hardly applicable to another instrument of the same type and still less to an instrument of a different type. Also, there is an element of luck in purchasing a glass atomizer or an instrument embodying one.

We have succeeded in constructing an atomizer of molded glass, brought to its final dimensions by grinding.[341] This method easily permits close tolerances, but it was discontinued, being time-consuming and costly. Others[764] have fused wires into glass, subsequently dissolving them out; this method provides glass nozzles of good uniformity.

Atomizers have been made also of machined brass, stainless steel, and even platinum and silver.[33] Plastics are now oftener used, even in direct atomizers (Section 33), which are in direct contact with the flame.[196] Tolerances can be held with thermoplastic materials by injection molding techniques.

24. *Classification of Atomizers*

Atomizers can be classified from various points of view.

(*a*) *According to the source of energy.* 1. By *nonpneumatic* atomizer we denote one that operates on external sources of energy; examples are electrostatic, electrolytic, chemical, ultrasonic, and centrifugal atomizers.

2. By *pneumatic* atomizer we mean one in which the atomization is done by a current of gas (generally compressed air or oxygen), used at the same time for supplying the flame.

(*b*) *According to the relation between the atomizer and the burner.* 1. A *direct* atomizer sends the spray directly into the flame without intermediate conduits between the atomizer and the burner port, which here form a single unit, an *atomizer-burner* (Section 33).

2. An *indirect* atomizer, on the other hand, is separated from the burner. The liquid is sprayed first into a spray chamber and the mist is conveyed to the burner through a tube.

(*c*) *According to the method of supplying liquid to the atomizer.* 1. A *suction* atomizer draws up the sample liquid from a container, such as a Petri dish, placed below it.

2. A *gravity-feed* atomizer receives the liquid from above, poured in through a funnel. Even here, suction may exceed gravity.

3. In a *controlled-flow* atomizer the liquid is supplied by positive and adjustable displacement, as by means of some type of syringe.

(*d*) *According to the design of the atomizer nozzle.* 1. *Angle* atomizers, with the two tubes set at an angle to each other.

2. *Concentric* atomizers, with annular orifices.

3. *Immersion* atomizers (injectors, Fig. 17d), etc.

Various other systematic distinctions might be listed in addition to these.

25. Nonpneumatic Atomizers

1. Chemical atomizers, for historical reasons, deserve first mention; they were used, in fact, by Kirchhoff and Bunsen. The sample is mixed with chemicals, such as zinc and hydrochloric acid, which evolve gases. The escaping gas bubbles entrain the liquid as a mist, which is carried with the air stream to the flame. Atomizers of this kind are still useful for qualitative analysis; but for quantitative analysis they did not prove workable because the atomization is too unreproducible.[125] The evolution of gas and the degree of atomization depend strongly upon the composition of the liquid. In the zinc–hydrochloric acid method, for instance, copper, bismuth, or tin will interfere, precipitating upon the zinc. Further, the added chemicals alter the composition and, during atomization, the concentration of the sample. Samples chemically atomized in this way can be introduced more simply by holding the flame obliquely over a test tube, crucible, or the like, containing the bubbling sample solution.[704]

2. Electrolytic atomizers. In these, a current is passed between two electrodes immersed in the sample, liberating gas. The rising bubbles as they burst generate droplets of sample, which pass into the air stream. These atomizers also have proved impractical, having the same disadvantages as chemical atomizers.

3. Centrifugal atomizers. An early attempt was made to disintegrate liquids mechanically by a rapidly spinning disk or the like.[125] But even very high-speed centrifugal atomizers have not succeeded in producing drops as fine as those from pneumatic atomizers.

4. Ultrasonic atomizers. If a liquid is dropped upon a rapidly vibrating quartz plate (at, say, 800 kc.), it is dispersed into very fine drops. Focusing the acoustic waves with a concave quartz plate or the like will enhance this atomizing effect. Unfortunately, the needed equipment is costly; also, it is not easy to feed the liquid to the quartz and carry the mist away at a uniform rate, or to avoid carry-over of the residue from one sample to the next.

5. Electrostatic atomizers.[683-684] The liquid sample is allowed to rise in a capillary separated by a few centimeters from a counter electrode charged to 5–15 kv. (Fig. 16). The very high field at the surface of the liquid produces electric forces which atomize the liquid. Very small samples can be handled in this way. Another advantage is that the walls of the spray chamber soon acquire a charge, which by electrostatic repulsion prevents later drops from condensing on the walls. A disadvantage is that the rate of atomization depends upon the dipole moment of the molecules of the liquid as well as upon viscosity and

other properties. Differences in dipole moment or spray rate will be observed between one sample and the next. Distilled water, in particular, behaves badly.

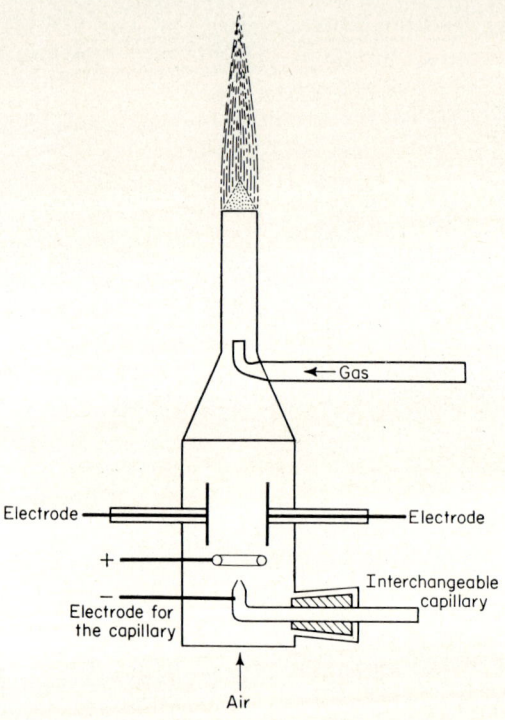

Fig. 16. Electrostatic atomizer with condenser plates for lengthening the paths of the droplets (from ref. 684).

In summary, nonpneumatic atomizers have the advantage that atomization and air or oxygen for the flame can be separately controlled. Another advantage is that the analyte concentration in the air can be greater than with a pneumatic atomizer, in which, for greater suction rate or finer drop size, more air flow is needed, diluting the analyte. On the other hand, the air flow in a pneumatic atomizer can be adjusted so that flow fluctuations have no effect on the emission. At this adjustment the effect of air flow on the flame temperature and the excitation balances its effect on the mist concentration and atomization efficiency. This cannot be done with nonpneumatic atomizers.

Except by ultrasonic and electrostatic atomization, it has not yet been possible to obtain drops as fine as those from a pneumatic atomizer. The fact that these last two methods have not yet been widely applied may be due in part to the reluctance of instrument manufacturers to supply the necessary accessories (ultrasonic generators, high-voltage transformers, etc.). Moreover, there appear to have been no studies to show whether the higher cost would be justified by better efficiency—smaller drops, lower sample consumption, or better excitation in the flame.

We next take up the more commonly used pneumatic atomizer.

26. *Pneumatic Atomizers and Spray Chambers*

In this section we take up the structure and operation of pneumatic atomizers. Following this we discuss the spray chamber and its influence on atomization and drop size.

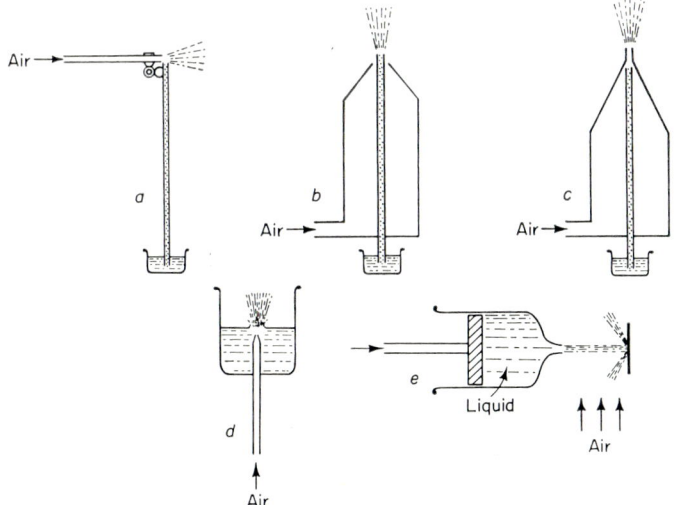

Fig. 17. Various types of atomizers (schematic). *a*, Angle atomizer; *b* and *c*, concentric atomizers without and with protruding air nozzle; *d*, injector; *e*, hydraulic atomizer.

Forms of atomizers. 1. Angle Atomizers (Fig. 17a). The liquid and the compressed air pass through two perpendicular tubes. Utilization of the air stream for atomization is not highly efficient in this arrangement, since part of the air can blow unused past the liquid capillary. The suction rate and fineness of atomization depend critically upon the relative position of the two tubes. Manufacturers usually provide means of adjusting the tubes with respect to each other for best atomization. Drops can condense on the air tube; whenever one reaches the forward edge, it is re-atomized. Since the walls may be contaminated by preceding samples, carry-over can then occur. This can be avoided by slanting the tube to prevent the drops from running toward the nozzle.

2. Concentric Atomizers (Figs. 17b, 17c). The compressed air is applied not from one side but from all sides concentrically about the suction capillary, through a cylindrical tube tapering toward the tip. The air stream is utilized more efficiently than in an angle atomizer, especially when the dimensions are properly adjusted (annular gap, angle of the cone or convergence of the air stream, positioning of the capillary tip within the nozzle orifice, etc.). For the same air pressure and air and liquid flow, a concentric atomizer produces smaller drops.

With a very small air or oxygen orifice and high pressure, the sharp air jet may not adequately contact all of the liquid, including the center of the column flowing up the capillary, which may not be atomized completely. The liquid should then be fed to a narrow annular orifice (see below) (ref. 347 and Fig. 22, page 124). This can be done by means of a channeled plug, which can take the form of an adjustable needle.[567] This needle valve permits the liquid flow to be easily adjusted nearly independently of the other atomizer parameters. By screwing it in and out, one can free obstructions during operation.

The air nozzle can extend beyond the end of the suction capillary as in Fig. 17c. The nozzle then acts like a Venturi tube. Lundegårdh earlier described a concentric atomizer in which the liquid went through the annular orifice and the air through the central capillary. This inverse arrangement is satisfactory, but the narrow annulus is more easily clogged than a smooth, round capillary bore by particles in the liquid. A spiral motion can be imparted to the air or oxygen stream before it reaches the atomizer nozzle, by tangential injection into the conical nozzle body; this improves the efficiency of atomization at the air–liquid interface.[341]

3. Injectors (Fig. 17d). A sharp air jet is passed through the sample liquid. The larger drops separate by falling back into the sample after the air bubbles burst. In this respect an injector is a kind of reflux atomizer (Section 28). Only the finest droplets reach the spray chamber and the flame. Despite this advantage injectors are hardly used any more because contamination from the preceding sample can be avoided only by thoroughly cleaning the atomizer vessel between samples. Moreover, evaporation of the solvent changes the concentration of the sample during prolonged atomization.

4. Hydraulic Atomizers (Fig. 17e). The sample liquid is squirted under pressure through a small orifice in a sharp jet against a barrier surface, where it is shattered into fine drops and carried off by the air or gas stream. The air or gas does not have to be compressed, so that no compressor is needed; but the liquid must be compressed. For this purpose the pressure of tap water may suffice. A drawback is that fine liquid orifices are more easily clogged than air orifices, and it is troublesome to avoid contamination by a preceding sample. Strictly speaking, this type of atomizer belongs to the nonpneumatic category.

The efficiency of atomizers of types 1 and 2 above can be appreciably improved by placing a barrier surface (a sphere) in the jet of spray at a suitable distance from the air nozzle. The high-speed droplets are further reduced in size on striking the surface. At high jet speeds pressure waves, shock waves, or ultrasonic vibrations may arise, as in a Hartmann pipe. These further disintegrate the drops. Sometimes the edges of annular orifices, gas nozzles, etc. (Fig. 17c), can function similarly.

In summary, concentric atomizers of type 2 (Fig. 17b and c) are at present most commonly used in commercial flame photometers. But angle atomizers (Fig. 17a) also are still popular because of their simple construction.

Functioning of the atomizer. Superficially the action of a pneumatic atomizer looks very simple. The liquid is sucked up a capillary and at the top is blown into little drops. On closer examination the atomization is seen to be a complex process depending on many factors. These factors may depend in turn on the temperature of the aspirated liquid and of the compressed air, and on the shapes and relative orientation of the nozzles. With indirect atomizers the size and shape of the spray chamber and the turbulence of the air stream also control the sizing of the droplets. It would carry us too far to discuss all of these factors, and so we limit ourselves to a few that are of practical importance for the common pneumatic atomizers, especially the effect of air pressure on the atomization; for the more fundamental aspects see Section 6.

For a given nozzle, change of the air pressure at the nozzle changes the air consumption rate, the liquid flow (suction) rate, and the fineness of atomization (the drop size distribution in the mist). We deal first with the drop size, assuming for simplicity that the suction rate is kept constant under varying air pressure by some means such as control of the length or bore of the capillary or use of a sample pump (Section 33).

Ordinarily, the suction rate would change with air pressure. But an increase of sample flow at a given pressure will coarsen the drops. If both pressure and suction rate are given free play, they will change together, and it becomes harder to generalize (see below).

The formula in Section 6 for the mean drop diameter, accurate only under ideal conditions, shows that the drop size depends on wv^2, w being the density of the air and v the velocity of the air stream. With rising pressure P, only v will increase, making the drops finer. But if P is more than 1.9 times the outside pressure, the air velocity in the throat of the nozzle equals the speed of sound while the diverging part of the air jet may attain supersonic speeds.[13] In this range the density w of the air jet rises proportionally to P, reducing the drop size for constant suction rate. The consumption of air (and indirectly, of gas) rises with pressure for a given nozzle. In the supersonic range the air consumption is theoretically proportional to P. But the higher air flow entails dilution of the aerosol (since the suction rate is assumed constant), opposing the effect of finer atomization. This dilution effect in the supersonic range can overbalance the enhancement of emission due to finer atomization on increase of pressure. Usually, however, intensities or detection limits, for constant suction rate, are better at moderately supersonic velocities (pressure somewhat above 0.9 atm. gage) than in the subsonic range.

Normally the suction rate, assumed constant above, will change (roughly proportionally) with P. In the following example we shall try to show what to expect, under simplifying assumptions. If the ratio of air or oxygen to fuel gas is held constant (keeping the flame temperature constant), the ratio of analyte concentration in the aerosol at the outlet of the spray chamber (see below), to analyte concentration in the flame, neglecting secondary effects, is constant. This ratio depends in fact only on the expansion of the mixture on combustion. If by raising P the flow of air and (by our assumption) of gas is doubled, and if the effective spray rate (suction rate times atomizer efficiency, the suction rate being given its freedom) is thereby tripled, then the intensity of the emission per unit volume of the flame rises by the factor 3/2, in view of the 2-fold dilution resulting from the doubling of air flow, provided that the working curve is straight.

There is another way in which high air velocities can enhance atomization. At very high speeds shock waves form on walls, corners, and protruding edges, and with a suitably shaped nozzle lead to ultrasonic vibrations. These shock waves, which are not necessarily periodic, are associated with supersonic speeds. The shock waves and ultrasonic vibrations can atomize liquid or further disrupt droplets already formed. This effect has been deliberately utilized in atomizer design.[498]

Under less restrictive assumptions, however, there are several reasons for preferring high air pressure at the atomizer. Many authors actually use pressures high enough to generate supersonic speeds.[33,37,65]

Lundegårdh[33] noted that with pressures of 6–7 atm. gage the sensitivities were 10–20 times as high as with low-pressure atomizers. To keep gas and air consumption within reasonable limits, the nozzle orifices must then be smaller. In simple angle atomizers this has the disadvantage that the thin air jet catches a smaller fraction of the aspirated liquid. In any case these slender nozzles are hard to make, especially if the atomizers are to be manufactured to close tolerances. Hence the atomizers of commercial flame photometers are not usually intended for pressures above 0.9 atm. gage.

The spray chamber. The mist produced by an indirect atomizer expands into a spray chamber, where the larger droplets fall out, pressure and density fluctuations of the aerosol due to atomization are smoothed, and the aerosol attains a uniform temperature.

1. Separation of Drops. The following factors control the separation: (*a*) The primary drop size distribution (see above and Section 6) and all factors that affect it. (*b*) Evaporation and recombination in the spray chamber, which alter the primary drop size distribution. (*c*) All factors that affect the evaporation, including the temperature of the air stream, the sample liquid, and the spray chamber walls (see below), and the size and shape of the spray chamber. (*d*) Gravity, and centrifugal force where the aerosol stream meets bends in the conduit, where the larger, heavier drops are unable to take the turn fast enough. (*e*) Finally, Brownian motion of very small drops.

Special spray chambers have been invented that separate the larger drops efficiently. Basically, the mist stream follows a circuitous path through a suitably shaped or baffled spray chamber in which centrifugal separation becomes effective and exchange between the mist and the moisture-covered walls can take place. For example, umbrella-shaped barrier surfaces have been installed in spray chambers,[513, 567] or the mist stream is passed through a glass spiral countercurrent to the draining liquid film.[245]

2. Damping of Pressure and Density Fluctuations. Fluctuations of aerosol density occurring in the spray chamber can be communicated to the flame and lead to fluctuation of the emission and an unsteady meter reading. This can be mitigated, as in Section 18 in connection with air compression, by increasing the time constant (resistance times capacitance) of the meter or by installing a buffer flask between the spray chamber and burner.

From ref. 65 we have

$$S_2 = (1 + C_2^2 \omega^2)^{1/2} \quad \text{where } C_2 = V_2 P_0 / QB$$

where S_2 is again the factor of stabilization, referring here to the fluctuation in the spray chamber. Then $1/S_2$ is the fraction to which a periodic disturbance is reduced after passing the buffer vessel; V_2 is its volume in liters, and the other quantities are as in Section 18.

3. Cooling Effects in the Spray Chamber. When air is released from a pressurized tank, it is cooled several degrees C. by the Joule–Thomson effect. This cooling is partly compensated in the hose connections by conduction of heat from outside. A slighter cooling, due to the same cause, occurs as the air expands from the atomizer nozzle into the spray chamber. Hence the air in the spray chamber, even without atomized liquid, will be a few degrees below room temperature.[65,513]

If the compressed air for atomization is supplied by a compressor, the air is warmed by the compression; this heat is partly lost in the pipes leading to the instrument. In any case, the temperature of the air jet in and immediately beyond the atomizer nozzle is considerably (more than 10°C.) below room temperature. This cooling is due to conversion of the disordered thermal energy of the air molecules in the nozzle to the directed kinetic energy of the air blast. In the spray chamber the air jet expands, and the kinetic energy again changes to heat. But when liquid is sprayed, a small part of the kinetic energy is converted to surface energy in the formation of many small drops. Under practical conditions this work corresponds to a cooling estimated at less than 0.1°C., and so the effect is negligible.

A much greater cooling of the air stream results from partial evaporation of the solvent in the mist droplets and on the walls. But this cooling lowers the rate of evaporation and hence the atomization efficiency (Section 6). If we assume that the initially dry air becomes saturated with moisture and that the walls conduct no heat, the temperature drop can be calculated at about 10°.[65] This cooling causes a downward drift of the emission which starts anew as each fresh sample is sprayed.[65,513] The walls of the chamber start to cool shortly after the spraying begins, until they reach equilibrium. The cool walls cool the air stream in the spray chamber, reducing the evaporation rate of the droplets and the efficiency of atomization. To avoid these effects, liquid should be sprayed continuously, even during the intervals between samples. These effects cannot, of course, occur with atomizer-burners (Section 33).

4. The Effect of Sample Temperature. The initial temperature of the aspirated liquid affects the emission with both indirect and direct atomizers.[139,233,289,591,661,760] The emission rises about 1% for 1° warming of the liquid. With indirect atomizers this is presumably due to the strong temperature dependence of the vapor pressure. With direct atomizers the situation is more complex; the dependence of emission on

sample temperature varies with the element and the concentration.[233]

5. The Effect of Air Temperature. If the air supplied to the atomizer is preheated, the emission rises considerably with indirect atomizers.[65,139] This effect must be due to an improvement of atomization efficiency resulting from faster evaporation of the mist droplets. On the other hand, if the (unheated) air is saturated with moisture before entering an indirect atomizer, the emission drops, because evaporation is inhibited.[139]

27. Introduction of Sample into the Atomizer

Various devices have been invented for carrying the sample to the atomizer, with a view to doing so in the quickest, safest, and most convenient way. Further, the liquid must pass to the atomizer at a uniform rate; the quantity aspirated per unit time should be as nearly as possible independent of the suction height (head) between the atomizer nozzle and the surface of the sample, and of changes in the properties of the liquid (surface tension, viscosity, density, etc.). The system should be resistant to partial or complete clogging; but if it clogs, it should be easy to clear. The extent to which the commoner devices satisfy these requirements will be discussed below.

The uniformity of sample flow affects the stability of emission, which will not be gone into here; see Sections 6 and 14. But note that the emission rises usually less rapidly than the suction rate, owing to compensating effects (drop size and flame temperature). Thus, although the effect of suction rate on emission is usually not large, still the sample flow rate must be as uniform as possible if the highest precision is to be achieved.

In gravity-feed atomizers the sample is poured into a funnel from above. A small preliminary portion may be poured in first to rinse traces of the preceding sample from the funnel, or to displace the distilled water that may have been poured through between samples. The actual sample for measurement is then poured in. This method thus has the drawback that extra time and sample are needed for cleaning the funnel. Besides, there is the risk that drops condensed in the spray chamber may run down the outside of the capillary tube and cause irregular atomization. For these reasons a suction atomizer seems preferable, with a capillary that draws sample from below.

With the suction atomizers most widely used, the sample is commonly placed below the atomizer in a shallow dish (Petri dish or watch glass). The container should be shallow, so that the liquid level will not drop much during the analysis, and the suction rate will remain nearly constant. Watch glasses have the advantage of permitting consumption of the entire sample, to the last drop. A Petri dish offers the same advantage if it is tilted somewhat under the suction capillary so that the lowest point of the dish lies just below the capillary inlet. Watch glasses tip easily on a flat surface, but they take less storage space than Petri dishes and are easier to clean. To avoid blocking the capillary inlet by pressing against the watch glass or dish, a notch should be filed in the lower edge of the inlet, or the inlet may be cut obliquely.[764] The bore of the capillary should not be much under 0.5–0.3 mm., or it will be too easily clogged.

The effect of density upon the feed rate can be avoided, for example, by designing the atomizer[661, 764] to bring the liquid surface to the same level as the atomizer tip (Fig. 18), or by using controlled (positive-displacement) liquid flow.[347] Since surface tension, viscosity, etc., change with temperature, it is best to work if possible in a thermostated room. The samples should be at room temperature; they should not be applied to the atomizer fresh from the refrigerator.

For volatile liquids such as acetone or gasoline (Sections 79 and 123), open Petri dishes or watch glasses are unsuitable because the liquid evaporates from them too readily. Flammable liquids in open dishes may also catch fire. For these applications sample dishes with ground glass lids such as weighing dishes should be used. Precautions necessary with flammable liquids are discussed in Section 123.

In many materials (especially of biological origin), flocculent particles, protein coagula, filaments, etc., can hardly be avoided; these appear in the sample solution and occasionally clog the capillary partly or completely. Serum, cements, and similar substances adhere to the capillary and gradually block it. Such suspended material should not be filtered off, especially from small samples, for filter papers, even if "ashless" and washed with acids, release traces of alkali metals, etc., while they may also adsorb part of the substance to be determined. It is better to centrifuge the sample. But even with these precautions obstructions will sometimes have to be removed. Hence it is important that the atomizer be built so that (a) clogging is unlikely in the first place, and (b) when it does occur it can be easily corrected.

If the atomizer has a straight, polished capillary bore (Fig. 21), plugging is unlikely as there are no constrictions in the capillary. Large suspended particles will lodge at the inlet, where they can be easily removed. Whatever enters the inlet is not likely to get stuck within the bore. But when this does happen, a steel wire, slightly smaller than the capillary bore, can be thrust through from below. With atomizer-burners (Fig. 21) the wire must not enter the flame, or it will fuse at the end and cannot be withdrawn. Some types of atomizer-burners[758] can be freed by covering the tip (capillary tip, oxygen nozzle, and gas nozzle) with one's finger, to force the compressed oxygen back through the capillary, blowing out obstructions.

Some atomizer designs have a troublesome constriction at the upper end, just below the point where air and liquid meet. Particles are easily wedged here and may be rather hard to remove. Usually it is necessary to turn the gas and air off, take the atomizer apart, and clean the thin, very delicate glass nozzle with a wire, a sharp water jet, or the like, or by sucking water backward through the capillary with a water pump, or by chemical means. A better design employs a readily exchangeable sieve,[513] glass frit,[680] or the like, at the inlet of the capillary. Such a filter, however, can transfer contamination from a preceding sample.

Since a constriction is unavoidable in many atomizers, it is better located at the inlet of the capillary than at the outlet. Some designs[513,514,764] have a means for blowing the atomizer out, for example, a lateral air inlet (branching from the compressed air tube) with a stopcock. If the capillary clogs, this cock is opened and air is blown through it.

At high concentrations of solute, the sample feed rate may be altered by deposition of salt crystals upon the atomizer nozzle. Some manufacturers[514] provide special means for cleaning the atomizer nozzle, such as an auxiliary nozzle for squirting a rinsing liquid through it. Such cleaning devices are easier to install in the simple angle atomizer than in the concentric atomizer. In difficult cases and for thorough cleaning the atomizer has to be taken apart and cleaned chemically, as with cleaning solution, if the material will endure it.

Uniform atomization is revealed by uniform drainage of liquid in the spray chamber and by constancy of sample consumption. Instead of measuring the consumption rate, one can time the interval between application of the sample to the capillary and the brightening of the flame; a change in this interval points to partial obstruction of the atomizer.

28. *Drainage of the Sample*

In indirect atomizers with spray chambers much of the sample runs down the walls unused. This liquid is usually drained by special means, unless it is allowed to return to the sample (see below for reflux atomizers). At the bottom of the spray chamber there may be a siphon-like glass tube. The trap prevents the air needed for maintaining the flame from escaping through the drain. As the aerosol experiences resistance in the narrow ports of the burner and in the connections to the burner, the spray chamber has an excess inside pressure of about 10–30 mm. of water. The trap permits maintenance of the level difference, h, needed to preserve this pressure (Fig. 18). At the upper bend there is an opening to prevent the liquid as it drains into the condensate receiver from emptying the trap by siphon action; or else the exit of the drain can be placed high enough to prevent this. Drainage must be steady; otherwise,

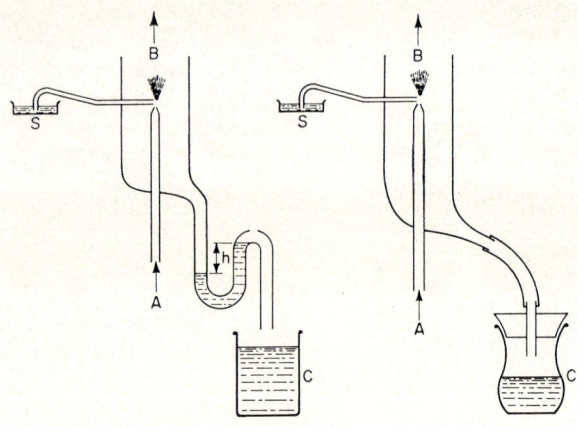

Fig. 18. Two arrangements for draining liquid from a spray chamber: (*left*) with siphon; (*right*) with closed drainage receiver. S, sample; C, condensate; A, compressed air inlet; h, head of liquid obstructing siphon; B, aerosol to burner.

variations in the height of the liquid column will be reflected in the pressure in the spray chamber and will affect the steadiness of the flame. We have found the trapped drain to have the following disadvantages:

1. At the start of a series of analyses there may be no liquid in the siphon, following a cleaning of the spray chamber, for instance. The air and mist then escape, and the flame when lit (if acetylene) is luminous and very sooty. The same thing happens if in hastily setting the regulator the air pressure is momentarily raised enough to expel the liquid from the trap.

2. The liquid may drain unsteadily, especially if the drain is too narrow and the walls are greasy.

We have used the device of Fig. 18 (*right*) to avoid these difficulties. The closed drainage receiver can be emptied during operation (while the flame is burning) by clamping the connecting tube during removal of the receiver. In this arrangement the rubber or cork stopper of the receiver must be tight, so that no air or aerosol can escape. The system can be made tight more easily by simply immersing the drain tube in a half-filled beaker. The liquid in the beaker above the outlet of the tube serves to seal it. In either case, the liquid should flow smoothly down the tube, the inside diameter of which should be large enough (8 mm.) for the liquid to flow down the walls without occluding the tube and trapping air bubbles, which would upset the uniformity of drainage.

Other special siphon arrangements or communicating vessels have been devised for this purpose. Some authors use a drain tube as in Fig. 18 (*left*), but normally keep it closed. The lower part of the spray chamber is large enough to hold the drainage from many samples; but when the atomizer jet is about to be submerged in the condensate, it is drained through the siphon or through a capillary tube with a stopcock. The stopcock can be dispensed with if the capillary is so designed as to provide effective closure of the spray chamber.[513]

Some designs, called *reflux atomizers*,[33,443] allow the condensate to return to the sample. The liquid is then subject to more than one atomization. An advantage is that less sample is needed, and it lasts longer, a matter of importance in taking spectrograms. The disadvantage is that on prolonged spraying the concentration changes. Moreover, a sample is easily contaminated by the preceding one. Either the atomizer must be taken apart, washed, and dried, or it must be rinsed repeatedly with the fresh solution. One must also take care that no drops of the preceding sample run back from the connecting tubes which carry off the mist. A further drawback is that the temperature of the sample falls during the continued spraying. Consequently, properties depending on the temperature, such as viscosity and surface tension, affect the results.

29. *Atomizers for Microanalysis*

In medical, biological, and similar applications, one often has to work with very small quantities of fluid. For this purpose special atomizers have been devised with very low suction rate or sample consumption. But an atomizer of normal suction rate is also useful for microanalysis provided that its other properties such as drop size permit lower detection limits than usual. A scarce sample can be diluted further with such an atomizer than with an ordinary one. A small quantity of sample will then last long enough for a flame analysis. Obviously, both approaches (low suction rate and low detection limit) can be combined.

The following measures, applicable to the atomizer, are useful for extending flame photometry to microanalysis.

(*a*) *Methods applicable to both indirect and direct atomizers.* 1. The flow rate is kept small by use of an especially small suction capillary. Such capillaries unfortunately clog more easily.

2. Another method of restricting the flow is to use long capillaries of normal diameter (about 0.3–0.5 mm.). The likelihood of clogging and the tendency for constituents to adhere are greater than with short capillaries, but less than with narrow ones.

3. Without narrowing or lengthening the capillary, the suction can be held within tolerable limits solely by proper form and positioning of the nozzles.[341]

4. Controlled introduction of the liquid, as with a motor-driven syringe, can keep the sample flow rate small, while avoiding difficulties from clogging.[347]

5. High-pressure atomizers[347] produce fine droplets which have little tendency to condense on the walls of a spray chamber and are more efficiently vaporized in the flame above a direct atomizer (Section 6).

(*b*) *Methods suitable only for indirect atomizers.* 1. In a reflux atomizer the liquid condensed on the walls runs back to the sample and is atomized repeatedly. The entire atomizer including the spray chamber has to be cleaned and dried after each sample. Moreover, the protracted exchange between air and sample solution changes the concentration and temperature of the solution. So far as we know, reflux atomizers are used only for photographic flame spectrography.

2. A heated spray chamber (Section 30) will diminish losses between the atomizer and the burner.

3. A nonpneumatic atomizer (for example, an electrostatic atomizer) may be used.[684]

4. The droplets can be electrostatically charged, to prevent recombination.

(*c*) *Direct atomizers.* The atomizer-burner (Section 33) can be modified as described under (*a*) 1–5. There is no loss between atomizer and burner top; the efficiency is always 100%. However, there are losses from incomplete or insufficiently rapid evaporation of the droplets in the flame; these losses depend on the design of the atomizer and on the pressure settings (Section 6).

Two or more of the above measures can be used jointly. In addition, there are measures applicable not to the atomizer but to other components of the flame photometer or to the preparation of the sample.

It is noteworthy that none of these methods of reducing sample consumption entails a proportional loss of flame intensity; the intensity diminishes much more slowly,[65,85,347] and with direct atomizers may even increase with diminishing sample flow (above a certain point), owing to the less marked cooling of the flame and downward shifting of the drop size spectrum. Recent advances in micro flame analysis have been notable. Lundegårdh, for instance, used at least 3 ml. of solution and generally about 20 ml.[33] Today 0.1–0.5 ml. of the ready-diluted solution will often suffice. Ordinary atomizers and any of the microatomizers named above can be equipped with contacts that signal the start and end of atomization electronically and feed the photocurrent to an integrating circuit during the interval.[255] These matters will all be discussed in greater detail in Section 89.

30. *The Heated Spray Chamber*

With indirect atomizers only a small fraction of the atomized sample, usually 1–3%, occasionally up to 12%,[65,341] reaches the flame. The rest is lost in the spray chamber and elsewhere. By heating the walls of the spray chamber externally, usually with an electric heating coil, to a temperature of from fifty to a few hundred degrees C., evaporation of the solvent in the spray is enhanced. Drops which would be lost by deposition on unheated walls mostly reach the flame with a heated spray chamber. Thus the efficiency of atomization rises. But an efficiency of 100% cannot be reached because even solid materials deposit on the walls. The heated spray chamber has the following disadvantages:

1. The interior of the spray chamber is invisible; the uniformity of atomization cannot be checked.
2. To avoid the drifts due to cooling of the walls during spraying and reheating between samples, heat transfer between the heating coil and the walls should be efficient. Glass, usually used for the chamber, is not a very good conductor. But the commoner metals, such as copper, corrode easily, and a cleaning problem arises.

For these reasons heated spray chambers are little used today.

31. *Injection of the Gas*

After the aerosol has been formed and the larger drops removed (Section 26), the fuel gas must be added before the mixture reaches the burner. The gas should be injected as uniformly as possible, so that the flame temperature and the excitation will remain steady (Sections 8 and 14). The choice of gas and air flow rates was discussed in Section 19, and the regulation of the flow in Section 18. Recall also the earlier remarks about standard nozzles. We are concerned here with the nozzle or gas-mixing jet, which is intended to assure uniform flow of gas to the flame and thorough mixing with the aerosol.

Constancy of flow. We assume that the pressure at the gas jet is sufficiently steady and properly adjusted. The uniformity of flow can then be spoiled by either a change in the properties of the nozzle (which is often a short, narrow capillary) by obstruction, corrosion, etc., or by a change of the pressure confronting the gas in the connections to the burner. We discuss these further below.

1. Nozzle Characteristics. The gas jet can be more or less completely clogged by foreign material carried by the gas (particles from gaskets, rubber tubes, etc.). But the flow resistance of the nozzle is more commonly altered by contamination by the aerosol as it flows by. The gas jet should therefore be so placed that sample residues cannot readily settle upon it. To this end, a protective tube, for example, can be placed about the jet itself, as in Fig. 8 (No. 10). Another way is to aim the jet downward, to keep the larger droplets from reaching the jet against gravity. Still another is to pass the gas stream, after emerging from the jet, through a porous body of large area and negligible flow resistance, such as a vessel filled with glass beads, before it mixes with the aerosol. Only the outer surfaces of the beads can become coated with residues, while the jet remains free.

2. Pressure. If there is a change in the pressure against which the gas must flow between spray chamber and burner top, the gas flow is indirectly affected. Such pressure changes can be caused by changes of the flow resistance in the connections to the burner, as from kinks in a rubber tube or obstructions or corrosion in the burner ports. A change in the air flow also affects the pressure and indirectly the gas flow. These effects can be minimized by keeping the flow resistance of the gas nozzle as high as possible, by making the nozzle orifice long and narrow. Higher gas pressure will then be required at the nozzle, and small pressure fluctuations beyond it will hardly affect the gas flow. However, the pressure cannot be increased indefinitely, especially of city gas, its pressure being rather low in the mains. A narrow metal nozzle also protects against flashback in case of error, as in lighting a flame above a burner without a flashback-proof cap when the gas hose still contains a gas–air mixture.

Mixing the gas. The gas should be so injected as to assure thorough mixing of all the components before combustion. This can be done by injecting the gas tangentially into a mixing chamber through which the aerosol is passing; the swirling promotes mixing. Vortical motion at the gas jet can be brought about also by enclosing the jet in a Venturi tube so that the sharp gas stream entrains the aerosol turbulently. Alternatively, the mixing can take place in the conduits leading to the burner, which may contain hollow glass balls, baffles, or the like, to aid the mixing. The eddying gas–aerosol mixture must be calmed again, that is, rendered laminar, to yield a quiet flame. This generally takes place in the burner cap (Section 32).

Change of fuel gas. It may happen that a flame photometer is to be used with different gases, according to the application, but it has no separate pressure gage for monitoring the gas flow. Interchangeable gas jets should then be available, unless the different gases are very much alike in those properties that determine the pressure drop, at the required flow rate, across the orifice.

BURNERS

32. *The Burner (with Indirect Atomizers)*

The mixture of mist, air or oxygen, and fuel gas is conveyed to a burner, above which the flame burns. There are also atomizer-burner combinations, in which the liquid is atomized and the oxygen and gas are mixed at the actual tip of the burner; these will be discussed in Section 33. Here we consider burners of the types used with indirect atomizers. Their requirements are listed below.

1. Suitable form, to provide a steady, reproducible flame. In most cases flow of the gas–air–mist mixture where it enters the flame should be as nearly laminar as possible.

2. A design that minimizes the possibility of flashing back, even on faulty manipulation of the gas and air.

3. Corrosion resistance and freedom from incrustation.

4. Readily interchangeable burner caps, to permit changing from one fuel gas to another that requires a different port size in the cap (see below). The cap should fit the burner loosely to avoid damage in case of flashback.

5. Easy lateral and vertical adjustability of the position of the burner, to permit peaking the emission intensity. Many interferences are sensitive to height in the flame (Section 98). Thus, by adjusting the height of observation (by moving the burner, the entrance beam being usually fixed), interferences can often be mitigated.

Manufacturers of flame photometers usually provide properly built burners. Stainless steel is often used. Shape and size of the burner are governed by the following factors:

1. The kind of gas. Burners intended for a gas mixture of low burning velocity, such as air–propane, cannot be used directly for one of high burning velocity, such as oxyhydrogen.

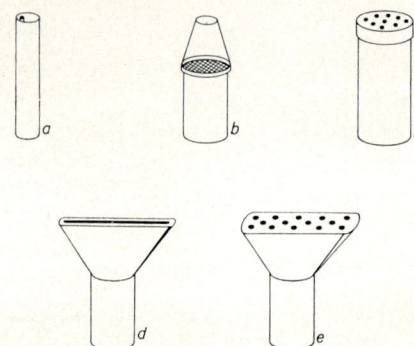

Fig. 19. Several types of burner. *a–c*, Round forms; *d–e*, elongated forms. *a*, Open tube as in a bunsen burner; *b*, with safety grille and constricted tip; *c*, Méker burner; *d*, fishtail burner; *e*, elongated Méker burner.

2. The desired size of the flame or gas flow rates. The speed at which the mixture issues from the burner port or ports must be about thrice the mean burning velocity. Flashback prevention imposes a maximal port diameter. These factors determine the number of orifices required in the cap.

3. Other factors, such as heat transfer from the burner cap to the surroundings, rate of aerosol output by the atomizer, and drop size distribution.

Types of burners. A few typical designs are shown in Fig. 19.

Open burners (the bunsen burner, Fig. 19a) are hardly used any more, since the flame easily flashes back in case of a blunder. Burners with auxiliary safety grilles (Fig. 19b) are also rare. Heat loss through the grille is supposed to cool the flame front below the ignition point in case the flame retreats into the grille, and thus prevent flashback into the burner barrel. An open burner may be justified in the uncommon case that the emission of an element in the inner cone is to be used for analysis. In a Méker burner (see below) the many small inner cones are so short that it is hard to focus the observation beam upon them. We return to this at the end of the section.

§ 32]   THE BURNER (WITH INDIRECT ATOMIZERS)   119

For premixed flames the *Méker burner* (Georges Méker, Paris, 1903) is most commonly used at present. This burner has a cap sitting loosely upon the burner tube; the grille of the cap, usually 1–5 mm. thick, contains many holes of 0.5–1.5 mm. diameter. The inner cone of the simple bunsen flame (Fig. 1, page 14) is thereby divided into many small primary combustion zones (Fig. 24). The thicker the grille is (the longer the holes are), the more nearly laminar will the flow be, and the more stable the inner cones and the entire flame. There are several other advantages:

1. More secondary air is drawn from the surroundings, helping to cool the burner cap.
2. The temperature in the flame above the inner cones is more uniform.
3. The individual inner cones burn more quietly than a single large one. The flame is hence steadier.
4. Like the safety grille mentioned above, the grille of the Méker burner, when properly designed, keeps the flame from flashing back past the burner cap.

It is hard to make a burner proof against flashback (assuming inept handling of oxygen and gas) when gas mixtures of very high burning velocity such as oxyacetylene and oxyhydrogen are to be used. It requires long, narrow ports and thermally conductive material in the burner cap, permitting adequate heat loss to the surroundings (sometimes artificial cooling is needed). But the long, narrow ports are easily clogged by deposits of sample residues from the aerosol–gas mixture. Méker burners are therefore seldom used with highly explosive mixtures, and open burners are of course never used. For such mixtures a burner in which the un-premixed gases come together directly at the base of the flame (Section 33) is more commonly used.

If, notwithstanding this, a Méker burner is to be used with premixed gases of high burning velocity, the holes in the cap can be somewhat larger than the above criteria indicate; the rate of clogging is then slower, but flashback is more likely, and safety precautions should be provided. For one thing, the loose cap offers some protection against damage of the apparatus, since it flies off in case of explosion. A foil-covered window (rupture disk) can be placed at the bottom of the burner close to the gas jet; it bursts in case of explosion.

Noncircular forms. Both open burners with or without a safety grille and Méker burners may have elongated tops (Figs. 19d, 19e). These flatten the flame longitudinally or transversely, and offer certain advantages:

1. Transverse widening of the flame (normal to the line of sight) reduces the flame thickness (Section 8). Self-absorption then sets in at higher concentration; the knee of the (logarithmic) working curve, other things being equal, lies higher, and the linear part extends to higher concentrations. The intensity will, however, be lower, a matter of importance at lower concentrations.

2. If the flame is widened longitudinally, it is thicker in the line of sight and provides somewhat higher intensity, but self-absorption starts at lower concentration.

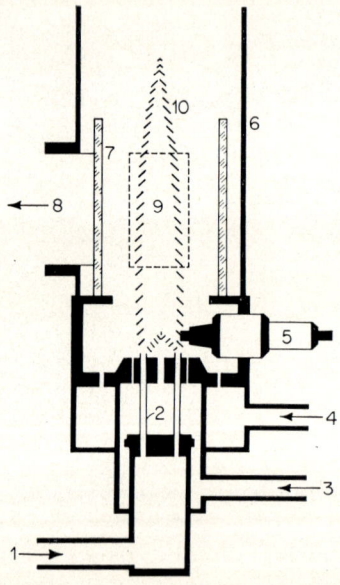

Fig. 20. Burner in a closed system (from ref. 513). 1, Fuel–air-mist mixture; 2, burner tubes corresponding to the orifices in the cap of a Méker burner; 3, supply tube for optional oxygen for raising flame temperature; 4, entry for filtered air for stabilizing the flame; 5, spark plug; 6, metal chimney; 7 glass cylinder; 8, beam window; 9, window for observing and adjusting the flame; 10, flame.

But these differences between flat and round flames are hardly significant. There is also the disadvantage that the central temperature of the flame is usually somewhat lower than that of a round flame, and the elongated flame entrains more external air and with it more dirt.

Intermediate forms. In addition to the burners described above for premixed laminar flames and those for un-premixed turbulent flames (Section 33), there are intermediate forms not readily classified in either group. Fig. 20 shows an example. When oxygen is added in this device, it is not premixed with either the aerosol or with the gas, unlike the air used in the burner. Many other variations exist, which we cannot discuss here.

Closed systems. The burners described above all burn in the open air, except for a chimney (Section 35) or other equipment surrounding them; that is, the ambient air has free access. This has the drawback that impurities can enter the flame from the atmosphere, raising the flame background, causing flashes of light from dust particles, etc. In some industrial areas where the air is very dusty, the contamination can make flame analysis impossible without special provisions. A flame will not burn stably in a fully enclosed space without access of air. Therefore, a closed system, while isolating the flame from the impure outer air, must be supplied with a stream of purified air or oxygen surrounding the flame, as in Fig. 20. For a premixed laminar flame, about twice as much purified air must be supplied around the flame as is consumed internally for atomization and primary combustion. For an un-premixed flame, even more purified air is needed. Purification of the air was discussed in Section 16.

Burners for separated flames. Certain lines, such as those of tin and molybdenum, and several bands are emitted only in the inner cone of the flame, or are perhaps stronger there than in the interconal zone, which is customarily used for analysis. To observe these emissions it is helpful to separate the inner cone from the diffusion flame. In the determination of nitrogen (Section 87) a separation of the flame will prevent reaction of the flame with atmospheric nitrogen. The separation can be done with special burners like those proposed by Teclu (1891), Smithells (1892), and others for chemical investigation of the flame. The technique depends on the difference in burning velocities between the first and second oxidations.[362, 664]

Cleaning. Any burner or burner cap must be cleaned occasionally, and oftener when more concentrated samples are sprayed. Salt deposits can be simply washed off. Organic substances should be cleaned off mechanically. A steel wire somewhat smaller than the hole is used for cleaning the ports in the grille of a Méker burner, by pushing it through each port in turn. The upper surface of the grille can be smoothed with fine emery paper.

33. *Atomizer-burners*

Fig. 21 shows an atomizer-burner. It comprises a straight capillary tube of palladium, reinforced over the lower part, concentrically surrounded by a cylindrical oxygen nozzle with a conically tapering tip.

Centering of the capillary in the oxygen nozzle is quite critical and is accomplished by means of centering screws. The oxygen stream produces suction and atomizes the sample. The oxygen nozzle is, in turn, concentrically surrounded by a nozzle for the gas (acetylene or hydrogen). Both the mixing of the gases and the atomization thus occur above the tip of the nozzles. The main part of the flame burns somewhat above the burner tip, not directly upon it, as the constituents must first mix. A small diffusion flame, however, will be observed just above the

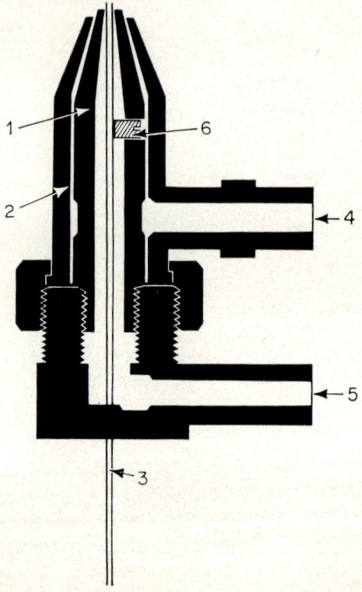

Fig. 21. Atomizer-burner (direct atomizer) for oxyhydrogen, by Beckman Instruments, Inc., Fullerton, Calif. 1, Oxygen nozzle; 2, hydrogen nozzle; 3, suction capillary; 4, hydrogen tube; 5, oxygen tube; 6, one of the three screws for centering the capillary in the tip of the oxygen nozzle.

fuel gas nozzle, where the issuing gas burns in the ambient air, creating a pilot flame to help maintain combustion of the turbulent, un-premixed flame. The burner tip remains fairly cool, since the main flame rides higher. The slight warming is due to radiation and the un-premixed pilot flame. Because of its turbulence, the atomizer-burner flame hisses noisily like a blowtorch, and the turbulence is intensified appreciably when the spray enters the flame. For this reason atomizer-burners are generally enclosed in housings with acoustic insulation (Section 63).

The atomizer-burner, also called direct atomizer, has the following advantages over the indirect atomizers discussed in Section 26:

1. There are no losses by condensation of spray in a spray chamber; all of the material enters the flame. The atomization efficiency is 100%. There are no interferences of the sort that result from varying condensation of drops in the chamber (Section 11). A gain of accuracy usually accrues.
2. It is safe to use flames of high burning velocity such as oxyhydrogen. The flame cannot flash back, since the gases first mix above the burner tip.
3. For these two reasons, the emission from a given concentration of analyte in the solution is higher than that from a premixed flame with indirect atomization.
4. Combustible substances such as gasoline can be safely sprayed.

There are also some disadvantages:

1. An interference of the sort that is sensitive to the drop size distribution of the spray (e.g., the effect of phosphorus or aluminum on alkaline earths) is more severe with an atomizer-burner (Section 98).
2. Although no material is lost in a spray chamber, there are other losses due to incomplete or insufficiently prompt evaporation of the larger drops in the flame (Section 6). The flame is more strongly cooled by the heavier load of liquid. Accordingly, an alteration of suction rate, as from slight obstruction of the capillary, changes not only the concentration of analyte in the flame but also the flame temperature and hence the excitation; to be sure, these usually have opposite effects.
3. Atomizer-burners are very sensitive to misalignment of the atomizer nozzle or displacement of the capillary, which puts the issuing oxygen stream askew and makes the flame lopsided.
4. Atomizer-burners are more liable to incrustation of the tip of the nozzle. This may distort the flame, decentering it in the optical path and leading to unpredictable losses of intensity.

There are also many other interferences that affect atomizer-burners in the same way as indirect atomizers with separate burners. An example is the effect of differences in viscosity and surface tension between samples on the reading. Here neither type of burner has an especial advantage over the other.

On the whole, when spectrophotometers are used, the advantages of atomizer-burners prevail. Attempts have been made to overcome or at

5*

least to mitigate some of the listed drawbacks by proper design and various innovations. Some of these are the following:

The cooling of the flame by the high flow rate of liquid can be lessened by reducing the liquid flow. This is done with a long or narrow suction capillary (Section 29), but such a capillary is more susceptible to clogging. This difficulty can be circumvented by controlled sample delivery (pumping). One scheme is to place the sample in a syringe and discharge it into the atomizer very slowly and uniformly with a synchronous motor and a driving mechanism. This device acts as a brake

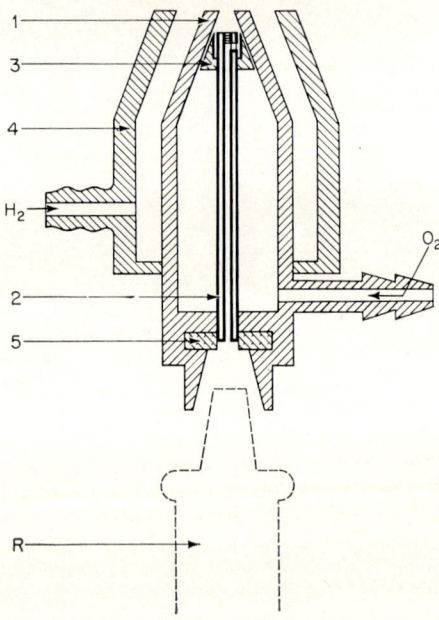

Fig. 22. Atomizer-burner for high oxygen pressure (5 atm.) and controlled sample feed using a motor-driven syringe R (from ref. 347). 1, Oxygen nozzle; 2, capillary; 3, conical centering piece with grooves for passage of oxygen; 4, hydrogen nozzle; 5, rubber gasket for connecting with the syringe. The liquid emerging from the tip of the capillary passes to a narrow annulus to increase the area of liquid coming into contact with the high-speed oxygen stream.

upon the sample flow, which would be faster without it, but it acts also as a positive-displacement pump in case of clogging. Fig. 22 shows such an arrangement. See also ref. 307a.

The loss of sample (mentioned in 2 above) from incomplete or slow evaporation of the larger drops in the flame can be lessened by reducing the drop size. Controlled, slow sample flow helps to reduce the drops. High pressure (with a small nozzle orifice) can also help a good deal in this direction (Section 6). The precise machine work required to make such a tiny, accurately centered orifice is, however, very exacting, and this recourse has not become popular.

Another method is now more commonly used for diminishing the drop size and the cooling of the atomizer-burner flame by the solvent. The customary solvent water is replaced by a combustible solvent of low surface tension, such as acetone, if the sample is compatible with it. The gain in sensitivity or improvement of the detection limit can amount to one or two orders of magnitude in favorable cases. We return to this in Sections 79 and 90.

34. *Size and Shape of the Flame*

The size and shape of the flame in a flame photometer affect the magnitude and reproducibility of the photocurrent yielded by the photodetector. These relationships will be dealt with in Sections 37, 43, and 64. The size and shape of the flame are also a simple and sure criterion, for the analyst, of normal behavior of the flame (which is the most important part of the whole instrument); they show whether it is keeping the adjustment established as optimal for the application at hand (Section 19).

The size and shape of the flame depend upon the following factors:

1. The nature of the fuel and oxidant.
2. The flow rates of fuel gas and air or oxygen, and their ratio.
3. The efficiency of mixing and the turbulence of the inflowing gases and of the flame.
4. The design and size of the burner, especially the number, size, and arrangement of the ports in the cap of a Méker burner (Section 32), the material of construction, its thermal conductivity, and the method of cooling it.
5. The kind and quantity of sample introduced; particularly, the kind and quantity of solvent (water, acetone, gasoline, etc.).
6. The drop size distribution arising on atomization, which especially affects the size and shape of an atomizer-burner flame.
7. Deposits of sample that may form on the burner tip or cap.

Some further details regarding these points are given below.

(*a*) *Gas and air flow*. If the gas and air (or oxygen) flow are simultaneously increased at constant ratio, the flame grows larger; if the speed of discharge from the burner port or ports (of a Méker burner) appreciably exceeds the burning velocity of the mixture, the flame blows off the burner, usually with marked turbulence and noise. But if the flow rates are too low in a premixed flame, it flashes back into the burner tube unless one of the protective measures of Section 32 is provided, in which case the flame finally goes out above the burner, even though some gas and oxygen continue to flow. Just before going out, the flame burns very close to the top of the burner, and it may heat the cap or burner tip. If this continues too long, the top of the burner can become incandescent.

If the total flow of gas plus oxidant is held within the range in which the flame can burn, while the ratio of the components is varied, the following phenomena are observed. As the ratio of gas to air (oxygen) flow is reduced, the flame grows more unsteady and turbulent and tends to blow off the burner. As the ratio is raised, the flame shows first a whitish color (chemiluminescence) at the edges and next, with carbonaceous gases, a stronger yellowish luminescence, first in the middle and then throughout the flame, due to incandescent carbon particles. This raises the flame background. The flame temperature (Section 14) changes with the gas/air ratio, and the line/background ratio does also, owing both to the temperature dependence of the line emission (Section 8) and to the change of flame background (Section 95) with gas/air ratio. When the chemiluminescence mentioned above appears in the edge of the flame (the outer combustion zone), it signifies that the gas/air ratio is too high for complete primary combustion in the inner cone and that by lowering the ratio the flame background can be reduced. The incandescence of carbonaceous flames appears first in the middle of the flame because the ambient air entering the edges promotes combustion there; but it spreads over the whole flame at very high gas/air ratios.

(*b*) *Turbulence.* The flow of a gas–air mixture in the ports of a Méker burner is generally laminar. Nevertheless, more or less turbulence can be noticed in the flame burning above, especially at the edges, and the turbulence increases with height. The marked turbulence at the tip of a premixed flame makes it pulsate or flicker, and mixes the hot flame gases with the surrounding air, eventually quenching the flame and limiting its height. The emission is measured, of course, not in the upper part of the flame, but usually within 2 cm. above the inner cones. The best height of observation is discussed in Sections 5, 6, 7, 8, and 98. The turbulence in the edge of a flame, increasing with height, shows a fairly regular pulsation in the lower part at about 30 c.p.s.,[65] which must be heeded in choosing a frequency for an a.c. method.

§ 34] SIZE AND SHAPE OF THE FLAME 127

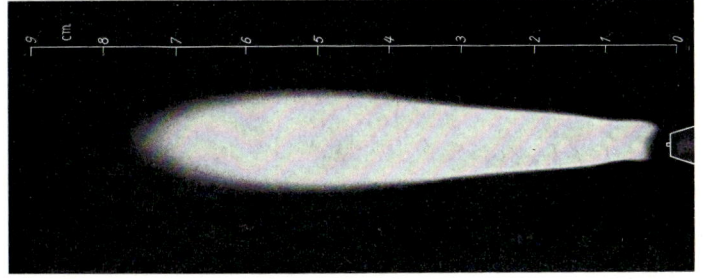

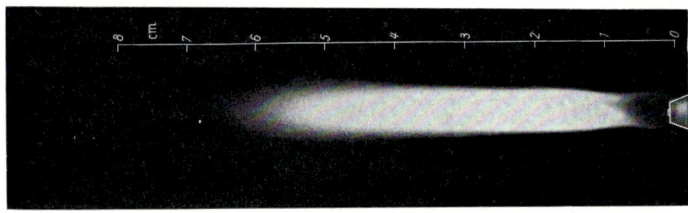

Fig. 25. Turbulent oxyhydrogen flame above an atomizer-burner. (*left*) Empty flame burning without addition of sample solution; (*right*) with sample spraying.

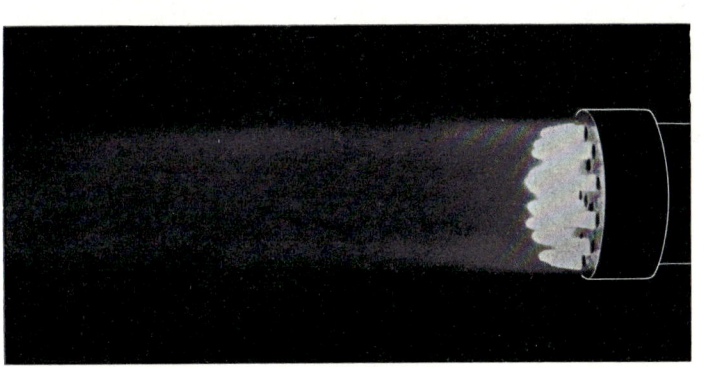

Fig. 24. Detail of the flame of Fig. 23. The luminous cones above the orifices of the Méker burner are clearly seen.

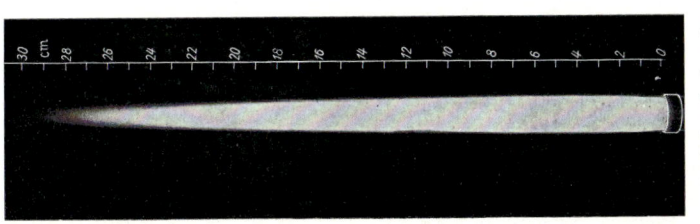

Fig. 23. Nearly laminar premixed flame (air–acetylene) above a Méker burner.

In an atomizer-burner with un-premixed gases (Fig. 21), the fuel and oxygen are turbulent as they enter the flame, which becomes more turbulent and therefore noisy. When liquid is sprayed, the turbulence increases, and the shape of the flame changes, as seen in Figs. 25a and 25b. Rather long and narrow and fairly quiet in the absence of spray, the flame becomes thicker and louder on spraying. Here, too, the best height of observation depends a great deal on the analyte and interferents.

(c) *Atomizer design.* In Section 32 we described various kinds of burner top (fishtail burner, etc.) and the differently shaped flames they produce. In atomizer-burners, the effect of drop size is seen on comparing burners having larger atomizer orifices and operating on low oxygen pressure (yielding coarser droplets) with burners having smaller orifices and using higher oxygen pressure (yielding finer droplets).

(d) *The sample.* We have mentioned the effect of the sprayed water on the turbulence and shape of the flame. If instead of water a combustible solution is atomized, the flame temperature rises provided there is enough oxygen, and the flame usually gets bigger, owing to the greater thermal expansion as well as the added fuel.

(e) *Drop size distribution.* The effect of the solvent was dealt with above.

(f) *Incrustation.* Uneven deposits of sample material on the burner tip can make the flame crooked or divide it into parts. This can be cured by cleaning the tip after shutting off the gas and oxygen.

The various factors outlined here (kind of gas, atomizer, etc.) and their effect on the size and shape of the flame must be viewed in relation to the effective aperture of the optical beam. These relations will be taken up briefly in the next sections; for more details see the special studies.[65]

35. *The Chimney*

The burner or flame is usually surrounded by a chimney or a burner housing made of metal with suitable windows, one for the optical beam, one for inspecting the flame, and perhaps others for inserting an electrical igniter, etc. The chimney is commonly of metal with glass windows, but sometimes it is made entirely of (heat-absorbing) glass or even quartz. The following are the functions of the chimney or burner housing.

1. To protect the flame from breezes such as result from the motion of people in the room.

2. To prevent accidental contact with the flame, which can cause an immediate third-degree burn.

3. To protect the observer from ultraviolet light, which may possibly cause inflammation of the conjunctiva.

4. To keep adjoining parts of the instrument from becoming overheated by the radiation from the flame.

5. To protect parts of the optical–electric system exposed to the flame (condenser lenses in filter instruments, entrance slits in monochromators) from contamination or corrosion, which can be especially harmful with direct atomizers.

6. Through proper design, to cut down the noise of a turbulent flame (Section 63).

7. To prevent contamination of the flame by dust from the outside air, when purified auxiliary air is supplied around the flame in a closed system (Section 32, Fig. 20).

8. Often, to help stabilize the flame; the turbulence in the outer and upper parts of the flame (Section 34) can be diminished under the proper conditions, but a chimney can also augment it.

In setting up a flame photometer, one should be careful to have no object vulnerable to heat, such as a lamp shade, above the chimney, and to prevent the room lighting or other extraneous light from entering the photometer through the chimney (cf. Section 76).

Optics and Electronics

36. *General*

It is the function of the optical system to collect the radiation from the steadiest part of the flame at the best height, disperse it spectrally, transmit only the desired spectral range, and collect this finally upon a radiation receiver (photodetector). The associated electronic system delivers the resulting photocurrent, perhaps after amplification, to a meter or indicator.

The beam is generally so chosen as to avoid the flickering tip and the unsteady edges of the mantle as well as the bright inner cone. Since the radiant flux from the part of the flame usually used for analysis is rather weak, especially at low concentrations of analyte, the limits of photometric measurement are often pressed, particularly with monochromators. The optics must therefore gather this inherently small radiant flux from the flame as efficiently as possible and transmit it with minimal loss to the photodetector. The ability of the optics to do this is characterized by a property called the luminosity.[318]

The *luminosity* defines the geometric transmission of radiant energy through an optical system, neglecting losses by reflection, absorption, diffusion, and diffraction. Its dimensions are cm.² sr. (sr. = steradian). If the radiance of the surface of a source (a flame) in w. cm.$^{-2}$ sr.$^{-1}$ is multiplied by the luminosity L in cm.² sr., the product is the radiant power in watts delivered to the receiving surface, e.g., the cathode of a phototube. It can be shown[318] that with complete imaging the luminosity is given approximately by

$$L = f_1 f_2 / a^2$$

where f_1 is the area of one radiating surface or transmitting aperture such as a lens or diaphragm opening, and f_2 is the next consecutive such area in the instrument (lens, photocathode, etc.), separated from the first by the distance a. The luminosity at any aperture in the instrument is proportional to the product of the cross-sectional area of the beam at this aperture and the solid angle within which usable radiation can pass through any point of this area. In a properly designed optical system the luminosity is the same everywhere in the instrument, except as it may be controlled by a variable iris diaphragm. If light losses due to absorption, reflection, diffusion, and diffraction are to be reckoned in, then L should be multiplied by the dimensionless optical transmittance T, between 0 and 1. In a good instrument T is little less than 1, apart from absorption in a filter.

The optics of a flame photometer must accordingly have a high luminosity. In this respect the two principal kinds of instrument—filter instruments (Sections 39–58) and spectrophotometers with monochromators (Sections 59–69)—differ from one another by orders of magnitude. After Sections 37 and 38, which apply to both filter and monochromator instruments, we shall separately discuss the filter instruments first. Figs. 26, 28, and 29 show such instruments. Table 15 in the Appendix summarizes all of the commercial flame photometers known to us and lists their chief technical data.

37. *The Optical Alignment of the Flame*

Most analysts will be working with ready-built commercial flame photometers, which may be assumed to be preadjusted except for certain adjustments to be discussed below.

Since most commercial instruments have to settle on a compromise among several mutually antagonistic requirements, a skilled experimenter can, by suitably adjusting or changing the existing arrangement, more effectively satisfy a particular requirement of special

§ 37]   THE OPTICAL ALIGNMENT OF THE FLAME   131

importance to him, while forgoing other desiderata. This is done by proper choice of fuel and oxidant flow (Section 19), or by one of the expedients described below. Incidentally, an instrument can be adapted to a special application by changing to an entirely different flame, which may require a different burner, atomizer, etc., from those supplied by

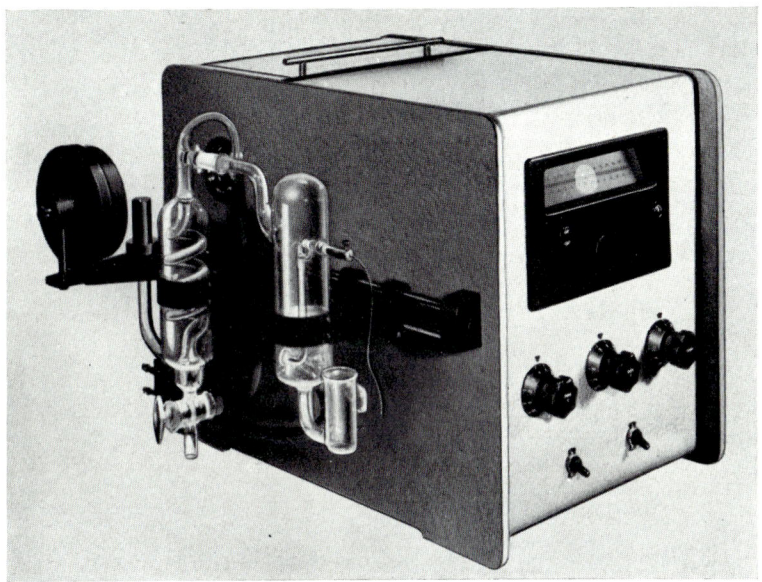

Fig. 26. The Schuhknecht flame photometer (ref. 245).

the manufacturer. Conversely, the atomizer and burner can be kept, while the optical system is replaced by another. In any such undertaking, the adjustments described below must be made.

The following adjustments are generally provided.

(a) A vertical adjustment of the burner with respect to the optical beam.

(b) A lateral adjustment of the burner with respect to the beam.

(c) Sometimes also a longitudinal adjustment to vary the distance of the burner from the optics.

(a) *Vertical adjustment.* Any part of the flame can be brought into the entrance beam (that is, the region of space within which the instrument can respond to radiation; in other words, its field of view). The entrance beam may be conical, or, with monochromators, shaped like a narrow truncated pyramid of small divergence. The flickering tip of a premixed flame and the more luminous inner cones should be excluded from the entrance beam by proper adjustment. If these parts are excluded, the fairly long premixed flame (Fig. 23) still offers a wide area within which observations can be made. The best height adjustment for a given case must be found by trial.

The following points are pertinent:

(1) The line or band intensity varies with height, but in a different way for each emission and each flame, as seen for example in Fig. 27. This behavior is determined by the dependence of molecular dissociation

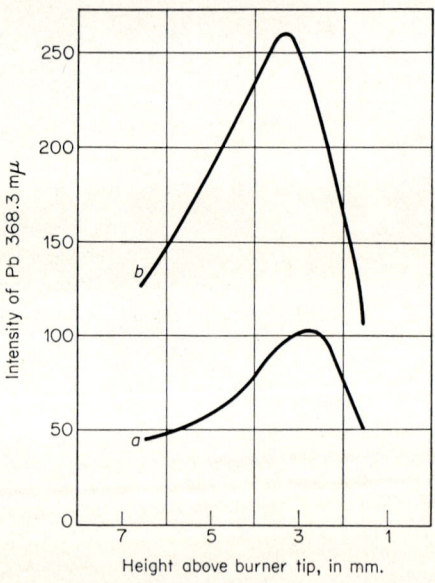

Fig. 27. Emission of the lead line 368.3 mμ as function of height in the flame (from ref. 300). *a*, Flame background; *b*, lead line plus background at a concentration of 2 ml. tetraethyllead per gallon of gasoline.

or association (e.g., hydroxide formation, Section 8), ionization, evaporation of solvent and solute (Section 6), temperature (excitation, Section 8), and other phenomena, on the height in the flame.

(2) In working close to a detection limit, the line or band intensity is not the only criterion, but the intensity of the flame background is important too. This likewise depends on height in the flame, but differently from line intensity. Fig. 27 also shows an example of this. The dependence of background on height depends in turn upon the

gases and their flow rates, the wavelength, and the solvent. If the ratio of line to background intensity is plotted against height, the best height adjustment corresponds to the peak of this curve.

(3) Interferences vary with the height (Section 98). If an interference is to be minimized, the best height adjustment will generally differ from that found as above.

In practice, then, a compromise will have to be found among these partly antagonistic requirements, determined largely by the relative importance of these requirements for the application.

The height adjustment is usually made simply by means of a bracket on the burner or its barrel, by which it is lifted or lowered. The manufacturer will have provided instructions. For measuring height effects, as in Fig. 27, a millimeter scale, for instance, can be attached to the burner.

(b) *Lateral adjustment.* If the flame is moved across the beam, the photocurrent will show a trapezoidal function with respect to lateral displacement. The sides of the trapezoid correspond to the edges of the flame, and the best adjustment will correspond to the center.

(c) *Longitudinal adjustment.* This is generally fixed by the manufacturer, so that the user is not likely to alter it. But if it can be changed, then the flame should normally be set close enough to the optics (but not unduly close) so that the entrance beam definitely excludes the unsteady sides of the flame. To find this distance, a ground-glass plate can be placed in the plane of the photodetector, with the effective width of the photosensitive surface marked upon it with crayon. The flame is then focused upon it with the condenser, and the image can be inspected. It should be somewhat wider than the photosensitive area; the longitudinal adjustment will establish this condition. The longitudinal adjustment of a flame with respect to a monochromator will be taken up in Section 64.

38. *The Mirror*

The total radiant flux emitted by a flame is not great, and only a small fraction of it enters the flame photometer. Filter photometers are therefore provided with condenser lenses having the largest possible aperture. With monochromators (Section 61), the situation is decidedly less favorable because the small entrance slit can accept only a much smaller part of the radiation from the flame. Hence, the detection limits (Section 90) are often fixed less by the optical resolution (or the ratio of line intensity to background) than by the ratio of net photocurrent to residual current (including dark current, Johnson noise, etc.; see Section 65). In such cases, a genuine gain in sensitivity can be achieved by increasing the light intensity, as with a mirror behind the flame, although the ratio of line to background may be almost unaffected by the mirror. By tilting about the horizontal axis, the mirror can also be used, instead of a burner height adjustment, for vertical selection of the area of the flame viewed by the optics. This can help a good deal, especially with elements hard to excite and sensitive to the height in the flame.[300]

Even with filter photometers a mirror is generally placed behind the flame to increase the light intensity, although the luminosity of such photometers can be very high. The gain provided by the mirror permits economy of sensitivity in the photoelectric measurement.

The backing mirror commonly images the flame upon itself; that is, the concave mirror creates an inverted image of the flame, of equal size, at the location of the flame. This might be expected to double the light intensity; but owing to self-absorption (Section 8) the gain is less, depending on the quality of the mirror and the concentration of analyte in the flame.

Conversely, instead of a backing mirror, a light trap or black absorbing surface may be placed behind the flame.[735a] It often suffices merely to turn the backing mirror aside, provided that the housing wall thus exposed to view is black. The purpose is to displace toward a higher concentration the onset of self-absorption, which increases the curvature of the working curve at high concentrations. But this entails an unavoidable loss of sensitivity at low concentrations.

FLAME PHOTOMETRY WITH FILTERS

Up to this point the discussion has been general, applying to all types of flame analysis and apparatus. In the sections to follow, we shall be concerned with flame photometry in the restricted sense, in which the light emitted by the flame is resolved by means of filters. A diagram of a filter photometer was given in Fig. 8. Later sections (59–75) will take up flame spectrophotometry and flame spectrography. The sequence of sections will correspond to the sequence followed by the light beam.

39. *Condensers*

It is possible to place filters and photodetectors directly in front of the flame, opposite the mirror; but this is seldom done, because the detector and filter would be heated. Usually a lens system is employed to form an image of the flame upon the photodetector, which is placed at a distance. Under these conditions the aim is to maximize the radiant

Fig. 28. Lange flame photometer, Model 6 (ref. 443).

power at the detector and to select the best area of the flame (Section 37). At the same time, the condensing lens system should send the rays through the filter as an approximately collimated (parallel) beam (Section 40); this is especially important with interference filters. High-quality lens systems of large aperture may achieve an aperture ratio of about f/1.

When the aperture ratio of the lens system, the width of the flame, and the photosensitive area are specified, there is still the question of the best magnification ratio. It can be shown, for example, that to a first approximation the photocurrent is maximal when, with sufficiently large photosensitive area, the flame is imaged on the photodetector at a magnification of unity.[67]

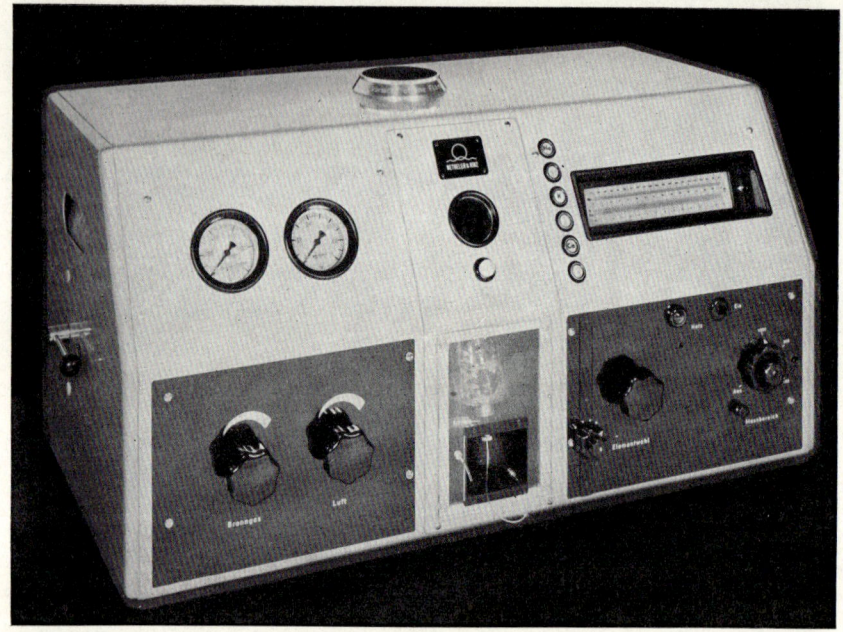

Fig. 29. Eppendorf flame photometer, by Netheler und Hinz, Hamburg (ref. 513).

40. *Filters*

The flame radiation admitted by the condenser system passes as an approximately parallel beam through optical filters whose purpose is to transmit the desired spectral line or band as fully as possible, while excluding the lines, bands, or continua of other elements and the flame background. Generally, there are various filter combinations for the isolation of different spectral lines; these can be introduced successively into the light path by means of a revolving plate or similar device. There may also be an arrangement for changing the amplifier sensitivity and preferably also the zero compensation, simultaneously with the filter, so that in shifting from one element to the next, the sensitivity and zero need not be readjusted. This arrangement is particularly helpful when the auxiliary internal-standard method (Section 52) is to be used. The transmitted light, in any case, next falls upon the radiation receiver (Section 42).

There are several ways of filtering out the desired spectral region, of which we name only the following as being the most important for flame photometric analysis: (a) liquid filters, (b) glass or gelatine filters, and (c) interference filters.

(a) *Liquid filters*. A glass cuvet filled with organic or inorganic colored solution is placed in the light path. The solution absorbs certain parts of

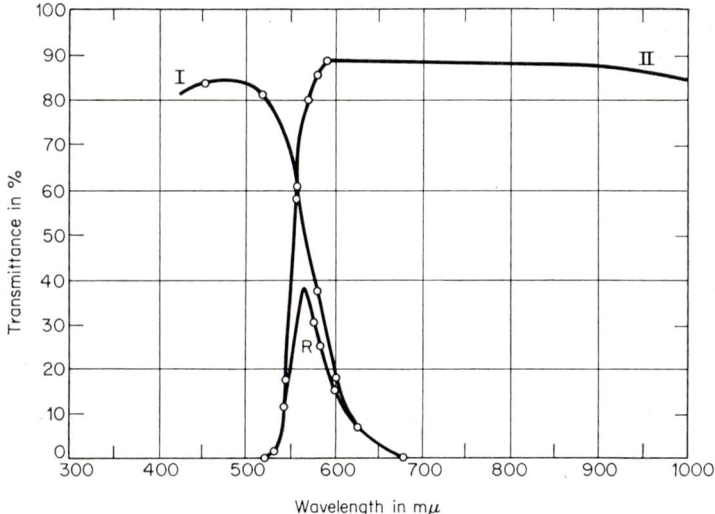

Fig. 30. Combination of two color filters for isolation of the yellow sodium line at 589 mμ (from ref. 224). I and II are the two component filters; R is the resultant transmission of the combination. I is Corning 9780; II is Corning 3482.

the spectrum more or less completely while transmitting other parts relatively well. The desired line or band can be more or less well separated by proper choice of the coloring agent and the thickness and concentration of the medium, or by superposition of several cuvets with different colored solutions. For practical reasons, glass cuvets with colored solutions are rarely used.

(b) *Solid filters*. It is better to have the colorant dissolved in glass or gelatine. Various spectral transmission curves can be obtained by combining filters and by varying the thickness. Fig. 30 shows a combination of two filters that isolates the yellow sodium line at 589 mμ fairly well. Filter I has maximal transmittance at about 500 mμ, and thus appears green. Filter II is uniformly transparent in the long-wave part of the spectrum, with a relatively steep absorption edge below 600 mμ and thus appears reddish or orange by transmitted light. If the filters are superposed, the resulting transmission curve, R, has a transmittance of about 20% near 589 mμ.

The efficiency of a filter can be characterized by the ratio of the desired transmittance at the line or band (in the example of Fig. 30, this is 20%) to the residual (undesired) transmittance in regions of the spectrum where the filter should be entirely opaque. The half-intensity bandwidth is also used as a measure of efficiency (see below and ref. 320). It indicates roughly the wavelength range within which the flame background is appreciably transmitted by the filter.

If the sample contains appreciable calcium along with the sodium, the undesired calcium light (bands of CaOH) at 554 and 616–630 mμ will also be transmitted by the filter of Fig. 30, though less strongly (in the example, the residual transmission is 20% at 554 mμ, 9% at 623 mμ). In other words, this combination is useful only when the sodium light greatly exceeds the interfering calcium light. In our example, a 1000-fold excess of sodium intensity is required to keep the calcium light below 0.1%. If this condition cannot be satisfied, better filters must be used, to absorb the calcium light more effectively; otherwise, errors will result from the type of interference called *cross-sensitivity* (Section 94), which can be allowed for only by troublesome corrections.

Spectral resolution cannot be easily improved by combining several color filters. The absorption edge can in any case be made steeper by greater thickness of the filter, but this reduces the peak transmittance, though not greatly, and imposes greater demands upon the sensitivity of the photodetector and the electronics (Sections 42 and 44). As shown below, this problem is better solved with interference filters.

For the sake of simplicity, we have discussed optical filtration up to this point as if the transmittance of the filter alone determined the resolution attainable. Actually, however, the spectral sensitivity of the photodetector (Section 42) must be taken into account.

This may be shown by an example. Suppose that the red potassium doublet at 767–770 mμ is to be isolated with filters. If the photodetector is one (such as a selenium photocell [Section 42]) which is scarcely sensitive to light beyond this wavelength (toward the infrared), then for isolation of the potassium doublet it suffices to use a filter having a sharp absorption edge (cut-off) just below this line (for example, the filter RG8 of Schott u. Gen., Mainz). But even then, a residual cross-sensitivity (Section 94) may exist with respect to the infrared sodium line at 819 mµ; this can be suppressed with an auxiliary heat-shielding filter, absorbing more strongly toward the infrared. But if a photo-detector is employed which has its maximal sensitivity in the infrared (Fig. 35), more effectual means will be needed for suppressing this unwanted sodium line, such as an interference filter (see below).

Strictly, we have to deal with the effective spectral sensitivity, that is, the product of the filter transmittance and the sensitivity of the photo-detector, expressed as a function of wavelength. The effective spectral sensitivities requisite or permissible at the wavelengths of interest are defined by the maximal tolerable cross-sensitivity and the intensity ratio of the interfering line to the analytical line. We cannot enter into particulars here, but the example used above will clarify the point. Here the infrared sodium line will not interfere if little sodium is present in the sample. But it would be hopeless to try to isolate the red potassium line with color filters if much rubidium were present, for it has two strong lines at 780 and 795 mµ. Fortunately, this element is rather rare.

It may happen that a desired line can be sufficiently well separated with color filters from all other emissions except one, viz., a strong line of another element. In this case it may be possible to suppress the interference with an ancillary barrier filter, which may itself be a color filter. The following is an example. Assume that calcium is to be determined flame-photometrically by the green CaOH band at 554 mµ in the presence of sodium. Adequate separation of this green band from the neighboring yellow sodium line at 589 mµ can be achieved only with the help of a barrier filter, for the long-wave cut-off of green color filters is fairly gradual. A suitable barrier filter for this case is a didymium-containing glass (for example, BG20 of Schott u. Gen.), which has an absorption band exactly at the location of the yellow sodium line (as well as bands in several other parts of the spectrum). This is shown in Fig. 31. The logarithmic scale of transmittance is better for judging the residual transmission in regions of interference. The filter used for calcium has three absorption peaks due to didymium. The chief transmission maximum lies in the green; subsidiary maxima occur in the orange and red. The latter do not interfere, as CaOH has bands here as well.

(c) *Interference filters*.* The simplest filters of this type, not widely used today, are constructed basically as follows. A thin, semitransparent layer of metal (generally silver) is evaporated on glass, and coated by evaporation with a spacing layer of transparent material (e.g., magnesium fluoride) with suitable index of refraction and thickness of the

---

* We wish to thank Dr. R. Schläfer of Mainz for many suggestions.

order of magnitude of a wavelength of light. Upon this a second metallic layer like the first is evaporated. A second plate of glass is finally cemented upon it for protection. In operation, the rays reflected from the upper and lower layers interfere with each other and with the transmitted ray. Depending on the path difference, which is determined by the thickness (more exactly, the thickness times the cosine of the angle

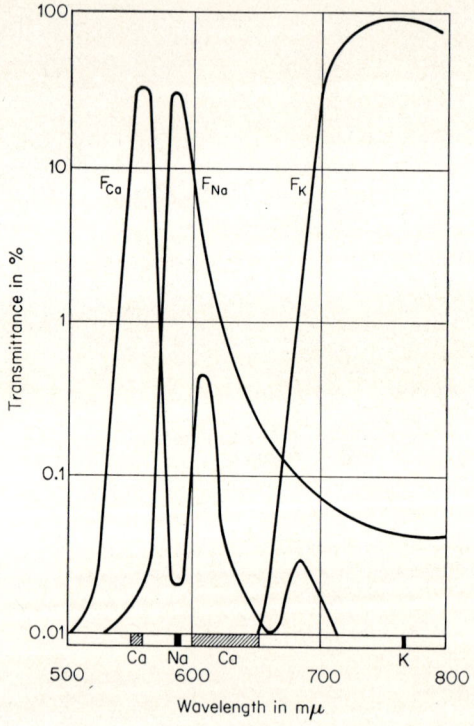

Fig. 31. Transmittance of a few color filter combinations for isolation of the lines Na 589 mμ, K 767 mμ, and the band CaOH 554 mμ (from ref. 524).

of incidence of the refracted ray), and the refractive index of the intermediate layer, certain parts of the spectrum are transmitted and other parts reflected. These simple interference filters have a maximal transmittance* T_m of about 40%, a half-intensity bandwidth HW (or simply *half-width*, the width of the transmission curve in mμ between the points where the transmittance is $\frac{1}{2}T_m$) of about 10 mμ, and a *tenth-width* (width between the points where the transmittance is

---

* The transmittance is the percentage of light transmitted, reflection losses at the surfaces of the filter being ignored.

$T_m/10$) of about 30 mµ. Since these filters have several maxima, they are combined with color filters to suppress the unwanted maxima. The color filter also lowers the filter "background" or stray transmission—the flat, extended transmission "tails" farther away in the spectrum.

This kind of interference filter is little used today, since several kinds of interference filters with better properties (steeper cut-off, lower stray transmission) have been developed specially for flame photometry.[620] These improved types include *double monochromatic filters*, consisting of two superposed simple monochromatic filters cemented together; *band-pass filters*, consisting of two directly coupled systems of layers (metallic layer, spacer, metallic layer, spacer, metallic layer); and *double band-pass filters*, consisting of two band-pass filters cemented together. Table 5 below lists typical transmission data for these four

TABLE 5. Transmittance of various types of interference filters for flame photometry (from the Jenaer Glaswerk, Schott u. Gen., Mainz.)

Peak wave-length	Type of filter	Relative transmittance at:			
		589	620	671	768 mµ
589	Monochromatic	1	.0022	1×10^{-4}	2.5×10^{-5}
	Double monochromatic	1	5×10^{-5}	$< 10^{-6}$	$< 10^{-6}$
	Band-pass	1	.01	1×10^{-4}	4×10^{-5}
	Double band-pass	1	3×10^{-5}	$< 10^{-6}$	$< 10^{-6}$
620	Monochromatic	$< 10^{-6}$	1	3×10^{-4}	1.5×10^{-5}
	Double monochromatic	$< 10^{-6}$	1	$< 10^{-6}$	$< 10^{-6}$
	Band-pass	$< 10^{-6}$	1	.0015	4×10^{-4}
	Double band-pass	$< 10^{-6}$	1	3×10^{-6}	$< 10^{-6}$
671	Monochromatic	$< 10^{-6}$	3×10^{-6}	1	8×10^{-5}
	Double monochromatic	$< 10^{-6}$	$< 10^{-6}$	1	$< 10^{-6}$
	Band-pass	$< 10^{-6}$	1.5×10^{-6}	1	1×10^{-4}
	Double band-pass	$< 10^{-6}$	$< 10^{-6}$	1	$< 10^{-6}$
768	Monochromatic	$< 10^{-6}$	$< 10^{-6}$	3×10^{-6}	1
	Double monochromatic	$< 10^{-6}$	$< 10^{-6}$	$< 10^{-6}$	1
	Band-pass	$< 10^{-6}$	$< 10^{-6}$	1.5×10^{-6}	1
	Double band-pass	$< 10^{-6}$	$< 10^{-6}$	$< 10^{-6}$	1

kinds of filters in relative units (referred to T_m) for each of four different wavelengths of interest in flame photometry (Na 589, Ca 620, Li 671, K 768 mµ). A 1 in the table denotes the maximal transmittance of the filter, while the other numbers show the relative residual transmittance of unwanted light at the other three wavelengths, without allowance for the spectral sensitivity of the photodetector. It is seen that the new double band-pass filters suppress the residual transmission to less than 1 p.p.m., so that they practically eliminate cross-sensitivities.

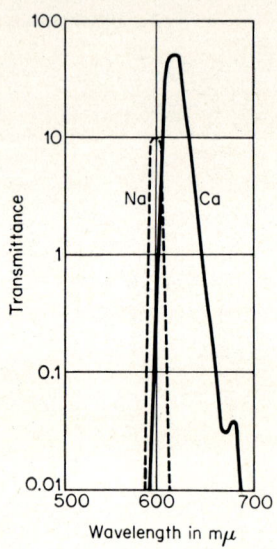

Fig. 32. Transmittance of two interference filters. ---- Double monochromatic filter for Na 589 mμ cemented to an OG 2 filter 2 mm. thick. —— Simple band-pass filter for Ca 620 mμ cemented to a BG 36 filter 2 mm. thick and an OG 3 filter 2 mm. thick (all filters are from the Jenaer Glaswerk of Schott u. Gen., Mainz). Note the logarithmic transmittance scale.

Fig. 32 shows the spectral transmission of a double monochromatic filter of recent type, combined with a 2 mm. OG2 filter, for sodium determination, and that of a simple band-pass filter for calcium at 620 mμ combined with a 2 mm. BG36 filter and a 2 mm. OG3 filter. Note the logarithmic scale of transmittance, which is more useful than a linear scale for showing large differences of transmittance.

Interference filters must be used in a nearly parallel beam, since the path difference and consequently the wavelength of peak transmission depend on the angle of incidence. However, in some of the newer filters the peak wavelength and other properties (half-width, etc.) have been made less sensitive to oblique incidence. These new filters can tolerate an aperture angle of 40°, and band-pass filters even 60°. Interference filters are nowadays available throughout the range from 215 to 2700 mμ.

We should mention also the *interference wedge* filter. This is a filter strip having a tapering intermediate layer, along which the wavelength of the transmitted light changes continuously and linearly, corresponding to the colors of the spectrum. As above, *monochromatic wedge* filters and *band-pass wedge* filters can be made. Such filters permit continuous adjustment of wavelength. A good interference wedge will yield a luminosity one or two orders of magnitude above that of a typical monochromator (Section 60), but its spectral purity is generally lower. The luminosity cannot, however, equal that of an interference filter, since the interference wedge must be used in conjunction with a slit.

The problem, discussed above for color filters, of barring an exceptionally strong interfering line or an entire spectral region, is better solved with an interference filter, since such a filter can be used also in reflection. It reflects all wavelengths except those which it transmits. By means of two similar interference filters, arranged in series as reflectors, any given line (transmitted by the filter) not too close to the region of interest may be removed so completely as not to interfere.

The practicing analyst will not wish to be concerned with the development of filter combinations according to the principles discussed above. For his application with its particular concentration ratios, we recommend a test by the methods of Section 94 of the efficiency of optical separation of the desired line from the interfering lines in his instrument (filter plus photodetector). If the results of this test are not satisfactory, perhaps a filter manufacturer can help him improve the situation, making use of his cross-sensitivity values.

41. *Diaphragms*

An adjustable diaphragm is often placed before or behind the filter for regulation of intensity. The function of this component will be considered along with the electronic control of intensity in Section 48. Sometimes a fixed diaphragm is employed to exclude light from the outer parts of the flame.

42. *Radiation Receivers*

The radiation receiver (variously called *photoreceiver* and *photodetector*) converts the luminous flux of the line or band isolated from the mixed radiation into an electric current proportional to the intensity, which is easily measured. For this purpose, many types of radiation receiver are available. We shall mention the most important, with their advantages and drawbacks, without inquiring further into their mode of operation.[28,58,148,250]

(a) *Barrier-layer cells* (*photovoltaic cells, photocells*). These generally consist of selenium, now also of silicon, deposited upon a metallic base and coated by evaporation with a thin, transparent metallic layer. This layer and the metallic base serve as the electrodes. Fig. 33 (upper part)

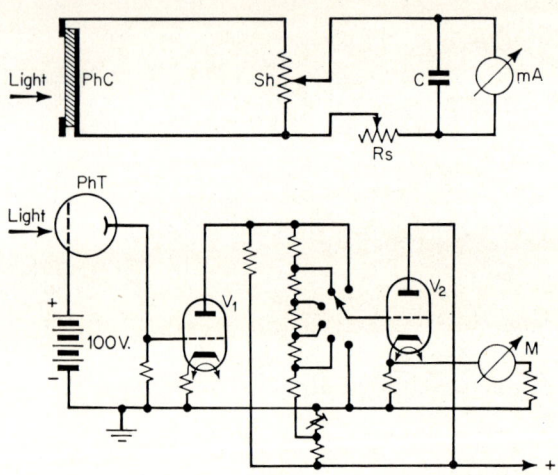

Fig. 33. Above: Circuit for a selenium photocell with intensity control and damping condenser. PhC, photocell; Sh, shunt; Rs, series resistor; C, condenser; mA, milliammeter or galvanometer. Below: Circuit for a phototube with vacuum-tube d.c. amplification. PhT, phototube, biased with a 100-volt battery; V_1, first amplifier tube with fixed grid resistor; V_2, second amplifier tube, at higher potential, with switchable grid resistor providing stepped sensitivity control; M, meter.

shows the circuit for a photocell. No auxiliary voltage is required for removing the electrons produced on illumination; the illuminated photocell operates as a voltage source. No dark current exists in the absence of radiation. Selenium photocells are small, rugged, and light. Besides, their sensitivity, about 0.1 amp. per watt of incident radiation, is quite good. The photosensitive area and the aperture angle within which radiation is received can be made as large as desired. Owing to their low internal resistance, photocells are insensitive to electrical interference in the vicinity. However, they have certain disadvantages. The current is proportional to the light only within a limited range, and only if the indicating meter has a sufficiently low internal resistance, not over 500 ohms. This nonlinearity can be removed by shunting with a low resistance, or by balancing the photocurrent potentiometrically. Because of their low internal resistance and the low load resistance necessary for linearity, amplification is impracticable. Selenium photocells age; they show fatigue and after-effects on prolonged illumination, and the zero returns to its original position only after a considerable interval of darkness. The photocurrents are temperature-sensitive, although recently cells with very low temperature coefficients have become available. Photocells offer a smaller choice of spectral sensitivity curves than the photodetectors listed below. Also, their total spectral range (300–800 mμ.) is narrower than that of the various kinds of vacuum phototubes taken together.

(b) *Vacuum photoelectric cells (phototubes)*. These consist of an evacuated glass or quartz envelope containing an evaporated photosensitive film of alkali metal and an insulated mesh or circular anode. Fig. 33 (lower part) shows the circuitry with amplification symbolized. The photoelectrons liberated from the photosensitive layer are drawn to the anode by a biasing voltage. If the voltage is strong enough to bring all the electrons to the anode, a saturation current results, which is proportional to the intensity of the light. The need of biasing voltage makes the circuitry more elaborate than that for a photocell. Moreover, owing to the low sensitivity of phototubes, the photocurrent must usually be amplified. The high-value load resistors required for this purpose are sometimes liable to insulation troubles. However, there are these advantages: generally good proportionality between current and light intensity (linearity), ease of amplification, and the availability of a variety of phototubes with different spectral sensitivity curves, most commonly the curves shown in Fig. 35. Gas-filled phototubes are not recommended, being only moderately linear and not particularly stable.

(c) *Multiplier phototubes (photomultipliers)*. These are vacuum phototubes with amplification through secondary electrons. They combine the advantages mentioned above with that of amplification within the tube itself, which, unlike electron-tube amplification, is practically free from noise. These tubes will yield currents up to 10^7 times that from a simple phototube. Accordingly, they need less amplification (or none) and a smaller load resistor, and cause less trouble with insulation. The amplification in the photomultiplier is easily adjusted by changing the voltage on the dynode stages (Fig. 34). Against these advantages there are the following drawbacks. A very stable direct-current power supply is required, of 500–2500 v., depending upon the type of photomultiplier and the desired amplification. The stability of this high-voltage supply must be an order of magnitude better than the required precision; for a precision of 1%, fluctuation of the stabilized voltage should not exceed 0.1%. Other drawbacks are the fatigue and saturation that appear at higher levels of illumination, and the sensitivity to electromagnetic interference. There is also the high cost of the photomultiplier and its requisite accessory equipment, the rectified high-stability voltage supply.

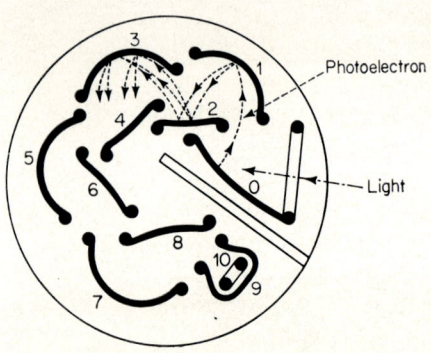

Fig. 34. Structure of a multiplier phototube (cross-section of an RCA 1P28). The heavy lines show the dynodes; the dotted lines show the paths of the primary and secondary electrons.

Amplification within the tube occurs in the following way: In the first stage, between the cathode and the first dynode plate (1 in Fig. 34), the photomultiplier operates like the vacuum phototube described above. The electrons striking the first dynode generate additional electrons, called secondary electrons, at this point. Depending on the voltage between the cathode and the first dynode, each arriving electron generates one to four secondary electrons. These are accelerated by the field between the first and second dynode, and on striking the second dynode liberate further electrons, so that the number of electrons leaving the second dynode can be as many as 16 for each original photoelectron. In each successive stage this operation is repeated. Thus photomultipliers with 9–11 stages can amplify up to 10^7 times.

The analyst will usually be working with a ready-built filter photometer equipped with a radiation receiver by the manufacturer. Sometimes there is a choice of two or more receivers having different spectral sensitivity characteristics (see below). The one best suited to the case at hand must then be selected. This point deserves further comment. To be universally applicable, the photoreceiver ought to be uniformly sensitive in all parts of the spectrum (a multi-octave receiver). But, on the other hand, to avoid cross-sensitivity (Section 94) the receiver should be sensitive only at the wavelength of measurement. In the extreme case, a photoreceiver of very selective sensitivity would need no filter or monochromator. However, neither of these two extremes can be realized with the photoreceivers under consideration here. Fig. 35 shows the spectral sensitivity curves of several practically important photomultiplier tubes. Phototubes are made with similar curves. It is usually preferable to operate as close as possible to the maximum of these curves.

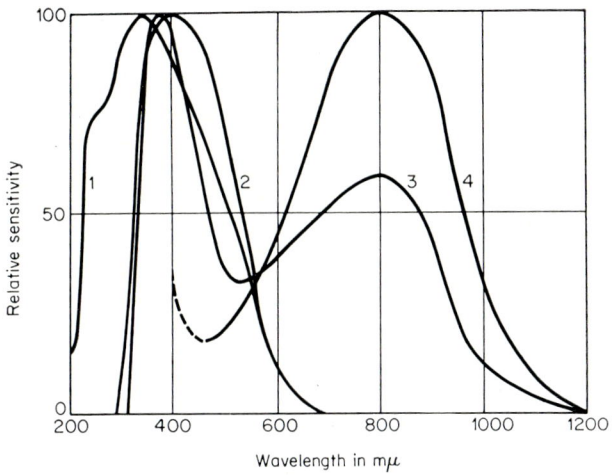

Fig. 35. Spectral sensitivity of several multiplier phototubes. 1, RCA 1P28; 2, RCA 1P21; 3, Maurer A/e; 4, RCA C7160.

The user of a filter photometer will often wish to increase its sensitivity. He can do this by substituting a more sensitive photodetector, such as a photomultiplier in place of a selenium photocell. In making this change, he has to keep in mind its effect upon the resolution required of the filter (Section 40); otherwise, troublesome cross-sensitivities may show up.

A photoreceiver should be so installed that no stray light can strike the photosensitive layer, which must also be protected from dust, combustion products, etc. Multiplier phototubes operated at highest sensitivity are advantageously suspended by springs to avoid the microphone effect caused by vibration. It is also advisable to employ a

6+C.A.F.P.

shield of magnetically soft metal (Mu-metal) against electromagnetic disturbances.

Insulation troubles with phototubes having high load resistance or high amplification can be alleviated with a desiccant in the phototube housing, such as indicating (blue) silica gel. This must be replenished occasionally. Hence, too, a flame photometer should be set up if possible in an area of low humidity.

All photodetectors show changes of sensitivity to some degree. These are more marked at the start, but after an hour or two become so gradual as scarcely to interfere in the usual mode of operation (Section 85) with regularly interposed standardizations. With photocells it is almost entirely a matter of fatigue, but this hardly occurs in vacuum phototubes. In multiplier phototubes positive or negative changes of sensitivity may be noticed at currents above 10^{-5} amp. These are due, however, less to changes in the photocathode than to changes of secondary electron emission from the last, most heavily loaded dynode, where saturation may occur. Currents of this magnitude, therefore, should be avoided. These drifts are mitigated by pre-exposure; that is, the photometer should be allowed to warm up before the start of the actual analysis by spraying a standard solution or the like for a time with the instrument turned on. A photomultiplier should be turned on several hours in advance, or else kept under voltage continually, day and night. Since it consumes no current when not illuminated and in the absence of a voltage divider, there is no drain on batteries, if used. A good way to prevent changes of sensitivity in a photomultiplier is to avoid high measuring currents by using low dynode voltage; this also minimizes difficulties from imperfect insulation. In the case of photocells, fatigue can be lessened by reducing the diaphragm aperture. The loss of photocurrent can then be compensated by (non-fatiguing) electronic amplification. The more serious errors can also be avoided by frequent calibration with the standard solutions and adjustment of the sensitivity (Section 85).

The effect of fatigue of photocells, etc., upon the precision of flame-photometric analytical results is usually overestimated. A large part of the effects sometimes blamed on the radiation receiver may reside actually in variable, temperature-dependent precipitation of the spray upon the walls of a spray chamber (Section 26), or in partial clogging of the atomizer.

The luminosity of the filter photometers under consideration is generally so great that the photocurrent of a phototube with or without electron multiplication is much larger than the dark current. Dark-current drift will then cause trouble only in red-sensitive photomultipliers, which have larger dark current. The problem of dark current and its drift, less critical for filter photometers, will be taken up in Section 65 in connection with flame spectrophotometry.

43. *Adjustment of the Radiation Receiver*

The photodetector should be so placed that the beam of light emerging from the optical system (the *exit beam*) covers the entire photosensitive area. If the detector is too close to the exit aperture, only a part of the photosensitive layer will be utilized, and the small illuminated zone is more likely to be overloaded. Both fatigue and the effect of mechanical vibration will then be more noticeable.[148] But if the detector is placed too far away, some of the light will pass the photosensitive surface unused. The lateral and vertical adjustments should be such as to maximize the signal from the exit beam. Since the exit beam is not homogeneous and the sensitivity of the photosensitive area is not uniform, the detector can be optimally positioned only by trial. This optimal position can vary with wavelength, but in practice this refinement is generally ignored. In a good instrument the photodetector has usually been correctly adjusted. The flame should be imaged upon the detector only when its surface is large and of roughly uniform sensitivity, as in the selenium photocell. On small photosensitive surfaces, as in the 931A and 1P28 photomultipliers, lateral displacement of a flame image will bring light from different parts of the flame into play; the flame should not be imaged upon these photodetectors.

44. *The Amplifier*

In many cases, the photocurrent from the photocell or photomultiplier goes directly to a sensitive indicating meter (a moving-coil galvanometer or, better, a mirror galvanometer) to provide a reading. If the sensitivity of the optical–electric system (comprising flame, condenser with mirror, filter, photodetector, and meter) is insufficient, the current or voltage can be enhanced (up to about a million-fold) by incorporating an amplifier between the detector and the meter. Amplification should not be carried so far as to reveal the statistical fluctuations (shot-effect noise) of the dark current or the electronic noise of the circuit elements (resistors, tubes). Manufacturers often equip their instruments with amplifiers only to avoid expensive, sensitive, cumbersome galvanometers.

A distinction is made between alternating-current and direct-current amplifiers. For the former, the light emitted by the flame has to be converted into alternating light. This may be done, for instance, with a rotating sector—a partly transparent, partly opaque disk, interrupting the beam periodically, for example, at 50 c.p.s. This method will eliminate any constant or stray light as well as the dark current of the photodetector, and it provides a steadier zero point than d.c. amplifiers, but the equipment is costlier.

An unsuitable amplifier can lead to errors of measurement:

1. The amplification and over-all sensitivity depend, in many designs, upon the condition of the amplifier—upon warming time, room temperature, condition and age of the tubes. Moreover, the amplification may be nonlinear; that is, the output current or voltage may depart from proportionality to the input current or voltage.

2. An a.c. amplifier generally introduces a hum or noise component into the measuring current, while a d.c. amplifier introduces an anode current. These have to be subtracted from the total current like a dark current or an unwanted component of flame photocurrent (Section 95), or else compensated in some way. These currents may not be constant, especially during warm-up of the amplifier. The dark-current noise and amplifier tube noise impair the detection limits more with d.c. than with a.c. amplifiers.

3. If the amplifier is poorly designed, the reading may be affected by line voltage fluctuations and other electric, magnetic, or mechanical disturbances, such as the microphone effect.

Most of these troubles become insignificant with proper design, as with a cathode-follower circuit or an a.c. balanced (push–pull) amplifier. But a given amplifier should be used for flame photometry only if tests show it to be satisfactory.

A good amplifier must have constant sensitivity and zero at sufficiently high amplification, good linearity, long-lived tubes and other components, and simple, accessible construction to facilitate repairs. It should have several steps of amplification; for instance, consecutive positions of the range switch might represent a factor of 2 in amplification. There should also be a fine adjustment to cover the gaps between these discrete steps and allow accurate balance. The sensitivity adjustment and the zero adjustment should be independent of each other. An adjustable time constant of the meter (Section 46) is very desirable; this can be had by suitable circuitry, such as a selection of capacitors. If little material is available for analysis (see Section 89 for micro methods), a quick response or short time constant of the meter will be wanted: for instance, an *RC* of about 0.5 sec.* But with quicker response the readings of the photometer will be less steady and the attainable precision consequently lower. When there is enough sample, a more sluggish indicating system (*RC* about 1–5 sec.) is preferable. It will then take longer to get a final reading on the meter, but this reading will be steadier. In any case, a skilled observer can estimate the mean reading of an unsteady needle quite accurately. There is also the practical matter of eliminating electrical interference by good grounding. Details of amplifier design cannot be covered in this book; see the literature.

45. *Power Supplies*

The amplifier requires heater current and anode voltage. The phototubes require a well stabilized d.c. biasing voltage, and multiplier phototubes require a very stable d.c. supply of 500–2500 v., depending on the type and the desired amplification. Requirements of stability are very stringent here, because voltage fluctuations are considerably magnified in the output owing to the multiplying effect. In a.c. amplifiers, hum in the rectified, stabilized high voltage must be kept low. These various voltages or currents cannot be taken directly from the line. There are different ways of providing the requisite power: (1) The heater currents can be taken from a storage battery, and the plate voltage from B batteries. A set of batteries in series provides the high voltage. (2) The needed voltages and currents can be taken from the line via converters, transformers, rectifiers, etc. These methods can be combined. Their advantages and drawbacks are as follows.

With battery operation, the storage battery must be kept charged, the acid checked, the clamps lubricated, the dry cells changed, etc. Charging is unnecessary when the battery is connected as a buffer across a charger. Analytical errors result from neglect (insufficient battery charge or failure to replace dry cells). But with freedom from the line, it becomes possible to carry out analyses directly in the field, as in agricultural investigations. There can be no errors from fluctuations of line voltage or frequency or from switching transients, etc.

With line operation there is, of course, the convenience of not having to tend batteries; the instrument merely needs to be plugged in. Line disturbances may upset the measurements, but their effect can be kept within tolerable limits by proper circuitry.

* This corresponds to an equilibration time (time required for 99.5% attainment of the final value) about five times as great, about 2.5 sec. in this instance.

46. *The Indicating Meter*

The simple, direct-reading filter photometers under discussion here generally have a moving-coil or a mirror galvanometer. On this, the zero lies at one end of the scale, so that the reading (in scale divisions) is a measure of the photocurrent and therefore of the concentration of the element being determined in the flame. (Exceptions: potentiometric circuits, Section 49; internal standardization, Section 51.) The term *direct-reading* may denote (a) readout by the deflection of a galvanometer or other directly indicating meter, as opposed to potentiometric (null-balance) readout, or (b) both direct and potentiometric readout on a single element, as opposed to any system of internal standardization.

The meter reading must be large enough to have some reading precision, and must also be proportional to the current and sufficiently reproducible. The length of the scale determines the reading precision and the possible range of concentration. If, for instance, the scale has 100 divisions and can be read to one-half division, then readings of less than 50 scale divisions are insufficient for a precision of 1%. In other words, the lowest and highest concentrations may not differ by more than a factor of 2 for 1% precision with this scale. But if the scale has 1000 divisions with the same legibility, a 10 times greater range of concentration can be embraced while retaining precision of 1%.

Modern galvanometers generally have a projected scale, like that of a rapid-weighing balance. Sometimes the ordinary scale supplied with the instrument can be replaced by a special precalibrated scale that permits direct reading of the concentration of analyte, for example, in milliequivalents per liter or milligrams percent in serum, or in percent CaO or MgO in cement or slag, provided the samples are properly prepared and the instrument properly adjusted.

It should be noted that the readings of most meters depend upon position; that is, it makes a difference whether the meter stands straight or tilted or lies on its side, owing to effects of friction, balance, etc., and the reading can depend also on the position of the zero. This is particularly true of mirror galvanometers. Further, distortion of the working curve can result from the fact that the angular displacement of the rotating mirror is projected upon a straight scale.

All of these meters are affected by certain disturbances, such as vibration, temperature fluctuations, or external magnetic fields (as from another nearby moving-coil meter). In an industrial laboratory it may be hard to mount a galvanometer vibration-free; the meter can be mounted on a bracket fixed to a supporting wall of the building, or an inertial mass flexibly suspended, such as a cast iron plate resting on tennis balls or soft rubber stoppers.

When electronic means are not employed for adjusting the zero, this is done mechanically, as by displacing the scale, the return spring of a moving-coil meter, or the supporting pivot of the galvanometer. If the zero point is changed too much, as in mechanical suppression of the dark current, the sensitivity may be affected, or hysteresis or nonlinearity may arise. The smaller the reading is, the greater are the percentage errors of reading a meter (both personal and real). It is desirable, especially with short scales, to adjust the sensitivity or to choose the concentrations (Section 79) so that the readings will lie in the upper (usually right-hand) third of the meter scale. With needle meters, errors of parallax must be avoided by viewing the scale perpendicularly above the needle. This is facilitated with a scale backed by a mirror (a mirror scale), together with a vertically flattened needle; in reading the scale, the needle is superposed on its reflection.

Meters must not be subjected to too heavy a current; in general, the meters mentioned here can be briefly overloaded to 10 times the full-scale current. Likewise, the photoreceivers will not endure too much illumination. In instruments with amplifiers, the meter furnished by the manufacturer usually cannot be overloaded, since the tubes act as a current limiter.

In the choice of a meter, the period and the external resistance are very important. The latter should be such that the meter will operate exactly in the limiting aperiodic state, i.e., so that it is critically damped, and neither oscillates (overshoots) nor moves too slowly. Short-period meters with a response time under 2 sec. are not advisable, unless the amplifier has a larger time constant. Otherwise, the reading will be unsteady, owing to the unavoidable inconstancy of the atomizer and flame. A very sluggish meter gives considerably more exact readings, since it averages the photocurrent over an extended time. A long time constant is advisable only when the light flashes caused by dust particles entering the flame are eliminated, as by a closed system (Section 32). Otherwise the signals from the flashes, which, though small, constitute an increment in the meter reading, take a long time to decay. But when only a small amount of sample is at hand, a slowly equilibrating meter cannot be used. In any case, long equilibration time means slower analyses (Section 91). A compromise must be sought between these requirements.

In null methods (Sections 49–51), a less sluggish meter should be used, with a period of about 1 sec. Resistors and/or capacitors in parallel with the meter will duly raise the time constant; several authors have used this scheme.

In simple photometers with selenium barrier-layer photocells without amplification, a galvanometer must be used. Its sensitivity must be about 10^{-9} amperes per 0.001 radian, and the internal resistance must not exceed 500 ohms. In instruments with phototubes and amplifiers, the required meter sensitivity is governed by the amplification and the required over-all sensitivity.

Before being put to use, an instrument, including the meter, should be tested for linearity. The test will show whether curvature of a working curve is real. If fictitious curvature due to nonlinearity is encountered, steps can be taken to eliminate it from the instrument.

The test can be made with a steady, sodium-containing flame or some other sufficiently constant source, such as a battery-operated incandescent lamp. The meter is made to read full scale by adjusting the sensitivity. If the light is now attenuated in an aperture plane (the filter is usually located in such a plane in commercial instruments) with a quantitatively adjustable rotating sector disk or with gray filters of known optical density, such as are used in colorimeters, the meter should give the correct reading corresponding to this attenuation. If it does not, then one or more of the components—photodetector, amplifier, or meter (or the subdivisions or slidewire of a potentiometer, Section 49)—is nonlinear.

A simpler test can be carried out by first taking a reading A_1 on a high concentration of sodium in the flame. A beam attenuator is then inserted at a suitable point as above, and the reading A_2 is taken; this attenuator can be of any kind, for example, a reasonably uniformly smoked glass plate or a not too dark photographic plate (exposed, developed, and fixed). The test is then repeated with a lower concentration of sodium, yielding an unattenuated reading A_3 and an attenuated one A_4. If the instrument is linear, $A_1/A_2 = A_3/A_4$.* An appreciable departure from this proportion shows that one or more of the components (detector, amplifier, meter) is nonlinear.

The faulty component can be identified as follows. The meter alone is first tested with a battery and a precision decade resistor box. It is made to read full scale; if the resistance is doubled, the meter reading should drop to half; if quadrupled, to one-quarter, etc. If the meter is linear, the amplifier plus meter can be tested by applying quantitatively variable voltage from a precision voltage divider. The meter must give

* Unfortunately, the converse is not necessarily true. If the instrument response is equal to any power of the light intensity ($R = I^n$, n any constant), the test will be passed.—*Translator*.

proportional readings; if it does not, the amplifier should be modified or replaced. If the amplifier and meter are linear but a nonlinearity is observed in the over-all instrument including the photodetector, then only the latter can be at fault.

These tests should be carried out with the various sensitivity settings likely to be used in practice, and should be repeated occasionally, perhaps once a year.

47. *Integration Methods*

We have hitherto assumed that the meter reading represents the momentary photocurrent, perhaps amplified. In the spectrographic methods to be considered later (Section 70), the principle is a different one. Here the radiant flux is integrated over an extended period in the form of increasing blackening of the plate. The plate blackening is a measure of the time integral of the intensity during the exposure. An integrating method has the advantage of being nearly unaffected by fluctuations of the excitation source (unsteadiness of the flame, atomizer, etc.). Integration can also be done electrically. The ideal case of complete integration is approximated to some extent by increasing the time constant or response time of the meter in a direct-reading method. There are electronic circuits that can integrate exactly. Most simply, the photocurrent charges a condenser during illumination. The final charge on the condenser measures the time integral of the photocurrent during illumination. There are many other kinds of integrating circuit, which we cannot discuss here.

In a recent technique used for micro- and ultramicroanalysis, the exposure time (integration time), instead of being arbitrarily chosen, commences when the sample first ascends the capillary tube and ends when the volumetrically measured sample (about 0.1 ml.) is consumed.[255] Various circuits can be used. For example, the liquid itself can serve as an electrolytic switch.[724] For this purpose, the capillary is covered with a plastic tube surrounded by a metal cylinder, and a voltage is imposed between the cylinder and the capillary. As soon as the liquid bridges the plastic, a current flows, starting the integration via a relay; when the liquid is consumed, the current is interrupted and the integration stops.

A basic drawback of all the above integration methods is that accidental flashes of light, caused by dust particles entering the flame from the surrounding air, produce photocurrent pulses that add to the total reading, which cannot afterward be corrected for them. This is

6*

true not only of sodium and potassium analyses (dust particles contain considerable amounts of these elements), but also, if to a slighter extent, of determinations at other wavelengths, for sodium and potassium emit a continuum as well (Section 9). But when instantaneous readings are being taken, these light flashes, causing brief jerks of the meter needle, can be ignored. Closed systems in which purified air is supplied relieve this difficulty, although not completely (Section 32).

For the owner of a recorder there is another way of integrating that is free from the above basic disadvantage. During atomization of a sample of measured volume, the photocurrent is recorded as a function of time. The difference between the areas under the sample and the blank recordings measures the integrated net emission, provided that the light flashes are deducted, and that the sample flow rate lies in the range in which the photocurrent is proportional to it. Many manufacturers (see ref. 513, for example) equip their flame photometers with means of attaching a recorder. The area can be determined with a planimeter or with a special integrating attachment on the recorder which reads directly on a counter or the like. In this method a correction for flashes based on the recorded curve can be deducted from the automatically integrated reading.

However, the above assumption of proportionality between flow rate and reading is seldom valid in practice. It can be realized, for example, by means of controlled sample pumping. More commonly, the flow rate is such as to yield peak intensity. The integration described above is then invalid. Recording is still useful, however, since it easily gives the mean intensity reading during spraying without the need of averaging many separate readings on a meter. Also, it serves to check the atomizer performance; samples of equal volume should be consumed in equal times, and the times are easily read from the record.

48. *Control of Sensitivity*

In practical flame photometry, it is useful to have one or more controls by which the reading on a standard solution of known concentration can be set to the same value each time. Several schemes are used for this purpose.

1. An adjustable (iris) diaphragm may be placed in the light path and set at about one-half or two-thirds open at the start of a series of measurements, so that in case the over-all sensitivity declines during prolonged operation, the diaphragm can be gradually opened to compensate this fatigue. To avoid varying the observed area of the flame

on opening or closing the diaphragm, the latter is placed at a point in the beam where there is no imaging of the flame, as immediately before or behind the filters or one of the lenses. A drawback is that with very high intensities the diaphragm must be nearly closed, and with the usual designs control is then very coarse.

2. A variable resistance may be placed in parallel and/or in series with the meter, to shunt part of the (amplified) photocurrent or to lower the voltage at the meter (Fig. 32). An undesirable consequence is that such a control usually affects the response time of the meter. With photocells, this arrangement is suspect because the external resistance of the circuit affects the characteristics of the photocell; but these effects may be disregarded when less accuracy is expected of the photometer.

3. If an amplifier is used, the gain can be easily altered, as by varying the grid leak of one or more tubes (Section 44).

4. The sample itself can be quantitatively diluted or concentrated (Section 79) to arrive at the desired range of sensitivity.

With any of these methods of control, a test should be made to determine whether change of the over-all sensitivity displaces the zero point of the scale or the flame background.

49. *Potentiometric Circuits*

Instead of being read out as a meter deflection in scale divisions, as described in the preceding sections, the (amplified) photocurrent passing to the meter can be balanced against an opposed, adjustable compensating current; that is, the auxiliary current is so adjusted as to make the meter read zero. This potentiometric method requires a sufficiently stable source of auxiliary voltage, such as a battery (B_2 in Fig. 36), and an adjustable resistance or potentiometer P, which provides a balancing current or voltage that can be adjusted by moving the slider on the potentiometer. The voltage taken off P is varied until the null meter N is balanced. The concentration is then read not on N but from the scale of the potentiometer P.

Fig. 36 shows the principle of the balancing circuit. Light strikes the phototube T, liberating photoelectrons from the cathode, which are drawn off as photocurrent by the biasing voltage B_1 (about 100 v. for phototubes without multiplication) through the resistors R_1 (10^6–10^{10} ohms) and R_2 (about 10 ohms). The resulting voltage drop in R_1 (with R_2) amounts to about 1 v. This is amplified and fed to the null meter, which goes off scale (say) to the right. By means of the auxiliary circuit, comprising battery B_2 and the variable resistor P, a balancing voltage is imposed across R_2 by adjusting P, which changes the voltage level of R_1 just enough to balance N again. The auxiliary voltage drop across R_2 is then equal and opposite to the voltage drop due to the photocurrent across R_1. The voltage drop across R_2 due to the photocurrent can be neglected because R_1 greatly exceeds R_2.

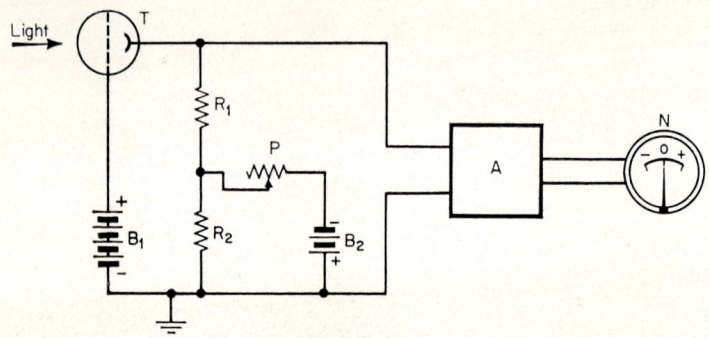

Fig. 36. Example of a schematic null-balancing potentiometric circuit (see text). T, phototube or photomultiplier; B_1, battery or sufficiently constant d.c. voltage supply for drawing off photoelectrons; R_1, load resistor of about 10^6–10^{10} ohms; R_2, zero-point resistor of about 10 ohms; P, precision potentiometer; B_2, battery for potentiometer; A, amplifier; N, null meter.

A practical circuit will have special provisions for switching the sensitivity and for compensating the dark current and flame background; we shall not enter upon these details here.

The following are basic advantages of the potentiometric circuit:

1. The attainable precision is generally better than in direct-reading methods, for good potentiometers depart less from linearity (not over 0.025%) than moving-coil galvanometers or the like. The reading precision is also good, especially when there is a choice of several (say, 10) values of the resistance.

2. Potentiometers are less easily disturbed than precision meters by vibration, tilting, thermal effects, magnetic fields, etc.

3. Inaccuracy of needle deflection or of the scale of the null meter N does not affect the measurement. The only requirement of the meter, aside from a minimal sensitivity, is that the null point be reproducibly indicated when after successful balancing no current flows through the meter.

4. The null meter can be more sensitive than a direct-reading meter,

since it does not matter if the deflection prior to balancing exceeds the range of the meter. A limit is imposed only by the fact that no meter should be overloaded with too heavy a current for an extended time. Overloading will be impossible if the amplifier limits the current to a certain maximum. A device for switching the meter sensitivity will help by permitting the circuit to be balanced first roughly in the "coarse" (low sensitivity) position of the switch, and then precisely in the "fine" (high sensitivity) position. The latter position may be engaged by pressing a button.

5. Similar remarks apply to the amplifier (if any), provided that the balancing is done on the input side of the amplifier. The amplifier then does not need to be at all linear or even of constant amplification. It is required only that zero voltage at the input should always correspond to zero current at the output of the amplifier.

The following are disadvantages of the potentiometric circuit:

1. The cost is higher because of the more elaborate circuitry (precision potentiometer, etc.).

2. Additional effort is required to watch the meter while balancing the potentiometer.

3. In taking readings on very small samples (micromethods, Section 89), the sample may be consumed before the potentiometer can be balanced.

4. For a beginner, the procedure is harder to keep track of; also, it is harder to diagnose difficulties than in a direct-reading instrument. Greater skill is therefore demanded of the user.

5. Scanning a spectrum (measuring the intensity while advancing the wavelength steadily) with a spectrophotometer (Sections 59 ff.) is very awkward or impossible with a null meter.

Manufacturers sometimes offer instruments permitting both direct reading and null balancing. The bridge can also be balanced automatically with a motor. Further, the measuring potentiometer can be combined with a printing roll which makes a printed record of the analytical results without intervention of the observer. A strip-chart recording potentiometer can be used in this way.

50. *Direct Multiple-Beam Methods*

When several elements are to be determined quantitatively in a sample solution by flame analysis, one of the above types of flame photometer can measure the elements in sequence. But if several optical systems are disposed around the flame, with separate filters, detectors, and meters, the desired elements can be measured simultaneously.[339, 548] This method has the following advantages over sequential measurement:

1. Time is saved by avoiding change of filters, adjustment of sensitivity and zero, and repeated spraying of standard solution.
2. Sample is economized since only enough is needed to yield the single set of readings.
3. The concentration ratios of the analytes are obtained more accurately, because irregularities of the atomizer and flame affect these ratios considerably less than in sequential measurement.
4. The sensitivities of the photoreceivers can be better adjusted to the problem at hand, because no compromises are needed among the competing demands that must be met when different elements are determined consecutively with the same receiver.
5. The several meters can be equipped with fixed scales calibrated in units of concentration. This obviates reference to working curves and change of scales.

There are the following disadvantages:

1. The instrumentation is costlier.
2. The intensity gained with backing mirrors must be forgone, since there is hardly space for mirrors in a multiple-beam instrument.
3. Multiple-beam flame photometers are not commercially available (see the index of manufacturers in the Appendix).*

51. *Internal Standardization*

The internal-standard method has also been called the indirect method or double-beam method. It is distinguished from direct methods (Sections 39–50) in that the intensity of a spectral line or band is not used directly as a measure of the concentration of the analyte, but instead the ratio of the intensities of two different lines is used as the measure. The one line is, as before, the analytical line, and the other the so-called *internal-standard line*. The latter is usually a line of an *internal-standard element* added in equal concentrations to the standard and sample solutions—generally with the diluent—and not present appreciably in the original sample. The internal-standard emission may also be a band of a suitable internal-standard element. Alternatively, the comparison standard may be an element already present in sufficiently constant concentration, usually a major constituent; in

*One is now available. See p. 430.

this case, no internal standard is added. Lundegårdh[33] introduced a further modification in using a suitable part of the flame background adjacent to the analytical line as internal standard. In flame-spectrographic methods of sufficient inherent precision, he recommended the deliberate addition of an internal standard only when disturbances of atomization might be expected.

Comparison with an internal-standard line diminishes or removes an assortment of errors which afflict the direct methods. In any given case, however, the suitability of the chosen auxiliary emission (line, band, or background) must be tested by determining the direction and magnitude of the interferences affecting the two emissions separately and in conjunction. The interferences should be as nearly as possible equal for the internal-standard and analytical emissions, but at all events at least of the same sign, so that the interferences may be largely cancelled by taking the ratio of analytical to internal-standard emission. Otherwise, the internal-standard method will give less accurate results than the direct method. A good many line pairs are known to react very similarly to interferences (see Section 55). Moreover, there are several types of error that will always be pretty well cancelled regardless of the choice of internal-standard substance or line (but excluding flame background). These include all errors due to irregularities of atomization and variable condensation of the liquid in the spray chamber of an indirect atomizer. Any residual errors may then be due, among other things, to the fact that the analytical line and the internal-standard line have working curves of different curvature; but these errors are sometimes reversed in sign.

Lithium is the internal-standard element most commonly used. As a matter of convenience, we shall speak of lithium instead of internal-standard element. But misgivings have lately arisen regarding this use of lithium, because its emission is more sensitive to location in the flame (Section 11) than that of other elements.[662, 696] Also, changes in the gas composition and the OH content of the flame have a much more marked effect on lithium than on sodium and potassium.

52. *Auxiliary Method of Internal Standardization*

Flame analysis by internal standardization requires special apparatus, to be discussed in the next section. Nevertheless, a simple flame photometer of the type described in Sections 39–49 can accommodate this method by the sequential measurement of analytical and internal-standard lines, as in the following example:

Suppose that a rather viscous sample is to be analyzed for potassium with the help of aqueous standard solutions. Because of the effects of viscosity upon the suction rate, drop size, and condensation, systematic errors will be likely if the viscous sample and aqueous standards are compared directly; specifically, the result for potassium may come out too low.

The error can be revealed by adding to the sample and standard equal quantities of lithium, for example, as lithium chloride at 20 p.p.m. of lithium; the apparent potassium concentrations of sample and standards are then directly measured with a filter transmitting the potassium line at 767 mμ, and next the lithium intensities are measured with a filter for the line at 671 mμ. If, now, the lithium intensity in the sample is smaller than in the standards, the ratio of these intensities is applied as a correction factor to the potassium value. This method has the advantage of requiring no special apparatus except the filter for the internal-standard line.

A modification of this method employing a more sophisticated evaluation is the following.[96] The photocurrents from the analytical and internal-standard lines are not compared, but rather the apparent concentration of internal-standard element in the sample is found from a working curve obtained with standard solutions. The ratio of apparent to true concentration of internal-standard element (lithium) in the sample is called the "concentration correction factor". This factor is then applied to the apparent concentration of the analyte, likewise found with the help of a working curve prepared from standards. This method, unlike the preceding one, gives accurate results when the working curves of the analyte and the internal-standard element have different curvatures. However, it would be difficult to make this method automatic (Section 91).

In routine work, these expedients lack convenience for the following reasons: (1) Sequential measurement of internal-standard and analytical lines (potassium and lithium in the above example) takes more time. (2) Consequently, more material is consumed, for liquid must be atomized during the entire analysis. (3) Additional time is required for calculating the quotients and concentration correction factors. (4) The results may still be open to question, as when the over-all sensitivity of the instrument changes between the measurements of the analytical and internal-standard lines. (5) Only systematic changes in atomization, such as those due to differences in viscosity and surface tension, are equalized; accidental irregularities of atomization are not offset.

Hence, long series of determinations by the internal-standard method are better carried out with special equipment which permits *simultaneous* measurement of both intensities (internal-standard and analytical) with automatic presentation of the quotient. Such instruments are described in the next section.

53. *The Optical–Electric System for Internal Standardization*

Two optical systems must be provided—two condensers, two sets of filters (one for the analytical and one for the internal-standard line), and (unless a switching arrangement is used) two photoreceivers. Fig. 37 shows the arrangement invented by Berry, Chappell, and Barnes. The light from the flame is separated into two beams, directed

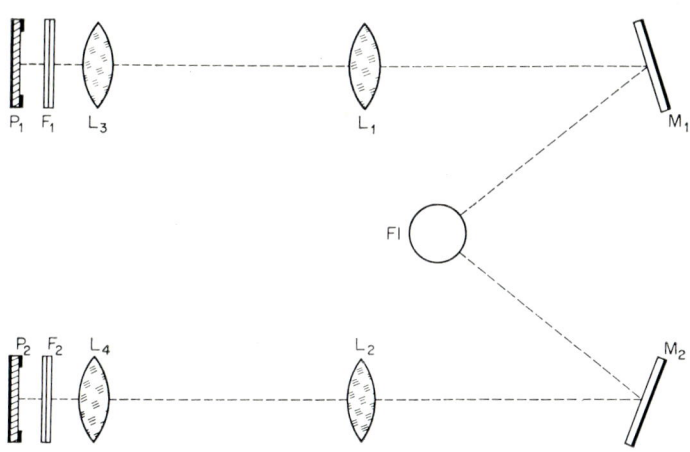

Fig. 37. Double-beam system of Berry, Chappell and Barnes.[140] Fl, flame; M_1 and M_2, mirrors; L_1 and L_2 condenser lenses; L_3 and L_4, field lenses; F_1 and F_2, two sets of filters; P_1 and P_2, photocells.

by the mirrors M_1 and M_2. The flame is focused by the lenses L_1 and L_2 upon the field lenses L_3 and L_4, respectively. A set of filters next isolates the analytical and internal-standard lines. For the one filter there is a choice of sodium (589 mμ) or potassium (767 mμ); the other is a fixed lithium filter (671 mμ).

The intensity quotient of analytical to internal-standard line can be obtained, e.g., from a bridge circuit. One branch contains a fixed resistance, while a second, symmetric branch contains a precision potentiometer, and the other two branches contain the two photocells.

When the null meter in the bridge is made to read zero by adjusting the potentiometer, the position of the slider (the reading on the potentiometer) measures the quotient of the two photocell voltages and hence the concentration of analyte.

Regarding the photometric system of Fig. 37, it should be added that the flickering of the flame is not necessarily directionally isotropic; compensation of the errors due to flicker will hence be incomplete. If flicker is to be eliminated completely by a two-beam technique, the photoreceivers must be arranged to view the flame from the same direction. This can be done with a semitransparent mirror. The single beam of radiation emitted by the flame is divided by the mirror into two partial beams. Each passes through a filter, one for the analytical and one for the internal-standard line, and thence to a photoreceiver for each.

Disadvantages of circuits with two photoreceivers: (1) If the sensitivities of the receivers change by different amounts during the measurement, the accuracy will suffer. (2) The varying dark currents of the receivers (except photovoltaic cells) likewise affect the accuracy.

Attempts have therefore been made to develop systems having two optical paths, as above, but only one photoreceiver. One such scheme is to modulate each beam with a different frequency and convert these to alternating current with a single photoreceiver; the a.c. signals can then be separated in the amplifiers. The two signals can also be modulated with the same frequency but 180° out of phase. One of the signals is varied with an optical attenuator until the reading on a synchronous phase-sensitive meter is exactly zero. The calibration of the attenuator directly gives the intensity ratio of analytical to internal-standard line.

54. *Multiple-Beam Internal Standardization*

In the preceding section we have described a method of balancing the intensity of a single analytical line (sodium or potassium), chosen at will, against that of a single internal-standard line (lithium). As with direct-reading flame photometers (Section 50), internal standardization can be extended to the simultaneous determination of several elements by means of multiple light paths. However, the several analytical line intensities are measured not directly but as ratios to the intensity of one or conceivably several properly chosen internal-standard lines.[344] As in Section 50, the instrument requires a multiplicity of light paths with different filters and photoreceivers for the several analytical lines and one or more filters and photoreceivers for the internal-standard line or lines. Each intensity quotient of analytical to internal-standard line can be obtained with a bridge circuit as in Section 53, one bridge being required for each analyte. If one internal-standard element is to serve simultaneously for several analytes, the (amplified) photocurrent from its line must be suitably distributed to the bridges of the several analytes.

The advantages of such an instrument are the same as those of a multiple-beam direct-reading flame photometer. There are also the general advantages peculiar to any internal-standard method (next section).

55. *Advantages and Drawbacks of Internal Standardization*

(*a*) *Advantages*. 1. For a given type of instrument, internal standardization yields appreciably better relative precision (reproducibility) than direct reading, by compensating the effects of irregularities of atomization, variable loss of liquid in the spray chamber, and flame flicker on the reading. If the highest precision is not required, the deficiencies of a simpler instrument, for instance one with an inferior atomizer, can be made up by internal standardization.

2. Systematic errors due to differences in viscosity and surface tension (Section 101) are considerably reduced.

3. Errors due to radiation interference are usually reduced, often to less than half that occurring in the direct method.

4. An instrument set up for internal standardization can generally also be used for direct reading.

(*b*) *Disadvantages*. 1. The multiple-beam system with its extra photodetectors, bridges with precision potentiometers, etc., is costlier. Moreover, the filters have to have appreciably higher selectivity, especially the one for the internal standard, to avoid errors due to cross-sensitivities (Section 94) afflicting the internal-standard line.

2. It takes more work to establish the method initially, since the standardizing and measuring procedures must take two concentrations and the associated interferences simultaneously into account.

3. Pipetting the lithium-containing diluent entails unavoidable errors. These cause larger errors in an internal-standard determination than the corresponding errors of dilution in the direct method, for they affect oppositely the concentrations of analyte and internal standard, and thus are magnified when the quotient is taken. This unfavorable effect of the dilution error is greatest for equal quantities of sample and

diluent (1:1 dilutions). Sometimes the addition of the lithium and the dilution are done separately,[269] in which case this effect occurs only with the addition of lithium, not with the subsequent dilution, which does not then have to be quantitative.

4. In unfavorably constituted cases, other sources of error can affect the analytical and internal-standard lines in opposite senses; an example is the change of flame temperature resulting from change of the ratio of gas to air and its unequal effects upon the excitation conditions of the two lines.

5. Considerable error can arise when the sample unexpectedly contains appreciable amounts of the internal standard.

Some further remarks are in order. The addition of the internal-standard element in equal concentrations to the individual sample solutions represents no additional work in case a dilution of the samples happens to be necessary. A lithium-containing diluent is then merely added, instead of distilled water.

Some line pairs useful for internal-standard analysis are listed in Table 6. Calcium is not very good as an internal-standard element, as it is a common impurity.

TABLE 6

Analytical line, mμ	Internal-standard line, mμ	References
Na 589	Li 671	269
K 767	Li 671	269
Ca 554	Li 671	130, 344
Ca 630	Li 671	340, 342, 344
Ca 554	Na 589	269
Li 671	K 767	269
Li 671	Co 352.9	164a
Cu 324.8	Ag 328.0	207
Ba	Li 671	309
Fe 386.0	Co 387.1	209

The reduction of error by internal standardization has been the subject of many investigations.[130, 131, 132, 140, 270] We mention two instances below.

1. The effect of varying *gas and air pressures* upon the result. Fig. 38 shows how the photocurrent varies with the gas pressure in the direct and internal-standard methods respectively. It is evident that

gas pressure has much less effect upon the result in the internal-standard method. The effect of inconstant air pressure is also diminished. In Fig. 38, the air pressure is shown as parameter; the set of curves is seen to be spaced more closely in the internal-standard method. At all events, the direct method would be in a very unfavorable range here, far from the maximum of the photocurrent–gas pressure curve (cf. Section 19).

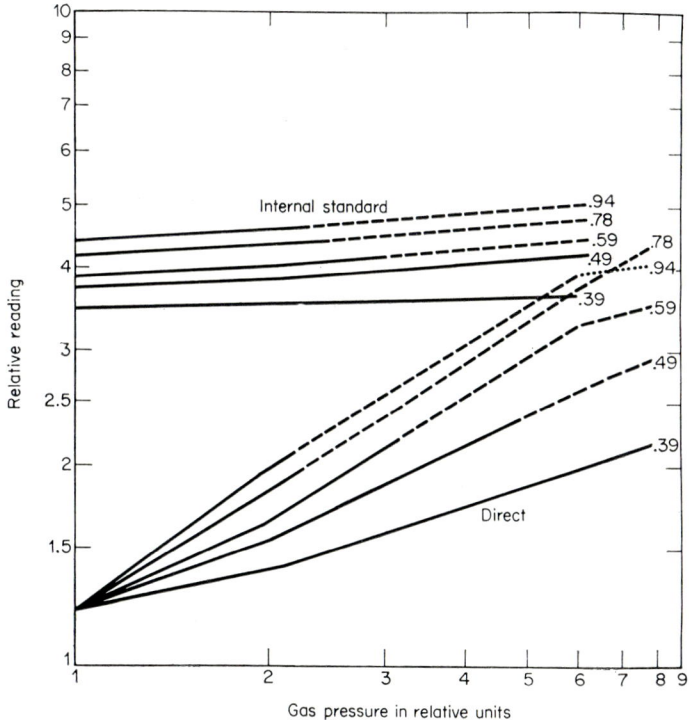

Fig. 38. Effect of gas pressure on the reading in the direct-reading method and in the internal-standard method. The air-inlet orifice is shown as parameter (from ref. 735a).

2. Reduction of *nonspecific radiation interference* (Sections 11 and 101). For example, sulfuric acid depresses the emission of sodium increasingly with rising concentration of acid, as shown in Fig. 39. It is clear that internal standardization can reduce this error to as little as one-tenth that of the direct method, although part of this effect can be ascribed to the normalizing action of lithium (Section 101).

56. Recording Methods with Filter Photometers

The recording of flame spectra is now quite common, but it cannot be done with a filter flame photometer unless it has an interference wedge. We return to this in Section 68. But even without an interference wedge, a recorder can be put to good use in several ways with a filter instrument. In these applications the reading is usually recorded as a function of time, the wavelength being fixed by the filter. In Section 47 we mentioned the use of a recorder for obtaining the time integral of the intensity of a line, which in the present case would be isolated

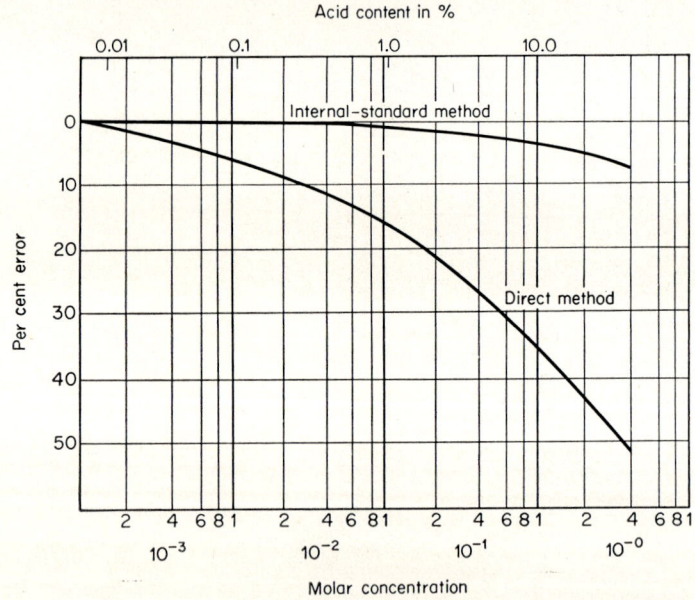

Fig. 39. Effect of sulfuric acid concentration upon the percentage error for sodium in the internal-standard and direct methods (from ref. 140).

by a filter. By this technique appreciably lower detection limits can be reached, a matter of importance for micro- and ultra-microanalysis.[255,585,724]

Another matter of interest is the variation of a concentration with time in the control of an industrial process. Thus, for instance, the sodium content of a solution flowing through a pipe can be continuously indicated by diverting a small fraction of the liquid through a flame photometer. The amplified photocurrent is then fed to a recorder. If the concentrations of several elements are needed, they can be recorded with the help of an appropriate multiple-beam optical system (Section 50) and a multiple-pen recorder using inks of different color. Such measurements can be made directly or potentiometrically (Section 49) or by internal standardization (Section 51). In the two latter cases the recorder must be a recording potentiometer which, at intervals, records that fraction of the auxiliary voltage necessary for null balance or the voltage due to the internal-standard line. The periodic calibrations (Section 85) necessary for controlling the sensitivity of the entire system can also be made automatically. Some recording potentiometers compare the auxiliary voltage with a standard cell at intervals such as 5 min. and automatically adjust the sensitivity accordingly. For the present purpose, comparison with the standard cell must be replaced by comparison with the photocurrent from a periodically atomized standard solution.

The recorder can be set up to activate an alarm when the concentration exceeds previously set limits either upward or downward. Automatic production control can also be effected by holding a stream concentration as constant as possible with a recorder-controller. This topic belongs to the field of control engineering.

57. *Multiple-Use Flame Photometers*

There are several types of flame photometer which, through interchange of parts, can be used not only for flame analysis but also as colorimeter, turbidimeter, conductimeter, ohmmeter, flowmeter, fluorimeter, reflectometer, thermometer, etc. Such instruments are generally built on the module (building-block) principle; that is, parts can be rearranged so as to convert the instrument to a different function. This design principle is based on the self-evident fact that the optical–electric measuring system of a flame photometer is useful also for other optical measurements such as colorimetry or nephelometry, while the electrical system can be set up with appropriate adjuncts for other electrical measurements, etc. Thus an assortment of instruments is obtained at a comparatively low price.

But certain drawbacks mitigate this advantage: (1) Only one of the instruments can be used at a time. (2) The instrument must be altered for every change of use, often with readjustment and probably recalibration. This demands more skill of the user. (3) If the apparatus is used for a long time for only one purpose, the unused parts can go astray. (4) The design of the modules entails compromises between efficiency of a given part and the functioning of the entire combination. In other words, a flame photometer of modular design is generally less efficient than an instrument designed specifically for flame photometry.

A multipurpose instrument can therefore be recommended only when financial means are restricted, and when one or another of the parts will be used only occasionally. The situation is different with spectrophotometers (Section 59). Here the main component, for example, a grating monochromator, is so costly that the use of separate attachments to convert the main instrument to different functions is justified.

58. *Limitations of Filter Photometers*

Interference filters have been developed to the point where filter flame photometers can now be fully competitive with the usual monochromator instruments in the visible range, at least for the determination of sodium, potassium, lithium, and calcium. For this purpose, filter instruments, which commonly use low-temperature flames and indirect atomizers, can be even better than a monochromator instrument, since certain interferences, especially ionization interference among the alkali metals (Sections 7 and 97), are less pronounced in a cooler flame, and the phosphate interference with calcium and other elements (Sections 7 and 98) is also less marked with an indirect atomizer. But if other elements are to be determined, such as strontium, barium, magnesium, or elements with lines in the ultraviolet, then filters give way to monochromators, which, however, owing to their low luminosity (Section 36), generally employ atomizer-burners with their hotter, turbulent flames of higher luminous output. Thus filter and monochromator instruments nicely supplement but cannot fully replace each other.

FLAME SPECTROPHOTOMETRY

59. *General*

In spectrophotometry, the spectral lines are isolated with a monochromator instead of filters. A monochromator is a spectroscope having an exit slit in the focal plane of the spectrum, where the photographic plate would be in a spectrograph (Section 70). This exit slit isolates from the spectrum a single, very narrow wavelength range—practically a "line". A polychromator is a similar instrument having several exit slits for the simultaneous isolation of several wavelengths.

The following are the advantages of flame spectrophotometry as compared with filter flame photometry (excluding interference wedges):

1. The instrument can be set for any application, that is, for any element determinable in the flame, by simply selecting the wavelength on the monochromator. Among several lines of the same element, the most suitable can be chosen (Section 82), whereas with filter instruments the choice is restricted to the few wavelengths transmitted by the filters provided with the instrument. Moreover, any region of the flame background can be selected and separately studied.

2. A line can be resolved from adjacent lines and flame background generally better with a large monochromator than with filter combinations. (A small monochromator is about equal to a good interference filter in the visible.) For example, the lines Mn 403.5 mμ and K 404.4 mμ cannot be separated with filters. Moreover, the spectral bandwidth of a monochromator can be varied continuously by varying the entrance and exit slit widths; this is not possible with filters. If the resolution is inadequate in an exceptional case (owing to stray light, for instance), it can be improved sufficiently for the most exacting needs with the help of filters (Section 40) or by the use of a double monochromator (two single monochromators in tandem). This improves the dispersion as well as the stray light.

3. The analytical wavelength range can be extended into the ultraviolet with monochromators having ultraviolet-transmitting optics (usually quartz). This can be done at best only imperfectly with filters, which are not available with sufficiently well-defined transmission bands in the ultraviolet. Ultraviolet monochromators hence permit determination of many additional elements and offer a larger choice of analytical lines. Such a choice is often very helpful in eliminating crosssensitivities (Section 94) or other interferences.

The following are disadvantages of flame spectrophotometry in comparison with filter flame photometry:

1. The apparatus is elaborate and costly, especially grating and quartz prism monochromators and double monochromators.

2. Monochromators transmit a smaller portion of the flame radiation than filters, because of the small entrance slit. For the measurement of such small radiant output, the most sensitive photoreceivers and very good electronics are needed. Since the flame radiation is inherently weak, the detection limits may be set less by the optical resolving power (governing the intensity ratio of line to background) than by the ratio of photocurrent to dark current (electron emission of the unilluminated photocathode) (Section 65).

3. Operation is more complex; in particular, location of the intensity maximum (peak of the spectral line) and choice of appropriate slit widths demand some care. These adjustments affect the coordinates of the working curves, the magnitude of the flame background (Section 95), and the cross-sensitivities. Spectrophotometry therefore requires more skill than filter photometry.

Since the advantages outweigh the drawbacks, flame spectrophotometry has lately gained popularity, especially in America. Simple flame attachments are connected to spectrophotometers, which have become established in a great variety of laboratories for absorption measurements and the like. A few such combinations will be described in Sections 60 and 63.

60. *Spectral Dispersion*

The light from the flame is dispersed spectrally with a prism or grating; accordingly, we distinguish prism instruments and grating instruments. The prism makes use of the fact that the refrangibility of light by the prism material (glass or quartz) varies with wavelength. Normally the refractive index rises toward shorter wavelength. The grating employs the wavelength-dependent interference of rays diffracted at the lines of the grating. Grating instruments ordinarily give higher dispersion (greater expansion of the spectrum). Since flame spectra are not rich in lines, most applications are satisfied by lower dispersion, which is more advantageously supplied by a prism.[298] Prism instruments are therefore commoner, and only these will be described here.

It must be admitted, however, that the number of grating instruments, though small at present, is gradually increasing (see the list of manufacturers in the Appendix). This is due to the need for better detection limits (provided by higher dispersion) in the analytical problems of nuclear physics (Section 117), biology and medicine (Section 115), etc. We therefore intercalate a few remarks on grating monochromators.

The angular dispersion (see below) of a grating is nearly constant with small angles of incidence and therefore excels that of a prism especially in the visible and infrared. Unlike that of prism instruments, the wavelength scale is nearly linear (equal spaces on the scale corresponding to equal wavelength intervals) everywhere in the spectrum, permitting more convenient and accurate interpolation. The image of a straight entrance slit in the plane of the exit slit is nearly straight with a grating;

this aids resolution. With a prism the slit image is more curved, the curvature depending on wavelength. The disadvantage of the older gratings, which spread the meager radiation of the flame among the spectra of differerent orders, has been removed in the new echelette gratings through their specially shaped (blazed) grooves. The blazing concentrates the energy largely into the spectrum of a single order. A comparison of grating and prism monochromators with respect to luminosity and optical transmittance is not a simple matter.[389,390,541] In the larger grating monochromators, lower transmittance (caused, e.g., by preliminary dispersion with a filter) or a lower luminosity can be compensated by widening the slits, without much loss of resolution.

Returning now to prism monochromators, we define the *dispersion*, D. This is the property of a prism by virtue of which it refracts light of different wavelengths to different degrees. It represents the dependence of the refractive index n of the prism material (glass, quartz, etc.) upon the wavelength; specifically,

$$D = dn/d\lambda$$

Values of D or n as functions of the wavelength λ for various materials have been collected in tables.[3,29] This property D (which is, incidentally, temperature-dependent) determines the so-called *angular dispersion* $d\theta/d\lambda$, which gives the angular difference between rays of different wavelength separated by the prisim. For the commonest case, in which the light traverses the prism symmetrically, the following equations hold:[28]

$$d\theta/d\lambda = (d\theta/dn)(dn/d\lambda) = D(d\theta/dn) = D(b/d)$$

where b is the length of the prism base (in prisms with autocollimation it is twice the base) and d is the diameter of the beam at the prism. This angular dispersion, together with the dimensions of the monochromator or the spectrograph, determines the spacing of the spectral lines on the photographic plate or in the plane of the exit slit of the monochromator. This spacing is called the *linear dispersion* $d\lambda/ds$. It is expressed in mμ/mm. or in A./mm. Under the above assumptions,

$$d\lambda/ds = 1/f(d\theta/d\lambda) = d/fbD$$

where f is the focal length of the imaging system, e.g., the focal length of the camera objective of a spectrograph.

The most important property of a monochromator or spectrograph is not the linear dispersion but the so-called (theoretical) *resolving power* R, which gives the least separation $\Delta\lambda$ of two lines at the wavelength λ which can be barely shown by the instrument to be separate. The value of the resolving power is given here without proof:[28]

$$R = \lambda/\Delta\lambda = b\, dn/d\lambda$$

This relation holds only if the prism is fully illuminated. Conversely, the formula gives the base length of the prism needed for resolving a given line pair. Sometimes several prisms are placed in series to attain

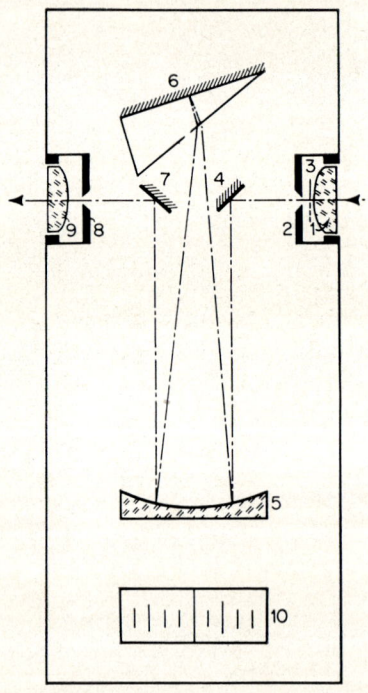

Fig. 40. Beam path in a monochromator with an autocollimating (Littrow) prism (schematic) (Carl Zeiss, Oberkochen). 1, Field lens; 2, entrance slit; 3, pivoted shutter; 4, reflecting prism; 5, concave mirror; 6, Littrow prism; 7, reflecting prism; 8, exit slit; 9, field lens; 10, optical wavelength indicator.

a greater base length b, or the light may be passed through the same prism several times. The resolving power of an actual instrument is generally poorer than the theoretical, owing to unavoidable departures from the ideal. However, the theoretical resolving power gives an indication of the efficiency of a monochromator or spectrograph.

Further, the luminosity L (for a definition see Section 36) is important in judging a monochromator for purposes of flame photometry. It is a coefficient which, multiplying the radiance of the source, yields the power received by the photodetector behind the monochromator. If radiance is measured in w./cm.^{2}sr. (sr. = steradians) and luminosity is in cm.^{2}sr., then the power comes out in watts. This relation is valid

Fig. 41. Model DU spectrophotometer by Beckman Instruments, Fullerton, Calif., with flame attachment.

only if light from the source fills the entrance beam fully and the detector receives the entire exit beam. In the interest of adequate radiant power, the height l and width s of the entrance slit of the monochromator should not be too small, aside from the question of proper positioning of the flame before the monochromator (Section 64). At all events, a limit is imposed upon s by other considerations (Section 62).

Specifically, we assume that, as is customary, the entrance and exit slits have the same height l and the same width s, that the focal lengths f of the collimating lenses or mirrors are equal, and that their diameters are large enough not to restrict the beam. The beam aperture is assumed to be limited by the prism. The luminosity L of the monochromator is then given [318] by:

$$L = slah/f^2$$

where h is the useful (beam-limiting) height of the prism and a is its width projected upon the plane normal to the beam axis. Alternatively,

$$L = \Delta\lambda R l h / f$$

where $\Delta\lambda$ is the spectral slit width (bandwidth corresponding to the width of the exit slit) and R is the resolving power as above. If S is the radiance of a monochromatic source and N the power received by the photodetector, then

$$N = LS$$

But for continuous light the radiance must be expressed in terms of bandwidth. If S_λ is the radiance per unit bandwidth (e.g., per mμ) at wavelength λ, then

$$N = LS_\lambda \Delta\lambda$$

In view of the various relations discussed above, this can also be written

$$N = S_\lambda \overline{\Delta\lambda}^2 R l h / f = S_\lambda \Delta\lambda s l a h / f^2$$
$$= S_\lambda (d\lambda/ds) s^2 l a h / f^2$$

The last formula is useful when the linear dispersion $d\lambda/ds$ is available. Incidentally, the last formula shows clearly that for a continuum N is proportional to s^2, while the earlier formula for monochromatic emission shows N proportional to s. We return to this in Section 62.

It is assumed in the above that no losses of light result from absorption, reflection, and diffraction in the optical components such as mirrors, lenses, or prisms.[318] If these losses are to be allowed for, the luminosity L should be multiplied by the dimensionless optical transmission factor T (between 0 and 1). T depends somewhat on λ. In good instruments T will not be much less than 1.

61. *Requirements of the Monochromator*

1. A monochromator should have adequate luminosity (Section 60) and good optical transmission.

2. The resolving power should be sufficient to permit separate measurement of close lines such as K 404.4 and Mn 403.5 mμ. High dispersion or resolving power also makes for high line-to-background ratio and low detection limits (Section 90). Data on separability and recorded spectrograms of certain lines useful in flame photometry have been published.[307,518]

3. Operation should be simple, and the wavelength and slit dials should be easily legible and free from parallax.

4. The wavelength should be readily adjustable. It should be possible to locate the maximum of a line exactly even with slits as narrow as a few hundredths of a millimeter.

5. The wavelength setting should remain constant. It should be free from backlash or play and independent of temperature, so that it will not drift off the line during an extended run. The prism material should have a low thermal coefficient of refractivity, or the monochromator should be compensated for temperature, or else the room should be thermostated, if high stability of the wavelength is expected.

6. A wide range of wavelength is needed. Instruments with glass optics are not useful much below 350 mμ owing to absorption of the short-wavelength part of the spectrum. Quartz monochromators are better in this respect, but they have only moderate dispersion in the visible.

7. Stray light must be low. The monochromator should pass very little interfering light of unwanted wavelengths along with the wavelength being measured.

In a (single) monochromator of good quality, the stray light is so low that the response drops by several powers of ten when the wavelength setting is moved off a line to the adjacent region. If a weak line is to be measured next to a very strong one, this stray light may be too great. It can be further reduced by insertion of suitable color filters or interference reflection filters for suppression of the interfering line, or by use of a double monochromator.

62. *The Slit Width*

In the operation of a monochromator, the entrance and exit slits must be adjusted properly. In the types commonly used for flame analysis, the entrance and exit slits are coupled so as to be equal for all width settings. Only this case will be considered below.

We assume the slit to be of good mechanical quality. Among other things, its edges must be smooth and parallel to each other and to the refracting edge of the prism. Further, the exit slit has to be curved so as to match the mean curvature of the image of the entrance slit as closely as possible. The two jaws of each slit must move with simultaneous and equal displacement, so that the central or mean wavelength of the transmitted band will not shift on changes of slit width.

In the case of a line, according to Section 60, when the slit width is changed, the radiant flux density (illumination) at the exit slit does not change, but the radiant power collected by the photoreceiver does—except with extremely narrow slits. Under the above conditions, therefore, the photocurrent from a line is approximately proportional to the slit width. More exactly, the photocurrent varies as $W^2/(W+\Delta W)$, where W is the slit width and ΔW the practical (effective) resolving power expressed in terms of slit width. In the Beckman DU spectrophotometer, for example, ΔW is about 0.03 mm.

But the response to a continuum such as the flame background behaves differently; the radiant power received varies as the square of the slit width. Consequently, the ratio of line to background, important in determining the detection limit, is approximately inversely proportional to the slit width; more exactly (see above), the ratio varies inversely as $W+\Delta W$. These different relations between the photocurrent and the slit width for lines and for continua, respectively, do not, however, cover all cases; narrow and wide bands show still different (intermediate) behavior. Further complexities arise if the slit jaws are not parallel; e.g., the quadratic relation between continuum response and slit reading will fail.

The following considerations favor narrow slits: (*a*) A higher ratio of line intensity to flame background and, hence, better detection limits (Section 90); (*b*) better separation of close lines. The following considerations favor wide slits: (*a*) the photocurrents are greater and more easily measured; (*b*) sensitivity of the wavelength and slit adjustments to shock and temperature is greatly reduced or eliminated.

A compromise has to be drawn. Fig. 42*a* shows an idealized recorder trace obtained by moving the exit slit slowly past the image of a line. The base of the isosceles triangle diminishes with the slit width. For defining the wavelength range transmitted by the monochromator, the dashed mid-line HIB is used, rather than the base of the triangle. This quantity, the half-intensity bandwidth (Section 40), is also called the *effective spectral bandwidth*. The actual recorder trace, as compared with the idealized one, is somewhat broadened and rounded at the corners (Fig. 42*b*), owing to errors in the lenses or mirrors used for imaging, diffuse reflection from the imperfect surfaces, inaccuracies in the prism surfaces, imperfect parallelism of the entrance and exit slits, inaccurate curvature of the exit slits, etc. A consequence of these optical defects, which are hardly avoidable, is that at very narrow slits the ratio of line to background no longer improves on narrowing the slit to the extent that it would with perfect optics.[300,408] In the Beckman DU monochromator (Fig. 41), on narrowing the slits, for instance, from 0.05 to 0.01 mm., the ratio of line to background increases only in the ratio 1:2 rather than 1:5. For simplicity, diffraction at very narrow slits is ignored here. While the ratio of line to background changes little at these narrow slits, the intensities diminish with the square of the slit width. Therefore,

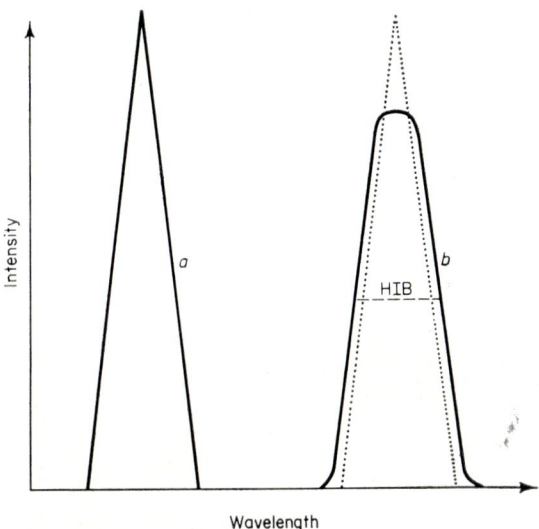

Fig. 42. *a*, Idealized recorder curve of a line, for equal entrance and exit slits; *b*, recorder curve actually obtained under these conditions. ---- Midline, the effective spectral slit width or half-intensity bandwidth (HIB).

the optimal slit settings must strike a compromise between high line-to-background ratio and measurability of the signal. This optimum lies in the region where the ratio begins to level off. In the Beckman DU it is about 0.02 to 0.03 mm. for the hotter flames.[152, 307]

63. *Flame Accessories for Spectrophotometers*

The monochromator employed for flame analysis is often part of a spectrophotometer used for absorption analysis. If the light source of the instrument is replaced by a flame accessory (Fig. 43) comprising atomizer, burner, etc., a flame spectrophotometer results. Such accessories or attachments include (1) a burner, usually combined with (2) an atomizer; (3) a sample changer; (4) a concave mirror; (5) a housing with mounting supports and means of adjustment; and (6) pressure gages and regulators (Section 18), generally separate from the rest of the accessory.

7+C.A.F.P.

Atomizer-burners are preferred because they yield higher radiance than indirect atomizers. Nearly always, very hot flames (oxyacetylene or oxyhydrogen) are used to provide good excitation for ultraviolet lines. Since these flames—in which the gas, oxygen, and liquid spray mix in the flame itself—burn turbulently, they hiss loudly. For damping the noise, the accessory is equipped with a housing lined with a sound-absorbing material. The same purpose is served to some extent by the chimney, composed of several rings, which also guards against contact.

Fig. 43. Flame attachment for the M4Q monochromator (Carl Zeiss, Oberkochen).

There are adjustments to permit placing the flame and mirror in the best position before the monochromator. The entire array should be so stable that no readjustment will be needed after changing the arrangement, as during an interval of use for absorptiometry.

64. *Alignment of the Flame with Respect to the Monochromator*

The same considerations are valid here as in Section 37. There are a few additional points.

1. With a horizontal optical axis, the lower end of the entrance slit must be about 5–8 mm. above the upper tip of the inner cone of the flame.

2. Lateral adjustment is also important; the vertical axis through the center of the flame (the flame axis) must intersect the horizontal

optical axis of the monochromator. To achieve this, the flame is adjusted laterally to peak the signal at the photoreceiver behind the exit slit.

3. The best distance of the flame from the entrance slit depends upon the over-all arrangement. In the simplest case, no auxiliary lenses or mirrors are used in front of the entrance slit, and the flame may be regarded as a point source, at least when sufficiently distant. If the flame is placed at a great distance r in front of the entrance slit, the energy entering this slit will be very small, by the inverse-square law. If the flame (assumed to present a uniformly luminous area) continuously approaches the slit, it will reach a distance s at which the flame just fills the solid angle embraced by the entrance beam of the monochromator. Beyond this point, further approach of the flame to the entrance slit should bring little further gain of intensity. Actually, though, the intensity does continue to rise, if less rapidly, because the inner region is usually more luminous in the turbulent atomizer-burner flames commonly used with monochromators. The flame should not be brought too close to the slit lest the monochromator be overheated, or the slits corroded by spray from the samples. Theoretically, thus, to a rough approximation, the distance s would seem to be the best. But in practice the flame is brought somewhat nearer the monochromator so as to increase the signal and exclude the cooler, unsteadier outer areas of the flame. These same purposes can be served by a backing mirror focusing a duly enlarged image of the flame on or near the slit. If the flame width is b (usually about 1 cm.), s is given by

$$b/s = d/f \quad \text{or} \quad s = bf/d$$

where f is the focal length of the collimator and d its effective diameter.

A field lens before the entrance slit in a commercial instrument will have been set by the manufacturer at the best distance between flame and slit. The lens serves to image the flame upon whichever aperture within the monochromator, in conjunction with the slits, determines the luminosity (Section 60). This aperture is generally the prism face.

65. *The Photoreceiver in a Monochromator*

The basic considerations of Section 42 apply here also. Owing to the small luminosity of monochromators and the low radiance of flames, the most sensitive multiplier phototubes are chiefly used. Above 600 mμ, however, vacuum phototubes are still often used.

Their sensitivity must be raised virtually to the theoretical limit. Several problems result—problems that can be neglected with filter photometers, which need less sensitivity in the photoreceiver.

When the amplification of the photomultiplier or of the amplifier or the sensitivity of the galvanometer is increased, a dark current becomes observable; that is, a photomultiplier under voltage yields a current in the absence of light. This dark current, like the photocurrent due to flame background (Section 95), is subtracted from the total photocurrent to obtain the net reading. The dark current would have no effect on the detection limits if it were constant; but it is not. The limits of measurement are reached when, on reduction of the light intensity, the photocurrent becomes so small that it is comparable with the fluctuations of the dark current. Since the dark current fluctuation depends on the magnitude of the dark current itself (see below), it is desirable to keep the dark current as small as possible, or to strive for a high ratio of photocurrent to dark current. These matters are elucidated below. First, as to the origin of the dark current.

(A) *Origin of the dark current of a photomultiplier.* The dark current comprises three parts: (1) leakage current through imperfect insulators, (2) thermionic emission of electrons from the cathode, and (3) currents due to nonpropagating gas discharges and optical feedback at high voltages. As the voltage V applied to a photomultiplier is continuously varied from a low to the highest tolerable voltage, the relative contributions of the three components of the dark current change rapidly. In the following discussion, the data in parentheses are given, by way of example, for the RCA 1P28 photomultiplier.[250]

1. Low Voltages. When the voltage is too low (below 350 v.), there is no secondary electron emission at the dynodes. Neither photocurrent nor dark current is observed. At higher V (350–500 v.), the dominant component is the leakage current through imperfect insulation between the dynode leads inside and outside the tube envelope. This leakage current is approximately proportional to the applied voltage, whereas the photocurrent as well as the thermionic component of the dark current increases exponentially according to

$$I = cV^a$$

where c is a constant and the exponent a is positive (7–8). In this range of low voltage, then, the ratio of net photocurrent to dark current increases with voltage (up to about 500 v.) in direct-current measurements. The fluctuations of the leakage current, dominant in this range, are decidedly slower than those of the thermionic current, and they interfere only in direct-current work. In alternating-current methods the leakage current does not interfere.

2. Intermediate Voltages. In this range (500–1000 v.), the component due to thermionic electrons emitted from the photocathode and to a certain extent also from the dynodes is dominant. These electrons are amplified in just the same way as the photoelectrons expelled from the cathode by light quanta. The above equation for photocurrent applies also to dark current. The ratio of photocurrent to dark current is thus nearly constant in this range. It does diminish somewhat with rising voltage, owing to the incipient appearance of the effects described in the next paragraph. The leakage current is negligible in this range, being only proportional to V, while the increase of thermionic current is more rapid than linear. The thermionic dark current exhibits both a rapid fluctuation (shot effect or shot noise; see below) and a slower variation (drift). The drift interferes in direct-current measurement and can be eliminated by alternating-current (chopped-beam) methods, while the effect of shot noise on the result can be kept small by use of a narrow-band amplifier or a sluggish meter, but it cannot be eliminated entirely. We return to these matters in paragraph B4 below.

3. High Voltage. At high dynode voltages (1000–1200 v.), in addition to the above two effects, pulsations due to momentary gas discharges and optical feedback are observed. These currents can considerably exceed the thermionic dark current, which is already quite large at these voltages. The dark current fluctuations that result are more rapid than those of the leakage current but slower than the shot noise, to be discussed below. These pulsations interfere in both direct-current and alternating-current operation. The ratio of photocurrent to dark current deteriorates rapidly with rising voltage in this range. In practical work the voltage is therefore generally kept to lower values.

(B) *Methods of reducing dark-current interference.* Various expedients, singly or in combination, can be used for maximizing the ratio of photo-current to dark current or of photocurrent to dark current noise (fluctuation). The following five recommendations are applicable whenever the dark current or its noise interferes in some way; that is, it is assumed here, unless otherwise indicated, that the dark current noise exceeds the other fluctuations determining the detection limits (fluctuations of flame background intensity, amplifier noise, etc.), and that it is large enough to be observable on the meter.

1. Proper Choice of Voltage. We concluded above that in the middle range of voltage the ratio of photocurrent to dark current is approximately constant and also maximal; at higher voltages (above about

1000 v.) and at lower voltages (below about 500 v.), the ratio diminishes owing to the rise of other interfering effects. It is therefore best to work in the middle range, preferably in the lower part (500–700 v.).

This choice of voltage is governed also by two other considerations:

(a) The method of measurement (direct or chopped-beam) and of amplification (direct-current or alternating-current). The gradual drift of the leakage and thermionic currents interferes in direct-current but not in alternating-current operation. It follows that an alternating-current method permits operation at a somewhat lower voltage V on the photomultiplier (down to about 400 v.).

(b) The other fluctuations fixing the detection limits. Thus, the flame background emission, owing perhaps to turbulence, may show fluctuations that exceed the dark current noise below a certain critical voltage V_f (e.g., 900 v.). Below V_f, then, the background fluctuation alone determines the detection limits, and there is no point to reducing V further in accordance with the above recommendation. In this case, V_f represents an optimum.

The dark current of a 1P28 averages about 10^{-9} amp. at the moderate voltages recommended above; in red-sensitive photomultipliers it is considerably greater, about 10^{-6} amp.* It rises by two decades at the higher voltages.

2. *Cooling.* The photomultiplier can be cooled with dry ice or even better with liquid air. This greatly reduces the dark current as well as its noise.

3. *Selection of the Photomultiplier.* The tube should be selected for high ratio of sensitivity to dark current. Photomultipliers of a given type can vary in this respect by a factor of more than 10, as well as in sensitivity at the red end of the spectrum.

4. *Chopped Beam and Alternating Current.* In alternating-current methods, the radiation to be measured can be interrupted with a rotating disk, giving light and dark intervals; the difference between the currents during the illuminated and dark phases of the phototube is the only thing that is amplified and measured. The direct-current component of the dark current and its slow variations are entirely eliminated. Only the short-period (higher-frequency) fluctuations still interfere; this noise component can be kept small by a suitable method of alternating-current amplification (see below for narrow-band amplification) or rectification or read-out.

To clarify this method we next examine the cause of the rapid statistical fluctuations (noise) of the dark current, assuming that the dynode

---

* The dark current is much lower in the newly developed red-sensitive photomultiplier FW-118 of the ITT Laboratories, Fort Wayne, Indiana.—*Translator.*

voltage of the photomultiplier is properly chosen (500–700 v., as above).

The dark current and photocurrent consist of a stream of electrons released from the photocathode by the agencies of heat and light, respectively. Each single electron is amplified in the succeeding dynode stages, so that the photocurrent and dark current consist, strictly, of single, random, statistically distributed, brief impulses of charge. Consequently the amplified dark current and photocurrent reveal a rapid fluctuation or noise, called shot effect. By virtue of a general statistical law, the absolute magnitude of this noise is proportional to the square root of the dark current or of the photocurrent plus dark current, that is, the square root of the primary current emitted from the photocathode; the relative noise (its ratio to the current) is inversely proportional to the square root of the current. A large signal is hence less affected by noise than a small one. This noise, like any other fluctuation, can be formally resolved into the sum of a great many small alternating currents of various frequencies and phases. In shot noise the frequencies f can cover a rather large range. In the dark-current noise of photomultipliers the frequency range extends far beyond a megacycle. An alternating-current amplifier with limited bandwidth passes only a part of this frequency range. This part, Δf, is deliberately kept small in a narrow-band amplifier; Δf is defined as the half-intensity bandwidth of the frequency transmission curve, in the same way as the bandwidth of a light filter (Section 40).

Two cases must be distinguished in the effect of the shot noise on the measurement:

(a) A narrow-band alternating-current amplifier is followed by a linear (or less often a square-law) rectifier and then a direct-current meter, a direct-current servomotor for a recorder, an integrating motor, or the like.

(b) An ordinary (not necessarily narrow-band) alternating-current amplifier is followed by a phase-sensitive rectifier (such as a ring demodulator) or a lock-in amplifier, followed in turn by a direct-current meter or direct-current recorder; or the alternating-current amplifier is followed directly by a phase-sensitive (synchronous) meter, servomotor or integrating motor.

In case (a) the mean amplitude of the transmitted alternating-current component of the dark-current noise displaces the zero. This dark reading is still overlaid by noise due to the statistical fluctuations of the amplitude of the transmitted alternating-current component of the dark current. With the more commonly used linear rectification, this zero displacement is proportional to $\sqrt{\Delta f}$, and the mean amplitude A_d is given approximately by

$$A_d = cSV \sqrt{2eGI_d(1/T)n/(n-1)}$$

where c is a constant of the order of magnitude of 1, e is the electron charge, n is the mean amplification factor for the secondary-electron amplification at a dynode, G is the over-all amplification factor of the photomultiplier, I_d is the mean dark current, S is the current sensitivity of the meter, V is the amplification factor of the alternating-current amplifier, and T is the period of the meter or the time constant of the RC circuit employed or the integration time. When Δf is taken equal to $1/T$ (it does not have to be), the zero displacement is about equal to the mean noise amplitude A_d. In practice it is difficult to make Δf less than 1 c.p.s. But T can easily be made to exceed 1 sec. (Sections 46 and 47), increasing the precision of reading but also increasing the consumption of sample. Notice that A_d is independent of the bandwidth Δf of the amplifier and that the zero displacement is independent of T.

The meter reading should be corrected for the zero displacement. Unfortunately, the dark reading cannot be simply subtracted from the total reading A_t to find the net signal reading, because, approximately,

$$A_t = \sqrt{A_n^2 + A_s^2}$$

where A_n is the noise reading and A_s is the signal reading; that is, light intensities yielding a signal of about the same magnitude as the noise no longer give a meter reading proportional to the intensity, even when the dark reading due to the rectified noise is electrically subtracted.[65,69] The departure from proportionality amounts to only about 0.5% when the signal is 10 times the noise reading, and at higher signal-to-noise ratios the departure from proportionality can be ignored. This loss of proportionality on reduction of the signal to the noise level in method (a) can be alleviated only by making Δf still smaller, so as to reduce the noise-to-signal ratio further; but there are practical lower limits to Δf.

In case (b), a phase-sensitive rectifier or meter is used, whose phase and frequency must correspond to those of the interrupted light beam. The zero displacement of case (a) does not occur here, and the reading remains strictly proportional to the signal down to the smallest signals. The dark-current noise and any amplifier noise cause only an irregular fluctuation of the zero point in case (b), the mean amplitude of the noise being given approximately by the above formula for A_d.[65,148] The same considerations still apply to the reduction of the zero fluctuations.

5. Counting Methods. Still greater sensitivity could be achieved by cooling the photomultiplier tube in liquid air and counting the individual amplified pulses of each electron cascade. Beam chopping could be combined with pulse counting by means of a two-channel counter for counting the pulses during the dark and light phases separately and subtracting the dark count. Such an arrangement does not seem to have been yet applied in flame spectrophotometry, perhaps because of practical difficulties. Since the quantum yield is not 100%, development of more efficient photoemissive surfaces could further improve the limiting sensitivity. Increase of the sensitivity of photoreceivers beyond this point (100% quantum efficiency) is theoretically impossible.

66. *Amplifiers, Meters, and Power Supplies*

The material covered in Sections 44–46 still applies basically, but for flame spectrophotometry the requirements of sensitivity are substantially more exacting. It is harder to keep the sensitivity and the zero sufficiently constant. We have already discussed the use of chopped light with narrow-band amplification and rectification (Section 65, (B) 4(*a*) or, better, with phase-sensitive rectification (4(*b*)), for effectively separating small photocurrents from the dark current and its noise current. It is also more important to filter out the line frequency hum.

67. *Multiple-beam Methods in Flame Spectrophotometry*

When the light emitted by a flame has been resolved into a spectrum, it is possible to collect at least the more important flame lines upon separate photoreceivers, to amplify them individually, and to measure them on separate meters. This system, in spectrophotometry with spark or arc, is known as "quantometry". The multiple monochromator is called a *polychromator*. More recently, such instruments have been described also for flame spectrophotometry.[160,645,710] One of them is shown in Fig. 44.

In one arrangement, two multiplier phototubes are placed close together, each behind one of two closely adjacent slits. Of this slit pair, one is set upon a line and the other upon the flame background next to the line. Since, for mechanical reasons, the slits cannot be placed very close together, a grating monochromator of high dispersion is employed, and owing to the thickness of the photomultiplier tube one of the beams must be deflected sideways by a mirror. The photomultipliers are chosen with approximately equal sensitivity; residual differences are removed by adjusting the dynode voltages (Section 42). The photocurrents from line plus background, and from the background only, are amplified separately. The currents from the two amplifiers are placed in opposition, so that the second is subtracted from the first.

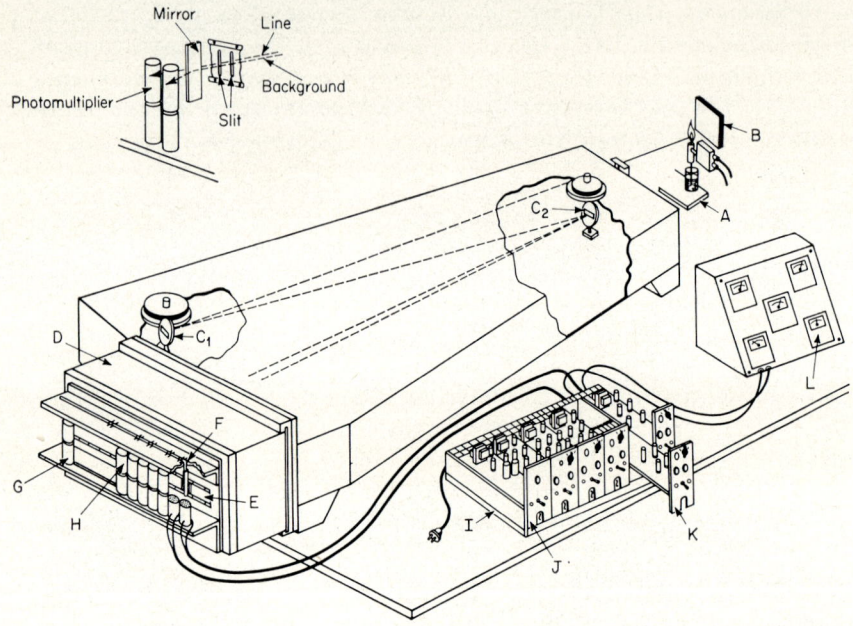

Fig. 44. Construction of a grating flame spectrophotometer for the simultaneous determination of several elements. A, Beckman atomizer-burner; B, backing mirror; C_1, collimating mirror; C_2, grating; D, housing for the photodetectors; E, plane of the exit slit; F, mirror; G, socket for a photomultiplier; H, photomultiplier; I, amplifier; J, low-voltage power supply; K, high-voltage power supply; L, microammeter for each analyte (from ref. 458).

The meter then shows the difference between line plus background and background, i.e., the line intensity alone. This device permits quantitative measurements when the line amounts to as little as one-seventh of the background. The instrument of Fig. 44 has five pairs of slits, photomultipliers, and amplifiers.

68. *Recording Methods in Flame Spectrophotometry*

Flame spectrophotometers permit the same techniques in recording, integrating, monitoring, and control, as were described in Section 56 for flame photometers with filters. Thus, the monochromator can be set upon a particular analytical line whose intensity is recorded with the passage of time in a sample of changing composition, and compared with intermittently interposed standard solutions. With a flame spectrophotometer, furthermore, the intensity can be scanned as a function of wavelength, yielding a recording of the spectrum of a given solution. Such procedures constitute a valuable extension of flame spectrography, to be discussed later; their advantages prevail also in flame spectrophotometry. Recording, moreover, is generally more exact than spectrography, because the measurement is made directly and not via the blackening of a photographic plate. But the consumption of sample is greater, because the spectrum must be scanned sequentially, wavelength by wavelength, whereas in spectrography the whole spectrum is recorded simultaneously.

There are other advantages of flame spectrophotometric recording:

1. Blank interferences, due to bands in the flame background, the effect of concomitants on the flame background, cross-sensitivities due to stray light, etc., can be recognized and eliminated more simply and surely by means of recording, along with a suitable method of graphical evaluation of the recorded spectrum (see Sections 94 and 95).[544]

2. Narrow slits can be used without trouble from creeping of the wavelength setting due to thermal effects, mechanical vibration, etc. In a recording, the peak of each line is shown unequivocally. Detection sensitivity is thereby improved.

3. With the help of systematic recording, the optimal operating conditions can be found more quickly and accurately, by experimenting with different gases, fuel/oxygen mixing ratios, heights of observation in the flame, slit widths, etc. To a large extent, written records in the form of tabulated data can be replaced by a properly annotated recording, which constitutes a complete record of the analytical measurement.

4. Recording always detects unexpected phenomena, such as a new interfering line in one particular sample in a routine series, which would be overlooked in the usual procedures.

5. In addition to time integration of a signal at fixed wavelength, discussed earlier, recording enables another kind of integration, viz., integration (by planimetry) of the area under a line profile in the recorded spectrum. This yields more reproducible results than momentary readings taken on the line without recording, especially for a weak shoulder on the side of a stronger line or band. In this way it has been

possible to determine 0.02 p.p.m. of sodium in the presence of 200 p.p.m. of potassium and 200 p.p.m. of calcium with an accuracy of 0.0003 p.p.m.[152]

6. Internal standardization (Section 52) can be applied to single-beam operation without additional equipment, by recording over both the analytical and the internal-standard line, and comparing the recorded intensities or the corresponding concentrations. Most of the advantages of internal standardization listed in Section 55 are valid for this method.

7. Various interferences such as displacement of the dissociation or ionization equilibria in the flame can be recognized more easily and surely, since several lines or bands of the analyte and concomitants are available.

8. Procedures for correcting interferences (Section 100), such as self-standardization or buffering, can be carried out more accurately by recording, since the effects of the interferents on the flame background, etc., can also be reliably eliminated at the same time.[544]

There are a few disadvantages:

1. Extra cost of the recorder, consumption of chart paper, and care of the equipment.

2. Higher sample consumption, since it takes some time to record a flame spectrum.

3. Irregularities of atomization, especially drift due to gradual clogging of the capillary, cause error in the internal-standard method described in 6 above and in other procedures, since the drift introduces changes of intensity that vary with the wavelength in a recording.

69. *Internal Standardization with a Spectrophotometer*

In connection with filter flame photometry, we considered in Sections 51 and 55 the advantages of double-beam methods entailing the determination of the intensity quotient of an analytical line and an internal-standard line. This method is applicable also to prism or grating instruments. In any case, a special spectrophotometer is generally required, having two exit slits, one for the analytical line and one for the internal-standard line. Fig. 45 shows the schematic arrangement of such an instrument. One slit is set permanently upon the internal-standard line, e.g., Li 671 mμ, while the other can be set upon any of various analytical lines. The quotient of the two photocurrents is read directly on the meter. These instruments have the same advantages

§ 69] INTERNAL STANDARDIZATION WITH A SPECTROPHOTOMETER   191

and drawbacks as internal-standard filter photometers (Section 55), and they enjoy the additional advantages of spectral dispersion (Section 59).

A simple monochromator with only one exit can be converted (aside from the procedure of Section 52) to internal-standard operation by the system shown in Fig. 46. The light from the flame is divided by a

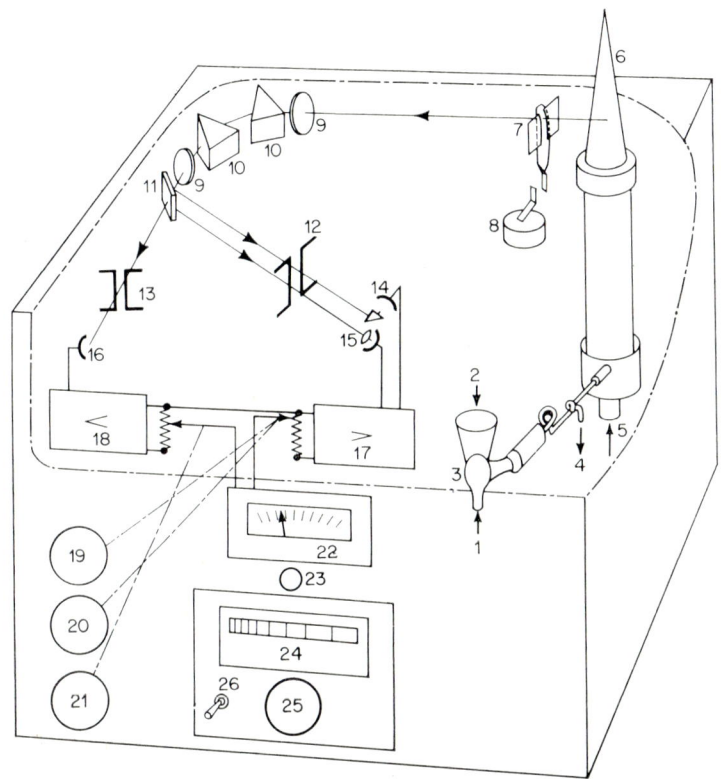

Fig. 45. Perkin-Elmer Model 52A flame photometer with internal standardization (schematic). 1, Compressed-air line; 2, sample funnel; 3, atomizer; 4, condensate drain; 5, fuel gas line; 6, flame; 7, entrance slit; 8, slit shutter drive (for a.c. operation); 9, collimating lenses; 10, prisms; 11, beam splitter (semitransparent mirror); 12, adjustable exit slit; 13, fixed exit slit for the internal-standard line; 14, red-sensitive phototube; 15, blue-sensitive phototube; 16, phototube for the internal-standard line; 17, amplifier for the analytical line; 18, amplifier for the internal-standard line; 19, fine gain control; 20, coarse gain control; 21, internal-standard gain control; 22, meter; 23, zero control; 24, wavelength scale; 25, wavelength control ("element selector"); 26, power switch.

semi-transparent mirror. Half the light passes through the monochromator, which has a phototube (not shown) behind the exit slit. With the wavelength dial of the monochromator, any of various analytical lines can be selected. The rest of the light encounters a filter combination transmitting only the internal-standard line (Li 671 mμ). The two photocurrents can be combined by means of a bridge

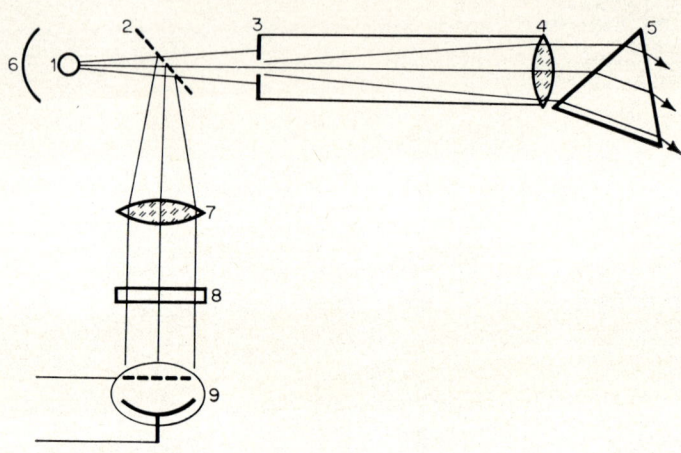

Fig. 46. Optics for internal-standard flame photometry using a single monochromator and a filter. 1, Flame; 2, half-silvered mirror; 3, entrance slit; 4, collimator lens; 5, monochromator prism; 6, backing mirror; 7, condenser; 8, filter; 9, phototube.

circuit to yield a quotient. But no such scheme seems to have been put to use. Internal standardization can be further extended to the simultaneous determination of several analytes, as in Section 54. For this a large spectrophotometer is required, having several exit slits, one for each analytical and internal-standard line, as described in Section 67. For each line pair the photocurrents are suitably opposed so as to form the quotient. Such an internal-standard polychromator system could be combined with filter photometers.

FLAME SPECTROGRAPHY

70. *General*

Unlike filter photometry or spectrophotometry with a monochromator, in which the spectral line intensities are measured with a photodetector, spectrography makes use of a photographic plate for receiving the

§ 70] FLAME SPECTROGRAPHY 193

spectrum. Exposure produces varying degrees of backening along the plate, depending upon the line intensities, the nature of the instrument, the type of plate, and the exposure time.

Advantages of flame spectrography as compared with flame photometry and spectrophotometry include the following:

1. Less dispersion is needed, because the resolving power is generally better exploited in a spectrograph than in a monochromator. The cost is therefore less. Also, photography requires less expensive equipment than photoelectric photometry.

2. Each analysis yields a photograph of the spectrum, or spectrogram, which can be preserved as a record. This is of value in many applications (legal matters, criminology, etc.). The spectrogram yields a quick qualitative view of the elements present in the sample, and this may suffice in many cases.

3. Many determinations can be made with a single exposure (using a step-sector disk; see 6 under the drawbacks), as the lines of the different elements are photographed all at the same time.

4. When many elements are present together in a sample, the labor required for a single analysis is less than when the elements are determined singly in sequence by one of the methods discussed earlier; but, owing to the relatively small number of lines in flame spectra, this condition seldom arises, i.e., the advantage of spectrography is less with the flame than with arc or spark. Moreover, multiple-beam spectrophotometry (Section 67) yields the same results, but it falls short of spectrography in that the slits for different photodetectors cannot be placed arbitrarily near each other.

5. In multiple analyses as above, the consumption of sample is lower.

6. The blackening of the plate directly yields the time integral of the illumination intensity during the exposure time at any given wavelength. The result is therefore unaffected by brief irregularities of atomization or burning.

7. Of the elements excitable in the flame and present in the sample, none is likely to be overlooked even though unexpected, in contrast with filter photometry and spectrophotometry (without recording).

8. Spectrography is adapted more easily when samples of wholly different composition are to be investigated.

The following are *drawbacks* of flame spectrography in comparison with flame photometry and spectrophotometry:

1. The operating cost is higher, owing to the use of plates or film.

At all events a good many spectra (up to 70) can be photographed on one plate, whereby the cost is spread out.

2. The result is not immediately forthcoming, but is delayed by the developing, fixing, washing, and drying of the plate, and the measuring and calculating of the optical density. This lengthier procedure is particularly burdensome when only one or a few elements are to be or can be determined. However, the procedure can be made automatic.

3. Spectrography is generally of lower precision, since the intensity is found not directly with a photodetector but circuitously via the density of the image on the plate or film. The spectral sensitivities of plates differ from those of photodetectors, a circumstance which can affect the choice of the line for analysis (Section 82).

4. Often the detection limits are poorer, because the flame background is raised by plate fog. But this drawback is balanced by losses in spectrophotometry due to mismatch of the exit slit curvature to that of the line, and the difficulty of attaining enough radiant flux at the exit slit. Hence, in a monochromator of dispersion equal to that of a competing spectrograph, the exit slit will have to be made larger than the theoretical limiting spectral bandwidth. But this lowers the ratio of line to background and consequently raises the detection limit (lowers the detection sensitivity). However, if we assume, as under advantage 1 above, that the dispersion of the spectrograph, in view of its better resolving power, is taken smaller than that of the monochromator, then the above effect of plate fog remains as a drawback of the spectrographic method.

5. The relation between line intensity and density on the plate is more complex than that between photocurrent and intensity in photometers or spectrophotometers. This makes the task of evaluation more troublesome.

6. It is desirable to operate not only in the linear region of the working curve (graph of intensity vs. concentration) but also in the middle, linear portion of the photographic *calibration curve* or *characteristic curve* (Section 75). Favorable densities lie between 0.4 and 1.0. When lines with very different intensities are to be measured, either the ends of the calibration curve (the regions of under- and over-exposure) will have to be used, or the exposure time must be varied. Several exposures must then be made with different exposure times, unless a step-sector disk* is used. But multiple exposure times naturally increase the consumption of sample, or entail the risk of error resulting from evaporation of the solution in a reflux atomizer (Section 25).

* See footnote in Section 75.

7. The integration of intensity during exposure includes light flashes caused by impurities entering the flame, and their effect cannot be deducted.

8. There are no photographic emulsions sensitive to the infrared beyond about 1000 mμ. For this region semiconductive photocells, thermocouples, etc., have to be used.

Historically, spectrographic methods are the earliest. With the increasing use of photometric and spectrophotometric methods, flame spectrography is waning. The above advantages 1, 2, 3, and 5 are

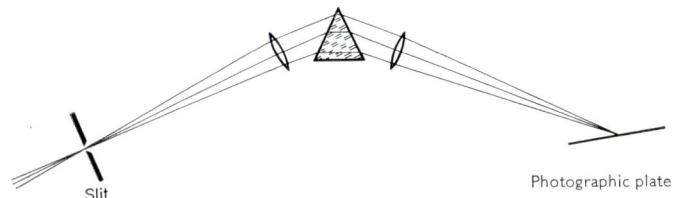

Fig. 47. Optics of a prism spectrograph (schematic).

rendered less cogent by the increasing use of flame-photometric recording. To be sure, strip-chart recording requires more sample than spectrography, which records the whole spectrum simultaneously.

Flame spectrography will therefore be discussed only briefly. The procedures are analogous to those with arc or spark. We refer to the literature.[26, 33, 37, 44]

71. *The Spectrograph*

The same considerations apply to spectrographs as to monochromators (Section 60). The statements in that section concerning dispersion and resolving power remain valid. As before, the light can be dispersed into a spectrum with a prism or grating; we shall confine ourselves here to prism instruments.

Fig. 47 shows the basic design of a prism spectrograph. A photographic plate or film lies in the plane in which the exit slit of a monochromator would lie. The important datum characterizing the efficiency of a spectrograph is the resolving power. In order to exploit fully the resolving capacity of the photographic emulsion, instrument manufacturers strive to push the resolving power of spectrographs to the theoretical limit. Monochromators, on the other hand, are usually intended for absorptiometry, for which, owing to the nature of absorption spectra with their wider bands, there is less need for high resolution.

For this reason and because of the difficulties mentioned at the end of Section 62, monochromators have lower practical resolving power than spectrographs of equal prism size.

In flame spectrography, as against arc or spark spectrography, there is the difficulty that the radiance of conventional flames is small. In principle, this can be compensated by a correspondingly greater exposure time. But the analysis then takes more time and much material is consumed during the prolonged atomization, or, if a reflux atomizer is used, the concentration of the sample solution changes during the exposure time. Therefore, for quantitative analysis, a spectrograph with a good relative (angular) aperture (ratio of lens diameter or limiting aperture to focal length, or f-number) should be used. In spectrographs, unlike monochromators with photoelectric receivers, it is a question not of the total emergent radiant flux, but rather of the intensity of illumination (radiant flux per unit area [320]) at the plate, because the photographic density is determined by the product of illumination intensity and exposure time. For lines, the illumination intensity is determined to a first approximation only by the relative aperture and not by the slit width or other parameters. However, owing to diffraction losses at very narrow slits and the small number of lines in flame spectra, somewhat wider slits may be used in flame than in arc or spark spectrography. For the same reasons it is inexpedient to use a camera with focal length larger than that of the collimator. If one wishes to enlarge the spectrum, one should do so subsequently with the developed plate. A linear dispersion of about 10 mμ/mm. at 400 mμ is generally adequate for flame spectrography.[37]

72. *The Photographic Plate*

Because flame spectra are not rich in lines, the plate does not need to be as fine-grained as for most other spectrochemical applications with spark, etc., but on the other hand it must have higher sensitivity owing to the weaker radiance of flames. Choice of spectral sensitivity will depend on the spectral region of interest. It is well to use the more popular commercial types of plate, as these are more likely to be fresh and are usually also more uniform in sensitivity. Storage longer than two months is usually inadvisable, for the background density becomes too uneven, even with good development. Old plates also show more emulsion fog and edge fog. Atmospheric conditions are important in storage; refrigeration will help to delay aging.

73. *The Photographic Process*

We cannot here examine the complex photographic process. Only a few points will be brought out. The density attained depends on the type of plate and developer, the concentration and temperature of the developer, and the development time. In spectrographic work these variables should always be held fixed, and the developer should never be used twice. To get comparable line densities and constant background density over the whole plate, it must be kept in motion during development, for instance, with a steady motor-driven rocker, or else brushed if necessary; otherwise the background will not be uniform. Duration, temperature, and agitation must be under control also in fixing, washing, and drying. Developing machines are now available for the spectrographic laboratory.

74. *The Exposure*

Obviously, the exposure time must be controlled very exactly, so as to keep the densities of the several spectra mutually comparable. For this purpose, electric shutters or timers are generally used. In addition, gas and air pressure, etc., must be reproducibly set (Sections 18 ff.). Each plate receives several calibrating spectra obtained from standard solutions of the same composition and known concentration, at least one at the start and one at the end of each series of exposures, to permit comparison of the densities of corresponding lines and to cancel variations of plate characteristics and errors resulting from differences in the development process. Because of errors of development, which are variable and hard to avoid, spectrograms on different plates taken under otherwise identical conditions cannot be accurately compared with each other. A step-sector disk in front of the slit enables the same line to be exposed at various densities simultaneously. In this connection, recall internal standardization (Section 51), originally used in spark spectrography. This method is, of course, applicable to flame spectrography. By its means exposure errors and variations of plate characteristics can be more fully compensated.

75. *Evaluation*

Evaluation of the exposed, developed, and fixed plate consists basically in a measurement of transparency with a densitometer. The transmittance T is the ratio of the intensity M of densitometer light transmitted at an exposed location on the plate to the intensity M_0

transmitted at an unexposed point: $T = M/M_0$. The density S is the negative logarithm of this quantity: $S = \log(M_0/M)$. The ratio M_0/M is given by the ratio A_0/A of the meter readings on the densitometer, A and M being, of course, proportional in a good instrument.

It is the relation between S and the intensity J of the spectral line or band that is of interest. If test exposures are made with various intensities J and various exposure times t, it is found that to a first approximation, S depends only on the product $Q = Jt$, or, more accurately, on Jt^p, where $p < 1$ is the so-called Schwarzschild exponent. S plotted against $\log Q$ or $\log Jt$ gives the *characteristic curve*, which is sigmoid, its form depending on the type of plate, the wavelength, the development, etc. For quantitative measurement, the steepest (central) part of the curve is used almost exclusively. Its slope is denoted by $\tan \alpha = \gamma$. This part of the curve, in which γ is nearly constant, generally extends from $S = 0.5$ to $S = 2$. If $\gamma < 1$, the plate is called soft; if $\gamma > 1$, it is called contrasty. To ensure that each spectral line, despite great variations of J, will lie within the linear part of the characteristic curve, a step-sector disk is used. This is a rotating sector placed before the slit plate, in which the shape of the opaque part of the sector varies the ratio of open to shut interval stepwise along the slit height. The exposure time, which must be long enough to accommodate the weakest line, is thus varied along the height of the slit. For the densitometry, that segment of the line is selected which falls within the best part of the characteristic curve. The relative intensity J is then calculated from the density S measured by the densitometer. As a rule, this is done not directly from the above equations, but by means of a suitable transformation, which yields a transformed density W. The equation transforming A into W can be so chosen in a given case than even over a fairly wide range of $\log Q$ the transformed density W shows a linear characteristic.[44, 180a, 407, 409] The simplest is the Seidel transformation:

$$W = \log(A_0/A - 1)$$

The line intensities thus obtained can be used in the same way as those measured directly by a phototube or other photodetector. But the precision is poorer, because the intensity is measured indirectly via the optical density. Basically, also, all of the corrective procedures given for flame spectrophotometry (Sections 96 ff.) can be applied to flame spectrography. The accuracy of densitometry is limited not only by errors in determining the ratio of line to background but also by the local variations of density due to the graininess of the emulsion. The effect of these random fluctuations of density diminishes as the area of uniform density available for measurement increases.

CHAPTER 4

ANALYTICAL TECHNIQUE

Use of Equipment

76. *The Laboratory*

The room in which flame analysis is to be conducted should be large enough, at least 60 cu. yd., and dry, dust-free and of uniform temperature. In a small room the residues from the evaporated spray will form a salt aerosol soon filling the room and raising the background reading on the analyte. In most instruments the flame aspirates the ambient air extensively;[632] hence an empty flame makes a good indicator of the purity of the room air. The room should also be large enough so that prolonged burning of the hot flame will not vitiate the atmosphere. The combustion produces carbon dioxide and water. Chemical operations, including preparation of the samples for analysis, should not be conducted in this room, lest the mechanical and electrical parts of the flame photometer should be damaged by acid fumes. Basement rooms are unsuitable, especially for working with propane; recall the safety precautions of Section 17, and the matter of chaining gas tanks to the wall. The electrical outlets should be equipped with means of grounding the flame photometer and the compressor, if used. A ventilating flue such as a funnel of 30 cm. diameter placed about 40 cm. above the flame with a draft conduit of 8–10 cm. diameter and a fan of 0.2–0.5 m.^{3}/min. capacity, is always advisable; it is indispensable when poisonous substances such as heavy metals are to be sprayed or analyzed, or when, for instance, trichloroacetic acid is used as solvent (Section 110), which yields obnoxious chlorine-containing fumes. Special purifiers for the laboratory air have been invented,[527] but are rather expensive. During and before the determination of potassium at low concentrations, smoking is not permissible in the room, since tobacco smoke contains much potassium. Normal ventilation of the room, as with an open window, is inadvisable, since it may cause fluctuations of temperature, which adversely affect the behavior of the instrument, as by altering the sensitivity of the photodetector, etc. Marked disturbance of the air such as is caused by opening a door should be avoided; it may upset the flame, even in the presence of a chimney. When solutions differing greatly in concentration are to be analyzed, it is well to start with the most dilute, as the air will be increasingly contaminated with combustion products. Further, the sample and standard solutions should be at nearly the same temperature (they can be kept in a refrigerator).

The working area should be as convenient and well fitted as possible. It should include a shock-proof mounting, such as a heavy table (Section 46) for sensitive galvanometers and for the monochromator, the wavelength setting of which is easily shifted by jarring. The surface of the table, at least below the atomizer, should be easy to clean and acid-proof. The shut-off valves for gas and oxygen should be within reach. Recall the recommendations in Section 17 for setting up the gas tanks. All meters including pressure gages and flowmeters should be visible at a glance from the working location. Proper lighting is essential. It should be sufficient for good vision but not so strong as to obscure the projected scale of a mirror galvanometer, or to introduce unnecessary stray light. The light should therefore not be placed above the chimney. There should be enough space to set out the sample and standard solutions, empty containers, and the notebook. Within reach there should be means for unclogging the atomizer and a waste container for excess liquid. Outside the room but near the entrance, there should be a suitable fire extinguisher, ready for use (Section 17(k)). In the analysis of combustible liquids, special precautions are in order (Section 123). If compressed-air flames (air–acetylene, air–propane, etc.) are to be operated quite often in the room, a compressor with its air and liquid filters should be permanently installed (Section 16). It should be placed in a soundproof box for damping the noise, but adequate cooling must be attended to. Better, the compressor and its accessories can be installed in an adjoining storeroom, basement, etc., with permanent air lines brought in to the flame photometer.

77. *Operating Procedure*

Instructions for use of the commercial types of flame photometer are furnished by the manufacturers. In this section we shall emphasize only a few basic matters, assuming that all components have been set up and adjusted successfully (Sections 37, 43, 64, etc.). The electronics should be turned on 15 min.–2 hr. before the start of the actual analysis, and a multiplier phototube should be warmed up even longer. About 10 min. before the start, the compressed air should be turned on and the pressure or flow rate set to the optimal value as previously determined

(Section 19). Not until then should the gas be turned on and the gas pressure or flow rate regulated. After opening the gas cock, one should wait about 10 sec. or long enough for the gas to displace air completely from the connections and to stabilize the reading of the manometer; the gas–air or gas–oxygen mixture may then be lit. Without this delay, a combustible mixture may remain in the connections at the time of ignition, and flash back. But with too long a delay the explosive mixture will be flowing into the room. This danger is especially great with hydrogen (which is odorless) if it escapes unnoticed for any length of time. If the flame flashes back into the burner tube, the gas should be shut off at once (and then the oxygen), heated apparatus allowed to cool, and the flame then relit according to the above sequence, but with greater flow rate. If the flame flashes back during normal operation, the cause may lie in leaky hose connections or the like.

With an indirect atomizer, after the flame is burning normally, a vessel of water or the proper solvent is placed beneath the suction capillary and allowed to spray while the burner is warming up. This helps to bring the walls of the spray chamber, burner connections, etc., to their equilibrium temperatures, which will differ from the room temperature. The solvent may have to be replenished during warm-up. An atomizer-burner (Section 33) can be started up immediately before the analysis, without having to warm up.

When all parts of the apparatus have reached their equilibrium working temperatures and the photoreceivers have warmed up, the mechanical zero of the galvanometer and the electrical zero of the amplifier are adjusted with the shutter closed. The desired wavelength is then set. In a filter flame photometer this is done by installing the proper filter (Section 40); in a flame spectrophotometer, by setting the wavelength dial (see Section 82 on selecting wavelengths) and the slit width (Section 62). The reference standard solution (Section 83) is next sprayed, and the wavelength setting of the monochromator is adjusted to maximize the reading. The sensitivity is then roughly adjusted to give the desired meter reading (say, 100 divisions). The blank (flame background plus cross-sensitivity, Sections 94 and 95) is now balanced by zeroing the meter while spraying a blank solution. This is followed by a final adjustment of the sensitivity. These last two steps, standardization and zeroing, are repeated. The analytical measurements can then begin. They should be interspersed with regularly spaced standardizations and zero checks. Whenever the sensitivity is readjusted upward, the zero should be checked again with a blank solution. For setting up internal standardization, see Section 53.

The working curve yields the concentration corresponding to each reading, unless the meter scale has been precalibrated in concentration units. The establishment of the working curve is discussed in Section 83. A source of error that must be allowed for is interelement interference (Sections 96–100). Following each set of 1–5 consecutive measurements, the reference standard solution (check solution) should be inserted to check the sensitivity and, if necessary, to restandardize it. The amplifier zero and the background compensation (blank zero) must also be checked from time to time. In a flame photometer with a selenium cell, the readjustment of the zero takes more time, as the cell exhibits hysteresis (Section 42). These standardizations and zero checks will be made oftener at the beginning of a long series of measurements than toward the end, for the meter readings will become steadier and more reproducible as the instrument continues running under constant conditions. The wavelength adjustment of a monochromator should be checked occasionally by spraying a standard solution, as it may be shifted by jarring or thermal effects.

Additional corrections will be needed for accurate analysis of solutions of complex composition, whose constituents (such as protein) may be hard to imitate in a standard solution, or in which radiation interferences or complex background effects must be allowed for. Further, the samples should be so diluted or concentrated as to bring them into the most favorable range of concentration (Section 79). The standard solutions must resemble the samples in composition and concentration. Impurities can clog the atomizer, and they can directly affect the result. Incipient clogging or gradual occlusion of the capillary, as by protein, can be detected in either indirect or direct atomizers by measuring the rate at which the liquid is aspirated, or by timing the interval between application of the liquid and brightening of the flame. A quick ear will detect clogging by the change in the sound of the spraying.

At the end of the analysis, the atomizer is rinsed with distilled water or solvent. The shut-down procedure is the reverse of starting: the gas is turned off first, then (after the flame is out) the compressed air or oxygen, and finally the electronics. The burner top and atomizer must be cleaned every so often, after perhaps every 100 analyses, or oftener with certain types of sample such as cement and unprepared serum. The optimal pressure or flow-rate setting (Section 19) should be checked about weekly, since they can drift owing to accumulation of debris (abraded bits of rubber tubing, gaskets, etc.) in the connections, especially in the nozzles. With skill one can recognize large changes of this kind from the altered flame color.

78. Development of a Procedure

A basic study of the analytical conditions must usually precede the routine application of an analytical method in flame photometry, unless prior experience with the same sample material and the same type of instrument can be put to use. For the original establishment of a procedure, the following preliminary steps will probably have to be taken.

1. In preliminary experiments of moderate accuracy, perhaps partly chemical, a determination is made of the elements present in the sample, their relation to the anions, etc., any other materials present (e.g., dextrose, urea), and their concentrations. For well-known and common analytical problems, published information can be used.

2. A procedure for preparation of the sample must be decided upon. Solids must usually be dissolved in acids. The type and quantity of the acid are important in preparing the standard solutions (Section 84). The chemical nature of the solid sample will usually indicate the best solvent or acid to use. If there is a choice of acids, nitric acid is generally preferred to hydrochloric; acid interference is usually least with nitric, larger with hydrochloric acid.[540] Organic substances can be either worked up directly and perhaps diluted (Section 114), or first ashed (Section 109).

Sometimes physical or chemical separations are necessary in preparing the sample. For instance, suspended particles may have to be centrifuged or filtered out, and the solid and dissolved portions prepared separately for analysis. Chemical separations such as group or selective precipitation, extraction, or ion exchange, may have to come first, to avoid interference from concomitants in the flame analysis or for other reasons to be dealt with later. The preparation usually concludes with a dilution, most simply with distilled water, sometimes with a special diluent such as an organic solvent, wetting agent, radiation buffer, or perhaps a solution of an internal-standard element (Section 51). We return to these matters in Sections 79, 88, 89, 100, and 101, and in the sections on applications.

3. The analytical line or band must be chosen. This is a very essential step, for a proper choice can eliminate at the outset many inconveniences in the subsequent routine determination, such as correcting for interfering cross-sensitivities or circumventing an unduly high flame background. In any case, this choice is largely influenced also by the type of instrument. With an efficient instrument well suited to the task, much preliminary work in setting up the method can be avoided, not to mention the considerable saving of time in the routine determinations.

Thus, a dilution in each analysis can be avoided by using a line that is suitable for the original range of concentration. The factors governing the choice of the line will be examined in Section 82. The index of flame lines, the spectrograms at the end of the book, and the detection limits (Section 90) provide data needed for the choice.

4. A decision has to be made between direct reading and internal standardization. For the latter, the internal-standard element and line must be properly selected (Section 55).

5. A blank and a series of standard solutions resembling the samples in composition and concentration are prepared (Sections 83 and 95).

6. The gas (Section 13) and the ratio of gas to air (Section 19) are selected, and the burner is aligned with the optics if necessary (Section 37).

7. Radiation interferences are tested as in Section 100. If they exceed the greatest permissible error, they must be circumvented by one of the methods listed in Section 96–100. The number of standard solutions may then have to be increased, as in the parametric method of removing interference (Section 97).

8. A suitable measuring and standardizing procedure is selected according to Section 85, and a suitable corrective method is adopted for removing possible sources of error (Sections 96–100).

9. An estimate is made, finally, of the magnitude of the errors that are present, as follows:

(*a*) The random error of the actual flame-photometric measurement is estimated on one and the same sample solution prepared for analysis (Section 102).

(*b*) The adscititious errors arising in the conversion of the sample into the final solution must also be estimated. For example, from the same batch of blood, a series of samples is transferred to centrifuge tubes and centrifuged, and each serum is separated from the clot, ashed if necessary, treated with the reagents (e.g., protein precipitant), diluted, and read on the flame photometer. The result shows the sum of random and preparative errors, including errors of dilution, contamination of containers, etc.

(*c*) Other errors due to inhomogeneity or temporal variation of the sample or alterations during sampling should be tested. For example, mixed soil samples (Section 105) are taken repeatedly from the same field and each is analyzed as in (*b*), or blood samples are taken repeatedly from the same patient and individually centrifuged, separated from the clot, and prepared for analysis as in (*b*). One has to keep in mind the "biological errors" that arise in the biological factor under study through the act of removing blood (Section 112). Similar difficulties will be encountered with other living subjects.

(d) Finally, systematic errors that may be present should be tested for by comparison of the flame analysis with another well-tested method (e.g., gravimetric) on the same material (Sections 103, 104). In this comparison the greater time consumed in the more tedious chemical method should be taken into account. We mention here also the self-standardization method (Section 100(b), paragraph 15) as a means of testing for errors.

In this initial development of the procedure, great pains should be taken to see that the routine determination based upon it is as simple as possible but still sufficiently accurate. Such a method can then be entrusted permanently to a technician.

79. The Solvent: Adjustment of the Concentration

Prior to the actual analysis, the sample solution must generally be diluted or concentrated. Whether and to what extent the concentration should be adjusted depend chiefly on the analytes and their concentrations, the type of instrument, the flame temperature, the atomizer, and the spectral line or band chosen. The choice of solvent depends on the type of sample or on the preceding chemical preparation as well as on the degree of purity in which the solvent is available. The dilution will usually be such as to permit use of the linear portion of the working curve. The convex portion at higher concentration (see Section 8 on self-absorption) can be avoided by dilution, while the concave part at low concentration (see Section 7 on ionization) can be avoided by concentrating the solution.

Table 7 lists recommended concentration ranges for several elements in aqueous solution for a commercial flame photometer[513] with indirect atomizer. The lower limit of the range is that at which the line or band intensity is 10 times that of the flame background in the spectral range of the filter, under the assumption that the working curve is not strongly concave.

If for any reason the sample solution ought not to be made more concentrated, the limit of the favorable range can be lowered under the following conditions:

1. The analyst will be willing to work with less precision, with curved working curves, or with lines less than 10 times as strong as the background. We assume throughout that the electronic sensitivity of the instrument is adequate for these measurements.

TABLE 7. Optimal concentration ranges for the determination of various elements in different flames, based on their working curves (from ref. 513).

A.	Air–propane	
	Na 589 mμ	1–10 mg./l.
	K 767	1–50
	Li 671	3–100
	Rb 780	50–250*
B.	Air–acetylene	
	Na 589	5–50
	K 767	50–250*
	Li 671	3–100
	Rb 780	200–500*
	Ca 623	20–200
C.	Air–oxygen–hydrogen	
	Na 589	5–100
	K 767	200–500*
	Li 671	5–50
	Rb 780	500–2000*
	Ca 623	100–10000
	Ca 423	1000–10000
	Mg 383	2000–10000
	Sr 461	500–1000
	Mn 403	500–1000

* These elements show noticeable ionization curvature of the working curve in the lower part and self-absorption curvature in the upper part of the listed ranges.

2. If the lower limit is set less by the ratio of line to background than by concavity of the working curve (this condition is marked by a * in Table 7), measures can be taken for straightening the curve at low concentrations (Section 10), and thus improving the precision in this range.

A concave working curve, unlike that of Fig. 50 (page 231), shows a steeper slope (>1) in logarithmic coordinates than a straight curve (slope $=1$). Hence, a given percentage error of the meter reading corresponds to a smaller percentage error of analyte concentration when the curve is concave than when it is straight. This is a marked advantage; nevertheless, we advise against the use of a concave curve for the following reasons:

(a) Many standard solutions are needed for establishing the working curve. Moreover, the samples may contain variable quantities of organic substances (such as protein in serum), which unpredictably affect the flame temperature and hence the ionization and concavity, preventing establishment of a reliable working curve.

(b) It is harder to interpolate between standardization points and, especially, to extrapolate beyond the highest point.

(c) When the samples vary considerably in concentration, the advantage of concavity can be turned to account only by so controlling the sensitivity as to avoid the limitations imposed by readability of the meter, zero drift of the amplifier, etc. But this is inconvenient.

Still, there will be instances in which a concave working curve offers real advantages over a straight one.

3. With a monochromator of adequate dispersion and good resolving power, affording better spectral isolation than the interference filters on which Table 7 is based, equally precise measurements are possible at lower concentrations provided the working curve is not concave. Table 10 in Section 90 gives an idea of the detection limits* available with the usual monochromators, which permit use of ultraviolet lines or bands that are difficult or impossible to isolate with interference filters but offer better sensitivity for many elements.

Conversely, the upper limits of the ranges of Table 7 can be raised under the conditions named below. Analysis at higher concentrations may be desired for avoiding dilution with its inevitable errors (see below) or for simultaneously determining elements at both high and low concentration in the same solution without separate dilutions for each element.

1. The convex part of a working curve can be used without great loss of precision (Section 82).

2. It may be possible to use another line of the analyte whose curve is sufficiently linear in the desired range.[349] A monochromator is ordinarily required for this.

High concentrations can cause incrustation and clogging of the atomizer and burner and will load the room air with the analyte.

The *solvent* oftenest used is distilled or double-distilled water. Water properly demineralized with an ion exchanger is entirely comparable in purity with good distilled water. For testing the purity of a given lot of distilled water intended as solvent for a flame analysis, the flame photometer itself is most convenient. The instrument is used at highest sensitivity at the analytical wavelength. For a preliminary qualitative

* Since the detection limits of Table 10 are defined as the concentrations at which the net intensity is only 1% of the flame background, they should be multiplied by 1000 for comparison with the lower limits of the ranges recommended in Table 7.

test, the readings on the available stocks of water are compared. The one giving the lowest reading (showing the least impurity) is adopted. Conductivity has also been proposed for this test.[49] Other solvents can be tested similarly.

The storage container for the water or solvent must contribute no noticeable quantity of electrolyte even over long periods; otherwise, an originally pure liquid will become contaminated. It must also be protected from dust. We return to the subject of containers in the next section.

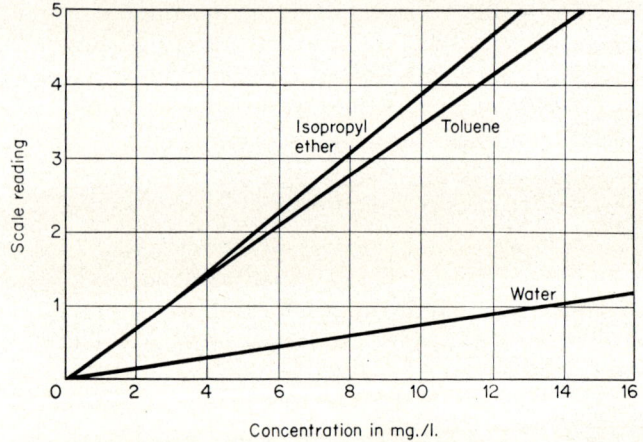

Fig. 48. Effect of various solvents on the intensity of Ba 870 mμ in a propane flame (from ref. 202).

Nonaqueous solvents are often used. They raise the emission typically fivefold, yielding better detection limits. Mixtures of alcohols or acetone with water are frequently recommended in the literature as solvents, but other solvents—gasoline, n-butanol, etc.—are successful in many applications.[65, 139, 140, 143, 182, 202, 228, 255, 268, 274, 298, 420, 537, 575, 661, 718] Fig. 48 shows, as an example, the effect of various solvents on the emission of barium (the background being subtracted). Organic solvents sometimes raise the flame background appreciably. A combustible solvent can form an explosive mixture in the spray chamber and connecting tubes of an indirect atomizer. Hence, the use of organic solvents for improving sensitivity is perhaps better confined to direct atomizers, which are not subject to this danger. Solutions in volatile solvents such as alcohol should not remain exposed, except briefly, to avoid errors in concentration due to evaporation (cf. Section 123).

The concentration of alcohol should be such that its enhancing effect upon the emission does not depend too critically on the concentration.[575] It should be kept in mind that many solutes may be less soluble in nonaqueous solvents.

Several investigators have tried to elucidate these solvent effects, for example, by comparing the viscosity, surface tension, and vapor pressure (volatility) of alcohol mixtures, etc., with those of water. These physical properties, among other things, govern the rate of transport of sample to the flame and hence the concentration of analyte in the flame and its emission (Section 6). The low surface tension and high vapor pressure of alcohols help atomization and evaporation, as shown by measurements of the atomization efficiency[139] and drop diameters.[182] These effects, which greatly favor the emission, outweigh the higher viscosity of alcohol–water mixtures, which diminishes the suction rate and would therefore lower the emission in the absence of the other effects.[139] These favorable effects of alcohol on the evaporation and the emission operate not only in the spray chamber of an indirect atomizer but also in a direct atomizer, especially one in which water drops evaporate incompletely in the flame (Section 6).

Many organic solvents change the flame temperature as well as the drop size. The considerable heat of combustion of such solvents as acetone has often been mentioned in the literature. Their heat of vaporization is also lower than that of water. However, it has been shown that, in contradiction to the usual assumption, the flame temperature prevailing when mixtures of water with acetone or propanol are sprayed can be either higher or lower than that prevailing with purely aqueous solutions.[71] This depends on whether the mixture is lean or rich. If it is rich, the combustion of the organic solvent is incomplete, and the liberated heat of combustion is apparently less than the heat required to raise the reaction products to the temperature of the flame.[65]

The effect of organic solvents on the flame temperature is especially hard to interpret in a direct atomizer, since it depends also on the factors affecting the sample flow rate and the evaporation. When an alcohol is sprayed, a small spherical diffusion flame forms around each droplet. The rate of evaporation of the liquid depends on many factors.[11] If evaporation is incomplete in the flame, equilibrium is not attained. The conditions in an atomizer-burner flame are rendered more complex by the strong cooling due to the entering solvent and by the change of size and shape of the flame, which is conspicuous when water is sprayed (Section 14, Figs. 25a, 25b).

From the above it is obvious that the solvent effect can vary a good deal with the conditions, and that experimental results obtained under one set of conditions are hardly valid for another. Thus, the effect of alcohol upon the atomization efficiency depends on the quality of the atomizer.[65] The question of practical importance, to what extent change of flow rate and change of temperature are separately responsible for the solvent effect, can be answered fairly easily by experiment. For example, the relative enhancements of the yellow sodium and the red potassium lines can be measured; if the enhancements, expressed in terms of concentration, are equal, then only a flow rate effect is present. But if temperature is the sole factor at work, then owing to the difference in excitation energy the relative enhancement of sodium should be about 1.3 times that of potassium, ionization and dissociation effects being left out of consideration.

Advantages and disadvantages of dilution are compared below.

Advantages:

1. Dilution permits greater accuracy when it brings the concentration into the linear part of the working curve. However, with the more easily ionized elements (alkali metals), the curve becomes concave at lower concentration. At still higher dilution, however, the curve can become linear again, owing to dominance of the ions contributed by the flame gases (Section 7).
2. There is less risk of incrustation and clogging at the atomizer, the burner tip, etc., by salt crystals or the like.
3. The room air is less rapidly contaminated.
4. Interferences are reduced (Section 96).
5. Differences in viscosity and surface tension between sample and standard are reduced, and their effect on the result is diminished. For instance, in serum analysis at higher dilution, the precipitation of protein with trichloroacetic acid can be dispensed with.
6. More sample is available. For a given initial quantity, high dilution permits multiple determinations; with very small initial quantities, the analysis may not even become possible until the material is diluted. For micro methods, see Section 89.

Disadvantages:

1. There are the unavoidable errors of pipetting.
2. There is greater risk of contamination from dirt or by diffusion from or adsorption on the walls of the container.
3. More work is needed in the preparative stage.

4. The purity required of the solvent becomes more critical with increasing dilution.*

5. Fluctuations in the electrical components of the instrument (shot effect, Johnson noise, etc.) become more noticeable at lower concentration.

6. Requirements of sensitivity of the instrument and purity of the atmosphere become more stringent.

Naturally, all dilutions have to be performed quantitatively, since an error of dilution creates an equivalent error in the analysis. The following points call for attention.

1. Because of the risk of concentration change from diffusion of alkalies out of the container walls or adsorption on the walls, all necessary dilutions should be made shortly (within a few hours) before the actual analysis. Since the diffusion is temperature-sensitive, the diluter solutions that are not to be used promptly should be kept refrigerated; but they must be brought to room temperature before the analysis.

2. For dilution use only good pipets and volumetric flasks, preferably officially calibrated (certified). Pipetting should be done with the greatest care, preferably with an automatic pipetter.

3. Volumetric apparatus such as pipets and flasks should be properly cleaned with chromic sulfuric acid or other suitable agent (Section 80), for if unreproducible quantities of liquid adhere as drops to the walls of a pipet, the best pipet and the greatest care are of no avail. For aqueous solutions the precision of a pipet can be appreciably improved[301, 303] by coating the interior with a silicone water-repellent, which eliminates the wetting of the wall that causes error, and permits more precise adjustment of the meniscus. However, the pipet must be recalibrated after coating with silicone.†

It can also happen that the analyte is at such low concentration that, even with every means of improving the detection limit, its lines or bands are too weak to be measured above the flame background or give too small a net reading owing to insufficient electronic sensitivity. If enough sample is available, it can be concentrated, for instance, by

---

* Our experience (in Germany) has shown that commercially packaged redistilled water is often worse for sodium, potassium and calcium analyses than regular distilled water. This might be owing to the fact that the twice-distilled water is generally prepared in glass apparatus and distributed in small, seldom-used glass containers, whereas ordinary distilled water is prepared in large quantities in metal equipment and distributed in large flasks.

† Directions for applying water-repellent films will be found in Bulletin 262C of Beckman Instruments, Inc., and elsewhere.

evaporating liquid constituents. In doing this, one must beware of contaminating the solution (as by diffusion of alkalies from a glass vessel), and the final product must not be too viscous. Biological materials are therefore often concentrated by ashing (Section 109). Other methods of concentration include chemical enrichment by precipitation, etc. We return to this in Sections 88 and 89.

80. *Containers and their Maintenance*

The ranges of concentration best for flame analysis are so low for the alkali and alkaline-earth metals that diffusion of alkali ions out of glass containers or adhering impurities affect the dilute solutions. This is especially true of new glass vessels or such as have been improperly cleaned (see below). Conversely, solute may become adsorbed on or diffuse into the walls.[420] Sometimes a slow precipitation can simulate this effect; for example, calcium can be lost by precipitation from a dilute, neutral solution by the carbon dioxide of the air. Loss to the walls is especially likely in polyethylene bottles when the solution contains a wetting agent (see below). In handling sample and standard solutions, therefore, special attention must be paid to the containers and the matter of cleaning them. The diffusion and adsorption phenomena become more critical the longer a dilute solution is to be kept.

Glass bottles with ground stoppers are recommended only for concentrated stock solutions. The friction of loosening a stopper liberates alkalies and alkaline earths, especially calcium. Bottles of hard glass with screw caps are better; the cap liner should consist not of cork but of rubber or, better, an elastic plastic disk. In particular, borosilicate glass is recommended for calcium analyses at low concentrations, or ordinary glassware should be coated with a water-repellent film (see below).

More recently, plastic containers (polyethylene, polystyrene) have come into extensive use. They have the great advantage that aqueous solutions do not wet the walls, and these containers are usually unbreakable. But wall effects are still not entirely absent, although they are usually a good deal smaller than with glass. Dirt on the walls can of course be as troublesome as with glass. Wetting agents, furthermore, favor exchange between the plastic and the solution. When this happens, alkali-containing plasticizer from the plastic can cause the same kind of contamination as glass, while plastic without plasticizer can absorb some of the ions in the solution (Schön, in the discussion of ref. 348). Equilibrium between wall and solution is reached after a sufficient period. If wetting agent must be present, the same plastic container should always be used for the same solution, and the contents should be renewed frequently at first. Plastic bottles have another drawback in that they cannot be used when certain organic solvents are present in the samples or standards. Also, very hot solutions should not be put into the bottles.

Despite these disadvantages, plastic bottles are an important advance in the storage of standards for flame photometry, and are to be recommended. Unfortunately, they are not available in all of the forms required for microanalysis. Hence glassware such as weighing dishes, watch glasses, and test tubes will often be used, sometimes with a silicone coating (see below).

The use of a silicone coating* is recommended for ordinary glassware, preferably having smooth surfaces and no sharp corners.[301, 303] The material is usually a solution of a substituted chlorosilane (methyltrichlorosilane, dimethyldichlorosilane, lauryltrichlorosilane, etc.), which is applied in diluted form as a thin film upon the well-cleaned and dried surface of the glass. The solvent is allowed to evaporate, and the film of silicone oil on the glass hydrolyzes on contact with water, liberating hydrogen chloride (from chlorosilanes) and leaving an adherent siloxane film on the surface.

Other types of silicone oil must be baked on. Silicone oil AK 350, for example, is applied as a 3% solution in benzene or a chlorinated hydrocarbon to the clean, degreased glass, the solvent is allowed to evaporate, and the glass is baked for 2 hr. at 300°C. or 6 hr. at 200°C.

Glassware coated with a baked-on silicone has advantages like those of plastic containers. The thin, unbaked films afford only water-repellency without chemical protection. This type of film will last (for years) only if properly applied, used, and cleaned. The manufacturers provide details. Excess silicone oil from the coated glassware must not be allowed to enter the atomizer capillary. The resulting water-repellency makes aspiration unreproducible or severely impedes it, and encourages gas dissolved in the sample or standard to deposit as bubbles within the capillary. These bubbles make atomization erratic. Conversely, the gas-releasing action of a siliconed surface can be turned to advantage if the samples and standards are allowed to stand for some time in a siliconed flask prior to the analysis, in order to deaerate them. In an emergency, glass vessels coated with paraffin can be used, but there is the risk that loosened particles of paraffin may get into the solution and clog the atomizer capillary.

* For example, Desicote, supplied by Beckman Instruments, Inc., Fullerton, Calif., or silicone oil AK 350 from Wacker-Chemie, Munich 22, Germany.

Even though the liquid in the bottle is pure, it can become contaminated by pouring it out over the dirty lip of the bottle. To avoid this, bottles containing solutions for flame photometry should be kept in a dust-free place, and the first drops of the poured liquid should be discarded.

The great susceptibility of flame analysis to traces of dirt can be forcefully demonstrated by two simple experiments:

1. Set the instrument at maximal sensitivity for sodium at 589 mμ and spray distilled water. Now dip the finger tip momentarily into another portion of about 5 ml. of the same distilled water and repeat. The flame photometer will show a distinct response. The same result is obtained by touching a watch glass, in which the pure distilled water is then placed and atomized.[143]

2. Set the instrument for maximal sensitivity to calcium and spray distilled water. Repeat with a sample of the same distilled water to which a drop of tap water has been added.

For cleaning glassware the following measures are recommended for the commonest applications: Used vessels should be rinsed with water as soon as possible. If necessary, the rinsing can be aided by gentle mechanical or chemical means (brushes, warm water, wetting agents). The glassware is next immersed in cleaning solution (chromic sulfuric acid) for 24 hr.; care should be taken that no bubbles prevent access of the solution to isolated spots on the glass surfaces, and that the solution is not exhausted (turning green). After the cleaning-solution bath the glassware is rinsed about five times with fresh tap water and then immediately with one or two portions of distilled water; it is then dried dust-free and stored for future use under dust-free conditions (in a closed cabinet).

New glassware may be steam cleaned or first cleaned with cleaning solution and then left for 8 or 10 days in distilled water, which should be changed every 2 days. This treatment removes any soluble alkalies. Glassware thus treated may be cleaned after subsequent use in the same way as above. Glassware that has been in use for some time and already repeatedly cleaned as above, and that is not further contaminated by the analytical solution, can be cleaned with an alkali-free wetting agent* and rinsed a few times with distilled water.

* For example, Arkopal N-060 or N-100, from Farbwerke Hoechst; BW$_2$, from the Chemische Werke Hüls; Foryl 100, from Fettchemie Düsseldorf; Non-Ion-Ox of Aloe Scientific, St. Louis 3, Mo.

§ 81]   FLAME SPECTRA OF THE ELEMENTS: ANALYTICAL EMISSIONS   215

We here give explicit warning against the following errors:

1. Drops of tap water must not be allowed to dry upon the containers, as it contains much calcium.

2. No polishes or scouring powders should be used for cleaning, as these may scratch the glass surface, and scratches favor the diffusion of calcium, sodium, potassium, etc., from the glass into the sample solution.

3. No soaps should be used, since they will form lime soaps. Detergents such as alkyl sulfonates are permissible.

4. Vessels should never be wiped with a cloth or the like, as this will inevitably transfer invisible dirt and easily introduce lint, which may lead to clogging of the atomizer capillary and burner jet.

5. Cork stoppers must not be used for closing flasks, for cork or particles of cork falling into the solution release much electrolyte and may likewise lead to irregularities of atomization.

If very dilute solutions are to be preserved without change of concentration for several months, even the above precautions for cleaning the glass vessels will not suffice. For this purpose the plastic containers mentioned earlier are recommended. Special cleaning methods are needed for these or for containers with a protective silicone coating; the manufacturer should be consulted.

The Flame Emission

81. *Flame Spectra of the Elements: the Principal Analytical Emissions**

The lines and bands that are useful for the determination of the elements are shown in the spectrograms and listed in the index of flame spectra at the end of the book, and in the table of detection limits

---

* Although the text of this Section is an essentially unrevised translation of the second German edition, Table 10 (Section 90) and Tables 13 and 14 (Appendix) have been brought more nearly up to date, at least up to 1962. For the sources of some of the more recent information, see the introduction to Table 13. The atlas of spectrograms also includes material too recent to have been heeded in this Section. For detailed wavelength tables and spectrograms showing the line spectra of the rare earths in the incandescent oxyacetylene flame, not included in the Appendix or Atlas of this book, see V. A. Fassel, R. H. Curry and R. N. Kniseley, *Spectrochim. Acta*, **18**, 1127–53 (1962). Similar extensive tables and spectrograms for vanadium, niobium, rhenium, titanium and molybdenum in the incandescent oxyacetylene flame are compiled in a paper by V. A. Fassel, R. B. Myers, and R. N. Kniseley, to appear in *Spectrochim. Acta*. This paper also shows a few flame lines of tantalum, tungsten, thorium, and uranium. A detailed tabulation and many spectrograms of the lines and bands (most of them chemiluminescent) of

(Section 90). Section 82 further deals with the choice of lines and bands in a general way, and Section 79 discusses the optimal ranges of concentration. In this section we cover some of the characteristic features of the flame spectra of the elements from the standpoint of the periodic system, not readily discerned from the tables mentioned above.

Group IA, alkali metals (*Li, Na, K, Rb, Cs*). These yield simple spectra with few strong lines. The resonance doublets of these elements (Li 670.8, Na 589.0-589.6, K 766·5-769.9, Rb 780.0-794.8, Cs 852.1-894.4 mμ) are the strongest lines that occur in the flame. The alkali metals are therefore the most amenable to flame analysis. We have already mentioned characteristics of the well-known lines of sodium (Sections 7, 8, 10), potassium (Sections 8, 10, 11), and rubidium (Section 7). Such bands of alkali-metal compounds as occur in the flame are of no analytical use. The alkali metals emit a continuum (Section 8), which acts as a blank interference (Section 95). At higher concentrations it may reduce the detection sensitivity for other elements simultaneously present.

The second members of the principal series (cf. Fig. 2) (Li 323.3, Na 330.2-330.3, K 404.4-404.7, Rb 420.2-421.6, Cs 455.5-459.3 mμ) are sometimes used for analysis, except for the weak lithium line. Temperature fluctuations have a stronger effect on these lines (Section 8). Common spectral interferences at these wavelengths include Mn 403.1-403.3-403.4 mμ with potassium, Ca 422.7 or Sr 421.6 mμ with rubidium, and Ba 455.4 or Sr 460.7 mμ with cesium. A good monochromator can alleviate most of these interferences. But in using the resonance lines, K 769.9 mμ can interfere with Rb 780.0 mμ unless resolution is adequate. Aside from this, Rb 780.0 mμ and Cs 852.1 mμ are favored because of their intensity, provided the instrument is sensitive enough in the near infrared.

For the most part, the alkali metals are comparatively insensitive to interferences from other elements, except such as change the ionization in the flame (Sections 7 and 97). It is noteworthy that the alkali metals show different degrees of sensitivity to height in the flame (Section 11), on account of the variation of concentration of the OH radical and H atom with height; the concentration is highest just above the inner cone. The tendency to form the hydroxide (which does not

---

Al, Ba, B, Cd, Ca, Cr, Cu, Fe, Pb, Mg, Mn, Mo, Ni, Si, Ag, Na, Sr, Sn, Ti, V, and Zn in the oxyhydrogen flame with naphtha as solvent has been compiled by B. E. Buell (Union Oil Co., Research Center, Brea, Calif.). The paper will be published, probably in condensed form, and any one interested in the complete data should write to Mr. Buell. For data on the emission of Zn, Cd, Hg, Sn, As, Sb, Bi, and Pt in the reaction zone of oxyacetylene, see the thesis by W. J. Carnes, University of Tennessee, Dec. 1961, and the paper by J. A. Dean and Carnes, *Analyst*, **87**, 743-7 (1962).—*Translator*.

emit) in the flame varies greatly among the alkali metals, following the decreasing order Li, Cs, Rb, K, Na; in the cooler hydrogen-containing flames lithium forms hydroxide so extensively that much of the element becomes unavailable for excitation in the flame, whereas with sodium this effect is negligible.[391] Hydroxide formation also depends strongly upon the composition of the flame gases[169] and the temperature.[394] Owing to this variable alkali hydroxide formation, the absolute intensities of the lines of the several metals for equal atomic concentrations are very diverse in the cooler flames, even with allowance for the different excitation potentials and transition probabilities; the intensity is greatest with sodium, least with lithium.[394]

One of us (Alkemade, unpublished) found that only 7% of the lithium is present as free atoms at a height of 6 cm. in a premixed air–acetylene flame, as deduced from the ratio of absolute intensities of the sodium and lithium lines, with the justifiable assumption that hardly any of the sodium is bound as molecules. Corrections were made for ionization and self-absorption. Another factor that can affect the height-dependence of emission is the ionization of the flame gases (Section 7); since it diminishes with height, the alkali emission should diminish correspondingly.

Self-absorption is marked with the resonance lines although least with lithium. It makes the working curves convex in the upper ranges. It is less severe for the second members of the principal series. The first members of the diffuse series (Li 610.4, Na 818.3-819.5, K 1169.0-1177.2 mμ) show little or no self-absorption since the transition is not to the ground state, and so the working curves are nearly straight. These effects as well as those of ionization (increasingly marked with K, Rb, Cs) have been discussed in Section 8.

Spectra of the alkali metals appear on Spectrograms 1–8.

Group IB (Cu, Ag, Au). Like the alkali metals, these elements yield simple spectra in the flame, but of lower intensity. The strongest features are the resonance doublets (Cu 324.8-327.4, Ag 328.1-338.3, Au 242.8-267.6 mμ), which are strongly self-absorbed and give very curved working curves. The gold lines, being of shorter wavelength, are much weaker than the copper and silver lines. Note the possibility of interference between copper and silver.

Copper emits bands in the green due to CuOH, CuO, and CuH, but the analytical value of these is slight. In cooler flames and in the presence of the appropriate halogen, copper yields the characteristic blue-green monohalide band spectra, still sometimes used for analysis (Section 87).

Spectra are found on Spectrograms 62 and 63.

Group IIA, alkaline-earth metals (Be, Mg, Ca, Sr, Ba, Ra). The flame determination of the alkaline earths is less easy than that of the alkali metals. The lines are less intense, and there are extensive bands of compounds such as CaOH and CaO. These bands are often stronger than the lines, depending on flame temperature and bandwidth of observation. Therefore, despite basic disadvantages (Section 82), the bands are often used for analysis. They do, at any rate, provide straighter working curves than the lines, but even the lines may show less convexity than those of the alkali metals.

Although the origin of these alkaline-earth bands, at least those in the visible, is not certain in every case (both oxide and hydroxide bands occur; see Sections 8 and 9), the greater portion of the alkaline-earth element is present in the flame as oxide.[393] Except for oxygen–cyanogen, the flame is not hot enough to dissociate these (hydr)oxides fully; their bands are very weak in the oxygen–cyanogen flame. The alkaline-earth metals are more sensitive than the alkalies to interferences from other elements such as aluminum and beryllium, or anions such as phosphate and sulfate. We return to this in Section 98.

The resonance lines (Be 234.9, Mg 285.2, Ca 422.7, Sr 460.7, Ba 553.6, Ra 482.6 mμ) are strong in conventional flames (except for beryllium; see below). In addition, the lines of the singly ionized atoms, homologous with the alkali-metal resonance doublets (Mg 279.6-280.3, Ca 393.4-396.8, Sr 407.8-421.6, Ba 455.4-493.4, Ra 381.4-468.2 mμ), appear in hot flames and, except for magnesium, become dominant in oxygen–cyanogen.

As with the alkali metals, introduction of alkaline-earth salts noticeably enhances the ionization of the flame. The relations are, however, much less simple than with the alkali metals, for in addition to ions such as Ca^+, positive and negative molecular ions such as $(BaOH)^+$ may occur.[695] Again as with the alkali metals, marked continua are contributed, especially by calcium, strontium, and barium.

Beryllium emits no lines in conventional flames but in oxygen–cyanogen its resonance line 234.9 mμ is strong and useful. In oxyhydrogen and oxyacetylene BeO bands at 471 and 510 mμ can be used for analysis at high concentrations. The working curve is nearly straight. There is appreciable continuous emission in the visible. Thus, beryllium would not be easy to determine in the usual flames. However, it can be determined indirectly (Section 87(b)). A spectrum in oxyhydrogen is shown on Spectrogram 10*.

Magnesium is hard to observe at its resonance line 285.2 mμ except with a hot flame and good monochromator, but it is exceptionally

* See also Spectrogram 9, showing the line spectrum in incandescent oxyacetylene.—*Translator.*

strong in oxygen–cyanogen, in which the nearby Mg^+ doublet also appears. The latter appears (weakly) in conventional flames chiefly in the reaction zone. The reaction zone also emits weak groups of lines at 278, 333, and 383 mμ. The MgOH bands in the near ultraviolet (360–400 mμ) are strong in hydrogen flames, and the chief peaks at 370 and 382 mμ are most commonly used for analysis. The MgO bands with chief heads at 500.7 and 520.6 mμ are much weaker. Spectra of magnesium appear on Spectrograms 11 and 12.

Calcium appears on nearly all long-exposed spectrograms at 422.7 mμ, revealing trace contamination. With suitable equipment and not too cool a flame, this line is good for analysis, but Rb 421.6 and Sr 421.6 mμ may rarely interfere. With filter photometers, however, the CaOH bands between 600 and 645 mμ, peaking at 622 mμ, or at 554 mμ are mostly used. In work with these calcium bands, especially with filters, separation of the sodium line at higher sodium concentrations can prove difficult. Band-pass filters with a wide spectral band (Section 40) will transmit relatively more of the background and sodium continuum. There are bands of CaO in the far red with maxima at 824 and 872 mμ, which can interfere with barium determinations. A spectrum is shown in Spectrogram 14.

Strontium and barium give characteristic spectra homologous with that of calcium, the lines and bands being shifted progressively to longer wavelengths. Whereas on a molar concentration basis the spectrum of strontium is about as intense as that of calcium, that of barium is weaker. The best SrO(H) band peaks are at 606 and 682 mμ, and those of barium are at 745 and 830–873 mμ. Strontium is most commonly determined at 460.7 or 682 mμ, and barium at 873 mμ. However, barium has a series of bands in the green, with many good peaks between 488 and 604 mμ, due to BaOH and BaO. The Ba resonance line 553.6 mμ is badly interfered with by calcium. The spark lines, 455.4 and 493.4 mμ, are more characteristic for barium in hot flames. High concentrations of sodium and potassium will create background interference. Spectra of strontium are found on Spectrograms 15–17, and of barium on Spectrogram 18.

Radium has a flame spectrum more like that of strontium than that of barium, but of lower intensity.

Group IIB (Zn, Cd, Hg). The resonance lines are Zn 213.9, Cd 228.8, Hg 185.0 mμ. Lying in the vacuum ultraviolet, the mercury line has not been reported in flames.* The zinc and cadmium lines are emitted

---

* J. B. Willis, *Anal. Chem.*, **34**, 614–17 (1962), has observed this line in absorption in an oxyhydrogen flame. Carbonaceous flames are opaque at this wavelength.—*Translator.*

weakly in conventional flames (best seen in oxyhydrogen, especially with an organic solvent). They are quite strong in oxygen–cyanogen. The intercombination lines, Zn 307.6, Cd 326.1, Hg 253.7 mµ, are noteworthy. The zinc line falls in the midst of the strongest OH bands and has not been used for analysis. All three lines are remarkable in that their absolute intensities diminish slightly as flame temperature rises. The mercury line is more intense in the inner cone of oxyacetylene. A number of weak zinc lines at 328, 346, 472, and 481 mµ have also been observed. Zinc emits a continuum in the visible, whose intensity increases rapidly with flame temperature. Spectra of zinc appear on Spectrogram 64; of cadmium, on Spectrogram 65; and of mercury, on Spectrogram 66.

Group IIIA, rare-earth metals, etc. (Sc, Y, lanthanides, actinides). Most of the rare earths emit characteristic if not very intense oxide (and perhaps hydroxide) band spectra, together with resonance and ion lines in the case of a few elements, especially europium, thulium and ytterbium. The periodicities appearing in these spectra (Fig. 49) reflect the periodicities in the structure of the outer electron shells. Cerium, not included in Fig. 49, emits bands at 480–495 mµ (Spectrogram 23). Both the bands and the lines can be used for analysis. However, if several of the elements are present together, their band systems may overlap; for quantitative analysis, special methods are then required, which will be discussed in Section 86. The flame spectra of the rare-earth elements are shown on Spectrograms 19–35.

The actinides for obvious reasons have been little studied. The analysis of these elements will be mentioned in Section 117. Uranium emits a weak continuum in the flame, due to its stable oxide.

Group IIIB (B, Al, Ga, In, Tl). Boron is the only nonmetal with a flame spectrum intense enough for easy analysis. It emits no lines, but the familiar green "fluctuation" bands, with principal peaks at 495, 518, 547, and 579 mµ, are quite distinctive. See Spectrogram 67.

The four metals yield the resonance doublets (Al 394.4-396.2, Ga 403.3-417.2, In 410.2-451.1, Tl 377.6-535.0 mµ). The indium and gallium lines in conventional flames have an intensity (for equal atomic concentration) approaching that of strontium; the aluminum and thallium lines are weaker. The aluminum lines are greatly enhanced (up to 1000-fold) and the detection limits improved by organic solvents or by chemical additives such as ethylenediaminetetraacetic acid[219a, 252] or oxine.[217] Hydrofluoric acid[431b] also improves the ratio of line and band emission to background for aluminum; cf. Section 88. Indirect methods of determination (Section 87) are also used, but the sensitivities are poorer.

§ 81] FLAME SPECTRA OF THE ELEMENTS: ANALYTICAL EMISSIONS 221

Aluminum has a band system, not very intense, between 435 and 510 mμ, with chief heads at 464.8, 484.2, 486.6, and 507.9 mμ, which can be used for analysis at moderate or high concentrations. Gallium, indium, and thallium normally emit no detectable flame bands, except for a weak system due to InO with principal head at 428.3 mμ.* For

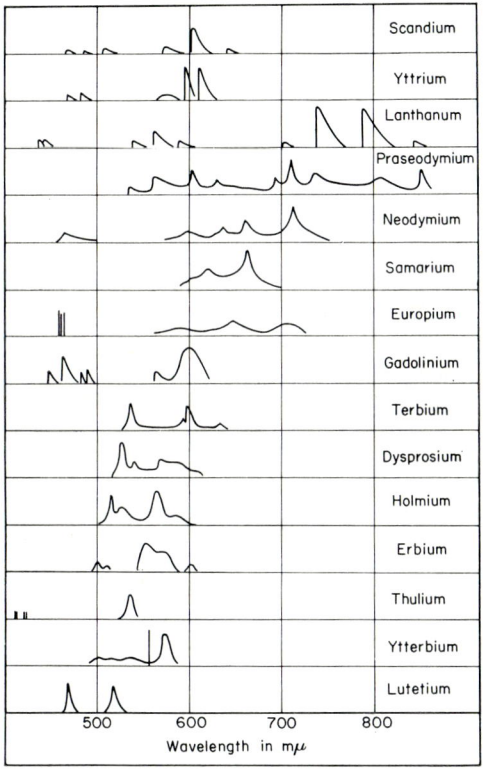

Fig. 49. The chief bands and lines of the rare-earth elements in air–acetylene (from ref. 546).

spectra of aluminum, see Spectrograms 68 and 69; of gallium, Spectrogram 70; of indium, Spectrograms 71 and 72; of thallium, Spectrogram 73.

Common interferences are Ca$^+$ 393.4-396.8 mμ with aluminum, and Mn 403.1-.3-.4 and K 404.4-.7 mμ with gallium. Magnesium and

* E. M. Bulewicz and T. M. Sugden, *Trans. Faraday Soc.*, **54**, 830–7 (1958), observed flame bands of InH and TlH.—*Translator.*

barium can interfere with the thallium lines. The resonance lines of gallium, indium, and thallium give straight working curves in the useful ranges of concentration; so also does the AlO band at 484 mμ.

Group IVA (Ti, Zr, Hf). In the usual flames these elements are present only as undissociated oxide, and emit no atomic lines. Titanium yields a rather complex band spectrum at 500–750 mμ. Some of the better band heads lie at 500.0, 516.8 and 545.0 mμ, and should be useful for analysis.* They are easily interfered with by many other elements (Ba, Fe, Mn, Cr, V, etc.) having band spectra in the green.

Zirconium yields chiefly a continuum, with weak bands of ZrO superposed at 564 and 574 mμ, of little use for analysis. However, it can be determined indirectly (Section 87). Hafnium has not been studied in the flame.

Group IVB (C, Si, Ge, Sn, Pb). The carbon line at 247.9 mμ is detectable in most carbonaceous flames, especially in the inner cone (Spectrograms 66 and 74). Bands of CN, CH, CO_2, CO and C_2 also appear in flames (Spectrograms 80, 83, 94 and 95; Fig. 6). Ordinarily these bands are accepted as part of the flame background. Obviously, carbon cannot be determined by flame photometry in a carbon-containing flame, but these lines and bands might be used for carbon determination in a noncarbonaceous flame, e.g., oxyhydrogen (Section 87). Silicon and germanium emit no lines or bands, except for weak bands of SiO in the shorter-wavelength ultraviolet.†

Tin and lead emit a few lines. The strongest are Pb 364.0, 368.3, 405.8 and Sn 284.0, 286.3, 300.9, 303.4 mμ. In the usual flames only the lead lines are strong enough to be useful for analysis (Spectrogram 82). The tin lines are much weaker; they are seen best in the inner reaction zone. However, in the oxygen–cyanogen flame the tin lines are about as intense as the lead lines and could well be used for analysis (Spectrogram 81).‡

In addition, lead emits a very weak band system of PbO, not very characteristic. The bands of tin, on the other hand, are stronger and have

---

* On the flame-photometric determination of titanium, see H. Goto and E. Sudo, *Bunseki Kagaku*, **10**, 456–62 (1961), and C. L. Chakrabarti, W. F. Pickering and C. L. Wilson, *Talanta*, **9**, 451–2 (1962).—*Translator.*

† Line spectra have recently been obtained in flames. See Spectrogram 75 for silicon and 76 for germanium.—*Translator.*

‡ For chemiluminescent line spectra of tin and lead see Spectrograms 78–80, 83 and 84. This tin emission has been put to practical use by W. Schmidt, *Beckman Report* (Beckman Instruments G.m.b.H., Munich), No. 4, pp. 6–7 (1961).—*Translator.*

sharp heads; they lie in the near ultraviolet and violet, and occur in both the inner cone and the mantle (Spectrogram 77). They are absent from the oxygen–cyanogen flame, in which all tin compounds dissociate.

Group VA (V, Nb, Ta). These elements, like those of *IVA*, form oxides stable in flames of moderate temperature and emit no lines. The band spectrum of VO (Spectrogram 36) is much like that of TiO in location and intensity, and subject to similar spectral interferences. The best band heads, useful for analysis, lie at 522.9, 547.0, and 573.7 mμ.*
In oxygen–cyanogen, however, the VO spectrum disappears, while a characteristic line spectrum of V and V⁺ takes its place (Spectrogram 37). The strongest triplet, 318.3-318.4-318.5 mμ, permits sensitive analysis.

Niobium emits only a continuum in hydrogen and acetylene flames. No information is available on tantalum.†

Group VB (N, P, As, Sb, Bi). Nitrogen produces a greenish continuum in the inner cone of flames containing free atomic oxygen,[392] which is due to the energy liberated in the reaction $NO + O \rightarrow NO_2$. The mantle of oxygen-containing flames also shows this continuum, though more weakly, owing to entrainment of atmospheric nitrogen. Sharp bands of NO occur in the air–hydrogen flame when nitric acid in the presence of alcohol is sprayed into it (Spectrogram 85).

Bands of PO in the short-wavelength ultraviolet appear with relatively good intensity in the air–hydrogen flame when alcohol is added (Spectrogram 86), but are weaker in the absence of such a solvent. Phosphorus also emits a continuum in the visible due to incandescent unvaporized phosphorus oxide.

Arsenic and antimony emit weak ultraviolet lines in the inner cone. Even in oxygen–cyanogen, As 228.8 mμ and Sb 231.1 mμ are quite weak and hardly suited for analysis. Arsenic and antimony yield band spectra in cooler flames, too weak to be useful for quantitative determination.‡

Bismuth is only slightly more easily excited. A line at 472.3 mμ is best seen in hydrogen flames (Spectrograms 89 and 90); it has a straight working curve. This line disappears in oxygen–cyanogen, which, on the other hand, emits the resonance line at 306.8 mμ with good intensity, standing out above the OH bands.

---

* On the flame-photometric determination of vanadium, see C. M. Stander, *Anal. Chem.*, 32, 1296–9 (1960), and E. Schöffmann and H. Malissa, *Mikrochim. Acta*, 1961, 319–22.—*Translator*.

† See the footnote at the beginning of this Section.

‡ Chemiluminescent line spectra of arsenic, antimony and bismuth are shown in Spectrograms 87, 88, and 91.—*Translator*.

Group VIA (Cr, Mo, W). Chromium has a sensitive flame spectrum, especially in organic solvents, with triplets at about 359, 427, and 521 mμ. The line at 425.4 mμ is one of the best for analysis. Only Ca 422.7 mμ may give difficulty if resolution is inadequate. The corresponding molybdenum triplet (379.8-386.4-390.3 mμ) is seen only weakly in the inner cone of acetylene flames, along with a still weaker triplet near 317 mμ. In oxygen–cyanogen the 380–390 mμ molybdenum triplet is well developed. This flame also excites many additional lines of chromium. For spectra of chromium see Spectrograms 38 and 39; of molybdenum, Spectrograms 40 and 41.

A band spectrum due to CrO in the green and red is emitted in the conventional flames. This disappears in the oxygen–cyanogen flame, owing to dissociation of the CrO. Nothing is known of the flame spectrum of tungsten.* Tungsten and molybdenum, however, can be determined indirectly; see Section 87.

Group VIB, chalcogens (O, S, Se, Te, Po). Although sulfur and selenium emit various kinds of flame spectra, these are not observed in analytical flame photometers.† Tellurium, however, emits an interesting band system in hydrogen flames, covering the near ultraviolet and violet. Peaks at 371.4, 388.4 and 400.7 mμ are among the most marked (see Spectrogram 93). Polonium has not been studied.

Group VIIA (Mn, Tc, Re). Manganese is very well suited to flame photometric determination, with its intense triplet at 403.1-403.3-403.4 mμ. The ultraviolet triplet 279.5-279.8-280.1 mμ is also emitted. Manganese has a well-developed oxide band spectrum (MnO), responsible for the green color of a manganese flame. The strongest heads (539.0, 558.6 mμ) are hardly used, owing to the greater convenience of the violet line. The intercombination lines 539.5, 543.3 mμ lie in the heart of the band system. There are other band systems 363–400 mμ and in the vicinity of 810 and 860 mμ. Spectra of manganese are shown on Spectrograms 42–44.

Technetium has not been studied in the flame. Rhenium is said to impart a green color to the flame, its line at 488.9 mμ perhaps appearing weakly.‡

Group VIIB, halogens (F, Cl, Br, I). No flame lines of these elements are known. Weak bands, hardly useful for practical analysis, have been observed under certain conditions, particularly an uncharacteristic

---

* See the footnote at the beginning of this Section.

† For the recently obtained spectrum of sulfur (CS), see Spectrogram 92. A chemiluminescent line spectrum can be obtained for tellurium.—*Translator*.

‡ See the footnote at the beginning of this Section, and Spectrogram 45.

band system of IO in the blue and green. Iodine might thus be determined (cf. Section 87).

Group VIIIA, iron and platinum metals (*Fe, Co, Ni; Ru, Rh, Pd; Os, Ir, Pt*). The first six of these elements emit rich line spectra. Spectral interference among them is very common, and high spectral resolution is needed for avoiding it. Osmium and iridium have not been investigated; their low volatility in conventional flames might limit the intensities of their spectra.* Above the point of saturation these metals would contribute a continuum from the incandescent aerosol. Platinum emits weakly, as its strongest lines are of shorter wavelength.

Iron has its strongest lines at 372.0, 373.7, 374.6, and 386.0 mμ, commonly used for analysis. The magnesium bands may interfere but can be allowed for by measurement of the background next to the lines (Section 95). Many visible lines occur at higher concentrations, while the inner cone of air–acetylene excites deep ultraviolet lines as far down as 247 mμ. In addition, iron shows a band spectrum (FeO) with the chief peaks in the green and yellow. See Spectrograms 46–52.

Cobalt has its strongest lines at 345.4 and 352.7 mμ. Other good lines are 340.5, 341.3, 350.2, and 387.3-387.4 mμ. Interferences are likely from Fe 387.9 mμ, Ni 341.5, 345.8, 346.2, 349.3, 352.5 mμ, Ru 349.9 mμ, Rh 350.3 mμ, and Pd 340.5 mμ. A good monochromator and narrow slits are needed to avoid these. Cobalt emits lines in the violet and a nearly structureless continuum in the green and red. See Spectrograms 53–57.

Nickel is best determined at 341.5, 345.8-346.2, 349.3, 351.5, 352.5, or 361.9 mμ. Lines likely to interfere include Co 341.3, 345.4, 346.3, 346.6, 350.6, 351.3, 352.7, 353.0 mμ, Ru 349.9 mμ, Pd 342.1, 351.7, 361.0 mμ. In the visible, nickel emits a weak band spectrum superposed on a continuum comparable with that of iron in intensity. See Spectrograms 46–48 and 58–60.

Ruthenium has a slightly simpler spectrum with strongest lines at 372.8 and 379.9 mμ. The former suffers interference from iron. A moderately strong ruthenium line at 569.9 mμ is noteworthy. Ruthenium appears to emit some bands in the vicinity of 600 mμ.

Rhodium emits most strongly at 369.2 mμ, at which it is not likely to suffer much interference from other elements. It has a good many other lines throughout the near ultraviolet and violet (see Spectrogram 61).

Palladium has strong and characteristic lines at 340.5, 361.0, and

* According to J. A. Dean, *Flame Photometry*, (McGraw-Hill, New York, 1960), p. 259, osmium and iridium emit no flame lines even in organic solvents.—*Translator.*

363.5 mμ. It is best determined in hydrogen flames. Working curves for these and other palladium lines are rather curved, 340.5 mμ showing the most self-absorption. Interference may be encountered from Co, Ni, and Cr. Pd 363.5 mμ is best for analysis, being the freest from interference. A weak ruthenium line at 363.5 mμ should be noted. In the visible, a continuum is observed in air–hydrogen throughout the blue and green.

Platinum has been reported to emit only its two principal lines, 265.9 and 306.5 mμ; they are weak, and the 306.5 mμ line falls squarely within one of the strongest OH bands.

82. *The Choice of a Line or Band*

In flame photometers with built-in filters the selection of an analytical line or band is up to the manufacturer. But with instruments having monochromators, interference wedges (Section 40), or interchangeable filters, the analyst has to decide upon the most suitable line or band among all those that are available (see Table 14 in the Appendix). This choice is governed by (*a*) the spectral sensitivity and resolution of the instrument, (*b*) the relative intensity of the line or band, (*c*) the shape of the working curve, (*d*) blank interference, (*e*) the question of line vs. band, (*f*) temperature effects, and (*g*) whether internal standardization is to be used. These points will now be discussed individually.

(*a*) *Spectral sensitivity and resolution.* It is obvious that only lines and bands can be used that lie within the spectral range of sensitivity of the photoelectric or photographic detecting system and that can be sufficiently well separated from interfering emissions. The limited range of transparency of available optical components, such as lenses of glass or quartz, and the spectral sensitivity of photoreceivers (or photographic emulsions) limit the spectral working range. The range of interference filters (with appropriate photoreceivers) is at present limited approximately to 360–1000 mμ. For the determination of sodium, then, only the lines 589 and 819 mμ are available. Glass color filter combinations have the disadvantage of poorer resolution, especially in the ultraviolet; they limit the spectral range further.

A glass prism monochromator with good resolving power can separate lines and bands in the range 350–1000 mμ better than color or interference filters. The commonest glass monochromators with low resolving power are about equal to interference filters. If the farther ultraviolet region is to be used, a quartz or grating monochromator is necessary. But a quartz monochromator has lower resolving power in the visible and especially in the near infrared than a glass monochromator of identical construction. With a quartz monochromator, for example, any of the lines 819, 589, and 330 mμ can be for sodium, provided suitable photoreceivers are available and the sample concentrations are adequate. Grating monochromators are equal to a quartz monochromator in spectral range, or better at long wavelengths. They have the advantage of uniform dispersion over the whole spectrum (Section 60). In resolving power, accordingly, they excel a quartz monochromator at longer wavelengths but may be poorer at shorter wavelengths.

The spectral sensitivity of the photodetector (or photographic plate) can likewise limit the spectral range (see Sections 42 and 72). The manufacturer usually selects the photodetector to fit the spectral transmission range of the optics.

(b) *Relative intensity of the line or band.* If several lines or bands are available in a given spectral range, usually the strongest will be selected. The quantity of interest is not the relative intensity alone but rather the product of the actual intensity and the effective spectral sensitivity of the instrument. It is this product that is measured by the instrument in terms of the scale reading when lines or bands are compared at fixed instrumental sensitivity. The effect of slit width will not be considered here; see Sections 62 and 95. The use of the strongest emission for the analysis has the advantage that stray light, filter background, and the associated cross-sensitivities (Section 94) are least troublesome, the precision of the reading is least affected by zero fluctuations and electronic noise in the amplifier and phototubes, and the effect of flame background is minimal (see paragraph (d) below). Also, the slit of a monochromator can be widened, to make the wavelength setting less critical; this worsens the line/background ratio, which, however, is unimportant for a strong line.

(c) *Shape of the working curve.* This varies with the line or band of a given element; the reasons were given in Section 10. A linear curve is most useful; linearity facilitates interpolation between calibration points (see the bracketing method, Section 85) and the daily adjustment of the slope of the curve.

A curved working curve can be straightened, in sections, by a suitable transformation or method of graphing. This is easily done for resonance lines having strong self-absorption at higher concentrations, by plotting on logarithmic coordinates (see the right-hand part of the curve in Fig. 50). Since the intensity is proportional to the square root of the concentration, the upper part of the curve has a constant slope of $\frac{1}{2}$. This part of the curve, however, though straightened by the logarithmic scale, has the disadvantage that errors of reading on the galvanometer scale due to zero fluctuation, amplifier noise, etc., have twice the effect upon concentration that they do in the steeper, linear part of the curve (left-hand part in Fig. 50). This is not true of changes of sample flow rate such as result from irregularity of atomization,[65] since the relative effect of these changes on the intensity reading equals the logarithmic slope of the working curve. For example, if the slope is $\frac{1}{2}$, a 1% change of flow rate causes a $\frac{1}{2}$% change of intensity (in the absence of compensatory effects), which in turn is interpreted as a 1% change of concentration.

The convex part of the working curve of a resonance line can be avoided by diluting sufficiently (cf. Fig. 50 and Section 79) or by going over to another line or band of the element that has a linear curve in the concentration range in question.[349] The beginning of the self-absorption curvature can be displaced to higher concentrations by obstructing the mirror behind the flame (Section 38). Curvature can develop at very high concentration or intensity through the effect of the solute on the flow rate of the sample (Section 11) or through nonlinearity of the photodetector.[65,361] Controlled sample pumping at the atomizer[307a,347] or, more simply, dilution will circumvent the flow rate effect. Dilution will also cure photodetector nonlinearity (Section 42); another expedient is to diaphragm the light at some suitable point (an aperture stop) in the beam.

Concavity of the working curve, as shown in Section 10, can arise from ionization of the analyte in the flame, which is most marked with easily ionized metals (alkali metals) in hot flames and at low concentrations. Unlike the self-absorption curvature, which occurs chiefly with resonance lines, ionization curvature affects all the lines and even bands of an element equally, since ionization reduces the concentration of the emitting atoms as well as molecules, and equilibrium can be assumed to exist between the atoms and the oxides or hydroxides (see Fig. 7, page 55, and Fig. 55, page 295). Ionization curvature can be lessened by using a cooler flame, concentrating the sample, or adding another easily ionized element to suppress the ionization of the analyte. When the flame gases themselves contain enough electrons, ionization curvature can be decreased also by higher dilution. The effectiveness of this measure for straightening a working curve will be the greater, the lower the concentration of electrons derived from the metal is in comparison with that of the flame gas electrons (Hollander and Alkemade, unpublished measurements). With laminar flames of the kind considered here, concavity can be reduced also by setting the beam lower in the flame,[65] where the electron concentration is higher. Finally, ionization curvature is much more sensitive than self-absorption curvature to flame temperature and hence to the mixing ratio as controlled by the gas pressures, and to addition of alcohol, etc.

(d) *Blank effects.* If the sample contains concomitants at high concentration, their effect on the blank or background in the filter photometer has to be considered in choosing the line or band (Sections 94 and 95). If the ratio of the intensity of neighboring lines, bands, or continua of the concomitants to that of the analytical emission is too high, it is better to go to another line or band of the element even though it is weaker. A high ratio of net intensity to blank is, however, also essential, not only for weaker emissions but also for diluter solutions. Detailed generalizations can hardly be laid down, since much depends on the resolution of the instrument, the type and temperature of the flame, etc.

As an example, strontium should be determined at 680 mμ in the presence of much barium and at 461 mμ in the presence of much calcium.[361,636]

(e) *Line vs. band.* The choice between a line and a band of a given element in the flame will generally favor the line, since it can be better separated from the flame background and other continua by narrowing the slit or improving the filter. But sometimes intensity favors the band. With filter photometers and cool flames, the green or the orange calcium band (554 or 623 mμ), for instance, will be preferred to the line at 423 mμ.[361,636] Alkali metals, however, emit only lines. On the other hand, some elements (boron, scandium, yttrium, lanthanum, etc.) emit only bands in the flame.* In a very hot flame (oxygen–cyanogen) an ion line can be stronger than an atom line, as with strontium.[304]

(f) *Temperature effects, etc.* Lines of higher excitation potential give less analytical precision than the resonance line of longest wavelength, since their intensities change more rapidly with flame temperature. For example, the red and ultraviolet lines of sodium are more sensitive to temperature than the resonance lines at 589 mμ. The relation between line intensity and temperature becomes more complex when the element is appreciably ionized in the flame or is partly combined as a molecule such as CaOH, since ionization and dissociation also depend on temperature. Other factors, such as transport or evaporation effects (Section 6), that change the total concentration of the analyte in the flame but not the temperature affect the analytical result found from any of the lines or bands equally. This is not necessarily true of ion lines, however, for the fraction of ionized atoms may vary only as the

* Cf. footnote to Section 81.

square root of the concentration of unionized atoms, in view of the Saha equation (Section 7).

(g) *Internal standardization.* In this method the choice of an emission is embarrassed by the need of considering two lines in their relation to each other. The analytical line should resemble the internal-standard line as closely as possible in excitation potential and in the shape of the working curve as governed by self-absorption. Ionization and incomplete dissociation can complicate the situation here too.

83. *Establishment of Working Curves*

In practical flame photometry, the existence of a generally applicable working curve for each element or each emission, permitting calibration of the meter scale in concentration units, would be most advantageous; such curves or scales could be purchased with the instrument, a set for each application. Unfortunately, this is impossible, for several reasons:

1. The presence of other elements, anions, etc., affects the position and form of the working curve of the analyte (Section 10). But these concomitants, such as protein or other organic material, vary with the application.

2. Even for one and the same application, the working curve determined by one investigator can be used by others only with caution, if at all, for the form of the curve and the background reading depend on the instrument and the various experimental conditions, such as the fuel/oxygen ratio, characteristics of the atomizer, and diluent used.

If the individual instruments of a given type all had the same characteristics, or even if standardized components (valves, burners, atomizers, etc.) could be used in all instruments, it would become more nearly possible to use the same working curve for each instrument, with strict attention to all experimental conditions. For the present, so long as this cannot be realized, it will remain necessary to prepare several series of standard solutions in setting up the procedure for any new application. Accordingly, we show only one working curve as an example (Fig. 50).

Fortunately, some manufacturers try to maintain high uniformity among their instruments and supply well-written instructions for the more usual applications. They sometimes supply ready-made standard solutions for the commonest determinations. But for the reasons mentioned above it is not safe to apply these instructions or working curves to other instruments.

A procedure for preparing the sample for analysis (Sections 78 and 79) has to be selected before setting up the working curve with the standards, since the composition of the standard solutions depends on the preparative method. For example, if a solid sample is to be dissolved in nitric acid, the standards should contain an equivalent

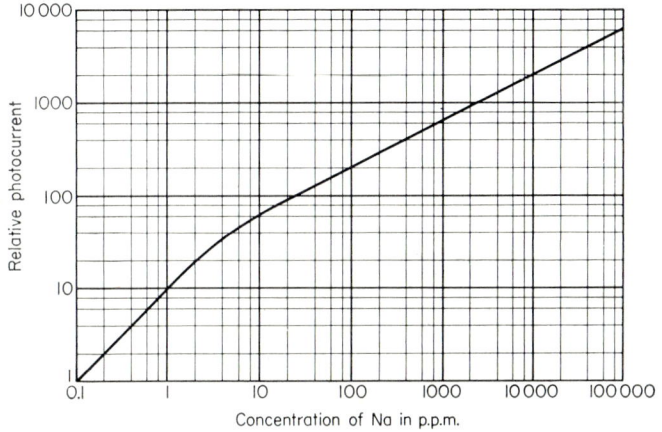

Fig. 50. Working curve for sodium at 589 mµ in logarithmic coordinates (from ref. 100).

concentration of nitric acid. Likewise, a method of dealing with interferences—for instance, the buffer method (Section 98)—must be chosen, since the working curves will depend on this too. In certain procedures, such as the bracketing method (Section 85) or self-standardization (Section 100), working curves can sometimes be entirely dispensed with; but work is not saved, for self-standardization is itself rather time-consuming.

Accordingly, the standard solutions used for determining the working curve should resemble the sample solutions as closely as possible, because the other constituents, such as acid radicals, affect the flame background and the curvature and slope of the working curve. But if, notwithstanding this, one elects to use simple standard solutions, one must test for errors that may arise from neglecting these effects (Sections 96 ff.). With sample solutions of very complex constitution, such as those of biological origin, it may be practically impossible to duplicate the essential constituents. As an example, proteins with their variable bonds to the alkaline-earth metals cannot be synthetically simulated. But errors from this source can be eliminated by subsequent corrections. We return to this point in Section 100. Basically, however, one should try to match the sample with the standard solutions as far as possible, since the work has to be done only once, whereas the correcting of the results has to be repeated indefinitely. How close the resemblance has to be, depends among other things upon the requisite accuracy.

We take first the case that only one element, say, sodium at 589 mμ, is to be measured, and that no great interferences are likely. A set of standard solutions will then be prepared with various known concentrations of sodium. These should contain the other concomitants of the sample in average concentrations. The number of standard solutions depends on the range of concentration expected and on the curvature of the working curve. The concentrations of these standards should not be in arithmetic series (e.g., 10, 20, 30, 40, 50, ... p.p.m. Na) but rather in geometric series, consecutive standards differing in concentration by a suitable factor such as $\sqrt{2}$, thus: 10, 14, 20, 28, 40, 57, 80, ... p.p.m. Na, so that the intervals are in a constant ratio. In the interests of a rational and exact pipetting technique—pipets are not available in all sizes—these ratios can be realized only approximately. Thus the following series of concentrations would be more practical: 10, 15, 20, 30, 40, 60, 80, ... p.p.m. Na.

When the expected ranges of concentration are small, as is often true in biological and medical applications (Section 111), an arithmetic series as in the first example above will suffice. If the curvature of the working curve is marked, the standard solutions will have to lie closer together in concentration, or else intermediate standards can be prepared later as needed. If the concentrations of the concomitants are very low in comparison with the analyte, they may sometimes be omitted from the standards.

One of these standard solutions should be prepared in larger quantity (a few liters). With this, the *reference standard* (or *control standard*), the sensitivity of the entire instrument is checked at regular intervals, e.g., after every four analyses (Section 85). Also, in establishing the working curves, the reference standard should be read periodically between measurements on the individual standards.

In addition, a blank solution is required for the adjustment or control of the zero point, to be discussed in Section 95. This is a solution containing all constituents of the sample except the analyte. If the concomitants have no effect on the background, they can sometimes be omitted from the blank solution; in this case, then, the zero can be

§ 83]   ESTABLISHMENT OF WORKING CURVES   233

established with the solvent alone, usually distilled water, but this simplification must be justified in each case by preliminary experiments.

It is debatable whether to use a reference standard equivalent in concentration to the average sample or one equivalent to the most concentrated sample likely to occur. Each has advantages and drawbacks. In the former method, there are the following advantages: only deviations from the mean concentration are measured; the commonest values, near this mean, are affected with only small errors; and only for extreme values will errors become serious, owing to poor reproducibility of the working curve, drift of the operating conditions, etc. The disadvantage is, of course, the lower precision at the highest concentrations. But if the reference standard is set at the top of the concentration range, the inverse difficulty exists. Moreover, contamination of the less concentrated samples by drops of the stronger standard clinging to the capillary inlet will be greater than in the former case.

When large ranges of concentration are to be covered with a single series of standards, several well-spaced reference standards should be prepared, with the usual standards between them. The analysis is then referred to the nearest reference standard. When the concentration range is large, curvature of the working curve must be heeded. Otherwise, one should keep to a smaller range of concentration, adjusting the individual samples by dilution (Section 79) so as to bring them into the calibrated range. To gage the dilution, preliminary semiquantitative trials are made with the flame photometer, or else a series of dilutions of each sample can be prepared at the outset.

The readings on the standard solutions, which establish the working curve, and the occasional intermediate calibration with the reference standard, are made by one of the methods of Section 85. Even with the bracketing technique, a working curve is advisable, to check the linearity. These calibrations are repeated for several days, their mean serving as basis for the working curve and the later routine determinations. The position of the working curve should be checked occasionally with a few control measurements, especially on changing the gas tank (Section 15).

Instead of drawing the working curve, one can draw a scale for the meter or for the potentiometer, showing the concentration directly. It is harder to prepare such a scale than a curve, but in routine analysis it obviates tedious reference to the working curve. However, with a scale there is the risk of forgetting to check it occasionally, or of tending to ignore departures actually detected.

When interferences are to be eliminated, as by the parametric method (Section 97), a good many more series of standard solutions may be needed. In this case, the concomitants cannot merely be added to the standards in average concentrations, but they must be varied as they vary in the sample; that is, in one series the concentration of one element is varied while the others are held constant at their mean values, and in the next series another element is varied while the rest, including the first, are kept constant at *their* mean values, etc. When the interferences are severe and the concentration ranges great, this procedure may have to be carried out not only for the average but also for the extreme concentrations. The result, in the case of two variables, is a family of working curves like that of Fig. 52.

When many interferences have to be taken into account at the same time, the requisite number of series of standards can exceed all bounds. To minimize the work of setting up the method, preliminary tests should be made to see whether the accuracy needed justifies consideration of all the interferences. The accuracy requirements themselves should be scrutinized; sometimes the demands are unreasonable. Further, background interferences, such as those resulting from filter transmission at the wrong wavelength, can perhaps be removed by increasing the optical resolution, as by use of additional filters (Section 40). In this event, much work is saved in preparing standards and in referring to the families of working curves.

Many interferences can be eliminated by chemical means, as will be discussed in Section 100. The standard solutions must naturally be adapted to such a procedure. They will then usually be simpler in composition and easier and quicker to prepare. But there will be a great deal of labor in the repeated chemical operations, such as precipitation, in the routine analysis of many samples.

Table 8 shows as an example how to choose the series of standards to be prepared. The preparation itself will be discussed in Section 84. We assume here that sodium, potassium, and calcium exhibit mutual interferences; this is true in hot flames only of the pair sodium–potassium and, with inadequate optical resolution, also of the pair sodium–calcium. We assume also the presence of certain other elements or radicals, viz., iron, copper, magnesium, and phosphorus, at low concentrations compared with the sodium, potassium, and calcium, so that they need be added only in constant, average concentrations. (We are thinking of blood serum.)

Standard solution No. 0 is also the reference standard. In Nos. 1–4 the sodium concentration is varied, in Nos. 5–8 the potassium concentration, and in Nos. 9–12 the calcium concentration; in each of these sequences the reference standard serves as the central member. The magnesium, copper, iron, and phosphorus, kept at constant concentration, are added as a concentrated mixed solution of all four com-

TABLE 8. Example of a series of standard solutions (simulated sera)

Standard no.[a]	Na	K	Ca	Mg	Cu	Fe	P
0	3300	200	100	20	1	1	18
1	2500	200	100	20	1	1	18
2	3000	200	100	20	1	1	18
3	3600	200	100	20	1	1	18
4	4000	200	100	20	1	1	18
5	3300	100	100	20	1	1	18
6	3300	150	100	20	1	1	18
7	3300	250	100	20	1	1	18
8	3300	300	100	20	1	1	18
9	3300	200	50	20	1	1	18
10	3300	200	75	20	1	1	18
11	3300	200	125	20	1	1	18
12	3300	200	150	20	1	1	18

[a] All concentrations are in mg./l.

ponents, to simplify the pipetting. The general technique of preparing such a series of standards will be covered in the next section.

The example merits some further remarks. The variable phosphorus content of serum affects the calcium values (Section 98), but in the example only the mean phosphorus content of normal serum is represented. In any case, it is not easy to estimate this mean, for serum contains not only inorganic phosphorus but also phosphorus combined with lipoids, etc. Further, the serum protein somewhat buffers the effect of phosphorus.[188,703] Magnesium, copper, and iron have practically no effect upon the determination of sodium, potassium, and calcium in serum, and they might as well be omitted. We have added them here only to allow the same standard solutions to be used for magnesium, copper, or iron determinations. When the optical resolution is adequate, the potassium and calcium can be varied together; this saves four standard solutions.

In this example, salt solutions are used as standards, the salt concentrations corresponding to those occurring in the animal or human body. But at these concentrations the solutions do not atomize well, being too viscous and containing too much sodium. Depending on the instrument and the element, the sample serum and standards (simulated sera) are therefore diluted 10-fold to 1000-fold.

This subsequent dilution of the standards can be avoided of course by preparing them in proper dilution in the first place. But the method outlined above, using concentrated stock standards (simulated sera), has certain advantages:

1. The same diluent can be used for the standards as for the samples, and it is possible to change the diluent or solvent, the buffering substance, or the internal-standard concentration. With dilute standards this is not possible without preparation of fresh standards.

2. Dilution errors of pipets, volumetric flasks, etc., are the same for samples and standards, and their effects cancel if the entire procedure is carried out with the same glassware.

3. Unduly low concentrations are avoided. The standards keep their nominal concentrations longer, because the effects of diffusion, adsorption, etc., are smaller at higher concentrations (Section 80). To be sure, more dilutions have to be performed.

Clearly all standardizations must be done under the same instrumental conditions (gas pressures or flow rates, etc.) as the measurements on the samples. If for any reason these conditions are changed, the working curves will have to be reestablished or at least their validity tested. It was pointed out in Section 19 that optimal operating conditions vary with the element.

Standardization and Measurement

84. *The Preparation of Standards*

The various standards including the blank or blanks must cover a range of concentration. Besides, their composition must be duly related to the composition of the samples. We discussed this in the preceding section. In the present section we deal in a general way with the preparation of standard solutions.

The analytical results depend critically upon the reliability of the standards, since flame analysis is a relative measurement. A concentration error in the standard due to faulty weighing, pipetting, etc., generally shows up in the analysis as an error of equal magnitude but opposite sign.

Starting materials must be of adequate purity, but reagent-grade chemicals are not always necessary. However, when a trace element is to be determined in a material, and the major component of the sample is to be added to the standards, this substance must be of the highest purity; except in rare cases, even a reagent-grade chemical will not suffice. The analyst will then have to purify it himself, or get a spectrochemically pure reagent.

The purity of the distilled water or other solvent must be considered. Ordinary distilled water is usually good enough. But when traces of alkali metals are to be determined, the solvent will have to be further purified, as by ion exchange or distillation in quartz or in silvered or tinned apparatus (see Section 79); this may be necessary with calcium or even boron. The containers for preparing and storing the standards (pipets, volumetric flasks, etc., of certified calibration if necessary) must be clean and properly selected (Section 80).

As starting materials, salts (often chlorides), oxides, and pure metals are commonly used. The moisture content must be definitely known; if necessary, the material should be dried for a time at 110°C. in an oven, but it must not be allowed to decompose. The chlorides of some elements such as calcium and lithium are hard to weigh dry, being very hygroscopic. For these elements another salt such as precipitated calcium carbonate or acetate is better; after weighing, it can be dissolved in a minimal quantity of dilute hydrochloric acid in the volumetric flask. The resulting chloride is then carefully dried in the flask to expel the carbon dioxide and the excess acid, and the flask is cooled and filled to the mark. The quantities weighed should not be too small; stronger stock solutions should be prepared first, from which the diluter standards can be prepared by quantitative dilution.

Dilute solutions such as one of calcium below 1 p.p.m. can be kept only a few days in ordinary containers (Section 80), and should not be prepared until just before the analysis. If, however, a dilute solution has to be kept for several months without noticeable change of concentration, then it should be kept in a plastic bottle in sufficient quantity—at least 250 ml., which, however, must not be depleted too far; the same bottle should always be used for the same solution.

Another risk is the development of algae or other microflora (from the air) in the standard solutions. In some types of standards they flourish as in a nutrient solution. The flora withdraw a portion of the material in solution (potassium, etc.) and concentrate it, forming a green deposit on the walls. This can be avoided with a little formalin or a few drops of carbon disulfide [182] in the distilled water; 5 ml. of 40% formalin per liter of water will do.[369] No detrimental effects on the analysis have been detected. Fungi can also enter the standard solutions. They make their presence known by a white, flocculent precipitate. Some of these species are not curbed by formalin or the like. In this case they may be killed by brief heating to nearly 100°C.

A further risk is that certain constituents of the standard may precipitate partially after a time; for example, the carbon dioxide of the air can precipitate calcium, but this is prevented by making the solution slightly acid (say, $0.1N$ in hydrochloric acid). A like amount of acid must then be added also to the sample with the diluent, to avoid an uncompensated error from the acid in the standard.

Sometimes constituents that react chemically are inadvertently included in a standard solution. For example, calcium chloride for the calcium standard and potassium sulfate for the potassium standard cannot be used in the same solution, for some of the calcium will precipitate as calcium sulfate, and the calcium sulfate remaining in solution gives less emission in the flame than calcium chloride. In dilute solutions such precipitations seldom occur. Further, the solubility of the precipitate depends on temperature, and so the concentration of the standard solution will vary with the room temperature.

A few firms* supply standard solutions for the commonest applications of flame photometry.

85. *The Technique of Measurement and Standardization*

The measurements and standardizations should be carried out in a uniform rhythm after a preliminary warm-up (Section 77), to avoid drift of sensitivity during the work. Recall the effect of the temperature of the spray chamber wall on the evaporation of the spray and hence on the rate of transport and the emission (Section 26), and the matter of photodetector fatigue and recovery, especially in photocells (Section 42). If the sequence of readings has to be interrupted, the atomizer should be connected by a rubber tube to a supply of distilled water or, better, of reference standard, to avoid changes in components sensitive to temperature or light during the pause. For the same reason, the interval between the start of atomization and the reading of the meter should always be the same. With atomizer-burners or internal-standard instruments, these precautions are hardly needed, provided the photodetectors are reasonably fatigue-proof.

The usual suction atomizers do not ordinarily have to be rinsed with distilled water before every sample; on the contrary, the water is likely to differ more in concentration from the sample than the samples differ among themselves, and so a carry-over of water to the next sample occasions a larger error than occurs without this precaution. Most atomizers of this kind clean themselves as the fresh sample rinses

* For example, Dade Reagents, Inc., Miami 35, Florida; E. Merck A.G., Darmstadt, Germany; and Netheler und Hinz G.m.b.H.[513]

§ 85] THE TECHNIQUE OF MEASUREMENT AND STANDARDIZATION   239

out the residues of the old. One need only make sure that when the new sample is applied, no large drops of the preceding sample cling to the capillary; they can be wiped off with suitable material if necessary. These residual drops cause the greater error, the more the consecutive solutions differ in concentration. This effect can be mitigated by a proper sequence, taking the samples and standards in order of increasing concentration; residual drops will then hardly matter. The drops can be more troublesome when the zero point is occasionally checked with the blank solution; it may be advisable to rinse with distilled water before the blank check and with the remains of an earlier sample between the zeroing and the restandardization. The sequence can be continued as before without intermediate rinses. In any case, a thorough rinse with distilled water is desirable at the end of a set, so that salt incrustations will not form on the atomizer, spray-chamber walls, burner, etc.

There are several ways of carrying out the measurements and standardization, all of which have advantages and drawbacks. The principal methods are described below. For the sake of simplicity, we assume that the flame background (Section 95) has been suitably eliminated.

(a) *The adjustment method*. The readings are taken in about the following sequence: reference standard (Section 83), sample 1, sample 2, reference, sample 3, sample 4, reference, sample 5, etc. Whenever the reference standard, after every 2–5 samples, is applied, the sensitivity is adjusted by means of a diaphragm or an electric control (Section 48). The frequency of adjustment depends on the required accuracy. This procedure is very simple and is often used, but it does not yield the highest accuracy (see below).

(b) *The quotient method*. The procedure is as above, but the sensitivity is not adjusted; instead, the photocurrent I_R of the reference standard is recorded each time. To each sample reading I_X on the photometer corresponds an interpolated reference standard value I_R, found by linear temporal interpolation between the two adjacent readings on the standard. The quotient I_X/I_R then measures the concentration. The readings can also be expressed as a percentage of the standard. In the same way, the working curve is prepared from the other standard solutions in association with the reference standard. Concentration is taken as abscissa and the I_X/I_R values as ordinate. An example is given in Table 9. Here RS denotes the reference standard. The table is drawn up and the first column filled before the readings begin. The second column receives the results—readings on a potentiometer

TABLE 9

Solution no.	Photometer reading, scale divisions	Interpolated I_R	I_X/I_R
RS	673		
56	428	672	0.637
57	759	671	1.083
RS	670		
58	314	670	0.469
59	869	670	1.297
RS	670		
60	944	669	1.411
61	460	668	0.689
RS	667		

or galvanometer. Columns three and four are calculated later (with a slide rule). The concentrations (added as a fifth column) may then be found from the working curve. If the range of concentration is very large, a logarithmic representation of the results as in Fig. 50 may be helpful; the logarithm of the quotient I_X/I_R is plotted against the logarithm of concentration.

A similar procedure is followed in setting up the working curve for the quotient method, but the measurements are repeated several times and averaged. It is then helpful to interchange columns and rows in Table 9; the readings, being in a column, are easier to add and average. The calibration can be presented in tabular form[172] instead of as a working curve, both in the quotient method and in other methods.

The quotient method is more exact than the adjustment method above, since the standard value changes more smoothly. Also, the I_R values reveal any instability of the instrument, as from partial clogging of the capillary. Doubtful measurements made at such times can be repeated later. This method is, however, more tedious than the adjustment method.

(c) *The bracketing method.* From a suitable series of standards, two are chosen by flame-photometric measurement with concentrations as close as possible to, but on either side of that, of the sample; these may be called the "bracketing" standards. The difference in concentration between the two standards should be kept as small as is practical, so that the working curve in this interval can be considered approximately straight. The concentration of the sample can then be determined with an error that becomes smaller, the smaller the interval between the standards and the straighter the working curve. Let the diluter standard have concentration c_1 and yield the reading I_1, and the other, concentration c_2 and reading I_2; let the concentration of the sample be c_x and the reading I_x. Then c_x is found from the known and measured values by

$$c_x = c_1 + \frac{I_x - I_1}{I_2 - I_1}(c_2 - c_1)$$

For highest accuracy, the following sequence is repeated several times: standard 1, sample, standard 2, sample, standard 1, etc. The value of c_x is found for each of both the upward and the downward sequences, i.e., with rising and with falling concentrations.

The method has these advantages:

1. The location of zero (scale reading for zero concentration) has little importance. One can work with a suppressed zero; that is, the reading for zero concentration can be entirely off-scale. The concentration range of interest can thus be magnified considerably. Indeed, when many repeated measurements are made by this method, the zero point—suppressed if possible—may be allowed to drift slowly during the determination.[505]

2. The same is true for gradual changes of sensitivity during the repeated measurements, or for any influences which affect the slope of the working curves of the sample and standards equally.

3. It is not necessary to set up a working curve but one may be desirable, for two reasons: (*i*) It should be ascertained, once for all, that the working curve is sufficiently straight between the two adjacent points of standardization; otherwise, additional standards should be intercalated. (*ii*) From the residual curvature thus found, it can be determined how large an error is committed in calculating the concentration by the above formula, i.e., what the departure is of the linearly interpolated value from the true value found with the curved working curve.

The method can be modified when the working curve is linear in the entire range with a known slope, k. The formula then becomes

$$c_x = c_1 + k(I_x - I_1) \quad \text{where} \quad k = (c_2 - c_1)/(I_2 - I_1)$$

In this case, only one standard solution of concentration c_1 not far from c_x need be measured. The choice of sign in the formula should be heeded.

86. *Measurements with Overlapping Bands or Lines*

Analyses sometimes have to be carried out with mixtures of elements, such as the rare-earth metals, emitting only band spectra, and it can happen that the bands more or less overlap (Fig. 51). For such cases a workable scheme has to be prescribed for making the measurements and standardizations. One is given below (from ref. 546); it can be used also when a band overlaps a line.

We assume that that part of the light intensity or photocurrent due to the flame background has already been accounted for in some suitable way (Section 95). Suppose element E_1 gives a recorder curve I as in

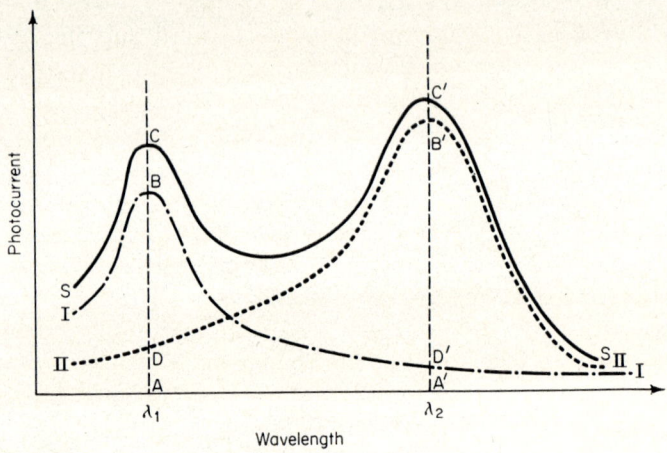

Fig. 51. The overlapping of two band spectra (schematic) (from ref. 546).

Fig. 51, while element E_2 gives curve II. In the mixture only the sum S of these two curves can be observed. Let curve I have a maximum at wavelength λ_1, and curve II at λ_2; the observed summation curve S will then have peaks at λ_1 and λ_2. The photocurrent AC measured at λ_1 comprises two parts, viz., AD due to E_2 and AB due to E_1; similarly, at λ_2, E_1 gives photocurrent $A'D'$ and E_2 gives $A'B'$. Then

$$AD = BC \quad \text{and} \quad A'D' = B'C'$$

Let us assume that the meter reading is proportional to concentration at all concentrations and at all wavelengths throughout the respective ranges of interest, for both band systems (for both elements), and that no interferences affect the intensities. Then, for E_1 and E_2,

$$AB/A'D' = K \quad \text{and} \quad A'B'/AD = K'$$

Further,

$$AC = AB + AD \quad \text{and} \quad A'C' = A'B' + A'D'$$

On substituting for AD and $A'D'$, we get

$$AC = AB + A'B'/K' \quad \text{and} \quad A'C' = A'B' + AB/K$$

§ 87] THE DETERMINATION OF NONMETALS: INDIRECT METHODS 243

The solution of these equations is

$$AB = \frac{K \cdot K' \cdot AC - K \cdot A'C'}{K \cdot K' - 1}$$

$$A'B' = \frac{K \cdot K' \cdot A'C' - K' \cdot AC}{K \cdot K' - 1}$$

Thus, it is necessary only to measure the ratios K and K' at the wavelengths λ_1 and λ_2 on each of the intensity curves I and II recorded in a preliminary experiment, and in the analysis to measure AC and $A'C'$ at λ_1 and λ_2, in order to find the photocurrents due to the separate portions AB and $A'B'$ of the emission. These calculated values depend on concentration in just the same way as do readings on the separate lines and bands.

There are two essential assumptions inherent in this procedure: (a) The reading is proportional to concentration up to the highest concentration of interest, at all wavelengths λ. (b) The sum spectrum S actually represents the sum of the separate spectra I and II, and interferences are absent. These assumptions have been tested and confirmed in particular cases,[546] and should be tested in other cases as they arise. The method permits recognition of other possible errors, such as inadequate elimination of flame background.

An analogous method of measurement and calculation can be given for the case of three or more bands, the measurements then being made at three or more wavelengths. With proper modification the method can accommodate the case of overlapping of one or more bands by lines or of a continuum of unknown magnitude by one or more bands or lines.

87. *The Determination of Nonmetals: Indirect Methods*

(a) *Determination of emitting nonmetals.** In the determination of metals, the flame bands of the metallic oxides or hydroxides are often

---

* Recently discovered flame emissions now permit the direct determination of several nonmetals. The line spectrum of silicon appears in the incandescent oxyacetylene flame. Distinctive band spectra of nitrogen and phosphorus (NO and PO) are well developed in air–hydrogen–acetylene or air–oxygen–hydrogen–acetylene flames, and the NO spectrum can be excited without interference from atmospheric nitrogen. As for sulfur, the CS spectrum is well excited in the air–oxygen–acetylene flame. The spectrum of selenium in the oxyhydrogen flame was described by W. N. Hartley, *Trans. Roy. Soc.*, **A185**, 161–212 (1894), but there is no recent work on this element. Tellurium and arsenic emit line spectra in air–hydrogen with alcohol, and arsenic is well excited in the inner cone of oxyacetylene (cf. footnote at the beginning of Section 81). For these new data and references, including data on band spectra useful for determining the halogens, see Tables 10 and 13 and the Atlas.—*Translator.*

9+C.A.F.P.

used. The alkaline earths and rare earths will be recalled. It is less commonly known that certain nonmetals also emit characteristic bands in the flame, which can be used in the same way for analysis.

1. Determination of Boron. Boron emits several characteristic bands in the near ultraviolet and visible (see Spectrogram 67).

2. Determination of Silicon. Bands of SiO have been reported in the ultraviolet.* [41]

3. Determination of Nitrogen. The cyanogen bands in an oxygen flame containing no nitrogen, especially the sequence starting at 388.3 mμ, can be used for the determination of organic nitrogen both in liquids and in gases.[362] It helps to separate the luminous inner cone from the outer diffusion flame (Section 32), because the CN bands in the inner cone are then independent of the atmospheric nitrogen, which can give rise to CN bands in the mantle. In this way nitrogen can be determined quantitatively in liquids and gases by the flame in normal room air. Instead of a flame separator, a closed burner (Section 32, Fig. 20) can be used, in which nitrogen-free oxygen flows around the flame in place of air, to eliminate the effect of atmospheric nitrogen on the flame emission. Instead of the CN bands, it might be possible to use the NH bands (336.0 mμ) or perhaps the NH_2 bands at 665.2 mμ and below[12] or the green NO_2 continuum in the inner cone[392] for the determination of nitrogen.

4. Determination of Phosphorus. Under the right conditions, phosphorus can be determined by the continuum emitted by unvaporized oxide particles. Moreover, in the air–hydrogen flame with an alcoholic solvent, the PO band in the short-wavelength ultraviolet can be observed. There are also indirect methods, to be discussed in part (b) below.

5. Determination of Arsenic. Ultraviolet bands of arsenic as well as a continuum in the visible region can be seen in various flames.

6. Determination of Sulfur. Bands of CS in the farther ultraviolet (256.7 mμ) and of SH (323.7 mμ) occur in organic flames, and SO bands are known at 316.5 and 327.1 mμ,[12] which might be used under favorable conditions for the determination of sulfur. However, indirect methods (differential methods, Section 88) are in use and are probably more convenient.

7. Determination of Tellurium. Tellurium also emits a series of

* These bands have been recorded in a spectrogram by H. Watanabe of Beckman Instruments, Inc., under favorable conditions in an oxyhydrogen flame at high silicon concentration. The flame-photometric determination of silicon needs further study.

§ 87]   THE DETERMINATION OF NONMETALS: INDIRECT METHODS   245

characteristic bands in the longer-wavelength ultraviolet and the visible.

8. Determination of Chlorine. If a known amount of copper is added to a flame, as by spraying copper sulfate solution, together with a compound containing chlorine but not copper, the flame will emit the bands of CuCl at 428.1, 435.4, and 443.4 mµ,[12,363,461] which can be used for the determination of chlorine. However, this method seems to have been less popular than the differential method for chlorine (see below). Other bands that might conceivably be used under special conditions for chlorine determination include those of CCl (286.2 mµ and below), Cl_2 (600.3 and 588.2 mµ), and ClO (384.1 mµ and above).[12]

9. Determination of Other Halogens. Like chlorine, fluorine has been determined by means of the CaF bands at 515 to 530 mµ or the SrF or BaF bands.[39] At higher concentrations it might be possible to use the bands of CF (247.9 mµ and below).[12] Fluorine can probably be determined more reliably by the differential method.

The other halogens might be similarly determinable under special conditions by their characteristic band spectra: iodine by means of the diffuse I_2 bands (342.5 mµ) in the inner cone of organic flames or the IO bands (448.8 mµ and above through the green);[12,526] bromine by means of the orange-red Br_2 system (654.6 mµ and below) or the ultraviolet system at 291.0 mµ (observed only in the inner cone of organic flames), or the bands of BrO.[12]

(b) Indirect methods for non-emitting elements. The effect of other elements or anions on the emission of an analyte will be discussed in detail in Sections 96–100. It has been found in some cases that the interference is proportional, within a certain range, to the concentration of the interferent; this is true of the interference of phosphorus with calcium (Fig. 57, page 302). This effect, though undesirable in the determination of an alkaline earth, can be put to use for the determination of phosphorus. Various other interferences such as those of sulfate and aluminum can be used similarly. The following are some examples of this indirect method.

1. Determination of Phosphorus.[83,229,704] For this purpose the sample is provided with a fixed concentration of calcium, usually by mixing it with a measured volume of calcium solution. Other alkaline-earth elements, which would act as buffers (Section 98), must be absent. The phosphorus concentration can then be found from the weakening of the calcium emission, provided it lies in the range where the interference is linear. The method has been tested for phosphoric acid between

0.0005 and 0.012 molar. The linear region can be attained by dilution or concentration or by proper choice of the calcium concentration. The accuracy is reported as ± 0.0002 molar. Interferences from other anions such as chloride have apparently not been investigated; such interference might be expected only if the interferent is at least as concentrated as the phosphate. Since other anions such as sulfate can cause a similar if weaker quenching of the emission of calcium, magnesium, etc., it may be desirable to separate these anions from the phosphate with an ion-exchange column before analysis. To the phosphate-containing effluent a known amount of calcium or strontium solution (say, 1000 p.p.m.) is pipetted. The calcium emission is then measured with and without the phosphate.

If the sample contains variable amounts of calcium and perhaps other alkaline earths to begin with, these can be removed by ion exchange against another element such as hydrogen, after which a known, fixed amount of calcium (or magnesium, etc.) is added for the analysis. Good results can be obtained, even in the presence of extraneous alkaline earths, by self-standardization; we discuss this further below.

Under proper experimental conditions, the method gives reasonably good precision and sensitivity. The conditions should be such as to maximize the effect of the phosphate on the calcium; this is contrary to the situation in a calcium determination, where the phosphate effect is not wanted. See the discussion of this in Section 98. Direct atomizers are better for phosphorus determination, giving stronger phosphorus interference than indirect atomizers.

2. Determination of Aluminum. Most flames are not hot enough to excite aluminum at its resonance line 396.1 mμ (excitation energy 3.3 e.v.) and its oxide at the blue and green bands sufficiently for determination at low and intermediate concentrations. The low emission is probably due also to incomplete evaporation of the aluminum oxide entering or forming in the flame. As with phosphorus, however, the formation of involatile Ca—Al(—O) or Sr—Al(—O) compounds, tying up a part of the otherwise excitable alkaline-earth metal and preventing it from emitting, can be utilized for the determination of aluminum. The depression of the emission of the added calcium or strontium is a usually linear measure of the aluminum concentration, provided the calcium concentration is not too low;[183,493,501,551,645,704] see also Fig. 58, page 303. Interfering alkaline earths present at the start can be separated as for phosphorus. It must be kept in mind that the effect of the aluminum on the calcium depends not only on the added calcium concentration but also on the concentrations of other alkaline earths, phosphorus, iron, nitrate, etc. Moreover, the calcium depression depends

on whether the aluminum is present as nitrate or as chloride, etc. In Section 98 we discuss the factors that influence the aluminum–calcium interference and hence the sensitivity of the indirect aluminum determination. As for phosphorus, the experimental conditions are chosen to maximize the aluminum effect.

Beryllium, molybdenum, selenium, tellurium, titanium, tungsten, vanadium, and other elements can be determined in the same way as aluminum, by their depression of the emission of strontium or other alkaline-earth element.[452, 453, 453a]

As indicated above, indirect methods will yield good results only when alkaline earths and ions interfering with the alkaline earths (except the analyte) are either absent from the sample or present in constant concentration. The method has been refined[183, 645] so as to measure and eliminate errors from this source by self-standardization. An accuracy of 4% can be attained.[493]

3. Determination of Organic Substances. Organic materials entering the flame enhance the emission of the elements, an effect which can be used for determining the organic substance.[228] The element whose intensity is to be measured must be present in constant concentration, and there must be no interferences from concomitants other than the substance to be determined.

We might mention here a more direct method of determining organic substances (more exactly, their carbon content) by flame photometry. The material is introduced into a carbon-free flame such as oxyhydrogen. Various carbon bands then appear: C_2 at 516.5 mμ and CH at 431.5 and 388.9 mμ (these are the strongest), CH_2O at 422.0 mμ, CO at 219.7 mμ, and HCO at 337.7 mμ.[12] These might be used for determining the carbon content of, for example, a solution of acetone in water.[228]

(c) *Differential methods.* This class of indirect methods, used for the determination of fluorine, chlorine, bromine, etc., will be covered in the next section.

88. *Combined Flame-Photometric and Chemical or Physicochemical Methods*

(a) *The differential determination of anions.* Certain elements and acid radicals such as chlorine and bromine can be directly detected in the flame only with difficulty if at all (cf. Sections 81, 87), but by a combination with chemical methods they can be determined with nearly the same accuracy as can other elements by direct methods.

For example, the *chloride* content of a solution can be found by adding a known quantity of silver nitrate in excess to a measured volume of the sample; silver chloride precipitates. In the supernatant solution, after centrifugation if needed, the residual silver concentration is determined in the flame, suitably by the line at 328.1 or 338.3 mμ.[33,302,488] The difference between this and the concentration that would be expected without chloride yields the chloride content of the sample. The sensitivity may be as good as 0.1 p.p.m. Cl[302] or even better.[488] Such a method is called a *differential* determination. The method can be extended to *bromide* and *iodide* simultaneously present, with ammonia for masking the lighter halogens.[488]

Similarly, *sulfate* can be determined by addition of strontium or barium chloride, the excess metal being determined in the supernatant with a flame photometer or spectrophotometer.[177] The addition of alcohol with a strontium salt has been suggested.[302] *Carbonate* can be determined by addition of calcium, and the use of zirconium for *phosphate* has been thought possible,[302] but the spectrum of zirconium in conventional flames is neither very sensitive nor specific. *Organic acids* have been determined by conversion to their sodium salts and measurement of the sodium content of a solution of the salt with the flame.[428] Each acid combines with a definite and characteristic percentage of sodium.* The method is not directly applicable to mixtures of such acids. Similar applications might be sought in other fields of organic chemistry.

(*b*) *Enrichment and separation*. Chemical enrichment or separation may be advisable or indispensable for one or more of the following reasons.

1. The initial concentration of the analyte may not be high enough for flame analysis. Enrichment provides a higher concentration in smaller volume. This procedure contrasts with the concentrating procedure discussed in Section 79, in that simple concentration leaves unchanged the ratio of line or band intensity of the analyte to that of the other constituents, while enrichment changes it substantially in favor of the analyte.

2. Enrichment can eliminate or sharply reduce interferences from possibly more concentrated concomitants.[61] We return to this in Section 100.

---

* This method is not, strictly, a differential method; it is more closely related to the substitution method (paragraph (*d*)). Another method of this kind is that in which sulfate is determined via barium sulfate not by measuring the excess barium but by determining the barium precipitated as sulfate. See D. C. Cullum and D. B. Thomas, *Analyst*, **84**, 113–16 (1959) and **85**, 688–9 (1960); I. Odler and J. Gebauer, *Chem. Zvesti*, **15**, 568–70 (1961).—*Translator*.

3. An organic extractant used as in 1 or 2 can intensify the emission (Section 79).

4. When the analyte is separated, the standard solutions can be very simple. They may have to contain only the analyte and the extractant. This advantage is important when the sample is complex.

By an intelligent choice of the working conditions, which depend on the nature of the sample, all four of the above objectives can be attained.

There are the following disadvantages:

5. Enrichment or separation must be carried out for each sample individually. The total labor may then equal that of a purely chemical analysis, and the time-saving advantage of flame photometry is more or less lost.

6. If a trace concentration is to be determined, there is always the risk of contamination through the reagents used for the enrichment or separation. Blank determinations must therefore always be carried out in parallel with the samples. In any case, the strictest cleanliness is requisite for any trace-element determination. It should be emphasized that the requirements of purity for chemicals used in flame photometry often differ greatly from those for ordinary chemical work. This is because some elements such as sodium and potassium are conspicuous in very low concentration at the wavelengths of their lines and even in the background, whereas other elements with higher excitation energy interfere less than in chemical methods.

We append some remarks on points 1 and 2.

1. Increase of the Initial Concentration. In separations, because of the spectral isolation of the radiations, it is less important to remove the interferents fully than to prevent any loss of the analyte and to make sure that none of it is added through the reagents. The best chemical method of separation will have to be determined in each case individually.

2. Separation of Interferents. Sometimes the separation is desired not for flame-photometric reasons but for other reasons—for instance, safety. Thus, radioactive materials must not be allowed to atomize and burn, contaminating the air. Examples of chemical separation of radioactive constituents are given in the literature. For instance, in the flame-photometric determination of lathanum in uranium, the latter was separated by ether extraction.*

* See J. F. Possidoni de Albinati, *Anales asoc. quim. arg.*, **43**, 106–20 (1955).—*Translator.*

A few examples of enrichment and separation methods follow.

(i) Precipitation. (a) The desired element can be precipitated along with any other elements that are present in not too high a concentration; or (b) the bulk of the interferent can be precipitated and the supernatant solution analyzed. The latter method is the less advisable, as a part of the desired element is generally lost through adsorption, occlusion, etc. In method (a) there must be no noticeable losses through residual solubility of the precipitate; otherwise, a different precipitation reaction must be chosen. The best precipitations are often those in which the metal ion forms a complex with a large organic molecule. If the organic compound added for this purpose is combined with another metallic ion such as sodium, this metal may also be excited in the flame; it must then be decided whether the presence of the carrier metal will interfere with the flame analysis.

There are reagents whose affinity for a given metal depends on the conditions of precipitation; this enables a degree of separation. Thus, 8-hydroxyquinoline (oxine) preferentially complexes the following metals at the pH values indicated below.

pH	Metal
> 5.4	Cd
8	Cr (III)
> 4.3	Co
> 5.3	Cu
6–8	Ga
9.4–12.7	Mg
4.6	Zn

Another well-known reagent for the enrichment of traces of heavy metals is sodium pyrrolidine dithiocarbamate. The following elements, detectable in the flame, are quantitatively complexed by it: Ag, Cu, Cd, Ga, In, Pb, Mn, Co, Fe, Pd. General directions for the concentration of trace elements with such reagents are available.[336] Another example is the determination of copper, nickel, and manganese in aluminum alloys[212] by precipitation with diethyldithiocarbamate and extraction with chloroform.

Aside from these complexing agents, which are applicable to whole groups of elements, relatively selective precipitants are sometimes recommended, such as oxalate for calcium[430,509] and similar reagents.[210,252]

§ 88] COMBINED FLAME-PHOTOMETRIC AND CHEMICAL METHODS   251

(*ii*) Evaporation. Many interferents such as excess nitric or hydrochloric acid can be removed by heating; others can be fumed off, for example, silica with hydrofluoric acid.

(*iii*) Extraction. If a sample solution is shaken with an immiscible solution or solvent (of different density) serving as extractant, certain substances more soluble in the extractant will pass from the sample to the extractant and concentrate in it, while other substances will remain preferentially in the sample solution. This phenomenon of extraction is useful for certain separations. Examples are the determination of lanthanum,[490] aluminum,[252] and iron.[174,208,435] Extraction methods usually improve the detection limits by one or two orders of magnitude. The ratio of solubilities of a substance in the sample solution and the extractant can be such that an essentially quantitative transfer is feasible. From the point of view of flame photometry, a solvent is especially valuable when, in addition, it improves the detection limits (Section 79). Extraction is not necessarily confined to transfer from one liquid phase to another. It may be carried out between a solid phase and a liquid, or vice versa. Easily fusible solids such as lead and borax can also be subjected to extraction.

(*iv*) Electrolytic separation. Electrolysis offers a further possibility for separation. Micro methods have been developed.

(*v*) Mechanical separation. If the substance to be analyzed is present as solid or colloidal particles suspended in a solution, it can be separated by centrifuging and dissolving. The use of filter paper for separating solids should be avoided because the paper, never entirely ashless, can introduce impurities and because it tends to adsorb ions. There is also the risk of obstructing the atomizer with fibers from the paper.

(*vi*) Selective adsorption. By means of ion exchangers, interferents can be replaced by substances having little or no interfering effect;[154,288,289,617] see Section 100. Conversely, the analyte can be adsorbed onto the resin to separate it from the concomitants, and subsequently eluted.

(*c*) *Complexing the analyte.* As will be explained in Section 100, interferences from concomitants can be removed or mitigated by chemical means; this will not be discussed here yet. But even under normal conditions, the analyte may undergo chemical changes such as oxidation in the flame, or even in the spray chamber, that are detrimental to the flame-photometric analysis. There may be no question of the influence of concomitants; the undesirable effect is often due simply to the oxygen of the air, components of the flame gases, or the high temperature.

9*

As an example, aluminum, on entering the flame or possibly during evaporation of the droplets in the spray chamber, may form involatile oxides, which fail to evaporate promptly, so that much of the material is not excited in the flame. If this formation of oxide (or other chemical change that depresses the emission) can be inhibited by means of a chemical additive, the detection sensitivity and sometimes the accuracy can be considerably improved. Little is yet known of the mechanisms of these reactions.

Ethylenediaminetetraacetic acid (EDTA) added to an aluminum solution can raise the intensity of the Al or AlO emission 100-fold.[252] 8-Hydroxyquinoline has a similar effect,[219a] but the enhancement seems to operate equally upon the lines, bands, and continuum of aluminum. If hydrofluoric acid is added, the band system is distinctly intensified while the continuum is weakened.[431b] It was shown in this case that the enhancement is due to the formation of AlF_3, which sublimes at 1200°C. and thus enters the flame in a readily excitable form.

Conversely, aluminum is sometimes removed chemically prior to the determination of other elements, particularly when it raises the flame background too much.[512] This will be discussed in Section 95.

(d) *The substitution method.* If the analyte is difficult or impossible to detect in the flame, it can be replaced (by means of ion exchange, for example) by another element easier to determine in the flame, such as sodium, provided that the exchanging element is present in negligible amount at the outset. Thus, beryllium might be replaced on an ion-exchange resin by sodium. The exchanging element must, of course, replace no element other than the analyte, and the exchange must be quantitative.

(e) *Flame-photometric titration.* In the preceding methods the determination of the analyte is made possible or the sensitivity improved by combining flame photometry with a chemical method. On the other hand, in flame-photometric titration the ordinary chemical method of titration is improved or simplified by combining it with flame photometry. For example, calcium can be determined by flame-photometric titration with phosphoric acid.[251] As the phosphoric acid is added to the calcium sample, there is at first a linear drop in the calcium emission, which is read at intervals during the titration. When a particular stoichiometric ratio (about 1:2; the value can vary with the flame photometer; see Section 98) is reached, the curve of emission vs. titrant shows a breakpoint, beyond which the calcium emission is practically constant. From the amount of phosphate needed to reach the breakpoint, the calcium concentration can be calculated. The initial sample must, of course, contain no phosphate or other substance interfering with calcium, such as aluminum or sulfate, or any other alkaline-earth elements (Section 87). The other alkaline earths can be individually titrated in the same way. Conversely, phosphate can be titrated by an alkaline earth such as barium, provided no alkaline earth is present initially.

89. *Micro and Ultramicro Methods*

Advances in various disciplines such as experimental animal and plant physiology have depended critically upon the development of suitable micro- and ultramicroanalytical methods. The flame analyses described up to this point generally require so little material or such low concentrations that in the usual chemical sense they have to be counted as micro methods. We list below some suggestions and some reported methods for still further decreasing the requirement of material.

A. The sample is specially prepared for the purpose, prior to the actual flame analysis. Examples are enrichment by precipitation, extraction, or ashing of organic material; see Section 88. Dilution may sometimes serve. An organic solvent (alcohol, acetone, etc.), which may be used at the same time for extraction, can enhance the emission 100-fold as compared with water; or it permits analysis at the same intensity and in the same measurement time with 1/100 the quantity of material; refer to Section 79. Particular attention must be paid to the cleanliness of the containers (Section 80). In extreme cases special technique, using ultramicropipets, a microscope, etc., may have to be resorted to; we cannot enter into this here (see, e.g., ref. 181).

B. The flame photometer is selected for its ability to handle minimal samples. The following factors come under consideration.

1. An atomizer with minimal suction rate is employed and the atomizer and spray chamber should have high efficiency (Sections 26 and 29); or the atomizer is designed for highly uniform introduction of minimal quantities into a flame.[584,585]

2. The instrument should permit choice of the best analytical line or band. See Sections 81, 82, and 90.

3. The gas mixture is chosen to yield a temperature best suited for exciting the line or band (Sections 13 and 14), and the gas and oxygen flow are optimized. Refer to Section 19 and to the measures for keeping the gas and air pure, discussed in Sections 16 and 26.

4. The dispersing element (filter or monochromator together with its slit adjustment: Section 62) is made to yield the best resolution,

that is, the narrowest spectral bandwidth compatible with adequate luminosity. See Sections 40 and 60.

5. All of the components of the instrument are optimally adjusted with respect to each other; see Sections 37 and 64.

6. Sensitive optical–electric equipment of high stability is employed. See especially Sections 65 and 66.

7. A reading or recording method is selected that yields either the most readings in a given time (since sample consumption increases with time), to provide the best average, or else a single mean value within a short interval. Refer to Sections 47 and 56 and to the two-slit method (Sections 67 and 96), which simultaneously measures the blank during the brief period of measurement and automatically subtracts the blank from the total reading.

These various schemes can be combined with each other. Some additional notes are in order:

Regarding B1, with microatomizers of low suction rate (below 1 ml./min.) the line or band intensity does not diminish proportionally to the suction rate; it falls more slowly, or may even rise at first as the suction rate diminishes, gradually falling again at very low suction rates.[65,140,270,347,513,545] Thus, a microatomizer replacing an ordinary one will save much sample without necessarily requiring higher concentration. Sometimes an ordinary atomizer can be converted to micro use by throttling the sample flow, as by a wire inserted into the capillary. But it is hard to go below 0.1 ml./min. or below 0.1 ml. sample volume, since carry-over of sample on the walls of the capillary, the capillary volume, and the spray chamber volume become inconveniently large in a relative sense, and with indirect atomizers the entering mist must attain equilibrium in the spray chamber. Direct atomizers (Section 33) permit a measurement time $\frac{1}{2}$ to $\frac{1}{5}$ of that required by indirect atomizers, since there is no spray chamber. A reflux atomizer, in which liquid condensing on the walls of the spray chamber is sprayed more than once, can save sample; but it has disadvantages (Section 28). Further, a heated spray chamber can significantly raise the efficiency of atomization, but this too has disadvantages (Section 30).

Regarding B4, higher dispersion in a monochromator for a given slit width (more exactly, illuminated slit area) proportionally diminishes the spectral bandwidth, improving the separation of line from background (Section 62) and the detection limit (Section 90) in the same ratio. In other words, for the same atomizer, photodetector, electronics, etc., a higher dispersion in the monochromator (with correspondingly better resolving power) permits detection of a correspondingly lower concentration. But since, unfortunately, the price rises with the resolving power (and not necessarily linearly), detection limits in microanalysis can be improved in this way only at higher cost. The inherent limitations on resolving power were discussed in Section 65.

Regarding B7, averaging by photoelectric recording (Section 47) yields better reproducibility, which aids detection sensitivity. For checking the uniformity of atomization speed, the time between the beginning and end of spraying, for a fixed sample volume, can be measured on the recorded curve. The uncertainty of measuring (and subtracting) the blank—for example, when a blank cannot be prepared synthetically—is avoided in the two-slit method (Sections 67 and 96). Recall the difficulty due to the contribution of light flashes from dust particles to an average value obtained electronically, as with capacitors; the average cannot be corrected for the flashes. The difficulty can be overcome by recording the signal (Section 56) and taking an average value from the record, omitting the flashes. It has been recently suggested that the integrating method can be improved by use of an electrolytic switch, which starts the integration when the liquid first contacts the capillary and stops it when the liquid is consumed.[255, 724] It seems likely, though, that unreproducible wetting of the switch might cause errors, and that variations in the integration time (due to variations in the spraying time for a fixed volume of sample) might not necessarily be compensated by proportional changes in the photocurrent being integrated (cf. B1 above). In any case, the integrated photocurrent from a blank solution can be subtracted electronically from the integral representing the total reading, to give the net reading.[255]

It should be borne in mind that if several elements are to be determined in a small sample, each must be present in the requisite minimal quantity; therefore, the requisite minimal quantity of sample increases with the number of analytes. Multichannel (multiple-beam) methods can obviate this difficulty, by measuring all the desired elements simultaneously in the same sample (Section 50).

90. *The Detection Limits*

We now take up the subject of the minimal concentration of analyte qualitatively detectable by a given flame photometer, spectrophotometer, or spectrograph, usually called the *detection limit*. The detection limits vary with the element (Table 10) and depend on the instrument, the experimental conditions, and the composition of the solution. These limits also fix the minimal concentration at which an element can be determined with specified accuracy; this concentration is called the *determination limit*. The detection and determination limits depend on various factors to be discussed below. Following this discussion, we give a definition of detection limit.

If the concentration of an element is diminished while the over-all sensitivity of the instrument is increased so as to keep the same meter reading, it will be found that the fraction of the reading representing the emission of the element grows smaller, while the fraction representing background increases. The fluctuation of the reading, relative to the net reading (total minus blank), also rises. With continued reduction of concentration and increase of sensitivity, the following effects arise and become dominant. The photocurrent or meter reading corresponding to the net emission of the element diminishes with respect to the background (blank) and the fluctuation of the reading; or, if the sensitivity cannot be increased indefinitely, the net reading at length diminishes until the reading error dominates. When the fluctuation or the reading error has increased to the point where it compares with the net reading, a quantitative determination is no longer possible, and when the fluctuation exceeds the net reading, even qualitative detection of the element becomes impossible. The detection limit has been overstepped. It can now be debated what definition to adopt for this limit of detectability. We shall not argue the merits and drawbacks of all possible definitions (see ref. 1, for example), but we recommend the definition given below, favoring ref. 410. We first present some preliminary concepts.

We have on hand various individual readings of a magnitude whose true value is usually unknown. The true value can be approximated statistically, on the basis of probability theory, by means of many measurements, and an examination of the errors of this approximation reveals a residual error that can never be eliminated. By error we mean the difference between the measured and the true value. However, since the true value is generally unknown, although it is to be ascertained by measurement, this definition of error has only theoretical meaning. The error includes two essentially distinct components—a systematic error causing a deviation, and a random error causing a scattering. We shall deal with these in Sections 102 and 103. For simplicity in defining the detection limit, we heed only the random error; that is, we assume that the measurements are "correct" but "imprecise". We further assume that the frequency distribution of the repeated readings about this mean value M has the form of a Gaussian or normal curve; this is usually true. The blank reading is considered to represent the fluctuating measurement obtained on atomizing the blank solution (Section 95). If the standard error (see Section 102 for its definition) of the blank reading is σ_B and the mean value of the blank reading is $\bar{x}_B$, then the detection limit $\bar{x} - \bar{x}_B$, in scale divisions, is defined by the equation

$$\bar{x} - \bar{x}_B = 3\sqrt{2}\sigma_B \cong 4\sigma_B,$$

or, in words: the difference between the mean analytical reading $\bar{x}$ (in scale divisions) and the mean blank reading $\bar{x}_B$ must at least equal a particular multiple $k\sqrt{2}$ times the standard error σ_B of the blank reading, if an element at a concentration giving the mean reading $\bar{x}$ is to be detected by means of a single pair of readings on the sample and blank. The value chosen for k depends on the desired statistical confidence of detection. If, as is customary, the qualitative detection is to be 99.7% certain, then, provided that several (more than 10) separate measurements are used for establishing $\bar{x}_B$ and σ_B, k should be 3, as in the above formula.

The following is a practical procedure for determining a detection limit defined as above. The instrument is optimally adjusted (see below), and the blank is prepared according to Section 95 and sprayed at least 10 times to give the readings $\bar{x}_B$. From these the mean $\bar{x}_B$ and the standard error σ_B are found according to Section 102, and the value of $\bar{x}$ is found from the above formula. Four standard solutions are then prepared (blank plus analyte) at four concentrations which should give $\frac{1}{2}$, 1, 2, and 4 times the quantity $3\sqrt{2}$ times the standard error of the blank; preliminary trials may be necessary. The mean meter readings on these solutions are $\bar{x}_\frac{1}{2}$, $\bar{x}_1$, $\bar{x}_2$, and $\bar{x}_4$. If $\bar{x}_B$ is subtracted from each of these, the net readings corresponding to the analyte are found, viz., $\bar{x}_{\frac{1}{2}A}$, $\bar{x}_{1A}$, $\bar{x}_{2A}$, and $\bar{x}_{4A}$. If these $\bar{x}_{nA}$ are plotted against the concentrations, a working curve is obtained; it should be reasonably straight in this range. A curvature might result from the use of an alternating-current system with linear rectifier (Section 65), in which the rectified noise component is appreciable. From the curve, the concentration c is found that corresponds to $\bar{x} - \bar{x}_B$ as given by the formula; c is the detection limit expressed in units of concentration. Actually, c and $\bar{x} - \bar{x}_B$ are not "limits" in a strict sense, but merely statements of probability. The coefficient of variation at the detection limit amounts to 33%, so that determinations carried out at the detection limit are very uncertain.

The response time of the meter also inheres in this definition, for the fluctuations are smaller with a more sluggish indicating system, other things being equal. As the speed of response can be easily changed

(Section 46), detection limits defined as above are hardly unique for a given instrument. Such detection limits should, therefore, be normalized to a standard response time, by multiplying them by the square root of the time constant. To avoid ambiguity, the resulting values should be called *normalized detection limits* or normalized limiting sensitivities. Values not normalized should be accompanied by a statement of the time constant. Normalization by the square root of the time constant is exactly valid only when the noise spectrum is white, i.e., when noise is independent of frequency. However, in flame photometry this proviso is only roughly satisfied.[65]

Regarding normalization by time constant, however, it must be pointed out that a given number of observations at a longer time constant naturally takes more total time and sample than the same number at a shorter time constant. If we consider detection limits attainable in a given total time of measurement, or with a given volume of sample, they turn out to be independent of time constant on a given instrument. Hence, the detection limits should be normalized to a standard total time of measurement, by dividing by the square root of the measurement time.

If the detection limit, properly measured and normalized for time of measurement, is multiplied by the sample volume required for the time of measurement, the product is the least absolute quantity of substance that can be detected under the normalized conditions. This can be as low as 10^{-10} g. for a good commercial instrument in a favorable case. Note that the sample volume depends on the properties of the atomizer, etc.; cf. Sections 29 and 89.

Another convenient datum is the least concentration that can be measured with a specified precision. We have called this the *determination limit*. The determination limit, like the detection limit, is affected by concomitant substances in the sample (see below). If, for instance, a precision of $\pm 2.5\%$ is required, then the determination limit is about 14 times the detection limit; since the precision is 33% at the detection limit by our definition, the detection limit must be multiplied by 33/2.5, or 14.

Sometimes overly optimistic detection limits are reported, resulting from the use of extrapolative methods of measurement or calculation that are not valid for the problem. An example will clarify this. Given an instrument with its optics and electronics, a reasonably steady flame, and sufficiently constant atomization. The wavelength is selected and a slit is chosen which is, for example, five times as wide as that assumed for the definition of detection limit (say, five times 0.05 mm. or 0.25 mm.). The dark current is balanced with the photodetector dark (the instrument is zeroed). Next, the shutter is opened, the solvent (generally distilled water) is sprayed, the sensitivity is adjusted, and the blank reading is taken, representing the flame and solvent only. Finally, solutions are sprayed which contain the analyte in various known concentrations in the same solvent. From these readings, by interpolation or extrapolation if necessary, that concentration is found at which the reading is exactly double the background reading; the net reading on the line then equals the background. If this concentration is divided by 100 and then by 5, because of the larger slit, i.e., by 500 altogether, a concentration is arrived at, apparently, at which the line amounts to 1% of background for a 0.05 mm. slit. If it is assumed that the reproducibility of repeated readings is $\pm 0.3\%$, the impression is conveyed that the concentration thus found can be distinguished with certainty from the background. But to justify this conclusion, the following conditions have to be fulfilled: (1) The line-to-background ratio must be inversely proportional to the slit width; and (2) with the reduction of signal (instrumental response) associated with reduction of concentration and slit width, the relative mean deviation of the reading ($\pm 0.3\%$ in the example) must remain the same as at higher signals.

However, these assumptions may be invalid. In the monochromator of the Beckman DU spectrophotometer, for instance, reduction of the slit width much below 0.05 mm. brings comparatively little further improvement of the line-to-background ratio, owing to unavoidable optical inadequacies (Section 62). Moreover, with very narrow slits (0.03 mm. and below) the relative precision of reading on this instrument diminishes owing to loss of signal, either absolutely, or relatively to resistor noise with the usual phototubes, or relatively to shot-effect noise with photomultipliers. Hence, in our example the reproducibility can no longer be taken as 0.3% but may become much worse at very narrow slits. To this must be added the fact that the net photocurrent may not be proportional to the concentration, owing to curvature of the working curve (cf. Fig. 7). In consequence of these effects, the detection limits attainable in practice may be appreciably poorer than those found by simple extrapolation. Hence, data for detection limits obtained without heeding these factors have little practical value. They can be used at best for purposes of intercomparison and as a rough indication of the sensitivity to be expected.

Table 10 (from ref. 307a) lists detection limits in mg./l. for the Beckman DU flame spectrophotometer. The definition differs from that above, but yields essentially the same values. The data are for aqueous solutions except in columns AHn, OHn, and OAn, where such data as are available for nonaqueous solutions are listed. No account is

TABLE 10. Best wavelengths and detection limits of the elements[a]

For multiple lines or bands (resolvable at narrow slits), a weighted mean wavelength is given. For bands with conspicuous heads, the wavelength of the head is given. The "Character" entry applies to all data in that line, except as indicated ahead of any specific detection limit entry. The detection limit is defined in the text. It is given for aqueous solutions in air–hydrogen (AH), oxyhydrogen (OH), oxyacetylene (OA), and oxycyanogen (OC); detection limits for the most favorable known nonaqueous solvent are listed under AHn, OHn, OAn; the solvent is named after the first of the set of data for that solvent.

a = arc line (neutral atom)
A = ethanol
AA = acetylacetone
ab = arc line on strong band
Ac = acetone
b = peak of a well-defined band
Bu = butanol (4% in water)
Bz = benzene
c = continuum; the wavelength represents the region of maximal intensity; if c follows a, b, r, etc., the latter is superposed on an at least equally strong continuum
C = chloroform
CT = carbon tetrachloride
d = double
EA = 1:2:1 ether–alcohol–water
f = intercombination line (transition between terms of different multiplicities)
G = gasoline
h = with sheath providing oxygen atmosphere
H = hexone (methyl isobutyl ketone)

HAP = 5:2:2:1 hexone–acetone–isopropanol–water
i = in the inner cone (reaction zone)
k = detection limits given for a good red-sensitive photomultiplier such as the Farnsworth 16PMI
K = kerosene
m = multiple band
M = methanol
N = naphtha
Oc = 2-octanone
p = resonance line
P = isopropanol
r = head of band degraded to longer wavelengths
s = spark line (singly ionized atom)
t = triple
v = head of band degraded to shorter wavelengths
w = wide or diffuse band
z = very rich oxyacetylene flame burning in oxygen atmosphere
— = known to be absent
( ) = denote doubtful or roughly estimated data

[a] Adapted from refs. 307 and 307a and corrected to mid-1962. —*Translator*.
[b] Seen in absorption against an effectively hotter band of OH.
[c] Obscured by C_2 bands.
[d] Obscured by CH bands.
[e] Obscured by CN bands.
[f] Calculated theoretically.
[g] Bands observed only with iodate ion.
[h] For a 1P28 photomultiplier with red filter.
[i] With oxygen added to the air and acetylene to the hydrogen.
[j] Ethyl silicate.
[k] With air added to the oxygen.
[l] In air–hydrogen–acetylene flame.
[m] In ethyl acetate.

Element	Wave-length	Character	AH	AHn	OH	OHn	OA	OAn	OC
Aluminum	394.40	p	pc 35		12	0.5 H	10	0.1 H	0.4
	396.15	p	pc 30		10	0.3	5	0.05	0.3
	467.2	r	rc 13		5	1	65	0.6	—
	484.2	r	7		3	0.2	35	0.15	—
	510.2	r	rc 10		5	1	55	0.6	—
Antimony	217.58	p		0.6 P				0.7 i H	50
	231.15	p	150	0.4			(100) i	1 i	
	252.85	a	1000	1.2			(100) i	1 i	
	259.81	f	400	1.4					
Arsenic	228.81	a		4 P				3 i H	100
	234.98	a		2.5			(100) i	2 i	300
	500	c	5						
Barium	455.40	s	c 1.3		1.3	0.5 Bu	2	0.1 N	0.1
	488	b	0.10		0.3	0.25	5	2	—
	493.41	sb	c 0.12		0.4	0.25	2	0.1	0.15
	513	b	0.07		0.3	0.2	3	2	—
	553.56	p	0.06		0.25	0.1	1	0.2	2
	830	b	k 0.05		0.3	0.4 Bz	0.7	(0.03)	
	873	b	k 0.12		0.3	0.4	0.7	(0.03)	
Beryllium	234.86	p	—		—		0.4 z	0.1 zM	1
	470.9	r	rc 50		15		40		—
	473.3	r			20		50		—

Detection limit (mg./l.) in: (column group header spanning OH through OC)

§ 90] THE DETECTION LIMITS 261

ANALYTICAL TECHNIQUE

Element	Wave-length	Character	\multicolumn{7}{c}{Detection limit (mg./l.) in:}						
			AH	AHn	OH	OHn	OA	OAn	OC
Bismuth	223.0	dpf		1.5 P			1000 i^b	7 i H	3
	306.77	p	50	50	600	2000 P	200		
	472.26	a	20	c	40				
Boron	494	db	2		0.3	0.2 Bu	2	1.2 M	—
	518.0	b	1.2		0.2	0.12	1.7	0.7	—
	547.6	b	0.9		0.16	0.11	0.7	0.6	—
	579	b	1.5		0.3	0.15	1	1	
Cadmium	228.80	p	50	0.3 P	10	0.6 P	40; 0.8 z	4 i H	0.5
	326.11	f	0.5	1.8	5	2 Ac	40	2 Ac	25
Calcium	393.37	s			0.4		0.3		0.02
	396.85	s			0.5		0.6		0.04
	422.67	p	0.04		0.01	0.006 Bu	0.04	0.01 EA	0.03
	554	b	0.02		0.006	0.004	0.06	0.015	1
	622	b	0.02		0.004	0.002	0.02	0.01	1
Cerium	481	mbc	c 0.3		2	2 Ac	c 20		
	494	mrc	c 0.25		1.5	2	c 15		
	550–600	c	0.15		1	1	15		
Cesium	455.54	p	0.5		0.4	0.3 Bu	30		300
	852.11 k	p	0.01		0.01	0.005	0.01		0.3
	894.35 k	p	0.03		0.02	0.015	0.02		1

§ 90] THE DETECTION LIMITS 263

Element	Wave-length	Character	AH	AHn	OH	OHn	OA	OAn	OC
Chromium	357.87	p	0.25	0.6 P	0.12	0.12 N	0.5; 0.2 h	0.010 H	0.2
	425.43	p	0.08	0.7	0.10	0.07 Ac	0.5; 0.2 h	0.011	0.08
	427.48	p	0.09	1.2	0.12	0.09	0.6; 0.3 h	0.018	0.09
	428.97	p	0.10	*a*	0.15	0.12	0.8; 0.4 h	0.03	0.12
	520.6	ta	ac 0.06		0.15	0.14 Bu	1	0.1	0.1
	579.4	rc	0.04		0.5	0.08	0.7		—
Cobalt	340.51	a	0.25		0.5	0.3 Bu	2		1.1
	341.25	dpf	0.25		0.4	0.3	3		1.1
	345.4	dap	0.22		0.4	0.2	1.5		0.2
	350.23	a	0.35		0.6	0.3	3		1.4
	352.8	tap	0.3		0.3	0.1 CT	1.5		1.0
	387.35	da	0.35		0.5	0.3 Bu	2.5		*e*
Copper	324.75	p	0.16	0.25 P	0.1	0.02 M	0.1	0.03 K	0.3
	327.40	p	0.25	0.4	0.1	0.02	0.1	0.03	0.5
	537	w	0.1		0.6	0.5 Bu	1.7		—
Dysprosium	526.3	r			0.15	0.12 H			
	540.4	mr			0.17	0.14			
	572.9	b			0.08	0.07			
	583.4	tb			0.08	0.07			

Detection limit (mg./l.) in:

264   ANALYTICAL TECHNIQUE

Element	Wave-length	Character	Detection limit (mg./l.) in:						
			AH	AHn	OH	OHn	OA	OAn	OC
Erbium	504 506.7 553	mb } r mb			0.3 0.2	0.18 H 0.12			
Europium	459.40 462.72 598 623 702	p p bc wc w			0.4 0.5 0.1 0.15 0.2	0.3 H 0.35 0.08 0.1 0.1	1 1 (1) (1) (1)		
Gadolinium	461.7 581 599 621	mr dvc drc mrc	(2) (1) (1)		0.6 0.2 0.12 0.15	0.35 H 0.08 0.04 0.04	(1) (0.5) (0.5)		
Gallium	403.30 417.21	p p	1 0.5		0.1 0.05	0.07 Bu 0.03	1 0.5		
Germanium	259.25 265.14	p dp	>1000	14 P 5	1000	400 P 300	700 i 250 i	400 iP 200 i	
Gold [f]	242.80 267.60	p p	100 30		10 5	0.2 H	9 6		

§ 90] THE DETECTION LIMITS 265

Element	Wave-length	Character	\multicolumn{7}{c}{Detection limit (mg./l.) in:}
			AH	AHn	OH	OHn	OA	OAn	OC
Holmium	515.7	r			0.2	0.2 H			
	527	mb			0.2	0.2			
	532.0	mr			0.2	0.2			
	566.0	mb			0.08	0.06			
Indium	410.18	p	0.07	0.2 P	0.05	0.03 Bu	0.2	0.09 h Ac	
	451.13	p	0.04	0.13	0.03	0.02	0.15	0.07 h	
Iodine [g]	484.5	rc			(10)				
	513.1	rc			(10)				
	530.8	rc			(10)				
Iron	302.06	p	2.5	2 P	4	0.2 H	7	0.1 AA	0.2
	371.99	p	0.12	0.9	0.25	0.1 CT	0.7; 0.5 h	0.15	0.5
	373.71	(d)ap	0.13	1.0	0.35	0.2 Bu	1.2; 0.7 h	0.2	0.7
	374.7	tpa	0.15	1.3	0.4	0.2	1.7; 1 h	0.15	1.0
	385.99	(d)p	0.15	1.1	0.3	0.2	0.9; 0.6 h	3 M	e
	564.7	rc	c 0.06		0.25	0.08	1.0	3	—
	581.9	rc	c 0.06		0.25	0.08	0.9		—
Lanthanum	438.4	mr	2		0.4	0.4 H	0.6	0.13 H	
	442.3	mr	2		0.4	0.3	0.6	0.12	
	560.1	dr	0.6		0.15	0.1	0.6	0.06	
	743 k	mr	0.3		0.5; 0.14 h	0.01	0.1; 0.05 h	0.08	
	792 k	mr	0.4		0.5; 0.14 h	0.01	0.1; 0.05 h	0.07	

266   ANALYTICAL TECHNIQUE

Element	Wave-length	Character	Detection limit (mg./l.) in:					
			AH	AHn	OH	OHn	OA	OAn	OC
Lead	261.42	a		2.5 M		0.9 iG			
	280.20	a		2.5		1.1 i			
	363.96	a	7	5 G	2	1.5	30; 10 h	2 G	4
	368.35	a	3	2.5	1	0.7	20; 5 h	1	1.5
	405.78	a	3	2.5	1	0.7	15; 3 h	0.1 H	1.5
Lithium	610.36	a	0.3		0.5	0.3 Bu	1; 0.17 h	0.25	0.4
	670.78[h]	p	0.001	0.007 P	0.0002	0.00015	0.001	0.0002	0.003
Lutetium	466.2	mr			0.3	0.13 H			
	517.0	mr			0.2	0.09			
Magnesium	285.21	p	0.1	0.5 P	0.1	0.04 Ac	0.06 z	0.05H	0.01
	370.2	b	0.02	0.17	0.1	0.1 EA	2		—
	381–383	mb	0.02	0.2	0.12	0.12	3		—
Manganese	403.2	tp	0.02		0.01	0.007 N	0.02	0.005	0.08
	539.0	r	0.10		0.20	0.13 Bu	1.2		50
	558.6	r	0.08		0.12	0.08	0.6		70
Mercury	253.65	f	6	4 P	30	6 Ac	8 i	2.5 i H	
Molybdenum	379.83	p	c 4	>100 P	c 12	1 iN	(20) i	0.5 HAP	3
	386.41	p	c 3.5	>100	c 12	1 i	(30) i	0.5	10
	390.30	p	c 3.5	>100	c 12	2 i	(40) i	0.5	0.9
	550–600	c	0.4		1		(3)		—

Element	Wave-length	Character	Detection limit (mg./l.) in:						
			AH	AHn	OH	OHn	OA	OAn	OC
Neodymium	661	mb			1	0.1 H	1		
	691	mrc			1	0.02	2		
	702 k	mbc			1	0.02	1		
	712 k	mbc			1	0.02	1		
Nickel	341.48	p	0.12	0.4 P	0.2	0.17 Bu	1; 0.4 h	0.25 H	0.2
	346.0	dpf	0.22	0.7	0.5	0.3	1.5; 0.8 h	0.4	0.5
	349.30	p	0.4	1.0	0.6	0.4	1.4; 0.8 h	0.5	0.5
	351.51	p	0.22	0.7	0.45	0.3	1.1; 0.7 h	0.35	0.5
	352.45	p	0.13	0.4	0.2	0.12	0.7; 0.3 h	0.17	0.3
	361.94	a	0.25	1.0	0.45	0.3	2; 0.8 h	0.5	0.5
Niobium	405.89	p	—		—		—	17 z A	
	550	c	2		10		30		
Nitrogen	214.9	dv	25^i	25 P					
	226.3	dv	25^i	25					
	236.3	dv	14^i	20					
	385.3	mv				60 iN			
Palladium	340.46	a	0.22		0.14	0.05 H	1		0.12^f
	360.95	a	0.35		0.22	0.11	1.5		0.2^f
	363.47	a	0.18		0.12	0.07	1		0.10^f

268 ANALYTICAL TECHNIQUE

Element	Wave-length	Character	Detection limit (mg./l.) in:						
			AH	AHn	OH	OHn	OA	OAn	OC
Phosphorus	238.3	dyb	25	1.2 P	10^l				
	246.4	dv	20	1.0	7^l				
	247.8	dv	25	1.3	10^l				
	520	c	0.3						
Platinum^f	265.95	p	90		13		20	15 i H	1.2
	306.47	p	40		10		15		1.5
Potassium	404.5	dp	0.3	10 P	0.15	0.1 Bu	1	1.7 h Ac	70
	766.49 k	p	0.001		0.0003	0.0002	0.0003		0.03
	769.90 k	p					0.0005		0.05
Praseodymium	576.3	rc			0.7	0.9 H			
	601.9	r			1	0.5			
	709.5 k	bc			(1)	0.2			
	735 k	drc			(1)	0.3			
Radium	381.44	s			(3)				
	482.59	p			(2)				
	627	b			(2)				
	665	b			(2)				
Rhenium	346.05	p	—	—	—	—	300 z	5 z A	

§ 90]   THE DETECTION LIMITS   269

Element	Wave-length	Character	\multicolumn{8}{c}{Detection limit (mg./l.) in:}							
			AH	AHn	OH	OHn	OA	OAn	OC
Rhodium	343.49	p	4		0.7	0.4 P	1.1	0.6 P	
	365.80	p	ac 1.7		0.9	0.7	1.8	1.1	
	369.24	(d)p	1.4		0.27	0.25	0.7	0.45	
	542.5	bc	c 0.25		0.7		10		
Rubidium	420.19	p	0.5		0.3	0.2 Bu	5		100
	780.02 k	p	0.003		0.003	0.002	0.005		0.1
	794.76 k	p	0.004		0.004	0.003	0.006		0.2
Ruthenium	349.89	p			2	0.3 M	3		
	372.75	dp			0.5	0.07	0.3		
	379.9	tp			1	0.1	0.4		
Samarium	614	mrc			0.25	0.12 H	5		
	624.3	drc			0.35	0.12	4		
	642	mbc			0.5	0.10	5		
	651.0	mrc			0.5	0.09	3		
Scandium	485.8	r			0.5	0.16 H			
	581.1	dr			0.25	0.07	2		
	607.3	tr			0.04	0.006	0.3		
	611.0	tr			0.05	0.008	0.4		
Silicon	241.4	r	500ʲ	>400 P		(100)ʲ		100 zM	
	248.7	r	1000ʲ	>400		(100)		60 z	
	251.61	p	400ʲ	>400			20 z	7 z	

270   ANALYTICAL TECHNIQUE

Element	Wave-length	Character	Detection limit (mg./l.) in:						
			AH	AHn	OH	OHn	OA	OAn	OC
Silver	328.07	p	0.10		0.1	0.1 EA	0.2	0.3 EA	0.4
	338.29	p	0.04		0.06	0.06	0.2	0.25	0.5
Sodium	330.3	dp	4		0.5	1 Bu	1	0.2	20
	589.2	dp	0.0003		0.0002	0.00015	0.0004	0.0001	0.005
	819 k	da			2	1	0.5		1
Strontium	407.77	s		> 2 P	0.3	0.2 Bu	0.6; 0.08 h	0.09 h Ac	0.03
	421.55	s			0.5	0.3	0.8; 0.12 h		0.06
	460.73	p	0.02	0.07	0.01	0.005 Ac	0.05; 0.015 h	0.012 h	0.07
	605.9	b	0.002	0.04	0.01	0.007 Bu	0.1		—
	666	w	0.02		0.015	0.01	0.25		—
	682	w	0.04		0.01	0.01	0.3		—
Sulfur	257.6	r	500^f				70^k	25 M^k	
	259.0	r	700^f				110^k	40^k	
Tellurium	238.5	da	1000	10 P			800 i	4 i H	
	371.4	drc	2.7		35		c 700		
	388.4	drc	2.7		40		c 600		
	400.7	drc	3		45		c 600		
	434.3	rc	3		45		c 600		
Terbium	535	mr			0.25	0.16 H			
	573	mbc			0.2	0.10	(1)		
	592.1	rc			0.12	0.06	(1)		
	597.9	drc			0.15	0.07			

§ 90] THE DETECTION LIMITS 271

Element	Wave-length	Character	Detection limit (mg./l.) in:						
			AH	AHn	OH	OHn	OA	OAn	OC
Thallium	377.57	p	0.2		0.1	0.06 Bu	1	0.3 Oc	1.7
	535.05	a	0.35		0.15	0.1	2	0.5	1.2
Thulium	490	mb			0.35	0.2 H			
	537	mb			0.3	0.16			
	542	mb			0.3	0.16			
Tin	235.48	p	—	0.5 M	—	7 iP	8 z	2 i H	
	242.95	a	1.5^l	0.5	1.7^k	7 i	7 z	1.7 i	
	270.65	p	120	0.6	800		6 i	6 z M	
	284.00	a	120	0.8	(700)		50 i		1.2
	303.41	p	25	0.6	200	1 iN	25 i		1.0
	358.5	r	3	20 P	11	45 Ac	50		—
	485	b	2	40	c 45				
Titanium	516.8	drc			0.25		7	7 P	
	545.0	drc			0.22		7	7	
	576.1	drc			0.22		7	7	
	713	mrc			0.5		1		
Uranium	550	c	0.7		2		(10)		

ANALYTICAL TECHNIQUE

Element	Wave-length	Character	Detection limit (mg./l.) in:						
			AH	AHn	OH	OHn	OA	OAn	OC
Vanadium	318.4	tp	—		—		40 z	0.5 z A	0.5
	437.92	a	—	ac 4 P	—			3 z	1.7
	440.7	da	—	ac 2.5	—			3.5 z	1.8
	522.9	drc	0.12	c 0.55	0.35	0.12 Bu	3.5	1 ^{m}	—
	547.0	drc	0.10	c 0.6	0.25	0.10	2.5	1 ^{m}	—
	573.7	drc	0.09	c 0.3	0.25	0.09	1.7		—
Ytterbium	398.80	p	0.15		0.4	0.08 H	(1)		
	498.1	b	0.05		0.2	0.2	(2)		
	532.5	b	0.04		0.13	0.13	(2)		
	555.65	pb	0.04		0.14	0.16	(1)		
	572.5	b	0.03		0.09	0.10	(1)		
Yttrium	481.8	r			0.35	0.2 H		0.15 H	
	599	mr			0.03	0.010	0.3	0.2	
	615	mr			0.035	0.013	0.3		
Zinc	213.86	p	250	60 P	160	10 Ac	600	80 i H	1.7
	481.05	a	c 30		c 170	16 iN	i	c 6 Ac	
	520–600	c	20		80	0.8 Ac		2.5	—
Zirconium	564	mrc			8		(100)		—
	574	mrc			8		(100)		—

taken of interelement effects (see Sections 96–100). These detection limits are calculated for an average instrument with standard atomizer-burner (Fig. 21) and an average 1P28 photomultiplier (Fig. 34) operated at the light level at which the shot-effect noise equals the flame flicker; the light level is controlled by the slits, and the dynode voltage is adjusted to give a convenient reading. The detection limit is then taken as the concentration of element in solution needed to give a net response amounting to 1% of the background under optimal conditions of focus and flame adjustment. In the red end of the spectrum, beyond 650–670 mμ, a Beckman red-sensitive phototube is assumed, except as indicated in the notes. For this phototube, the amplifier sensitivity must be such that full scale corresponds to a drop of at least 100 mv. across a load resistor of 10 000 megohms; the slit is then chosen to make the background read 100 mv., or it is set at 1 mm. (an arbitrary limit, to avoid excessive bandwidth), whichever would be smaller, and the detection limit is the concentration giving a net response of 1 mv. The slit width implied in this definition,* for the 1P28, is typically 0.1–0.2 mm. for oxyhydrogen, 0.02–0.05 mm. for oxyacetylene (but smaller with nonaqueous solvents in each case) and 0.01–0.02 mm. for oxycyanogen; but it will be 1 mm. beyond 670 mμ unless a red-sensitive photomultiplier is used, which permits narrower slits.

As seen in this table, aside from the alkali metals, most of the detection limits lie between 0.01 and 10 mg./l. In this range, about 55 of the elements can be detected under the indicated experimental conditions, half of them between 0.01 and 0.1 mg./l. and half between 0.1 and 10 mg./l. The alkali metals, Ca, Sr, Mn, and Sc have detection limits of 0.01 mg./l. or below, while N, S, I, and Re are listed with comparatively poor limits of 10 mg./l. or more.

We enquire next how the detection limits of the table change with the instrumental or procedural conditions. The question is hard to answer in a general way, since many variables enter, and they are sometimes unpredictable or interrelated with others. Since, as we have seen, the detection limits depend essentially on the fluctuations of the reading, anything that affects the fluctuation of the reading will affect the detection limits. But since these fluctuations depend on the flame flicker, anything that alters the flicker of the flame emission will affect the detection limits. The importance of the operator's skill should not be underrated. A practiced observer can accurately average the reading

* A more recent theoretical analysis has shown that, for lines, greater shot-effect noise can be tolerated without loss of detection sensitivity, permitting narrower slits and a considerable gain of resolution, especially in those cases where the above criterion would dictate a rather wide slit (> 0.1 mm.).—*Translator*.

of a wavering meter needle, while a beginner is annoyed by the fluctuation and is likely to take the reading at a moment when it happens to be quieter than usual—but that will be at the extreme values, which are least representative of the reading. In the following, then, we limit the discussion to qualitative statements, with references to other sections.

1. The *adjustment of the flame* with respect to the aperture of the beam received by the monochromator or the filter photometer (cf. Sections 37, 43 and 64) and the adjustment of the backing mirror behind the flame have an important effect on the steadiness of both the flame background emission and the net analyte emission and hence on the detection limit. For instance, if the instrument sees too much of the unsteady outer or upper parts of the flame, the greater unsteadiness of the reading makes the detection limit worse. By means of special mirror systems, a larger portion of the flame radiation emitted in all directions can be gathered by the optical system (Section 38). In such a situation the detection limit is improved for two reasons: (*a*) The total intensity increases, so that electronic noise is relatively less. (*b*) The flame radiation is integrated over a larger volume of the flame and larger solid angle; except for the adverse effect of coherence between the fluctuations of different parts of the beam, this makes for a steadier total emission than that from a small volume of the flame or limited aperture (solid angle). Point (*a*) above is not necessarily related to improvement of the line-to-background ratio. If this ratio remains constant when the total intensity increases, the detection limit will improve nevertheless.

2. The *gas/oxygen ratio* or gas/air ratio (Section 19) affects the flame temperature and its stability; these in turn affect the emission and its stability (Section 8) as well as other factors that influence the emission, such as dissociation and ionization (Section 7). The unavoidable fluctuations of the regulated gas and oxygen pressure (cf. Sections 14, 18 and 19) cause greater instability of the blank and analytical emissions (and these are not affected to the same degree) when the gas/oxygen ratio is suboptimal; at the same time, the unavoidable fluctuation of temperature, through its action on the evaporation of the spray, makes the emission less stable and adversely affects the detection limits. The detection limits can be improved by burning the flame in a protective atmosphere of clean oxygen instead of in the air (Section 32). This raises the flame temperature, intensifying the emission, and it improves the stability by eliminating the flashes from dust particles (see below).

3. *Influence of the atomizer.* The properties of the atomizer—its

suction rate, the uniformity of atomization, and the drop size distribution (Sections 6, 22–31)—govern losses between atomizer and burner (transport effects) and, in both direct and indirect atomizers, the evaporation of the droplets and solid particles in the flame and thus the intensity and stability of the emission.

4. *Concomitant elements* and other substances in the sample (or sometimes in the room air; see Section 93) generally worsen the detection limits, owing to decrease in the line-to-background ratio resulting from cross-sensitivities (Section 94), from increase of background by continuum and band emission (Sections 8 and 95) of the concomitants, from adverse effects of viscosity and surface tension on the atomization (Sections 6 and 101), and from slowing of spray evaporation in the spray chamber and the flame (Sections 6 and 7). These and other effects generally also increase the flicker, worsening the detection limits. For the same reason it is very important that the solvent should be free from extraneous elements, particularly the analyte.

5. The *optical–electric sensitivity* should be high for good detection sensitivity; this includes luminosity (Section 61), optimal transmittance (Sections 59, 60), sensitivity of the photoreceiver (Section 42), amplification (Section 44), and sensitivity of the meter (Section 46), together with minimal noise level (Section 65). But (contrary to a rather commonly held notion) increasing the sensitivity alone will not yield indefinitely better detection limits; refer to the definition above. Sometimes, to be sure, an increase in the luminosity of the optical system or in the electronic amplification can alone provide better detection sensitivity, when the limit is set not by the emission flicker or the electronic noise but, for example, by the precision with which a small meter deflection can be read or, in the short-wavelength ultraviolet, by poor transmittance of the optical components. In the latter case some gain can be realized by increasing the luminosity only, without increasing the dispersion of a monochromator or decreasing the bandwidth of a filter.

6. The *spectral bandwidth* (Sections 40 and 61) has a substantial effect on the detection limits. Except at very narrow slits, the limits improve (for lines and narrow, line-like bands) as the bandwidth diminishes, other parameters such as luminosity being held constant. But this requires an efficient monochromator. Such a reduction of spectral bandwidth or gain in resolving power, at constant or, if possible, greater luminosity, offers the chief means of appreciably improving the detection limits listed in Table 10. It is a costly approach, to be sure. The situation is different with wider bands; reducing the spectral bandwidth brings no gain in detection sensitivity unless it becomes possible to resolve the individual lines of the band. Indeed, for a given

monochromator the detection limit for a wide band even improves as the slit width or spectral bandwidth is widened, since this raises the radiant flux through the instrument and thereby reduces the relative contribution of electronic noise to the fluctuation. Filter instruments generally have a greater (less favorable) spectral bandwidth and hence poorer detection sensitivity. Typical values will be found in ref. 513.

7. The *measuring procedure* has a large effect on the detection limits.[584] This includes the question of internal standardization (Section 51), choice of time constant of the meter (Section 46), method of integration (Section 47), and recording with averaging (Section 56). Recall the remarks regarding normalization of the detection limit. In short, anything that reduces the standard error of the blank reading reacts favorably on the detection limits. However, most methods of reducing the fluctuation demand more time and more sample (Section 89). A recording and integrating procedure, nevertheless, gives a better average or integral of the instantaneous readings; the gain in precision and in detection limit may amount to a factor of ten.

91. *The Speed of Flame Analysis*

Analysis by flame photometry takes little time compared with chemical methods of analysis, apart from the initial adaptation of the method to the application at hand, which includes testing of the best experimental conditions, preparation of standards and working curves, etc. We next discuss the factors affecting the time requirement, pointing out possibilities for reducing it further. Saving time becomes the more important, the greater the expected number of routine determinations is.

1. The *preparation* of the sample for analysis—sampling the material and working it up for the actual flame analysis (Sections 78, 79, 87, 88) —takes the more time, the more complicated the procedure is. In the simplest case the sample is directly aspirated by the flame photometer. This is often possible in water analysis. The next simplest case is that of mere dilution prior to analysis. On the other hand, chemical manipulations such as precipitation take a good deal longer. The extent to which time will be saved by more efficient preparation can be judged only in each individual case. Refer to the sections mentioned above and to the many published applications in which such experience is described. It is possible to automate the repetitious preparative manipulations along with the entire flame-photometric procedure in routine analysis, for example, by means of an electronically controlled proportioning pump.[80]

2. The time required depends not only on the sample preparation but also on the *instrument* and the *organization of the work*—sample handling, bookkeeping, discarding used sample, calculating the results (use of working curves, etc.), and applying corrections to the results. We list below those aspects of instrument and method that influence the analytical speed, with a few comments on organization.

(*a*) *Sample changing* technique is important for speed. With the help of a sample changer (Fig. 43, page 180, for example), many sample containers, perhaps as many as 40, can be placed simultaneously on a rotating table and analyzed in rapid succession. The used containers can meanwhile be replaced by new samples without interrupting the sequence. If the personnel preparing the samples, applying and removing the containers, and operating the flame photometer are well organized and coordinated, then the instrument is being used most effectively. Otherwise, in case of a heavy work load, several flame photometers will have to be set up in parallel.

(*b*) The *reading delay* between application of the sample to the atomizer and the attainment of the final meter indication affects the analytical speed. It is briefer with direct than with indirect atomizers, since the latter entail a delay for equilibration in the spray chamber.

(*c*) The *time constant* of the optical–electric system also governs the time required for attainment of the final meter reading; see Sections 46 and 65. With a potentiometric circuit (Section 49), the time needed for balancing the slidewire usually exceeds that needed for reading the meter of a direct-reading instrument, because the act of balancing adds to the delay inherent in the response time of the null meter. With internal standardization (Section 51), except in the auxiliary procedure (Section 52), speed is usually greater, since the balancing of the potentiometer can get under way before equilibrium is established; the concentration ratio of analyte to internal-standard element in the mist in the spray chamber is the same before equilibration as after, so that the measurement can be started early. A multichannel flame photometer for the simultaneous determination of several elements can save both time and sample material.

Basically, there are no limits to automation of the entire procedure, by which the time for an analysis can be considerably shortened. The Quantometer is a good example. The flame background and dark current can be subtracted by automatic means. A potentiometer can be balanced against a reference voltage or against an internal-standard signal with a motor, yielding the ratio automatically. Corrections for interferences could be calculated electronically; the final result could then be printed directly upon a paper strip or the like. Flame photometers or flame spectrophotometers with such conveniences have not yet made their appearance, but it seems likely that there will be further development in this direction.

Recording generally takes more time and material than the taking of a single reading at a given wavelength. The speed of scanning a spectrum is governed by the response time of the recorder; if the speed is too great, the pen fails to reach the peaks of the lines. For faster recording of a spectrum, an oscilloscope can be used with a monochromator in which either the prism or a slit is vibrated over the desired wavelength range (e.g., ref. 343a). An alternating voltage synchronized with the vibration frequency is applied to the horizontal plate of the oscilloscope.

The Errors of Flame Analysis

92. *Introduction*

In Section 11, following the discussion of fundamentals (Sections 5–10), we surveyed the various kinds of interference and offered a classification that we think suitable for our purposes. Sections 93–104, to follow, will deal in detail with some of these interferences, methods of recognizing the errors that they cause, and practical schemes for eliminating them.

Section 93 discusses errors from impurities in the gases or the atmosphere; Sections 94 and 95, blank interferences due to cross-sensitivity and background effects; Sections 96–100, specific radiation interferences from concomitants; and Section 101, nonspecific radiation interferences. Sections 102–104 cover the mathematical basis of the estimation of error, required for the objective comparison of different methods.

Before starting with the topic of impurities in the air, we discuss first some general points of view applicable to the succeeding sections.

The more exact a result is to be, the more numerous are the sources of error that have to be considered. The preliminary labor of working out the method as well as the repetitive work of the routine determinations, therefore, become much easier when less accuracy is demanded. How much accuracy is required in a given case can be decided only from the particulars of that case. These questions will reappear in the later sections on applications.

The most easily recognized errors are those due to instrumental instability (gas pressure drift, flame flicker, clogging of the atomizer, zero inconstancy, changes of sensitivity in the detector, etc.). One need only insert a reference standard regularly after every two to five analyses, to check the sensitivity and perhaps also, with a blank solution, the zero point of the instrument. Errors of this type affect the reproducibility of the readings, defined in Section 102. The total instrumental error, including the contribution from the flame, should be distributed about equally among the individual sources of error; thus, there is little point in pushing the precision of one part of the apparatus, such as a good potentiometer, to $\pm 0.025\%$, while the indirect atomizer being used permits a reproducibility of only $\pm 1\%$ owing to erratic condensation of drops in the spray chamber (on the effect of temperature see Section 26).

93. *Impurity of the Gases and the Ambient Air*

Compressed air for the flame may contain dust, oil droplets from the compressor, water, etc., which can falsify the readings or at least worsen the detection limits (Section 90). The use of a filter in the compressed-air line was mentioned in Section 16. The fuel can also be impure. Hydrogen, for example, may contain potassium, iron, and a few other elements. But even with clean gases, flames are very sensitive to contaminants in the surrounding air, unless a closed burner (Section 26) is used. Any flame, of course, entrains ambient air, the purity of which affects the flame emission.

The light flashes from sodium-containing dust particles can not only upset the determination of sodium but also raise the flame background over the whole wavelength range, for sodium and potassium, like many other elements, emit a continuum (Section 8). The light flashes can thus unexpectedly interfere at other wavelengths; moreover, they can exert this effect through cross-sensitivity (Section 94). The flame background can also be raised by the flashes from coarse, unevaporated, incandescent particles of involatile substance.

Cigarette smoke can be very disturbing in the determination of potassium. Purification of the room air is possible[527] but expensive. In industrial areas marked contamination of the air, as from soap powder, can severely embarrass flame analysis. Further, the flame discharges in the form of dust the salts entering it. During a series of measurements, the room air may thus become contaminated, especially when the sample and standard solutions are rather concentrated. This can be avoided by installing a funnel and fume duct over the flame, with an exhaust fan to vent the fumes outdoors. This is more important for a smaller room. For the same reason, a sequence of measurements should proceed from the diluter to the more concentrated solutions. With the older flame photometers having an open flame, the increasing pollution of the room air can sometimes be seen from the flame color. Newer types generally have an opaque chimney; for these it may be helpful to set up separately a bunsen burner with a nonluminous flame to serve as a sensitive indicator of the purity of the air. It has been reported that on days of dust storms, flame analysis cannot be carried out. Similarly, strong sea breezes near the seacoast may cause trouble from salt particles in the air.[143]

A. BLANK INTERFERENCES*

94. *Cross-Sensitivity*

Ideally, a filter or monochromator should transmit only the line or band to be measured. Actually, however, a range of wavelengths is always transmitted, and even in remote parts of the spectrum the transmittance is not zero, but has a positive if generally very small value. Because of this unwanted transmission of light of other wavelengths, a filter intended for a given wavelength, such as the sodium line, or a monochromator set upon a line will transmit light of another element, such as potassium. This interference by an extraneous emission is called *cross-sensitivity*.[632] The efficiency of optical isolation is thus characterized by the cross-sensitivity, which should be as small as possible. Unfortunately, instrument manufacturers generally do not quote figures for cross-sensitivities, although they would be a good criterion of the quality of the instrument. In monochromators this residual transmittance results from stray light; in interference filters it may result from defects in the evaporated layer. Stray light can also occur in filter instruments if the filter and detector are not properly installed. Stray light in filter photometers or faulty transmittance of filters is especially serious when very weak lines, as from trace elements, have to be measured alongside very strong lines, as of the major component. For this reason, when the requirements are stringent, two monochromators can be placed in series to make a double monochromator, or several interference filters can be superimposed, or a filter or some other means of

* Blank interferences are called *spectral interference* by J. A. Dean (*Flame Photometry*, McGraw-Hill, New York, 1960) and many others. Dean tabulates many data on spectral interference (largely background interference, cross-sensitivity being kept low by use of a monochromator), giving the *specificity factor* or ratio of readings (at the analytical wavelength) of analyte and interferent at equal concentrations. Dr. Herrmann emphasizes that cross-sensitivity and background interference (the two kinds of blank interference) are rigorously distinguished in the present book.—*Translator*.

preliminary dispersion be combined with a monochromator. But such arrangements are expensive and not always available. When necessary, the interfering stray light can be eliminated by a suitable method of measurement and standardization. In any case, an effort should be made to diminish the cross-sensitivity by optical means to a negligible magnitude.

In establishing a method, suspected cross-sensitivities are tested as follows. Suppose sodium is to be measured; the line Na 589 mμ is set on the monochromator, or, in the case of a photometer, the sodium filter is placed in the light beam, and the sensitivity of the optical–electric system is adjusted to permit detecting the lowest sodium concentration of interest. To do this, of course, a preliminary experiment is needed. Next, the highest likely concentration of the interferent to be tested, say, potassium, is introduced into the flame and the photocurrent observed at the sodium wavelength; in this trial only the interferent is present. If the meter shows no response, the cross-sensitivity can be ignored in subsequent routine determinations in this application. If, however, a deflection is observed in this experiment, it does not necessarily follow that a cross-sensitivity is present. Some other kind of interference may be responsible; we return to this later.

In Table 11, listing cross-sensitivities, the interferent and analyte are assumed to be present in equal weight concentrations. The cross-sensitivity is expressed as a percentage of the reading on the analyte. The values will depend on the flame temperature, the method of spectral isolation, and the composition of the solution.

If the test described above shows an appreciable response, the cause should be determined as exactly as possible. The observed interferent response is not necessarily always caused by inadequate spectral separation; this can be checked most simply as follows. Place in the beam an auxiliary filter sufficiently transparent (90%) at the line of the analyte (e.g., 589 mμ) but less transparent (say, 20%) at the line of the interferent (767 mμ). If the sodium reading drops proportionally to the transparency of the filter at 589 mμ (in this example, by 10%), cross-sensitivity is absent. But if the drop is greater (exceeding 10% here), cross-sensitivity may be present.

If such a filter is unavailable, the following trial can be made. The interferent reading at the wavelength of the analyte is measured as a function of interferent concentration. This curve is compared with the working curve of the interferent at the wavelength presumed to be responsible for the cross-sensitivity, generally the strongest line (in our example, K 767 mμ). If the two curves have the same shape—perhaps the same sigmoid curvature or the same bend point (cf. Fig. 50, page 231)—at the same concentrations, then cross-sensitivity may be assumed present.

If these tests for cross-sensitivity are negative but the potassium still interferes with the sodium, then the following possibilities should be considered.

TABLE 11. Cross-sensitivities for the commonest interferents as observed in the flame photometer of Fig. 29[513] with air–propane flame and interference filters.

	Filter for:				
Interferent	Li 670 mµ	Na 589 mµ	K 770 mµ	Ca 620 mµ	Rb 780 mµ
Lithium	100	0.11	0.04	0.87	0.01
Sodium	0.03	100	0.06	0.87	0.01
Potassium	0.02	0.03	100	0.23	280
Calcium	0.16	0.20	0.01	100	0.01
Rubidium	0.02	0.05	1.70	0.25	100
Cesium	0.02	0.01	0.05	0.24	0.05
Magnesium	0.01	0.01	0.01	0.06	0.01
Strontium	8.50	0.05	0.01	4.05	0.01
Barium	0.07	0.02	0.62	0.83	0.27
Aluminum	0.01	—	0.01	0.01	0.01
Manganese	0.03	0.14	0.29	0.98	0.07
Iron	0.05	0.06	0.07	2.12	0.05
Copper	0.01	0.01	0.01	0.19	0.01

1. The chemicals employed may not be pure enough. A potassium salt may contain traces of sodium. This can be checked by using chemicals of different origin and purity, by measurement of the presumed impurity at another wavelength, by a chemical trace analysis, or by other means.

2. The supposed cross-sensitivity may result from a marked increase of the flame background in the neighborhood of the line. A recording of this region of the spectrum (e.g., the vicinity of the sodium line) with and without addition of the interferent (potassium), using a double monochromator, will reveal such an effect; or the information may be available from the literature. Potassium is known to raise the background (Section 8), but this effect should be noticeable only at low sodium and high potassium concentrations. A monochromator can be used to distinguish between possibilities 1 and 2 also by noting the dependence of the interfering intensity on slit width. In case 1, the relation is linear; in case 2, quadratic (cf. Sections 62 and 95).

3. The increase in the sodium reading (for example) may be due to the

specific radiation interference of potassium (Section 97). This often happens with these two elements. To test whether and to what extent cross-sensitivity is participating, an instrument nearly free from stray light can be used to measure or record the line and background intensity or to measure the interfering intensity as a function of slit width, as in case 2.

4. The interfering response may arise if the line lies within a band of the interferent, such as the CaOH band at 623 or 554 mμ. Or, perhaps, two closely adjoining spectral lines of the two elements cannot be separated by filters. This may be tested in the same way as in 2. In cases 2 and 4, the ratio of the analyte response to interferent response (potassium or calcium here) can be improved by using two or more superposed filters in a photometer, or narrower slits in a monochromator, or else a monochromator of greater dispersion. These expedients generally increase the requirement of sensitivity in the optical–electric system. But if two bands, such as the CaOH band at 623 mμ and the orange SrOH band, occupy the same region, then even good resolution is of no avail. Unless one of the calibration or correction methods given below and in Sections 86 and 98 is used, either the interferent (strontium) must be removed chemically (Section 88), or else another wavelength (for instance, the green CaOH band at 554 mμ) must be used where the interference (from strontium) is less or absent.

If a cross-sensitivity persists despite the optical improvements mentioned above, or if these means are not available, the interference can be circumvented by a suitable (parametric) calibrating procedure, in which the interferent (potassium) and the element sought (sodium) are both determined in each sample. Further, working curves (for sodium) are prepared for various concentrations of the interferent (potassium). In each determination, one uses the working curve corresponding to the concentration of interferent found, or interpolates between working curves. As an example, Fig. 52 shows various calcium working curves obtained with a filter appreciably transmitting sodium and potassium light. The sodium and potassium concentrations appear as parameter.

This method is to be distinguished from the parametric method of Section 97 for the elimination of specific radiation interferences. Cross-sensitivity introduces an additive increment of response independent of the sought value, which can be eliminated by subtraction. But specific interferences are usually multiplicative, and the factors may be greater or less than 1. Such factors are allowed for not by subtraction but by division. The parametric curves of Fig. 52 should be contrasted with those of Fig. 55; in the former, the working curves are parallel, whereas in the latter they radiate from a common point.

The magnitude of a cross-sensitivity sometimes depends significantly upon the flame temperature. For example, if the infrared sodium line at 819 mμ interferes with a potassium determination using a filter for 767 mμ, this interference is considerably smaller in a cooler flame (for example, propane instead of acetylene), since excitation of the sodium line requires higher temperatures, its excitation potential being higher[65] (cf. Fig. 2).

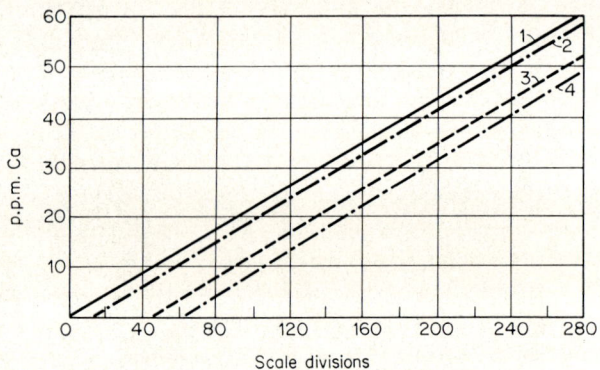

Fig. 52. Calcium working curves in the presence of stray sodium light passed by the filter (from ref. 128). 1, Pure calcium solution; 2, the same with 800 p.p.m. K; 3, with 800 p.p.m. Na; 4, with 800 p.p.m. Na and 800 p.p.m. K.

When the cross-sensitivity is not large and the concentration of the interferent or composition of the matrix is fairly constant, it suffices to simulate this small, constant interference in the standard solutions. In the example of Fig. 52, a constant, average concentration of sodium is added to each of the different calcium standard solutions—the "pre-cancellation" method. If the concentration of the interferent is only roughly constant, a combination of the pre-cancellation method and the parametric method described above is recommended, as follows. Standard solutions containing the average concentration of interferent are used for the routine determinations, and by means of standards with higher and lower interferent concentrations the magnitude of the analytical error is gaged. If the concentration of interferent in the sample departs appreciably from the average, a suitable correction is applied to the result. Which method is adopted in practice depends greatly upon the accuracy needed.

A more serious cross-sensitivity, as from the CaOH bands at 554 or

623 mμ in the determination of sodium at 589 mμ, can, as mentioned, be removed by a suitable method such as the parametric. Despite this correction, however, unsuspected errors may still result from specific radiation interferences (interelement effects). Suppose that a low sodium content is to be determined in the presence of much calcium, a problem which arises, for instance, in the cement industry (Section 122). Then if cross-sensitivity of the sodium filter for calcium light is eliminated numerically, it can happen that some other effect, e.g., an interelement effect due to silicon, that directly affects only the calcium, will affect also the sodium indirectly via the cross-sensitivity.[224] Theoretically, this can be allowed for by means of further corrections, but it is clear from these remarks that an optical elimination of the cross-sensitivity is always preferable. It must be kept in mind that the cross-sensitivity can depend not only on the concentration of the interferent but also sometimes on that of the analyte itself, as when the latter affects the emission of the interferent.

With solutions of complex composition, the preliminary test for cross-sensitivity described earlier has to be made in the same way for all the concomitant elements present in the sample, except those for which the cross-sensitivity is already known to be insignificant from other preliminary trials or from reliable information from the manufacturer. Standardization and correction methods analogous to those above can be prescribed for eliminating the effects of two or more interfering elements. These methods are too complicated to be presented here. Here, again, it it is important to use only optical equipment that provides good spectral purity—low cross-sensitivity.

At the end of the next section we shall show how cross-sensitivity and flame background can be eliminated together in one step.

95. *Elimination of the Background*

For simplicity we assume first that interfering light from other elements (cross-sensitivity) is absent or suitably eliminated. Later in this section we deal with the case in which both cross-sensitivity and background are to be eliminated at the same time. These are the two main types of blank interference (Section 11).

Any spectral apparatus, however good, transmits a range of wavelength (Section 62). Consequently, the instrument sees not only the analytical line but also a portion of the background. The background can be reduced but never wholly eliminated by improving the resolution. In this respect, background effects differ basically from cross-sensitivity. As discussed in Sections 8, 9, and 11, the background is composed of the radiation (bands and continua) of the flame itself and of any superposed radiation (usually continua, sometimes bands or weak lines) from concomitant solutes and solvent sprayed into the flame. These two components of the background, from the flame and from the concomitants, are unfortunately not constant but depend on the concentrations of the concomitants and are affected by different interferences in different ways. Hence special means are needed to eliminate this variable interfering radiation. But even when the interference is constant, the radiation is not steady, and this unsteadiness determines the detection limits (Section 90). We are concerned here only with the mean radiation intensity, not the fluctuation. The background radiation depends to some extent both directly and indirectly also on the concentration of the analyte itself. This component of the background, however, is usually considered not as an interfering emission but as part of the emission of the analyte.

We next discuss practical methods of eliminating the background, with the assumption that the instrumental response is linear. Three cases can be distinguished.

1. The line intensity is so high that the flame background can be ignored; this is the simplest case.

2. The flame background is small relatively to the line but not entirely negligible. The background can then generally be considered constant and subtracted from the total signal. There are several ways of doing this, viz.:

a. The flame background is measured by atomizing a blank solution, that is, a solution containing all the constituents of the sample including the solvent, except the analyte. This blank reading is subtracted from the total readings on the standard and the sample, or else the zero point of the meter can be displaced by an amount equal to the blank reading; this is called suppressing the zero. Instruments with amplifiers or bridge circuits generally have means for electronically adjusting the zero and thereby suppressing the background response. This method assumes the availability of chemicals for the blank that contain negligible amounts of the element sought; it assumes also that the other components of the sample can be well simulated and that they have a sufficiently constant influence on the background. In simple cases it suffices to atomize the solvent alone, often distilled water. This simplification must be justified by test in each instance.

b. The working curve is prepared with standard solutions containing all essential constituents of the sample (Section 83), without measurement or subtraction of the flame background. Such a working curve will not pass through the origin of the coordinate system but shows a parallel displacement by the amount of the flame background, as in Fig. 52. The reading on the sample can then be at once compared with those on the standards, and the concentration deduced without prior subtraction. The same assumptions must be made as in (*a*) regarding simulation of the sample solution. Working curves prepared as above for other applications having different zero points cannot be used. The points of the working curve and the blank should be checked fre-

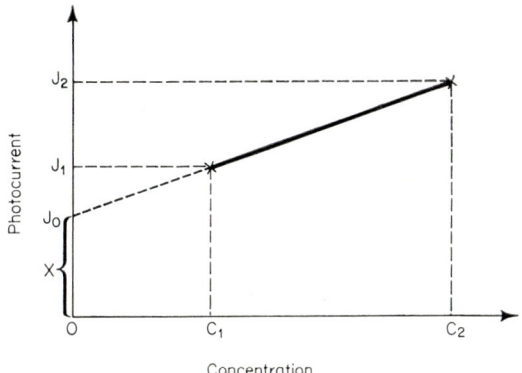

Fig. 53. Determination of an unknown flame background by extrapolation.

quently between determinations. This is very important, for changes in atomization due to obstruction of the atomizer, etc., have a larger effect on the analytical line than on the background, and therefore alter the ratio of line to background and displace the working curve (Fig. 53), thus throwing the result off.

c. The readings (photocurrents) are taken for two standard solutions of composition similar to that of the sample, containing different known concentrations of the analyte; let the standard of concentration c_1 yield photocurrent I_1, and the standard of concentration c_2 the photocurrent I_2. Then the background response x is obtained (Fig. 53) by extending the line through the two measured points to the left until it intersects the ordinate axis at I_0. Instead of using a graph, the background response x (distance from the origin O to I_0) can be found from

$$x = I_1 - c_1(I_2 - I_1)/(c_2 - c_1)$$

or

$$x = I_2 - c_2(I_2 - I_1)/(c_2 - c_1)$$

It has to be assumed here that the working curve is sufficiently linear in this interval; the assumptions made above also apply. By use of several calibration points, the method can be extended to nonlinear working curves.

It should be emphasized again that the background cannot be read on the empty flame, for the background emission changes on entry of a solution or solvent such as water, owing to its effect on flame temperature, etc. The background may be further changed if the sample contains concomitants that emit an appreciable continuum or bands. At all events, it is often possible in practice to make out with the background reading obtained by spraying the solvent alone, usually water. This reading is called the "water reading" or simply the *blank*.

3. The line intensity is comparable with the background response, and the latter may vary from one sample to another. This happens with measurements close to the detection limit when the sample contains substances such as the alkali metals, certain other metals, phosphoric acid, or organic substances, which raise the background by chemiluminescence, by changing the temperature or the concentration of a molecule such as CO_2, or by thermal radiation of unevaporated particles (Sections 8 and 11). The background then varies with the concentration of these substances. Besides these additive background effects, there may be more complex interferences which either raise or lower the individual contributions to the background, as described in Sections 97 and 98. Detailed investigations have not been reported.

Inadequate allowance for the background and its variation can lead to considerable analytical error in this case. Let us assume for simplicity that a monochromator (or spectrograph) is available. A few typical cases are shown in Fig. 54. In (*a*), the flame background needs to be measured only next to the line and subtracted from the total reading on line plus background. The measurement of the background close to the line usually includes part of the skirt of the line, which, strictly, should not be counted part of the background. But it has been shown[692] that subtraction of this analyte-specific part of the background does no harm, since it is proportional to the total emission of the analytical line. Hence measurement next to the line for background subtraction does not require complete spectral resolution.

In (*b*), the background varies more or less linearly with wavelength; in this case, it is measured at equal distances on both sides of the line and the mean of these two readings is subtracted from the total reading. In (*c*), this method is valid only if the slit can be reduced sufficiently to make the base width of the measured line small enough (Section 62). None of these methods is applicable to case (*d*), where by mischance a

§ 95] ELIMINATION OF THE BACKGROUND

band peak of the flame background coincides with the line. For this case the following procedure is recommended. Measure the blank at λ_0, the wavelength of interest, and by recording the spectrum of the background or by pointwise measurements at various wavelengths, locate a wavelength λ_1 where the blank reading is the same as at λ_0 or else larger or smaller by a constant factor. In the analysis, read the line plus background in the sample at λ_0 and the background at λ_1, multiplying it if necessary by the correction factor to find the background at λ_0; then subtract this value from the total reading. In this method, of course,

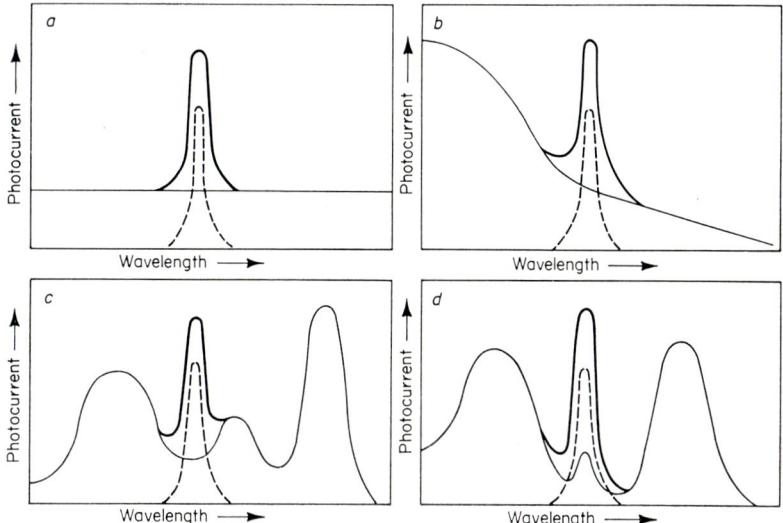

Fig. 54. The overlapping of a line and flame background (schematic). ——— Flame background alone; ---- line alone; ——— line plus background (sum). a, Line above a constant background; b, line above a smoothly sloping background; c, line above an irregular background; d, line superposed on a weak band.

the background measured with the blank must be affected by interferents in the same way as the background in the samples. This should be tested in each case. The procedure can be automated by suitable instrumentation.[160,457,458]

For completeness we should mention one other method, the slit-variation method. It was pointed out in Section 62 that the responses to a line and a continuum do not vary in the same way with the entrance and exit slit widths. If the total reading is plotted as a function of slit width, the contributions of line and background can be separately evaluated. The method is of use only when the background is sufficiently uniform in the immediate neighborhood of the line; but, owing to the band structure of the flame emission, this requirement is not often met.

A filter photometer does not generally permit setting the wavelength for a reading adjacent to the line, or varying the slit. Background effects are therefore more difficult to cope with, and poorer detection limits will usually have to be accepted. In favorable cases one can proceed as in 3, provided that a filter is available for the photometer that transmits only background at a suitable wavelength. This may be a filter intended for the determination of an element which does not happen to occur in the solutions.

A useful way of expressing the background intensity registered by a given instrument is to give the extrapolated concentration of analyte that yields the same photocurrent as the background itself. The background is thus expressed in equivalent concentrations.[361] These data are, of course, highly dependent upon the transmission characteristics of the optical system (filter transmittance curve, effective bandwidth, etc.) and upon the operating conditions (gas mixture, solvent, etc.).

We take up next the case in which both cross-sensitivity (Section 94) and variable background have to be considered, assuming first that the line intensity is high compared with the blank intensity (background plus cross-sensitivity) embraced within the range of transmission of the optics. It is then usually possible to work with an average blank value, obtained by the methods described for eliminating the background in cases 2a–2c; these methods are suitable for eliminating the blank (background plus cross-sensitivity) in the present case under the above assumption.

The situation is more difficult when the background, cross-sensitivity, and line or band intensity are all of similar magnitude. First thought should be given to the possibility of improving the spectral separation sufficiently to effect a substantial reduction of the blank, especially the cross-sensitivity. Recall the use of a blocking filter (Section 94) for suppressing cross-sensitivity. If this succeeds, the situation reduces to the simpler case described earlier. If not, the procedure described under 3 for coping with background can be used, background being now interpreted as blank. In the most difficult case, 3(d), the method described there may not eliminate all error, because the correction factor may not be the same for background as for cross-sensitivity; that is, if the interferent concentration changes, the wavelength dependence of the relative contributions of background and cross-sensitivity to the blank will change. Thus, three unknowns—line, background, and cross-sensitivity—have to be dealt with simultaneously. In order to find the true magnitude of the blank interference, which is to be deducted from the reading (line plus blank), the following procedure is suggested.

Take readings at three preselected wavelengths: λ_0, the wavelength of the analytical line or band, and λ_1 and λ_2, somewhat above and below, so chosen as to offer a large difference in the ratio B/C (background over cross-sensitivity). This can be judged by varying the interferent concentration or by means of blocking filters (see below). For instance, on approach to a strong line of an interferent, C is likely to rise faster than B. Then for determination of the net line intensity L, we have the three readings:

$$I_0 = L + B_0 + C_0 \quad \text{at } \lambda_0$$
$$I_1 = B_1 + C_1 \quad \text{at } \lambda_1$$
$$I_2 = B_2 + C_2 \quad \text{at } \lambda_2$$

If we make the usually valid assumption, as in Section 86, that the relative values of B at different wavelengths are independent of the interferent concentration, and if we make the same assumption for C, then

$$B_0/B_1 = b_{01}; \quad B_1/B_2 = b_{12};$$
$$C_0/C_1 = c_{01}; \quad C_1/C_2 = c_{12}$$

These ratios b and c can be found experimentally (see below). By inserting them in the first equations, we obtain a solution for L:

$$L = I_0 - \frac{b_{01}b_{12} - c_{01}c_{12}}{b_{12} - c_{12}} I_1 + b_{12}c_{12} \frac{b_{01} - c_{01}}{b_{12} - c_{12}} I_2$$

For practical calculation, the factors containing only b and c can be expressed as single numbers, giving a simple expression both for L and for the blank at λ_0. B and C at λ_1 and λ_2 can be found by reading their sums on spraying the blank solution. The interferent lines are then suppressed by one or more blocking filters to find B alone. Thence, the C, b, and c values can be found; they will be applicable to all subsequent analyses of the same kind under the given assumptions.

Many other methods for eliminating background have been devised (for example, refs. 206, 513 and 544), but they cannot be described here.

Special difficulties attend the elimination of background when it is infeasible to prepare a blank solution. An example is the determination of an impurity (lanthanum) in an element (praseodymium) which itself considerably raises the flame background by an unknown amount, but is unavailable in sufficient purity for preparation of a blank. In this case, the slope of the working curve can be found empirically by adding known increments of lanthanum, but the intersection of the curve with the ordinate axis (Fig. 53) remains unknown. In other words, a family of parallel working curves can be drawn without an easy way of deciding which is the right one. One sample should then be analyzed chemically to obtain one pair of coordinates (concentration, intensity) needed for fixing a point on the true working curve. Readings should be taken on this reference sample at intervals for purposes of calibration.

The work of numerical subtraction of the background, as above, perhaps coupled with subtraction of the cross-sensitivity, can be performed by a suitable automatic device. An example [457] was shown in Fig. 44, in which a large grating instrument is provided with a pair of slits for the determination of each element, one of the slits being set upon the line to be measured, and the other upon the background next to the line. By adjustment of the dynode voltages of the photomultipliers, their sensitivities are made equal. The amplified photocurrents from each pair of photomultipliers are then placed in opposition, so that the meter in the circuit registers only the difference, line plus background minus background, i.e., the line signal alone. This system permits quantitative measurements even when the line is only one-seventh as strong as the background.

Instead of these physical methods of eliminating background, a chemical method can be used. For example, a concomitant that raises the background to an exceptionally large and variable extent can be removed by precipitation.[512] The best chemical approach can be selected only in each case individually. In this connection consult Sections 88 and 100.

B. INTERFERENCES FROM CONCOMITANTS

96. *Introduction*

The blank interferences discussed in the last two sections consist only of an added, unwanted radiation. In Sections 97–101 we shall consider only those interferences (and methods for their elimination) in which the emission of the analyte itself is affected. Sections 97–100 cover radiation interferences that are specific for certain elements and concomitants (cf. Section 11), while Section 101 covers certain nonspecific interferences and methods of removing them.

For simplicity, we assume in Sections 97–101 that the filter or monochromator has sufficient resolution to avoid cross-sensitivity (Section

94), and that background interference is absent or eliminated. These simplifying assumptions should lead to no errors. In Section 97, for instance, we discuss the specific mutual radiation interferences between sodium and potassium, neglecting blank (background) effects. But, as is well known, sodium and potassium raise the background (Section 8), causing error or simulating additional radiation interference. In other words, not only the corrections of Section 97 but also those described in Section 95 must be applied to the result. The reader must keep this in mind throughout the following sections.

We also assume in Sections 97–100 that the properties of the atomizer, the separation of aerosol droplets, the gas and air flow to the flame, and the sensitivity of the optical–electric system remain constant. Thus we confine ourselves to those effects that are due to the presence of other substances in the solution, especially other elements (anions and cations). As in Section 11, we distinguish on the one hand specific interferences, among which ionization, dissociation, and vaporization interferences are more or less well known at present (Sections 97–100), and on the other hand nonspecific transport interferences (Section 101), including such effects as those of protein and alcohol on the surface tension and viscosity.

Specific radiation interferences due to change of flame temperature will not be discussed here. Those that are due to the presence of a combustible substance such as acetone have already been considered in Section 79. With indirect atomizers, noncombustible substances have usually little effect on the flame temperature; with direct atomizers, the effect can be significant but is not easily excluded (Section 7; cf. ref. 569a). For this reason we assume in the following that the flame temperature is constant.

In the strictest sense, transport interferences such as drop size interference cannot be separated from the more specific interferences or from the blank interferences. For one thing, they affect the processes in the flame and, through them, the flame background; for another, every change of concentration entails a change of drop size. But we cannot go into this matter in these sections (91–100).

Some aspects of the interelement effects and other interferences in the flame had been studied rather early.[125] In later years these results often went unheeded, perhaps because the low precision of the photographic methods of flame spectrography did not permit detection of these small effects. With improving precision additional interelement effects were continually discovered. We discuss below the errors caused by interelement effects (radiation interferences) and the means of circumventing them.

Specific radiation interference obstructs flame analysis worse than blank interference does, especially when high accuracy is needed for complex samples. Certainly by no means all radiation interferences are yet known; but simple tests will quickly detect specific radiation interference and even estimate its degree (Section 100). In this way, results obtained by a simple method can be corrected, but the procedure then becomes more complicated. The best-known examples of radiation interference will next be presented together with means of alleviation which apply as well to other cases as to those discussed specifically.

97. *Specific Radiation Interferences with the Alkali Metals*

Among the specific radiation interferences from concomitants discussed in Section 11, by far the most important is the mutual ionization interference of the alkali metals. Together with methods of correcting them, they will be considered in detail in the present section for the case of sodium and potassium.

Dissociation interferences from halides such as HCl or KCl can also be observed in the flame-photometric determination of the alkali metals, but usually only when the halide is present in considerable excess.[67,263] Vaporization interference, as from aluminum, phosphate, or sulfate, is much weaker with the alkali than the alkaline-earth metals (Section 98) and may be expected only at high concentrations of interferent.

For a constant potassium concentration and fixed experimental conditions (atomization, flame, etc.), the intensity of the potassium doublet at 767–770 mμ increases with the sodium content of the sample. This very common enhancement becomes distinctly noticeable in the hotter flames, such as air–acetylene, and at low analyte concentrations. Figs. 7 (page 55) and 55 show a set of potassium working curves for various sodium concentrations. The following facts are apparent:

1. For a fixed sodium concentration, the enhancement depends on the potassium concentration; therefore, it cannot be accounted for by an additive blank interference.

2. For a given sodium concentration, the relative enhancement is less for higher than for lower potassium concentrations.

3. For a given potassium concentration, the enhancement does not increase proportionally with the sodium concentration, but appears to approach a saturation value (Fig. 7).

4. In the presence of sodium, the concave working curve of potassium is straightened (Section 10).

Similar interelement effects can be observed with the other alkali metals, and they are the more marked, the lower the ionization potential of the metal is. All of these effects can be explained at least qualitatively by ionization, as shown in Sections 7 and 11.

As can be seen from Fig. 55, the ionization interference can be described in certain concentration ranges to a sufficiently good approximation by starting with an original working curve (or straight line) in

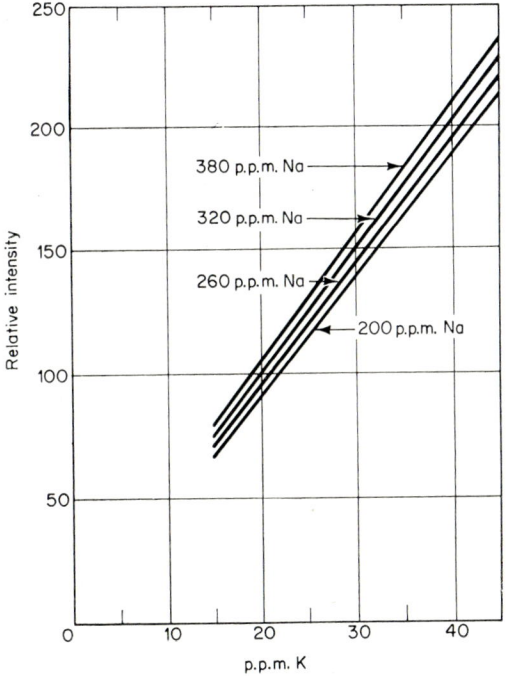

Fig. 55. Potassium working curves (767 mµ) in the presence of various concentrations of sodium in an air–acetylene flame (from ref. 661).

the usual linear coordinates, and multiplying the ordinates by a factor depending on the concentration of interferent. If the same information is transferred to logarithmic coordinates as in Fig. 50 (page 231) the working curve will have a different shape, and these factors will have the effect of displacing the curve parallel to itself. In the most general case, the interference can be described by two displacements of the curve, one parallel to the concentration axis and one parallel to the intensity axis ("log-log translation rule"). In ref. 274 the displacements are given for certain pairs of elements under certain experimental conditions. These displacement factors cannot, however, be applied to a case in which the flame temperature is different or several interferents are present together. Certain interferences, such as that of phosphoric acid on the alkaline earths, cannot be represented by parallel displacements of this kind.

If mutual ionization interference between two elements is suspected in a practical analysis, it can be tested as follows. Readings are taken on two solutions containing the elements separately at the appropriate concentrations, and on a third containing them together. If the readings on the mixture exceed those on the separate solutions (blanks being corrected for), then a mutual ionization interference is probably present. For the practical elimination of ionization interference, the following methods are available.

1. Use of an Optimal Flame. The enhancing effects of the alkali metals upon each other can be substantially reduced most simply by using a flame of low temperature, such as air–propane, in which ionization is hardly noticeable. But if less easily excited metals such as copper are to be determined at the same time, or if no cooler flame is available, the hotter flame will have to be used. Various schemes can then be employed for correcting the analytical result for the interference. These are discussed below for the case of sodium and potassium.

2. Parametric Method. As in Fig. 55, a series of standards is prepared with various sodium concentrations. These yield a family of potassium working curves with sodium concentration as parameter. Readings on both potassium and sodium are then taken, the latter at a sodium line along with sodium standards. The working curve corresponding to the sodium content found is then used for the estimation of potassium. If necessary, one can interpolate between two potassium working curves (two different sodium concentrations). This method is useful when the sodium concentration is likely to vary over a wide range. It is relevant that this divergent, fanlike array of potassium working curves is temperature-dependent; the fan spreads out with rising flame temperature. Moreover, in the practical use of these parametric families of curves and in some of the other methods of correction described below (except the simulation and normalization methods), the curves depend strongly on the ratio of fuel to oxygen and, with direct atomizers, also on the sample flow rate. They are therefore also shifted by partial occlusion of the atomizer, since alteration of any of these variables affects the flame temperature and hence the magnitude of the interferences, the coordinates of the parametric set of curves, and the

magnitude of the corrections to be applied. The divergence of the curves depends further on the nature of the anion.[675]

If high accuracy is desired and if the potassium concentration of the samples is quite variable, it becomes necessary in using the parametric method to take into account also the inverse effect of potassium on the sodium. In other words, it is necessary to allow for the effect of the measured potassium concentration upon the sodium before using this corrected sodium concentration for correcting the potassium reading. The true potassium concentration is thus found by successive approximations. Fortunately, in many applications (biological and medical), the effect of potassium on sodium is less important than that of sodium on potassium, and the corrections of higher order can be ignored.

3. Simulation Method. The sodium content of the sample is determined first, and a potassium standard is then prepared, containing the sodium concentration found. This method is useful when there are many samples whose sodium concentrations vary only slightly.

4. Correction Method. When the sodium varies somewhat more than in 3, a combination of 2 and 3 is recommended. A potassium working curve is established for an average sodium concentration with a suitable set of standards, and this curve is used directly. When the sodium content departs sufficiently from the mean, a correction is applied to the potassium values; the correction is taken from a graph established with the help of a series of potassium working curves with sodium concentrations varying over a small range.

5. Self-standardization. If the concomitants causing an approximately multiplicative ionization interference cannot be separately determined, or if they are too difficult to simulate in the standards, then the slope of the working curve, which is unknown because of the interference, can be found empirically by self-standardization, also called standard addition, which is essentially a recovery test.[275] A known quantity of the analyte is added to the sample, and the increase in the reading gives the slope of the working curve, at the particular concentration tested, in the presence of the interference. The test is carried out similarly with samples containing the analyte at other concentrations. The working curve can thus be approximated piecemeal by line segments, provided that samples with very little analyte are available. The method works only if the concomitants are present in the same concentrations or at least have the same effect on the analyte in all the samples. If it can be shown by preliminary trials that the multiplicative interference factor is the same at all concentrations of the analyte, then a single determination of the slope, coupled with knowledge of the working curve in the absence of interference, will suffice to establish

the working curve in the presence of the interference. With additional slope determinations, the method can be refined into the so-called double-coefficient method.[276] It must be made clear that this and similar corrective methods are inapplicable to determinations of calcium and other alkaline-earth metals when, as is common with these elements, the interference cannot be represented by a multiplicative factor (see Section 98).

6. Normalization and Saturation Methods. *a.* Normalization. If the range of concentration of the interferent is moderate, ionization interference can be largely eliminated by raising the concentration of free electrons in the flame to a higher, practically constant, *normalized* level. This can be done by adding a suitable concentration of another element, a *normalizer*, such as cesium[263] or barium as chloride;[332] if the ionization interference amounts to only a few percent, it suffices to raise the electron concentration about 10-fold. Even sodium can then be used as normalizer.

It has been found that other elements at sufficiently high concentration, such as aluminum and especially lead, can lessen the mutual interference between sodium and potassium[71,691] Even for these additives, the effect can be explained by a rise of electron density due to the added metal, which ionizes usually to a much greater degree than what corresponds to equilibrium.[71,697]

b. Saturation. If, however, the concentration of interferent (say, sodium) varies more widely, normalization must be carried farther, in the limiting case to the point where ionization of the analyte (potassium) is largely suppressed. The sodium concentration, however great, will then no longer affect the potassium emission. For this reason the method is called *saturation*.

Experimentally, the onset of saturation can be readily detected by plotting the enhancement of the potassium emission as a function of the quantity of added sodium or cesium. The potassium emission approaches a saturation value asymptotically, blank interferences being ignored or corrected in accordance with our earlier premises (cf. Sections 94 and 95).

However, saturation as a method of eliminating ionization interference—specifically, by adding a large excess of sodium—has several drawbacks: (*i*) Much sodium salt is usually needed for saturation. This easily leads to incrustation of the atomizer and burner top. (*ii*) If the analyte is at low concentration, error will result if it occurs as impurity in the added sodium salt, and blank interference (cross-sensitivity and background enhancement) causes trouble. (*iii*) The air is rapidly contaminated with sodium, making it hard to conduct subsequent sodium determinations in the same room. (*iv*) When sodium is used for normalization or saturation, sodium cannot be determined in the same sample.

A similar method of normalization can be used also for other kinds of interference (Section 101).

7. *Optimization of the Concentration.* The sample is diluted or concentrated. The sodium–potassium interference in a given flame can sometimes be markedly reduced by choice of the proper concentration. Ionization and ionization interference diminish with rising concentration of the analyte in the flame. However, there are practical limits to the concentration, and certain other errors are enhanced at high concentration (Section 101).

If the flame gases themselves supply an appreciable concentration of electrons (Section 7), ionization interference with the alkali metals can be reduced also by diluting the solution. The electrons from the flame gases then dominate those from the metals and effectively normalize them (method 6*a*). The favorable effect of dilution upon mutual interferences among the alkali metals reported in several papers[640] can be explained in this way.[65] There are also limits to dilution; the detection limits set one, and these depend on the instrumental conditions (sensitivity, resolution, etc.).

8. *Optimization of the Height of Observation.* The dependence of flame temperature and flame electron concentration on the height above the burner results in a dependence of the ionization interference on height. The optimal height of observation is determined experimentally in each case by measuring the magnitude of the interference at various heights to find the height at which it is least.

In conclusion, we should mention that internal standardization (Section 51) can mitigate but, in principle, never fully compensate ionization interference. The residual error will be least when the ionization potential of the internal-standard element is closest to that of the analyte, as when the elements are neighbors in the same group of the periodic table. The internal-standard element, moreover, actually reduces the interference owing to its normalizing effect. Similarly, excitation interference (Section 11) is mitigated by internal standardization most effectively when the excitation potentials of the analytical and internal-standard lines are closest. Incidentally, all of the above procedures for the relief of ionization interference, described for direct flame photometry, are applicable to the residual errors of internal-standard flame photometry.

It must be realized, in applying these various methods for correcting interferences, that the difficulties increase rapidly with the degree of accuracy demanded. For example, in the parametric method repeated approximations are necessary for high accuracy. In practice, therefore, one should consider the possibility of making out with less accuracy, before undertaking the troublesome task of compensating interferences.

98. *Specific Radiation Interferences with the Alkaline-earth Metals*

The flame lines of the alkaline-earth metals and the bands of the alkaline-earth molecules commonly used for measurement are more susceptible than those of the alkali metals to interference from anions and various cations such as aluminum and iron. On the other hand, the mutual interferences of the alkaline earths are generally considerably less severe than those of the alkali metals, especially in hot flames. This is due to the higher ionization potentials of the alkaline-earth metals. Interference from organic substances is generally not substantially greater with the alkaline-earth than with the alkali metals.[182, 202, 274]

In this section we discuss the strong, specific depression of the alkaline-earth emission by acids such as sulfuric and phosphoric and by a few elements such as aluminum. Calcium, which has been carefully investigated, will be used as example; but the discussion is probably valid for the other alkaline earths. Following a description of the typical depressing effects, we present an explanation of them, which has been worked out only very recently. After that, we list methods of eliminating these interferences in flame analysis. Some of these methods are more widely applicable, and can be used for other types of interference.

(*a*) *The effect of phosphate on calcium.* Fig. 56 shows the concentration-dependent effects of several anions on the CaOH band at 554 mμ at constant calcium concentration. These data are for an oxyacetylene atomizer-burner.[83] Phosphoric, sulfuric, and arsenic acids cause a strong, characteristic depression. This depression of calcium by phosphorus, always in much the same form, has been observed also by many others[37, 70, 229, 274, 276a, 400, 550, 582a, 637] under various experimental conditions; the depression by phosphorus occurs also with the other alkaline earths and with the lines as well as the bands.

As seen in Fig. 56, the interference by phosphorus (and sulfur) is linear with interferent concentration up to where the atomic ratio of interferent to calcium is roughly 1. Beyond this point there is no further effect (see also Fig. 57). There is thus a constant final (plateau) value for the interference of phosphorus with calcium. Interestingly, the P/Ca ratio at the bend point or knee is generally independent of the calcium concentration and the height of observation in the flame, although the relative level of the plateau depends on these factors (Fig. 57). With a given atomizer, the depression diminishes with rising flame temperature, while the P/Ca ratio at the knee remains unaffected (cf. refs. 70, 258, 637). Still, different authors give somewhat different values for the P/Ca ratio at the knee (from 0.38 to 1.0).

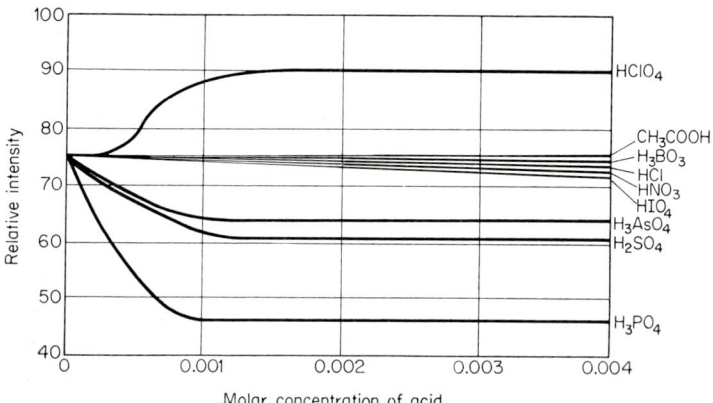

Fig. 56. Effect of various acids as function of concentration upon the emission of calcium at 554 mμ at a constant calcium concentration of 0.001 M (CaCl$_2$); oxyacetylene flame with an atomizer-burner (from ref. 83).

The increase of the relative depression of calcium by phosphorus with rising calcium concentration, for a given height of observation, as seen in Fig. 57, has been observed also by other authors.[83, 274, 400, 450, 637] It follows that the calcium working curves should show a more or less marked convexity in the presence of phosphorus; this has been actually observed.[70] Hence, the depression of calcium by phosphorus cannot in general be represented or corrected by a constant depression factor independent of the calcium concentration.

There is evidence that in the turbulent flames of atomizer-burners the phosphorus-calcium interference curve is less clear cut and more rounded than in laminar flames with indirect atomizers. An actual plateau may then be lacking.[458, 582a] The calcium depression in an atomizer-burner flame is also generally distinctly stronger than with an indirect atomizer, even when the atomizer-burner flame is the hotter.[258, 513, 637]

If a series of solutions is prepared with constant calcium and sulfate concentrations but increasing phosphate, there will be, at sufficiently high phosphate content, a plateau whose level is independent of the sulfate concentration;[83] the same is true for constant phosphate and varying sulfate. Since the phosphate effect on calcium exceeds the sulfate effect, the more marked depression of the calcium reading by phosphate can be somewhat diminished by a high concentration of sulfate. It thus appears that the depressive effects of sulfate and phosphate are neither additive nor multiplicative with respect to each other. If either sulfate or phosphate is added in excess, the effect of the other is suppressed.

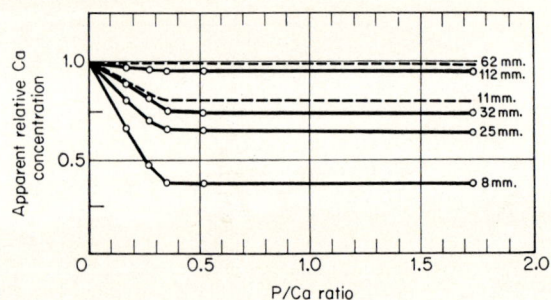

Fig. 57. Apparent calcium concentration as function of the atomic ratio P/Ca; the parameter is the height of observation above the inner cone in the laminar air–acetylene flame, with indirect atomization (from ref. 70). —— 8.3 meq./l. Ca; ---- 0.55 meq./l. Ca.

At very high phosphate concentrations—rare in practice—the calcium reading can rise markedly and even considerably exceed the reading in the absence of phosphate (cf. ref. 229). This enhancement can be ascribed to the incandescence of incompletely vaporized particles of phosphorus pentoxide or the like.[228] This effect is unrelated to the depression (cf. also ref. 276a), and will be ignored in the following. (In another investigation,[719] however, no increase of the flame background by phosphorus in the vicinity of the Ca 423 mµ line was observed.) This enhancement by high phosphate concentrations does not occur in the presence of alkali metals;[229] this is not necessarily at variance with the above observation,[228] since the rise of background might pass undetected in the presence of the relatively more intense background emission of the alkali metal.

This depression by phosphorus is characteristic of all the alkaline earths. It is usually absent or weak with the alkali metals.[228,229,274,289,458]

(b) *The effect of aluminum on calcium.* The interferences of aluminum salts with the alkaline-earth emission are related to those of phosphate and sulfate.[37,68,183,253,276a,633] The aluminum interference, like the phosphate, is much stronger with alkaline-earth than with alkali metals (cf. refs. 156, 633). Iron, chromium, silicon, beryllium, molybdenum, and tungsten can have a similar effect on the alkaline-earth emission (cf. refs. 253, 263, 453, 637, 645).

There is a distinct difference between the effect of aluminum nitrate on the one hand and that of aluminum halides on the other. With the nitrate, the calcium emission on the plateau is low or very low compared with the undepressed emission. This remains true at greater heights of observation in the flame. But with an excess of aluminum chloride, the calcium emission is significantly higher, especially at

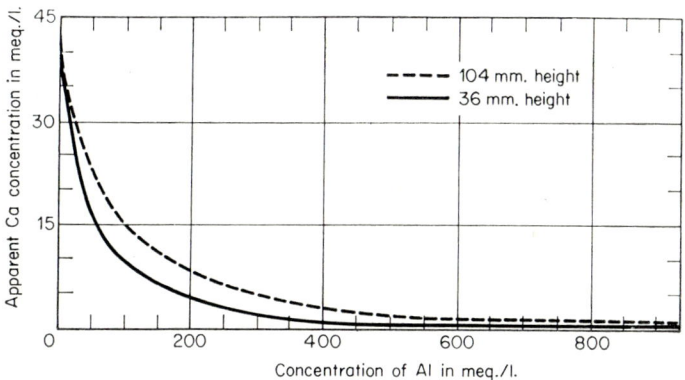

Fig. 58. Apparent calcium concentration as function of the aluminum concentration for two different heights of observation in the laminar air–acetylene flame with indirect atomization. The Ca concentration was 41.7 meq./l. (from ref. 70).

greater heights (cf. refs. 59, 183, 370a, 582a, 637, 645). As in the phosphate effect, both nitrate and chloride of aluminum cause the initial rapid, linear fall of calcium emission with rising aluminum concentration (see also the discussion of indirect methods for determining phosphorus and aluminum in Section 87 (b).) As seen in Fig. 58, the interference curve for aluminum nitrate lacks the sharp knee.

When the molar calcium concentration sufficiently exceeds the molar aluminum concentration (as nitrate), then for fixed aluminum concentration the apparent absolute loss of calcium is practically independent of the calcium concentration itself, while with rising aluminum but fixed calcium concentration, the apparent loss of calcium increases proportionally.[68] Hence, the molar calcium loss per mole of aluminum appears constant and is roughly unity (compare the P/Ca ratio at the knee of the phosphate–calcium interference curve). In the presence of a constant but relatively low concentration of aluminum, the calcium working curve runs practically parallel to the curve in the absence of aluminum.[68] As in the phosphate effect, then, the depression of calcium by aluminum cannot be described by a constant depression factor independent of the calcium concentration. With both phosphorus and aluminum, the magnitude of the interference is the same for a line (423 mµ) as for a band (623 mµ).[276a, 637]

(c) *Interpretation.* The enhancing effect of perchloric acid on calcium (Fig. 56), which is absent with alkali metals, has not been explained, so far as we know (cf. refs. 228, 274).* But the characteristic depression of calcium and similar elements by phosphorus and aluminum may now be regarded as pretty well accounted for. We take up the phosphorus effect first.

The constant P/Ca ratio at the knee led many authors[37, 458, 572, 680] to the hypothesis that the depression must be related to the formation of some non-emitting complex between calcium and phosphorus having a constant atomic ratio. The observed ratios suggest compounds such as $Ca_2P_2O_7$, $Ca_3(PO_4)_2$, or Ca_3P_2 (cf. refs. 83, 680). The formation of such complexes should diminish the concentration of calcium atoms and molecules capable of emitting in the flame.

The question then arises whether the P–Ca complex occurs in the flame as vapor or in solid or liquid form. Solid particles might form from the aerosol entering the flame under the action of the flame heat. If the complex vaporizes, it must be assumed to have a high dissociation energy. If a condensed phase persists, involatility of the complex may account for the loss of free calcium species.[458]

A first clue to the solution of this problem is offered by the observation that no calcium depression occurs when calcium and excess phosphoric acid are fed to the flame simultaneously by two separate, parallel atomizers.[70, 276a, 562a, 637] According to the latest views, these phenomena can be interpreted by the following assumptions, presented somewhat schematically, regarding the interference of phosphorus on calcium in the absence of other concomitants. It should be kept in mind, meanwhile, that a greater height in the laminar flame corresponds to a longer residence time of the particles in the flame. Following ref. 70, we distinguish three separate phases:

First Phase. Before entering the flame, each aerosol droplet yields a

* J. H. Gibson, W. E. L. Grossman and W. D. Cooke (International Conference on Spectroscopy, University of Maryland, June, 1962) find that the enhancement of calcium by perchloric acid is due entirely to an increase in concentration of calcium atoms, and suggest that the evaporating droplets, containing concentrated perchloric acid at the last moment, explode and thus improve the efficiency of atomization.—*Translator.*

solid particle, in which calcium is combined with phosphorus in a constant atomic ratio. If enough phosphorus is present, practically all the calcium is combined with it.

Second Phase. Assume phosphorus to be present in excess. In the hot flame the excess phosphorus (the part not combined with the calcium) evaporates promptly; that is, it is stripped from the particle low in the flame. The particle then contains only calcium combined with phosphorus, probably in association with oxygen. The size of the particle depends on the calcium concentration but not on the phosphorus concentration of the initial solution. If calcium had been present in excess, the converse situation might prevail.

Third Phase. These Ca–P(–O) particles should evaporate completely at the temperature of the flame. But owing to the low volatility of the material of the particle (cf. also ref. 458), the aerosol persists for some time in the flame, steadily giving off calcium by evaporation. The free calcium content of the flame hence increases with the height of observation, i.e., with residence time in the flame. Since the phosphate effect is absent in the experiment with dual atomizers, it must be assumed that in the vapor phase the calcium is not combined with the phosphorus. Accordingly, if the phosphorus and calcium are sprayed together with a single atomizer, all the calcium should be liberated at sufficient height in the flame. The rate of release of calcium depends on the rate of evaporation of the Ca–P particle. This in turn depends on the size of the original particle and on its average temperature[637] while in the flame.

These hypotheses now easily explain the knee in the phosphorus–calcium interference curve and the plateau different from zero. The atomic ratio at the knee corresponds to the ratio in the Ca–P complex. The plateau beyond the knee (at higher phosphorus concentrations) is due to the fact that the excess phosphorus is quickly stripped from the particle, so that it has no further effect on the rate of evaporation of the calcium from the Ca–P(–O) particle higher in the flame. The height and temperature effects noted above are accounted for by the limited volatility of the particle. It also becomes understandable why, at constant height of observation, the relative depression of calcium by phosphorus is smaller at lower calcium concentration in the solution; in the presence of excess phosphorus, when the calcium concentration is lower, the evaporation rate is *relatively* higher, since the Ca–P particles are then smaller.*

* The *absolute* evaporation rate should be proportional to the surface of the particle. For a sphere, it varies as the square of the radius, while the mass varies as the cube, i.e., faster. The increase of vapor pressure with diminishing particle radius can also help the evaporation rate.

These hypotheses can also explain the phenomena that appear on treating a phosphorus-containing calcium solution with sulfuric acid or another alkaline-earth metal as releasing agent (see below). Apparently the sulfate displaces the phosphate from combination with the calcium, forming an involatile (sulfur) compound with the calcium. Similarly, in the presence of a large amount of another alkaline earth such as strontium, most of the phosphorus combines with it, leaving very little free to combine with the calcium.

These considerations show also that the depression of calcium by phosphorus should depend on the drop size distribution of the atomizer. If the spray is very fine, then for a given calcium concentration the Ca–P(–O) particles in the flame are correspondingly small. The vaporization rate and the rate of release of calcium are then relatively high. Hence, for a given height of observation, the calcium emission is relatively higher (referred to the calcium emission without phosphorus) than with a coarser spray. This accounts for the observation with commercial flame photometers[258,513] that the phosphate effect on calcium can be diminished by inserting a barrier surface in the spray chamber, which removes the larger droplets. Our hypothesis also explains the fact, mentioned earlier, that the depression of calcium by phosphorus is stronger in a flame with a direct than with an indirect atomizer, even when the former is hotter (cf. ref. 680). This difference can be accounted for in terms of the mean temperature reached by the Ca–P particles in the flames commonly used with these two types of atomizers:[637] with the direct atomizer the residence time of the particles is much shorter owing to the higher speed of emergence of the flame gases.

Since the interference does not occur unless calcium and phosphorus are present jointly in the same aerosol drop, phosphine as impurity in commercial acetylene should not affect the calcium emission.

The observations on the depression of calcium by aluminum, described above, point similarly to the formation of an involatile Al–Ca(–O) complex. As in the case of phosphorus, the effect of aluminum on calcium disappears when the two elements are supplied to the flame through separate atomizers (refs. 68, 276a, 562a, 637, and the discussion of ref. 59). The hypothesis is further supported by the dependence of the interference on height[68,637] and by evidence for the low volatility of calcium aluminate complexes.[458]

The difference between the chloride and nitrate of aluminum in the magnitude of their interferences, mentioned earlier, can be accounted for by the stronger oxidizing power of the nitrate ion and the low volatility of aluminum oxide. When the atomic concentration of aluminum exceeds that of calcium, the excess aluminum not bound to calcium is strongly oxidized. Unlike the much less stable phosphorus oxide, the aluminum oxide vaporizes only slightly at the temperature of the flame, so that the Al–Ca compound is presumably encapsulated in solid aluminum oxide, which impedes the evaporation of the Al–Ca compound and the release of calcium in the flame; this effect increases with the relative excess of aluminum and with the particle size (i.e., with the absolute concentration of aluminum in the solution). In fact, a sufficient excess of aluminum will wholly suppress the calcium emission. The shape of the aluminum–calcium interference curves and the effect of height in the flame (Fig. 58) can be qualitatively interpreted by these considerations.

The absence of a sharp knee in the aluminum nitrate interference curve makes it hard to judge the atomic Al/Ca ratio in the complex. The ratio might be deduced from the slope of the initially linear drop in the interference curve, which, however, depends on the height of observation because of the evaporation of the complex. The slope might permit a decision among $CaAl_2O_4$, $Ca_3Al_2O_6$, etc.[458]

This presentation still leaves unanswered several questions regarding the interference of phosphorus, aluminum, etc., on calcium. In particular, the compositions of the assumed complexes have not been unequivocally ascertained. Also, the variation of the P/Ca and Al/Ca ratios as reported by different observers calls for further investigation. Observations of a more direct nature than the optical ones used hitherto are needed to elucidate these matters.

We should mention that a divergent hypothesis for the interference of aluminum, phosphorus, and sulfur on the alkaline earths has recently been proposed.[582a] On the basis of observations of the amorphous structure of the solid particles formed during the very brief period of evaporation of the mist droplets in the flame, delayed evaporation of the Ca–Al particles, as assumed in the hypotheses described above, is not believed critical. For instance, the effect of phosphorus on calcium is thought to be related to the kinetics of formation of CaO from the calcium (meta)phosphate molecule in the *gas* phase. Owing to the brief residence time of the analyte in the flame, the comparatively slow formation of oxide remains incomplete, accounting for the observed depression. This hypothesis is actually capable of explaining the results of the dual-atomizer experiment and the height dependence of the depression. But it does not seem able to account for the effects of the fineness of atomization and the calcium concentration in solution (in the presence of excess phosphorus) on the magnitude of the depression. To explain these phenomena, additional hypotheses would be required (cf. ref. 582a).

11+C.A.F.P.

(*d*) *Elimination of the interferences.* We return now to practical flame photometry and to the question of removing or correcting the phosphorus and aluminum interferences. The methods to be considered are applicable also to the other alkaline-earth metals and to other acids (such as sulfuric) or other interfering metals (such as iron). In the discussion, we show first how to minimize the depression of the calcium light by proper choice of the instrumental and experimental factors. Secondly, we give methods of eliminating the errors in the determination of calcium that are due to any residual depression, and the errors that exist under fixed instrumental conditions.

1. The depression of calcium by phosphorus can be diminished, sometimes eliminated, by the following means:

(*i*) Higher dilution of the phosphorus-containing calcium solution (see Fig. 57).
(*ii*) Observation at greater height in the flame (Fig. 57; cf. ref. 276a).
(*iii*) Use of a hotter flame (cf. refs. 70, 258, 637). If, contrary to the recommendations of Section 19, the flame is operated below its temperature maximum or off the emission maximum of the analyte, the interference from phosphorus, aluminum, etc., is greater than on the maximum. Proper adjustment of the gas/oxygen ratio will then reduce the interference. We emphasize that the corrective methods given below are valid only when the gas/oxygen ratio is the same for the analyses as for the preliminary standardization. Otherwise, the interferences and the corrections determined in the standardization will not be correct in the analyses.[350a]
(*iv*) Refinement of the spray, as by means of a throttle in the atomizer capillary, higher pressure, a finer atomizer, removal of the larger drops by a suitable spray chamber with barrier surfaces, a longer tube between spray chamber and burner, etc. Direct atomizers are disadvantageous both because of their limited possibilities in this regard, and because the mean residence time is shorter and the mean temperature attained by the particles in the flame is apparently lower than with indirect atomizers.

Methods (*iii*) and (*iv*) can be recommended also for diminishing the depression of calcium by aluminum (cf. ref. 68, 513, 637, 692). In the presence of aluminum nitrate, (*ii*) is also applicable, but with only moderate success (Fig. 58; cf. ref. 276a). Method (*i*) also seems useful for aluminum.[692]

If, as is to be expected, equilibrium exists between calcium atoms and calcium-containing molecules in the flame, the calcium depression should be the same for lines and bands,[370a,637,754] provided the depression is calculated in terms of equivalent loss of calcium *concentration*. Owing to self-absorption with atomic lines at higher concentrations (cf. ref. 361), the depression, expressed as a loss of *emission*, may be smaller for the line. Hence, the interference of phosphorus or aluminum with calcium cannot be substantially improved by choice of a different emission.

2. We next describe several methods, some of which have a wider applicability, for eliminating the analytical error due to the depression of calcium by phosphorus or aluminum, for given instrumental conditions and sample concentration.

The Parallel Method. This is a saturation method in which the constancy of the interference of phosphorus with calcium within a certain range of concentration is put to practical use by adding a sufficient quantity of the interferent, e.g., in the form of $(NH_4)_2HPO_4$, to put the phosphate concentration into this range. The situation is then represented by the horizontal part of the interference curve.[33,130,450,616] This method is distinguished from the saturation method of Section 97 in that relatively little material is added. It is successful, of course, only when the interference curve has a horizontal portion. The effects of sulfate and aluminum chloride but not aluminum nitrate are of this kind; see Figs. 56 and 58. The interference of zirconium with barium can also be eliminated in this way.[253]

Releasing Agents. As shown above, the effect of phosphorus on calcium can be diminished or entirely suppressed by adding an element, such as strontium or magnesium, which is also interfered with by phosphorus. The calcium is thus "released" from the interference by the releasing agent (strontium, etc.). The method is also called *buffering*: the phosphate interference is said to be buffered by the added element, called a buffer. Conversely, in the determination of strontium, calcium salts can be used as releasing agent against the phosphate interference. Interference by aluminum and silicon can also be well buffered by strontium chloride.[637] Anions can also serve as releasing agents. Iron and especially lanthanum, which emits no interfering light, are effective for releasing calcium from interference by phosphorus and sulfur.[754]

Naturally, the materials added must contain none of the element sought in appreciable concentration. One should also watch for interferences from lines or bands of the buffer; the strontium bands can interfere with the orange calcium bands, in which case a different line or band of calcium, such as 423 or 554 mμ, should be used. Further, especially when working close to a detection limit, one should select a releasing agent that raises the flame background little if at all.

The Comparison Method. The following procedure can be used instead of the buffer method. A different alkaline-earth element, such as strontium, is added in known concentration, and the effect of the interferent on the strontium is employed to correct the calcium value correspondingly. The principle is like that of internal standardization (lithium would not, of course, be used here). The buffering action reduces the interference with each of the two elements. The concentrations of both alkaline earths must lie in the steep, linear part of the interference curve, and other interferents must be absent. See ref. 645.

Several of the methods discussed in Sections 97 and 100 can be used for eliminating these interferences. In particular, the interferent is often removed chemically. In our opinion, chemical methods are worth considering only if the analysis does not have to be repeated too often (in view of the total labor involved in the chemical manipulations) and if the separation (precipitation, etc.) is simple to carry out and does not upset the subsequent flame-photometric determination.

Recent investigations have shown that interferences with the alkaline-earth elements, especially those of phosphorus and sulfur, can be largely suppressed by addition of ethylenediaminetetraacetic acid. This method seems notable for its simplicity.[213, 350, 453, 454, 748]

99. *Catalog of Specific Interferences*

For compendious presentation it would be useful to have a table listing all specific interferences in order of their magnitude. This is infeasible at present because the subject is very large, incompletely investigated, and afflicted with too many variables, especially instrumental variables. If we consider only the 50 or so elements excitable in the flame, there are basically about 50^2 or 2500 possible binary interelement effects. But we must add interferences from other elements such as chlorine that are not excited, from radicals such as sulfate and phosphate, from organic materials such as protein and urea, etc. There is a further complication in that the interferences of several components acting together are not additive or multiplicative but may behave far less simply.[83] These effects depend also on the flame temperature and hence upon the gases employed, their mixing ratio, the solvent, the concentrations, etc. Any one developing a new method should, therefore, himself investigate the magnitude of these effects in his particular case (Section 100).

§ 99]   CATALOG OF SPECIFIC INTERFERENCES   311

In the interest of standardizing the analytical procedures, a standardization of instrument models and a high degree of interchangeability among individual instruments of a given make would be very desirable, so that results obtained with one instrument would be valid for another. But despite the unfavorable current situation in this regard, it is already possible to derive from published work indications as to the interferences to be expected in a given case. The qualitative results published on specific interferences are assembled in Table 12; we have included no quantitative data, because the variables named above usually prevent such data from being transferred from one instrument to another. This catalog is presented with reservations, since in some cases we could not clearly establish that the reported interference was specific and not due to a cross-sensitivity inherent in the instrument (Section 94), insufficient compensation of the background (Section 95), or some other effect such as differences of viscosity or surface tension (Section 101).

Table 12 includes only a selection of the published information, primarily the more recent work. The list is confined to cases in which only one analyte and one interferent are involved. Interferences in ternary and higher systems are not included. To save space, interferents of different chemical composition but similar action are listed together; for example, the interferents P_2O_5, H_3PO_4, $(NH_4)_2HPO_4$, etc., are listed as P; HCl, NH_4Cl, etc., are listed as Cl. The many alcohols and other organic compounds are not listed separately, but are comprised under "org.".

TABLE 12. Index of specific interferences

The analyte and interferent are listed alphabetically by symbol. Legend: d, depression of the emission of the analyte by the interferent; e, enhancement. A combination, e.g., dde, means that in general (e.g., at low and medium interferent concentrations) there is a depression, but sometimes (e.g., at higher concentrations) an enhancement. A dash — means that under the experimental conditions there was no marked interference; a blank signifies that no unambiguous relation could be established or that none could be deduced from the publication. Parentheses () denote an insignificant effect; org. = organic substances such as alcohol.

Table 12

Analyte	Interferent	Interference	Reference	Analyte	Interferent	Interference	Reference
Ag	Cl	d	228	B	F	e	431b
	ClO₄	d	228, 582		Fe	e	206
	Mg		587		K	d	206
	NO₃	d	228, 582		Li	d	206
	Na	e	587		Mg	—	206
	P	e	228		Mn	e	206
	S	e	228, 582		NO₃	e	206, 228
	org.	eed	228, 582, 587		Na	e	206
Al	Ca	d	431b		Ni	e	206, 268
	Cl	e	431b		P	e	206, 228
	ClO₄	d	431b		Pb	(e)	206
	F	ee	431b		S	de	206, 228, 580
	K		334		Sr	—	206
	Na		334		Zn	d	206
	P	dd	431b		org.	eed	206, 228, 431a, 580
	S	dd	431b	Ba	Al	d	253, 274, 633, 647
	org.	eed	219a, 252, 333, 431a		B		274
B	Al	e	206		CO₃		581
	Br		206		Ca	e	274, 636, 704
	Ca	d	206		Cl	—d	228, 274, 647
	Cd	—	206		ClO₄	—d	228, 647
	Cl	d	228, 580		Cu	e	253
	ClO₄	e	228, 580		F	e	581
	Co	e	206		Fe	de	253, 647
	Cs	e	206		K	e	647
	Cr	e	206		Mg	e	274, 647
	Cu	e	206		Mn	e	647
					NO₃	d	228

§ 99] CATALOG OF SPECIFIC INTERFERENCES

Analyte	Interferent	Interference	Reference
Ba	Na	e	274, 647
	Ni	e	253
	P	dde	228, 274, 704
	Rb		268b
	S	d	228
	SiF$_6$	e	581
	Sr		274, 636, 647
	Zr	d	253
	org.	eed	228, 274, 647
Ca	Al	d	68, 78, 183, 217, 274, 276a, 370a, 433, 501, 513, 544, 550, 551, 616, 632, 633, 637, 645
	As	d	83, 452, 644
	B	e	83, 274, 276a
	Ba	d	274, 632
	Be		452
	CO$_3$		369
	Cd	e	632
	Cl	dde	83, 189, 228, 239, 274, 369, 616, 680, 703
	ClO$_4$	eed	83, 228, 274, 688, 703
	Cr	d	632, 645
	Fe	d	183, 294, 433, 544, 551, 616, 645
	IO$_4$	d	83
	K	d	189, 430, 433, 550, 616, 632, 680, 731

Analyte	Interferent	Interference	Reference
Ca	Li	d	632
	Mg	ed	189, 274, 433, 543, 616, 632, 688
	Mn	d	433, 513, 550, 632
	Mo		452
	NO$_3$	d	83, 90, 228, 239, 274, 369, 616, 680, 703
	Na	d	189, 274, 430, 433, 616, 632, 680, 731
	P	dde	37, 61, 70, 78, 83, 228, 229, 239, 274, 276a, 294, 369, 400, 419, 430, 433, 435, 450, 452, 513, 544, 550, 582a, 616, 637, 680, 686, 704, 754
	Rb		268b
	S	d	61, 83, 189, 228, 239, 274, 276a, 369, 452, 510, 613, 616, 680
	Se		452
	Si	d	224
	SiO$_2$		637
	Sr	e	189, 274, 544, 637, 688
	Te	d	452
	Ti	d	452, 544, 765
	TiF$_6$	d	452
	V		452
	W		452

Analyte	Interferent	Interference	Reference
Ca	Zn	d	632
	Zr	d	253, 452
	org.	eed	83, 228, 259, 274, 369, 653, 703
Co	Fe	d	174
	S	e	174
Cr	Al	d	544
	Cl	e	228
	ClO_4	d	228
	Fe	d	174, 544
	NO_3	d	228
	P	e	228, 544
	S	e	174, 228
	org.	eed	228, 569a
Cs	Ca		274
	Cl		274
	K	e	274, 572
	Na		274
	Rb	e	572, 575
	S		274
	org.	eed	274
Cu	Al	e	146, 207
	Ba	d	146
	Ca		146
	Cd	—d	146, 207
	Cl	dde	146, 228, 574, 579

Analyte	Interferent	Interference	Reference
Cu	ClO_4	ed	146, 228, 574, 579
	Co	d	146, 207
	Cr	—e	146, 207
	Cs	e	207
	Fe	e	146, 207
	K	—e	146, 207
	Mg	edd	146, 207
	Mn	—d	146, 207
	NO_3	de	146, 228
	Na	—d	146, 207, 345
	Ni	e	146, 207
	P	e	228
	Pb	—	207
	S	e	146, 207, 228, 574, 579
	Sn	dee	146
	Zn	—e	207
	org.	eed	145, 146, 228, 574, 579
Fe	Ca		460
	Na		460
Ga	Cu	e	487
	org.	eed	146a
In	Al		453a
	Mo		453a
	P	d	453a
	S	d	453a, 485
	Te	d	453a

Analyte	Interferent	Interference	Reference
In	V	eed	453a
	org.		146a
K	Al	d	335, 537, 632, 716, 742, 760
	B		716
	Ba	de	537, 632, 760
	Be	d	506
	Br	d	197
	CO$_3$	e	289
	Ca	de	274, 289, 537, 632, 706, 731, 742, 760
	Cd	d	632
	Cl	d	197, 214, 228, 274, 572, 589, 742
	ClO$_4$	e	228, 274, 572
	Co	d	537
	Cr	d	632
	Cs	e	537, 572
	Cu	d	537
	Fe	d	335, 537, 716, 742
	Li	d	214, 335, 537, 632, 742
	Mg	d	289, 537, 632, 716, 742
	Mn	d	537, 550, 632, 760
	Mo	d	537
	NO$_3$	—d	197, 228, 289, 742
	Na	eed	64, 67, 134, 143, 274, 289, 332, 334, 354, 537, 632, 661, 691, 731, 742, 760

Analyte	Interferent	Interference	Reference
K	Ni	d	537
	P	d	197, 228, 229, 274, 289, 550, 572
	Pb	de	537
	Rb	e	537, 572
	ReO$_4$	d	482
	S	d	197, 228, 289, 572, 742
	Sr	de	537, 632
	Te	de	537
	Ti		716
	W	d	537
	Zn	d	537, 632, 760
	org.	dee	96, 228, 274, 289, 653
La	Al	d	490
	Ba	d	490
	Ca	—e	490
	Ce		490
	Cl		490
	ClO$_4$	—d	490
	Cr	ed	490
	Cu	dd	490
	F		490
	Fe	e	490
	K		490
	Li	d	490
	Mg	d	490
	Na		490
	Nd	d	490

316 ANALYTICAL TECHNIQUE

Analyte	Interferent	Interference	Reference
La	P	dd	490
	Pr	—e	490
	Rb	—	490
	S	d	490
	Th	ed	490
	Ti	d	490
	U	—	490
	Zr	d	490
	org.	eed	489
Li	Al	d	166, 214, 232, 679, 699, 760
	B	—	238
	Ba	de	632, 760
	Be	d	506
	Ca	de	214, 274, 632, 760
	Cd	d	632
	Cl	dde	214, 228, 274, 572
	ClO$_4$	e	228
	Cr	d	632
	Cs		579
	Fe	d	214
	K	eed	166, 274, 579, 632, 699, 760
	Mg	d	214, 632, 679
	Mn	de	632, 760
	NO$_3$	d	228
	Na	eed	274, 579, 632, 760
	P	dde	228, 274, 572, 699

Analyte	Interferent	Interference	Reference
Li	Rb	d	699
	S		166, 228, 274, 572, 699, 760
	Ti		214
	Zn	d	632, 760
	org.	eed	96, 228, 264, 274, 718, 760
Mg	Al	d	544
	Ca	de	498, 544, 676
	Cl	d	228, 578
	ClO$_4$	e	228, 578
	Fe	e	550
	K	e	550
	Mn	e	550
	NO$_3$	—d	228
	Na	e	346, 460
	P	dde	33, 544, 550, 704
	S	d	228, 259, 578
	Ti	d	544
	org.	eed	228, 259, 578, 611a
Mn	Al	de	544
	Ca	e	544
	Cl	ed	544
	Cr	e	544
	Fe	de	174, 544
	Mg	e	544
	Na	e	544
	P	d	544, 704

§99] CATALOG OF SPECIFIC INTERFERENCES

Analyte	Interferent	Interference	Reference
Mn	S	d	174, 544
	Ti	d	544
Na	Al	d	64, 65, 67, 156, 163, 232, 334, 335, 367, 537, 632, 716, 742, 760
	B	d	156, 274, 537, 716
	Ba	eed	274, 537, 632, 760
	Be	d	506
	Br	d	197
	Ca	eed	274, 537, 632, 672, 680, 716, 731, 742, 760
	Cd	d	632
	Cl	d	134, 156, 197, 214, 228, 233, 263, 274, 537, 572, 589
	ClO$_4$	d	228, 274, 572
	Co	d	537
	Cr	dde	537, 632
	Cs	d	233, 537
	Cu	d	537
	Fe	ed	335, 537, 716, 742
	K	eed	143, 233, 263, 274, 332, 334, 537, 550, 632, 653, 661, 691, 731, 742, 760
	Li	dde	156, 214, 233, 335, 537, 632, 742, 760
	Mg	d	274, 335, 537, 632, 716, 742
	Mn	dde	537, 632, 760
	Mo	d	537
	NO$_3$	d	197, 228, 274, 537, 589
	Ni	d	537
	P	d	134, 197, 228, 229, 274, 537, 572, 589
	Pb	d	537
	Rb	d	233, 537
	S	d	134, 197, 228, 274, 537, 572, 589, 760
	Sr	e	537, 632
	Ti	d	716
	Tl	d	537
	W	d	537
	Zn	d	537, 632, 760
	org.	eed	96, 134, 228, 274, 589, 653
Ni	B	e	268
	Co	e	174
	Cr	e	174
	Fe	e	174
	Mn	e	174
	S	e	174
	org.	eed	268
P	Ca	e	163
	I	—	163

318

Analyte	Interferent	Interference	Reference
P	NO_3	—	163
	Na	e	163
	S	—	163
	org.	eed	163
Rb	Cl	e	274
	Cs	e	232, 233, 572, 575
	K	e	232, 233, 274, 572
	Li	e	232, 274
	Na	e	232, 233, 274
	org.	eed	274
Sr	Al	dd	217, 228, 370a, 501, 537, 633, 645
	As		452
	Ba	e	213, 216, 582b, 636
	Be	d	228, 452, 453
	Ca	e	213, 216, 274, 314a, 370a, 537, 582b, 636, 700
	Cl	de	215, 228, 537
	ClO_4	e	228
	Cr	d	228
	Fe	d	228, 253, 700
	K	ed	213, 216, 370a
	Li	d	213, 216
	Mg	e	213, 216, 700
	Mn		228
	Mo	d	452

Analyte	Interferent	Interference	Reference
Sr	NO_3	—	215, 228, 537
	Na	d	213, 216, 700
	P	d	33, 130, 215, 228, 452, 537
	Rb		268b
	S	d	215, 228, 452, 537
	Se	d	452
	Te	d	452
	Ti	d	452, 765
	TiF_6	d	452
	V	d	452
	W	d	452
	Zr	d	452
	org.	eed	228, 314a
Tl	Al	d	453a
	Cl	d	228
	ClO_4	d	228
	Mo		453a
	NO_3	—	228
	Na	e	377
	P		228, 453a
	Rb		268b
	S	d	228, 453a
	Te		453a
	V		453a
	org.	eed	146a, 228
V	org.	eed	431a

100. *Detection and Elimination of Specific Interferences*

In Sections 97 and 98 we have listed several methods of recognizing and eliminating or correcting specific interferences. In this section we present some additional methods of more general applicability. They can also be used in some cases for nonspecific interferences (Section 101). We take up first the recognition and then the elimination of interferences.

(a) *The detection of specific interferences.* 1. A method of recognizing ionization interference was given in Section 97. It can be generalized to include other specific interferences. Readings are first taken on a set of standard solutions, each containing one of the elements present in the sample and detectable by flame photometry. A mixture of these elements is next prepared, in the proportions corresponding to the composition of the sample, and readings are again taken on each element, the blank being subtracted. If the readings on the mixture differ by varying factors from those expected on the basis of the standard readings, specific interferences are at work. But if they are all larger or smaller then the expected readings by an approximately constant factor, then nonspecific interferences are present, perhaps a difference in viscosity or density between the mixture and the separate standards. A definite conclusion can be reached only if various lines or bands of the different elements are measured repeatedly in the mixture and the standards, and if the observed differences distinctly exceed the random error of the method (Section 102). We assume again that blank interferences have been suitably eliminated. These two assumptions (regarding blank interferences and statistical significance of the observations) are retained throughout this section.

2. If only one element is of interest, the above procedure is unduly elaborate; if the instrument is a filter photometer which can handle only one element, the procedure is impossible. The following procedure[298, 306] can then be used. A working curve is established with a series of standards containing only the one element, and the analyte in the sample is tentatively determined by means of this curve. The sample is then diluted to twice the volume and the analyte redetermined. If it appears to have half the former concentration, it is unlikely that specific interferences are affecting the emission or the slope of the working curve. To test this conclusion further, the dilution and analysis can be repeated.

3. Instead of the method of successive dilution just described, a doubling method (admixture method or self-standardization) can be applied.[532] An amount of the analyte equal to that found in the tentative determination is added to the sample, and a redetermination is

made with the help of the working curve. In the absence of interferences, twice the original concentration will be found.

Methods 2 and 3 can be used together. Since some kinds of interference remain practically constant in certain ranges of interferent concentration and might thus escape detection by these methods, it is advisable to extend the tests over a wide range of concentration, preferably over several decades.

(b) *The elimination or correction of specific interferences.* If the interference is not very large, a moderate reduction may suffice. A systematic error due to specific interference can be considered practically eliminated when it does not exceed the random error of the analysis. We list below various instrumental and procedural methods of reducing (and often practically eliminating) specific interferences.*

1. Use of a *cooler flame*, as for ionization interference (Sections 7 and 97).

2. Use of a *hotter flame*, as for dissociation or vaporization interference (Sections 7 and 98).

3. Use of an *atomizer* or atomizer-burner best suited to the application in terms of drop size distribution or of drop separation (ref. 513, and Section 98).

4. Optimization of the *observation height* in the flame (Sections 37, 97, and 98).

5. Optimization of the *concentration* of the sample for a given flame temperature (ref. 263, and Sections 79 and 97).

6. Use of the best *solvent* (Section 79), which in turn affects the flame temperature, the drop size distribution, and the optimal height.

7. *Internal standardization* (Section 51). This reduces a specific interference only with proper choice of the internal-standard element; the interferent should react upon it and the analyte at least in the same direction and preferably to the same extent. For a different analyte, a different internal-standard element at proper concentration generally will be needed.

We next list again several methods given in Sections 97 and 98 for the special applications discussed there; these can often be applied to other specific radiation interferences.

8. The *parametric method* (Section 97) is applicable quite generally, not alone to ionization interference. However, it is troublesome to carry out when more than one interferent has to be considered.

---

* One other method that should be included in the following list is the method of successive dilutions with extrapolation to infinite dilution, described in ref. 306.—*Translator.*

9. The *simulation method* (Section 97) also is generally applicable, whenever it is not too hard to simulate the sample adequately. This method is recommended when the interferents in the sample remain approximately constant in composition and concentration.

10. The *correction method* (Section 97). The same remarks apply as to 8 and 9.

11. The *normalization* and *saturation methods* (Section 97) are applicable whenever, as in ionization interference, the interference approaches an asymptote with rising interferent concentration.

12. The *parallel method* (Section 98) is applicable whenever at least part of the interference curve is parallel to the interferent concentration axis. There will then be a concentration range within which the interference is independent of the interferent concentration.

13. The *buffer method* (Section 98) (use of releasing agents) differs from the two preceding in that a third element is added (but not measured), instead of the interferent itself, for the purpose of reducing the interference. This method is applicable whenever a buffer element (releasing agent) can be found that releases the analyte from the interference. It is particularly useful for magnesium determinations, since the compounds of magnesium with phosphorus, aluminum, titanium, etc., are considerably less stable in the flame than those of the releasing agents calcium, strontium, and barium with these interferents. Hence, lower concentrations of releasing agent are needed for magnesium than for calcium, etc.[544]

14. The *dilution method* can be used with interferences that diminish in relative magnitude on dilution, like that of phosphorus on calcium. Sufficient dilution then practically eliminates the interference, provided that the analytical precision does not fall below that required, through approach to the detection limit (Section 90). Dilution is effective with interferences of the type discussed in Section 98. However, it must be pointed out that not all interferences diminish with dilution. For example, ionization interferences increase with dilution, apart from the effect of the flame electrons (Section 97) at very low concentrations.

15. *Numerical* or *graphical correction*, using logarithmic translation (see Section 97), will eliminate multiplicative interferences. Other interferences besides those due to ionization can be described with reasonable accuracy within certain concentration ranges by a multiplicative factor. The graphical method has an advantage over methods 16a and 17 in that the working curve does not have to be linear.

16. *Self-standardization*[370, 602a] was mentioned at the start of this section as a means of detecting specific interferences. In certain cases,

duly adapted, it can be used for quantitative evaluation of the interference and for correction of the analytical result (see also Section 97). Method 16a below applies to an interference which, in the concentration range of interest, can be described as a constant, concentration-independent factor, the working curve being linear. Method 16b is more general in scope, applicable when the intensity I can be expressed by $I = kC^m$, where k and m are constants which, however, depend on the interferent concentration, and C is the concentration of the analyte; the working curve is thus not linear. Method 16a yields the slope of the working curve as affected by the interference, and from it the analytical result. The limitations of this method prevent it from being used for interferences of the sort discussed in Section 98, but on the other hand it can be used for nonspecific interferences (Section 101). However, the more general method 16b is applicable to the alkaline-earth interferences of Section 98.

16a. Measurements are taken at the wavelength λ_0 on the analytical line and at λ_1 next to the line, to permit subtraction of the background. We assume the background equal at λ_0 and λ_1. Let A be the reading on the sample and $A + Z$ that on the sample containing a suitable additional quantity of the analyte (added as salt). Subscripts refer to the two wavelengths. The slope of the working curve is given by

$$\frac{C_a + C_z}{C_z} = \frac{(A+Z)_0 - (A+Z)_1}{(A+Z)_0 - (A+Z)_1 - (A_0 - A_1)}$$

where C_z is the added concentration. The original concentration of the analyte is then

$$C_a = C_z \frac{A_0 - A_1}{[(A+Z)_0 - (A+Z)_1] - (A_0 - A_1)}$$

This is based on the assumption of a linear working curve. However, by means of a correction, the calculation can be applied to a curved working curve,[316] provided the interference can still be expressed as a constant, concentration-independent factor.

16b. This method of self-standardization applies to a working curve of the form $I = kC^m$, k and m being unknown constants depending on the interferent concentration.[140a] Logarithmic coordinates are used. At least three readings are taken, one on the original sample and one each on portions of this sample after addition of at least two different quantities of the analyte, selected to raise the concentration by amounts not too different from the expected original concentration. The intensities are then plotted on doubly logarithmic paper against the total concentrations, adopting for the moment a guessed or expected original concentration. The three (or more) points will generally not lie on a line. The assumed original concentration is now adjusted to make the curve straight. This gives the true original concentration of the sample, and the straightened curve is the true working curve. This method has been used successfully in arc spectrography.[62,243,409,600] It can also be carried out numerically.[140a]

17. The *doubling method* is a special case of self-standardization. The added concentration C_z is so chosen that the net reading (with blank subtracted) is exactly doubled. To do this, various additions must be tried in order to find the right one. If the conditions of method 16a obtain, C_z thus found equals the sought concentration C_a. As compared with 16a, this method avoids calculation and utilizes a range of concentration that affords the best accuracy in self-standardization. But it has the drawback of requiring repeated trials, which consume time and sample material.

The dilution that supervenes in 16 and 17 must be allowed for, or else each portion must be diluted by the same amount, to compensate for the dilution caused by the addition of analyte. Self-standardization may thus be carried out at fixed or variable volume. The errors of self-standardization, which is often used (under the term *recovery*) in chemistry, have been the subject of several studies.[140a,243,316] We caution against self-standardization when adsorbing substances such as protein are present in the sample.

18. The *substitution method*.[544] In methods 16a and 17, but not in 16b, it was assumed that the interference, expressed as a factor, was independent of the concentration of the analyte; 16a and 17 are therefore not applicable to the interferences of Section 98. This limitation is avoided by the substitution method, which we explain below, using ionization interference as an example.

As in the buffer and comparison methods, a foreign element is added, not in large amount but at a concentration comparable with that of the analyte. The added element must meet these requirements: (*i*) Its radiation must not interfere directly with that of the analyte, through either cross-sensitivity or band overlap. (*ii*) It must react to the interference in the same way as the analyte does.

In the case of potassium interfering with sodium, the ionization constant J_X of the added third element should be close to that of the sodium, J_{Na}. If in fact $J_X = J_{Na}$, it can be shown by the Saha equation (Section 7) that for a given concentration of interferent (potassium), the electron concentration in the flame remains fixed so long as the sum of the molar concentrations of sodium and the third element is constant. Then, likewise by the Saha equation, the degree of ionization of the analyte (sodium) is constant, and the working curve for sodium will be linear, apart from self-absorption. Self-standardization can then be used for eliminating the potassium interference. The potassium concentration need not be known and can vary from one sample to the next.

The procedure is as follows. Prepare a series of solutions from the original sample (in which the sodium and potassium concentrations are unknown) by adding to it both sodium and the third element in such amounts that the sum of their molar concentrations remains fixed at some arbitrary value, which should be of the same order as the original sodium concentration. The sodium emission is plotted against the added sodium concentration for the series. The abscissa of the intersection of the extended line with the concentration axis is the negative of the original sodium concentration.

The method is exact only under ideal conditions. Generally the ionization potentials of analyte and foreign element cannot be equal. The two elements may also enter side reactions, such as hydroxide formation, to different degrees. If the ionization potentials are unequal, the admixtures should be such that the sum of the molar analyte concentration plus a certain multiple or fraction F of the molar concentration of the third element is constant. This factor F, which depends on the interferent concentration, can be determined experimentally by finding a set of pairs of concentrations that yield a straight working curve. This requires many trials, consuming time and sample.

To avoid this inconvenience, F can be calculated in comparatively simple cases. It must be assumed that no side reactions such as hydroxide formation interfere. The formula is $F = (J_{Na}/e + 1)/(J_{Na}/e + J_{Na}/J_X)$, where e is the constant concentration of electrons. Unless $J_X = J_{Na}$ (when $F = 1$), F depends on e and hence on the analyte and interferent concentrations. But if, when the ionization is very strong, J_{Na} greatly exceeds e, and if J_{Na}/J_X is at most of the order of 1, F is practically independent of the concentrations of interferent and analyte and equal to 1. The same result follows if the ionization is weak, so that J_{Na} is much less than e, and if J_{Na}/J_X is at least of the order of 1.

Similarly, with more complex interferences, as of phosphorus with calcium, another alkaline-earth element such as strontium or barium can be added in such quantities as to produce a linear working curve of the analyte. In this case no factor like F can be specified, since the interferences of phosphorus, aluminum, titanium, etc., cannot be expressed in as simple form over extended ranges as ionization interference can (recall the bend-point in the phosphorus–calcium interference, Section 98). Only an experimental approach is possible. Similar considerations apply when several interferents are present together.

Other advantages of the substitution method are that it can handle nonspecific interferences as well, and that, at least in the types of interference mentioned above, the emission of the analyte is enhanced by addition of the third element.

19. *Chemical removal* of the interferent is at present often resorted to with complex samples. In routine determinations it has the disadvantage of greater total working time, but the time can be kept within reason when the chemical procedure is simple and there are not too many samples. Thus, the interference of aluminum with the alkaline earths can be eliminated by precipitating the aluminum with pyridine or urotropine,[293] or by precipitating the calcium as oxalate, washing it free from interferents, and redissolving it.[508] Phosphorus is removed by ion exchange.[36, 61, 174, 289, 337, 550] Also, the enrichment and extraction methods of Section 88 can be used for separating interferents.[174, 208]

There is another method that might be included here. When a little alkali metal is to be determined in the presence of much alkaline-earth metal, the background of alkaline-earth bands can interfere. This interference can be removed "chemically" by adding a sufficient amount of aluminum to the sample.[633] The aluminum, as described in Section 98, suppresses the alkaline-earth emission almost completely and thus practically eliminates the interference. Unlike other chemical methods, this one exploits a reaction that occurs only in the flame.

In closing, we call attention to the possibility of eliminating sufficiently well-defined interferences by means of a computer or data-handling system. The concentration of interferent could be monitored by a second optical beam. This signal would control, for example, a motor-driven potentiometer applying a correction factor to the analytical reading in the manner of an analog computer. So far as we know, the correction of interference has not yet been automated in this way.*

101. *Nonspecific Interferences*

As shown in Section 11, concomitants can give rise to nonspecific interference with the analytical line or band by altering the rate of transport of analyte to the flame and the rate of evaporation of the droplets. The evaporation effect in the flame is likely to be especially pronounced with direct atomizers, whereas with indirect atomizers

---

* Instruments that monitor an interferent (calcium) and automatically correct for it have now been described by Z. A. Kel'tseva, *Tsement*, **1958**, No. 2, pp. 19–21; A. Hinde, U.K. Atomic Energy Authority, Report PG–147 (1961); and D. N. Ivanov, *Zavodskaya Lab.*, **27**, 297–9 (1961).—*Translator.*

transport interference is more marked. The reasons for this difference are the following. With a direct atomizer, the larger drops that enter the flame evaporate incompletely. Slight changes of the evaporation characteristics of the sample alter the supply of analyte in a form excitable in the flame. Such effects are called evaporation interference. With an indirect atomizer, on the contrary, this effect does not appear in the flame, because the small amount of liquid that reaches it is very finely divided and evaporates fully.

Instead, a change of the evaporation characteristics alters the fraction of liquid separated in the spray chamber. This effect is called transport interference. Both these interferences are due to the same cause, viz., changes in the physical properties of the solution sprayed (density, surface tension, viscosity, vapor pressure), resulting from the presence of concomitant salts, acids, or organic substances in variable concentration along with the analyte. These properties influence the suction rate, the atomization, and the evaporation of the spray (Section 6).

Several schemes, some already discussed, are used for minimizing these interferences. They include the following:

1. The differences in the physical properties of the sample and standard solutions can be reduced by adding to the standards, in appropriate amounts, such substances as glycerine, gelatine, or agar-agar.[502] These additions must be as nearly as possible free from the analyte and elements which might cause blank or radiation interference.

2. The sample and standards can be normalized by adding a salt (lithium chloride), an acid (sulfuric), or an organic medium (see above) in excess, [241,263,563] to equalize the atomization and vaporization characteristics; in contrast with the preceding method, the normalizing agents are added not merely to the standards, but in excess to both standards and sample. In this way the nonspecific interferences from other variable concomitants of unknown concentration can be held small or suppressed entirely (compare the normalization method in Section 97).

3. Since these interferences diminish in relative magnitude with diminishing interferent concentration, dilution will reduce them (cf. Section 79).

4. Since nonspecific interferences act equally on all elements at a given concentration and on a given element at all concentrations (Section 11), they are approximately multiplicative;[274,275] that is, they can be described by a constant factor. Internal standardization will therefore very nicely compensate them; moreover, the lithium chloride

(for example) that is added as internal standard has something of a normalizing action as well (see 2 above).

5. Nonspecific interferences are less pronounced with direct than with indirect atomizers.[563] With a direct atomizer having controlled sample flow (Section 33), the interferences are still smaller, but effects acting on the evaporation rate of the mist droplets still remain.

These sources of error and their effects are unfortunately very complex and depend strongly upon the characteristics of the instrument, especially of the atomizer. Hence no universal correction formulas can be given, for instance, for eliminating the effect of surface tension or viscosity, even though these may be measured in advance. We have earlier spoken of the complex relation between flame emission and suction rate, among other factors (Section 6). It is, however, possible to develop empirical correction formulas for special applications with a specific instrument.[545]

C. PRECISION AND ACCURACY

102. *The Random Error*

If the concentration of an element is repeatedly determined with a flame photometer by comparison with a standard solution, the experimental conditions being held as constant as possible, the same result will not be obtained each time; instead, the measurements will be more or less spread out. This scattering is called the random error. In addition, all of the readings, including their mean, can be affected with a systematic departure from the usually unknown "true" value. Systematic errors will be taken up in the next section; the causes of random error will be discussed in the present section. Consideration of the possible causes often permits recognition of the chief contributory factors in a practical case, and suggests methods of reducing the error. We shall also show how to express the scattering objectively by means of a statistic.

(*a*) *Causes of random error.* 1. Inhomogeneity of the sample or standard solutions. This is commoner with biological materials.

2. Nonuniformity of temperature in the solutions or differences of temperature among solutions, such as might result from holding test tubes in the hand.

3. Change of the suction height as the liquid is withdrawn from the sample container.

4. Partial obstruction of the suction capillary.

5. Formation and dislodgment of deposits on the atomizer tip.

6. Irregularity of atomization due to fluctuations of pressure in the compressed-air supply or at the regulating valve, or to pressure surges in the spray chamber.

7. Variations in separation and evaporation of the drops in the spray chamber of an indirect atomizer, due to temperature changes, etc., as from variation of the time interval between application of the sample and the taking of the reading.

8. Inhomogeneity of the combustible mixture due to incomplete mixing of gas and air, and irregularity of gas flow through the regulating valve.

9. Formation of deposits on the burner cap.

10. Variations in drop and particle evaporation due to fluctuations in the drop size distribution (see above) and in the flame temperature (as from gas pressure fluctuations).

11. Emission fluctuations due to the effects of the above variables on the atomic concentration of analyte in the flame at the height of observation, and due to variation of temperature, which in turn are due to various causes (irregularities of gas flow, of liquid flow in direct atomizers, etc.).

12. Fluctuation of the background radiation.

13. Zero-point fluctuation of the optical–electric system.

14. Variation of sensitivity of the system.

15. Errors of standardization due to carry-over of drops of sample adhering to the capillary into the standard or blank, and the resulting shifts in the full-scale value and the zero point (suppressed blank reading), respectively.

16. Errors of reading the meter.

17. Errors of reading a working curve or of calculating a correction for interference.

(b) Calculation of the standard deviation as a measure of the random error. The most frequently occurring value (the *mode*) is normally the same as the (rounded-off) average value M (or *mean*) of all the values, for long series of readings (over 10). This mean, M, is found by dividing the sum of the individual values c_i by the number of observations n:

$$M = \Sigma c_i/n, \qquad i = 1,2,3,\ldots, n$$

This is also called the *arithmetic mean*. The greater the departure of a value from the mean M, the less frequently does it occur in a set of values. Widely deviant values are hardly ever found. If the frequency of occurrence of the individual values within equal intervals is plotted against the value itself (the midpoint of the interval), a roughly bell-shaped curve is usually obtained; this is the symmetric *Gaussian (normal) distribution*.* The maximum of the curve lies at M. The quantity σ, a measure of the scattering about the mean, is very important in judging an analytical method; it is given by

$$\sigma = \sqrt{\Sigma(M-c_i)^2/(n-1)}$$

Here $(M-c_i)$ is the deviation of an individual measured value (of the concentration) from the mean value; σ might be called the mean error of an individual measurement. This term would be ambiguous because, as will be seen in the next section, systematic departures from the true value may be superposed upon this error. Accordingly, σ is better called the *standard deviation* (often *standard error*) in this connection. The quantity 2σ then corresponds to what the practical analytical chemist calls *precision*.

If σ is divided by $\sqrt{n}$, we get the *standard error of the mean*, σ_M, a measure of the uncertainty of the mean value M:

$$\sigma_M = \sigma/\sqrt{n} = \sqrt{\Sigma(M-c_i)^2/n(n-1)}.$$

This shows the variation to be expected in the mean M when several series of measurements are carried out with equal numbers of observations. Thus, σ_M is useful in calculating the error of self-standardization (Section 100), in which the means of sets of replicate measurements are employed.

But σ is the most important quantity for judging the relative precision or reproducibility of a method. It is usually quoted not in absolute terms as above, e.g., 10.4 ± 0.6, but rather as a percentage of the mean M, thus: $10.4 \pm 6\%$; specifically,

$$\sigma(\%) = 100\ \sigma(\text{abs})/M$$

This is called the *coefficient of variation*.

Unlike systematic errors, the random errors of reproducibility cannot be allowed for by any method of calculation. The magnitude of σ directly shows the general behavior of the error distribution. Thus, in a set of repeated measurements of a quantity, 2/3 of the errors lie within $\pm \sigma$, 1/20 of them exceed 2σ, and only about 1/300 or 0.3% of them exceed 3σ. However, as mentioned in Section 90, there is a certain subjective element in the value of σ as thus obtained from sequences of

* Departures from a Gaussian distribution may result from rounding off the readings to whole scale divisions when the distribution covers only a few scale divisions. To avoid this difficulty, the readings should not be rounded off.

readings, since the judgement of a "mean" meter reading depends on the observer's skill. An experienced observer can integrate well while watching an unsteady needle, whereas a beginner is likely to weight the extreme excursions too heavily.

For testing whether the results of two series of measurements are to be regarded as equal within the experimental uncertainty, or whether the difference of their means is large enough to be considered "statistically certain," the limit 3σ is often used as criterion.* Compare Section 90.

In direct flame methods (without internal standardization), σ may lie between 0.2% and 5%, depending on the type of sample, flame photometer and application, and the skill of the operator. Internal standardization can reduce these errors by fully one-half.

It is worth stressing that the relative error of flame methods remains constant over wide ranges of concentration, unlike that of chemical methods. Hence, with respect to their accuracy, flame methods gain increasing advantage over chemical methods at lower concentrations—of course, with the limitation that the concentration cannot come too near the detection limit, for the coefficient of variation then increases.

For judging the attainable precision and for establishing the concentration difference that can barely be distinguished with "certainty," it is necessary to make repeated measurements and to calculate σ as above. In taking repeated readings on the same sample, it is easy to fall prey to autosuggestion—an unconscious tendency to make a reading on an erratically moving spot of light, needle, etc., the same or nearly the same as the preceding reading. The standard error then comes out too small. This can be cured by distributing the replicate measurements used for error determination among the routine measurements without identifying the samples, so that they are not recognized at the time. The result becomes even more objective if the readings are taken over several days and, if possible, by different observers independently but under the same experimental conditions. Autosuggestion can be avoided by allowing an observer to carry out blind reproducibility measurements on two solutions contained in bottles of similar appearance, differing in concentration by an amount about equal to the expected scatter (perhaps 1%). An assistant submits the bottles in a random sequence without the observer's knowing which bottle each is. The observer calls the readings to the assistant, who records them in two columns corresponding to the two samples.

* In America, a difference of 2σ is commonly used as a criterion of statistical significance; such a difference is then called "significant", "probably significant", or "significant at the 95% confidence level".—*Translator.*

103. Systematic Errors

The standard error σ characterizes the reproducibility of a replicate measurement. But it is no criterion of the agreement between the mean found by the (flame) method in question and the mean found by some other suitable method such as a good chemical method. The departure of the mean M defined in the preceding section from the absolute, "true" value M_t, to the extent that it is not accounted for by the random error $\sigma/\sqrt{N}$, is called the *systematic error*. We assume for simplicity that it is practicable to find M_t by means of a "reliable" chemical method.

The following are causes of the systematic error.

1. Errors in the standard solutions, as from moisture in the chemicals, errors of weighing, impurities, alkali contamination from glass vessels, etc.

2. Dilution errors in preparing the samples for analysis—pipet errors, etc.

3. Effects of concomitant substances not simulated in the standard solutions; that is, partial or complete neglect of specific interferences (Sections 96ff), cross-sensitivity (Section 94), background (Section 95), etc.

4. Differences in surface tension and viscosity between the sample and standard solutions (Section 101), or differences in temperature, pH, etc.

The possibility of determining the "true" value with a reliable chemical method is often moot. In difficult cases, the results obtained with different chemical methods may differ considerably from one another. As an example, the *mean* calcium concentration in human serum as found by different chemical methods varies from 8.8 to 13.2 mg. Ca per 100 ml., depending on the method.[588] In this and similar cases even the chemical methods, when they have not been critically clarified, require a certain rather empirical standardization,[410] which can be made by accepting one method as the norm, or by making available to all laboratories standard samples of definitely known composition.

References in the literature to systematic errors are not so numerous or reliable, since they are harder to measure than random errors. In comparing flame-photometric or -spectrophotometric mean values with those of other analytical methods, one must allow for the systematic departures of the latter from the true value. It may be idle to debate which of the two methods has contributed most of the observed departure. Moreover, in many practical problems a systematic error is inconsequential, as when only the variation of concentration among samples is of interest, the value of the standard against which these variations are measured being unimportant; besides, the standard can be adjusted to any need by transformation of the scale or the working curve. However, if inadequate methods are used for the determination of trace elements, the systematic error can be larger and possibly significant.

104. *The Aggregate Error of Measurement*

A more serious matter is the fact that radiation interferences, cross-sensitivity, etc., can vary among samples of different composition; this happens when the interferents vary in concentration among the samples, and are not determined. But in such cases it is still quite possible for the random error to be small and the reproducibility of replicate determinations on a single sample to be very good, because of good experimental conditions and the fact that in the series of replicates the interferents are constant. It is further possible that the apparent systematic error will likewise be small, as shown by the difference of the mean values obtained by two methods of determination (flame and chemical), perhaps because the several interferences happen to cancel each other, or perhaps because the chemical method used for comparison has by chance the same systematic error. Thus, despite small random and apparent systematic errors, the results can be afflicted with appreciable errors not included under either of the types defined above. Errors of this kind, together with the random and systematic errors, constitute the *aggregate error* of measurement. Section 100 listed methods of detecting and compensating such errors.

CHAPTER 5

APPLICATIONS

The practicing analyst, always pressed for time, wants simple, reliable "recipes". If he finds one, he will often use it uncritically, in order to get the analytical result, which alone interests him. This desire for ready-made procedures is certainly understandable. But it cannot be satisfied within the limits of this book.

There are several reasons for this. The number of types of sample material analyzed by flame photometry amounted, at the time of the first edition of this book, to about 700, in each of which, on the average, about two elements were determined by the flame. These alone represent 1400 procedures. Meanwhile, the number of published flame-photometric procedures may have doubled, coming to perhaps 3000. The procedure depends not only on the nature, mode of combination, and concentration of the analyte, interferents, and other concomitants, but also, for a given element in a given material, on the type of flame photometer. For instance, ionization interferences may be small or negligible (Sections 7, 14 and 97), if the flame is cool; but if it is hotter, they must be removed or corrected by adaptation of the procedure or by calculation. Another instrumental characteristic strongly influencing the method is the type of atomizer—direct or indirect—and its drop size distribution and drop-separating power. This characteristic intimately affects correction or removal of the interference of phosphorus with calcium, for example. Since there are about 30 firms making flame photometers, and some manufacture more than one type (Table 15), there must be about 50 kinds of commercial flame photometer at present. Accordingly, with the usually unwarranted assumption that all instruments of a given make are sufficiently alike, and not counting homemade instruments, we should be dealing in principle with 50 × 3000 or 150 000 procedures. But we cannot stop here. Any instrument can be set up in many different ways, depending on the choice of line or band, adjustment of height of observation in the flame, adjustment of gas/oxygen ratio, etc., all of which affect the procedure, as for instance in the matter of applying background corrections. If the essential operating parameters of an instrument are divided into 10 sets of values, we come out with a million and a half possible detailed procedures. In addition to these instrumental variables, there is a host of other factors that are left, within limits, to the choice of the analyst, such as sample preparation and choice of solvent or dilution, which affect the procedure. Thus with the same instrument, sample, analyte, and adjustments, the procedure can be worked out differently by different chemists. This source of variability, which must not be underestimated, brings the number of possible procedural distinctions to astronomical magnitude.

Confronted by this situation, we considered limiting ourselves to procedures for the commonest materials using widely distributed makes of instrument with prescribed settings, and, from among other published procedures, selecting a few apparently reliable ones for recommendation. The following reasons oppose this:

1. A procedure tested for a particular instrument with given settings is often used uncritically with other settings or even another type of instrument, and failure is then unjustifiably blamed on the method or on the author. Indeed, errors can arise because a published instrumental setting such as a pressure may not be optimal in another flame photometer of the same make; the same setting may yield a different gas/oxygen ratio and flame temperature, altering ionization interference, etc. Refer to Sections 19 and 97.

2. We have no information on the distribution of the various makes of flame photometer, and it varies with the country and in the course of time.

3. The authors of this book, being a physicist and a physical chemist, would find it very difficult to prescribe procedures in other branches of science, because the viewpoints of flame photometry are not the only ones of importance. Special precautions are needed in taking the sample, such as soil or blood, and in preserving and preparing it for analysis. Therefore, recommendations could be made only with the help of an expert in the field in question. In any case, workers in these several fields of application should check and if necessary revise their procedures, which are often of long standing, in the light of the new knowledge of flame photometry (for example, interference mechanisms).

Accordingly, we shall give no detailed procedures. Instead, we shall discuss the several fields of application from the purely flame-photometric point of view, with references to the extensive (though selected) bibliography at the end of the book.

Agricultural Applications

105. *Agriculture and Botany*

Flame analysis was first applied to agriculture and allied fields in the work of the plant physiologist, Lundegårdh. His work will therefore be summarized first. He was concerned on the one hand with the soil constituents absorbable by plants (agricultural chemistry, pedology, fertilization, agriculture, etc.) and on the other hand with the relation between plant growth and the composition of the soil together with the environment—climate, humidity, temperature, and insolation—(plant physiology). The nutrient content of the soil influences the composition of the plant tissues, viz., root, stem, leaf, ear, etc., which differ in their reactions. Naturally, there are connections with other fields, such as zoology, animal husbandry, veterinary medicine, nutrition, agricultural chemistry, mineralogy, and plant breeding. The practical significance of these studies is obvious: determination of the fertilizer requirements of the soil, increase of yield, improvement of quality and quantity of vegetable food and indirectly of animal food such as meat and milk. The chief purpose of fertilization remains, as always, the alleviation of deficits (Liebig).

The phenomena of plant metabolism, e.g., the antagonism between potassium and calcium, the mass-action law, and Donnan equilibrium, are like those of animal metabolism. Plants absorb food through the roots, generally as ions, and the growth of the plant depends on the concentrations and interrelations of the nutrients, chiefly potassium, calcium, phosphorus, and nitrogen. There are also the essential trace elements—magnesium, manganese, copper, and others known to be important; but the part they play in the plant organism is far from fully understood.

The concentrations of the elements in the plant depend on the type of soil and its nutrient content, as well as climatic factors. The relation to soil composition is quite complex, since among other things the ratios of the ions to each other in the soil (ion antagonism, ion interference) are significant. Consequently, Lundegårdh usually studied fertilizer requirements not by chemical analysis of soil extracts (see below) but by determination of the nutrients in the plants grown on the soil, using his technique of "leaf analysis". Since little earlier work of this kind had been done by chemical methods, he concerned himself at first with three analyses for characterizing the composition of the soil, viz., determination of the nutrient content of (1) the subsoil, (2) the soil humus or surface soil, and (3) the tissues of the plant. Lundegårdh called this the "triple analysis". But he ultimately concluded that this triple analysis could be replaced by the simpler leaf analysis alone. The latter reflects the concentrations of the ions in the soil that can be taken up by the plant ("mirror analysis"). For this purpose, a total analysis of the ripe plant has little value, for the buds, ears, fruits, etc., show surprisingly constant composition. The leaves, however, are suitable, although the time at which the leaf is taken is important, as the concentrations change with age.

Lundegårdh analyzed the ashed leaves for potassium and calcium by flame spectrography, using an air–acetylene flame. Later, he employed an automatic "robot" which supplied the sample, made the exposure, and developed, fixed and washed the film;[34] naturally, a less elaborate method suffices. For his occasional soil analyses, he used a citric acid extract. Triple analysis is still used in pedological investigations.

Under the impetus of Lundegårdh's work, many others applied flame methods to the investigation of problems in plant physiology and agriculture. In Germany this was furthered by the availability of the simple filter photometers of Schuhknecht–Waibel and of Zeiss, and flame photometry has since been used extensively for soil and plant investigations.

Meanwhile, flame photometry has become firmly ensconced in laboratory practice for such routine work as the determination of fertilization requirements of soils. In this particular case a proper method of sampling the soil is essential; average samples of the surface and subsoil must be taken from various parts of a field in a suitable manner and well mixed. Preparation of these samples for flame analysis must be such as to include all and only those constituents of the soil that are capable of being assimilated by plants. We cannot enter further into these matters here; see, for example, ref. 50. Plants, also, are often subjected to routine analysis. We mention the classic seedling method of Neubauer. In this type of analysis, unlike that of soils, special ashing methods are an essential part of the procedure.

Besides these numerous routine methods, investigations of difficult problems with the aid of specially developed flame-(spectro)photometric methods are being reported in increasing number. An example is the behavior of trace elements in the complex physiological processes in plants. In these studies, the interferences in flame photometry must be heeded and as far as possible eliminated under the conditions prevailing. This research, of course, benefits other disciplines; advances of general interest were mentioned in the preceding section.

To help the agricultural chemist locate the special literature, we list here a selection of references:

General works on method: 1, 22, 33, 34, 39, 50, 73, 87, 107, 121, 144, 165, 196, 245, 254, 256, 267, 273, 281, 293, 294, 317, 337, 401, 426, 429, 430, 434, 443, 467, 468, 478, 479, 480, 513, 521, 523, 539, 549, 552, 570, 589, 590, 591, 594, 595, 611a, 614, 615, 617, 617a, 617b, 625, 649, 574, 680, 707, 730, 732, 759.

Interferences in the determination of calcium: 61, 258, 551, 616, 644, 680.

Ion exchange for the avoidance of interference: 61, 288, 289, 617a.

Studies of trace elements: 415, 416, 465, 466, 550, 630.

106. Foods and Nutrition

Nutritional science somewhat overlaps veterinary medicine, agriculture, botany, etc. There will also be interrelations with other disciplines; for instance, poisoning in man or animals may be a matter for criminology. A few applications are covered below.

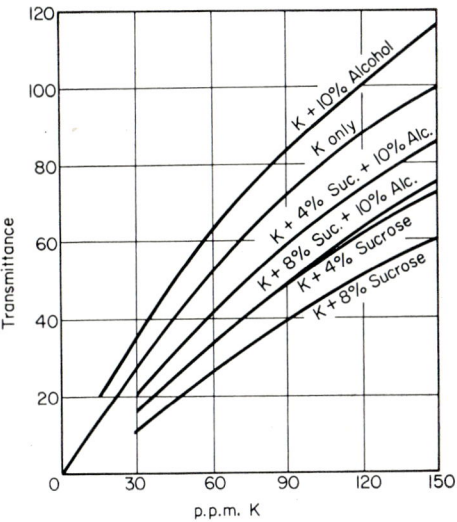

Fig. 59. Effect of sugar and alcohol on the potassium working curve (from ref. 96).

Butter, milk, etc., when they contain various amounts of solid phase, must usually be ashed before analysis, so that no solid material will clog the atomizer. However, fresh milk or the like can be atomized directly. Fruit juices, soft drinks, and alcoholic beverages, especially fruit liqueurs, should be ashed before analysis when good precision is required, as they contain interfering constituents, such as alcohol of varying concentration, which will markedly affect the flame background. Alcohol can cause a spuriously high sodium or potassium reading, unless the same content of alcohol is used in the standards. However, complete simulation of the sample by the standard is difficult, as all other components (sugar, etc.) have to be considered too. Hence ashing is advisable.

As an example, Fig. 59 shows the effect of sugar and alcohol on the potassium working curve.

Instead of ashing or the simulation method, internal standardization (Section 51) in a somewhat modified form[96] has had some success in investigations of this type. As described in Section 52, equal concentrations of lithium are mixed with the sample and standard solutions. The apparent lithium concentration in the sample is found by comparison with the standard and reference to a working curve. The concentration

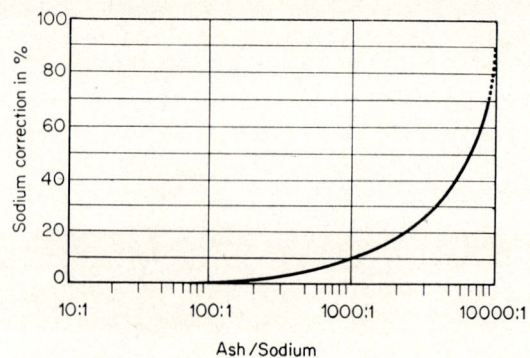

Fig. 60. Correction curve for sodium at very low ratios of sodium to total ash (from ref. 143).

correction factor (ratio of apparent to true lithium concentration in the sample) is applied to the apparent concentration of analyte in the sample. The method works even at higher concentrations where the working curves of lithium and potassium are not linear. This procedure obviates the preliminary ashing.

Almost all products of animal or vegetable origin contain much sodium. Hence, the determination of sodium is easy, and interference from other elements (Section 97) in these materials is infrequent. However, the sodium content of certain plant products, especially fruits, grain, nuts, etc., is exceptionally small relative to the total ash, and in these interference of other elements with the sodium determination is likely.[143] It is worth repeating that the sodium line 589 mμ in such cases requires to be especially well separated from the neighboring calcium bands.

The bean *Phaseolus vulgaris* is typical of plants with low sodium content;[143] the ash consists essentially of potassium, calcium, magnesium, phosphate, and sulfate, with only traces of sodium. Here we should expect considerable errors from specific interferences, especially ionization interference of potassium with sodium. The errors are easily eliminated by measuring the enhancement by self-standardization, as described in Sections 97 and 100. If the composition of the rest of the ash is reasonably constant, an empirical correction curve can be derived; that is, the ratio of apparent sodium to total ash is found first, and the curve then yields the correction to be applied to this sodium value. Fig. 60 shows such a correction curve.

The calcium and magnesium content of milk is of interest, because it affects stability of the casein suspension. Small concentrations of iron and copper are important in sugar sirups, because these metals degrade vitamin C. They should therefore be removed if beverages containing vitamin C are to be sweetened with the sirup. The copper content must be kept below 5 p.p.m., the iron below 15 p.p.m.[302]

For additional special applications, see refs. 22, 34, 39, 50, 91, 103, 107, 151, 165, 190, 203a, 312, 323, 401, 413, 566, 639, 753.

Medicine and Biology

107. *General*

The value of the speed and sensitivity of flame analysis for problems in biology and medicine has been apparent since the beginnings of spectrochemistry. Nearly every pioneer of flame analysis applied his methods, although often developed for other purposes, to biological problems. Thus, Lundegårdh analyzed blood serum, milk, tissue extracts, etc., and noted the advantages of adding hydrochloric acid to serum and similar liquids.[33] Jansen, Heyes, and Richter also applied their method to medical problems. Flame photometry developed similarly, although considerably later, in the United States. Here, too, the pioneers, such as Barnes and his collaborators,[90] worked with a view to medical and biological applications, or arrived at the flame method under the impetus of such problems. The work of these forerunners did not at first, however, have much influence on laboratory practice.

Only after 1948 did flame analysis find wider use in the medical laboratories of Europe and other lands. In Germany this development was sparked by the publication of Belke and Dierkesmann.[128] The number of publications quickly climbed to several hundred. Today sodium and potassium are determined in body fluids almost exclusively by flame analysis. We limit ourselves here to works dealing with method. Owing to the high speed of flame analysis it became possible for the first time to study large assemblages of material entailing long series of measurements yielding statistically reliable conclusions.[149] Besides, the consumption of material is generally much less than in chemical micro-methods; this has enabled studies requiring repeated sampling on children and small living animals.[93] Only by this method can sufficiently quick determinations be made in contingencies endangering life, such as marked potassium deficiency, and flame analysis has now provided the first insight into the incidence of such pathological disturbances.

Many papers on medical applications give practical information for the use of one or another type of flame photometer in the analysis of a given element in a given body fluid. It is not yet possible to prescribe uniform techniques, since, aside from the choice of method, the technique will depend largely upon the instrument available, as shown earlier.

However, it seems doubtful that even future working technique can be made uniform, for the requirements imposed upon a flame photometer by the individual problems are altogether too various. We discuss this in the following sections.

The following are general works dealing with the relation of flame photometry to medicine and associated problems: refs. 21, 36, 49, 81, 92, 93, 94, 95, 105, 123, 128, 131, 133, 134, 135, 137, 139, 149, 153, 181, 184, 279, 339–350a, 354, 422, 446, 449, 507, 510, 584, 585, 588, 609, 626, 651, 673, 687, 720, 747, 757.

108. *Characteristics of Biological and Medical Sample Material*

The analytical material encountered in biological and medical laboratories has certain characteristics favorable for flame analysis. But it often contains interfering constituents such as protein or sugar, seldom met in other applications, which complicate the procedure. The following peculiarities should be noted.

1. Unlike the samples of metallurgy, mining, etc., the materials are mostly *liquids* (blood, serum, urine, juices, perspiration, saliva, etc.), an advantage in flame photometry. Usually only liquids can be taken from a living subject (man or animal) without risk to life or health and without other drawbacks such as scarring after excision of a specimen. Sometimes, of course, the composition of solid bodies (bones, teeth, etc.) is of interest, but these cases are exceptional and occur chiefly in the examination of cadavers. We shall discuss mainly clinical analyses on living subjects.

2. The elements occur in fairly *constant concentration* in each body fluid, in contrast with analytical problems in industry. Even in urine, disregarding the water content, the relative concentrations of the elements vary only within moderate limits. This fact entails advantages and drawbacks for flame analysis: (*a*) *Precision* must be better than that needed for the commonest problems in industry and research, so that the small biological fluctuations that occur can be recognized with certainty. Of what use, for instance, would the usual $\pm 2\%$ precision be with concentrations (of serum sodium) which normally vary only $\pm 1\%$ in an experimental subject or animal? (*b*) On the other hand, the small *range of variation* is helpful in automating the procedure, which can then be handled by a technician; this is more difficult when the concentrations vary markedly.

3. Regarding the *composition* of body fluids, many of the elements in animals and man are not detectable, so that the spectrum has few lines, facilitating flame analysis; and there are few interferences from concomitants.

4. There are the following difficulties in this type of analysis: (*a*) Certain *precautionary measures* are needed to avoid communication of disease in handling infectious material. (*b*) Many body fluids do not keep well on *storage*. For instance, if 24-hour urine specimens are to be analyzed, sedimentation of urates during the unavoidable delay causes difficulty. Many body fluids become altered even during sampling; for example, the potassium content of serum is thus altered, so that, in principle, the determination of serum potassium *in vivo* might seem impossible. We return to this in Section 112. (*c*) The mode of *bonding* of certain analytes, for example, that of calcium to the serum protein, is very complex and variable. These conditions can hardly be adequately simulated in standard solutions, and many errors may result (see Section 114). (*d*) Many *concomitants* such as protein and bilirubin have a considerable effect on the surface tension and viscosity, and the variation of their concentration leads to additional interference.

109. *Ashing and Extraction*

Many biological fluids can be directly atomized and analyzed after appropriate dilution, without other treatment. This simple preparation is a great advantage for flame analysis. But with tissues, bones, feces, etc., a preliminary ashing is necessary. Also, body fluids will sometimes be ashed, as when the initial concentration of the analyte is insufficient, or when the interference of protein, sugar, etc., must be avoided. We discuss here some aspects of ashing and extraction in the preparation of samples for flame analysis; but we cannot list all known methods. For further information see the compendia and handbooks.[20, 21, etc.]

Before ashing, the moisture should first be determined; after drying, the sample is weighed in a weighing dish. Liquids can be pipetted quantitatively and dried. The actual ashing is not undertaken until after drying. A preliminary careful drying prior to the actual ashing helps to prevent loss of material by spurting and foaming during the first stages of the ashing.

Drying is most simply done at temperatures of 80–100°C. Tissue sections and the like are best dried in vacuum; the material may be warmed to speed the drying. Caution is needed to avoid loss of material by spurting. For surest results, freeze-drying is recommended, but it takes time and special equipment. If freeze-drying is not resorted to, a single-stage (gas-ballast) pump is satisfactory together with a drying agent. The evacuated vapors must be vented, for the odor of decomposing protein is disagreeable. The pump oil must be occasionally renewed. The weighing, drying, and second weighing are most easily done in weighing bottles that can be closed with a ground lid, to prevent evaporation of liquid before the first weighing and, later, contamination with dust.

The actual ashing can be either dry or wet. Dry ashing without use of oxidizing agents has the advantage of adding no impurities via the reagents. It is best carried out in quartz dishes that are not too small; quartz is preferred because it gives off no alkalies and is proof against thermal shock. The dishes should have at least six times the volume of the material to be ashed. Platinum dishes can also be used.

1. Dry Ashing.[143] This is usually done in a controlled electric muffle furnace at 550°C. Heating should be gradual, so that evolution of gas will be slow and puffing minimized. The material should remain in the oven until a grey ash is obtained. The temperature must on no account exceed 700°C. lest volatile substances vaporize; nor may it be too low, or ashing will be incomplete or take too long. After ashing, the sample is allowed to cool in the oven, an excess of concentrated reagent-grade hydrochloric acid is added to the ash, and the excess acid is evaporated by careful heating. The residue is then reheated to 550°C. in the oven until the powder is white; generally this takes 1–2 hours. But if a white powder is not obtained after many hours of heating at 550°C., it should be cooled again; this time pure distilled 25% nitric acid or similar oxidant is added, and the material is dried again and heated a third time to 550°C. Any remaining color will then be due to iron and cannot be removed. After cooling, a few drops of pure concentrated hydrochloric

acid are added and the solution is diluted quantitatively with water or other suitable solvent. With direct atomizers, dilution with acetone or the like will improve the detection limits.

The solutions should not be filtered even with ash-free paper, which may still yield traces of ash, easily detectable by the flame; furthermore, the paper fibers adsorb many substances. Suspended particles are best removed by centrifuging, after which the clear supernatant is carefully pipetted off. But if filter paper is used nevertheless, it must not be touched, for sweat and dirt on the fingers contain much sodium and potassium. Moreover, the ash must not be stirred with a glass rod or the like, for the friction may transfer alkalies from the rod to the sample.

2. Wet Ashing. There are many methods of wet ashing; we shall limit ourselves to describing one that uses nitric acid exclusively.[20, 21]

The material to be ashed (perhaps a fluid such as serum) is treated in a Kjeldahl flask with concentrated reagent-grade nitric acid and allowed to stand a few hours. For liquid samples an equal volume of nitric acid is taken, and for solids about 5–10 parts. After this digestion, intended to prevent foaming later, the mixture is heated carefully in a hood. As soon as evolution of the brown nitrous fumes has ceased, the material is cooled and treated again carefully with nitric acid, and reheated; this cycle is repeated until the residue is no longer dark. The last residues of nitric acid must be fumed off very thoroughly. After cooling, the residue is taken up in dilute hydrochloric acid or the like, as for dry ashing.

3. Extraction. Several authors [269, 720] have proposed to replace the tedious ashing process by extraction of the desired constituents. For this purpose the biological material to be analyzed is comminuted, weighed, perhaps also dried, and treated for several hours with at least 10 parts by weight of $0.75N$ nitric acid or sometimes merely boiled with water. The supernatant solution is diluted 5–20 times for the actual determination. The standard solutions must contain the same concentration of nitric acid. How well the analytical results of extraction agree with those of ashing, does not seem to have been investigated.

110. *The Removal of Protein*

Some authors recommend removal of protein before the actual flame analysis, usually when an atomizer-burner is to be used. There are the following advantages:

1. Removal prevents interference by coagulated protein, clogging and fouling of the atomizer, and crooked burning of the flame from deposits on the burner tip.

2. The viscosity of the protein-free sample solutions is closer to that of the aqueous standard solutions. Dilution need not be so great, and the flame emission will be stronger and more easily measured.

3. There will be no effect of the protein on the temperature, the excitation of the elements in the flame, and the background and blank.

Disadvantages include the following:

1. There is more work in preparing for the analysis.

2. There is the risk that the reagents, such as trichloroacetic acid, may introduce impurities into the sample. In any case, such errors are largely compensated by adding the same amount of reagent to the standards and the blank.

3. The interference of phosphorus with calcium is diminished by protein.[187] When the protein is precipitated, the variable phosphorus content of the serum exercises a stronger interference.

4. Certain elements such as calcium and copper are more or less strongly bound to the protein, partly by adsorption, partly chemically. On removal of the protein, some of the analyte may be coprecipitated and lost for analysis.[475a, 588] This effect can be kept small by prior acidification. Nevertheless, the analytical results come out lower than when the protein is not removed; but with uniform technique they can be validly compared among themselves. Removal of protein takes less time than ashing. However, ashing (especially dry ashing) avoids drawbacks 2 and 4 of deproteinization.

Protein is removed generally with trichloroacetic acid, sometimes with sodium tungstate or other reagents. The standard solutions must contain matching concentrations of trichloroacetic acid, sodium tungstate, etc. Trichloroacetic acid offers the additional advantage, like ethylenediaminetetraacetic acid (Section 100), of diminishing or removing certain interferences (see also ref. 661).

111. *The Determination of Sodium in Body Fluids*

Most body fluids, tissue extracts, etc., on introduction into a flame yield an intense yellow color, showing that the light of the sodium doublet 589 mμ preponderates and that the element is present in appreciable concentrations. Owing to this dominance the filtration or spectral separation of the Na 589 mμ line is easy in these applications. Indeed, it may be possible to do without spectral separation. Also, it is not hard to keep the containers and additives such as distilled water sufficiently free from sodium, except in ultramicromethods. Interferences from other elements, owing to the dominance of sodium, are *a priori* improbable. It is possible to make out with standard solutions containing only sodium, but with hot flames they should contain a constant amount of potassium; cooler flames are preferable, as they avoid ionization interference (Section 97).

Several items are worth noting. Dilutions should be such that the linear part of the working curve can be used (Section 79). The fairly high dilution needed for this (typically 1:1000) generally obviates the need of deproteinization or ashing. Analyses can be made also in the curved region of the working curve of Na 589 mμ, if certain disadvantages can be tolerated (Section 79).

Since the sodium doublet at 819 mμ has a straight working curve at higher concentrations, it can be employed without high dilution. The line at 330 mμ, requiring a spectrophotometer for isolation, is usable.[349] If the sample is not to be diluted greatly, some authors [64,184] recommend deproteinization with trichloroacetic acid (and addition of isopropyl alcohol) for sodium and potassium analyses, especially with atomizer-burners. This will avoid interference from the protein, as through incrustation of the atomizer tip at high concentrations (cf. Section 110); but, naturally, it will add to the time needed for preparing the sample (serum) for analysis.

The following works give details on the determination of sodium in body fluids: refs. 64, 248, 269, 272, 339, 340, 342, 349, 354, 355, 360, 364, 418, 419, 420, 436, 500, 524, 531, 533, 545, 611, 613, 661, 663, 708, 741, 750.

112. *The Determination of Potassium in Body Fluids*

Most biological fluids contain less potassium than sodium. Even so, the flame determination of potassium is simple, aside from the interferences of biological origin mentioned below. Generally the potassium doublet at 767 mμ is used. In hot flames the potassium readings are affected by sodium, and so cooler flames are preferable. Otherwise the error in the potassium reading resulting from the sodium must be compensated by one of the methods of Sections 97 and 100. The parametric method is commonly used for this purpose. A series of standard solutions suitable for the analysis of serum was listed in Section 83. Standards for serum are now available commercially.*

* For example, Dade Reagents, Inc., Miami, Fla.; E. Merck A.-G., Darmstadt, Germany; and ref. 513.

If the sample for potassium analysis is taken improperly, appreciable errors may result, which are likely to be blamed on the flame photometry; these sources of error therefore merit discussion. Incidentally, errors of this type were mostly first recognized, and their causes clarified, through the flame method. These biological interferences will be explained here using the potassium determination as an example.

The concentration of potassium in the blood corpuscles and especially in thrombocytes[542] is much higher than in serum or plasma. Between these corpuscles and the serum there is a continual exchange of potassium and an antagonistic exchange of sodium. The concentration gradient is maintained by the life processes, demanding a suitable blood sugar concentration, body temperature, etc. When a specimen is taken, potassium very quickly begins to diffuse from the corpuscles into the serum or plasma,[136] owing to the decline of blood sugar concentration, change of pH and temperature, etc. This effect falsifies the *in vivo* potassium concentration; only an artifact is measured. A proper technique of blood sampling (Section 113) will keep this error small and nearly constant. Basically, however, it is impossible, by any of the methods mentioned here, to determine the potassium actually present *in vivo*. The diffusion that begins at once on taking the blood depends strongly on the temperature, and so the potassium value will vary with the temperatures at which the blood was taken and further processed into serum.[94,136] The external conditions should therefore be held constant in such investigations.

A rise of potassium content can also result from hemolysis in varying degree.[136] Hemolysis and diffusion cannot strictly be distinguished. Absorption spectroscopy shows, by the hemoglobin absorption bands, that nearly every serum is more or less hemolytic. This hemolysis is rarely intravital (e.g., in hemolytic jaundice). However, it is usually induced by the technique of sampling and the manipulations of preparing the serum.

It might seem possible to correct a potassium value raised by hemolysis, by simultaneous determination of hemoglobin in the serum; but there is a fallacy in that the diffusion of potassium from the cells is not necessarily directly related to the release of hemoglobin. In any case, noticeably hemolyzed samples can be marked, to avoid errors in evaluating the potassium concentrations.

The possibility must also be heeded that the original potassium concentration may be falsified even *in vivo* through psychological effects or the like prior to the actual taking of the blood specimen. Anxiety of the subject before puncture by the hypodermic needle and the pain itself play a part here;[95] the emotion causes the serum potas-

§ 113] DETERMINATION OF SERUM ELECTROLYTES 347

sium to decline abruptly. This can be easily demonstrated by filling a sequence of vials with successive blood samples, marked with the time of removal, and centrifuging and analyzing them. During the first 20 seconds of blood withdrawal the potassium falls very distinctly by several mg. per 100 ml., and then returns slowly to the initial value. Neither the initial nor the final value of such a series yields (not counting diffusion, discussed above) the true concentration present *in vivo*, for a certain amount of time is needed to fill the first vial, during which the serum potassium is changing, while the final value is further falsified by loss of blood. Moreover, the series may be interrupted if the hypodermic needle used for withdrawal of blood becomes clogged with coagula. This drop in potassium concentration can be eliminated by complete narcosis, and sometimes by local anesthesia of the area to be punctured,[95] but it will still not be certain that the potassium values thus found (apart from diffusion) have not been altered by the narcotic or the anesthetic. Incidentally a kind of acclimation occurs, for after repeated puncture the concentration drop diminishes.[406] For practical purposes we conclude that if blood is to be taken for various laboratory examinations, only the first sample should be used for potassium analysis. In comparative experiments the time between puncture and filling of the first tube should be about the same.

Obviously every precaution should be taken regarding cleanliness in the withdrawal of blood, and every manipulation should be avoided that might cause hemolysis. The next section presents a sampling technique which takes all these matters into account, including some that are important for calcium analysis in serum (Section 114).

The following publications on the determination of potassium in body fluids are recommended: refs. 64, 95, 132, 136, 248, 269, 272, 313, 339, 340, 342, 354, 355, 360, 364, 406, 418, 419, 420, 436, 500, 524, 531, 533, 542, 545, 611, 613, 654, 661, 671, 708, 741, 750.

113. *Blood Sampling for Determination of Serum Electrolytes*

1. Blood is taken in the morning from the patient, still in bed and fasting 8 hours. (Take note of daily rhythm, seasonal variation of ion concentrations, effect of eating, effect of motion and breathing, effect of oxygen saturation of the blood, effect of therapeutics such as calcium injections or the like.)

2. The blood needle should be large, with a polished inner surface, clean and dry. The vein should be punctured shortly after being blocked, and the tourniquet removed shortly before withdrawal of the needle.

3. The first 0.5–1 ml. of blood should be drawn with a 1 or 2 ml. syringe and discarded. The syringe is then removed from the needle remaining in the vein, and the blood emerging from the needle allowed to flow slowly into a clean, dry centrifuge tube held beneath it. If the plasma is to be analyzed, the centrifuge tube should be coated with silicone (Section 80). If blood is taken at the same time for other purposes, then only the first specimen should be used for the potassium or calcium determination. Aside from that, the time taken to fill a centrifuge tube should be as uniform as possible; otherwise, the drop in potassium during the removal may have a noticeable effect.

4. If there is need of haste, the plasma can be analyzed. For this purpose, the blood sample is mixed with an anticoagulant such as heparin or ethylenediaminetetraacetic acid (EDTA) and then centrifuged at once. Note that the normal values for plasma and serum are somewhat different.[542] The standards and blank must contain the proper amount of anticoagulant; this is especially important in calcium determinations when EDTA is added. When there is less need of haste, we recommend analyzing the serum, since diluted serum atomizes better than plasma at equal dilution. The coagulation time during which the serum stands over the blood clot must always be the same (about $\frac{1}{2}$–1 hour), because the potassium value rises on protracted standing owing to diffusion out of the corpuscles. This diffusion can be inhibited for a while by adding fructose to the freshly sampled blood;[136] an equivalent quantity of fructose should then be added to the standards and blank. In handling the tubes of blood, one should avoid shaking (impact causes hemolysis), changes of temperature (the blood must not be placed in a refrigerator or at an open window), exposure to sunlight (ultraviolet light causes hemolysis), contamination by dust, etc.

5. After centrifugation, sometimes double centrifugation, the sample must be immediately pipetted quantitatively into a clean, dry tube. Otherwise, potassium will diffuse on continued standing from the blood clot into the supernatant serum. A clean, dry transfer pipet should be used for taking this sample. The storage tube should be tightly closed with a rubber stopper. In this condition the serum can be kept in a refrigerator until the actual analysis (at most 1 week with sodium and potassium analyses). If this pipetting is not done directly after the centrifugation, stratification may appear on prolonged standing, which can lead to considerable error especially in calcium analyses.[491] Even shaking a serum which has stood for some time over the clot before pipetting cannot completely reverse this unmixing phenomenon. In calcium analyses a delay of more than 1 day is inadvisable even with the above precautions.

During the entire procedure, contamination of every kind must be avoided, as from dried drops of tap-water (containing calcium), dust, cellulose fibers, or particles of cork, which can lead to clogging of the atomizer or liberation of calcium. The whole procedure, from withdrawal of the blood, through centrifugation, to pipetting, should always be carried out at as uniform a temperature as possible, for the potassium interchange with the erythrocytes depends strongly on temperature.[94,136]

The reproducibility of the electrolyte concentrations, especially potassium, will depend, further, on the state of excitement of the patient (fear and pain of the puncture, sight of blood, etc.) Vegetatively unstable patients will hence show greater variations among samples. Even the personality of the physician is not without effect.

114. *The Determination of Calcium in Body Fluids*

The determination of sodium and potassium by flame photometry or flame spectrophotometry is now widespread owing to its simplicity, speed, accuracy, and low consumption of material. But thus far the determination of calcium by the flame has met with only partial favor, although clinicians, among others, are not less interested in it than in the alkali metals. There are several reasons why the flame determination of calcium in body fluids has lagged:

1. The determination of calcium by the flame is more difficult, as alkaline-earth metals are harder to excite, and the emission intensities are therefore considerably weaker; the calcium line at 423 mμ is not strong in the conventional air–acetylene flame. Accordingly, the CaO(H) bands at 554 or 623 mμ are generally used for detecting calcium in these flames. The band at 623 mμ is preferred to the green band, being easier to separate from the sodium line at 589 mμ. With the hotter oxy-acetylene flame, the line at 423 mμ is often used.

2. Calcium is less abundant in the body. For this reason and the preceding one, many flame photometers have insufficient sensitivity for the determination of calcium in diluted body fluids.

3. The relatively strong CaO(H) bands at 554 and 623 mμ are close to the very strong sodium doublet at 589 mμ and therefore hard to separate with filters (Section 40). Even monochromators may pass an appreciable fraction of stray sodium light, because of its dominance. Moreover, the flame background at the CaO(H) bands is raised by the sodium. Calcium determinations may consequently be afflicted with cross-sensitivities (Section 94) and background interferences (Section

95), which complicate the procedure or cause error if the method is inadequate. In hot flames the weak lines of sodium at 568.5 and 615.7 mµ can interfere.

4. Conventional flame temperatures are not high enough for complete evaporation of aerosol particles containing certain calcium compounds such as the Ca–P(–O) complex. A portion of the calcium is thereby lost to analysis (Section 98). Hence interferences from concomitants are more numerous and severe with calcium than with sodium and potassium.

5. Calcium is only partly ionized in many body fluids, particularly in serum. Most of it is bound in a very complex manner to the serum protein, in part chemically, in part adsorptively. These variable complexes cannot be simulated with standard solutions. Hence, depending on the type and amount of protein and its calcium compounds, more or less serious errors may be expected unless suitable precautionary measures are taken (see below). Further, the serum protein, owing to its content of carbon and hydrogen, acts rather like alcohol on the flame temperature and drop size distribution (Section 79). On prolonged standing (more than 1 day), the serum calcium tends to stratify, and cannot be completely rehomogenized by shaking prior to pipetting.[491] This drawback applies, of course, also to chemical determinations of calcium, in which, however, it is less marked because of the larger samples taken.

6. Chemical determinations of calcium are easier and more accurate than those of sodium and potassium and are therefore still used more commonly.

In view of the need for a simple, rapid method for calcium, much effort is being spent toward overcoming these difficulties. Some of these studies are described below.

(*a*) Preliminary Ashing. Clearly, the difficulties might be removed by first ashing enough of the body fluid, e.g., serum (Section 109). The interfering protein compounds are thus destroyed and a uniform anion is introduced, such as nitrate or chloride, which is easily simulated by standards, and the concentration is raised by redissolving the residue with a smaller amount of liquid. This procedure, properly carried out, doubtless yields accurate results, but much time is needed for the ashing, whereby the advantage of speed in the flame analysis is lost. Besides, the sodium, potassium, etc., remain, and so blank interferences must still be allowed for in the flame photometry. Incidentally, ashing has also been used with chemical determinations, to avoid interference from precipitated protein during titration.

(b) Chemical Separation and Flame Photometry. It has been variously suggested[185,502,508,509,622] that the calcium be precipitated chemically (as by oxalic acid), the precipitate washed and taken up with hydrochloric acid or the like, and the calcium determined by the flame. The advantages are the same as those of ashing, but in comparison with the purely chemical method (precipitation by oxalic acid followed by titration) there is little advantage in speed. The method is, however, quicker than that involving ashing.

An advantage of this chemical and flame method over method (a) is that sodium and potassium do not interfere even with inefficient calcium filters, as they are almost quantitatively removed. A small amount remaining in the precipitate does no harm. The standard solutions in this method need contain only calcium salts. An advantage over the chemical method (precipitation with oxalic acid and titration of the oxalate) is that coprecipitated magnesium oxalate does not interfere with the analysis, as it does not contribute to the calcium reading in the subsequent flame photometry. Likewise, concomitant reducing substances, such as residual precipitated protein, will not interfere, as they do in titration.

A disadvantage is the incomplete precipitation of calcium bound to protein, especially when the serum has stood for some time and protein has coagulated.[188,475a] This trouble can be avoided to a certain extent by first acidifying the serum in order to free the calcium from the protein. The protein is first precipitated (see below), e.g., with trichloroacetic acid. The supernatant calcium-containing solution is pipetted off and the calcium then precipitated as above. Protein removal by this method entails, of course, extra work. Moreover, some of the calcium may be coprecipitated with the protein despite the acidification.

(c) Precipitation of Protein. Errors caused by differences of surface tension and viscosity, or by complexing of the calcium by protein, can be kept small by first precipitating the protein (Section 110). Apart from the loss of time, the main disadvantage is that some of the calcium is coprecipitated with the protein. The calcium values therefore come out lower than when the protein is not removed.[588] However, replicate determinations show that the calcium value is more reproducible on deproteinized serum than on untreated serum (see below).

(d) Improved Methods Avoiding Treatment of the Serum. Since ashing, deproteinization, and precipitation prior to flame photometry can introduce errors and also make extra work, efforts have been directed to using the serum in its natural composition for flame analysis with the help of other refinements requiring less time. The methods discussed below are useful not only for calcium but also for other determinations.

Use can be made of an efficient instrument with photomultiplier, hot flame, and good optical resolution, offering sufficient sensitivity for calcium in highly diluted body fluids. Various interferences, including the effect of protein on the flame temperature and of phosphorus on the calcium emission, are small with an indirect atomizer in which the larger drops are well separated (cf. Section 98). With such an instrument, the method can be as simple as for sodium and potassium and without substantially greater errors, provided an adequate procedure is used, with blanks for compensating the sodium effect on the background (Section 95), and fairly high dilution (1:20 to 1:50) to avoid errors due to viscosity and surface tension differences between sample and standards and to keep the effect of phosphorus on calcium small.

The situation is more complex with a direct atomizer, which sprays more ballast liquid into the flame, so that differences in composition between sample and standard, such as might be due to omission of protein from the standards, have a much greater effect on flame temperature, drop evaporation, dissociation, and other factors. When greater accuracy is required, it is better to precipitate the protein in advance, to separate the calcium, or to ash the entire sample (see above). If these tedious operations are to be avoided, the errors can be kept small by the following expedients:

(i) A hot flame (oxyacetylene) provides better excitation and vaporization.

(ii) The flame temperature is maximized, for the same reasons (cf. Section 19).

(iii) An atomizer with low suction rate and high oxygen pressure produces small drops.

(iv) An instrument with good optical resolution (narrow spectral bandwidth) and low stray light will keep the variable background low and suppress cross-sensitivity. The calcium line at 423 mμ is then best.

(v) The sample is diluted further, preferably with organic solvent, which raises the emission by raising the temperature and/or reducing the drop size (Section 79). This diminishes the effects of organic constituents of the body fluid on the flame temperature and of surface-active agents like bilirubin on the drop size.

(vi) By means of suitable additives such as EDTA in the diluent, certain interferences, especially that of phosphorus on calcium, can be kept small.[350, 350a] If an additive such as a wetting agent is used, there is the risk that its effect on the calcium reading in the sample will differ from that in the standards, depending on the presence of other concomitants that may not have been included in the standards. For example, the addition of a wetting agent to serum samples and standards[513] and the addition of phosphorus[187] have been successful under the proper conditions. Strontium and other buffers (releasing agents) can also keep the errors small; see Section 98.

(*vii*) The standards can be made to simulate the body fluid more closely by adding such substances as gelatine, albumen, fructose, and urea.[500] The solutions may not be stable over long periods. Stabilized synthetic standards containing protein are commercially available for some applications (see the footnote in Section 112). Even in these, however, the calcium is not combined with the protein, and this can cause error. (The magnitudes of all these errors depend greatly on the experimental conditions.) Body fluids with known (chemically determined) analyte concentrations can also be used as standards.

(*viii*) Internal standardization is particularly recommended for viscous material and higher concentrations (Section 101).

(*ix*) Corrections can be determined separately, as by self-standardization, and applied to the results.

(*x*) Proper technique of sampling the body fluid (Section 113) and preparing it for flame photometry will reduce or eliminate various difficulties, such as stratification or unmixing.

In conclusion, we list the following references dealing with the determination of calcium in body fluids: 49, 86, 130, 138, 185, 187, 188, 221, 237, 246, 257, 269, 282, 315a, 337a, 339, 340, 342, 350, 350a, 354, 355, 364, 369, 370, 411, 419, 420, 451, 459, 474, 475a, 491, 502, 504, 508, 509, 524, 545, 564, 570, 596, 596a, 602a, 613, 617b, 621, 622, 646, 703, 708, 719, 746.

115. *The Determination of Other Elements in Biological Materials*

Since flame photometry takes less time and material than the corresponding chemical methods, the advantages of being able to extend this elegant method to the determination of other elements beside sodium, potassium, and calcium in body fluids, tissue extracts, etc., have long been apparent. But difficulties exist: (*a*) these other elements of interest for flame analysis are appreciably harder to excite; and (*b*) they occur in considerably lower concentrations than sodium, potassium, and calcium in tissues or body fluids.

These difficulties are multiplicative. The sensitivity and resolving power of the most commonly used commercial flame photometers and flame spectrophotometers are inadequate for the purpose. Hence, efficient spectrophotometers of various special designs have been built, usually with laboratory equipment,[345,346,710,712] to overcome these difficulties. The considerations of Section 89 apply here.

If such apparatus is not available, the difficulties can be met by starting with a larger sample, and ashing it or concentrating the analytes by precipitating and redissolving them in a smaller volume of liquid or acid for the flame analysis (Section 88). But this approach is tedious and the size of the sample precludes its widespread use. Very hot flames or absorption flame photometry might solve some of the problems.*
These developments are still in a state of flux.

We offer the following references:

Determination of magnesium: 49, 138, 153, 203b, 237, 346, 411, 509, 652, 700a, 708.
Determination of copper: 97, 345.
Determination of strontium: 321, 710.

Geology, Industry and Other Fields

116. *Water Analysis*

The simplest types of flame-photometric analysis are those of natural and industrial waters, as the concentrations of elements normally present lie in the range favorable for flame work. Dilution, ashing, etc., are thus unnecessary. But sometimes the analyte must be concentrated (Section 79) or chemically enriched (Section 88), when the initial concentration in the water is not high enough for flame determination. If the water is heavily contaminated, the larger particulate matter must be removed before analysis to avoid clogging the atomizer. The preparation of standard solutions is relatively simple, because, owing to the low concentration, interferents have little effect. It is therefore not surprising that water analyses were among the first to be carried out by flame photometry: in 1860 Kirchhoff and Bunsen discovered the element cesium in mineral waters.

Analyses are made both of natural spring, river, and sea water and of industrial waters such as distilled water, condensate, rinse waters, boiler feed water, and waste water. The deposits yielded by such waters are also analyzed[640]; these are normally dissolved for a flame analysis.

* Absorption flame photometry is, in fact, doing this very well indeed. Two recent examples are the determination of magnesium in serum, urine and food by R. Herrmann and W. Lang, *Z. ges. exptl. Med.*, **135**, 569–82 (1962), and the determination of lead, mercury, bismuth, nickel, zinc and cadmium by J. B. Willis, *Anal. Chem.*, **34**, 614–7 (1962).—*Translator*.

A classification by these categories is impractical, for industrial waters affect the composition of stream water, and vice versa. Moreover, there are connections with botany and zoology, since the growth of aquatic plants and animals depends on the mineral content of the water. There are also connections with mineralogy, for seeping rain water dissolves portions of the rock on its way to a spring, and so the composition of the spring water yields inferences regarding its path through the ground.

Sometimes a rare element such as lithium is added deliberately to natural waters to trace their route. This scheme is of interest to sanitarians, among others, who are concerned with the supply of domestic water. Curative mineral waters are also analyzed by flame photometry. A few examples will be given from among these various applications.

The strontium content of sea water of different oceans at various depths has been studied. The strontium line at 460.7 mμ is measured, and the flame background next to the line is subtracted. There is the related problem of coping with radioactive waste, containing radiostrontium, etc., which is often disposed of in the sea. The relatively high magnesium content of sea water may cause background interference in flame photometry.

With regard to industrial water, there is the matter of testing the quality of water purified by ion exchange, which has extensively displaced the costlier method of distillation. The residual concentrations of sodium, potassium, and calcium are often determined by flame analysis, which saves much time. The same is true of control in plants using any other method for treating water; it is important to determine when the purification medium, such as an ion-exchange resin, needs regeneration.

In irrigation for agriculture or gardening, the correct metal content is vital for success: on the one hand, irrigation water must not contain too much solute, such as sodium salts, lest clogging or corrosion should interrupt operations of extended duration; on the other hand, there is an upper limit to the concentration of certain constituents such as boron which are easily taken up by grain, etc., and can lead to toxic effects.[309] However, soils must have a certain minimal content of boron to prevent boron deficiency in plants.

Boiler feed-water installations generally use pretreated water, differing in composition from the untreated water used for cooling the condenser. It has to be very pure, to avoid corrosion. If the sodium and potassium concentrations of the condensate are monitored, any leakage occurring in the condenser is easily detected. The composition of the deposits in a boiler give a clue to means of avoiding or at least minimizing them; they greatly lower the efficiency of the boiler and condenser, as the layers of salts reduce heat transfer through the metallic walls. The corrosive action of the water on boilers, tubes, heat exchangers, etc., is also of interest. Such corrosion, which may progress swiftly, can cause boiler explosions. The chlorine content of water can serve as an index of its corrosiveness. The chlorine can be determined flame-photometrically by the methods of Sections 87 and 88. The sulfate content of waters has also been determined by flame photometry.[177]

We list the following references: 73, 89, 189, 201, 309, 359, 363, 383, 460, 484, 570, 624, 640, 659, 693, 731, 740, 751, 755, 756.

117. *Physics, Nuclear Physics, and Electronics*

Flame processes have naturally aroused the interest of physicists and physical chemists. The results of physicochemical investigation that are important for the practice of flame photometry were presented by way of introduction in Sections 5–11. The development of means for keeping flames constant has been valuable also for experimental and theoretical studies, as in the measurement of excitation energies, transition probabilities, and equilibria of dissociation and ionization of molecules and atoms. The observation of interferences in the flame, though they are sometimes of small magnitude, has led to many theoretical investigations (Sections 7, 97, and 98), which have helped to clarify the complex processes in the flame (Section 5) and shed light also on those in arcs and sparks.[302]

Flame methods are ideal for carrying out experimental and theoretical studies of solubility, adsorption, corrosion, etc.—in short, any phenomenon for following which rapid and accurate liquid analysis is needed.[302]

There are other ways in which flame analysis can be useful in physical experiments. In nuclear fission, the uranium concentration might be determined by the flame in preparing the pure starting material; the uranium can be separated by ether extraction of the nitrate. Atomization of the ether extract into a hydrogen flame would be a simple and rapid method of determination, although the uranium emission is nonspecific.[302] The high sensitivity of the flame for alkali metals can be used in analyzing fission products, including their stable end-products, which are not detectable by radioactivity. Similarly, impurities in the starting materials, especially uranium, can be determined by flame spectrophotometry.[441a] A special method for the determination of sodium impurity in uranium has been reported. As uranium appreciably raises the flame background, self-standardization (Section 100) is used for the determination: a known quantity of sodium is added to the initial solution, in order to find the concentration originally present. The method is useful in the range from 2 to 5000 parts of sodium per million parts of uranium.[405]

A complication in the spectrochemical or flame-photometric analysis of nuclear materials, aside from the hazard of toxicity (discussed below), is the isotope shift of lines and bands. Thus, quantitative analysis of boron-10 solutions cannot be carried out with standards of normal boron.[162]*

In all research with radioactive material, it is necessary to avoid contamination of the air and risk to personnel. There are the following methods: (a) The radioactive material, if not of direct interest, is separated chemically. (b) The flame is operated in a closed burner, as in Fig. 20 (p. 120), the combustion products being purified by a filter before being vented to the atmosphere.[414] (c) Micro- or ultramicromethods (Section 89) are used, not only because the quantity of available sample may be minute, but also, in the interest of safety, to reduce the amount of material passing through the filter or entering the atmosphere.

In phosphors such as zinc cadmium sulfide the slightest impurities are important.[205,587] Flame analysis can serve to determine these impurities, some of which may be purposely added. One such method is used for sodium in the range 0.0001–0.001% and potassium at 0.001–0.02%. Zinc cadmium sulfide free from sodium and potassium is used for preparing standards with known additions of these elements.[205] More generally, the testing of highly purified metals or compounds for traces of impurities, as in the determination of physical constants of pure materials, is a good field for spectral analysis and, for the more easily excited elements, for flame photometry. The trace impurities of interest in semiconductor research (solid-state physics) are often, however, at such low concentrations as to be undetectable spectrochemically. Mass spectrometry is better suited to this analytical problem.

A final example of the use of flame analysis in applied physics is the study of the evaporation of barium, strontium, and calcium from the surface of an oxide cathode in a vacuum tube.[253,575] The evaporated material is caught on a quartz plate mounted in the vacuum tube. At the end of the test the condensate is dissolved and analyzed by flame photometry. Interferences from the components of the cathode coating as well as the supporting metal have to be allowed for.

Other references: 162, 205, 240, 253, 414, 441a, 540, 575, 587, 718.

---

* The isotope shifts in the green bands of BO_2 are beautifully shown in the paper by J. W. C. Johns, *Can. J. Phys.*, **39**, 1738–68 (1961).—*Translator*.

118. *Chemical Industry and Pharmacy*

Flame analysis with its high speed, specificity and sensitivity is applied in chemical industry and pharmacy to so many and such various problems that we can give no brief summary, especially since many of these analytical methods are unpublished. These applications include general analytical studies of raw materials, intermediates, and finished products in development and plant laboratories and in well-established chemical production processes. For example, certain process controls can be handled more or less automatically by a flame photometer, the readings of which serve to regulate a concentration during the process so as to optimize the operating conditions. For this purpose a small fraction of the process liquid flowing continually through a pipe can be diverted to the atomizer of a flame photometer. A recorder with the flame photometer then effects industrial process control for long periods; also, alarms can be activated if the permissible range of concentration is overstepped. In this way the processes in cation-exchange columns have been followed automatically for long periods.[528]

Of these and similar applications of flame photometry we can mention here only a few published examples. Purity tests of chemical products can be expedited by flame photometry, especially when it is a matter of detecting the alkali and alkaline-earth metals in the presence of each other or in any material. In such purity tests there are usually specified limits of concentration which must not be exceeded. Some of these tests can be carried out easily only by flame photometry.

In many industrial reactions it is necessary to remove unwanted sodium compounds from a precipitate by washing, as with water. A flame photometer will give the residual sodium content after each washing and quickly show whether the precipitate is pure enough or needs further washing. Also, in boiler feed-water installations the sodium content is important (Section 116).

For economical manufacture of phosphoric acid by the wet method, the by-product calcium sulfate must be filtered out of the acid as completely and reliably as possible. For this purpose the crystallization of the calcium sulfate is controlled for easy filtration. The crystal growth depends chiefly on the ratio of calcium to sulfate ion in the starting solution. Control of this ratio by the chemical method would be very tedious, because the oxalate must be precipitated twice, since iron, aluminum, and phosphate interfere with the analysis. A flame analysis cuts the time from several hours to a few minutes. But since aluminum and phosphate affect also the flame determination (Section 98), it is advantageous to use strontium as an internal standard. This method yields a precision entirely comparable with that of the chemical method.[205]

An example from the fertilizer industry is the preparation of potassium fertilizer salts by flotation. Here the flame photometer is well suited for following the content of potassium and other additives.

An example from the pharmaceutical industry is the preparation of nutrient solutions for the cultivation of antibiotics, in which the concentrations of calcium, magnesium, potassium, and sodium have to be controlled throughout the entire process.[302] Flame methods are convenient also for determining the purity of pharmaceuticals.

If an organic acid is suitably converted to the sodium salt, the salt will contain a characteristic percentage of sodium.[428] The sodium content, determined by flame photometry, will then help to identify the organic acid. But the method is not applicable to diluted acids or to mixtures of acids.

Industrial vapors, waste gases, smoke, etc., can be analyzed for alkalies by flame photometry.[302]

If a heavy metal is to be extracted, for instance, with dithizone in chloroform, it can be determined directly in the chloroform extract by flame spectrophotometry.[208] This technique could also be used for concentrating less easily excited heavy metals, to make them amenable to determination by flame methods (Section 88). In these applications, when chemical enrichment is resorted to, the availability of sample is usually a less limiting factor than in biological and medical applications. Flame methods have secured a firm place also in the paper industry,[106,400,543] the varnish industry,[200] industrial waste processing,[73] and many other applications.

The following are some additional references:

Chemical industry: 73, 106, 108, 154, 155, 200, 207, 252, 327a, 400, 528, 543, 702.

Pharmacy: 159, 314, 351, 383, 463.

119. *Geology and Mineralogy*

In the analysis of geological materials such as minerals, ores, clays, mineral waters, and petroleum, flame photometry can simplify the procedures. Since these are the raw materials for many branches of industry such as ceramics, glass, cement, and oil, analysis of these materials is important. Here we shall consider only geology and mineralogy, continuing in the following sections with the analytical problems of these industries.

In the early days of flame analysis, Kirchhoff and Bunsen were concerned with geological materials and discovered cesium in mineral waters (Section 4). Lundegårdh[33] also was active in mineral analysis, since the composition of minerals is closely related to agricultural chemistry, in which he worked.

Except for mineral waters (Section 116), the material for a flame analysis will generally have to be worked up, that is, brought into solution. The procedure will vary according to the composition of the material and the desired analysis. Minerals vary greatly in composition—much more, for instance, than biological and medical materials. Accordingly, the method of preparing the sample will depend on the physical properties and chemical composition of the individual sample. Directions for preparing various materials for flame analysis are given in the literature (ref. 532 and the references listed below). The flame-photometric procedure itself also depends greatly on the composition of the sample; the presence and relative concentrations of concomitant elements govern the use of various additives and corrective methods for avoiding interference (Sections 97–100). Since a rock cannot be analyzed directly but must be brought into solution (sulfuric and hydrofluoric acids are commonly used for decomposing it), the question of the time required for the preparative chemical manipulations is less important than it is, for instance, in medicine. Thus, interferents are commonly separated chemically before the actual flame analysis (for example, aluminum, iron, and silica are removed by treatment with ammonium chloride and calcium carbonate), so that these manipulations cannot be reckoned as impediments (cf. Section 88). The chemical procedure can sometimes be so arranged that the analytes are isolated, the interferents being automatically removed without additional steps.

On the other hand, various purely flame-photometric methods have been employed in special cases for eliminating interferences due to concomitants. Thus, in the determination of calcium in phosphate–carbonate–silicate rocks, the interferences of phosphorus and aluminum on the calcium are buffered by adding a large amount of magnesium. This method is relatively quick and the errors are only about 2%.[433]

Other pertinent points will be covered in the following sections on the glass, ceramic, cement and other industries. Here we append a selection of references on flame-photometric analyses and methods in geology and mineralogy: 1, 166, 247, 277, 309, 314b, 315, 317, 365, 433, 501, 516, 532, 562, 735, 736, 737, 742.

120. *The Glass Industry*

Physical properties such as viscosity which are important in the processing of a glass melt depend on the composition of the melt and therefore basically on the content of sodium and potassium. In many types of glass these constituents must be determined quite accurately, to permit reliable prediction of the behavior of the melt during further processing. A deviation of $\pm 0.2\%$ in the Na_2O content can result in appreciable changes in the glass melt. Larger negative deviations can lead, for instance, to stoppage of the processing machinery or to tears in the finished product. The sodium and potassium must be determined promptly in the fresh melt, before it can be released for further processing. Flame methods are here ideally suited for rapid analysis of many samples, and they are now solidly ensconced in glass laboratories.

The resolving power of the photometer employed must be high in this application, because plate glass contains traces of K_2O which must be determined along with about 15% of Na_2O and 10% of CaO.

The glass can be prepared for analysis somewhat more simply than minerals (Section 119), since glass powder, finely ground with an emery disk, is easily soluble in hydrofluoric and sulfuric acids, in hydrofluoric and perchloric acids, or (for sodium, potassium, and aluminum determinations) in hydrofluoric and oxalic acids[331,333] and it generally contains only traces of aluminum and titanium. When hydrofluoric and oxalic acids are used, the calcium, which might interfere, is separated at the same time. Owing to the low concentrations of aluminum and titanium, one can omit the precipitation with ammonia and addition of acid that are commonly employed with minerals. Sometimes, for relative measurements, it suffices to grind the glass powder in water.

For standardization, mixtures of known concentration in ratios similar to those of the sample are prepared as above. Recall that aluminum can interfere in the determination of calcium (Section 98). The mutual interference of sodium and potassium can be buffered by addition of barium chloride[332] (Section 97 and 100). If aluminum itself is to be determined, an organic solvent, especially n-butanol, added to the solution, will intensify its emission.[333] In these determinations, the varying concentrations of sodium, potassium, calcium, and iron must be kept in mind as possible sources of interference.

We refer to the following publications: refs. 102, 108, 142, 164, 193, 220, 260, 326, 328, 333, 335, 604, 605, 706, 743, 761, 762, 763.

121. *The Ceramic Industry*

There are many possible applications for the flame photometer in the laboratories of the ceramic industry. The alkali metals, for instance, are principal constituents of pastes and glazes, of slip, soda, and waterglass. Other materials such as clays, kaolins, or titanium dioxide contain little alkali, which, however, must be known, as the alkali metals affect the properties of the finished products. In fire clay the alkali content largely determines the degree of fire resistance and temperature of use, as the alkali acts as a flux at elevated temperatures. There is also interest in zonal analysis of used firebrick, e.g., glass-pot brick. In ceramic insulators the alkali content along with other ingredients determines the insulating properties. In the porcelain industry the pastes are made of feldspar, quartz and kaolin; the alkali content of the mixture is calculated as feldspar. These alkali values affect the refractoriness and shrinkage (ratio of size of the finished object to that of the starting form).

Ceramic products are prepared for flame analysis in the same way as minerals (Section 119), viz., with hydrofluoric acid, which removes silica at the same time. If fired products containing alumina are to be analyzed for calcium, the calcium may not dissolve entirely during the acid decomposition, presumably because of the presence of very resistant calcium aluminates. Such materials are more accurately analyzed for calcium gravimetrically after being decomposed by a potassium pyrosulfate fusion.

On this subject, see refs. 78, 88, 150, 209, 214, 217, 218, 219, 219a, 309, 330, 334, 368a, 445, 650, 716. On the flame-photometric analysis of refractories, see refs. 295, 592, 618, 619, 623.

122. *The Cement Industry*

The structure and mechanical properties of cements after setting are closely related to their composition. Several of the constituents of interest in this connection, especially magnesium, sodium, potassium, and calcium, can be determined in the raw material and the finished product considerably faster by flame photometry than by chemical methods. In the direct flame analysis of cements, there may be difficulties from incrustation or clogging of the atomizer and burner. Some preventive means are the following: (*a*) the lowest feasible concentrations of cement solution should be used; (*b*) a rinse with distilled water should follow each determination; (*c*) the samples should be prepared chemically so as to inhibit incrustation, as by adding hydrochloric acid to both samples and standards.

The magnesium content of cement has an important effect upon its strength and setting time.[302] Magnesium determinations can be carried out suitably at 371 mμ following precipitation of silicon, aluminum, and iron with ammonium carbonate.[192] Calcium markedly raises the flame background, owing to its preponderance; hence, unless the standards and blank contain the proper amount of calcium, the calcium should be determined flame-spectrophotometrically to permit correcting the background (Section 95). The interference of sulfate, chloride, and manganese on magnesium must be heeded. The calcium background continuum can interfere also in the determination of other elements.

The concentrations of sodium, potassium, lithium, and calcium in cements are important. The first three elements are prejudicial to quality. The determination of the sodium is beset with certain difficulties, as there is but little sodium (up to about 1%) along with much calcium. In the usual measurement with a glass filter or normal interference filter for isolating the sodium line, much calcium light from the CaO(H) bands in the green (554 mμ) and orange (623 mμ) will pass the filter, owing to the preponderance of calcium. Theoretically, these errors can be compensated by cross-sensitivity corrections (Section 94). But then interferences, as from aluminum, acting on the calcium may affect the sodium analysis via the cross-sensitivity of the sodium filter to calcium. For this analysis, therefore, a multilayer filter with small half-intensity bandwidth or a monochromator with sufficient dispersion and narrow slits is advisable. The calcium can perhaps then be omitted from the sodium standards. We should add that calcium is less susceptible to many interferences in a hotter flame. Conversely, the calcium background interference is smaller in a cooler flame. The interference of calcium with sodium can be suppressed by adding sufficient aluminum, suitably as nitrate, to the sample and standards.[633,634] Aluminum so greatly suppresses the calcium emission (Section 98) that it no longer interferes.

In most gravimetric methods of determining sodium, lithium is included in the result. Flame photometry has the advantage of permitting the separate determination of lithium along with the other alkali metals. In lithium determinations with instruments of limited resolution, calcium and strontium must be included in the standards at the proper concentrations, as these metals emit hydroxide bands that raise the background in the neighborhood of the lithium line at 671 mμ.[473]

See also the following: refs. 74, 98, 222, 224, 225, 226, 231, 244, 266, 308, 310, 473, 565, 686, 745.

123. *Petroleum and Fuels*

The atomization of combustible materials such as gasoline, benzene, or oils diluted with naphtha in flame photometers with indirect atomizers (Section 26) is somewhat risky, since the spray is explosive and can damage the apparatus if the flame flashes back. There is also the likelihood of fractionation, especially in heated spray chambers (Section 30). As an expedient, therefore, such materials have been evaporated and ashed, the residue being taken up with water or dilute acid for the flame analysis. But the development of the direct atomizer (Section 33) has made possible the safe analysis of combustible materials without prior evaporation and ashing. Naturally, some precautions will be needed (see below).

After it had been demonstrated that flame-photometric analysis of combustible liquids could be simple, rapid, and safe, this application became increasingly popular. The elements lead,[300, 302, 307a, 402, 661a] manganese,[661a] boron,[168] and copper[403] have been determined in petroleum fuels. These additives have important characteristics, such as antiknock quality and corrosiveness. Since the analysis of gasoline is typical of these applications, we shall describe it here, along with the appropriate safety precautions.

The oxyhydrogen flame is most commonly used for gasoline analysis, but other flames such as air–hydrogen can also be used. The fuel pressure or flow rate can be kept smaller than usual, since the gasoline spray contributes fuel (cf. Section 79). Since gasoline has lower viscosity and surface tension than water, the atomizer should have a narrow or long capillary, to restrict the flow and prevent flooding of the flame with gasoline. To keep the capillary clean, acetone should be sprayed between measurements.[661a]

The following safety measures are recommended for avoidance of accidents or fires. The sample cup when in position under the atomizer must be covered, e.g., with a punctured disk of neoprene slipped up over the inlet capillary; the disk has a second hole to permit equalization of pressure. This cover prevents burning and evaporation of the sample. As precaution against emergency, tongs should be handy for removing a sample cup that may have caught fire. The fire can be quenched by covering it with a beaker. The supplies of gasoline should be stored away from the flame photometer. A proper technique (see below) ensures that samples and standards remain exposed no longer than absolutely necessary, to prevent change of concentration by evaporation.

For best accuracy the following procedure is recommended. Have the sample and one or two standards at hand in closed bottles, and an open waste bottle; set the corresponding 5-ml. sample cups beside them on a sheet of paper. For transfer use a 1-ml. dropper, kept in a glass cylinder padded at the bottom with absorbent material. Open the sample bottle and transfer gasoline with the dropper to the sample cup to rinse it, pour into the waste bottle, and blow out the last drop in the tip of the dropper. Then transfer more of the same sample to the rinsed cup. Return the dropper at once to its cylinder, raise the sample cup under the atomizer, and read the spectrophotometer. Pour the rest of the sample into the waste bottle and invert the sample cup on the paper; close the sample bottle and record the reading. Repeat this procedure for the next sample or standard. The time required for one such operation is 35 seconds. Ordinary washing or wiping of the sample cups will not suffice.

Difficulties in the direct flame analysis of gasoline result primarily from differences among the base stocks. The volatile hydrocarbons, occurring in variable concentrations, that compose the gasoline, have different viscosities, surface tensions, boiling points, and heats of combustion, and different effects on the flame temperature and on the flame background at the analytical line. It is impracticable to simulate the composition of the sample in the standards or blank. Errors due to these differences can be minimized or eliminated by the following means:

1. Internal standardization (Sections 51 and 101) will compensate differences in surface tension or viscosity, but is less useful for differences of temperature or background.

2. Self-standardization (Section 100) has been used, but the frequent pipetting of volatile liquids is inconvenient and susceptible to errors of evaporation.

3. The sample flow rate can be controlled (ref. 307a; cf. Section 33). However, the necessary equipment is not yet commercially available.*

4. The sample, standards, and blank are diluted with a uniform diluent such as isooctane. This diminishes the differences among the solutions and base-stock errors.[402, 481] However, dilution reduces the accuracy available for an element when its concentration comes too close to the detection limit.

Besides gasoline, flame analysis is used for lubricants, motor oils, and the like. The lubricating and corrosive properties of new oil can be predicted, to a considerable extent, from its composition. Certain metals in lubricating oil markedly affect its corrosiveness toward metals. For example, copper salts present in oil enhance oxidation and lead to the formation of sludge. Spectrochemical studies of corroded steels provide evidence regarding the mechanism of this corrosion. The metals found in used oil show which constituents of an alloy are attacked

* The new Chaffee–Keyes flame photometer (Table 15) incorporates sample feed rate control.—*Translator.*

by the oil, and which are unaffected. Such studies are valuable in the effort to improve existing lubricants. Fig. 61 shows a flame-spectrophotometric recording of a diluted lubricating oil.

Procedural details in oil analysis are similar to those described above for gasoline. The following are a few examples of applications. Special methods have been developed for the determination of calcium in lubricating oils; the oil is first diluted with naphtha. Certain constituents of crude oil, such as iron and nickel, destroy the catalysts used

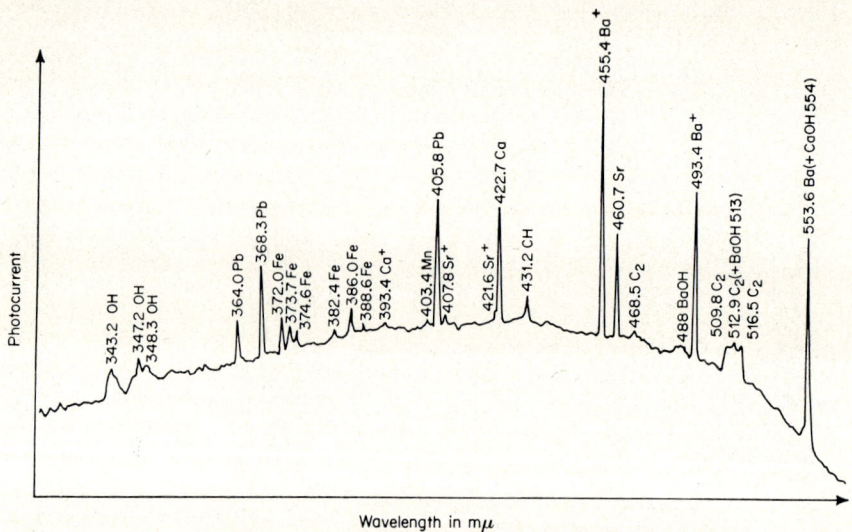

Fig. 61. Flame spectrum of a (diluted) used motor oil. Aside from a few bands in the flame background and the spectra of the additives, there are lines of impurities, indicative of the origin and quality of the oil, and of corrosion occurring during operation (from ref. 311).

in processing the oil. These metals, some of which can be detected by flame analysis, should be removed before processing.[265] The presence of sodium in steam engine oil can point to leakage of cooling water, or the like. In compressor residues, the metals iron, nickel, chromium, lead, copper, magnesium, and manganese are of interest and can be detected by flame photometry.[302]

Since coal is a raw material for the synthesis of hydrocarbon fuels, the elements present in various kinds of coal are of interest. We return to this in the next section.

There are the following additional references on petroleum fuels and oil: 75, 168, 265, 300, 302, 307a, 402, 403, 412, 481, 496, 661a.

124. The Metallurgical Industries: Analysis of Metals and Slags

Although metallurgical materials, being chiefly solids, are often analyzed directly by spark or arc, flame photometry can be very helpful in special studies. Analyses are performed on ores (chalcopyrite, bauxite), fluxes (calcium carbonate), intermediate products (pig iron, crude copper, spelter), final products (alloyed steels, aluminum–magnesium alloys, brass, bronze), and waste products such as slags. Slags of suitable composition can serve in turn as raw material for other industries (cement, Thomas meal). Analyses are also carried out on fuel ash,[635] exhaust gases, waste water (Section 116), etc.

The chemical composition of the ore from a given deposit and of the associated rocks shows whether it can be worked profitably, indicates the properties of the crude metal that it will yield, and suggests suitable methods of beneficiation (e.g., flotation) and extraction of the metal (roasting, smelting with suitable fluxes, electrolysis). The physical and chemical properties (melting point, density, tensility, hardness, elasticity, corrosion resistance) of the intermediate and end products depend on the concentrations of foreign elements. Apart from the effects of composition, the properties can be controlled within limits by various treatments (mechanical working, heat treatment, aging), which are of less interest from the point of view of flame photometry. Naturally, there are various interconnections with mineralogy and geology (Section 119) and with all branches of industry having to do with the processing of metals and slags.

We cite below a few of the many metallurgical applications of flame photometry.

In cast iron, the carbon, silicon, manganese, phosphorus, sulfur, and titanium are usually determined chemically or spectrochemically with the spark. Sometimes also calcium, magnesium, sodium, and lithium are determined with the flame. According to one such method[435] the metal is dissolved in hydrochloric acid and the insolubles filtered off. Most of the iron is extracted from the solution with ether. After further treatment for removal of ether and insoluble material, the aqueous solution is analyzed directly with a flame spectrophotometer against suitable standard solutions.

In the above example the unwanted iron was removed by extraction. Conversely, if iron is to be determined at low concentration in another metal, it can be concentrated by this same expedient. Thus, to determine iron in an aluminum alloy,[208] the iron is extracted with acetylacetone, in which it is more soluble than the other components of the alloy. Many interferences are avoided in this way, and the combustible solvent adds to the size and energy of the flame. Iron can be determined thus in the presence of large amounts of magnesium, calcium, and aluminum.

Other elements are also of interest in aluminum alloys, for instance, sodium, for which a special method has been developed.[161] The sample is dissolved in a mixture of hydrochloric and nitric acids, lithium chloride is added as internal standard, and the solution is diluted with distilled water and methyl alcohol.

A method has been worked out for the determination of lithium in magnesium–lithium–aluminum alloys.[679] The alloy is dissolved in hydrochloric acid, suitably diluted, and analyzed for lithium at 671 mμ. The interferences of magnesium and aluminum on lithium must be allowed for. Conversely, there is a method for determining residual sodium and potassium in purified lithium metal.[380]

The copper content of commercial alloys can be determined by flame photometry.[207] In this method a known concentration of silver can be added, the silver line at 328.1 mμ serving as internal-standard line. In complex copper–zinc alloys containing lead, manganese, nickel, etc., the concomitants can interfere; absorption flame photometry has proved valuable for analyzing these alloys, since it avoids the interferences.[353]

In blast furnace slags, calcium and magnesium are of interest. An addition of strontium will mitigate various interferences, such as that of silica on calcium.[59] To eliminate this interference entirely, the silica can be fumed off at the start with hydrofluoric acid, or filtered off, or left in solution while the residual error is corrected numerically; the last method is particularly recommended for slag analysis.

We list the following references:

Smelting: 59, 126, 156, 161, 167, 179, 210, 212, 213, 231, 239.

Metallurgy: 41a, 161, 268, 280a, 322, 324, 327, 367, 379, 435, 485, 487, 506, 512, 561, 679, 699, 725, 733, 743.

Slag analysis: 379, 388, 515, 583, 677, 692, 715.

125. *Criminology*

The criminologist is often confronted with the problem of identifying materials, marks of tools or weapons, etc., found at the site of a crime, with other materials or objects found upon a suspect. He may also have to determine the origin of traces of pigment, oil, or the like, in the clothing of a prisoner in connection with an alibi. Objective proof of the origin of a suspicious trace can be valuable evidence regarding a fire, traffic accident, etc. For this purpose spectrochemical methods, along with microphotography, x-ray analysis, gas chromatography, etc., are ideal, because they require so little material. Flame photometry excels for the determination of the alkali and alkaline-earth metals and is sometimes used for other metals.

In proceedings involving murder by poison, it is necessary to establish the cause of death with certainty. Flame-spectrographic determinations of thallium and lead can be useful here, perhaps after preliminary concentration of the metal.

A somewhat different example is the following. Much of the illegal traffic in narcotics is in opium, which is of interest directly as such, and also as the basis of a number of other preparations. In order to control the international opium traffic, the geographical origin of the material must be established, as an aid to suppressing the illegal commerce at its source. For this purpose the chemical and physical properties of opium from all source countries have been thoroughly investigated, and the ash has been analyzed flame-photometrically. With the help of these data the composition of the ash points to the origin of the opium.[91]

APPENDIX

Introduction to Tables 13 and 14

These tables list intensities and wavelengths of 3400 lines and bands of 145 atoms, ions, and molecules representing 70 elements that can be determined by flame photometry. The data have been compiled by the translator, largely from original spectrograms. Nearly every intensity value has been measured directly from recorded spectrograms or estimated from photographic spectrograms (usually with the help of a step-sector disk). About 400 spectrograms (200 recorded and 200 photographic) have been used in this work. Most of the wavelengths have been scrutinized and many have been corroborated from grating spectrograms. Not a few are new measurements or revisions of earlier wavelengths.

Still, these tables are incomplete and preliminary. Many original sources in obscure or old journals or books have not been examined; not all of the available spectrograms have been fully measured; and there exist many pertinent spectrograms that the translator has not seen. Moreover, except for a few later corrections, the work on these tables was completed before the end of 1961. The flood of new data on flame spectra in 1962 has been such that, were these new data to be incorporated, the tables would be nearly doubled in bulk. In the footnote at the beginning of Section 81 were mentioned the chief of these recent contributions—by Fassel, Curry, Myers, Kniseley and Buell.

These tables are intended as an aid to analytical flame spectrophotometry. Accordingly, many weak bands in the richer band spectra and most of the detailed structure of polyatomic bands are ignored. Rotational lines are listed for only a few bands of open structure (CH, CuH, OH). Many molecules emitting known flame bands [24, 66, 77]* are omitted as having (as yet) little analytical utility, and others are omitted owing to lack of information or time to collate the data—e.g., TlH,[7] MgH,[6a, 41] and the compounds of gold and rhenium that emit flame bands. Hartley's paper [41] was helpful for several elements, but his data on Sb, Bi, Se, and Ag have not proved decipherable in terms of modern wavelengths. There is clearly much room for new descriptions of neglected spectra known since the last century.

For supplementary information on the spectra in various flames and solvents and for more precise data on spectrochemical intensities or sensitivities, these tables should be used in conjunction with the briefer but broader tables by the translator [30, 31, 32, 34a, 35] as well as Table 10 of this book. The table in the *Handbook of Analytical Chemistry* [35] and the revised Beckman Bulletin 753-B, now in preparation (cf. ref. 32), are the fullest and most nearly up to date. The joint use of one of these briefer tables with Tables 13 and 14 and the Atlas at the end of the book will best serve the purposes of selecting analytical wavelengths, gaging spec-

* References in this section are to the special bibliography appended below.

APPENDIX

tral interferences, identifying elements, and estimating concentrations in semi-quantitative flame analysis.

Sources of Information

The table and bibliography following this Introduction show the sources used in compiling Tables 13 and 14. The meaning of each of the columns of this table is described below.

Species Identity. Where no reference is given for the identity of the species, it may be regarded as well known. Many of the molecules (e.g., HoO) are uncertain, there being little evidence to support the identity; some are entirely doubtful (e.g., EuOH). It is possible, consequently, that certain spectra now ascribed to a single species (e.g., BiO) are actually emitted by more than one molecule, and that in other cases spectra ascribed to different emitters (e.g., RaO, RaOH) may belong to the same one. Reference to one of Moore's works[60, 61, 62] for an atom signifies that her data were used to establish the character of the transition corresponding to each line. The excitation potentials in the Russian wavelength tables[91] have been helpful. Certain errors and omissions in these tables have been corrected with the help of Moore's later data. For example, the tin lines 2151, 2210, 2246, and 2483, assigned to Sn II by the standard tables,[40, 91] were found, by fitting them to Moore's energy levels,[62] to belong in fact to the neutral tin atom.*

Spectrograms Examined. The third column lists the sources of the spectrograms utilized for identifying and selecting the lines and bands and estimating their intensities. Roughly half the spectrograms employed were obtained at Beckman Instruments; whereas these were weighted more heavily for the intensity measurements, grating spectograms obtained chiefly from other laboratories were used for the more detailed wavelength measurements. A majority of these spectrograms are unpublished.

Wavelength Identification. The fourth column lists authorities for the occurrence and authorities for the wavelengths of the lines and bands. Except to the extent indicated by references to sources[5, 6, 9, 21, 37, 41, 42, 53, 54, 56, 59, 67, 71, 72, 75, 78] alleging or revealing the occurrence of given lines or bands in a flame (and for these the wavelengths have always been checked and often corrected), all identifications are on the responsibility of the translator and based on inspection of the spectrograms of column three. In a few cases[11, 38, 50, 68] spectrograms are mentioned that were used for auxiliary verification of the occurrence of certain debatable emissions. Authorities for the actual wavelengths of recognized lines or bands include standard compendia,[22, 23, 24, 40, 60, 61, 62, 66, 77] special articles on certain elements,[10, 39, 49, 51, 57, 65, 70, 74, 88, 89] and certain older works[21, 73, 76] when nothing better could be found. If the column is blank, the wavelengths were taken from standard works.[40, 66, 77, 91]

Wavelength Measurements. The fifth column shows the sources of the spectrograms from which new or revised wavelengths were measured. Unless otherwise noted, an entry in the fifth column indicates that many, most, or all of the

---

* Dr. Marvin Margoshes pointed out the case of Sn 2151; an improved value for its wavelength is given in the new tables by Meggers *et al.*[56a] Had these tables been available earlier, the task of identifying and characterizing lines could have been greatly simplified.

tabulated wavelengths are new measurements. It should be mentioned that many wavelengths had been measured prior to acquisition of the very fine atlas of molecular spectra by Gatterer et al.;[23] with the help of this atlas, many of the earlier data were corrected, replaced, or expanded. For some of the less well-defined spectra, however, wavelengths of this atlas were not accepted uncritically, but checked by measurement from the spectrograms of the atlas itself or from Curry's spectrograms[11] or, occasionally, from other sources.

Intensities. The last two columns list the spectrograms or other sources used for making the measurements or estimates of intensity. For many species, several sources, given under "intensity normalization", were compared for selection of a best set of intensities. If the last column is blank, the individual intensities are those of the normalized table. Entries in the last column indicate the sources of some or all of the individual data, supported, as the case may be, by the general level of intensity arrived at by the normalization. In a few cases (AsO r, BaCl, Cd^+, Ra species, SrCl r, and the blue SrO bands) it was not possible to relate the relative intensities to the conditions of analytical flame photometry; although internally consistent within each species, the absolute level of these intensities is merely guessed.

Acknowledgments. The translator wishes to acknowledge the aid of friends and colleagues, whose contributions were essential to the completion of these tables of flame spectra. Those who supplied spectrograms used in this work, either unpublished or in advance of publication, include A. W. Berger, Bruce E. Buell, Ronald H. Curry, Velmer A. Fassel, Roland Herrmann, Oscar Menis (in cooperation with Theodore C. Rains and John A. Dean), and Wolfgang Schmidt. Thanks are due also to J. W. Robinson for prepublication data and to Radu Mavrodineanu and Maurice Pinta for reprints containing spectrograms. Miss Jean E. Kinnear and Hideo Watanabe prepared many spectrograms in the laboratories of Beckman Instruments, Inc., that have yielded a great deal of basic information. Detailed acknowledgments are made also in the list of references.

Description of the Tables

Table 13 lists the flame spectra alphabetically by emitting species, and Table 14 is an index according to wavelength.

Character. At the head of the list for each species are given the character of the spectrum and the best flame and solvent. For neutral atoms all lines are assumed to be non-resonance, non-intercombination lines unless marked with p or f, respectively, or with u to signify that assignment to neutral atom or ion is uncertain. Atom lines that are not p, f, or u, are marked a in Table 14. Ion lines are not further classified in Table 13 and are marked s in Table 14. For molecules, all bands are of the character indicated at the start of the species unless marked otherwise with a notation that either adds to (d, t, m) or replaces the predesignated character.

Flame and Solvent. For each species, the best known analytical flame (i.e., the flame offering the best sensitivity) and the best known solvent are given in parentheses. A comma separates flame from solvent; a semicolon separates alternative flame–solvent combinations. If no solvent is indicated it may be assumed to be water, there being no information on other solvents. A flame or flame–solvent notation following any individual wavelength entry applies to that entry only. In a few cases special solutions are indicated; e.g., the CuCl bands, regarded as representing an emission of copper, appear in perchloryl fluoride–hydrogen or oxyacetylene–chloroform flames; but regarded as representing an emission of chlorine for the purpose of flame-photometric determination of chlorine, the bands appear best in air–hydrogen with deliberately added copper in solution.

Background Spectra. A distinction is made between analytical emissions and background emissions. A selection of the latter has been included as a guide to background spectral interferences and because background bands are sometimes taken for analytical bands. In fact, bands that arise from the background in one situation may serve as an analytical emission in another. Emissions that are always or sometimes to be considered background emissions are denoted by a q. Thus, the NO bands are analytical bands in air–oxygen–hydrogen–acetylene (AOHA) but background bands in oxycyanogen (OC q).

Wavelengths. These have been changed from millimicrons to angstroms in view of the increasing use of high-dispersion instruments. Lines and ultraviolet bands are quoted to the nearest 0.1 A and visible and infrared bands usually only to the nearest A. Many of the bands are not sharp enough to justify inclusion of the decimal; authors often disagree on the wavelength; the labor required to measure or confirm all band wavelengths to 0.1 A would have introduced much delay; and the additional precision would probably be of little help to the analyst. Diffuse bands are given only to the nearest 5 or 10 A, as shown by placing the insignificant digits in subscript form. As for lines, although many are known to 0.001 A, a surprisingly large number, owing to diffuseness, complexity, or disagreement among authors, cannot be given precisely to 0.01 A, and the additional precision is not yet useful in flame photometry. Indeed, a few lines in the table are given only to the nearest angstrom, owing to lack of better information in the sources consulted.

Spectral Resolution. The resolution of the table has been matched, more or less, to that of a good but small grating monochromator. Bands separated by no more than 2 A are generally counted as one and marked with a d, t, or m; the weighted mean wavelength is given. The extent to which this is done depends on the profile of the band. In short, bands are listed separately that are likely to be detected separately. Sometimes individual intensity estimates are not available for each band of a resolved group, and such bands are bracketed. The criterion for separating lines is stricter; two lines are listed as one only if separated by less than a few tenths of an angstrom. In Table 14, bracketed lines and bands are listed separately unless they are contiguous in the wavelength sequence.

Intensities. It has not yet proved practical to reduce the intensities to a single uniform, absolute scale. In view of the large variety of flames available for analytical flame spectrophotometry, ranging from air–gas to oxycyanogen, the range of excitation conditions is enormous, the more so since many emissions are not thermal in origin but chemiluminescent. The same line (e.g., Be 2349) may be entirely absent (undetectable, at any rate) or intense in one and the same flame (oxyacetylene) with different ratios of fuel to oxygen. The gross discrepancies in the published data on oxycyanogen are a good example of this. A homogeneous scale of intensities like that of Meggers, Corliss, and Scribner[56a] for arc spectra is therefore virtually out of the question for flame spectrophotometry unless the tables are made much more elaborate. Accordingly, after the intensity data for each species were carefully surveyed, a reference level was selected so as to achieve a rough normalization of the intensity scale against the general level of the flame background as normalizing factor. This is a precarious procedure, however, and comparisons of intensities of different emitters in different flames should not be trusted to better than an order of magnitude. However, the intensities are internally self-consistent for each species, and, as far as possible, for different species in the same flame and solvent. Here intercomparisons ought to be valid within a factor of 2 or 3 for different species and better than 2 within a single species. An attempt has been made to compensate for the spectral sensitivity curves of photodetectors and emulsions. The standard of intensity is thus roughly constant throughout the range 2000–12 000 A. The data represent the net intensity of line or band above the subjacent continuum of the element, if any. In complex band spectra, however, it is not easy to decide how much of the total emission to ascribe to the bands proper; there might be a wide range of opinion on this point.

The Wavelength Index. Table 14 lists the wavelength, the character, the emitting species, the best flame, and the intensity. In the interests of compactness, and since Table 14 is an index to Table 13, certain refinements of Table 13 are ignored in Table 14. In particular, ions and metallic oxides and hydroxides (and molecules of uncertain nature) are designated in Table 14 only by the symbol of the element; the character designation (*a*, *s*, *r*, *b*, etc.) will make it clear enough under which species in Table 13 the emission will be found. Solvents are not specifically shown in Table 14 unless they are essential (e.g., OA, C). If the solvent is water, no indication is given; if it is anything else, an *n* is appended to the flame symbol.

Sources of Information for the Tables of Flame Spectra

Species	Species Identity	Spectrograms Examined	Wavelength Identifications	Wavelength Measurements	Intensity Normalization	Detailed Intensity Estimates
Ag		27, 28			25, 50, 87	
Al		18, 30, 80, 87	6[a], 72[a]		18, 25, 26, 30, 80, 87	
AlO		17, 18, 30, 56, 80, 87	23, 66		17, 18, 25, 26, 30, 56, 80, 87	30
As	61	33	(53)			33
AsO v	66	33	66			33
AsO r	66		(41)			66
Au			53, 54, 56			25[b]
B			6, 75			75
BO$_2$	48, (37)	2, 4, 17, 30, 80, 84	37	26[c], 30, 84	4, 17, 25, 26, 30, 37, 80	
Ba	60	13, 38, 50, 56, 87	50[a]		13, 25, 78, 87	38, 50, 72, 78
Ba$^+$		13, 38, 50, 56, 87	50[a]		13, 25, 78, 87	50, 56, 78
BaCl	66	38				38
BaF	66		9, 67			9, 67
BaO		13, 23, 38, 50, 87	23		13, 25, 80, 87	23, 38, 50, 56, 87

APPENDIX

Species	Species Identity	Spectrograms Examined	Wavelength Identifications	Wavelength Measurements	Intensity length Normalization	Detailed Intensity Estimates
BaOH	24	13, 38, 87		25, 38, 87	12, 13, 25, 56, 80, 87	25, 87
Be		33				33
BeO		30	23		30, 32, 36	30
Bi	62	30, 33	41^a, (53)			30, 33
BiO	63, 66, 77	30	66	30		30
C		25, 85				
C_2 x	66	27		27		27
C_2 v		23, 25, 27				27, 33
CH		24, 25, 33	24, 66	24		25, 33
CN v		23, 27, 38	23, 66	27^d	5	27
CN r		27	24, 66	27^d		27
CO		33	66			33
CS		33	66	33^d		33
Ca		25, 50	9, 50, 78		9, 25, 50, 78	
Ca$^+$		25, 50			25, 50, 87	
CaCl mr			65, 66			9
CaCl r, v		38, 67	66, 77		9, 38, 67	
CaF		66, 67	66, 77		9, 47, 66, 67	
CaO	24	34, 50, 56, 87			34, 56, 87	50^a
CaOH low	24	25, 30, 56, 87		25, 32	25, 30, 80, 87, 90	
CaOH high		38, 50	(23), (24), (66), (77)	38, 50		38, 50
Cd	(62)	85	6^a			27, 29, 33, 72, 84
Cd$^+$			6			
Ce			75			75
Ce$^+$			75			75
CeO		1, 23, 30, 56	23, (88)		23, 25, 30	
ClCu	see CuCl					
Co	60, 61	27, 30, 38, 50, 52, 56, 87	(38), (50), (61)	30^a	15, 30, 80, 81, 87, 90	30, (50)
CoO	e	30		30		30
Cr		3, 25, 27, 30, 33^a, 50, 56, 87			(5), 25, 30, 87, 90	3, 25, 27, 33^a, 50^a, 56^a, 78^a
CrO		9, 23, 25, 80, 87	23		25, 80, 87	23^f, 87
Cs		8, 25, 38, 80, 86, 87	21^a, 40, 51, 38^a, 73		8, 25, 40, 80, 86, 87	25, 38^a, 73, 87
Cu		38, 80, 87			25, 80, 87	33^a
CuCl	66	84	24, 66		(9)	84
CuF	66, 77		66, 77		9, (67)	
CuH	66	38, 84, 87	66	38		87
CuO	24	80, 87	23, 66			23, 87

376   APPENDIX

Species	Species Identity	Spectrograms Examined	Wavelength Identifications	Wavelength Measurements	Intensity Normalization	Detailed Intensity Estimates
CuOH	24	38, 84, 87	(66)	87	80, 87	25, 87
Dy		11, 52, 68, 71, 83	11, 22, 71			11
DyO	23	11, 19, 23, 52, 68, 69, 71, 83	23		19, 23, 68, 71, 83	
ErO	23	19, 23, 68, 69, 71, 83	23		23, 68, 71, 83	
Eu	22, 60	11, 23, 68, 71, 83	22, 60		11, 68, 71, 83	
Eu⁺	22, 60	11, 23, 68, 71, 83	22, 60		11, 68, 71	
EuOH	e	11, 19, 23, 68, 71, 83	77	11, 23, 68, 71, 83	68, 71, 83	
FCa	See CaF					
Fe (AH)	61	33, 55, 56			33, 55, 56	
Fe (AAi)	60, 61	27, 55, 56	(78)		55, 56	27ª
Fe (OH)	60	2, 25, 27, 30, 38, 42, 50, 80, 87	(78)		15, 16, 17, 25, 30, 78, 80, 81, 87, 90	27ª
FeO		2, 23, 25, 30, 56, 80, 87	23	23ᵈ	15, 16, 17, 23ᶠ, 25, 30, 56ᶠ, 80, 87, 90	
Ga		52, 56			7, 56, 78	
Gd		11, 22, 23	22			11
GdO	23	1, 11, 23, 38, 52, 68, 69, 71, 83	23, (66), (68), (76)		19, 68, 71, 83	11, 23
Ge	61	33				33
HO	see OH					
H₂O		25	24, 66			25
Hg		85	59		25, 33	
HoO	23	11, 19, 23, 38, 68, 69, 71, 83	23, (77)		68, 71, 83	11, 23
IO		64	66		64, 66	
In		30, 56, 80	(78)		28, 30	
InO	89	30	89			30
K	62	2, 5, 25, 38	40, 62ᵇ, 73, 74		5, 25, 38, 80, 87	
LaO v		11, 23, 30, 57	23		11, 23, 30	30
LaO r		11, 23, 30, 34, 56, 57, 58, 68, 71, 80, 83	23		25, 30, 34, 57, 58, 71	11, 30, 34ᶠ
Li		30, 38, 56, 80, 87			9, 13, 14, 25, 30, 80	30, 56ª, 72ª

APPENDIX

Species	Species Identity	Spectrograms Examined	Wavelength Identifications	Wavelength Measurements	Intensity Normalization	Detailed Intensity Estimates
LuO	23	1, 11, 19, 23, 68, 69, 71, 83	22, 23, 71[d], (77)	11[d]	11, 22, 23, 68, 71, 77, 83	
Mg (AAi)		56			33, 56, 60, 61	27[a]
Mg (OH)		9, 25, 42, 44, 87			25, 42, 87	
Mg+		27, 30, 56				27
MgCl		9, 67	66, 77		9, 67	
MgF		9, 67	66, 67		9, 67	
MgO	24	2, 17, 23, 42, 80, 87	23		25, 66, 80, 87	
MgOH	24	2, 15, 25, 30, 42, 56, 80, 87		30, 42, 87	25, 30, 50, 84, 87	87
Mn	60	2, 27, 38, 50, 90			2, 5, 27, 38, 50, 60, 78	
Mn+		27				27
MnF			66			9
MnO		2, 23, 80, 87	23		2, 8, 80, 87, 90	(23), 87
MnOH b	64a	38, 56, 80, 87, 90, 92		38		87
MnOH w	e	56, 87		56	56, 87	
Mo		25, 27, 30, 56			5, 33, 56	56
NC	see CN					
NH		25, 27	66			27
NO		27, 33	66			27, 33
Na		4, 25, 38, 44	60		4, 25, 38, 80, 90	72[a]
NdO	23	1, 11, 19, 23, 56, 68, 69, 71, 83	23, 66, 68, 71, 76, 77	1, 11, 56, 68, 71, 83	11, 23, 68, 71, 83	11, 23[f]
Ni (AH)	61	33	61			33
Ni (OH)	60	25, 27, 30, 50, 80, 84, 87, 90	(50), 60		30, 50, 78, 80, 84, 87	
NiO		87, 92	66		87, 92	
OH		11, 25, 27, 30, 46	66	11, 25, 30, 46		25, 30
Os			(41)			75
PO		33	66			33
Pb	62	33, 87			25, 32, 33, 87	33
Pb+	62	33	(6)			33
PbO		38, 92	23, 24, 66, 77	38	(24), 38, (77), 92	
Pd		13, 52, 56, 85			13, 25, 31, 52, 56, 85	

378   APPENDIX

Species	Species Identity	Spectrograms Examined	Wavelength Identifications	Wavelength Measurements	Intensity Normalization	Detailed Intensity Estimates
PrO b	23	71	71			71
PrO r	23	11, 23, 68, 71, 83	23, (88)	11[d]	68, 71	11, 23[f]
Pt			53, 56			25[b], 27[b]
Ra		45			21, 45	
Ra+		45			21, 45	
RaO	[e]		21			21
RaOH	[e]	45	21	45	21, 45	
Rb		8, 25, 38, 56, 80, 87	40, 73	38[a]	8, 25, 38, 73, 86, 87	
Re		33				33
Rh	56	25, 32, 52, 56	(54)		25, 52, 56	25
RhO	[e]	25		25		25
Ru	56, 60	13, 52, 56			13, 52, 56	
SC	See CS					
Sb	62	33	(53)			33
SbO	81a	33		33		33
ScO		1, 11, 23, 52, 68, 71, 83	22, 23, 57	11, 23	68, 71, 83	11, 23[f], 71, 83
Si		33	(6)			33
SiO		25, 33, 66, 84	66		25, 33, 66, 84	
Sm	(60)	11, 68, 69	22, (71)		56, 68, 71, 83	11
SmO	23	11, 23, 56, 68, 71, 83	23, (68), (71), (76)	11	56, 68, 71, 83	11, 23[f], 68[f]
Sn	62	25, 27, 30, 33, 55, 56, 85	6[a]			33, 56[a]
SnH		25				25
SnO r		25, 38, 56, 85	10		5, 25, 38, 56, 85	25, 38[d]
SnO b	[e]	25, 30, 85		25, 85		25
Sr	60	30, 38, 50, 80, 87	9, 42[a]		9, 32, 38, 78	
Sr+	60	30, 38, 50, 87, 90	6[a], 9		9, 32, 78, 80, 90	
SrCl r		38	66			38
SrCl v		38	77		21, 38, 67, 77	
SrF			77		9, 67	
SrH	23	23, 56	23, 66			56
SrO r(OG)		38	66			38
SrO v	23	23, 38	23		23, 38	
SrO r(AA)	23	23, 56	23		23, 56	
Sr₂O₂	24	23, 30, 38, 87		23, (38)	25, 30, 87	
SrOH	24	23, 30, 38, 56, 80, 87		23, 30, 38	25, 30, 38, 80	30, (38), (80)

APPENDIX

Species	Species Identity	Spectrograms Examined	Wavelength Identifications	Wavelength Measurements	Intensity Normalization	Detailed Intensity Estimates
Ta	91		75			75
Ta+	62		75			75
TbO	23	11, 19, 23, 68, 69, 71, 83	23	11^d	68, 71, 83	11, 23
Te	40		33			33
TeO	39, 70	25, 33	39, 70			25, (39), (70)
Ti			6, 75			75
TiO		23, 90	23		25, 32, 90	23, 25
Tl		8, 38, 52, 56, 80, 85, 90	52, 56, 72^a, 78		8, 25, 38, 50, 56, 72, 78, 80, 82, 85	
Tm	60	11, 71	68, 71		11, 68	
TmO	23	11, 23, 71, 83	23	11^d	11, 71, 83	
V	60	27, 30				27
V+	60	27, 30				27
VO		13, 23, 49, 80, 85	23, (40), 49^f	23^d	13, 25, 80, 85, 90	23^f, 85
W	60		75			75
YO		1, 11, 23, 38, 52, 68, 69, 71, 83	23	11^d	13, 68, 71, 83	11, 23, 71, 83
Yb		11, 30, 68, 71, 83	(11)		25, 30, 71, 83	30
Yb+	60	11, 30				30
YbO b, w	^{e}	23, 30	(71)	23, 30		30
YbO r	23	23, 30	23, (71)	(30)		30
YbOH	^{e}	11, 30, 68, 71, 83	(71), (76)	30, 68^d	30, 71, 83	25
Zn	60, 61	25, 30	5, 6, 56, 72			25^a, 27, 30
Zr	60		75			75
Zr+	61		75			75
ZrO		23	23		26, 84	26

^{a} Certain (usually) weaker lines or bands.
^{b} Calculated.
^{c} B^{10} isotope.
^{d} Only a few bands.
^{e} Merely surmised; no evidence.
^{f} Beyond about 700 mμ.
Parentheses ( ) denote auxiliary information.

References

1. Berger, A. W., Powers, T. F., and Maher, J., "Flame spectra of some rare earths", Electrochem. Soc., San Francisco, May 1956. Manuscript was made available through courtesy of A. W. Berger.

2. Boycks, E. C., "Instrumental studies in flame spectrophotometry", Ph. D. Thesis, Univ. Wisconsin, 1955.
3. Bryan, H. A., and Dean, J. A., *Anal. Chem.*, **29**, 1289–1292 (1957).
4. Buell, B. E., *Anal. Chem.*, **30**, 1514–1517 (1958).
5. Buell, B. E., "A study of flame photometric determinations of elements", Union Oil Co., Brea, Calif., Report dated Sept. 18, 1958. Copy was made available through courtesy of author.
6. Buell, B. E., Personal communications, 1960–1961. Information contained in part in *Anal. Chem.*, **34**, 635–640 (1962).
6a. Bugrim, E. D., Lyutyi, A. I., and Rossikhin, V. S., *Optika i Spektroskopiya*, **10**, 801–804 (1961).
7. Bulewicz, E. M., and Sugden, T. M., *Trans. Faraday Soc.*, **54**, 830–837 (1958).
8. Chirnside, R. C., *J. Soc. Glass Technology*, **43**, 5T–29T (1959).
9. Collier, H. E., Jr., "The mechanism of spectral excitation of metallic ions by a new high-temperature source", Ph. D. Thesis, Lehigh Univ., 1955; Univ. Microfilms, Ann Arbor, Mich., Publ. 13027. Available through courtesy of R. H. Curry.
10. Connelly, F. C., *Proc. Phys. Soc. (London)*, **45**, 780–791 (1933).
11. Curry, R. H., Iowa State Univ., Personal communications, 1960–1961, and sets of photographic and recorded spectrograms of the rare earths.
12. Curtis, G. W., Knauer, H. E., and Hunter, L. E., "The effect of organic solvents on the flame photometric emission of certain elements", pp. 67–76 of *Symposium on Flame Photometry*, Am. Soc. Testing Materials, Spec. Tech. Publ. 116 (1951).
13. Dean, J. A., *Flame Photometry*, McGraw-Hill, New York, 1960; supplemented by personal communications.
14. Dean, J. A., "Enhancement effects in flame photometry", Eastern Anal. Symposium, New York, Nov. 1960.
15. Dean, J. A., and Burger, J. C., Jr., *Anal. Chem.*, **27**, 1052–1055 (1955).
16. Dean, J. A., and Lady, J. H., *Anal. Chem.*, **27**, 1533–1536 (1955).
17. Dean, J. A., and Thompson, C., *Anal. Chem.*, **27**, 42–46 (1955).
18. Eshelman, H. C., Dean, J. A., Menis, O., and Rains, T. C., *Anal. Chem.*, **31**, 183–187 (1959).
19. Fassel, V. A., *Anal. Chem.*, **32**, No. 11, pp. 19A–46A (Oct. 1960).
20. Fassel, V. A., Curry, R. H., and Kniseley, R. N., "Flame spectra of rare earth elements and their analytical application", Am. Chem. Soc., Chicago, Sept. 1961; *Spectrochim. Acta*, **18**, 1127-53 (1962). Some of the data were made available by Curry (ref. 11).
21. Formánek, J., *Die qualitative Spektralanalyse anorganischer und organischer Körper*, 2nd ed., Mückenberger, Berlin, 1905.
22. Gatterer, A., Junkes, J., and Frodl, V., *Spektren der seltenen Erden*, Vatican City, 1945.
23. Gatterer, A., Junkes, J., Salpeter, E. W., and Rosen, B., *Molecular Spectra of Metallic Oxides*, Vatican City, 1957.
24. Gaydon, A. G., *The Spectroscopy of Flames*, Wiley, New York, 1957.
25. Gilbert, P. T., Jr., Unpublished work at Beckman Instruments, Inc., 1948–1961.
26. Gilbert, P. T., Jr., Unpublished work at North American Aviation, Inc., Downey, Calif., 1952–1955.

27. Gilbert, P. T., Jr., "Oxycyanogen flame photometry", Pittsburgh Conference on Anal. Chem. and Appl. Spectroscopy, Mar. 1958; report printed by Beckman Instruments, Inc., 1958.
28. Gilbert, P. T., Jr., *Spectrochim. Acta*, **12**, 397–400 (1958).
29. Gilbert, P. T., Jr., *Anal. Chem.*, **31**, 110–114 (1959).
30. Gilbert, P. T., Jr., "Analytical flame photometry: new developments", pp. 73–156 in *Symposium on Spectroscopy*, Am. Soc. Testing Materials, Spec. Tech. Publ. 269 (1960).
31. Gilbert, P. T., Jr., "Flame photometry. Less familiar elements", pp. 346–362 in *Encyclopedia of Spectroscopy*, edited by G. L. Clark; Reinhold, New York, 1960.
32. Gilbert, P. T., Jr., "Flame spectra of the elements," 2nd ed., Bulletin 753-A, Beckman Instruments, Inc., 1961.
33. Gilbert, P. T., Jr., "Chemiluminescent flame spectrophotometry", Pittsburgh Conference on Anal. Chem. and Appl. Spectroscopy, Mar. 1961; pp. 171-215 in *Proceedings of the Xth Colloquium Spectroscopicum Internationale*, E. R. Lippincott, and M. Margoshes, editors, Spartan Books, Washington, D.C., 1962.
34. Gilbert, P. T., Jr., Seminar, Soc. Appl. Spectroscopy, Chicago, May, 1961; Pittsburgh Conference on Anal. Chem. and Applied Spectroscopy, Mar. 1962.
34a. Gilbert, P. T., Jr., "Flame spectra of the elements", pp. 2804–33 of *Handbook of Chemistry and Physics*, 43rd edition, C. D. Hodgman, R. C. Weast, and S. M. Selby, editors, Chemical Rubber Publishing Co., Cleveland, 1961; and pp. 2852–81 of 44th edition, 1962.
35. Gilbert, P. T., Jr., "Flame photometry", pp. 6–197 to 6–227 in *Handbook of Analytical Chemistry*, L. Meites, editor, McGraw-Hill, New York, 1962.
36. Gramont, A. de, *Compt. rend.*, **153**, 318 (1911).
37. Grenat, H., Herman, L., Herman, R., and Weniger, S., *Rev. opt.*, **38**, 134–137 (1959).
38. Hagenbach, A., and Konen, H., *Atlas der Emissionsspektren der meisten Elemente*, Fischer, Jena, 1905.
39. Haranath, P. B. V., Rao, P. T., and Sivaramamurty, V., *Z. Physik*, **155**, 507–517 (1959).
40. Harrison, G. R., *Massachusetts Institute of Technology Wavelength Tables*, Wiley, New York, 1939.
41. Hartley, W. N., *Phil. Trans. Roy. Soc. London*, **A185**, 161–212 (1894).
42. Herrmann, R., *Flammenphotometrie*, Springer, Berlin, 1956.
44. Herrmann, R., and Alkemade, C. T. J., *Flammenphotometrie*, 2nd ed., Springer, Berlin, 1960.
45. Hofmann, K. A., *Lehrbuch der anorganischen Chemie*, 2nd ed., Vieweg u. Sohn, Braunschweig, 1919.
46. Hornbeck, G. A., and Herman, R. C., *Ind. Eng. Chem.*, **43**, 2739–2757 (1951).
47. Ishida, R., *J. Chem. Soc. Japan, Pure Chem. Sect.*, **77**, 241 (1956); *Repts. Govt. Chem. Ind. Research Inst. Tokyo*, **51**, 337–338 (1956).
48. Kaskan, W. E., and Millikan, R. C., *J. Chem. Phys.*, **32**, 1273–1274 (1960).
49. Keenan, P. C., and Schroeder, L. W., *Astrophys. J.*, **115**, 82–88 (1952).
50. Kinnear, J. E., "Report on emission spectrography with the Beckman oxyacetylene flame", Beckman Instruments, Inc., Report dated May 27, 1955; also unpublished data and many photographic spectrograms.
51. Kleiman, H., *J. Opt. Soc. Am.*, **52**, 441–447 (1962).

52. Lundegårdh, H., *Die quantitative Spektralanalyse der Elemente*, Fischer, Jena; Part I, 1929; Part II, 1934.
53. Lundegårdh, H., *Lantbruks-Högskol. Ann.*, **3**, 49–97 (1936).
54. Lundegårdh, H., *Metallwirtschaft*, **17**, 1222–1226 (1938).
55. Mavrodineanu, R., "Quelques considérations sur l'analyse quantitative par la flamme", pp. 15–29 in the Proceedings of the Colloquium Spectroscopicum Internationale VIII, Lucerne, Sept. 1959; published by H. R. Sauerländer, Aarau, Switzerland, 1960.
56. Mavrodineanu, R., and Boiteux, H., *L'analyse spectrale quantitative par la flamme*, Masson, Paris, 1954.
56a. Meggers, W. F., Corliss, C. H., and Scribner, B. F., *Tables of Spectral-Line Intensities*, Natl. Bur. Standards Monograph 32; Part I, 1962; Part II, 1961.
57. Meggers, W. F., and Wheeler, J. A., *Bur. Standards J. Research*, **6**, 239–275 (1931).
58. Menis, O., Rains, T. C., and Dean, J. A., *Anal. Chem.*, **31**, 187–191 (1959).
59. Mitscherlich, A., *Ann. Physik*, **121**, 459–488 (1864); quoted by Ramage (ref. 72) and by Hartley (ref. 41).
60. Moore, C. E., *A Multiplet Table of Astrophysical Interest*, revised ed., Natl. Bur. Standards Tech. Note 36, Washington, D.C., 1959; reprinted from edition of 1945.
61. Moore, C. E., *An Ultraviolet Multiplet Table*, Natl. Bur. Standards Circular 488; Section 1, 1950; Section 2, 1952; Sections 3, 4 and 5, 1962.
62. Moore, C. E., *Atomic Energy Levels*, Natl. Bur. Standards Circular 467; Vol. I, 1949; Vol. II, 1952; Vol. III, 1957.
63. Morgan, F., *Phys. Rev.*, **49**, 41–50 (1936).
64. Ohyagi, Y., *Sci. of Light (Tokyo)*, **7**, 14–22 (1958).
64a. Padley, P. J., and Sugden, T. M., *Trans. Faraday Soc.*, **55**, 2054–2061 (1959).
65. Parker, A. E., *Phys. Rev.*, **47**, 349–358 (1935).
66. Pearse, R. W. B., and Gaydon, A. G., *The Identification of Molecular Spectra*, 2nd ed., Wiley, New York, 1950.
67. Pennsalt Chemicals Corp., Philadelphia, "PF perchloryl fluoride flame photometry", Bull. dsE-1819; undated; received 1958.
68. Pinta, M., "Contribution a l'étude des spectres de flamme. Applications analytiques", Thesis, Univ. Paris, 1952. Copy was made available through courtesy of author.
69. Pinta, M., *J. Rech. Centre Natl. recherche Sci. Labs. Bellevue (Paris)*, No. 21, pp. 260–270 (1952).
70. Purbrick, R. L., *J. Chem. Phys.*, **30**, 962–963 (1959).
71. Rains, T. C., House, H. P., and Menis, O., *Anal. Chim. Acta*, **22**, 315–327 (1960). Manuscript and enlarged spectrograms were made available prior to publication through courtesy of O. Menis.
72. Ramage, H., *Proc. Roy. Soc. (London)*, **70**, 1–27 (1902).
73. Ramage, H., *Proc. Roy. Soc. (London)*, **70**, 303–312 (1902).
74. Risberg, P., *Arkiv Fysik*, **10**, 583–606 (1956).
75. Robinson, J. W., *Anal. Chem.*, **33**, 1226–1230 (1961).
76. Rodden, C. J., and Plantinga, O. S., *Phys. Rev.*, **45**, 280–281 (1934).
77. Rosen, B., editor, "Données spectroscopiques concernant les molécules diatomiques", Vol. 4 of *Tables de constantes et données numériques*, Hermann, Paris, 1951.

78. Rusanov, A. K., and Il'yasova, N. V., *Atlas Plamennykh, Dugovykh i Iskrovykh Spektrov Elementov* ("Atlas of flame, arc, and spark spectra of the elements"), Gosgeoltekhizdat, Moscow, 1958.
80. Schmidt, W., Forschungsinstitut der Feuerfest-Industrie, Bonn (now with Beckman Instruments G.m.b.H., Munich), copies of recorded flame spectrograms of many elements.
81. Schöffmann, E., and Malissa, H., *Mikrochim, Acta*, **1961**, 319–322.
81a. Shimauchi, M., *Sci. of Light (Tokyo)*, **9**, 109–133 (1960).
82. Stavinoha, W. B., and Nash, J. B., *Anal. Chem.*, **32**, 1695–1697 (1960).
83. Tremmel, C. G., "Flame spectrometric determination of lanthanum in rare earth mixtures", Master's thesis, Iowa State Univ., 1958. Copy was made available through courtesy of V. A. Fassel, who had also submitted spectrograms from the thesis prior to its completion.
84. Watanabe, H., Beckman Instruments, Inc., unpublished spectrograms and data, 1955.
85. Watanabe, H., Spectrograms published with his permission in Herrmann and Alkemade (ref. 44).
86. Watanabe, H., "Rubidium and cesium flame spectrograms", Technical Report 405, Beckman Instruments, Inc., Mar. 1955.
87. Watanabe, H., and Kendall, K. K., *Appl. Spectroscopy*, **9**, 132–140 (1955); the original spectrograms were made available by H. Watanabe.
88. Watson, W. W., *Phys. Rev.*, **53**, 639–642 (1938).
89. Watson, W. W., and Shambon, A., *Phys. Rev.*, **50**, 607–609 (1936).
90. Whisman, M., and Eccleston, B. H., *Anal. Chem.*, **27**, 1861–1869 (1955).
91. Zaĭdel', A. N., Prokof'ev, V. K., and Raĭskiĭ, S. M., *Tablitsy Spektral'nykh Liniĭ*, Gosudarst. Izdatel'stvo Tekh.-Teoret. Lit., Moscow, 1952; International edition (Saidel, Prokofiev, and Raiski), *Tables of Spectrum Lines*, VEB Verlag Technik, Berlin, 1955.
92. Zeiss, Carl, Oberkochen, West Germany, "Flammenspektren", Bestellnummer A 50-812 (1961).

Abbreviations used in Tables 13 and 14

Characters:

    a = line of the neutral atom not p, f, or u.
    b = peak of a narrow, headless band.
    d = double
    f = intersystem combination (intercombination) line.
    m = multiple (more than three components).
    p = resonance line (transition to ground state or a state within a few tenths ev of the ground state).
    r = head of a band degraded to longer wavelengths.
    s = line of singly ionized atom.
    t = triple.
    u = unclassified (as to neutral atom or ion).
    v = head of a band degraded to shorter wavelengths.
    w = peak of a wide or diffuse band.
    x = sharp peak of or in a headless band, formed of bunched rotational lines.
    y = rotational line or group of a few rotational lines.

Flames:

    AA = air–acetylene.
    AG = air–gas.
    AH = air–hydrogen.
    AOA = air–oxygen–hydrogen.
    AOHA = air–oxygen–hydrogen–acetylene.
    FH = fluorine–hydrogen.
    i = inner cone or reaction zone of flame.
    OA = oxyacetylene.
    OAz = fuel-rich (incandescent) oxyacetylene.
    OC = oxycyanogen.
    OG = oxygen–gas.
    OH = oxyhydrogen.
    PH = perchloryl fluoride-hydrogen

Solvents:

    AA = acetylacetone.
    Ac = acetone.
    Al = alcohol.
    Bu = butanol (4% in water).
    C = chloroform.
    Cu = copper solution.
    D = dimethylformamide.
    G = gasoline.
    H = hexone (methyl isobutyl ketone).
    Hp = heptane.
    IO_3^- = iodate solution.
    M = methanol.
    n = nonaqueous solvent (Table 14).
    N = naphtha.
    P = isopropanol.
    q = background emission.
    W = water.

TABLE 13. Flame spectra of the elements alphabetically by emitting species

Ag (AH, W)

3280.7 p	6000
3382.9 p	7000

Al

2568.0 p OH, H	?	
3082.2 p OH	(1)	
3092.7 p OH	(1)	
3944.0 p OA, H	100	
3961.5 p OA, H	180	

AlO r (OA, H)

4353	1
4374	1
4394	1
4471	2
4494	4
4516	5
4538	5
4558	4
4576	3
4648	15
4672	20
4695	18
4716	13
4736	9
4754	6
4842	65
4866	50
4888	26
5079	19
5102	24
5123	24
5143	18
5161	15
5177	11
5191	8
5337	2
5358	3
5377	4
5394	4
5410	4
5423	3

As (AH, P)

2288.1	17
2349.8	30
2381.2 f	2.5
2437.2 f	(1)
2456.5 f	6
2492.9 f	6
2780.2	5

AsO v (AH, P)

2437.3 ⎫ 2438.5 ⎭	10
2503.6 ⎫ 2504.7 ⎭	20
2569.7 ⎫ 2570.9 ⎭	20

AsO r

3105.6 ⎫ 3106.8 ⎭	6
3170.6 ⎫ 3172.4 ⎭	9
3207.8 ⎫ 3209.1 ⎭	6
3277.2 ⎫ 3279.1 ⎭	10

Au (OH; AA)

2428.0 p	5
2676.0 p	10
2748.3	2

B (OC, W)

2496.8 p ⎫ 2497.7 p ⎭	4

BO$_2$ b (OH, Bu)

408_0	9
418_0	6
437_0	12
453_0	30
471_5	90
493_0	165
495_0	155
5180	300
5476	350
579_0	140
603_0	25
620_0	20
639_0	20
682_0	5
721_0	?
775_0	?
822_0	?

In $B^{10}O_2$ the wavelengths are 10–15 Å lower.

Ba (AA; OH)

3501.1 p	2
3889.3 u	1
5535.6 p	200
5777.7	1
5971.7 ⎫ 5972.8 u ⎭	1
5997.1	1
6019.5	4
6063.1	4
6110.8	12
6450.9	4
6482.9 u	4
6498.8	50
6527.3	25
6595.3	25
7060.0	4

Ba$^+$ (OA, H)

4554.0	200
4934.1	100
6141.7	10
6496.9	50

BaCl (OG)

5066 v	2
5138 r, v	20
5167 r	4
5240 ⎫ 5244 ⎭ r	30
5321 v	10

BaF r (PH; FH)

4951	70
5001	50
5125 t	8

BaO r (AH, W)

4524	3
4538	1
4579	4
4622	7
4639	5
4664	3
4681	10
4723	5
4742	10
4786	15
4831	(10)
4850	50
4965	30
5012	(10)
5086	100
5214	70
5349	90

5367	(10)	4733	5	4382	8
5420	(5)	4755	3	4679	35
5455	20	4776	2	4685	35
5492	70	4795	1.3	4698	43
5510	10	4812	0.9	4715	43
5602	10	4828	0.6	4737	46
5644	40	5055	1.4	5098	50
5659	5	5076	1.4	5129	80
5673	10	5095	1.3	5165	110
5701	40	5112	1.1	5470	10
5758	15	5128	0.8	5502	13
5805	20	5141	0.4	5541	16
5864	40			5585	23
5887	10	**Bi** (AH, P)		5635	18
5976	(5)	2061.7 p	0.8		
5984	10	2110.3 p	0.6	**CH** y (OH, G q)	
6040	50	2228.3 p	17		
6100	20	2230.6 f	30	3872 r	7
6111	10	2276.6 f	7	3889 r	6
6163	10	2400.9	0.6	4237	5
6225	15	2524.5	0.8	4243	6
6291	25	2627.9	8	4249	7
6423	15	2898.0	35	4256	8
6493	25	2938.3	(10)	4262	9
6563	10	2989.0 f	(10)	4268	10
6634	1	2993.3	(5)	4274	11
6783	50	3024.6 f	(5)	4280	12
6857	40	3067.7 p	400	4286	13
6931	60	4722.6 AH, W	10	4292	15
7098	80			4297	20
		BiO (?) b (AH, W)		4315 v	60
BaOH b (AH, W)				4324 x	8
488_0	120	4394	0.7	4334	6
502_0	30	4424	0.6	4339	7
513_0	140	4542	0.2	4343	8
524_0	80	4619	0.1	4348	9
745_0 w	45	$5564^{(a)}$ r	0.3	4352	10
830_0	200	5639 r	0.1		
873_0	230	574_0	0.1	**CN** v (OHi, N; OC q)	
		$6037^{(a)}$ r	0.1	(Violet)	
Be (OAz, M)		$^{(a)}$ Definite BiO bands; origin of others doubtful.			
2348.6 p	2000			3560 b	0.26
2494.7	4			3583.9 ⎫	
2650.6 m	15			3585.9 ⎭	0.82
3321.3	15	**C** (OA q)		3590.4	0.54
		2478.6	—	3850.9	2.6
BeO r (OH, W)				3854.7	2.7
4427	0.3	**C₂** x (OA q; OC q)		3861.9	3.2
4452	0.4	(Mulliken)		3871.4	4.0
4475	0.3			3883.4	3.9
4496	0.2	2310	4	4152	0.41
4517	0.2	2322	5	4158	0.43
4535	0.1	**C₂** v (OH, G q) (Swan)		4168	0.50
4553	0.03			4181	0.60
4709	9	4365	15	4197	0.71
		4371	12	4216	0.68

APPENDIX

CN tr (OC q) (Red)

Strongest heads at low dispersion:

695_0	0.21
711_0	0.31
728_0	0.30
744_0	0.22
789_0	0.90
808_0	1.0
828_0	0.70
850_0	0.40
917_0	2.4
941_0	1.6
958_0	0.7
1095_0	0.11

CO r (OHi, G q) (Fourth positive)

2089.9	1
2113.1	1
2128.3	1
2150.2	2
2173.0	5
2196.8	5
2215.8	2
2221.5	4
2238.3	2
2247.2	2
2261.7	4
2273.9	2
2286.1	5
2311.5	5
2337.9	5

CS r (AOA, M)

2507.3	1.0
2511.2	0.8
2523.2	1.2
2530.0	1.0
2538.7	1.3
2549.5	0.9
2555.8	1.1
2572.7	0.9
2575.6	5
2589.6	3.2
2605.9	1.6
2621.6	1.0
2662.6	1.1
2677.0	1.3
2693.2	1.3

Ca (OH; AA; OA)

4226.7 p	10 000

4283.0	4
4302.5	15
4318.7	3
4425.4	1
4435.0 ⎫ 4435.7 ⎭	10
4454.8 ⎫ 4455.9 ⎭	15
4585.9	8

Ca⁺ (OA)

3933.7	300
3968.5	150

CaCl mr (OA, C)

3728	5
3775	8
3828	12
3882	13

CaCl (PH; OA, C)

5810 r	250
5934 r	500
6185 v	500
6212 v	500

CaF dr (PH; FH)

5146	3
5292 ⎫ 5298 ⎭	150

CaF (FH)

5830 r	50
6064 v	150
6285 v	50

CaO r (OA)

4084	1
4105	1
6625 ⎫ 6639 ⎭	100
6956 ⎫ 6959 ⎭	10
6984	10
7318 ⎫ 7326 ⎭	30
7716 ⎫ 7721 ⎭	70
8153	100
8652	150
9229	200
9807 v	10
9835 v	10

CaOH b (OH)

Peaks at low dispersion:

544_0	100
554_0	5000
572_0	200
602_0	2000
622_0	10 000
644_0	1500

CaOH x (OH)

Peaks at high dispersion:

5434	100
5462	100
5511	100
5515 r	200
5521	200
5526	200
5533 r	600
5539 r	1200
5540	800
5541	800
5542	1200
5544	800
5546	800
5547	1000
5548	600
5554 v	800
5557	400
5559	400
5562	200
5652	100
5697	100
5719	100
5813	100
5935	100
5965 r	200
5971	200
5986 v	400
6003 r	400
6006 v	600
6022 v	400
6029	300
6035 v	400
6049	300
6068 v	200
6076 v	200
6088 r	200
6092 v	200
6098 v	400
6115	200

Cd (OH)

2288.0 p OC	200
2980.6 OH, N	?

3261.1 f AH	200	2346.2 f	0.06	2764.2	0.3
4678.2	1	2352.9	0.17	2766.4	0.1
4799.9	2	2355.6 p	0.11	2775.0	0.1
5085.8	4	2358.2 p ⎫	0.07	2987.2 p	9
		2358.7 p ⎭		2989.6 f	9
Cd⁺ (OH, N)		2365.1 p	0.2	3000.5 f	1
2144.4	?	2369.7	0.10	3013.6 p	4
		2371.9 f	0.08	3017.5 p	9
Ce (OC, AA)		2380.5 p	0.25	3034.4 f	1
		2384.9 p	0.3	3042.5 p	2
5523.0	0.6	2392.0 p	0.2	3044.0 p	36
		2402.2 f	0.35	3048.9 p	7
Ce⁺ (OC, W)		2407.3 p	1.1	3072.3 p	15
3560.8	0.2	2411.6 p	0.9	3082.6 p	14
		2414.5 p ⎫	1.1	3086.8 p	16
		2415.3 p ⎭		3089.6 p	10
CeO (OH, Ac)		2419.1	0.03	3121.4 p	7
4612	10	2421.7	0.03	2137.3 p	10
4621 r ⎫		2424.9 p	1·0	3139.9 p	5
4624 r ⎬	6	2429.2 p	0.03	3147.1 p	13
4627 ⎭		2432.2 p	0.85	3158.8 p	10
4684 r	12	2436.7 p	0.75	3219.2 f	3
4695	10	2439.0 p	0.7	3237.0 f	7
4704	5	2456.2	0.07	3334.1	26
4792	15	2460.8 p	0.08	3337.2 f	1.4
4799	20	2464.6 p	0.05	3354.4	23
4807	25	2467.7 p	0.08	3367.1	24
4815 r	20	2470.3	0.08	3370.3 f	2.6
4846	8	2473.9 p	0.06	3385.2	25
4863 r ⎫		2476.6	0.12	3388.2	28
4873 r ⎬	6	2483.6	0.07	3395.4 f	66
4882 r		2495.6	0.03	3405.1	360
4891 ⎭		2504.5	0.1	3409.2	140
4925 r ⎫		2506.9	0.07	3412.3 f ⎫	380
4931 r		2511.0	0.4	3412.6 p ⎭	
4934 r		2517.9	0.2	3417.2	70
4936 r ⎬	20	2521.4 p	1.2	3431.6 p	190
4941 r		2529.0 p	0.9	3433.0	100
4947 r		2530.1	0.1	3442.9 p ⎫	280
4951 r ⎭		2532.2	0.12	3443.6 ⎭	
		2536.0 p	0.9	3449.2 ⎫	290
ClCu: See CuCl		2544.3 p	0.7	3449.4 ⎭	
		2548.3	0.06	3453.5	480
Co (OH)		2553.0	0.1	3455.2 p	200
2276.5	0.06	2556.8	0.04	3462.8	170
2287.8	0.06	2561.3	0.3	3465.8 p	340
2295.2 f	0.09	2567.3 p	0.5	3474.0 p	270
2296.0 f ⎫	0.05	2572.2	0.1	3483.4	42
2296.7 ⎭		2573.5	0.5	3489.4	110
2304.0 ⎫	0.06	2590.6	0.03	3491.3 p	60
2304.2 p ⎭		2622.4	0.1	3495.7	130
2309.0 p	0.22	2650.3	0.2	3502.3	500
2316.8 p	0.05	2679.8	0.05	3506.3	290
2323.1 p	0.16	2685.3	0.1	3510.4 p	280
2336.0 p	0.11	2695.8	0.1	3512.6	300
2338.7 p	0.18	2761.4	0.07	3513.5 p	370

3518.3	50	3974.7	13	3605.3 p	500
3520.1 p	60	3979.5 f	48	3636.6 OC	2.5
3521.6 f	180	3987.1	4.4	3749.0 OC	1.5
3523.4	150	3991.7	10	3886.8	5
3526.8 p	460	3995.3 f	290	3902.9 OC	0.3
3529.0 p	150	3997.9 f	100	3908.8 OC	0.8
3529.8	300	4011.0 f	0.8	3919.2 OC	1.1
3533.4 p	130	4020.9	33	3983.9 OC	0.5
3543.3	0.7	4027.0 f	15	4254.3 p	900
3550.6 p	58	4045.4 f	13	4274.8 p	700
3552.7 p	6	4057.2 f	12	4289.7 p	500
3558.8 f	5	4066.4 f	11	4295.8 ⎫	
3560.9	50	4076.1	21	4297.0 ⎬ OC	3
3565.0 f	45	4082.6	1	4299.7 ⎭	
3569.4	180	4086.3	2	4307.5 u OC	4
3575.0 ⎫	340	4092.4	50	4320.6 OC	4
3575.4 p ⎭		4110.5	21	4337.6	2
3584.8 p ⎫	90	4118.8	200	4339.4	10
3585.2 ⎭		4121.3	320	4344.5	10
3587.2	120	4132.2 f	1.1	4351.8	10
3594.9 p	190	4190.7 f	33	4359.6 OC	7
3602.1 p	140	4198.4 f	1	4371.3 OC	2
3605.4 f	30	4207.6	1	4496.9	2
3625.0	3	4234.0 f	6	4546.0 OC	0.8
3627.8 f	60	4252.3 f	15	4646.2	5
3631.4 p	42	4268.0 f	2.2	4651.3 ⎫	5
3647.7 p	28	4285.8 f	6	4652.2 ⎭	
3652.5 p	39	4303.2 f	2	5204.5 ⎫	
3657.0	3	4580.1 f	2	5206.0 ⎬	300
3683.0	0.8	4727.9 f	7	5208.4 ⎭	
3693.1 f	0.6	5483.3	5	5247.6 OC	1.5
3704.1	15			5264.2	5
3708.8	0.6	**CoO** (?) (OH)		5296.7	5
3732.4	2.3	563_5 w	100	5298.3	10
3745.5	23			5345.8	15
3749.9	0.8	**Cr** (OC; OA, H)		5348.3	10
3808.1	5			5409.8	20
3816.3	1	2364.7 p ⎫			
3842.0	55	2365.9 p ⎬ AH, P	2	**CrO** r (OH, Bu)	
3845.5	270	2366.8 p ⎭			
3851.0	4	2971.1 AAi	0.2	5168	10
3861.2	19	2975.5 AAi	0.2	5229	20
3873.1	700	2980.8 AAi	0.2	5293	10
3874.0	400	2988.6 OC	1	5356	20
3876.8	40	3014.8 d ⎫ OC	1.5	5417	30
3881.9	400	3015.2 ⎭		5564	40
3894.1	270	3017.6 ⎫		5623	35
3895.0	(100)	3018.5 ⎬ OC	1.5	5795	70
3906.3	12	3018.8 ⎭		5852	50
3909.9 f	54	3020.7 ⎫ OC	1.5	6052	100
3922.8 f	1.7	3021.6 ⎭		6394	100
3936.0 f	56	3024.4 OC	0.5	6452	50
3941.7	39	3030.2 AAi	1	6772	40
3945.3	9	3053.9 AAi	0.5	6830	70
3952.9	19	3578.7 p	1000	6896	60
3957.9	14	3593.5 p	800	7190	10

7249	20	6628.7	2	4441	7
7312	30	6723.3	15	4452	10
7375	30	6824.7	2	4464	11
7714	20	6870.5	2	4648 r	8
7777	30	6973.3	15		
7843	40	6983.5	7	**CuO** r (OH)	
7910	40	8015.7	5	6045	10
8339	40	8079.0	15	6060 d	50
8360	40	8521.1 p	10 000	6147	50
8386	40	8761.4	40	6162 m	50
8417	40	8943.5 p	5000		
8449	40			**CuOH** b (AH)	
		Cu (OH, M)		493_0 w	60
Cs (OH)		2492.1 p AH, P	20	505_0 w	70
3288.6 p	0.05	3247.5 p	10 000	524_0	110
3313.2 p	0.1	3274.0 p	6500	530_0	110
3347.4 p	0.2	5105.5	50	537_0 w	120
3348.7 p	0.1	5700.2	5		
3398.1 p	0.3	5782.1	10	**Dy** (OH)	
3400.0 p	0.2			4186.8	12
3476.9 p	0.5	**CuCl** r (PH; OA, C)		4211.7	15
3480.1 p	0.2	(For Cl: AH, Cu soln.)		4577.8	10
3611.5 p	1	4120	1		
3617.4 p	0.3	4188	2	**DyO** b (OH, H)	
3876.2 p	5	4211	2	442_0	10
3888.6 p	1	4259	5	4535 dr	20
4555.3 p	150	4281	10	4570 dr	30
4593.2 p	40	4333	10	4574 dr	35
5152.7	0.3	4354	20	4598	30
5196.7	0.5	4412	7	4775	15
5209.5 d	0.5	4434	15	5136 r	120
5256.6	1	4494	4	5140 dv	190
5301.4	0.5	4516	5	5150 r	210
5340.9	3	4882	3	5193 v	240
5350.4	1			5204	240
5406.7	1	**CuF** r (PH; FH)		5248 m	300
5413.9 d	2	4932	14	5263 r	600
5465.9	5	5061	2.5	5273 r	600
5502.9	4	5686	0.5	5285 r	400
5568.4	1			5296 r	300
5573.7	1	**CuH** y (OH)		5315 d	300
5635.2	7	4005 r	5	5333 d	300
5663.7	8	4280 r	30	5349 dr	350
5745.7	1	4328 r	11	5404 mr	500
5838.8	2	4348	10	5454 d	450
5845.1	9	4356	9	5493	500
5847.6	2	4364	8	5544 d	400
6010.5	10	4372	7	5563	350
6034.1	5	4380	9	5694 t	600
6213.1	10	4389	9	5705 t	700
6217.6	2	4399	9	5729	1000
6354.6	10	4409	9	574_0 m	600
6432.0	2	4419	7	5834 t	1000
6472.6	2	4430	8	5856 dr	800
6586.5	10				

5868	700	5628 r	300	2191.8 p	0.4
5884 d	600	5634 v	250	2200.4 p	1
5921 r	400	5645 v	200	2292.5 p	1
5940 dr	300	5664 v	200	2297.8 p	3
5979 dr	300	5674	250	2299.2 p	1
6006 tv	500	5684	150	2373.6 p	1
6027 d	200	5693	150	2457.6	2
6043 d	300	5706 d	100	2462.6 p	8
6050 r	300	5716	100	2465.1	1
6058 d	300	5723 v	100	2468.9	1
6078 dr	500	5730 r	50	2472.9 p	11
6084 d	500	5737	50	2474.8	1
6121 d	300	5751	50	2479.8 p	8
				2483.3 p	60

ErO b (OH, H)

Eu (OH, H)

				2488.1 p	40
				2490.6 p	40
4988	30	4594.0 p	170	2496.5	1
5002 r	30	4627.2 p	150	2501.1 p	5
5009 r	80	4661.9 p	120	2510.8 p	9
5034 v	150	5645.8	2	2518.1 p	9
5041 v	150	5765.2	6	2522.8 p	35
5050	150	6018.2	10	2524.3 p	5
5067 r	300	6864.6	10	2527.4 p	20
5073	200			2529.1 p	8
5078 r	200			2535.6 p	5
5081 v	200	**Eu$^+$** (OH, H)		2541.0 p	7
5089 d	150	4129.7	5	2546.0 p	6
5100 r	100	4205.0	5	2549.6 p	6
5135 v	70	4435.6	1	2584.5	4
5151 v	70			2599.6 AAi	1
5167 v	50	**EuOH** (?) (OH, H)		2606.8 AAi	3
5176 d	50			2623.5 AAi	2
5180 d	60	Bands are diffuse; red ones finely fluted.		2635.8 AAi	1
5184 d	50			2667.9 f AAi	1
5372 v	15	474_0 w	8	2679.1 AAi	2
540_0 m	10	520_0 w	4	2690.1 f AAi	1
5421 d	20	553_0 b	20	2706.6 AAi	1
5430 r	40	571_0 b	50	2719.0 p	35
5438 d	50	598_0 b	300	2720.9 p	20
5446 d	150	623_0 w	600	2723.6 p	7
5453 d	200	628_0 b	500	2728.0	2
5459	250	647_0 w	800	2733.6	7
5470	200	655_0 w	800	2737.3 p	7
5480 dr	250	662_0 w	700	2742.4 p	12
5498 d	300	684_0 w	1200	2744.1 p AAi	3
5513 m	400	702_0 w	1200	2750.1 p	7
5528	500			2753.7 ⎫	
5534 r	400	**FCa:** See **CaF**		2754.0 ⎬ AAi	2
5545 d	500			2754.4 ⎭	
5552	500	(For F: OA, Ca soln.)		2756.3 p	3
5557	400			2761.8 ⎫	3
5563 r	500	**Fe** (AH, P; AAi)		2762.0 ⎭	
5568 dv	400			2767.5	4
5581 d	350	2166.8 p	1	2772.1 p AAi	2
5596 d	500	2178.1 p	1	2778.2 AAi	2
5613 v	500	2186.5 p	1	2788.1	12

Fe (AAi)

2813.3	30
2823.3	10
2825.6	10
2832.4	15
2844.0	30
2851.8	15
2936.9 p	25
2941.3 p OC	2
2947.9 p	30
2953.9 p	20
2957.4 p	15
2965.3 p AA	20
2966.9 p	60
2970.1 p	30
2973.2 p	35
2981.4 f	30
2983.6 p	40
2994.4 p	25
3001.0 p	60
3008.1 p	40
3020.6 dp ⎫ 3021.1 p ⎭	100
3025.8 p	40
3031.6	25
3037.4 p	40
3042.7	30
3047.6 p	40
3057.4 OC	20
3059.1 p	70

Fe (OH, Bu; OC)

2973.2 p	2.5
2981.4 f OC	1.8
2983.6 p	5
2994.4 p	5
3001.0 p	2.5
3008.1 p	2.5
3020.6 dp ⎫ 3021.1 p ⎭	35
3025.8 p OC	4
3037.4 p	2.5
3047.6 p	5
3057.4 OC	2.5
3059.1 p OC	6
3440.6 p ⎫ 3441.0 p ⎭	200
3443.9 p OC	2
3465.9 p	25
3475.5 p	55
3476.7 p	10
3490.6 p	60
3497.8 p	10
3513.8 f	7
3521.3 f	3

3526.0 p	3
3558.5 f	3
3565.4 f	15
3570.1 f	40
3581.2	100
3587.0	7
3608.9	18
3618.8	20
3631.5	25
3647.8	25
3679.9 p	35
3683.1 p	3
3687.5	11
3705.6 p	80
3707.8 p	5
3709.2	15
3719.9 p	900
3722.6 p	120
3727.6	4
3733.3 p	70
3734.9	160
3737.1 p	550
3745.6 p	350
3748.3 p	130
3749.5	130
3758.2	40
3763.8	25
3767.2	15
3787.9	3
3795.0	6
3798.5	6
3799.5	6
3813.0 f	5
3815.8	10
3820.4	110
3824.4 p	110
3825.9	80
3834.2	30
3840.4 ⎫ 3841.1 ⎭	20
3850.0	4
3856.4 p	200
3859.9 p	800
3865.5	5
3872.5	7
3878.6 p	110
3886.3 p	270
3895.7 p	50
3899.7 p	80
3906.5 p	12
3917.2 OA	20
3920.3 p	50
3922.9 p	80
3927.9 p	80
3930.3 p	100
3969.3	5

4005.2	5
4045.8	20
4063.6	8
4071.7	6
4132.1	2
4143.9	2
4202.0	3
4216.2 f	5
4250.8 OC	0.7
4260.5 OC	0.7
4271.8	12
4294.1 f	6
4307.9	12
4325.8	11
4375.9 f	12
4383.5 f	35
4404.8 f	11
4415.1 f	3
4427.3 f	8
4442.3	3
4461.7 f	5
4482.2 f	5
5110.4 f	7
5167.5	15
5269.5	40
5328.0	30
5371.5	15
5405.8	5
5429.7	5

FeO r (OH, Bu)

4478	10
4544	4
4604	4
4659	6
4730	6
4929	10
5290	6
5382	30
5430	30
5532	30
5583	35
5614	90
5647	100
5679	90
5790	50
5808	20
5819	100
5868	200
5903	200
5975	200
6095 b	150
6109 b	150
6181	200
6219	300

763_0 w	40
779_0 w	60
792_0 w	90
811_0 mb	100
834_0 w	80
858_0 w	100

Ga (OH, Bu)

2874.2 p	10
2943.6 p	20
2944.2 p	10
4033.0 p	1000
4172.1 p	2000

Gd

5911.4	100

GdO r (OH, H)

4463	30
4481	30
4499	20
4617 m	300
4634 m	120
4653	40
4671	15
4717	10
4735	15
4756	10
4798	50
4817	60
4835	40
4856	15
4893 m	150
4910 m	100
4928 m	50
4949	10
5100 m	20
5117 t	30
5136	30
5247 v	30
5259 v	40
5277 v	70
5294 v	100
5303 v	100
5327	120
5336	120
5346	120
5356	120
5366	120
5377	120
5383	120
5405	150
5416	120
5427	120
5439	120
5450 d	130
5462	130
5469	130
5476 d	120
5481	120
5488	120
5653 d	40
5664	100
5681 d	400
5699 d	300
5716	200
5734 d	100
5746	150
5756	150
5771	250
5788	250
5807 v	600
5819 v	600
5832 v	500
5844 v	400
5858 db	400
5862 b	400
5905 b	400
5911 b	500
5918	500
5927	800
5942 d	600
5956 t	500
5972 d	400
5988 b	700
6001 db	800
6007 b	500
6019 v	400
6035 v	300
6044 tb	500
6064 db	600
6081 tb	1000
6099 d	700
6108	800
6115	500
6132	1000
6136	800
6152	1000
6156 t	1000
6172	1000
6176	1000
6183	1500
6201 d	2000
6212	2000
6221 d	2000
6232	1500
6242	1500
6245	800
6252	1000
6263	800
6267	600
6273	800
6284	500
6290	300
6294	500
6306	400
6316	200
6329	200
6338	200
6352	150

Ge (AH, P)

2498.0 f	0.5
2592.5 p	7
2651.2 p ⎫ 2651.6 p ⎭	30
2691.3 p	7
2709.6 p	15
2754.6 p	15
3039.1	10
3269.5 f	5

HO: See **OH**.

H₂O r (OH q)

6919	1
7165	3
7299	1
8097	8
8916	15
8974	10
9129	12
9183	8
9277	120
9333	90
9440	60
9485	50
9559	50
9610	60
9669	80

Hg

2536.5 f AH, P	25
4358.4 OH	weak
5460.7 OH	weak

HoO b (OH, H)

5069 d	100
5092	100
5097	100
5105 r	200
5119 r	200
5126 r	150
5132 r	200

394   APPENDIX

5142 dv	200	4587	1.0	4786.5	0.5
5148 r	200	4694	0.9	4791.0	1
5157 r	500	4845	1.0	4799.8	1
5162 r	200	4964	0.6	4804.3	2
5168 r	200	5131	0.9	4849.9	1
5186	200	5209	0.5	4856.1	2
5190 r	200	5308	1.0	4863.5	1.5
5196 r	150	5495	0.9	4869.8	3
5211 r	150	5533	0.7	4942.0	2
5251 v	300	5730	0.8	4950.8	3
5258	200			4956.0	3
5263 d	300	**In** (OH, Bu)		4965.0	6
5270 v	200			5084.2	4
5275 dv	300	2710.3 p	0.4	5097.1	6
5286 dv	300	2753.9	0.5	5099.2	7
5299	200	2932.6 p	1.0	5112.2	12
5304	200	3039.4 p	13	5323.2	7
5315 v	200	3256.1 p	120	5339.7	12
5320 dr	300	3258.6 p	40	5343.0	8
5326 r	200	4101.8 p	2000	5359.5	15
5333 r	300	4511.3 p	3000	5782.6	12
5346 r	250			5802.0	25
5359 r	250	**InO** r (OH)		5812.5	8
5380 t	150			5832.1	15
5565 dr	200	4126.4	0.1	6911.3	20
5584 r	400	4134.8	0.1	6939.0	40
5592 d	400	4142.5	0.1	7664.9 p	400 000
5609 d	400	4154.9	0.15	7699.0 p	200 000
5626 r	600	4165.3	0.2	8503.5 ⎫	
5635 v	500	4224.1	0.1	8505.2 ⎭	4
5643 tr	500	4233.2	0.15	8902.2 ⎫	
5656 v	700	4243.8	0.3	8904.0 ⎭	6
5660	1000	4257.6	0.45	9595.6 ⎫	
5672 v	700	4270.3	1.0	9597.8 ⎭	15
5677 d	600	4282.6 d	1.7	11 021 d	40
5688 v	700	4301.5	0.55	11 690.2	2000
5696 v	1000			11 769 ⎫	
5705	600	**K** (OH, Bu)		11 773 ⎭	3000
5711 v	600	2992.2 p	0.5	12 432	1000
5713 r	700	3034.8 p	1	12 522	1000
5729 dr	700	3102.0 p ⎫			
5737 d	500	3102.2 p ⎭	2	**LaO** v (OH)	
5752 v	400	3217.0 p	2	3556.4	1
5804 v	400	3217.5 p	1	3561.0	2
5820 v	600	3446.7 p	20	3566.1	4
5831 v	600	3447.7 p	10	3604.4	5
5835 v	600	4044.1 p	500	3608.1	5
5849 v	600	4047.2 p	250	3611.4	5
5854 v	400	4638.5	(1)	3614.7	4
5865 v	300	4641.6 p ⎫		3620.1	1
5880	400	4642.2 p ⎭	2	3622.6	1
		4740.9	(0.2)	3709.7	2
IO r (OH, Iodate)		4744.3	(0.5)		
4356	0.9	4753.9	0.5	**LaO** r (OH, H)	
4488	1.0	4757.4	1	4356 b	20

APPENDIX 395

4372	100	5977	100	8406	5
4376	100	6005	70	8454	20
4380	200	6033	50	8490	30
4384	250	6061	20	8527	35
4388	200	6188	20	8564	30
4392	100	6216	40	8601	25
4396	50	6245	40	8638	20
4400	50	6274	30	8677	15
4405	40	6302	20	8715	10
4410	30	6331	15	8754	5
4418	300	6360	10		
4423	300			**Li** (OH)	
4428	300	**LaO** r (OH, H)		2741.3 p	(2)
4433	250	6995	10	3232.6 p	30
4438	200	7011	20	3794.7	0.3
4443	100	7025	15	3915.0	1.2
4448	70	7041	30	4132.2	4
4453	50	7055	15	4273.3	0.7
4458	30	7071	35	4602.9	25
4464	20	7085	15	4972.0	6
4581	5	7101	35	6103.6	500
4585	6	7116	15	6707.8 p	700 000
4589	7	7132	35	8126.5	10
4594	7	7147	15		
4598	5	7163	30	**LuO** r (OH, H)	
4603	3	7179	10	4094	50
		7194	25	4096	70
LaO dr (OH, H)		7211	10	4107	30
5179	6	7225	15	4239	15
5204	11	7257	10	4242	25
5228	16	7380	150	4250	40
5253	18	7404	600	4253	20
5278	17	7434	600	4455	7
5303	13	7465	500	4489	7
5328	8	7497	350	4500	10
5354	6	7507	150	4509	20
5381	110	7528	250	4512	10
5407	160	7539	100	4521	12
5432	160	7560	180	4526	12
5458	140	7592	120	4533	20
5483	90	7625	80	4540	8
5509	60	7665	60	4561	10
5535	35	7702	30	4570	7
5561	15	7738	20	4575	10
5588	5	7877	170	4591	10
5601	700	7911	900	4644 v?	30
5627	600	7945	850	4654	60
5654	350	7980	600	4660	80
5680	200	8015	400	4662	1000
5707	80	8050	250	4672 d	250
5732	30	8086	150	4684	400
5868	100	8122	100	4696	200
5895	150	8159	60	4708	100
5922	200	8196	25	4721	30
5950	150	8233	15	4726	20

4735	35	**MgCl** v (PH; OA, C)		4034.5 p	3000
4749	20			4041.4	70
4764	15	3711 m	50	4048.8	35
4780	7	3776 m	150	4055.5	70
4992 b	30	3845 m	50	4058.9	20
5004	40			4063.5	25
5019	50	**MgF** dv (PH; FH)		4079.2 ⎫	50
5035	50	3503	12	4079.4 ⎭	
5041	60	3594	150	4082.9	40
5058 b	70	3684	10	4083.6	40
5120	150			4235.1 ⎫	10
5161	550	**MgO** v (OH, Bu)		4235.3 ⎭	
5170	700			4451.6	25
5186 b	350	4963	50	4462.0	15
5202 b	250	4975	70	4605.4	70
5206	170	4986	80	4754.0	80
5218	250	4997	80	4761.5	30
5233	100	5007	100	4762.4	30
5256	70	5206	10	4765.9	30
5403	30			4766.4	30
5449	60	**MgOH** b (AH, W)		4783.4	100
5464	60	3624	130	4823.5	100
5487	40	3702	480	5394.7 f	200
5971 ⎫		3719	390	5432.5 f	100
5993 ⎬	500	3729	330		
6014 ⎭		3767	260	**Mn**+ (OC)	
6688 ⎫		3784	270	2576.1	10
6749 ⎬	500	3791	290	2593.7	10
6810 ⎭		3807	410	2605.7	10
		3814	410		
Mg (AAi)		3823	420	**MnF** v (FH)	
		3834	420	Heads of close sequences:	
2776.7	2	3846	330	3360	6
2778.3	2	3877	220	3439	18
2779.8	3	3912	210	3518	20
2781.4	2			3595	12
2783.0	2	**Mn** (OA)		3671	8
2852.1 p OC	10 000	2794.8 p	75	3746	6
3329.9	3	2798.3 p	55		
3332.2	4	2801.1 p	40	**MnO** r (OH, Bu)	
3336.7	5	3531.8 ⎫		4976	30
3829.4	10	3532.0 ⎬	15	5013	45
3832.3	20	3532.1 ⎭		5051	40
3838.3	25	3547.8 ⎫		5159	80
		3548.0 ⎬	20	5192	130
Mg (OH)		3548.2 ⎭		5229	110
4571.1 f	5	3569.5 ⎫		5267	80
5167.3	10	3569.8 ⎬	25	5360	300
5172.7	20	3570.0 ⎭		5390	370
5183.6	40	3806.7	30	5424	220
		3823.5	35	5449	100
Mg+ (OC)		3834.4	30	5586	700
2795.5	500	4018.1	30	5610	500
2802.7	250	4030.8 p	5000	5860	300
		4033.1 p	4000		

5881	700	2440.0	1.0	544_0 w	40
5910	300	2447.0	1.2	5478	50
5944	150	2471.1	3.0	5681	60
6176	200	2478.7	1.5	5726	60
6204	100	2550.0	1.2	5971 r	200
6237	100	2559.0	1.2	5974 r	200
		2587.5	2.0	5990	250

MnOH b (AH; OG; AA)

Peaks in a nearly continuous band:
364_0
369_5
374_0
380_0
384_0
389_0 } 100
393_0
395_5
397_5
402_0
404_5
407_0

MnOH (?) w (OH; AA)

810_0	400
855_0	400

Mo (OA, H; AAi)

3132.6 p	10
3158.2 p	8
3170.3 p	8
3194.0 p	15
3208.8 p	3
3798.3 p	200
3864.1 p	170
3903.0 p	140

NC: See **CN**

(For N: OHi, N)

NH x (OC, W q)

3360	5
3370	1

NO v (AOHA; OC q)

2149.1	0.5
2154.9	0.4
2239.4	0.3
2245.4	0.3
2262.8	1.5
2269.4	0.9
2363.3	3.0
2370.2	1.5

2440.0	1.0
2447.0	1.2
2471.1	3.0
2478.7	1.5
2550.0	1.2
2559.0	1.2
2587.5	2.0
2595.7	1.2
2671.4	0.8
2680.0	0.6
2713.2	1.0
2722.2	0.5

Na (OH, Bu)

2680.4 dp	(1)
2852.8 p ⎫	5
2853.0 p ⎭	
3302.3 p	150
3303.0 p	70
4390.1	0.5
4393.4	1
4494.2	1
4497.7	2
4541.7	0.2
4545.2	0.5
4664.8	2
4668.6	4
4747.9	0.3
4751.8	0.7
4978.5	10
4982.8	20
5148.8	1.5
5153.4	3
5682.6	40
5688.2	80
5890.0 p	800 000
5895.9 p	400 000
6154.2	10
6160.7	20
8183.3	1000
8194.8	2000

NdO b (OH, H)

4414	10
4511	10
4619	20
4629	15
4645	10
4769	5
483_0 w	5
4959	5
5138	15
5250	20
5313	50
5327	30

544_0 w	40
5478	50
5681	60
5726	60
5971 r	200
5974 r	200
5990	250
5998 d	200
6002 v	200
6014 v	150
6027 r	150
6053	150
6133 v	100
6180	100
6219	150
6247	120
6269 r	200
6281 v	250
6287 r	250
6297 r	200
6313	200
6332	200
6336 r	200
6349 r	400
6354	400
6361	300
6370 d	500
6382 r	400
6389	400
6404	400
6410 r	400
6425 r	500
6442 r	500
6448 v	500
6453 v	400
6464 r	500
6493	500
6500 dv	1000
6518 d	500
6542 v	500
6564	500
6580 v	1500
6582 r	1500
6586 d	2000
6598 r	2000
6601 r	1500
6608 d	2000
6612 r	1500
6621	1200
6624	1200
6630 r	1500
6634 r	1000
6638	1000
6648 r	1200
6652 r	900
6655	1000

6664 r	800	7792 r	250	3366.2 f	25
6680 r	900	7996 v	400	3369.6 p	300
6683 r	700	8004 v	400	3372.0 p	25
6694	600	8216 v	200	3374.2 f	25
6748 r	400	8380	100	3380.6	200
6774 r	300	8661 d	100	3380.9 p	20
6819 v	250			3391.0 p	200
6831 v	250	**Ni** (AH, P)		3393.0 p	500
6839	200			3409.6 p	10
6846	200	2290.0 p	9	3413.5 p	30
6857 r	300	2300.8 p	3	3413.9 f	20
6880 v	300	2311.0 p	35	3414.8 p	1500
6887 d	400	2312.3 p	25	3423.7 p	120
6897 r	600	2314.0 p	27	3433.6 p	400
6908 r	800	2317.2 p	15	3437.3 f	150
6918 r	1000	2320.0 p	45	3446.3 p	600
6931 t	1000	2321.4 p	25	3452.9 f	150
6940 d	1200	2325.8 p	20	3458.5 p	600
6945 d	1200	2330.0 p	8	3461.7 f	1000
6957 r	800	2337.5 p	6	3467.5 f	15
6970	1500	2345.5 p	15	3469.5 f	15
6992 v	1500	2360.6 p	1	3472.5 p	300
6997 r	1500	2362.1 p	1	3483.8 p	60
7011 r	2000	2401.8 p	1	3485.9 f	10
7021 d	1500	2419.3 p	1	3493.0 p	1000
7027	1500	2943.9 p	15	3500.9 p	60
7033 r	2000	2981.6 p	20	3502.6 f	5
7056 r	1500	2984.1 p	15	3507.7 f	5
7064	1500	2992.6 p	25	3510.3 p	250
7092 d	2000	2994.5 f	25	3513.9 f	15
7100 r	1200	3002.5 p	60	3515.1 p	1000
7115 d	2000	3003.6 p	50	3519.8 p	30
7126 r	1500	3012.0	50	3523.4 f	10
7133 r	1500	3019.1 p	20	3524.5 p	2000
7158 r	1500	3031.9 p	10	3528.0 p	15
7172 tr	1500	3037.9 p	50	3548.2 p	30
7186 r	1000	3050.8 p	70	3551.5 p	5
7213 r	500	3054.3 p	40	3561.8 f	5
7227	800	3057.6 p	50	3566.4	400
7236 d	400	3064.6 p	(20)	3571.9 p	160
7245 r	300	3080.8 p	15	3587.9 f	10
7255 r	250	3101.6 p	50	3597.7 p	100
7268 r	200	3114.1 f	10	3602.3 r	15
7285 r	250	3134.1 p	50	3609.3 f	20
7318 v	200	3233.0 p	70	3610.5 p	240
7343 v	250	3234.6 p	15	3612.7 p	40
7405 r	250	3243.1 f	40	3619.4	900
7411 r	400	3248.5 p	10	3624.7 f	15
7420 r	250	3271.1 f	5	3662.0 f	3
7440 t	300	3282.7 p	5	3664.1 p	12
7467 d	300	3286.9 p	5	3669.2 f	5
7498 r	250	3315.7 f	40	3670.4 p	12
7529 r	200	3320.3 f	25	3674.1 f	15
7556 r	200	3322.3 f	12	3688.4 p	10
7592 r	200	3361.6 p	30	3693.9 f	4
7622 r	200	3365.8 f	15	3722.5 p	8

APPENDIX 399

3736.8 f	10
3739.2 f	5
3749.0 f	4
3775.6 f	50
3778.1 f	1
3783.5 f	50
3793.6 p	2
3807.1 f	100
3831.7 f	10
3858.3 f	200
3913.0 f	2
3972.2 f	2
3973.6 f	8

NiO r (OH, Bu)

4751	4
4877	10
4890	6
4939	4
5007	10
5024	25
5098	5
5175	20
5323	5
5408	10

OH x, y (OH q)

Prominent peaks:

2811.3 r	50
2816.0 r	40
2829.0 r	65
2831.7	45
2838.7	50
2875.3 r	45
2892.7 r	50
2908	40
2919	32
2945.2 r	23
3063.7 r	650
3067.8 r	750
3081.7	400
3089.9	800
3096	600
3099.5	550
3102.1	550
3106.0	600
3112.3	450
3117.8	500
3122.6	550
3128.3	400
3147.5	450
3169.6	250
3428.1 r	13
3432.1 r	15

3472.1 r	20
3484	13

Os (OC, W)

3301.6 p	0.2

PO dv (AH, P)

The following are P heads; the Q heads lie 1.3 A lower.

2288.2	4
2294.9	8
2301.7	8
2306.9	8
2313.7	4
2320.6	4
2367.3	30
2375.2	50
2379.9	40
2383.5	50
2387.9	35
2396.3	30
2454.6	70
2464.2	140
2477.9	100
2518.7	50
2529.4	70
2540.4	80
2555.0	50
2595.7	35

Pb (OHi, G)

2170.0 p	4
2237.4	0.3
2246.9	2
2332.4	0.2
2388.8	0.1
2393.8	4
2399.6	0.05
2401.9 f	3
2411.7	0.6
2443.8	1.7
2446.2	3
2476.4	9
2577.3 f	7
2613.7 } 2614.2 }	80
2663.2	25
2802.0	90
2823.2	10
2833.1 p	35
2873.3	20
3572.7	25
3639.6	110

3683.5	250
3739.9 f	25
4057.8	350

Pb$^+$ (OHi, G)

2203.5	0.01

PbO r (OH)

3878	0.5
3910	0.5
3955	1
3988	0.5
4038	0.5
4146	2
4229	3
4317	4
4410	5
4454	4
4509	4
4554	6
4658	6
4706	5
4749	5
4817	7
4851	6
4984	8
5022	7
5138	7
5162	7
5258	6
5331	5
5354	5
5405	4
5459	6
5618	4
5678	6
5911	5
6161	3
6251	4
6342	2
6428	2

Pd (OH, H)

2447.9 p OA	(10)
2476.4 f OA	(10)
2763.1 f OA	20
3242.7	450
3258.8 f	30
3287.2	30
3302.1	60
3373.0	150
3380.7	60
3404.6	1600
3421.2	400
3433.4	60

3441.4	100	6036	150	7827	300
3460.8	400	6042	100	7866	500
3481.2	100	6055	150	7873	300
3516.9	450	6098	100	7922	400
3553.1	180	6130	70	7986	500
3571.2	90	6135	70	8048	400
3609.5	1100	6163	70	8076	400
3634.7	2000	6259	120	8110	500
3690.3 f	30	6263	70	8173	300
3718.9	260	6271	150	8184	300
3799.2	240	6279	150	8234 d	250
3832.3	30	6298	120	8386	200
3894.2 f	40	6318	60	8392	100
3958.6 f	40	6336	50	8427	200
4213.0 f	80	6352	60	8432	150
		6357	60	8477	150
PrO b (OH, H)		6363	70	8489	600
416_0	5	6377	60	8494	300
438_0	5	6458	70	8555	400
452_0	10	6474	250	8622	300
460_0	7	6494	150		
467_0	5	6514	100	**Pt** (OC; AAi)	
473_0	2	6589	100	2440.1	(1)
482_0	5	6609	80	2487.2	(1)
492_0	7	6631	60	2659.5 p	10
502_0	3	6722	100	3064.7 p	50
		6822	120		
PrO r (OH, H)		6851	70	**Ra**	
5137	15	6882	100	4825.9 p	50
5157	20	6893	100		
5288	10	6924	250	**Ra**$^+$	
5352	35	6930	200	3814.4	30
5367 d	25	6950	200	4682.3	10
5378	10	6987	200		
5597	50	6991 v	200	**RaO** (?) b	
5612	50	7018	200	4405	10
5628	30	7039	300	4592	10
5645	15	7095 b	700	4718	10
5675	20	7120	500	4750	10
5691	70	7129	500		
5705	40	7138	500	**RaOH** (?) b	
5720	35	7279	300	520_0 w	10
5731	40	7320	500	602_0 w	10
5749	30	7327	300	6210	20
5763	70	7343	400	6247	20
5777	30	7350	300	6269	50
5791	20	7376	500	6285	20
5942	40	7401	300	6329	20
5963	40	7413	300	6349	20
5972	50	7431 d	400	6653	50
5983	50	7485	400		
5991	35	7539	250	**Rb** (OH, Bu)	
6005 d	50	7663	600	3228.0 p	(0.3)
6019	250	7668	400	3229.1 p	(0.3)
6026	120	7708	400		
		7714	400		

APPENDIX 401

3348.7 p	(1)	3412.3 u	1	**RhO** (?) (OH)	
3350.9 p	(1)	3434.9 p	270	5425 b	10
3587.1 p	9	3455.2	1		
3591.6 p	3	3462.0	32	**Ru** (OH, M)	
4201.9 p	350	3470.7	20		
4215.6 p	140	3474.8	25	3428.3 p	80
4967	(0.02)	3478.9	17	3436.7 p	60
4983	0.02	3498.7 f	10	3498.9 p	200
5017	0.03	3502.5 p	110	3589.2	(10)
5023	0.05	3507.3	25	3593.0	40
5037	0.1	3528.0 p	100	3596.2 p	40
5075.8	0.1	3538.1	3	3634.9 p	30
5089.0	0.1	3543.9	4	3661.4 f	120
5132	0.02	3549.5	13	3719.3 u	10
5150.7	0.3	3570.2	9	3726.9 p ⎫	1500
5164.9	0.2	3583.1 p	57	3728.0 p ⎭	
5195.3	0.5	3596.2 ⎫	70	3730.4 p	10
5234.0	0.03	3597.1 f ⎭		3742.3	30
5260.0	1	3612.5	19	3760.0	30
5322.4	0.2	3626.6	3	3786.1	30
5362.6	2	3639.5	0.5	3790.5 p	30
5390.6	1	3658.0 p	115	3798.1 ⎫	
5431.5	4	3666.2	14	3798.9 p ⎬	1000
5578.8	1	3690.7 p	100	3799.3 p ⎭	
5648.1	6	3692.4 p	400	3925.9 f	20
5653.7	3	3700.9 p	100	4080.6 f	20
5724.4	10	3713.0	6	4112.7	(30)
6070.8	20	3735.3	1	4197.6 ⎫	
6159.6	25	3748.2	2	4198.9 f ⎬	(10)
6206.3	30	3765.1 f	17	4199.9 ⎭	
6298.3	50	3788.5	5	4206.0	(10)
6306.3	5	3793.2	18	4554.5	(10)
7618.9	(50)	3799.3	34	4584.4 f	(10)
7757.7	(50)	3806.8	14	5699.0	50
7800.2 p	50 000	3818.2	3		
7947.6 p	25 000	3822.3	13	**SC**: See **CS**	
		3828.5	5		
Re (OAz, M)		3833.9	6	**Sb** (AH, P)	
		3856.5	66		
3451.8 p	2	3877.3	2	2068.4 p	8
3460.5 p	8	3912.8	1.5	2127.5 f	1
3464.7 p	4	3922.2	1	2139.8	2
		3934.2 f	16	2145.0	1.5
Rh (OH; OA, P)		3942.7	1.5	2175.9 p	80
		3958.9	18	2179.3	6
3271.6	3	4082.8 f	4	2208.5	2
3280.6	10	4121.7	4	2220.8	2
3283.6 f	15	4128.9	6	2224.9	2
3323.1 f	42	4135.3 f	16	2311.5 p	200
3338.5	1	4196.5 f	1	2445.5	8
3344.2	0.5	4211.1 f	37	2528.5	180
3359.9	1	4288.7	6	2598.1 f	280
3362.2	0.5	4374.8 f	65	2769.9 f	40
3368.4	1	4528.7 f	3		
3372.3	5	4675.0	4	**SbO** (AH, W)	
3396.8 p	100	4745.1	2	2565 mb	1
3399.7	1			2574 mr	1

ScO r (OH, H)

4502	8
4536	17
4571	17
4607	13
4673	65
4707	70
4742	50
4779	20
4858	200
4894	40
5097	60
5134	50
5171	30
5209	15
5357	8
5395	12
5435	10
5724	25
5737	100
5761	25
5765	80
5773	150
5776	120
5798	80
5802	100
5811 d	150
5837	60
5840	100
5849 d	150
5878	120
5888	150
5918	100
5928	100
5959	60
5969 d	80
6002	60
6017	600
6036	10 000
6055	800
6064	5000
6073	5000
6079	8000
6092	1000
6102	3000
6110	2500
6116	5000
6140	1000
6149	800
6154	1000
6181	200
6188	400
6193	600
6220	150
6229	120
6233	150
6262	70
6272	40
6275	70
6318	30
6326	30
6408	120
6420	30
6424	30
6437	60
6446	120
6458	100
6462 d	30
6476	80
6485	150
6496	150
6503	40
6518	60
6526	100
6535	120
6558	100
6567	80
6576	100
6591	20
6601	60
6610	70
6618	70
6645	40
6654	50
6661	60
6691	30
6701	40
6706	40
6738	30
6748	30
6752	30
6787	30
6800 d	20
6824	20
6864	40
6906	40
6922	30
6937	30
6963	40
6991	50
7005	50
7026	30
7036	50
7082	40
7094	40
7276	(50)

Si (OAz, M)

2506.9 p	8
2514.3 p	5
2516.1 p	22
2519.2 p	5
2524.1 p	8
2528.5 p	8
2881.6	(10)

SiO r (OAz, M)

2236.3	1
2277.2	1
2298.9	2
2342.4	3
2344.3	6
2364.5	4
2387.9	2
2413.8	8
2436.3	1
2459.0	2
2481.9	2
2486.8	8
2509.9	2
2563.8	6
2587.1	6
2644.8	3
2669.0	6
2693.7	5
2755.0	4
2780.5	4

Sm (AA; OA)

4419.3	50
4441.8 ⎫ 4442.3 ⎭	50
4581.7	50
4670.8	30
4688.7	20
4716.1 ⎫ 4717.1 ⎭	50
4728.4	30
4760.3	50
4783.1	40
4785.9	20
4841.7	30
4883.8 ⎫ 4884.0 ⎭	100
5200.6	50
5251.9	15
5271.4	50
5403.7 ⎫ 5405.2 ⎭	20
5516.1 p	20
5626.0	10
5659.9 p	10
5867.8 ⎫ 5868.6 ⎭	10

APPENDIX

SmO r (OH, H)

4546 b	20
5681	15
5698	15
5729	15
5746	15
5808 v	30
5820 mb	60
5840 db	50
5857	50
5924 mb	100
5943 mb	100
5947 mb	120
5962	80
5972	150
5981	120
5988 d	200
5993	80
6019	120
6035 b	200
6047	150
6065 b	120
6071	250
6079 mb	100
6089	150
6113	150
6132	300
6138	250
6149 db	250
6165	200
6168	200
6195	250
6200	200
6229	200
6238 db	200
6243 db	250
6256 tb	200
6273 b	150
6328 b	150
6335 v	150
6339	400
6348	200
6353	200
6361 v	200
6374	250
6379 db	250
6388 b	300
6391	300
6396 b	300
6406 db	300
6421	300
6425 v	300
6439	250
6450 b	300
6456 d	300

6469	250
6474 db	300
6478	400
6485 d	600
6510 trv	2000
6525	400
6532 m	1200
6540 db	300
6557	700
6565	200
6580	300
6585	250
6600 b	300
6619	300
6634	150
6647 d	250
6651 v	250
6656	250
6669 v	250
668_0 mb	200
6700 v	250
6716 b	250
6728	200
6744 v	200
6751	200
6796	200
6808	200
6820	150
6872	100
7155 b	30
7229 b	30
731_0 w	30
759_0 w	30

Sn (AH, P)

2113.9 p	0.5
2148.7 p	0.5
2151.4	0.5
2194.5	2
2199.3 f	5
2209.7	6
2231.7	0.5
2246.1 p	35
2268.9	20
2286.7 f	4
2317.2 f	4
2334.8 p	20
2354.8 p	100
2380.7 p	2
2408.2	1
2421.7	15
2429.5	120
2483.4	13
2495.7 f	4
2546.6 f	70

2571.6 f	10
2594.4	5
2661.2 f	25
2706.5 p	350
2761.8 f AAi	1
2779.8 f	8
2785.0 f AAi	2
2813.6 f AAi	4
2840.0	1000
2850.6 f	25
2863.3 p	500
2913.5 f OH, N	weak
3009.1 p	350
3034.1 p	800
3175.0	1200
3262.3	750
3330.6 f	100
3801.0 f	200
4524.7	40

SnH r (AH, W)

6095	10

SnO r (AH, W)

3205.8	5
3229.8	3
3262.4	20
3291.8	4
3323.5	16
3344.7	7
3351.4	5
3374.7	3
3381.7	4
3388.3	13
3406.9	6
3415.8	11
3444.6	6
3484.5	20
3512.9	7
3530.0	3
3542.4	4
3574.5	3
3585.4	20
3614.8	7
3677.7	1
3691.4	11
3721.2	11
3752.3	5
3779.3	4
3802.7	7
3817	6
3833.2	7
3864.9	13
3877.6	10

14+C.A.F.P.

3884.2	9	5521.8	1	4565	2
3899.3	8	5534.8	1	4693	1
3919.5	4				
3947.6	(7)	**Sr**$^+$ (OH, Bu)		**SrO** v (OG)	
3951.0	11	3464.5	?	6076 d	25
3978.7	8	4077.7	500	6085 m	20
3983.9	12	4215.5	400	6090	15
4019	5	4305.4	1	6095 t	10
4026	5			6100 d	7
4079	13	**SrCl** r (OG)		6109 d	7
4109	12	3894.0	4	6114	5
4122	12	3918.3	4		
4144	8	3937.1	4	**SrO** r (AA)	
4174	3	3961.6	4	7501	5
4218	8	3983.4	4	7523	7
4240	9	4009.4	4	7542	10
4262	12			7558	10
4303	7	**SrCl** v (AG; PH)		7873	20
4343	1	6359	20	7882	25
4365	3	6362	10	7902	30
4379	1	6614	20	8258	100
4411	6	6745	10	8272	100
4429	5			8700	70
4452	7	**SrF** (FH; PH)			
4499	4	5633 ⎫		**Sr$_2$O$_2$** b (AH, W)	
4522	3	5637 ⎬ r	5	5943	500
4541	2	5641 ⎭		5970	500
4570	3	5772 ⎫			
4591	3	5775 ⎪		**SrOH** b (AH, W; OH, Bu)	
4612	4	5778 ⎬ r	100	6035	3000
4634	4	5779 ⎪		6059	7000
4661	4	5782 ⎪		624$_0$ w	150
		5785 ⎭		646$_5$ w	700
SnO (?) b (AH, W)		6283 r ⎫		659$_0$ w	1500
Complex bands of doubtful origin; principal peaks:		6306 mv ⎭	30	666$_0$ w	5000
		6395 r ⎫		672$_0$ w	4000
		6419 mv ⎭	30	682$_0$ w	7000
485$_0$	40	6512 mv	200	704$_0$ w	500
496$_0$	30	6633 mv	200		
516$_0$	13	Bands marked mv are heads of close sequences.		**Ta** (OC, Hp)	
563$_0$	12			4661.1	1.2
		SrH v (AA)			
Sr (OH, Ac; OA, D)		7347	100	**Ta**$^+$ (OC, Hp)	
4607.3 p	10 000			2432.7	0.6
4722.3	2.5	**SrO** r (OG)			
4741.9	2	4167	2	**TbO** r (OH, H)	
4811.9	5.5	4281	3	4473	50
4832.1	8.5	4303	3	4571	50
4876.3	6.5	4346	3	4590 b	30
4962.3	6	4400	2	4607	60
5238.6 u	1	4421	2	4619 b	50
5256.9 u	1.2	4523	2	4752	15
2480.9	1.7	4544	2	4780	25
5504.2	(0.5)				

APPENDIX 405

4792	15	6050 v	200	4205	13
5270	10	6056	400	4268	13
5327	100	6068	500	4318	4
5340 d	350	6071 d	350	4343	18
5351 d	300	6078 m	600	4383	5
5357	120	6091 db	300	4408	10
5363	70	6098 d	250	4460	3
5369	300	6103 b	250	4487	16
5373	120	6112 b	250	4556	10
5379	100	6122 v	200	4616	2
5387	100	6126	150	4640	10
5396	60	6134	150	4699	7
5415	200	6140	200	4711	7
5428	40	6245	100	4774	3
5443	200	6292	70	4802	5
5457 v	100	6341	70	4863	9
5467	100	6349	250	4941	6
5471 b	70	6395	100	4975	2
5550	100	6436	150	5006	2
5558	120	6450 v	120	5035	6
5566	120	6460	150	5118	5
5575	150	6485 v	70	5139	6
5593	100	6496	50	5280	4
5601	150			5504	5
5635	200	**Te** (AH, P)		5668	2
5639	400	2383.2	5		
5645	120	2385.8	7	**Ti** (OC, Hp)	
5664 d	200			3635.5 p OH, N	?
5681	150	**TeO** r (AH, W)		4981.7	4
5699	150			4991.1	5
5702	300	3345.0	2	4999.5	5
5727	300	3382.7	5		
5755 v	150	3422.4	4	**TiO** r (OH)	
5768 v	150	3463.8	3	4806 d	20
5802	250	3507.8	2	4849	15
5815	150	3516.7	4	4956 d	40
5821	120	3560.5	6	5000	30
5839	250	3606.8	9	5168 d	70
5857	500	3661.5	10	5450 d	80
5865	250	3710.8	7	5598	(50)
5876	300	3718	6	5761 d	(20)
5902	250	3767	6	6215 ⎫	
5909	250	3777	6	6222 ⎭	(20)
5921	1000	3819.7	7	6681 ⎫	
5925	400	3835	7	6714 ⎥	
5933 v	250	3879	9	6747 ⎥	30
5940	900	3890.7	7	6781 ⎭	
5960 db	300	3940.6	7	7057 d ⎫	
5966 tb	250	3952	8	7090 d ⎥	
5979 d	900	4003	8	7127 d ⎥	40
5999 b	400	4010	9	7159 ⎭	
6014 b	200	4060	5		
6022 b	200	4075	11	**Tl** (OH, Bu)	
6026	170	4125	6	2767.9 p	10
6036 t	200	4136	10	2918.3	10
6045	300	4185	3		

2921.5	3	5640 r	200	3133.3	0.2
3229.8	10	5684	200	3190.7	0.6
3519.2	60	5749	150	3267.7	1.0
3529.4	15			3271.1	1
3775.7 p	1000	**V** (OC)		3276.1	(0.5)
5350.5	700	3043.1 p ⎤			
Tm (AA; OA, Al)		3043.6 p ⎦	1.0	**VO** r (OH, Bu)	
4094.2	12	3052.2 p	0.3	4813	5
4105.8	11	3053.6 p	2.0	4858	9
4187.6	12	3056.3 p	2.6	4904	12
4203.7	8	3060.5 p	4.0	4951	5
4359.9 p	6	3066.4 p	3.0	5011	17
4386.4 p	4	3073.8 p	0.3	5057	30
4733.3 u	7	3183.4 p ⎤		5105	25
5675.8	6	3184.0 p ⎬	23	5229 d	50
		3185.4 p ⎦		5276	50
TmO b (OH, H)		3193.9	0.15	5325	35
4814	200	3198.0 p	1.2	5470 d	90
4892 r	300	3202.4 p	1.8	5517	60
4898 d	250	3207.4 p	0.5	5737 d	90
4903 d	200	3263.2	0.3	5838	40
4913 m	150	4330.0 p ⎤	0.4	5889	30
4938 d	250	4332.8 p ⎦		6087 d	50
4946	200	4341.0 p	0.4	6478 d	(20)
4962 v	150	4352.9 p	0.6	6531	(20)
5226	200	4379.2	6	6952	10
5257 r	200	4384.7 p	5	6979 b	10
5282 v	150	4390.0	3	7011	20
5294	200	4395.2	2	7038 b	20
5302 dv	250	4400.6	1	7070	20
5313 dv	300	4406.6 ⎤	5.5	7108 b	20
5329 dr	350	4408.5 ⎦		7374 d	5
5339 v	250	4441.7	0.5	7394	10
5347 r	400	4444.2	0.5	7405	10
5352	400	4460.3	1.7	7418	10
5361	300	4594.1 p	0.2	7434 d	15
5365	300	4827.5 p ⎤		7454	15
5374 r	350	4831.6 p ⎬	1.2	7472	15
5396	300	4832.4 p ⎦		7492	15
5401 r	300	4851.5 p	0.8	7517 b	15
5415	400	4864.7 p	1.3	7534	10
5425	350	4875.5 p	2.0	7851	5
5434	350	4881.6 p	2.0	7865	10
5473 r	400	5698.5 ⎤		7897	15
5488	300	5703.6 ⎬	0.5	7919	15
5500 r	400	5707.0 ⎦		7940	20
5520 r	300			7961	20
5524 r	400	**V**$^+$ (OC)		7973	20
5531	400	3093.1	1.8	800_0 b	20
5558 r	400	3102.3	1.7	8538	5
5566 t	350	3110.7	2.0	8574 d	5
5572 r	350	3118.4	1.3	8597	7
5578 r	300	3121.1	0.2	8605	7
5589 r	350	3125.3	1.4	8625	10
		3130.3	0.8	8642	10

8668	10	6108	300	4738 r	80
8702	10	6115	300	4749 r	80
		6127	200	4760 r	90
W (OC, W)		6132	7000	4765 r	100
4680.5 f	10	6148	5000	4778 mr	120
		6165	3000	4824 r	100
YO r (OH, H)		6182	2000	4840 r	120
4496	10	6200	1200	5438 dr	60
4526	15	6218	800		
4562	15	6237	500	**YbOH** (?) b (AH, W)	
4604	12	6251	150	4850	250
4650 d	60	6275	250	4981	500
4677 d	60	6295	100	5174 w	350
4707 d	50	6316	60	5325	700
4745 d	30	6338	40	555_0	600
4818 d	250	6352	40	5725	900
4842 d	150	6370	40	587_0	250
4870	50	6387	40	602_0 w	120
5025 d	40	6406	30	622_0 w	100
5050 d	60	6424	30		
5078 d	60	6468	50	**Zn** (OH, N)	
5112	50	6485	60	2138.6 p OC	50
5153	20	6501	70	2608.6	(0.3)
5275 d	15	6518	80	2684.2	(0.5)
5304 d	15	6536	60	2712.5	(0.5)
5698	30	6554	60	3075.9 f AH	1
5714	50	6573	50	3282.3	(1)
5731	60	6611	40	3302.8 d	(1)
5747	60	6631	30	3345.0 ⎫	
5764	60	6719	50	3345.6 ⎭	(1)
5783	40			4680.1	2
5800	50	**Yb** (OH, H)		4722.2	3
5819	40	3464.4	70	4810.5	6
5838	30	3988.0 p	700	6362.3	(1)
5842	50	5556.5 p	250		
5859	60			**Zr** (OC, Hp)	
5876	60	**Yb$^+$** (OH)		4687.8	2
5894	60	3289.4	2		
5912	50	3694.2	15	**Zr$^+$** (OC, Hp)	
5931	40			2571.4	6
5939	100	**YbO** (OH)			
5956	200	3880 b	7	**ZrO** r (OH)	
5972	6000	415_0 w	3	5629 ⎫	
5988	5000	419_0 w	4	5634 ⎪	
6004	3000	4442 b	3	5658 ⎬	5
6020	2000	4476 b	9	5663 ⎭	
6037	1200	4518 b	10	5718 ⎫	
6054	800	4574 b	3	5724 ⎪	
6073	300	4590 b	4	5748 ⎬	5
6089	300	4716 r	50	5754 ⎭	
6097	300				

408 APPENDIX

TABLE 14. Index of flame spectra by wavelength

λ					λ					λ				
2061.7 p	Bi	AHn	0.8		2288.2 dv	P	AHn	4		2370.2 v	NO	AOHA	1.5	
2068.4 p	Sb	AHn	8		2290.0 p	Ni	AHn	9		2371.9 f	Co	OH	0.08	
2089.9 r	CO	OHinq	1		2292.5 p	Fe	AHn	1		2373.6 p	Fe	AHn	1	
2110.3 p	Bi	AHn	0.6		2294.9 dv	P	AHn	8		2375.2 dv	P	AHn	50	
2113.1 r	CO	OHinq	1		2295.2 f	Co	OH	0.09		2379.9 dv	P	AHn	40	
2113.9 p	Sn	AHn	0.5		2296.0 f	Co	OH	0.05		2380.5 p	Co	OH	0.25	
2127.5 f	Sb	AHn	1		2296.7 a					2380.7 p	Sn	AHn	2	
2128.3 r	CO	OHinq	1		2297.8 p	Fe	AHn	3		2381.2 f	As	AHn	2.5	
2138.6 p	Zn	OC	50		2298.9 r	Si	OAzn	2		2383.2 a	Te	AHn	5	
2139.8 a	Sb	AHn	2		2299.2 p	Fe	AHn	1		2383.5 dv	P	AHn	50	
2144.4 s	Cd	OHn	?		2300.8 p	Ni	AHn	3		2384.9 p	Co	OH	0.3	
2145.0 a	Sb	AHn	1.5		2301.7 dv	P	AHn	8		2385.8 a	Te	AHn	7	
2148.7 p	Sn	AHn	0.5		2304.0 a	Co	OH	0.06		2387.9 dv	P	AHn	35	
2149.1 v	NO	AOHA	0.5		2304.2 p					2387.9 r	Si	OAzn	2	
2150.2 r	CO	OHinq	2		2306.9 dv	P	AHn	8		2388.8 a	Pb	OHin	0.1	
2151.4 a	Sn	AHn	0.5		2309.0 p	Co	OH	0.22		2392.0 p	Co	OH	0.2	
2154.9 v	NO	AOHA	0.4		2310 x	C₂	OAq	4		2393.8 a	Pb	OHin	4	
2166.8 p	Fe	AHn	1		2311.0 p	Ni	AHn	35		2396.3 dv	P	AHn	30	
2170.0 p	Pb	OHin	4		2311.5 r	CO	OHinq	5		2399.6 a	Pb	OHin	0.05	
2173.0 r	CO	OHinq	5		2311.5 p	Sb	AHn	200		2400.9 a	Bi	AHn	0.6	
2175.9 p	Sb	AHn	80		2312.3 p	Ni	AHn	25		2401.8 p	Ni	AHn	1	
2178.1 p	Fe	AHn	1		2313.7 dv	P	AHn	4		2401.9 f	Pb	OHin	3	
2179.3 a	Sb	AHn	6		2314.0 p	Ni	AHn	27		2402.2 f	Co	OH	0.35	
2186.5 p	Fe	AHn	1		2316.8 p	Co	OH	0.05		2407.3 p	Co	OH	1.1	
2191.8 p	Fe	AHn	0.4		2317.2 p	Ni	AHn	15		2408.2 a	Sn	AHn	1	
2194.5 a	Sn	AHn	2		2317.2 f	Sn	AHn	4		2411.6 p	Co	OH	0.9	
2196.8 r	CO	OHinq	5		2320.0 p	Ni	AHn	45		2411.7 a	Pb	OHin	0.6	
2199.3 r	Sn	AHn	5		2320.6 dv	P	AHn	4		2413.8 r	Si	OAzn	8	
2200.4 p	Fe	AHn	1		2321.4 p	Ni	AHn	25		2414.5 p	Co	OH	1.1	
2203.5 s	Pb	OHin	0.01		2322 x	C₂	OAq	5		2415.3 p				
2208.5 a	Sb	AHn	2		2323.1 p	Co	OH	0.16		2419.1 a	Co	OH	0.03	
2209.7 a	Sn	AHn	6		2325.8 p	Ni	AHn	20		2419.3 p	Ni	AHn	1	
2215.8 r	CO	OHinq	2		2330.0 p	Ni	AHn	8		2421.7 a	Co	OH	0.03	
2220.8 a	Sb	AHn	2		2332.4 a	Pb	OHin	0.2		2421.7 a	Sn	AHn	15	
2221.5 r	CO	OHinq	4		2334.8 p	Sn	AHn	20		2424.9 p	Co	OH	1	
2224.9 a	Sb	AHn	2		2336.0 p	Co	OH	0.11		2428.0 p	Au	OH	5	
2228.3 p	Bi	AHn	17		2337.5 p	Ni	AHn	6		2429.2 p	Co	OH	0.03	
2230.6 f	Bi	AHn	30		2337.9 r	CO	OHinq	5		2429.5 a	Sn	AHn	120	
2231.7 a	Sn	AHn	0.5		2338.7 p	Co	OH	0.18		2432.2 p	Co	OH	0.8	
2236.3 r	Si	OAzn	1		2342.4 r	Si	OAzn	3		2432.7 s	Ta	OCn	0.6	
2237.4 a	Pb	OHin	0.3		2344.3 r	Si	OAzn	6		2436.3 r	Si	OAzn	1	
2238.3 r	CO	OHinq	2		2345.5 p	Ni	AHn	15		2436.7 p	Co	OH	0.7	
2239.4 v	NO	AOHA	0.3		2346.2 f	Co	OH	0.06		2437.2 f	As	AHn	1	
2245.4 v	NO	AOHA	0.3		2348.6 p	Be	OAzn	2000		2437.3 v	As	AHn	10	
2246.1 p	Sn	AHn	35		2349.8 a	As	AHn	30		2438.5 v				
2246.9 a	Pb	OHin	2		2352.9 a	Co	OH	0.17		2439.0 p	Co	OH	0.7	
2247.2 r	CO	OHinq	2		2354.8 p	Sn	AHn	100		2440.0 v	NO	AOHA	1	
2261.7 r	CO	OHinq	4		2355.6 p	Co	OH	0.11		2440.1 a	Pt	OC	1	
2262.8 v	NO	AOHA	1.5		2358.2 p	Co	OH	0.07		2443.8 a	Pb	OHin	1.7	
2268.9 a	Sn	AHn	20		2358.7 p					2445.5 a	Sb	AHn	8	
2269.4 v	NO	AOHA	0.9		2360.6 p	Ni	AHn	1		2446.2 a	Pb	OHin	3	
2273.9 r	CO	OHinq	2		2362.1 p	Ni	AHn	1		2447.0 v	NO	AOHA	1.2	
2276.5 a	Co	OH	0.06		2363.3 v	NO	AOHA	3		2447.9 p	Pd	OA	10	
2276.6 f	Bi	AHn	7		2364.5 r	Si	OAzn	4		2454.6 dv	P	AHn	70	
2277.2 r	Si	OAzn	1		2364.7 p	Cr:	see 2365.9			2456.2 a	Co	OH	0.07	
2286.1 r	CO	OHinq	5		2365.1 p	Co	OH	0.2		2456.5 f	As	AHn	6	
2286.7 f	Sn	AHn	4		2365.9 p	Cr	AHn	2		2457.6 a	Fe	AHn	2	
2287.8 a	Co	OH	0.06		2366.8 p					2459.0 r	Si	OAzn	2	
2288.0 p	Cd	OC	200		2367.3 dv	P	AHn	30		2460.8 p	Co	OH	0.08	
2288.1 a	As	AHn	17		2369.7 a	Co	OH	0.1		2462.6 p	Fe	AHn	8	

APPENDIX 409

2464.2 dv	P	AHn	140	2530.0 r	CS	AOAn	1	2651.2 p	Ge	AHn	30
2464.6 p	Co	OH	0.05	2530.1 a	Co	OH	0.1	2651.6 p			
2465.1 a	Fe	AHn	1	2532.2 a	Co	OH	0.12	2659.5 p	Pt	OC	10
2467.7 p	Co	OH	0.08	2535.6 p	Fe	AHn	5	2661.2 f	Sn	AHn	25
2468.9 a	Fe	AHn	1	2536.0 p	Co	OH	0.9	2662.6 r	CS	AOAn	1.1
2470.3 a	Co	OH	0.08	2536.5 f	Hg	AHn	25	2663.2 a	Pb	OHin	25
2471.1 v	NO	AOHA	3	2538.7 r	CS	AOAn	1.3	2667.9 f	Fe	AAi	1
2472.9 p	Fe	AHn	11	2540.4 dv	P	AHn	80	2669.0 r	Si	OAzn	6
2473.9 p	Co	OH	0.06	2541.0 p	Fe	AHn	7	2671.4 v	NO	AOHA	0.8
2474.8 a	Fe	AHn	1	2544.3 p	Co	OH	0.7	2676.0 p	Au	OH	10
2476.4 a	Pb	OHin	9	2546.0 p	Fe	AHn	6	2677.0 r	CS	AOAn	1.3
2476.4 f	Pd	OA	10	2546.6 f	Sn	AHn	70	2679.1 a	Fe	AAi	2
2476.6 a	Co	OH	0.12	2548.3 a	Co	OH	0.06	2679.8 a	Co	OH	0.05
2477.9 dv	P	AHn	100	2549.5 r	CS	AOAn	0.9	2680.0 v	NO	AOHA	0.6
2478.6 a	C	OAq		2549.6 p	Fe	AHn	6	2680.4 dp	Na	OHn	1
2478.7 v	NO	AOHA	1.5	2550.0 v	NO	AOHA	1.2	2684.2 a	Zn	OHn	0.5
2479.8 p	Fe	AHn	8	2553.0 a	Co	OH	0.1	2685.3 a	Co	OH	0.1
2481.9 r	Si	OAzn	2	2555.0 dv	P	AHn	50	2690.1 f	Fe	AAi	1
2483.3 p	Fe	AHn	60	2555.8 r	CS	AOAn	1.1	2691.3 p	Ge	AHn	7
2483.4 a	Sn	AHn	13	2556.8 a	Co	OH	0.04	2693.2 r	CS	AOAn	1.3
2483.6 a	Co	OH	0.07	2559.0 v	NO	AOHA	1.2	2693.7 r	Si	OAzn	5
2486.8 r	Si	OAzn	8	2561.3 a	Co	OH	0.3	2695.8 a	Co	OH	0.1
2487.2 a	Pt	OC	1	2563.8 r	Si	OAzn	6	2706.5 p	Sn	AHn	350
2488.1 p	Fe	AHn	40	2565 mb	Sb	AH	1	2706.6 a	Fe	AAi	1
2490.6 p	Fe	AHn	40	2567.3 p	Co	OH	0.5	2709.6 p	Ge	AHn	15
2492.1 p	Cu	AHn	20	2568.0 p	Al	OHn	?	2710.3 p	In	OHn	0.4
2492.9 f	As	AHn	6	2569.7 v	As	AHn	20	2712.5 a	Zn	OHn	0.5
2494.7 a	Be	OAzn	4	2570.9 v				2713.2 v	NO	AOHA	1
2495.6 a	Co	OH	0.03	2571.4 s	Zr	OCn	6	2719.0 p	Fe	AHn	35
2495.7 f	Sn	AHn	4	2571.6 f	Sn	AHn	10	2720.9 p	Fe	AHn	20
2495.5 a	Fe	AHn	1	2572.2 a	Co	OH	0.1	2722.2 v	NO	AOHA	0.5
2496.8 p	B	OC	4	2572.7 r	CS	AOAn	0.9	2723.6 p	Fe	AHn	7
2497.7 p				2573.5 a	Co	OH	0.5	2728.0 a	Fe	AHn	2
2498.0 f	Ge	AHn	0.5	2574 mre	Sb	AH	1	2733.6 a	Fe	AAi	7
2501.1 p	Fe	AHn	5	2575.6 r	CS	AOAn	5	2737.3 p	Fe	AHn	7
2503.6 dv	As	AHn	20	2576.1 s	Mn	OC	10	2741.3 p	Li	OH	2
2504.5 a	Co	OH	0.1	2577.3 f	Pb	OHin	7	2742.4 p	Fe	AHn	12
2504.7 v	As:	see 2503.6		2584.5 a	Fe	AHn	4	2744.1 p	Fe	AAi	3
2506.9 a	Co	OH	0.07	2587.1 r	Si	OAzn	6	2748.3 a	Au	OH	2
2506.9 p	Si	OAzn	8	2587.5 v	NO	AOHA	2	2750.1 p	Fe	AHn	7
2507.3 r	CS	AOAn	1	2589.6 r	CS	AOAn	3	2753.7 a	Fe:	see 2754.0	
2509.9 r	Si	OAzn	2	2590.6 a	Co	OH	0.03	2753.9 a	In	OHn	0.5
2510.8 p	Fe	AHn	9	2592.5 p	Ge	AHn	7	2754.0 a	Fe	AAi	2
2511.0 a	Co	OH	0.4	2593.7 s	Mn	OC	10	2754.4 a			
2511.2 r	CS	AOAn	0.8	2594.4 a	Sn	AHn	5	2754.6 p	Ge	AHn	15
2514.3 p	Si	OAzn	5	2595.7 v	NO	AOHA	1.2	2755.0 r	Si	OAzn	4
2516.1 p	Si	OAzn	22	2595.7 dv	P	AHn	35	2756.3 p	Fe	AHn	3
2517.9 a	Co	OH	0.2	2598.1 f	Sb	AHn	280	2761.4 a	Co	OH	0.07
2518.1 p	Fe	AHn	9	2599.6 a	Fe	AAi	1	2761.8 f	Sn	AAi	1
2518.7 dv	P	AHn	50	2605.7 s	Mn	OC	10	2761.8 a	Fe	AHn	3
2519.2 p	Si	OAzn	5	2605.9 r	CS	AOAn	1.6	2762.0 a			
2521.4 p	Co	OH	1.2	2606.8 a	Fe	AAi	3	2763.1 f	Pd	OA	20
2522.8 p	Fe	AHn	35	2608.6 a	Zn	OHn	0.3	2764.2 a	Co	OH	0.3
2523.2 r	CS	AOAn	1.2	2613.7 a	Pb	OHin	80	2766.4 a	Co	OH	0.1
2524.1 p	Si	OAzn	8	2614.2 a				2767.5 a	Fe	AHn	4
2524.3 p	Fe	AHn	5	2621.6 r	CS	AOAn	1	2767.9 p	Tl	OHn	10
2524.5 a	Bi	AHn	0.8	2622.4 a	Co	OH	0.1	2769.9 f	Sb	AHn	40
2527.4 p	Fe	AHn	20	2623.5 a	Fe	AAi	2	2772.1 p	Fe	AAi	2
2528.5 a	Sb	AHn	180	2627.9 a	Bi	AHn	8	2775.0 a	Co	OH	0.1
2528.5 p	Si	OAzn	8	2635.8 a	Fe	AAi	1	2776.7 a	Mg	AAi	2
2529.0 p	Co	OH	0.9	2644.8 r	Si	OAzn	3	2778.2 a	Fe	AAi	2
2529.1 p	Fe	AHn	8	2650.3 a	Co	OH	0.2	2778.3 a	Mg	AAi	2
2529.4 dv	P	AHn	70	2650.6 ma	Be	OAzn	15	2779.8 a	Mg	AAi	3

APPENDIX

2779.8 f	Sn	AHn	8	2980.8 a	Cr	AAi	0.2	3060.5 p	V	OC	4
2780.2 a	As	AHn	5	2981.4 f	Fe	AAi	30	3063.7 r	OH	OHq	650
2780.5 r	Si	OAzn	4	2981.6 p	Ni	OHn	20	3064.6 p	Ni	OHn	20
2781.4 a	Mg	AAi	2	2983.6 p	Fe	AAi	40	3064.7 p	Pt	OC	50
2783.0 a	Mg	AAi	2	2984.1 p	Ni	OHn	15	3066.4 p	V	OC	3
2785.0 f	Sn	AAi	2	2987.2 p	Co	OH	9	3067.7 p	Bi	AHn	400
2788.1 a	Fe	AHn	12	2988.6 a	Cr	OC	1	3067.8 r	OH	OHq	750
2794.8 p	Mn	OA	75	2989.0 f	Bi	AHn	10	3072.3 p	Co	OH	15
2795.5 s	Mg	OC	500	2989.6 f	Co	OH	9	3073.8 p	V	OC	0.3
2798.3 p	Mn	OA	55	2992.2 p	K	OHn	0.5	3075.9 f	Zn	AH	1
2801.1 p	Mn	OA	40	2992.6 p	Ni	OHn	25	3080.8 p	Ni	OHn	15
2802.0 a	Pb	OHin	90	2993.3 a	Bi	AHn	5	3081.7 x	OH	OHq	400
2802.7 s	Mg	OC	250	2994.4 p	Fe	AAi	25	3082.2 p	Al	OH	1
2811.3 r	OH	OHq	50	2994.5 f	Ni	OHn	25	3082.6 p	Co	OH	14
2813.3 a	Fe	AAi	30	3000.5 f	Co	OH	1	3086.8 p	Co	OH	16
2813.6 f	Sn	AAi	4	3001.0 p	Fe	AAi	60	3089.6 p	Co	OH	10
2816.0 r	OH	OHq	40	3002.5 p	Ni	OHn	60	3089.9 x	OH	OHq	800
2823.2 a	Pb	OHin	10	3003.6 p	Ni	OHn	50	3092.7 p	Al	OH	1
2823.3 a	Fe	AAi	10	3008.1 p	Fe	OHn	2.5	3093.1 s	V	OC	1.8
2825.6 a	Fe	AAi	10	3009.1 p	Sn	AHn	350	3096 x	OH	OHq	600
2829.0 r	OH	OHq	65	3012.0 a	Ni	OHn	50	3099.5 x	OH	OHq	550
2831.7 x	OH	OHq	45	3013.6 p	Co	OH	4	3101.6 p	Ni	OHn	50
2832.4 a	Fe	AAi	15	3014.8 da ⎱ Cr		OC	1.5	3102.0 p	K	OHn	2
2833.1 p	Pb	OHin	35	3015.2 a ⎰				3102.1 x	OH	OHq	550
2838.7 x	OH	OHq	50	3017.5 p	Co	OH	9	3102.2 p	K:	see 3102.0	
2840.0 a	Sn	AHn	1000	3017.6 a ⎱				3102.3 s	V	OC	1.7
2844.0 a	Fe	AAi	30	3018.5 a ⎰ Cr		OC	1.5	3105.6 r	As:	see 3106.8	
2850.6 f	Sn	AHn	25	3018.8 a ⎰				3106.0 x	OH	OHq	600
2851.8 a	Fe	AAi	15	3019.1 p	Ni	OHn	20	3106.8 r	As		6
2852.1 p	Mg	OC	10 000	3020.6 dp	Fe	OHn	35	3110.7 s	V	OC	2
2852.8 p ⎱ Na		OHn	5	3020.7 a	Cr:	see 3021.6		3112.3 x	OH	OHq	450
2853.0 p ⎰				3021.1 p	Fe:	see 3020.6		3114.1 f	Ni	OHn	10
2863.3 p	Sn	AHn	500	3021.6 a	Cr	OC	1.5	3117.8 x	OH	OHq	500
2873.3 a	Pb	OHin	20	3024.4 a	Cr	OC	0.5	3118.4 s	V	OC	1.3
2874.2 p	Ga	OHn	10	3024.6 f	Bi	AHn	5	3121.1 s	V	OC	0.2
2875.3 r	OH	OHq	45	3025.8 p	Fe	AAi	40	3121.4 p	Co	OH	7
2881.6 p	Si	OAzn	10	3030.2 a	Cr	AAi	1	3122.6 x	OH	OHq	550
2892.7 r	OH	OHq	50	3031.6 a	Fe	AAi	25	3125.3 s	V	OC	1.4
2898.0 a	Bi	AHn	35	3031.9 p	Ni	OHn	10	3128.3 x	OH	OHq	400
2908 x	OH	OHq	40	3034.1 p	Sn	AHn	800	3130.3 s	V	OC	0.8
2913.5 f	Sn	OHn	weak	3034.4 f	Co	OH	1	3132.6 p	Mo	OAn	10
2918.3 a	Tl	OHn	10	3034.8 p	K	OHn	1	3133.3 s	V	OC	0.2
2919 x	OH	OHq	32	3037.4 p	Fe	OHn	2.5	3134.1 p	Ni	OHn	50
2921.5 a	Tl	OHn	3	3037.9 p	Ni	OHn	50	3137.3 p	Co	OH	10
2932.6 p	In	OHn	1	3039.1 a	Ge	AHn	10	3139.9 p	Co	OH	5
2936.9 p	Fe	AAi	25	3039.4 p	In	OHn	13	3147.1 p	Co	OH	13
2938.3 a	Bi	AHn	10	3042.5 p	Co	OH	2	3147.5 x	OH	OHq	450
2941.3 p	Fe	OC	2	3042.7 a	Fe	AAi	30	3158.2 p	Mo	OAn	8
2943.6 p	Ga	OHn	20	3043.1 p ⎱		OC	1	3158.8 p	Co	OH	10
2943.9 p	Ni	OHn	15	3043.6 p ⎰ V				3169.6 x	OH	OHq	250
2944.2 p	Ga	OHn	10	3044.0 p	Co	OH	36	3170.3 p	Mo	OAn	8
2945.2 r	OH	OHq	23	3047.6 p	Fe	OHn	5	3170.6 r ⎱ As			9
2947.9 p	Fe	AAi	30	3048.9 p	Co	OH	7	3172.4 r ⎰			
2953.9 p	Fe	AAi	20	3050.8 p	Ni	OHn	70	3175.0 a	Sn	AHn	1200
2957.4 p	Fe	AAi	15	3052.2 p	V	OC	0.3	3183.4 p ⎱			
2965.3 p	Fe	AAi	20	3053.6 p	V	OC	2	3184.0 p ⎱ V		OC	23
2966.9 p	Fe	AAi	60	3053.9 a	Cr	AAi	0.5	3185.4 p ⎰			
2970.1 p	Fe	AAi	30	3054.3 p	Ni	OHn	40	3190.7 s	V	OC	0.6
2971.1 a	Cr	AAi	0.2	3056.3 p	V	OC	2.6	3193.9 a	V	OC	0.15
2973.2 p	Fe	AAi	35	3057.4 a	Fe	OC	2.5	3194.0 p	Mo	OAn	15
2975.5 a	Cr	AAi	0.2	3057.6 p	Ni	OHn	50	3198.0 p	V	OC	1.2
2980.6 a	Cd	OHn	?	3059.1 p	Fe	AAi	70	3202.4 p	V	OC	1.8

APPENDIX

3205.8 r	Sn	AH	5
3207.4 p	V	OC	0.5
3207.8 r	As		6
3208.8 p	Mo	OAn	3
3209.1 r	As:	see 3207.8	
3217.0 p	K	OHn	2
3217.5 p	K	OHn	1
3219.2 f	Co	OH	3
3228.0 p	Rb	OHn	0.3
3229.1 p	Rb	OHn	0.3
3229.8 r	Sn	AH	3
3229.8 a	Tl	OHn	10
3232.6 p	Li	OH	30
3233.0 p	Ni	OHn	70
3234.6 p	Ni	OHn	15
3237.0 f	Co	OH	7
3242.7 a	Pd	OHn	450
3243.1 f	Ni	OHn	40
3247.5 p	Cu	OHn	10 000
3248.5 p	Ni	OHn	10
3256.1 p	In	OHn	120
3258.6 p	In	OHn	40
3258.8 f	Pd	OHn	30
3261.1 f	Cd	AH	200
3262.3 a	Sn	AHn	750
3262.4 r	Sn	AH	20
3263.2 a	V	OC	0.3
3267.7 s	V	OC	1
3269.5 f	Ge	AHn	5
3271.1 f	Ni	OHn	5
3271.1 s	V	OC	1
3271.6 a	Rh	OH	3
3274.0 p	Cu	OHn	6500
3276.1 s	V	OC	0.5
3277.2 r ⎫ 3279.1 r ⎭	As		10
3280.6 a	Rh	OH	10
3280.7 p	Ag	AH	6000
3282.3 a	Zn	OHn	1
3282.7 p	Ni	OHn	5
3283.6 f	Rh	OH	15
3286.9 p	Ni	OHn	5
3287.2 a	Pd	OHn	30
3288.6 p	Cs	OH	0.05
3289.4 s	Yb	OHn	2
3291.8 r	Sn	AH	4
3301.6 p	Os	OC	0.2
3302.1 a	Pd	OHn	60
3302.3 p	Na	OHn	150
3302.8 da	Zn	OHn	1
3303.0 p	Na	OHn	70
3313.2 p	Cs	OH	0.1
3315.7 f	Ni	OHn	40
3320.3 f	Ni	OHn	25
3321.3 a	Be	OAzn	15
3322.3 f	Ni	OHn	12
3323.1 f	Rh	OH	40
3323.5 r	Sn	AH	16
3329.9 a	Mg	AAi	3
3330.6 f	Sn	AHn	100
3332.2 a	Mg	AAi	4
3334.1 a	Co	OH	25
3336.7 a	Mg	AAi	5
3337.2 f	Co	OH	1.4
3338.5 a	Rh	OH	1
3344.2 a	Rh	OH	0.5
3344.7 r	Sn	AH	7
3345.0 r	Te	AH	2
3345.0 a ⎫ 3345.6 a ⎭	Zn	OHn	1
3347.4 p	Cs	OH	0.2
3348.7 p	Cs	OH	0.1
3348.7 p	Rb	OHn	1
3350.9 p	Rb	OHn	1
3351.4 r	Sn	AH	5
3354.4 a	Co	OH	23
3359.9 a	Rh	OH	1
3360 v	MnF	FH	6
3360 x	NH	OCq	5
3361.6 p	Ni	OHn	30
3362.2 a	Rh	OH	0.5
3365.8 f	Ni	OHn	15
3366.2 f	Ni	OHn	25
3367.1 a	Co	OH	25
3368.4 a	Rh	OH	1
3369.6 p	Ni	OHn	300
3370 x	NH	OCq	1
3370.3 f	Co	OH	2.5
3372.0 p	Ni	OHn	25
3372.3 a	Rh	OH	5
3373.0 a	Pd	OHn	150
3374.2 f	Ni	OHn	25
3374.7 r	Sn	AH	3
3380.6 a	Ni	OHn	200
3380.7 a	Pd	OHn	60
3380.9 p	Ni	OHn	20
3381.7 r	Sn	AH	4
3382.7 r	Te	AH	5
3382.9 p	Ag	AH	7000
3385.2 a	Co	OH	25
3388.2 a	Co	OH	28
3388.3 r	Sn	AH	13
3391.0 p	Ni	OHn	200
3393.0 p	Ni	OHn	500
3395.4 f	Co	OH	65
3396.8 p	Rh	OH	100
3398.1 p	Cs	OH	0.3
3399.7 a	Rh	OH	1
3400.0 p	Cs	OH	0.2
3404.6 a	Pd	OHn	1600
3405.1 a	Co	OH	360
3406.9 r	Sn	AH	6
3409.2 a	Co	OH	140
3409.6 p	Ni	OHn	10
3412.3 u	Rh	OH	1
3412.3 f ⎫ 3412.6 p ⎭	Co	OH	380
3413.5 p	Ni	OHn	30
3413.9 f	Ni	OHn	20
3414.8 p	Ni	OHn	1500
3415.8 r	Sn	AH	11
3417.2 a	Co	OH	70
3421.2 a	Pd	OHn	400
3422.4 r	Te	AH	4
3423.7 p	Ni	OHn	120
3428.1 r	OH	OHq	13
3428.3 p	Ru	OHn	80
3431.6 p	Co	OH	190
3432.1 r	OH	OHq	15
3433.0 a	Co	OH	100
3433.4 a	Pd	OHn	60
3433.6 p	Ni	OHn	400
3434.9 p	Rh	OH	270
3436.7 p	Ru	OHn	60
3437.3 f	Ni	OHn	150
3439 v	MnF	FH	18
3440.6 p ⎫ 3441.0 p ⎭	Fe	OHn	200
3441.4 a	Pd	OHn	100
3442.9 p ⎫ 3443.6 a ⎭	Co	OH	280
3443.9 p	Fe	OC	2
3444.6 r	Sn	AH	6
3446.3 p	Ni	OHn	600
3446.7 p	K	OHn	20
3447.7 p	K	OHn	10
3449.2 a ⎫ 3449.4 a ⎭	Co	OH	290
3451.8 p	Re	OAzn	2
3452.9 f	Ni	OHn	150
3453.5 a	Co	OH	480
3455.2 p	Co	OH	200
3455.2 a	Rh	OH	1
3458.5 p	Ni	OHn	600
3460.5 p	Re	OAzn	8
3460.8 a	Pd	OHn	400
3461.7 f	Ni	OHn	1000
3462.0 a	Rh	OH	30
3462.8 a	Co	OH	170
3463.8 r	Te	AH	3
3464.4 a	Yb	OHn	70
3464.5 s	Sr	OHn	?
3464.7 p	Re	OAzn	4
3465.8 p	Co	OH	340
3465.9 p	Fe	OHn	25
3467.5 f	Ni	OHn	15
3469.5 f	Ni	OHn	15
3470.7 a	Rh	OH	20
3472.1 r	OH	OHq	20
3472.5 p	Ni	OHn	300
3474.0 p	Co	OH	270
3474.8 a	Rh	OH	25
3475.5 p	Fe	OHn	55
3476.7 p	Fe	OHn	10
3476.9 p	Cs	OH	0.5
3478.9 a	Rh	OH	17
3480.1 p	Cs	OH	0.2
3481.2 a	Pd	OHn	100
3483.4 a	Co	OH	40
3483.8 p	Ni	OHn	60
3484 x	OH	OHq	13
3484.5 r	Sn	AH	20
3485.9 f	Ni	OHn	10
3489.4 a	Co	OH	110
3490.6 p	Fe	OHn	60
3491.3 p	Co	OH	60

APPENDIX

3493.0 p	Ni	OHn	1000
3495.7 a	Co	OH	130
3497.8 p	Fe	OHn	10
3498.7 f	Rh	OH	10
3498.9 p	Ru	OHn	200
3500.9 p	Ni	OHn	60
3501.1 p	Ba	OH	2
3502.3 a	Co	OH	500
3502.5 p	Rh	OH	110
3502.6 f	Ni	OHn	5
3503 dv	MgF	PH	12
3506.3 a	Co	OH	290
3507.3 a	Rh	OH	25
3507.7 f	Ni	OHn	5
3507.8 r	Te	AH	2
3510.3 p	Ni	OHn	250
3510.4 p	Co	OH	280
3512.6 a	Co	OH	300
3512.9 r	Sn	AH	7
3513.5 p	Co	OH	370
3513.8 f	Fe	OHn	7
3513.9 f	Ni	OH	15
3515.1 p	Ni	OHn	1000
3516.7 r	Te	AH	4
3516.9 a	Pd	OHn	450
3518 v	MnF	FH	20
3518.3 a	Co	OH	50
3519.2 a	Tl	OHn	60
3519.8 p	Ni	OHn	30
3520.1 p	Co	OH	60
3521.3 f	Fe	OHn	3
3521.6 f	Co	OH	180
3523.4 a	Co	OH	150
3523.4 f	Ni	OHn	10
3524.5 p	Ni	OHn	2000
3526.0 p	Fe	OHn	3
3526.8 p	Co	OH	460
3528.0 p	Ni	OHn	15
3528.0 p	Rh	OH	100
3529.0 p	Co	OH	150
3529.4 a	Tl	OHn	15
3529.8 a	Co	OH	300
3530.0 r	Sn	AH	3
3531.8 a }			
3532.0 a } Mn	OA		15
3532.1 a }			
3533.4 p	Co	OH	130
3538.1 a	Rh	OH	3
3542.4 r	Sn	AH	4
3543.3 a	Co	OH	0.7
3543.9 a	Rh	OH	4
3547.8 a }			
3548.0 a } Mn	OA		20
3548.2 a }			
3548.2 p	Ni	OHn	30
3549.5 f	Rh	OH	13
3550.6 p	Co	OH	60
3551.5 p	Ni	OHn	5
3552.7 p	Co	OH	6
3553.1 a	Pd	OHn	180
3556.4 v	La	OH	1
3558.5 f	Fe	OHn	3
3558.8 f	Co	OH	5
3560 b	CN	OHin	0.25
3560.5 r	Te	AH	6
3560.8 s	Ce	OC	0.2
3560.9 a	Co	OH	50
3561.0 v	La	OH	2
3561.8 f	Ni	OHn	5
3565.0 f	Co	OH	45
3565.4 f	Fe	OHn	15
3566.1 v	La	OH	4
3566.4 a	Ni	OHn	400
3569.4 a	Co	OH	180
3569.5 a }			
3569.8 a } Mn	OA		25
3570.0 a }			
3570.1 f	Fe	OHn	40
3570.2 a	Rh	OH	9
3571.2 a	Pd	OH	90
3571.9 p	Ni	OHn	160
3572.7 a	Pb	OHin	25
3574.5 r	Sn	AH	3
3575.0 a } Co	OH		340
3575.4 p }			
3578.7 p	Cr	OAn	1000
3581.2 a	Fe	OHn	100
3583.1 p	Rh	OH	55
3583.9 v	CN:	see 3585.9	
3584.8 p } Co	OH		90
3585.2 a }			
3585.4 r	Sn	AH	20
3585.9 v	CN	OHin	0.8
3587.0 a	Fe	OHn	7
3587.1 p	Rb	OHn	9
3587.2 a	Co	OH	120
3587.9 f	Ni	OHn	10
3589.2 a	Ru	OHn	10
3590.4 v	CN	OHin	0.5
3591.6 p	Rb	OHn	3
3593.0 a	Ru	OHn	40
3593.5 p	Cr	OAn	800
3594 dv	MgF	PH	150
3594.9 p	Co	OH	190
3595 v	MnF	FH	12
3596.2 p	Ru	OHn	40
3596.2 a } Rh	OH		70
3597.1 f }			
3597.7 p	Ni	OHn	100
3602.1 p	Co	OH	140
3602.3 f	Ni	OHn	15
3604.4 v	La	OH	5
3605.3 p	Cr	OAn	500
3605.4 f	Co	OH	30
3606.8 r	Te	AH	9
3608.1 v	La	OH	5
3608.9 a	Fe	OHn	18
3609.3 f	Ni	OHn	20
3609.5 a	Pd	OHn	1100
3610.5 p	Ni	OHn	240
3611.4 v	La	OH	5
3611.5 p	Cs	OH	1
3612.5 a	Rh	OH	19
3612.7 p	Ni	OHn	40
3614.7 v	La	OH	4
3614.8 r	Sn	AH	7
3617.4 p	Cs	OH	0.3
3618.8 a	Fe	OHn	20
3619.4 a	Ni	OHn	900
3620.1 v	La	OH	1
3622.6 v	La	OH	1
3624 b	Mg	AH	130
3624.7 f	Ni	OHn	15
3625.0 a	Co	OH	3
3626.6 a	Rh	OH	3
3627.8 f	Co	OH	60
3631.4 p	Co	OH	40
3631.5 a	Fe	OHn	25
3634.7 a	Pd	OHn	2000
3634.9 p	Ru	OHn	30
3635.5 p	Ti	OHn	?
3636.6 a	Cr	OC	2.5
3639.5 a	Rh	OH	0.5
3639.6 a	Pb	OHin	110
364_0 b	Mn	AH	100
3647.7 p	Co	OH	28
3647.8 a	Fe	OHn	25
3652.5 p	Co	OH	40
3657.0 a	Co	OH	3
3658.0 p	Rh	OH	115
3661.4 f	Ru	OHn	120
3661.5 r	Te	AH	10
3662.0 f	Ni	OHn	3
3664.1 p	Ni	OHn	12
3666.2 a	Rh	OH	14
3669.2 f	Ni	OHn	5
3670.4 p	Ni	OHn	12
3671 v	MnF	FH	8
3674.1 f	Ni	OHn	15
3677.7 r	Sn	AH	1
3679.9 p	Fe	OHn	35
3683.0 a	Co	OH	0.8
3683.1 p	Fe	OHn	3
3683.5 a	Pb	OHin	250
3684 dv	MgF	PH	10
3687.5 a	Fe	OHn	11
3688.4 p	Ni	OHn	10
3690.3 f	Pd	OHn	30
3690.7 p	Rh	OH	100
3691.4 r	Sn	AH	11
3692.4 p	Rh	OH	400
3693.1 f	Co	OH	0.6
3693.9 f	Ni	OHn	4
3694.2 s	Yb	OHn	15
369_5 b	Mn	AH	100
3700.9 p	Rh	OH	100
3702 b	Mg	AH	480
3704.1 a	Co	OH	15
3705.6 p	Fe	OHn	80
3707.8 p	Fe	OHn	5
3708.8 a	Co	OH	0.6
3709.2 a	Fe	OHn	15
3709.7 v	La	OH	2
3710.8 r	Te	AH	7
3711 mv	MgCl	OA, C	50
3713.0 a	Rh	OH	6

APPENDIX 413

3718	r	Te	AH	6	3798.9 p	Ru	OHn	1000	3872.5 a	Fe	OHn	7
3718.9 a	Pd	OHn		260	3799.2 a	Pd	OHn	240	3873.1 a	Co	OH	700
3719	b	Mg	AH	390	3799.3 a	Rh	OH	35	3874.0 a	Co	OH	400
3719.3 u	Ru	OHn		10	3799.3 p	Ru:		see 3798.9	3876.2 p	Cs	OH	5
3719.9 p	Fe	OHn		900	3799.5 a	Fe	OHn	6	3876.8 a	Co	OH	40
3721.2 r	Sn	AH		11	380_0 b	Mn	AH	100	3877 b	Mg	AH	220
3722.5 p	Ni	OHn		8	3801.0 f	Sn	AHn	200	3877.3 a	Rh	OH	2
3722.6 p	Fe	OHn		120	3802.7 r	Sn	AH	7	3877.6 r	Sn	AH	10
3726.9 p	Ru	OHn		1500	3806.7 a	Mn	OA	30	3878 r	Pb	OH	0.5
3727.6 a	Fe	OHn		4	3806.8 a	Rh	OH	14	3878.6 p	Fe	OHn	110
3728	mr	CaCl	OA, C	5	3807 b	Mg	AH	410	3879 r	Te	AH	9
3728.0 p	Ru:		see	3726.9	3807.1 f	Ni	OHn	100	3880 b	Yb	OH	7
3729	b	Mg	AH	330	3808.1 a	Co	OH	5	3881.9 a	Co	OH	100
3730.4 p	Ru	OHn		10	3813.0 f	Fe	OHn	5	3882 mr	CaCl	OA, C	13
3732.4 a	Co	OH		2.3	3814 b	Mg	AH	410	3883.4 v	CN	OHin	4
3733.3 p	Fe	OHn		70	3814.4 s	Ra		30	3884.2 r	Sn	AH	9
3734.9 a	Fe	OHn		160	3815.8 a	Fe	OHn	10	3886.3 p	Fe	OHn	270
3735.3 a	Rh	OH		1	3816.3 a	Co	OH	1	3886.8 a	Cr	OAn	5
3736.8 f	Ni	OHn		10	3817 r	Sn	AH	6	3888.6 p	Cs	OH	1
3737.1 p	Fe	OHn		550	3818.2 a	Rh	OH	3	3889 r	CH	OHnq	6
3739.2 f	Ni	OHn		5	3819.7 r	Te	AH	7	3889.3 u	Ba	OH	1
3739.9 f	Pb	OHin		25	3820.4 a	Fe	OHn	110	389_0 b	Mn	AH	100
374_0 b	Mn	AH		100	3822.3 a	Rh	OH	13	3890.7 r	Te	AH	7
3742.3 a	Ru	OHn		30	3823 b	Mg	AH	420	3894.0 r	SrCl	OG	4
3745.5 a	Co	OH		23	3823.5 a	Mn	OA	35	3894.1 a	Co	OH	270
3745.6 p	Fe	OHn		350	3824.4 p	Fe	OHn	110	3894.2 f	Pd	OHn	40
3746	v	MnF	FH	6	3825.9 a	Fe	OHn	80	3895.0 a	Co	OH	100
3748.2 a	Rh	OH		2	3828 mr	CaCl	OA, C	12	3895.7 p	Fe	OHn	50
3748.3 p	Fe	OHn		130	3828.5 a	Rh	OH	5	3899.3 r	Sn	AH	8
3749.0 a	Cr	OC		1.5	3829.4 a	Mg	AAi	10	3899.7 p	Fe	OHn	80
3749.0 f	Ni	OHn		4	3831.7 f	Ni	OHn	10	3902.9 a	Cr	OC	0.3
3749.5 a	Fe	OHn		130	3832.3 a	Mg	AAi	20	3903.0 p	Mo	OAn	140
3749.9 a	Co	OH		0.8	3832.3 a	Pd	OHn	30	3906.3 a	Co	OH	12
3752.3 r	Sn	AH		5	3833.2 r	Sn	AH	7	3906.5 p	Fe	OHn	12
3758.2 a	Fe	OHn		40	3833.9 a	Rh	OH	6	3908.8 a	Cr	OC	0.8
3760.0 a	Ru	OHn		30	3834 b	Mg	AH	420	3909.9 f	Co	OH	55
3763.8 a	Fe	OHn		25	3834.2 a	Fe	OHn	30	3910 r	Pb	OH	0.5
3765.1 f	Rh	OH		17	3834.4 a	Mn	OA	30	3912 b	Mg	AH	210
3767	b	Mg	AH	260	3835 r	Te	AH	7	3912.8 a	Rh	OH	1.5
3767	r	Te	AH	6	3838.3 a	Mg	AAi	25	3913.0 f	Ni	OHn	2
3767.2 a	Fe	OHn		15	384_0 b	Mn	AH	100	3915.0 a	Li	OH	1.2
3775	mr	CaCl	OA, C	8	3840.4 a } Fe		OHn	20	3917.2 a	Fe	OA	20
3775.6 f	Ni	OHn		50	3841.1 a }				3918.3 r	SrCl	OG	4
3775.7 p	Tl	OHn		1000	3842.0 a	Co	OH	55	3919.2 a	Cr	OC	1.1
3776	mv	MgCl	OA, C	150	3845 mv	MgCl	OA, C	50	3919.5 r	Sn	AH	4
3777	r	Te	AH	6	3845.5 a	Co	OH	270	3920.3 p	Fe	OHn	50
3778.1 f	Ni	OHn		1	3846 b	Mg	AH	330	3922.2 a	Rh	OH	1
3779.3 r	Sn	AH		4	3850.0 a	Fe	OHn	4	3922.8 f	Co	OH	1.7
3783.5 f	Ni	OHn		50	3850.9 v	CN	OHin	2.6	3922.9 p	Fe	OHn	80
3784	b	Mg	AH	270	3851.0 a	Co	OH	4	3925.9 f	Ru	OH	20
3786.1 a	Ru	OHn		30	3854.7 v	CN	OHin	2.7	3927.9 p	Fe	OHn	80
3787.9 a	Fe	OHn		3	3856.4 p	Fe	OHn	200	393_0 b	Mn	AH	100
3788.5 a	Rh	OH		5	3856.5 a	Rh	OH	65	3930.3 p	Fe	OHn	100
3790.5 p	Ru	OHn		30	3858.3 f	Ni	OHn	200	3933.7 s	Ca	OA	300
3791	b	Mg	AH	290	3859.9 p	Fe	OHn	800	3934.2 f	Rh	OH	16
3793.2 a	Rh	OH		18	3861.2 a	Co	OH	20	3936.0 f	Co	OH	55
3793.6 p	Ni	OHn		2	3861.9 v	CN	OHin	3.2	3937.1 r	SrCl	OG	4
3794.7 a	Li	OH		0.3	3864.1 p	Mo	OAn	170	3940.6 r	Te	AH	7
3795.0 a	Fe	OHn		6	3864.9 r	Sn	AH	13	3941.7 a	Co	OH	40
3798.1 a	Ru:		see	3798.9	3865.5 a	Fe	OHn	5	3942.7 a	Rh	OH	1.5
3798.3 p	Mo	OAn		200	3871.4 v	CN	OHin	4	3944.0 p	Al	OAn	100
3798.5 a	Fe	OHn		6	3872 r	CH	OHnq	7	3945.3 a	Co	OH	9

3947.6 r	Sn	AH	7	4075 r	Te	AH	11	4196.5 f	Rh	OH	1
3951.0 r	Sn	AH	11	4076.1 a	Co	OH	21	4197 v	CN	OHin	0.7
3952 r	Te	AH	8	4077.7 s	Sr	OHn	500	4197.6 a	Ru:	see 4198.9
3952.9 a	Co	OH	19	4079 r	Sn	AH	13	4198.4 f	Co	OH	1
395_5 b	Mn	AH	100	4079.2 a } 4079.4 a }	Mn	OA	50	4198.9 f } 4199.9 a }	Ru	OHn	10
3955 r	Pb	OH	1								
3957.9 a	Co	OH	14	408_0 b	B	OHn	9	4201.9 p	Rb	OHn	350
3958.6 f	Pd	OHn	40	4080.6 f	Ru	OHn	20	4202.0 a	Fe	OHn	3
3958.9 a	Rh	OH	18	4082.6 a	Co	OH	1	4203.7 a	Tm	OAn	8
3961.5 p	Al	OAn	180	4082.8 f	Rh	OH	4	4205 r	Te	AH	13
3961.6 r	SrCl	OG	4	4082.9 a	Mn	OA	40	4205.0 s	Eu	OHn	5
3968.5 s	Ca	OA	150	4083.6 a	Mn	OA	40	4206.0 a	Ru	OHn	10
3969.3 a	Fe	OHn	5	4084 r	Ca	OA	1	4207.6 a	Co	OH	1
3972.2 f	Ni	OHn	2	4086.3 a	Co	OH	2	4211 r	CuCl	AH, Cu	2
3973.6 f	Ni	OHn	8	4092.4 a	Co	OH	50	4211.1 f	Rh	OH	35
3974.7 a	Co	OH	13	4094 r	Lu	OHn	50	4211.7 a	Pd	OII	15
397_5 b	Mn	AH	100	4094.2 a	Tm	OAn	12	4213.0 f	Pd	OHn	80
3978.7 r	Sn	AH	8	4096 r	Lu	OHn	70	4215.5 s	Sr	OHn	400
3979.5 f	Co	OH	50	4101.8 p	In	OHn	2000	4215.6 p	Rb	OHn	140
3983.4 r	SrCl	OG	4	4105 r	Ca	OA	1	4216 v	CN	OHin	0.7
3983.9 a	Cr	OC	0.5	4105.8 a	Tm	OAn	11	4216.2 f	Fe	OHn	5
3983.9 r	Sn	AH	12	4107 r	Lu	OHn	30	4218 r	Sn	AH	8
3987.1 a	Co	OH	4.5	4109 r	Sn	AH	12	4224.1 r	In	OH	0.1
3988 r	Pb	OH	0.5	4110.5 a	Co	OH	21	4226.7 p	Ca	OH	10 000
3988.0 p	Yb	OHn	700	4112.7 a	Ru	OHn	30	4229 r	Pb	OH	3
3991.7 a	Co	OH	10	4118.8 a	Co	OH	200	4233.2 r	In	OH	0.15
3995.3 f	Co	OH	290	4120 r	CuCl	AH, Cu	1	4234.0 f	Co	OH	6
3997.9 f	Co	OH	100	4121.3 a	Co	OH	320	4235.1 a } 4235.3 a }	Mn	OA	10
4003 r	Te	AH	8	4121.7 a	Rh	OH	4				
4005 r	CuH	OH	5	4122 r	Sn	AH	12	4237 y	CH	OHnq	5
4005.2 a	Fe	OHn	5	4125 r	Te	AH	6	4239 r	Lu	OHn	15
4009.4 r	SrCl	OG	4	4126.4 r	In	OH	0.1	4240 r	Sn	AH	9
4010 r	Te	AH	9	4128.9 a	Rh	OH	6	4242 r	Lu	OHn	25
4011.0 f	Co	OH	0.8	4129.7 s	Eu	OHn	5	4243 y	CH	OHnq	6
4018.1 a	Mn	OA	30	4132.1 a	Fe	OHn	2	4243.8 r	In	OH	0.3
4019 r	Sn	AH	5	4132.2 f	Co	OH	1.1	4249 y	CH	OHnq	7
402_0 b	Mn	AH	100	4132.2 a	Li	OH	4	4250 r	Lu	OHn	40
4020.9 a	Co	OH	33	4134.8 r	In	OH	0.1	4250.8 a	Fe	OC	0.7
4026 r	Sn	AH	5	4135.3 f	Rh	OH	16	4252.3 f	Co	OH	15
4027.0 f	Co	OH	15	4136 r	Te	AH	10	4253 r	Lu	OHn	20
4030.8 p	Mn	OA	5000	4142.5 r	In	OH	0.1	4254.3 p	Cr	OAn	900
4033.0 p	Ga	OHn	1000	4143.9 a	Fe	OHn	2	4256 y	CN	OHnq	8
4033.1 p	Mn	OA	4000	4144 r	Sn	AH	8	4257.6 r	In	OH	0.45
4034.5 p	Mn	OA	3000	4146 r	Pb	OH	2	4259 r	CuCl	AH, Cu	5
4038 r	Pb	OH	0.5	415_0 w	Yb	OH	3	4260.5 a	Fe	OC	0.7
4041.4 a	Mn	OA	70	4152 v	CN	OHin	0.4	4262 y	CH	OHnq	9
4044.1 p	K	OHn	500	4154.9 r	In	OH	0.15	4262 r	Sn	AH	12
404_5 b	Mn	AH	100	4158 v	CN	OHin	0.4	4268 y	CH	OHnq	10
4045.4 f	Co	OH	13	416_0 b	Pr	OHn	5	4268 r	Te	AH	13
4045.8 a	Fe	OHn	20	4165.3 r	In	OH	0.2	4268.0 f	Co	OH	2.2
4047.2 p	K	OHn	250	4167 r	Sr	OG	2	4270.3 r	In	OH	1
4048.8 a	Mn	OA	35	4168 v	CN	OHin	0.5	4271.8 a	Fe	OHn	12
4055.5 a	Mn	OA	70	4172.1 p	Ga	OHn	2000	4273.3 a	Li	OH	0.7
4057.2 f	Co	OH	12	4174 r	Sn	AH	3	4274 y	CH	OHnq	11
4057.8 a	Pb	OHin	350	418_0 b	B	OHn	6	4274.8 p	Cr	OAn	700
4058.9 a	Mn	OA	20	4181 v	CN	OHin	0.6	4280 y	CH	OHnq	12
4060 r	Te	AH	5	4185 r	Te	AH	3	4280 r	CuH	OH	30
4063.5 a	Mn	OA	25	4186.8 a	Dy	OH	12	4281 r	CuCl	AH, Cu	10
4063.6 a	Fe	OHn	8	4187.6 a	Tm	OAn	12	4281 r	Sr	OG	3
4066.4 f	Co	OH	11	4188 r	CuCl	AH, Cu	2	4282.6 dr	In	OH	1.7
407_0 b	Mn	AH	100	419_0 w	Yb	OH	4	4283.0 a	Ca	OH	4
4071.7 a	Fe	OHn	6	4190.7 f	Co	OH	33	4285.8 f	Co	OH	6

APPENDIX

4286 y	CH	OHnq	13
4288.7 a	Rh	OH	6
4289.7 p	Cr	OAn	500
4292 y	CH	OHnq	15
4292.1 f	Fe	OHn	6
4295.8 a	Cr:	see 4297.0	
4297 y	CH	OHnq	20
4297.0 a } 4299.7 a }	Cr	OC	3
4301.5 r	In	OH	0.55
4302.5 a	Ca	OH	15
4303 r	Sn	AH	7
4303 r	Sr	OG	3
4303.2 f	Co	OH	2
4305.4 s	Sr	OHn	1
4307.5 u	Cr	OC	4
4307.9 a	Fe	OHn	12
4315 v	CH	OHnq	60
4317 r	Pb	OH	4
4318 r	Te	AH	4
4318.7 a	Ca	OH	3
4320.6 a	Cr	OC	4
4324 x	CH	OHnq	8
4325.8 a	Fe	OHn	11
4328 r	CuH	OH	11
4330.0 p } 4332.8 p }	V	OC	0.4
4333 r	CuCl	AH, Cu	10
4334 y	CH	OHnq	6
4337.6 a	Cr	OAn	2
4339 y	CH	OHnq	7
4339.4 a	Cr	OAn	10
4341.0 p	V	OC	.40
4343 y	CH	OHnq	8
4343 r	Sn	AH	1
4343 r	Te	AH	18
4344.5 a	Cr	OAn	10
4346 r	Sr	OG	3
4348 y	CH	OHnq	9
4348 y	CuH	OH	10
4351.8 a	Cr	OAn	10
4352 y	CH	OHnq	10
4352.9 p	V	OC	0.6
4353 r	Al	OAn	1
4354 r	CuCl	AH, Cu	20
4356 y	CuH	OH	9
4356 r	IO	OH, IO_3^-	0.9
4356 b	La	OHn	20
4358.4 a	Hg	OH	weak
4359.6 a	Cr	OC	7
4359.9 p	Tm	OAn	6
4364 y	CuH	OH	8
4365 v	C_2	OHnq	15
4365 r	Sn	AH	3
437_0 b	B	OHn	12
4371 v	C_2	OHnq	12
4371.3 a	Cr	OC	2
4372 y	CuH	OH	7
4372 r	La	OHn	100
4374 r	Al	OAn	1
4374.8 f	Rh	OH	65
4375.9 f	Fe	OHn	12
4376 r	La	OHn	100
4379 r	Sn	AH	1
4379.2 a	V	OC	6
438_0 b	Pr	OHn	5
4380 y	CuH	OH	9
4380 r	La	OHn	200
4382 v	C_2	OHnq	8
4383 r	Te	AH	5
4383.5 f	Fe	OHn	35
4384 r	La	OHn	250
4384.7 p	V	OC	5
4386.4 p	Tm	OAn	4
4388 r	La	OHn	200
4389 y	CuH	OH	9
4390.0 a	V	OC	3
4390.1 a	Na	OHn	0.5
4392 r	La	OHn	100
4393.4 a	Na	OHn	1
4394 r	Al	OAn	1
4394 b	Bi	AH	0.7
4395.2 a	V	OC	2
4396 r	La	OHn	50
4399 y	CuH	OH	9
4400 r	La	OHn	50
4400 r	Sr	OG	2
4400.6 a	V	OC	1
4404.8 f	Fe	OHn	11
4405 r	La	OHn	40
4405 b	Ra		10
4406.6 a	V	OC	5.5
4408 r	Te	AH	10
4408.5 a	V:	see 4406.6	
4409 y	CuH	OH	9
4410 r	La	OHn	30
4410 r	Pb	OH	5
4411 r	Sn	AH	6
4412 r	CuCl	AH, Cu	7
4414 b	Nd	OHn	10
4415.1 f	Fe	OHn	3
4418 r	La	OHn	300
4419 y	CuH	OH	7
4419.3 a	Sm	AA	50
442_0 b	Dy	OHn	10
4421 r	Sr	OG	2
4423 r	La	OHn	300
4424 b	Bi	AH	0.6
4425.4 a	Ca	OH	1
4427 r	Be	OH	0.3
4427.3 f	Fe	OHn	8
4428 r	La	OHn	300
4429 r	Sn	AH	5
4430 y	CuH	OH	8
4433 r	La	OHn	250
4434 r	CuCl	AH, Cu	15
4435.0 a	Ca	OH	10
4435.6 s	Eu	OHn	1
4435.7 a	Ca:	see 4435.0	
4438 r	La	OHn	200
4441 y	CuH	OH	7
4441.7 a	V	OC	0.5
4441.8 a	Sm	AA	50
4442 b	Yb	OH	3
4442.3 a	Fe	OHn	3
4442.3 a	Sm:	see 4441.8	
4443 r	La	OHn	100
4444.2 a	V	OC	0.5
4448 r	La	OHn	70
4451.6 a	Mn	OA	25
4452 r	Be	OH	0.4
4452 y	CuH	OH	10
4452 r	Sn	AH	7
4453 r	La	OHn	50
4454 r	Pb	OH	4
4454.8 a	Ca	OH	15
4455 r	Lu	OHn	7
4455.9 a	Ca:	see 4454.8	
4458 r	La	OHn	30
4460 r	Te	AH	3
4460.3 a	V	OC	1.7
4461.7 f	Fe	OHn	5
4462.0 a	Mn	OA	15
4463 r	Gd	OHn	30
4464 y	CuH	OH	11
4464 r	La	OHn	20
4471 r	Al	OAn	2
4473 r	Tb	OHn	50
4475 r	Be	OH	0.3
4476 b	Yb	OH	9
4478 r	Fe	OHn	10
4481 r	Gd	OHn	30
4482.2 f	Fe	OHn	5
4487 r	Te	AH	16
4488 r	IO	OH, IO_3^-	1
4489 r	Lu	OHn	7
4494 r	Al	OAn	4
4494 r	CuCl	AH, Cu	4
4494.2 a	Na	OHn	1
4496 r	Be	OH	0.2
4496 r	Y	OHn	10
4496.9 a	Cr	OAn	2
4497.7 a	Na	OHn	2
4499 r	Gd	OHn	20
4499 r	Sn	AH	4
4500 r	Lu	OHn	10
4502 r	Sc	OHn	8
4509 r	Lu	OHn	20
4509 r	Pb	OH	4
4511 b	Nd	OHn	10
4511.3 p	In	OHn	3000
4512 r	Lu	OHn	10
4516 r	Al	OAn	5
4516 r	CuCl	AH, Cu	5
4517 r	Be	OH	0.2
4518 b	Yb	OH	10
452_0 b	Pr	OHn	10
4521 r	Lu	OHn	12
4522 r	Sn	AH	3
4523 r	Sr	OG	2
4524 r	Ba	AH	3
4524.7 a	Sn	AH	40
4526 r	Lu	OHn	12
4526 r	Y	OHn	15
4528.7 f	Rh	OH	3
453_0 b	B	OHn	30

APPENDIX

4533	r	Lu	OHn	20	4607	r	Sc	OHn	13	4693	r	Sr	OG	1
4535	r	Be	OH	0.1	4607	r	Tb	OHn	60	4694	r	IO	OH, IO_3^-	0.9
4535	dr	Dy	OHn	20	4607.3	p	Sr	OHn	10 000	4695	r	Al	OAn	18
4536	r	Sc	OHn	17	4612	b	Ce	OHn	10	4695	b	Ce	OHn	10
4538	r	Al	OAn	5	4612	r	Sn	AH	4	4696	r	Lu	OHn	200
4538	r	Ba	AH	1	4616	r	Te	AH	2	4698	v	C_2	OHnq	45
4540	r	Lu	OHn	8	4617	mr	Gd	OHn	300	4699	r	Te	AH	7
4541	r	Sn	AH	2	4619	b	Bi	AH	0.1	4704	b	Ce	OHn	5
4541.7	a	Na	OHn	0.2	4619	b	Nd	OHn	20	4706	r	Pb	OH	5
4542	b	Bi	AH	0.2	4619	b	Tb	OHn	50	4707	r	Sc	OHn	70
4544	r	Fe	OHn	4	4621	r	Ce:		see 4624	4707	dr	Y	OHn	50
4544	r	Sr	OG	2	4622	r	Ba	AH	7	4708	r	Lu	OHn	100
4545.2	a	Na	OHn	0.5	4624	r	}Ce	OHn	6	4709	r	Be	OH	9
4546	b	Sm	OHn	20	4627	b				4711	r	Te	AH	7
4546.0	a	Cr	OC	0.8	4627.2	p	Eu	OHn	150	171_5	b	B	OHn	90
4553	r	Be	OH	0.03	4629	b	Nd	OHn	15	4715	v	C_2	OHnq	45
4554	r	Pb	OH	6	4634	mr	Gd	OHn	120	4716	r	Al	OAn	13
4554.0	s	Ba	OAn	200	4634	r	Sn	AH	4	4716	r	Yb	OH	50
4554.5	a	Ru	OHn	10	4638.5	a	K	OHn	1	4716.1	a	Sm	AA	50
4555.3	p	Cs	OH	150	4639	r	Ba	AH	5	4717	r	Gd	OHn	10
4556	r	Te	AH	10	4640	r	Te	AH	10	4717.1	a	Sm:		see 4716.1
4558	r	Al	OAn	4	4641.6	p	}K	OHn	2	4718	b	Ra		10
4561	r	Lu	OHn	10	4642.2	p				4721	r	Lu	OHn	30
4562	r	Y	OHn	15	4644	v	Lu	OHn	30	4722.2	a	Zn	OHn	3
4565	r	Sr	OG	2	4645	b	Nd	OHn	10	4722.3	a	Sr	OHn	2.5
4570	dr	Dy	OHn	30	4646.2	a	Cr	OAn	5	4722.6	a	Bi	AH	10
4570	r	Lu	OHn	7	4648	r	Al	OAn	15	4723	r	Ba	AH	5
4570	r	Sn	AH	3	4648	r	CuH	OH	8	4726	r	Lu	OHn	20
4571	r	Sc	OHn	17	4650	dr	Y	OHn	60	4727.9	f	Co	OH	7
4571	r	Tb	OHn	50	4651.3	a	}Cr	OAn	5	4728.4	a	Sm	AA	30
4571.1	f	Mg	OH	5	4652.2	a				473_0	b	Pr	OHn	2
4574	dr	Dy	OHn	35	4653	r	Gd	OHn	40	4730	r	Fe	OHn	6
4574	b	Yb	OH	3	4654	r	Lu	OHn	60	4733	r	Be	OH	5
4575	r	Lu	OHn	10	4658	r	Pb	OH	6	4733.3	u	Tm	OAn	7
4576	r	Al	OAn	3	4659	r	Fe	OHn	6	4735	r	Gd	OHn	15
4577.8	a	Dy	OHn	10	4660	r	Lu	OHn	80	4735	r	Lu	OHn	35
4579	r	Ba	AH	4	4661	r	Sn	AH	4	4736	r	Al	OAn	9
4580.1	f	Co	OH	2	4661.1	a	Ta	OCn	1.2	4737	v	C_2	OHnq	45
4581	r	La	OHn	5	4661.9	p	Eu	OHn	120	4738	r	Yb	OH	80
4581.7	a	Sm	AA	50	4662	r	Lu	OHn	1000	474_0	w	Eu	OHn	8
4584.4	f	Ru	OHn	10	4664	r	Ba	AH	3	4740.9	a	K	OHn	0.2
4585	r	La	OHn	6	4664.8	a	Na	OHn	2	4741.9	a	Sr	OHn	2
4585.9	a	Ca	OH	8	4668.6	a	Na	OHn	4	4742	r	Ba	AH	10
4587	r	IO	OH, IO_3^-	1	467_0	b	Pr	OHn	5	4742	r	Sc	OHn	50
4589	r	La	OHn	7	4670.8	a	Sm	AA	30	4744.3	a	K	OHn	0.5
4590	b	Tb	OHn	30	4671	r	Gd	OHn	15	4745	dr	Y	OHn	30
4590	b	Yb	OH	4	4672	r	Al	OAn	20	4745.1	a	Rh	OH	2
4591	r	Lu	OHn	10	4672	dr	Lu	OHn	250	4747.9	a	Na	OHn	0.3
4591	r	Sn	AH	3	4673	r	Sc	OHn	65	4749	r	Lu	OHn	20
4592	b	Ra		10	4675.0	a	Rh	OH	4	4749	r	Pb	OH	5
4593.2	p	Cs	OH	40	4677	dr	Y	OHn	60	4749	r	Yb	OH	80
4594	r	La	OHn	7	4678.2	a	Cd	OH	1	4750	b	Ra		10
4594.0	p	Eu	OHn	170	4679	v	C_2	OHnq	35	4751	r	Ni	OHn	4
4594.1	p	V	OC	0.2	4680.1	a	Zn	OHn	2	4751.8	a	Na	OHn	0.7
4598	b	Dy	OHn	30	4680.5	f	W	OC	10	4752	r	Tb	OHn	15
4598	r	La	OHn	5	4681	r	Ba	AH	10	4753.9	a	K	OHn	0.5
460_0	b	Pr	OHn	7	4682.3	s	Ra		10	4754	r	Al	OAn	6
4602.9	a	Li	OH	25	4684	r	Ce	OHn	12	4754.0	a	Mn	OA	80
4603	r	La	OHn	3	4684	r	Lu	OHn	400	4755	r	Be	OH	3
4604	r	Fe	OHn	4	4685	v	C_2	OHnq	35	4756	r	Gd	OHn	10
4604	r	Y	OHn	12	4687.8	a	Zr	OCn	2	4757.4	a	K	OHn	1
4605.4	a	Mn	OA	70	4688.7	a	Sm	AA	20	4760	r	Yb	OH	90

APPENDIX

4760.3 a	Sm	AA	50		4850 r	Ba	AH	50		4959 b	Nd	OHn	5
4761.5 a	Mn	OA	30		4850 b	Yb	AH	250		496_0 b	Sn	AH	30
4762.4 a	Mn	OA	30		4851 r	Pb	OH	6		4962 v	Tm	OHn	150
4764 r	Lu	OHn	15		4851.5 p	V	OC	0.8		4962.3 a	Sr	OHn	6
4765 r	Yb	OH	100		4856 r	Gd	OHn	15		4963 v	Mg	OHn	50
4765.9 a	Mn	OA	30		4856.1 a	K	OHn	2		4964 r	IO	OH, IO_3^-	0.6
4766.4 a	Mn	OA	30		4858 r	Sc	OHn	200		4965 r	Ba	AH	30
4769 b	Nd	OHn	5		4858 r	V	OHn	9		4965.0 a	K	OHn	6
4774 r	Te	AH	3		4863 r	Ce	OHn	6		4967 a	Rb	OHn	0.02
4775 b	Dy	OHn	15		4863 r	Te	AH	9		4972.0 a	Li	OH	6
4776 r	Be	OH	2		4863.5 a	K	OHn	1.5		4975 v	Mg	OHn	70
4778 mr	Yb	OH	120		4864.7 p	V	OC	1.3		4975 r	Te	AH	2
4779 r	Sc	OHn	20		4866 r	Al	OAn	50		4976 r	Mn	OHn	30
4780 r	Lu	OHn	7		4869.8 a	K	OHn	3		4978.5 a	Na	OHn	10
4780 r	Tb	OHn	25		4870 r	Y	OHn	50		4981 b	Yb	AH	500
4783.1 a	Sm	AA	40		4873 r	Ce	OHn	6		4981.7 a	Ti	OCn	4
4783.4 a	Mn	OA	100		4875.5 p	V	OC	2		4982.8 a	Na	OHn	20
4785.9 a	Sm	AA	20		4876.3 a	Sr	OHn	6.5		4983 a	Rb	OHn	0.02
4786 r	Ba	AH	15		4877 r	Ni	OHn	10		4984 r	Pb	OH	8
4786.5 a	K	OHn	0.5		488_0 b	Ba	AH	120		4986 v	Mg	OHn	80
4791.0 a	K	OHn	1		4881.6 p	V	OC	2		4988 b	Er	OHn	30
4792 b	Ce	OHn	15		4882 r	CuCl	AH, Cu	3		4991.1 a	Ti	OCn	5
4792 r	Tb	OHn	15		4882 r	Ce	OHn	6		4992 b	Lu	OHn	30
4795 r	Be	OH	1.3		4883.8 a ⎱ Sm	AA	100			4997 v	Mg	OHn	80
4798 r	Gd	OHn	50		4884.0 a ⎰					4999.5 a	Ti	OCn	5
4799 b	Ce	OHn	20		4888 r	Al	OAn	25		5000 r	Ti	OH	30
4799.8 a	K	OHn	1		4890 r	Ni	OHn	6		5001 r	BaF	PH	50
4799.9 a	Cd	OH	2		4891 b	Ce	OHn	6		5002 r	Er	OHn	30
4802 r	Te	AH	5		4892 r	Tm	OHn	300		5004 r	Lu	OHn	40
4804.3 a	K	OHn	2		4893 mr	Gd	OHn	150		5006 r	Te	AH	2
4806 dr	Ti	OH	20		4894 r	Sc	OHn	40		5007 v	Mg	OHn	100
4807 b	Ce	OHn	25		4898 db	Tm	OHn	250		5007 r	Ni	OHn	10
4810.5 a	Zn	OHn	6		4903 db	Tm	OHn	200		5009 r	Er	OHn	80
4811.9 a	Sr	OHn	5.5		4904 r	V	OHn	12		5011 r	V	OHn	17
4812 r	Be	OH	0.9		4910 mr	Gd	OHn	100		5012 r	Ba	AH	10
4813 r	V	OHn	5		4913 mb	Tm	OHn	150		5013 r	Mn	OHn	45
4814 b	Tm	OHn	200		492_0 b	Pr	OHn	7		5017 a	Rb	OHn	0.03
4815 r	Ce	OHn	20		4925 r	Ce	OHn	20		5019 r	Lu	OHn	50
4817 r	Gd	OHn	60		4928 mr	Gd	OHn	50		502_0 b	Ba	AH	30
4817 r	Pb	OH	7		4929 r	Fe	OHn	10		502_0 b	Pr	OHn	3
4818 dr	Y	OHn	250		493_0 b	B	OHn	165		5022 r	Pb	OH	7
482_0 b	Pr	OHn	5		493_0 w	Cu	AH	60		5023 a	Rb	OHn	0.05
4823.5 a	Mn	OA	100		4931 r	Ce	OHn	20		5024 r	Ni	OHn	25
4824 r	Yb	OH	100		4932 r	CuF	PH	14		5025 dr	Y	OHn	40
4825.9 p	Ra		50		4934 r	Ce	OHn	20		5034 v	Er	OHn	150
4827.5 p	V:	see 4831.6			4934.1 s	Ba	OAn	100		5035 r	Lu	OHn	50
4828 r	Be	OH	0.6		4936 r	Ce	OHn	20		5035 r	Te	AH	6
483_0 w	Nd	OHn	5		4938 db	Tm	OHn	250		5037 a	Rb	OHn	0.1
4831 r	Ba	AH	10		4939 r	Ni	OHn	4		5041 v	Er	OHn	150
4831.6 p	V	OC	1.2		4941 r	Ce	OHn	20		5041 r	Lu	OHn	60
4832.1 a	Sr	OHn	8.5		4941 r	Te	AH	6		505_0 w	Cu	AH	70
4832.4 p	V:	see 4831.6			4942.0 a	K	OHn	2		5050 b	Er	OHn	150
4835 r	Gd	OHn	40		4946 b	Tm	OHn	200		5050 dr	Y	OHn	60
4840 r	Yb	OH	120		4947 r	Ce	OHn	20		5051 r	Mn	OHn	40
4841.7 a	Sm	AA	30		4949 r	Gd	OHn	10		5055 r	Be	OH	1.4
4842 r	Al	OAn	65		495_0 b	B	OHn	155		5057 r	V	OHn	30
4842 dr	Y	OHn	150		4950.8 a	K	OHn	3		5058 b	Lu	OHn	70
4845 r	IO	OH, IO_3^-	1		4951 r	BaF	PH	70		5061 r	CuF	PH	2.5
4846 b	Ce	OHn	8		4951 r	Ce	OHn	20		5066 v	BaCl	OG	2
4849 r	Ti	OH	15		4951 r	V	OHn	5		5067 r	Er	OHn	300
4849.9 a	K	OHn	1		4956 dr	Ti	OH	40		5069 db	Ho	OHn	100
485_0 b	Sn	AH	40		4956.0 a	K	OHn	3		5073 b	Er	OHn	200

APPENDIX

5075.8 a	Rb	OHn	0.1	5153 r	Y	OHn	20	5233 r	Lu	OHn	100
5076 r	Be	OH	1.4	5153.4 a	Na	OHn	3	5234.0 a	Rb	OHn	0.03
5078 r	Er	OHn	200	5157 r	Ho	OHn	500	5238.6 u	Sr	OHn	1
5078 dr	Y	OHn	60	5157 r	Pr	OHn	20	524_0 b	Ba	AH	80
5079 r	Al	OAn	19	5159 r	Mn	OHn	80	524_0 b	Cu	AH	110
5081 v	Er	OHn	200	516_0 b	Sn	AH	13	5240 r	BaCl	OG	30
5084.2 a	K	OHn	4	5161 r	Al	OAn	15	5244 r			
5085.8 a	Cd	OH	4	5161 r	Lu	OHn	550	5247 v	Gd	OHn	30
5086 r	Ba	AH	100	5162 r	Ho	OHn	200	5247.6 a	Cr	OC	1.5
5089 db	Er	OHn	150	5162 r	Pb	OH	7	5248 mb	Dy	OHn	300
5089.0 a	Rb	OHn	0.1	5164.9 a	Rb	OHn	0.2	5250 b	Nd	OHn	20
5092 b	Ho	OHn	100	5165 v	C_2	OHnq	110	5251 v	Ho	OHn	300
5095 r	Be	OH	1.3	5167 r	BaCl	OG	4	5251.9 a	Sm	AA	15
5097 b	Ho	OHn	100	5167 v	Er	OHn	50	5253 dr	La	OHn	18
5097 r	Sc	OHn	60	5167.3 a	Mg	OH	10	5256 r	Lu	OHn	70
5097.1 a	K	OHn	6	5167.5 a	Fe	OHn	15	5256.6 a	Cs	OH	1
5098 v	C_2	OHnq	50	5168 r	Cr	OHn	10	5256.9 u	Sr	OHn	1.2
5098 r	Ni	OHn	5	5168 r	Ho	OHn	200	5257 r	Tm	OHn	200
5099.2 a	K	OHn	7	5168 dr	Ti	OH	70	5258 b	Ho	OHn	200
5100 r	Er	OHn	100	5170 r	Lu	OHn	700	5258 r	Pb	OH	6
5100 mr	Gd	OHn	20	5171 r	Sc	OHn	30	5259 v	Gd	OHn	40
5102 r	Al	OAn	25	5172.7 a	Mg	OH	20	5260.0 a	Rb	OHn	1
5105 r	Ho	OHn	200	5174 b	Yb	AH	350	5263 r	Dy	OHn	600
5105 r	V	OHn	25	5175 r	Ni	OHn	20	5363 db	Ho	OHn	300
5105.5 a	Cu	OHn	50	5176 db	Er	OHn	50	5264.2 a	Cr	OAn	5
5110.4 f	Fe	OHn	7	5177 r	Al	OAn	11	5267 r	Mn	OHn	80
5112 r	Be	OH	1.1	5179 dr	La	OHn	6	5269.5 a	Fe	OHn	40
5112 r	Y	OHn	50	5180 b	B	OHn	300	5270 v	Ho	OHn	200
5112.2 a	K	OHn	12	5180 db	Er	OHn	60	5270 r	Tb	OHn	10
5117 tr	Gd	OHn	30	5183.6 a	Mg	OH	40	5271.4 a	Sm	AA	50
5118 r	Te	AH	5	5184 db	Er	OHn	50	5273 r	Dy	OHn	600
5119 r	Ho	OHn	200	5186 b	Ho	OHn	200	5275 dv	Ho	OHn	300
5120 r	Lu	OHn	150	5186 b	Lu	OHn	350	5275 dr	Y	OHn	15
5123 r	Al	OAn	25	5190 r	Ho	OHn	200	5276 r	V	OHn	50
5125 tr	BaF	PH	8	5191 r	Al	OAn	8	5277 v	Gd	OHn	70
5126 r	Ho	OHn	150	5192 r	Mn	OHn	130	5278 dr	La	OHn	17
5128 r	Be	OH	0.8	5193 v	Dy	OHn	240	5280 r	Te	AH	4
5129 v	C_2	OHnq	80	5195.3 a	Rb	OHn	0.5	5282 v	Tm	OHn	150
513_0 b	Ba	AH	140	5196 r	Ho	OHn	150	5285 r	Dy	OHn	400
5131 r	IO	OH, IO_3^-	0.9	5196.7 a	Cs	OH	0.5	5286 dv	Ho	OHn	300
5132 r	Ho	OHn	200	520_0 w	Eu	OHn	4	5288 r	Pr	OHn	10
5132 a	Rb	OHn	0.02	520_0 w	Ra		10	5290 r	Fe	OHn	6
5134 r	Sc	OHn	50	5200.6 a	Sm	AA	50	5292 dr	CaF	PH	150
5135 v	Er	OHn	70	5202 b	Lu	OHn	250	5293 r	Cr	OHn	10
5136 r	Dy	OHn	120	5204 b	Dy	OHn	240	5294 v	Gd	OHn	100
5136 r	Gd	OHn	30	5204 dr	La	OHn	11	5294 b	Tm	OHn	200
5137 r	Pr	OHn	15	5204.5 a	Cr:	see 5206.0		5296 r	Dy	OHn	300
5138 r,v	BaCl	OG	20	5206 r	Lu	OHn	170	5296.7 a	Cr	OAn	5
5138 b	Nd	OHn	15	5206 v	Mg	OHn	10	5298 dr	CaF:	see 5292	
5138 r	Pb	OH	7	5206.0 a / 5208.4 a	Cr	OAn	300	5298.3 a	Cr	OAn	10
5139 r	Te	AH	6					5299 b	Ho	OHn	200
5140 dv	Dy	OHn	190	5209 r	IO	OH, IO_3^-	0.5	530_0 b	Cu	AH	110
5141 r	Be	OH	0.4	5209 r	Sc	OHn	15	5301.4 a	Cs	OH	0.5
5142 dv	Ho	OHn	200	5209.5 da	Cs	OH	0.5	5302 dv	Tm	OHn	250
5143 r	Al	OAn	18	5211 r	Ho	OHn	150	5303 v	Gd	OHn	100
5146 dr	CaF	PH	3	5214 r	Ba	AH	70	5303 dr	La	OHn	13
5148 r	Ho	OHn	200	5218 r	Lu	OHn	250	5304 b	Ho	OHn	200
5148.8 a	Na	OHn	1.5	5226 b	Tm	OHn	200	5304 dr	Y	OHn	15
5150 r	Dy	OHn	210	5228 dr	La	OHn	16	5308 r	IO	OH, IO_3^-	1
5150.7 a	Rb	OHn	0.3	5229 r	Cr	OHn	20	5313 b	Nd	OHn	50
5151 v	Er	OHn	70	5229 r	Mn	OHn	110	5313 dv	Tm	OHn	300
5152.7 a	Cs	OH	0.3	5229 dr	V	OHn	50	5315 db	Dy	OHn	300

5315	v	Ho	OHn	200	5378	r	Pr	OHn	10	5454	db	Dy	OHn	450
5320	dr	Ho	OHn	300	5379	r	Tb	OHn	100	5455	r	Ba	AH	20
5321	v	BaCl	OG	10	5380	tb	Ho	OHn	150	5457	v	Tb	OHn	100
5322.4	a	Rb	OHn	0.2	5381	dr	La	OHn	110	5458	dr	La	OHn	140
5323	r	Ni	OHn	5	5382	r	Fe	OHn	30	5459	b	Er	OHn	250
5323.2	a	K	OHn	7	5383	r	Gd	OHn	120	5459	r	Pb	OH	6
5325	r	V	OHn	35	5387	r	Tb	OHn	100	5460.7	a	Hg	OH	weak
5325	b	Yb	AH	700	5390	r	Mn	OHn	370	5462	x	Ca	OH	100
5326	r	Ho	OHn	200	5390.6	a	Rb	OHn	1	5462	r	Gd	OHn	130
5327	r	Gd	OHn	120	5394	r	Al	OAn	4	5464	r	Lu	OHn	60
5327	b	Nd	OHn	30	5394.7	f	Mn	OA	200	5465.9	a	Cs	OH	5
5327	r	Tb	OHn	100	5395	r	Sc	OHn	12	5467	r	Tb	OHn	100
5328	dr	La	OHn	8	5396	r	Tb	OHn	60	5469	r	Gd	OHn	130
5328.0	a	Fe	OHn	30	5396	b	Tm	OHn	300	5470	v	C_2	OHnq	10
5329	dr	Tm	OHn	350	540_0	mb	Er	OHn	10	5470	b	Er	OHn	200
5331	r	Pb	OH	5	5401	r	Tm	OHn	300	5470	dr	V	OHn	90
5333	db	Dy	OHn	300	5403	r	Lu	OHn	30	5471	b	Tb	OHn	70
5333	r	Ho	OHn	300	5403.7	a	Sm	AA	20	5473	r	Tm	OHn	400
5336	r	Gd	OHn	120	5404	mr	Dy	OHn	500	5476	b	B	OHn	350
5337	r	Al	OAn	2	5405	r	Gd	OHn	150	5476	dr	Gd	OHn	120
5339	v	Tm	OHn	250	5405	r	Pb	OH	4	5478	b	Nd	OHn	50
5339.7	a	K	OHn	12	5405.2	a	Sm:		see 5403.7	5480	dr	Er	OHn	250
5340	dr	Tb	OHn	350	5405.8	a	Fe	OHn	5	5480.9	a	Sr	OHn	1.7
5340.9	a	Cs	OH	3	5406.7	a	Cs	OH	1	5481	r	Gd	OHn	120
5343.0	a	K	OHn	8	5407	dr	La	OHn	160	5483	dr	La	OHn	90
5345.8	a	Cr	OAn	15	5408	r	Ni	OHn	10	5483.3	a	Co	OH	5
5346	r	Gd	OHn	120	5409.8	a	Cr	OAn	20	5487	r	Lu	OHn	40
5346	r	Ho	OHn	250	5410	r	Al	OAn	4	5488	r	Gd	OHn	120
5347	r	Tm	OHn	400	5413.9	a	Cs	OH	2	5488	b	Tm	OHn	300
5348.3	a	Cr	OAn	10	5415	r	Tb	OHn	200	5492	r	Ba	AH	70
5349	r	Ba	AH	90	5415	b	Tm	OHn	400	5493	b	Dy	OHn	500
5349	dr	Dy	OHn	350	5416	r	Gd	OHn	120	5495	r	IO	OH, IO_3^-	0.9
5350.4	a	Cs	OH	1	5417	r	Cr	OHn	30	5498	db	Er	OHn	300
5350.5	a	Tl	OH	700	5420	r	Ba	AH	5	5500	r	Tm	OHn	400
5351	dr	Tb	OHn	300	5421	db	Er	OHn	20	5502	v	C_2	OHnq	13
5352	r	Pr	OHn	35	5423	r	Al	OAn	3	5502.9	a	Cs	OH	4
5352	b	Tm	OHn	400	5424	r	Mn	OHn	200	5504	r	Te	AH	5
5354	dr	La	OHn	6	5425	b	Rh	OH	10	5504.2	a	Sr	OHn	0.5
5354	r	Pb	OH	5	5425	b	Tm	OHn	350	5509	dr	La	OHn	60
5356	r	Cr	OHn	20	5427	r	Gd	OHn	120	5510	r	Ba	AH	10
5356	r	Gd	OHn	20	5428	r	Tb	OHn	40	5511	x	Ca	OH	100
5357	r	Sc	OHn	8	5429.7	a	Fe	OHn	5	5513	mb	Er	OHn	400
5357	r	Tb	OHn	120	5430	r	Er	OHn	40	5515	r	Ca	OH	200
5358	r	Al	OAn	3	5430	r	Fe	OHn	30	5516.1	p	Sm	AA	20
5359	r	Ho	OHn	250	5431.5	a	Rb	OHn	4	5517	r	V	OHn	60
5359.5	a	K	OHn	15	5432	dr	La	OHn	160	5520	r	Tm	OHn	300
5360	r	Mn	OHn	300	5432.5	f	Mn	OA	100	5521	x	Ca	OH	200
5361	b	Tm	OHn	300	5434	x	Ca	OH	100	5521.8	a	Sr	OHn	1
5362.6	a	Rb	OHn	2	5434	b	Tm	OHn	350	5523.0	a	Ce	OCn	0.6
5363	r	Tb	OHn	70	5435	r	Sc	OHn	10	5524	r	Tm	OHn	400
5365	b	Tm	OHn	300	5438	db	Er	OHn	50	5526	x	Ca	OH	200
5366	r	Gd	OHn	120	5438	dr	Yb	OH	60	5528	b	Er	OHn	500
5367	r	Ba	AH	10	5439	r	Gd	OHn	120	553_0	b	Eu	OHn	20
5367	dr	Pr	OHn	25	544_0	b	Ca	OH	100	5531	b	Tm	OHn	400
5369	r	Tb	OHn	300	544_0	w	Nd	OHn	40	5532	r	Fe	OHn	300
537_0	w	Cu	AH	120	5443	r	Tb	OHn	200	5533	r	Ca	OH	600
5371.5	a	Fe	OHn	15	5446	db	Er	OHn	150	5533	r	IO	OH, IO_3^-	0.7
5372	v	Er	OHn	15	5449	r	Lu	OHn	60	5534	r	Er	OHn	400
5373	r	Tb	OHn	120	5449	r	Mn	OHn	100	5534.8	a	Sr	OH	1
5374	r	Tm	OHn	350	5450	dr	Gd	OHn	130	5535	dr	La	OHn	35
5377	r	Al	OAn	4	5450	dr	Ti	OH	80	5535.6	p	Ba	OH	200
5377	r	Gd	OHn	120	5453	db	Er	OHn	200	5539	r	Ca	OH	1200

554_0	b	Ca	OH	5000	5628	r	Er	OHn	300	5696	v	Ho	OHn	1000
5540	x	Ca	OH	800	5628	r	Pr	OHn	30	5697	x	Ca	OH	100
5541	v	C_2	OHnq	16	5629	r	Zr	OH	5	5698	r	Sm	OHn	15
5541	x	Ca	OH	800	563_0	b	Sn	AH	12	5698	r	Y	OHn	30
5542	x	Ca	OH	1200	5633	r	SrF:	see 5637		5698.5 a	V:	see 5703.6	
5544	x	Ca	OH	800	5634	v	Er	OHn	250	5699	dr	Gd	OHn	300
5544	db	Dy	OHn	400	5634	r	Zr	OH	5	5699	r	Tb	OHn	150
5545	db	Er	OHn	500	563_5	w	Co	OH	100	5699.0 a	Ru	OHn	50
5546	x	Ca	OH	800	5635	v	C_2	OHnq	18	5700.2 a	Cu	OHn	5
5547	x	Ca	OH	1000	5635	v	Ho	OHn	500	5701	r	Ba	AH	40
5548	x	Yb	OH	600	5635	r	Tb	OHn	200	5702	r	Tb	OHn	300
555_0	b	Yb	AH	600	5635.2 a	Cs	OH	7	5703.6 a	V	OC	0.5
5550	r	Tb	OHn	100	5637	r	SrF	FH	5	5705	tb	Dy	OHn	700
5552	b	Er	OHn	500	5639	r	Bi	AH	0.1	5705	b	Ho	OHn	600
5554	v	Ca	OH	800	5639	r	Tb	OHn	400	5705	r	Pr	OHn	40
5556.5 p	Yb	OHn	250	5640	r	Tm	OHn	200	5706	db	Er	OHn	100
5557	x	Ca	OH	400	5641	r	SrF	FH	5	5707	r	La	OHn	80
5557	b	Er	OHn	400	5643	tr	Ho	OHn	500	5707.0 a	V:	see 5703.6
5558	r	Tb	OHn	120	5644	r	Ba	AH	40	571_0	b	Eu	OHn	50
5558	r	Tm	OHn	400	5645	v	Er	OHn	200	5711	v	Ho	OHn	600
5559	x	Ca	OH	400	5645	r	Pr	OHn	15	5713	r	Ho	OHn	700
5561	dr	La	OHn	15	5645	r	Tb	OHn	120	5714	r	Y	OHn	50
5562	x	Ca	OH	200	5645.8 a	Eu	OHn	2	5716	b	Er	OHn	100
5563	b	Dy	OHn	350	5647	r	Fe	OHn	100	5716	r	Gd	OHn	200
5563	r	Er	OHn	500	5648.1 a	Rb	OH	6	5718	r	Zr	OH	5
5564	r	Bi	AH	0.3	5652	x	Ca	OH	100	5719	x	Ca	OH	100
5564	r	Cr	OHn	40	5653	dr	Gd	OHn	40	572_0	b	Ca	OH	200
5565	dr	Ho	OHn	200	5653.7 a	Rb	OHn	3	5720	r	Pr	OHn	35
5566	r	Tb	OHn	120	5654	dr	La	OHn	350	5723	v	Er	OHn	100
5566	tb	Tm	OHn	350	5656	v	Ho	OHn	700	5724	r	Sc	OHn	25
5568	dv	Er	OHn	400	5658	r	Zr	OH	5	5724	r	Zr	OH	5
5568.4 a	Cs	OH	1	5659	r	Ba	AH	5	5724.4 a	Rb	OHn	10
5572	r	Tm	OHn	350	5659.9 p	Sm	AA	10	5725	b	Yb	AH	600
5573.7 a	Cs	OH	1	5660	b	Ho	OHn	1000	5726	b	Nd	OHn	60
5575	r	Tb	OHn	150	5663	r	Zr	OH	5	5727	r	Tb	OHn	300
5578	r	Tm	OHn	300	5663.7 a	Cs	OH	8	5729	b	Dy	OHn	1000
5578.8 a	Rb	OHn	1	5664	v	Er	OHn	200	5729	dr	Ho	OHn	700
5581	db	Er	OHn	350	5664	r	Gd	OHn	100	5729	r	Sm	OHn	15
5583	r	Fe	OHn	35	5664	dr	Tb	OHn	200	5730	r	Er	OHn	50
5584	r	Ho	OHn	400	5668	r	Te	AH	2	5730	r	IO	OH, IO_3^-	0.8
5585	v	C_2	OHnq	23	5672	v	Ho	OHn	700	5731	r	Pr	OHn	40
5586	r	Mn	OHn	700	5673	r	Ba	AH	10	5731	r	Y	OHn	60
5588	dr	La	OHn	5	5674	b	Er	OHn	250	5732	dr	La	OHn	30
5589	r	Tm	OHn	350	5675	r	Pr	OHn	20	5734	dr	Gd	OHn	100
5592	db	Ho	OHn	400	5675.8 a	Tm	OAn	6	5737	b	Er	OHn	50
5593	r	Tb	OHn	100	5677	db	Gd	OHn	600	5737	db	Ho	OHn	500
5596	db	Er	OHn	500	5678	r	Pb	OH	6	5737	r	Sc	OHn	100
5597	r	Pr	OHn	50	5679	r	Fe	OHn	90	5737	dr	V	OHn	90
5598	r	Ti	OH	50	5680	dr	La	OHn	200	574_0	b	Bi	AH	0.1
5601	dr	La	OHn	700	5681	dr	Gd	OHn	400	574_0	mb	Dy	OHn	600
5601	r	Tb	OHn	150	5681	b	Nd	OHn	60	5745.7 a	Cs	OH	1
5602	r	Ba	AH	10	5681	r	Sm	OHn	15	5746	r	Gd	OHn	150
5609	db	Ho	OHn	400	5681	r	Tb	OHn	150	5746	r	Sm	OHn	15
5610	r	Mn	OHn	500	5682.6 a	Na	OHn	40	5747	r	Y	OHn	60
5612	r	Pr	OHn	50	5684	b	Er	OHn	150	5748	r	Zr	OH	5
5613	v	Er	OHn	500	5684	b	Tm	OHn	200	5749	r	Pr	OHn	30
5614	r	Fe	OHn	90	5686	r	CuF	PH	0.5	5749	b	Tm	OHn	150
5618	r	Pb	OH	4	5688	v	Ho	OHn	700	5751	b	Er	OHn	50
5623	r	Cr	OHn	35	5688.2 a	Na	OHn	80	5752	v	Ho	OHn	400
5626	r	Ho	OHn	600	5691	r	Pr	OHn	70	5754	r	Zr	OH	5
5626.0 a	Sm	AA	10	5693	b	Er	OHn	150	5755	v	Tb	OHn	150
5627	dr	La	OHn	600	5694	tb	Dy	OHn	600	5756	r	Gd	OHn	150

APPENDIX

5758	r	Ba	AH	15	5844	v	Gd	OHn	400	5942	r	Pr	OHn	40
5761	r	Sc	OHn	25	5845.1	a	Cs	OH	9	5943	mb	Sm	OHn	100
5761	dr	Ti	OH	20	5847.6	a	Cs	OH	2	5943	b	Sr	AH	500
5763	r	Pr	OHn	70	5849	v	Ho	OHn	600	5944	r	Mn	OHn	150
5764	r	Y	OHn	60	5849	dr	Sc	OHn	150	5947	mb	Sm	OHn	120
5765	r	Sc	OHn	80	5852	r	Cr	OHn	50	5950	dr	La	OHn	150
5765.2	a	Eu	OHn	6	5854	v	Ho	OHn	400	5956	tr	Gd	OHn	500
5768	v	Tb	OHn	150	5856	dr	Dy	OHn	800	5956	r	Y	OHn	200
5771	r	Gd	OHn	250	5857	r	Sm	OHn	50	5959	r	Sc	OHn	60
5772	r	SrF	FH	100	5857	r	Tb	OHn	500	5960	db	Tb	OHn	300
5773	r	Sc	OHn	150	5858	db	Gd	OHn	400	5962	r	Sm	OHn	80
5775	r	SrF:	see 5772		5859	r	Y	OHn	60	5963	r	Pr	OHn	40
5776	r	Sc	OHn	120	5860	r	Mn	OHn	300	5965	r	Ca	OH	200
5777	r	Pr	OHn	30	5862	b	Gd	OHn	400	5966	tb	Tb	OHn	250
5777.7	a	Ba	OH	1	5864	r	Ba	AH	40	5969	dr	Sc	OHn	80
5778	r				5865	v	Ho	OHn	300	5970	b	Sr	AH	500
5779	r	SrF:	see 5772		5865	r	Tb	OHn	250	5971	x	Ca	OH	200
5782	r				5867.8	a	Sm	AA	10	5971	r	Lu:	see 5993	
5782.1	a	Cu	OHn	10	5868	b	Dy	OHn	700	5971	r	Nd	OHn	200
5782.6	a	K	OHn	12	5868	r	Fe	OHn	200	5971.7	a	Ba	OH	1
5783	r	Y	OHn	40	5868	dr	La	OHn	100	5972	dr	Gd	OHn	400
5785	r	SrF:	see 5772		5868.6	a	Sm:	see 5867.8		5972	r	Pr	OHn	50
5788	r	Gd	OHn	250	587$_0$	b	Yb	AH	250	5972	r	Sm	OHn	150
579$_0$	b	B	OHn	140	5876	r	Tb	OHn	300	5972	r	Y	OHn	6000
5790	r	Fe	OHn	50	5876	r	Y	OHn	60	5972.8	u	Ba:	see 5971.7	
5791	r	Pr	OHn	20	5878	r	Sc	OHn	120	5974	r	Nd	OHn	200
5795	r	Cr	OHn	70	5880	b	Ho	OHn	400	5975	r	Fe	OHn	200
5798	r	Sc	OHn	80	5881	r	Mn	OHn	700	5976	r	Ba	AH	5
5800	r	Y	OHn	50	5884	db	Dy	OHn	600	5977	dr	La	OHn	100
5802	r	Sc	OHn	100	5887	r	Ba	AH	10	5979	dr	Dy	OHn	300
5802	r	Tb	OHn	250	5888	r	Sc	OHn	150	5979	dr	Tb	OHn	900
5802.0	a	K	OHn	25	5889	r	V	OHn	30	598$_0$	b	Eu	OHn	300
5804	v	Ho	OHn	400	5890.0	p	Na	OHn	800 000	5981	r	Sm	OHn	120
5805	r	Ba	AH	20	5894	r	Y	OHn	60	5983	r	Pr	OHn	50
5807	v	Gd	OHn	600	5895	dr	La	OHn	150	5984	r	Ba	AH	10
5808	r	Fe	OHn	20	5895.9	p	Na	OHn	400 000	5986	v	Ca	OH	400
5808	v	Sm	OHn	30	5902	r	Tb	OHn	250	5988	b	Gd	OHn	700
5810	r	CaCl	OA, C	250	5903	r	Fe	OHn	200	5988	dr	Sm	OHn	200
5811	dr	Sc	OHn	150	5905	b	Gd	OHn	400	5988	r	Y	OHn	5000
5812.5	a	K	OHn	8	5909	r	Tb	OHn	250	5990	b	Nd	OHn	250
5813	x	Ca	OH	100	5910	r	Mn	OHn	300	5991	r	Pr	OHn	35
5815	r	Tb	OHn	150	5911	b	Gd	OHn	500	5993	r	Lu	OHn	500
5819	r	Fe	OHn	100	5911	r	Pb	OH	5	5993	r	Sm	OHn	80
5819	v	Gd	OHn	600	5911.4	a	Gd	OA	1000	5997.1	a	Ba	OH	1
5819	r	Y	OHn	40	5912	r	Y	OHn	50	5998	db	Nd	OHn	200
5820	v	Ho	OHn	600	5918	r	Gd	OHn	500	5999	b	Tb	OHn	400
5820	mb	Sm	OHn	60	5918	r	Sc	OHn	100	6001	db	Gd	OHn	800
5821	r	Tb	OHn	120	5921	r	Dy	OHn	400	6002	v	Nd	OHn	200
5830	r	CaF	FH	50	5921	r	Tb	OHn	1000	6002	r	Sc	OHn	60
5831	v	Ho	OHn	600	5922	dr	La	OHn	200	6003	r	Ca	OH	400
5832	v	Gd	OHn	500	5924	mb	Sm	OHn	100	6004	r	Y	OHn	3000
5832.1	a	K	OHn	15	5925	r	Tb	OHn	400	6005	dr	La	OHn	70
5834	tb	Dy	OHn	1000	5927	r	Gd	OHn	800	6005	dr	Pr	OHn	50
5835	v	Ho	OHn	600	5928	r	Sc	OHn	100	6006	v	Ca	OH	600
5837	r	Sc	OHn	60	5931	r	Y	OHn	40	6006	tv	Dy	OHn	500
5838	r	V	OHn	40	5933	v	Tb	OHn	250	6007	b	Gd	OHn	500
5838	r	Y	OHn	30	5934	r	CaCl	OA, C	500	6010.5	a	Cs	OH	10
5838.8	a	Cs	OH	2	5935	x	Ca	OH	100	6014	r	Lu:	see 5993	
5839	r	Tb	OHn	250	5939	r	Y	OHn	100	6014	v	Nd	OHn	150
5840	r	Sc	OHn	100	5940	dr	Dy	OHn	300	6014	b	Tb	OHn	200
5840	db	Sm	OHn	50	5940	r	Tb	OHn	900	6017	r	Sc	OHn	600
5842	r	Y	OHn	50	5942	dr	Gd	OHn	600	6018.2	a	Eu	OHn	10

APPENDIX

6019	v	Gd	OHn	400	6078	mr Tb	OHn	600
6019	r	Pr	OHn	250	6079	r Sc	OHn	8000
6019	r	Sm	OHn	120	6079	mb Sm	OHn	100
6019.5 a		Ba	OH	4	6081	tb Gd	OHn	1000
602_0	b	Ca	OH	2000	6084	db Dy	OHn	500
602_0	w	Ra		10	6085	mv Sr	OG	20
602_0	w	Yb	AH	120	6087	dr V	OHn	50
6020	r	Y	OHn	2000	6088	r Ca	OH	200
6022	v	Ca	OH	400	6089	r Sm	OHn	150
6022	b	Tb	OHn	200	6089	r Y	OHn	300
6026	r	Pr	OHn	120	6090	v Sr	OG	15
6026	r	Tb	OHn	170	6091	db Tb	OHn	300
6027	db	Dy	OHn	200	6092	v Ca	OH	200
6027	r	Nd	OHn	150	6092	r Sc	OHn	1000
6029	x	Ca	OH	300	6095	b Fe	OHn	150
603_0	b	B	OHn	25	6095	r SnH	AH	10
6033	dr	La	OHn	50	6095	tv Sr	OG	10
6034.1 a		Cs	OH	5	6097	r Y	OHn	300
6035	v	Ca	OH	400	6098	v Ca	OH	400
6035	v	Gd	OHn	300	6098	r Pr	OHn	100
6035	b	Sm	OHn	200	6098	dr Tb	OHn	250
6035	b	Sr	AH	3000	6099	dr Gd	OHn	700
6036	r	Pr	OHn	150	6100	r Ba	AH	20
6036	r	Sc	OHn	10 000	6100	dv Sr	OG	7
6036	tr	Tb	OHn	200	6102	r Sc	OHn	3000
6037	r	Bi	AH	0.1	6103	b Tb	OHn	250
6037	r	Y	OHn	1200	6103.6 a	Li	OH	500
6040	r	Ba	AH	50	6108	r Gd	OHn	800
6042	r	Pr	OHn	100	6108	r Y	OHn	300
6043	db	Dy	OHn	300	6109	b Fe	OHn	150
6044	tb	Gd	OHn	500	6109	dv Sr	OG	7
6045	r	Cu	OH	10	6110	r Sc	OHn	2500
6045	r	Tb	OHn	300	6110.8 a	Ba	OH	12
6047	r	Sm	OHn	150	6111	r Ba	AH	10
6049	x	Ca	OH	300	6112	b Tb	OHn	250
6050	r	Dy	OHn	300	6113	r Sm	OHn	150
6050	v	Tb	OHn	200	6114	v Sr	OG	5
6052	r	Cr	OHn	100	6115	x Ca	OH	200
6053	b	Nd	OHn	150	6115	r Gd	OHn	500
6054	r	Y	OHn	800	6115	r Y	OHn	300
6055	r	Pr	OHn	150	6116	r Sc	OHn	5000
6055	r	Sc	OHn	800	6121	db Dy	OHn	300
6056	r	Tb	OHn	400	6122	v Tb	OHn	200
6058	db	Dy	OHn	300	6126	r Tb	OHn	150
6059	b	Sr	AH	7000	6127	r Y	OHn	200
6060	dr	Cu	OH	50	6130	r Pr	OHn	70
6061	dr	La	OHn	20	6132	r Gd	OHn	1000
6063.1 a		Ba	OH	4	6132	r Sm	OHn	300
6064	v	CaF	FH	150	6132	r Y	OHn	7000
6064	db	Gd	OHn	600	6133	v Nd	OHn	100
6064	r	Sc	OHn	5000	6134	r Tb	OHn	150
6065	b	Sm	OHn	120	6135	r Pr	OHn	70
6068	v	Ca	OH	200	6136	r Gd	OHn	800
6068	r	Tb	OHn	500	6138	r Sm	OHn	250
6070.8 a		Rb	OHn	20	6140	r Sc	OHn	1000
6071	r	Sm	OHn	250	6140	r Tb	OHn	200
6071	dr	Tb	OHn	350	6141.7 s	Ba	OAn	10
6073	r	Sc	OHn	5000	6147	r Cu	OH	50
6073	r	Y	OHn	300	6148	r Y	OHn	5000
6076	v	Ca	OH	200	6149	r Sc	OHn	800
6076	dv	Sr	OG	25	6149	db Sm	OHn	250
6078	dr	Dy	OHn	500	6152	r Gd	OHn	1000

6154	r	Sc	OHn	1000
6154.2 a	Na	OHn	10
6156	tr	Gd	OHn	1000
6159.6 a	Rb	OHn	25
6160.7 a	Na	OHn	20
6161	r	Pb	OH	3
6162	mr	Cu	OH	50
6163	r	Ba	AH	10
6163	r	Pr	OHn	70
6165	r	Sm	OHn	200
6165	r	Y	OHn	3000
6168	r	Sm	OHn	200
6172	r	Gd	OHn	1000
6176	r	Gd	OHn	1000
6176	r	Mn	OHn	200
6180	b	Nd	OHn	100
6181	r	Fe	OHn	200
6181	r	Sc	OHn	200
6182	r	Y	OHn	2000
6183	r	Gd	OHn	1500
6185	v	CaCl	OA, C	500
6188	dr	La	OHn	20
6188	r	Sc	OHn	400
6193	r	Sc	OHn	600
6195	r	Sm	OHn	250
620_0	b	B	OHn	20
6200	r	Sm	OHn	200
6200	r	Y	OHn	1200
6201	dr	Gd	OHn	2000
6204	r	Mn	OHn	100
6206.3 a	Rb	OHn	30
6210	b	Ra		20
6212	v	CaCl	OA, C	500
6212	r	Gd	OHn	2000
6213.1 a	Cs	OH	10
6215	r	Ti	OH	20
6216	dr	La	OHn	40
6217.6 a	Cs	OH	2
6218	r	Y	OHn	800
6219	r	Fe	OHn	300
6219	b	Nd	OHn	150
622_0	b	Ca	OH	10 000
622_0	w	Yb	AH	100
6220	r	Sc	OHn	150
6221	dr	Gd	OHn	2000
6222	r	Ti:		see 6215
6225	r	Ba	AH	15
6229	r	Sc	OHn	120
6229	r	Sm	OHn	200
623_0	w	Eu	OHn	600
6232	r	Gd	OHn	1500
6233	r	Sc	OHn	150
6237	r	Mn	OHn	100
6237	r	Y	OHn	500
6238	db	Sm	OHn	200
624_0	w	Sr	AH	150
6242	r	Gd	OHn	1500
6243	db	Sm	OHn	250
6245	r	Gd	OHn	800
6245	dr	La	OHn	40
6245	r	Tb	OHn	100
6247	b	Nd	OHn	120

APPENDIX

6247	b	Ra		20	6352	r	Pr	OHn	60	646_5	w	Sr	AH	700
6251	r	Pb	OH	4	6352	r	Y	OHn	40	6468	r	Y	OHn	50
6251	r	Y	OHn	150	6353	r	Sm	OHn	200	6469	r	Sm	OHn	250
6252	r	Gd	OHn	1000	6354	b	Nd	OHn	400	647_0	w	Eu	OHn	800
6256	tb	Sm	OHn	200	6354.6	a	Cs	OH	10	6472.6	a	Cs	OH	2
6259	r	Pr	OHn	120	6357	r	Pr	OHn	60	6474	r	Pr	OHn	250
6262	r	Sc	OHn	70	6359	v	SrCl	PH	20	6474	db	Sm	OHn	300
6263	r	Gd	OHn	800	6360	dr	La	OHn	10	6476	r	Sc	OHn	80
6263	r	Pr	OHn	70	6361	b	Nd	OHn	300	6478	r	Sm	OHn	400
6267	r	Gd	OHn	600	6361	v	Sm	OHn	200	6478	dr	V	OHn	20
6269	r	Nd	OHn	200	6362	r	Sc	OHn	30	6482.9	u	Ba	OH	4
6269	b	Ra		50	6362	v	SrCl	PH	10	6485	r	Sc	OHn	150
6271	r	Pr	OHn	150	6362.3	a	Zn	OHn	1	6485	dr	Sm	OHn	600
6272	r	Sc	OHn	40	6363	r	Pr	OHn	70	6485	v	Tb	OHn	70
6273	r	Gd	OHn	800	6370	db	Nd	OHn	500	6485	r	Y	OHn	60
6273	b	Sm	OHn	150	6370	r	Y	OHn	40	6493	r	Ba	AH	25
6274	dr	La	OHn	30	6374	r	Sm	OHn	250	6493	b	Nd	OHn	500
6275	r	Sc	OHn	70	6377	r	Pr	OHn	60	6494	r	Pr	OHn	150
6275	r	Y	OHn	250	6379	db	Sm	OHn	250	6496	r	Sc	OHn	150
6279	r	Pr	OHn	150	6382	r	Nd	OHn	400	6496	r	Tb	OHn	50
628_0	b	Eu	OHn	500	6387	r	Y	OHn	40	6496.9	s	Ba	OAn	50
6281	v	Nd	OHn	250	6388	b	Sm	OHn	300	6498.8	a	Ba	OH	50
6283	r	SrF	FH	30	6389	b	Nd	OHn	400	6500	dv	Nd	OHn	1000
6284	r	Gd	OHn	500	639_0	b	B	OHn	20	6501	r	Y	OHn	70
6285	v	CaF	FH	50	6391	r	Sm	OHn	300	6503	r	Sc	OHn	40
6285	b	Ra		20	6394	r	Cr	OHn	100	6510	trv	Sm	OHn	2000
6287	r	Nd	OHn	250	6395	r	SrF	FH	30	6512	mv	SrF	FH	200
6290	r	Gd	OHn	300	6395	r	Tb	OHn	100	6514	r	Pr	OHn	100
6291	r	Ba	AH	25	6396	b	Sm	OHn	300	6518	db	Nd	OHn	500
6292	r	Tb	OHn	70	6404	b	Nd	OHn	400	6518	r	Sc	OHn	60
6294	r	Gd	OHn	500	6406	db	Sm	OHn	300	6518	r	Y	OHn	80
6295	r	Y	OHn	100	6406	r	Y	OHn	30	6525	r	Sm	OHn	400
6297	r	Nd	OHn	200	6408	r	Sc	OHn	120	6526	r	Sc	OHn	100
6298	r	Pr	OHn	120	6410	r	Nd	OHn	400	6527.3	a	Ba	OH	25
6298.3	a	Rb	OHn	50	6419	mv	SrF:	see 6395		6531	r	V	OHn	20
6302	dr	La	OHn	20	6420	r	Sc	OHn	30	6532	mr	Sm	OHn	1200
6306	r	Gd	OHn	400	6421	r	Sm	OHn	300	6535	r	Sc	OHn	120
6306	mv	SrF:	see 6283		6423	r	Ba	AH	15	6536	r	Y	OHn	60
6306.3	a	Rb	OHn	5	6424	r	Sc	OHn	30	6540	db	Sm	OHn	300
6313	b	Nd	OHn	200	6424	r	Y	OHn	30	6542	v	Nd	OHn	500
6316	r	Gd	OHn	200	6425	r	Nd	OHn	500	655_0	w	Eu	OHn	800
6316	r	Y	OHn	60	6425	v	Sm	OHn	300	6554	r	Y	OHn	60
6318	r	Pr	OHn	60	6428	r	Pb	OH	2	6557	r	Sm	OHn	700
6318	r	Sc	OHn	30	6432.0	a	Cs	OH	2	6558	r	Sc	OHn	100
6328	b	Sm	OHn	150	6436	r	Tb	OHn	150	6563	r	Ba	AH	10
6329	r	Gd	OHn	200	6437	r	Sc	OHn	60	6564	b	Nd	OHn	500
6329	b	Ra		20	6439	r	Sm	OHn	250	6565	r	Sm	OHn	200
6331	dr	La	OHn	15	644_0	b	Ca	OH	1500	6567	r	Sc	OHn	80
6332	b	Nd	OHn	200	6442	r	Nd	OHn	500	6573	r	Y	OHn	50
6335	v	Sm	OHn	150	6446	r	Sc	OHn	120	6576	r	Sc	OHn	100
6336	r	Nd	OHn	200	6448	v	Nd	OHn	500	6580	v	Nd	OHn	1500
6336	r	Pr	OHn	50	6450	b	Sm	OHn	300	6580	r	Sm	OHn	300
6338	r	Gd	OHn	200	6450	v	Tb	OHn	120	6582	r	Nd	OHn	1500
6338	r	Y	OHn	40	6450.9	a	Ba	OH	4	6585	r	Sm	OHn	250
6339	r	Sm	OHn	400	6452	r	Cr	OHn	50	6586	db	Nd	OHn	2000
6341	r	Tb	OHn	70	6453	v	Nd	OHn	400	6586.5	a	Cs	OH	10
6342	r	Pb	OH	2	6456	dr	Sm	OHn	300	6589	r	Pr	OHn	100
6348	r	Sm	OHn	200	6458	r	Pr	OHn	70	659_0	w	Sr	AH	1500
6349	r	Nd	OHn	400	6458	r	Sc	OHn	100	6591	r	Sc	OHn	20
6349	b	Ra		20	6460	r	Tb	OHn	150	6595.3	a	Ba	OH	25
6349	r	Tb	OHn	250	6462	dr	Sc	OHn	30	6598	r	Nd	OHn	2000
6352	r	Gd	OHn	150	6464	r	Nd	OHn	500	6600	b	Sm	OHn	300

6601	r	Nd	OHn	1500	6751	r	Sm	OHn	200	6987	r	Pr	OHn	200
6601	r	Sc	OHn	60	6752	r	Sc	OHn	30	6991	v	Pr	OHn	200
6608	db	Nd	OHn	2000	6772	r	Cr	OHn	40	6991	r	Sc	OHn	50
6609	r	Pr	OHn	80	6774	r	Nd	OHn	300	6992	v	Nd	OHn	1500
6610	r	Sc	OHn	70	6781	r	Ti	OH	30	6995	r	La	OHn	10
6611	r	Y	OHn	40	6783	r	Ba	AH	50	6997	r	Nd	OHn	1500
6612	r	Nd	OHn	1500	6787	r	Sc	OHn	30	7005	r	Sc	OHn	50
6614	v	SrCl	PH	20	6796	r	Sm	OHn	200	7011	r	La	OHn	20
6618	r	Sc	OHn	70	6800	dr	Sc	OHn	20	7011	r	Nd	OHn	2000
6619	r	Sm	OHn	300	6808	r	Sm	OHn	200	7011	r	V	OHn	20
662_0	w	Eu	OHn	700	6810	r	Lu	OHn	500	7018	r	Pr	OHn	200
6621	b	Nd	OHn	1200	6819	v	Nd	OHn	250	702_0	w	Eu	OHn	1200
6624	b	Nd	OHn	1200	682_0	b	B	OHn	5	7021	db	Nd	OHn	1500
6625	r	Ca	OA	100	682_0	w	Sr	AH	7000	7025	r	La	OHn	15
6628.7	a	Cs	OH	2	6820	r	Sm	OHn	150	7026	r	Sc	OHn	30
6630	r	Nd	OHn	1500	6822	r	Pr	OHn	120	7027	b	Nd	OHn	1500
6631	r	Pr	OHn	60	6824	r	Sc	OHn	20	7033	r	Nd	OHn	2000
6631	r	Y	OHn	30	6824.7	a	Cs	OH	2	7036	r	Sc	OHn	50
6633	mv	SrF	FH	200	6830	r	Cr	OHn	70	7038	b	V	OHn	20
6634	r	Ba	AH	1	6831	v	Nd	OHn	250	7039	r	Pr	OHn	300
6634	r	Nd	OHn	1000	6839	b	Nd	OHn	200	704_0	w	Sr	AH	500
6634	r	Sm	OHn	150	684_0	w	Eu	OHn	1200	7041	r	La	OHn	30
6638	b	Nd	OHn	1000	6846	b	Nd	OHn	200	7055	r	La	OHn	15
6639	r	Ca:		see 6625	6851	r	Pr	OHn	70	7056	r	Nd	OHn	1500
6645	r	Sc	OHn	40	6857	r	Ba	AH	40	7057	dr	Ti	OH	40
6647	dr	Sm	OHn	250	6857	r	Nd	OHn	300	7060.0	a	Ba	OH	4
6648	r	Nd	OHn	1200	6864	r	Sc	OHn	40	7064	b	Nd	OHn	1500
6651	v	Sm	OHn	250	6864.6	a	Eu	OHn	10	7070	r	V	OHn	20
6652	r	Nd	OHn	900	6870.5	a	Cs	OH	2	7071	r	La	OHn	35
6653	b	Ra		50	6872	r	Sm	OHn	100	7082	r	Sc	OHn	40
6654	r	Sc	OHn	50	6880	v	Nd	OHn	300	7085	r	La	OHn	15
6655	b	Nd	OHn	1000	6882	r	Pr	OHn	100	7090	dr	Ti	OH	40
6656	r	Sm	OHn	250	6887	db	Nd	OHn	400	7092	db	Nd	OHn	2000
666_0	w	Sr	AH	5000	6893	r	Pr	OHn	100	7094	r	Sc	OHn	40
6661	r	Sc	OHn	60	6896	r	Cr	OHn	60	7095	b	Pr	OHn	700
6664	r	Nd	OHn	800	6897	r	Nd	OHn	600	7098	r	Ba	AH	80
6669	v	Sm	OHn	250	6906	r	Sc	OHn	40	7100	r	Nd	OHn	1200
668_0	mb	Sm	OHn	200	6908	r	Nd	OHn	800	7101	r	La	OHn	35
6680	r	Nd	OHn	900	6911.3	a	K	OHn	20	7108	b	V	OHn	20
6681	r	Ti	OH	30	6918	r	Nd	OHn	1000	711_0	tr	CN	OCq	0.3
6683	r	Nd	OHn	700	6919	r	H_2O	OHq	1	7115	db	Nd	OHn	2000
6688	r	Lu	OHn	500	6922	r	Sc	OHn	30	7116	r	La	OHn	15
6691	r	Sc	OHn	30	6924	r	Pr	OHn	250	7120	r	Pr	OHn	500
6694	b	Nd	OHn	600	6930	r	Pr	OHn	200	7126	r	Nd	OHn	1500
6700	v	Sm	OHn	250	6931	r	Ba	AH	60	7127	dr	Ti	OH	40
6701	r	Sc	OHn	40	6931	tb	Nd	OHn	1000	7129	r	Pr	OHn	500
6706	r	Sc	OHn	40	6937	r	Sc	OHn	30	7132	r	La	OHn	35
6707.8	p	Li	OH	700 000	6939.0	a	K	OHn	40	7133	r	Nd	OHn	1500
6714	r	Ti	OH	30	6940	db	Nd	OHn	1200	7138	r	Pr	OHn	500
6716	b	Sm	OHn	250	6945	db	Nd	OHn	1200	7147	r	La	OHn	15
6719	r	Y	OHn	50	695_0	tr	CN	OCq	0.2	7155	b	Sm	OHn	30
672_0	w	Sr	AH	4000	6950	r	Pr	OHn	200	7158	r	Nd	OHn	1500
6722	r	Pr	OHn	100	6952	r	V	OHn	10	7159	r	Ti	OH	40
6723.3	a	Cs	OH	15	6956	r	Ca	OA	10	7163	r	La	OHn	30
6728	r	Sm	OHn	200	6957	r	Nd	OHn	800	7165	r	H_2O	OHq	3
6738	r	Sc	OHn	30	6959	r	Ca:		see 6956	7172	tr	Nd	OHn	1500
6744	v	Sm	OHn	200	6963	r	Sc	OHn	40	7179	r	La	OHn	10
6745	v	SrCl	PH	10	6970	b	Nd	OHn	1500	7186	r	Nd	OHn	1000
6747	r	Ti	OH	30	6973.3	a	Cs	OH	15	7190	r	Cr	OHn	10
6748	r	Nd	OHn	400	6979	b	V	OHn	10	7194	r	La	OHn	25
6748	r	Sc	OHn	30	6983.5	a	Cs	OH	7	721_0	b	B	OHn	?
6749	r	Lu	OHn	500	6984	r	Ca	OA	10	7211	r	La	OHn	10

APPENDIX

7213	r	Nd	OHn	500	7542	r	Sr	AA	10	808_0	tr	CN	OCq	1
7225	r	La	OHn	15	7556	r	Nd	OHn	200	8086	r	La	OHn	150
7227	b	Nd	OHn	800	7558	r	Sr	AA	10	8097	r	H_2O	OHq	8
7229	b	Sm	OHn	30	7560	r	La	OHn	180	810_0	w	Mn	OH	400
7236	db	Nd	OHn	400	759_0	w	Sm	OHn	30	811_0	mb	Fe	OHn	100
7245	r	Nd	OHn	300	7592	r	La	OHn	120	8110	r	Pr	OHn	500
7249	r	Cr	OHn	20	7592	r	Nd	OHn	200	8122	r	La	OHn	100
7255	r	Nd	OHn	250	7618.9 a		Rb	OHn	50	8126.5 a		Li	OH	10
7257	r	La	OHn	10	7622	r	Nd	OHn	200	8153	r	Ca	OA	100
7268	r	Nd	OHn	200	7625	r	La	OHn	80	8159	r	La	OHn	60
7276	r	Sc	OHn	50	763_0	w	Fe	OHn	40	8173	r	Pr	OHn	300
7279	r	Pr	OHn	300	7663	r	Pr	OHn	600	8183.3 a		Na	OHn	1000
728_0	tr	CN	OCq	0.3	7664.9 p		K	OHn	400 000	8184	r	Pr	OHn	300
7285	r	Nd	OHn	250	7665	r	La	OHn	60	8194.8 a		Na	OHn	2000
7299	r	H_2O	OHq	1	7668	r	Pr	OHn	400	8196	r	La	OHn	25
731_0	w	Sm	OHn	30	7699.0 p		K	OHn	200 000	8216	v	Nd	OHn	200
7312	r	Cr	OHn	30	7702	r	La	OHn	30	822_0	b	B	OHn	?
7318	r	Ca	OA	30	7708	r	Pr	OHn	400	8233	r	La	OHn	15
7318	v	Nd	OHn	200	7714	r	Cr	OHn	20	8234	dr	Pr	OHn	250
7320	r	Pr	OHn	500	7714	r	Pr	OHn	400	8258	r	Sr	AA	100
7326	r	Ca:	see 7318		7716	r	Ca	OA	70	8272	r	Sr	AA	100
7327	r	Pr	OHn	300	7721	r				828_0	tr	CN	OCq	0.7
7343	v	Nd	OHn	250	7738	r	La	OHn	20	830_0	b	Ba	AH	200
7343	r	Pr	OHn	400	775_0	b	B	OHn	?	8339	r	Cr	OHn	40
7347	v	SrH	AA	100	7757.7 a		Rb	OHn	50	834_0	w	Fe	OHn	80
7350	r	Pr	OHn	300	7777	r	Cr	OHn	30	8360	r	Cr	OHn	40
7374	dr	V	OHn	5	779_0	w	Fe	OHn	60	8380	b	Nd	OHn	100
7375	r	Cr	OHn	30	7792	r	Nd	OHn	250	8386	r	Cr	OHn	40
7376	r	Pr	OHn	500	7800.2 p		Rb	OHn	50 000	8386	r	Pr	OHn	200
7380	r	La	OHn	150	7827	r	Pr	OHn	300	8392	r	Pr	OHn	100
7394	r	V	OHn	10	7843	r	Cr	OHn	40	8406	r	La	OHn	5
7401	r	Pr	OHn	300	7851	r	V	OHn	5	8417	r	Cr	OHn	40
7404	r	La	OHn	600	7865	r	V	OHn	10	8427	r	Pr	OHn	200
7405	r	Nd	OHn	250	7866	r	Pr	OHn	500	8432	r	Pr	OHn	150
7405	r	V	OHn	10	7873	r	Pr	OHn	300	8449	r	Cr	OHn	40
7411	r	Nd	OHn	400	7873	r	Sr	AA	20	8454	r	La	OHn	20
7413	r	Pr	OHn	300	7877	r	La	OHn	170	8477	r	Pr	OHn	150
7418	r	V	OHn	10	7882	r	Sr	AA	25	8489	r	Pr	OHn	600
7420	r	Nd	OHn	250	789_0	tr	CN	OCq	0.9	8490	r	La	OHn	30
7431	dr	Pr	OHn	400	7897	r	V	OHn	15	8494	r	Pr	OHn	300
7434	r	La	OHn	600	7902	r	Sr	AA	30	850_0	tr	CN	OCq	0.4
7434	dr	V	OHn	15	7910	r	Cr	OHn	40	8503.5 a		K	OHn	4
744_0	tr	CN	OCq	0.2	7911	r	La	OHn	900	8505.2 a				
7440	tb	Ba	OHn	300	7919	r	V	OHn	15	8521.1 p		Cs	OH	10 000
745_0	w	Ba	AH	45	792_0	w	Fe	OHn	90	8527	r	La	OHn	35
7454	r	V	OHn	15	7922	r	Pr	OHn	400	8538	r	V	OHn	5
7465	r	La	OHn	500	7940	r	V	OIIn	20	855_0	w	Mn	OH	400
7467	db	Nd	OHn	300	7945	r	La	OHn	850	8555	r	Pr	OHn	400
7472	r	V	OHn	15	7947.6 p		Rb	OHn	25 000	8564	r	La	OHn	30
7485	r	Pr	OHn	400	7961	r	V	OHn	20	8574	dr	V	OHn	5
7492	r	V	OHn	15	7973	r	V	OHn	20	858_0	w	Fe	OHn	100
7497	r	La	OHn	350	7980	r	La	OHn	600	8597	r	V	OHn	7
7498	r	Nd	OHn	250	7986	r	Pr	OHn	500	8601	r	La	OHn	25
7501	r	Sr	AA	5	7996	v	Nd	OHn	400	8605	r	V	OHn	7
7507	r	La	OHn	150	800_0	b	V	OHn	20	8622	r	Pr	OHn	300
7517	b	V	OHn	15	8004	v	Nd	OHn	400	8625	r	V	OHn	10
7523	r	Sr	AA	7	8015	r	La	OHn	400	8638	r	La	OHn	20
7528	r	La	OHn	250	8015.7 a		Cs	OH	5	8642	r	V	OHn	10
7529	r	Nd	OHn	200	8048	r	Pr	OHn	400	8652	r	Ca	OA	150
7534	r	V	OHn	10	8050	r	La	OHn	250	8661	db	Nd	OHn	100
7539	r	La	OHn	100	8076	r	Pr	OHn	400	8668	r	V	OHn	10
7539	r	Pr	OHn	250	8079.0 a		Cs	OH	15	8677	r	La	OHn	15

8700	r	Sr	AA	70	917_0	tr	CN	OCq	2.4	9610	r	H_2O	OHq	60
8702	r	V	OHn	10	9183	r	H_2O	OHq	8	9669	r	H_2O	OHq	80
8715	r	La	OHn	10	9229	r	Ca	OA	200	9807	v	Ca	OA	10
873_0	b	Ba	AH	230	9277	r	H_2O	OHq	120	9835	v	Ca	OA	10
8754	r	La	OHn	5	9333	r	H_2O	OHq	90	1095_0	tr	CN	OCq	0.1
8761.4 a		Cs	OH	40	941_0	tr	CN	OCq	1.6	11021	da	K	OHn	40
8902.2 a, 8904.0 a		K	OHn	6	9440	r	H_2O	OHq	60	11690.2 a		K	OHn	2000
					9485	r	H_2O	OHq	50	11769 a, 11773 a		K	OHn	3000
8916	r	H_2O	OHq	15	9559	r	H_2O	OHq	50					
8943.5 p		Cs	OH	5000	958_0	tr	CN	OCq	0.7	12432	a	K	OHn	1000
8974	r	H_2O	OHq	10	9595.6 a, 9597.8 a		K	OHn	15	12522	a	K	OHn	1000
9129	r	H_2O	OHq	12										

APPENDIX

TABLE 15. Manufacturers of flame photometers and flame spectrophotometers

Status of 1961-62. No claim is made for completeness. The data have been assembled from manufacturers' brochures and similar material.

Abbreviations: AA = air-acetylene; AB = air-butane; AG = air-city gas; AH = air-hydrogen; AP = air-propane; Att. = attachment; Dir. = direct; Em. = emission flame photometry; Fil. = filter; Gr. = grating monochromator; Ind. = indirect; I.S. = internal standard; I.W. = interference wedge; OA = oxyacetylene; OB = oxygen-butane; OG = oxygen-city gas; OH = oxyhydrogen; OP = oxygen-propane; PC = photoconductive cell; PM = photomultiplier; Pot. = potentiometric; PP = photographic plate; Pr. = prism monochromator; PT = phototube; PV = photovoltaic cell; Rec. = recording.

No.	Manufacturer and Address	Model	Method	Dispersion	Flame	Atomizer	Detector	Amplifier	Readout	Notes
1	Advanced Instruments, Inc., Newton Highlands 61, Mass.		Em.	Fil.	AG	Ind.	PC	No	I.S.	(a)
2	Applied Physics Corp., Monrovia, Calif.	14 + Att.	Em.	Pr.-Gr.	AH, OH, OA	Dir	PM, PC	a.c.	Rec.	
3	Baird-Atomic, Inc., Cambridge 38, Mass.	KY-1	Em.	Fil.	AG, AP, AB	Ind.	PM	d.c.	Dir., I.S.	(a), (h)
4	Baldwin Instruments Co. Ltd., Dartford, Kent, England		Em.	Fil.	AA	Ind.	PV	Yes	Dir., I.S.	
5	Beckman Instruments, Inc., Fullerton, Calif.;	DU + 9200	Em.	Pr.	AH, OH, OA	Dir.	PT, PM	d.c.	Pot., Rec.	(w)
	Beckman Instruments G.m.b.H., Frankfurter Ring 115, Munich 45	DK + 92490	Em.	Pr.	same	Dir.	PM, PC	a.c.	Rec.	
		B + 9125	Em.	Pr.	same	Dir.	PT, PM	d.c.	Dir.	
		DB + 28260	Em.	Pr.	same	Dir.	PM	a.c.	Dir., Rec.	
		C + Att.	Em.	Fil.		Dir.	PT	d.c.	Dir., Rec.	
6	Coleman Instruments, Inc., Maywood, Ill.	21	Em.	Fil.	OG, OP, OB	Dir.	PT	d.c.	Dir.	(b)

428   APPENDIX

No.	Manufacturer and Address	Model	Method	Dispersion	Flame	Atomizer	Detector	Amplifier	Readout	Notes
7	Electronest, Ing. Anton Nest, Saarbrücken-Ottenhausen; Sté. Électronest, S. A. R. L., Forbach, Moselle, France	STA 58 MED 58 SPE 58	Em. Em. Em.	Fil. Fil. Fil.	AA AA OA, OH	Ind. Ind. Ind.	PV, PT PV, PT PV, PT	No No No	Dir. Dir. Dir.	(c) (d)
8	Électro-Synthèse, 47, Rue Barrault, Paris (13e)	Ph.F. 56	Em.	Fil.	AP, AB, AG	Ind.	PV	No	Dir., I.S.	
9	Evans Electroselenium Ltd., Halstead, Essex, England	EEL	Em.	Fil.	AG, AP, AB	Ind.	PV	No	Dir.	(e)
10	A. Gallenkamp and Co. Ltd., Sun St, London, E.C.2	Mark II	Em.	Fil.	AG, AP, AB	(Ind.)	(PV)	(No)	Dir.	
11	Gelman Instrument Co., Chelsea, Mich.	Gelman-Evans	Em.	Fil.	AG	Ind.	PV	No	Dir.	
12	Hilger and Watts Ltd., 98 St. Pancras Way, London, N.W.1; Hilger and Watts France, 43, Rue Beaubourg, Paris (3e)	H 700+H 868 H 700+H 909	Em. Abs.	Pr. Pr.	OH AG	Ind. Ind.	PT, PM PT, PM	d.c. d.c.	Pot, Rec. Pot.	(f)
13	Hitachi Ltd., Ibaraki, Japan	FPF-2 EPU-2A + H-2	Em. Em.	Fil. Pr.	OH, OA	Dir.	PT, PM	d.c.	Dir. Pot, Rec.	
14	Jarrell-Ash Co., Newtonville 60, Mass.	82000 +82500 82000 + Att.	Em. Abs.	Gr. Gr.	OH, OA AH, OH, OA	Dir. Dir.	PM PM	d.c. d.c.	Rec. Rec.	(g)

APPENDIX 429

15	Jobin-Yvon, Arcueil (Seine), France; Établissements Jouan, 113, Blvd. St.-Germain, Paris (6e)	Jouan	Em.	Fil.	AG, AP, AB	Ind.	PV	No	Dir.	(h), (j)
		Research model	Em.	Gr.	OA	Dir.	PT	a.c.	Dir.	
16	Keyes Scientific Co., Inc., 122 Hampshire St., Cambridge, Mass.	Chaffee–Keyes Research Model 60	Em.	Fil.	OH, OA, OP	Dir.	PM	a.c.	Pot., Rec., Dir.	(l), (r), (u)
17	P. J. Kipp en Zonen, N.V. v/h, Voorstr. 67, Delft, Holland	53	Em.	Fil.	AP, AA	Ind.	PV	No	Dir.	
18	Dr. B. Lange, Hermannstr. 14, Berlin-Zehlendorf; Agent: Epic, Inc., 150 Nassau St, New York 38	2	Em.	Fil.	AG, AP, AA	Ind.	PV	No	Dir. (+ I.S.)	(i)
		6	Em.	Fil.	AG	Ind.	PV	d.c.	Dir.	
19	Marius, Utrecht, Holland	Industrial model 5	Em.	Fil.	AG, AP	Ind.	PC	No	Dir.	(h), (i), (y)
			Em.	Fil.	AP	Ind.	PT	a.c.	Dir., I.S., Rec.	(b)
20	Metrohm A.-G., Herisau, Switzerland	E1006, E1007	Em.	Pr.			PT, PM	a.c.		
21	National Instrument Laboratories, Inc., 828 Evarts St., N.E., Washington 18, D.C.	N.I.L. (Patwin FC)	Em.	Fil.	AG, AP	Ind.	PV	No	I.S.	(h)
22	Netheler und Hinz G.m.b.H., Hamburg-Wellingsbüttel; Agent: Brinkman Instruments, Inc., Great Neck, L.I., N.Y.	Eppendorf	Em.	Fil.	AP, AA, AH	Ind.	PM	d.c.	Dir., Rec.	(a), (k)
		Eppendorf II	Em.	Fil.	AP	Ind.	PT	d.c.	Dir.	(l)
23	Gebrüder Netzsch Maschinenfabrik, Selb, Bavaria		Em.	I.W.	AA	Ind.	PM	d.c.	Pot.	(m)
24	North American Philips Co., Inc., Mt. Vernon, N.Y.	12130 Norelco	Em.	Fil.	AG, AP	Ind.	PV	No	Pot., I.S.	

No.	Manufacturer and Address	Model	Method	Dispersion	Flame	Atomizer	Detector	Amplifier	Readout	Notes
25	Optica S.p.a., Via Calatafimi 7, Milan;	CF-4 + 6-1070	Em.	Gr.	OA, OH, AH	Dir.	PM	d.c., a.c.	Pot., Rec.	
	Optica United Kingdom Ltd., Gateshead-on-Tyne, 11;	B-2 or S-2	Em.	Gr.	same	Dir.	PP		Pot.	
		CF-4 + ATA	Abs.	Gr.	AA	Ind.	PM	d.c., a.c.	Pot., Rec.	
	Optica, Inc., 828 Evarts St., N.E., Washington 18, D.C.;	AT-6	Abs.	Pr.	AA	Ind.	PT	d.c.	Pot.	(n)
	Optica France S.r.l., 2, Rue Chabanais, Paris									
26	Optische Werke, Jena, Germany	III	Em.	Fil.	AA	Ind.	PV	No	Dir.	(o)
27	The Perkin-Elmer Corp., Norwalk, Conn.;	146	Em.	Pr.	AG, AA, AP	Ind.	PT	a.c.	Pot., I.S.	
	Perkin-Elmer and Co. G.m.b.H., Überlingen/See (Bodensee) Germany	202 + Att.	Em.	Pr.	OH, OA	Dir.	PM	a.c.	Rec.	(p)
		350 + Att.	Em.	Pr.	OH, OA	Dir.	PM, PC	a.c.	Rec.	(q)
		214	Abs.	Gr.	AA	Ind.	PM	a.c.	Dir., Rec.	
28	Process and Instruments Corp., 17 Stone Ave., Brooklyn 33	1 B	Em.	Fil.	AP	Ind.	PV	No	Dir., I.S.	(a)
			Em.	Fil.	AG, AP	Ind.	PC	No	Pot., I.S.	(v)
29	Research and Control Instruments, Inc., Woburn, Mass.		Abs.	Gr.	AG	Ind.	PM		Pot.	
30	Sigrist und Weiss A.-G., Falkenstr. 23, Zürich 1/8	U P3/I	Em.	Fil.	AP, AA	Ind., Dir.	PT, PM	a.c.	Pot.	(r)
31	Technicon Controls, Inc., Chauncey, N.Y.	AutoAnalyzer + Att.	Em.	Fil.	OP	Dir.	PC	d.c.	Rec., I.S.	(s)
		Multichannel	Em.							(x)

APPENDIX 431

32	Unicam Instruments Ltd., Arbury Works, Cambridge; Agent: Wilkens-Anderson Co., 4525 W. Division St., Chicago 51	SP.900	Em.	Pr.	AA, AP, AH	Ind.	PM	a.c.	Dir., Rec.
33	Waters Associates, Framingham, Mass.	Process	Em.	Gr.	OH	Dir.	PM	a.c.	Dir., Rec. (t)
34	Carl Zeiss, Oberkochen, Württemberg, Germany	III	Em.	Fil.	AG	Ind.	PV	No	Dir.
		PF 5	Em.	Fil.	AA, AP	Ind.	PM	d.c.	Dir.
		PMQ II + Att.	Em.	Pr.	OH, OA	Dir.	PM	a.c.	Dir., Rec.
		M4Q + Att.	Em.		OA	Dir.	PM		Dir.

Notes

(a) Supplies standard solutions.
(b) Offers sample-handling system.
(c) Designed by Schuhknecht.
(d) Phototube can be cooled.
(e) Offers a recording sodium flame monitor.
(f) Un-premixed flame; cooled burner.
(g) Triple burner and multiple-pass optical system; lamp turret.
(h) Built-in air compressor.
(i) Burns air-gasoline also.
(j) Heated spray chamber.
(k) Flame can be burned with added oxygen.
(l) Atomization is done by the fuel gas.
(m) Can be used as colorimeter also.
(n) Double beam; lamp turret; adaptable to digital readout.
(o) Formerly the Zeiss Model III.
(p) Double monochromator.
(q) Double beam; per cent absorption shown on counter.
(r) Can read continuously.
(s) Thermally compensated photocells. Automatic sample treatment and pumping. Can record sodium and potassium simultaneously.
(t) Continuously monitors several process streams.
(u) Null balance against adjustable tungsten lamp; voltage across lamp is read. Sample pumped by syringe; automatic water rinse.
(v) Assortment of hollow-cathode lamps placed before fixed entrance slits; all beams emerge through single exit slit.
(w) With burner sheath attached, can burn any fuel gas with oxygen or air-oxygen.
(x) Determines Na, K, Ca, Mg, Sr (or any other selection) simultaneously, with background compensation.
(y) Has attachment for colorimetry.

BIBLIOGRAPHY

The point of view governing the selection of literature was stated in the Preface. An asterisk following the reference number means that this publication contains many literature citations, which are recommended for further search or study.

In this translation, the journal abbreviations are those of *Chemical Abstracts*. All references accessible to the translator either directly or through abstracts or other bibliographies have been carefully checked in order to eliminate errors. The remaining references have been checked by Dr. Herrmann and Dr. Alkemade. To aid the reader, the *Chemical Abstracts* reference (given as *CA*, followed by the volume and column number) has been appended. In a few cases references to other abstracts journals are given when a *CA* abstract could not be found. To provide more information in the available space, titles of journal articles have been replaced by a compact statement of the contents of the article. In this statement, the abbreviations used in the abstracts of *CA* are employed, together with the following special abbreviations:

   f.p. = flame photometry, flame photometer, flame-photometric.
   f.s. = flame spectrophotometry, etc.
    d. = determination (of).
   intf. = interference.
  int. std. = internal standard.
   ionizn. = ionization.

Example: "Al, P intf. in f.s. d. Ca, Sr in glass, slag" = "interference of aluminum and phosphorus (i.e., phosphate) in the flame-spectrophotometric determination of calcium and strontium in glass and slag."

Books

1. Burriel-Martí, F., and J. Ramírez-Muñoz, *Flame Photometry, a Manual of Methods and Applications*, Elsevier, Amsterdam, 1957.
1a. Cannon, C. G. (editor), *Electronics for Spectroscopists*, Interscience, New York, 1960.
2. Cordes, H., *Propan-Ratgeber*, Drägerwerk, Lübeck, 1957.
3. d'Ans, J., and E. Lax, *Taschenbuch für Chemiker und Physiker*, 2nd ed., Springer, Berlin, 1949.
4. Delahay, P., *Instrumental Analysis*, Macmillan, New York, 1957.
5. Finkelnburg, W., *Einführung in die Atomphysik*, 6th ed., Springer, Berlin, 1958.
6. Flügge, S. (editor), *Handbuch der Physik*, Vol. XXVIII: "Spektroskopie II", Springer, Berlin, 1957.
6a. Flügge, S. (editor), *Handbuch der Physik*, Vol. VIII/1: "Strömungsmechanik", Springer, Berlin, 1959.

7. Gardiner, K. W., "Flame Photometry", p. 219 in W. G. Berl (editor), *Physical Methods in Chemical Analysis*, Vol. III, Academic Press, New York, 1956.
8. Gatterer, A., J. Junkes, and V. Frodl, *Spektren der seltenen Erden*, Vatican City, 1945.
9. Gatterer, A., J. Junkes, E. W. Salpeter, and B. Rosen, *Molecular Spectra of Metallic Oxides*, Vatican City, 1957.
10. Gaydon, A. G., *Dissociation Energies and Spectra of Diatomic Molecules*, 2nd ed., Chapman and Hall, London, 1953.
11. Gaydon, A. G., and H. G. Wolfhard, *Flames, Their Structure, Radiation and Temperature*, Chapman and Hall, London, 1953; 2nd ed., Macmillan, New York, 1960.
12. Gaydon, A. G., *The Spectroscopy of Flames*, Wiley, New York, 1957.
13. Geiger, H., and K. Scheel, *Handbuch der Physik*, Vol. VII: "Mechanik der flüssigen und gasförmigen Körper", Springer, Berlin, 1927.
14. Gorbach, G., *Mikrochemisches Praktikum aus Anleitungen für die chemische Laboratoriumspraxis*, Springer, Berlin, 1956.
15. Grotrian, W., *Graphische Darstellung der Spektren von Atomen und Ionen mit ein, zwei und drei Valenzelektronen*, Springer, Berlin, 1928.
16. Hald, P. M., "Determinations with the Flame Photometer", p. 79 in M. B. Visscher (editor), *Methods in Medical Research*, Vol. IV, Year Book Publishers, Chicago, 1951.
17. Hanle, W., "Anregung der Spektren", Vol. IX, Part 3, of A. Eucken and K. L. Wolf (editors), *Hand- und Jahrbuch der chemischen Physik*, Akademische Verlagsgesellschaft, Leipzig, 1936.
18. Harley, J. H., and S. E. Wiberley, *Instrumental Analysis*, Wiley, New York, 1954.
19. Herzberg, G., *Molecular Spectra and Molecular Structure, I. Spectra of Diatomic Molecules*, 2nd ed., Van Nostrand, New York, 1950.
20. Hinsberg, K., and K. Lang, *Medizinische Chemie*, Urban und Schwarzenberg, Munich, Berlin, Vienna, 1957.
21. Hoppe-Seyler, D. F., and W. Thierfelder, *Handbuch der physiologisch- und pathologisch-chemischen Analysen*, Springer, Berlin, 1959.
22. Humphries, E. C., "Mineral Components and Ash Analysis; the Flame Photometer", p. 474 in K. Paech and M. V. Tracey (editors), *Moderne Methoden der Pflanzenanalyse*, Vol. I, Springer, Berlin, 1956.
23. Ivenskaya, N. D., *Spektral'nyĭ analiz*, Gosudarstvennoe Nauchno-Tekhnicheskoe Izdatel'stvo Mashinostroitel'noĭ Literatury, Moscow, 1955.
24. Iwantscheff, G., *Das Dithizon und seine Anwendung in der Mikro- und Spurenanalyse*, Verlag Chemie, Berlin, 1958.
25. Jost, W., *Explosions- und Verbrennungsvorgänge in Gasen*, Springer, Berlin, 1939; Edwards, Ann Arbor, 1944.
26. Kayser, H., or H. Kayser and H. Konen, *Handbuch der Spektroskopie*, Vols. I–VIII, Hirzel, Leipzig, 1900–1932. Many references on flame photometry prior to 1900.
27. Kibisov, G. I., Kh. E. Sterin, and T. O. Vreden-Kobetskaya, *Spektral'nyĭ analiz*, Izdatel'stvo Akademii Nauk S.S.S.R., Moscow, 1955.
28. Kortüm, G., *Kolorimetrie, Photometrie und Spektrometrie*, Springer, Berlin, 1955.
29. Landolt, H., and R. Börnstein, *Physikalisch-chemische Tabellen*, 6th ed., Vol. I, part 2, Springer, Berlin, 1951.

30. Lange, B., *Kolorimetrische Analyse*, 5th ed., Verlag Chemie, Weinheim, 1956.
31. Lewis, B., and G. von Elbe, *Combustions, Flames and Explosions of Gases*, Academic Press, New York, 1951; 2nd ed., 1961.
32. Lewis, B., R. N. Pease, and H. S. Taylor (editors), *Combustion Processes*, Princeton Univ. Press, 1956.
33. Lundegårdh, H., *Die quantitative Spektralanalyse der Elemente*, Fischer, Jena; Part I, 1929; Part II, 1934.
34. Lundegårdh, H., *Die Blattanalyse*, Fischer, Jena, 1945; translation by R. L. Mitchell, *Leaf Analysis*, Hilger and Watts, London, 1951.
35. Mache, H., *Die Physik der Verbrennungserscheinungen*, Veit, Leipzig, 1919.
36. Margoshes, M., and B. L. Vallee, "Flame Photometry and Spectrometry. Principles and Applications", p. 353, in D. Glick (editor), *Methods of Biochemical Analysis*, Vol. III, Interscience, New York, 1956.
37. Mavrodineanu, R., and H. Boiteux, *L'analyse spectrale quantitative par la flamme*, Masson, Paris, 1954.
37a. Meggers, W. F., *Periodic Chart of the Atoms* (and *Key*), Welsh Scientific Co., Chicago, 1959.
38. Mitchell, A. C. G., and M. W. Zemansky, *Resonance Radiation and Excited Atoms*, Macmillan, New York, 1934. Reprinted 1961 by Cambridge University Press.
39. Mitchell, R. L. *The Spectrographic Analysis of Soils, Plants, and Related Materials*, Commonwealth Bureau of Soil Science, Technical Communication No. 44, Harpenden, England, 1948. *CA* 43, 1136.
40. Moritz, H., *Spektrochemische Betriebsanalyse*, 2nd ed., Enke, Stuttgart, 1956.
41. Pearse, R. W. B., and A. G. Gaydon, *The Identification of Molecular Spectra*, 2nd ed., Wiley, New York, 1950.
41a. Poluektov, N. S., *Rapid Flame-Photometric Analytical Methods in Nonferrous Metallurgy* (in Russian), Gosudarstvennoe Nauchno-Tekhnicheskoe Izdatel'stvo Khimicheskoĭ Literature, Moscow, 1958.
41b. Poluektov, N. S., *Metody analiza po fotometrii plameni*, Gosudarstvennoe Nauchno-Tekhnicheskoe Izdatel'stvo Khimicheskoĭ Literature, Moscow, 1959.
41c. Rusanov, A. K., and N. V. Il'yasova, *Atlas plamennykh, dugovykh i iskrovykh spektrov elementov*, Gosudarstvennoe Nauchno-Tekhnicheskoe Izdatel'stvo Literatury po Geologii i Okhrane Nedr, Moscow, 1958.
42. Scheichl, L., *Brandlehre und chemischer Brandschutz*, 2nd ed., Hüthig, Heidelberg, 1958.
43. Scheller, H., *Einführung in die angewandte spektrochemische Analyse*, 3rd ed., VEB Verlag Technik, Berlin, 1960.
44. Seith, W., and K. Ruthardt, *Chemische Spektralanalyse*, 5th ed. by W. Rollwagen, Springer, Berlin, 1958.
45. Sherman, J., "Emission Spectrography", p. 330 in W. G. Berl (editor), *Physical Methods in Chemical Analysis*, Academic Press, New York, 1950.
46. Simon, H., and R. Suhrmann, *Der lichtelektrische Effekt und seine Anwendungen*, 2nd ed., Springer, Berlin, 1958.
47. Sommerfeld, A. J. W., *Atombau und Spektrallinien*, 7th ed., Vieweg, Braunschweig, 1951.
48. Stuart, H. A., *Molekülstruktur. Bestimmung von Molekülstrukturen mit physikalischen Methoden*. Hirschwaldsche Buchhandlung, Berlin, 1934.
49. Teloh, H. A., *Clinical Flame Photometry*, Thomas, Springfield, Ill., 1959.

50. Thun, R., Rudolf Herrmann, and E. Knickmann, *Handbuch der landwirtschaftlichen Versuchs- und Untersuchungsmethodik (Methodenbuch)*, Vol. I, *Die Untersuchung von Böden*, 3rd ed., Neumann, Berlin, 1955.
51. Unsöld, A., *Physik der Sternatmosphären*, 2nd ed., Springer, Berlin, 1955.
52. Vallee, B. L., "Flame Spectrometry", p. 229 in J. H. Yoe and H. J. Koch, Jr. (editors), *Trace Analysis*, Wiley, New York, 1957.
53. West, W. (editor), *Chemical Applications of Spectroscopy*, Interscience, New York, 1957.
54. Westphal, W. H., *Physik*, Springer, Berlin, 1947; 18th/19th ed., 1956.
55. Whytlaw-Gray, R., and H. S. Patterson, *Smoke, a Study of Aerial Disperse Systems*, St. Martin's, New York, 1932.
56. Willard, H. H., L. L. Merritt, Jr., and J. A. Dean, *Instrumental Methods of Analysis*, 2nd ed., Van Nostrand, New York, 1951; 3rd ed., Van Nostrand, Princeton, 1958.
57. Zaĭdel', A. N., V. K. Prokof'ev, and S. M. Raĭskiĭ, *Tablitsy Spektral'nykh Liniĭ*, Gosudarstvennoe Izdatel'stvo Tekhniko-Teoreticheskoĭ Literatury, Moscow, 1952; Saidel, A. N., W. K. Prokofjew, and S. M. Raiski, *Spektraltabellen* (International ed.), VEB Verlag Technik, Berlin, 1955.
58. Zworykin, V. K., *Photoelektrische Zellen*, Springer, Berlin, 1958.

JOURNAL ARTICLES, ETC.

59. Abresch, K., and W. Dobner, *Arch. Eisenhüttenw.*, **29**, 25 (1958). *CA*, **52**, 6060. F.s.d. Ca, Mg in ferrous metallurgy.
60. Acetylenverordnung (Rules for handling acetylene), Bonn, 1947.
61.* Adams, F., and R. D. Rouse, *Soil Sci.*, **83**, 305 (1957). *CA*, **51**, 18424. Ion exchange to remove anion intf. with Ca.
62. Addink, N. W. H., *Mikrochim. Acta*, **1955**, 703. *CA*, **49**, 11483. Self-standardization in spectrography.
63. Alkemade, C. T. J., *Physica*, **18**, 933 (1952). *CA*, **47**, 6761. Ionization of Na, K in air–acetylene f.
64. Alkemade, C. T. J., J. Smit, and J. C. M. Hattinga-Verschure, *Biochem. et Biophys. Acta*, **8**, 562 (1952). *CA*, **47**, 168. New f.p. for d. Na, K in serum.
65.* Alkemade, C. T. J., *A Contribution to the Development and Understanding of Flame Photometry*, Thesis, University of Utrecht, 1954; published by Excelsior, The Hague. *Sci. Abstr.*, **A57**, No. 9035.
66. Alkemade, C. T. J., and J. M. W. Milatz, *J. Opt. Soc. Am.*, **45**, 583 (1955). *CA*, **49**, 12966. Double-beam absorption f.p.
67.* Alkemade, C. T. J., *Proc. Colloquium Spectroscopicum Internationale VI* (*Spectrochim. Acta*, **1957**), p. 7. *CA*, **54**, 152. Intf. (ionization, self-absorption, dissociation, compound formation).
68. Alkemade, C. T. J., and M. E. J. Jeuken, *Z. anal. Chem.*, **158**, 401 (1957). *CA*, **52**, 3532. Intf. Al with Ca.
69. Alkemade, C. T. J., and W. J. Lavèn, *Appl. Sci. Research*, **B6**, 337 (1957). *Sci. Abstr.*, **61B**, No. 1283. Null instrument for d. small a.c. currents.
70. Alkemade, C. T. J., and M. H. Voorhuis, *Z. anal. Chem.*, **163**, 91 (1958). *CA*, **53**, 1984. Intf. P with Ca.
71. Alkemade, C. T. J., in H. Guyer (editor), *VIII Colloquium Spectroscopicum Internationale* (Lucerne, 1959), Sauerländer, Aarau, Switzerland, 1960. Org. solvent effects; intf. mechanisms; K continuum.
72. Allan, J. E., *Analyst*, **83**, 466 (1958). *CA*, **53**, 125. Absorption f.s.d. Mg, etc.

72a. Allan, J. E., *Spectrochim. Acta*, **1959**, 800. *CA*, **54**, 5339. Absorption f.s.d. Fe, Mn.
73. American Public Health Association, American Water Works Association, and Federation of Sewage and Industrial Wastes Associations, *Standard Methods for the Examination of Water, Sewage, and Industrial Wastes*, 10th ed., American Public Health Association, New York, 1955; f.p. methods on pp. 172, 191.
74. American Society for Testing Materials, *Methods for Emission Spectrochemical Analysis*, Philadelphia, 1953; method C228-49T for f.p.d. Na, K in portland cement.
75. *Am. Soc. Testing Materials, Proc.*, **53**, 372 (1953). F.s. or f.p. d. Na in residual fuel oil.
76. American Society for Testing Materials, *Methods for Emission Spectrochemical Analysis*, 2nd ed., Philadelphia, 1957.
77. Angot, J., *Mikrochim. Acta*, **1959**, 346. *CA*, **54**, 24123. F.p. d. traces Na in K salts.
78. Anonymous, *Brit. Clayworker*, **64**, No. 761 (Sept. 1955). F.p.d. alkali metals in clay.
79. Same as 622a.
80. Anonymous, *Anal. Chem.*, **31**, No. 11, 111A (Nov. 1959). Announcement of Technicon f.p.
81. Archibald, R. M., *Anal. Chem.*, **27**, 677 (1955). Review of clinical chemistry, mentioning f.p.
81a. Avni, R., and C. T. J. Alkemade, *Mikrochim. Acta*, **1960**, 460. *CA*, **55**, 18426. Enhancement of Na by org. solvents.
82. Baird-Atomic, Inc. (formerly Baird Associates, Inc.), Cambridge, Mass., *Better Analysis*, Newsletter, **4** (1954). Survey on f.p. Also Bulletin 37. F.p. for Na, K.
83. Baker, G. L., and L. H. Johnson, *Anal. Chem.*, **26**, 465 (1954). *CA*, **48**, 6839. Intf. anions on Ca.
84. Baker, M. R., and B. L. Vallee, *J. Opt. Soc. Am.*, **45**, 773 (1955). *Sci. Abstr.*, **58A**, No 8586. O_2–C_2N_2 f.p.
85. Baker, M. R., K. Fuwa, R. E. Thiers, and B. L. Vallee, *J. Opt. Soc. Am.*, **48**, 576 (1958). *CA*, **53**, 3873. Emission as function of sample flow in O_2–C_2N_2.
85a. Baker, M. R., and B. L. Vallee, *Anal. Chem.*, **31**, 2036 (1959). *CA*, **54**, 6388. Theory of emission as function of sample flow in O_2–C_2N_2, etc.
86. Baker, R. W. R., *Biochem. J.*, **59**, 566 (1955). *CA*, **49**, 9076. F.p. d. Ca in serum.
87. Baldwin Instrument Co. Ltd., Dartford, Kent, England. F.p.
88.* Banerjee, S. K., *Indian Ceram.*, **3**, 165 (1956). *CA*, **52**, 14118. F.p. in ceramics.
89. Barnes, H., *Analyst*, **80**, 573 (1955). *CA*, **49**, 13831. Review on anal. of sea water.
90. Barnes, R. B., D. Richardson, J. W. Berry, and R. L. Hood, *Ind. Eng. Chem., Anal. Ed.*, **17**, 605 (1945). *CA*, **40**, 25. F.p. in general, and a new filter instrument.
91. Bartlett, J. C., C. G. Farmilo, and L. I. Pugsley, Pittsburg Conference on Anal. Chem. and Appl. Spectroscopy, Mar. 1955. Abstr. in *Anal. Chem.*, **27**, 320 (1955). F.p. anal. of opium.
92. Baumann, R., and R. Herrmann, *Z. ges. exptl. Med.*, **120**, 172 (1953). *CA*, **47**, 7021. F.p. d. electrolytes in capillary serum.

93. Baumann, R., and R. Herrmann, *Z. ges. exptl. Med.*, **124**, 404 (1954). Routine micro-d. Na, K, Ca in rat serum.
94. Baumann, R., and R. Herrmann, *Ärztl. Wochschr.*, **9**, 902 (1954). *CA*, **49**, 1906. Changes in blood Na, K, Ca due to temp.
95. Baumann, R., R. Herrmann, and R. Metzger, *Ärztl. Wochschr.*, **9**, 204 (1954). *CA*, **48**, 7098. Errors in d. Na, K, Ca in serum.
96. Bauserman, H. M., and R. R. Cerney, Jr., *Anal. Chem.*, **25**, 1821 (1953). *CA*, **48**, 3845. Int.-std. f.s. d. Na, K in sugar–alcohol solutions.
97. Beauchene, R. E., A. D. Berneking, W. G. Schrenk, H. L. Mitchell, and R. E. Silker, *J. Biol. Chem.*, **214**, 731 (1955). *CA*, **49**, 10799. F.s. d. amino N via equiv. Cu.
98. Becker, F., *Zement-Kalk-Gips*, **4**, 93 (1951). *Chem. Zentr.* **1951**, II, 2651. F.s. anal. cements, etc.
99. Beckman Instruments, Inc., Fullerton, Calif., Bulletin 167F (1950). Describes Model 9200 f.s. Replaced by Bulletin 303.
100. Beckman Instruments, Inc., Bulletin 259 (1951). Instructions, Model 9200 f.s. Replaced by Bulletin 334.
101. Beckman Instruments, Inc., *Beckman Bulletin*, No. 10, p. 3 (1953). F.s. d. Pb in gasoline.
102. Beckman Instruments, Inc., Application data sheet B-14-C (1952). F.s. d. Na, K in glass.
103. Beckman Instruments, Inc., Application data sheet DU-4-F (1953). F.s. d. Fe, Cu in citrus beverage.
104. Beckman Instruments, Inc., Application data sheet DU-6-C (1953). F.s. d. Mg in cement.
105. Beckman Instruments, Inc., Application data sheet DU-12-B (1953). F.s. d. Na, K in serum.
106. Beckman Instruments, Inc., Application data sheet DU-23-P (1954). F.s. d. Na, K in paper pulp.
107. Beckman Instruments, Inc., Application data sheet DU-24-A (1954). F.s. d. Cu, Fe, Mn in plant nutrient solns.
108. Beckman Instruments, Inc., Application data sheet ID 516 (1956) (International Division). Anal. of glass and raw materials.
109. Beckman Instruments, Inc., *Beckman Bulletin*, No. 7, p. 4 (1952). Model 9200 f.s.
110. Beckman Instruments, Inc., *Beckman Bulletin*, No. 7, p. 6 (1952). Detection limits and wavelengths for Beckman f.s.
111. Beckman Instruments, Inc., *Beckman Bulletin*, No. 16, p. 1 (1955). Model 4100 f.p. for Na, K.
112. Same as 99.
113. Same as 99.
114. Beckman Instruments, Inc., Bulletin 211A (1949). Servicing instructions for Model 10300 f.s. (obsolete).
115. Beckman Instruments, Inc., Bulletin 278 (1951). Instructions for Model B f.s. Replaced by Bulletin 334.
116. Beckman Instruments, Inc., Bulletin 325 (1956). Describes DU and B f.s.
117. Beckman Instruments, Inc., Data sheet 3 (1952). Wavelengths and detection limits for f.s. Replaced by Bulletin 753A (1961).
118. Anonymous, *Laboratory*, **20**, 32 (1950). Beckman f.s.
119. Anonymous, *Laboratory*, **21**, 74 (1951). Beckman f.s.

120. Beckman Instruments, Inc., Application data sheet DU-42-C (1956). Removal of anion intf. in f.s.
121. Beckman Instruments, Inc., Application data sheet DU-45-A (1955). F.s. in agronomy.
122. Beckman Instruments, Inc., Application data sheet DU-47-M (1955). Qual. anal. by f.s.
123. Beckman Instruments, Inc., Application data sheet 41-60-B (1956). F.p. d. Na, K in serum.
124. Beckman Instruments, Inc., Application data sheet 41-61-B (1956). F.p. d. Na, K in urine.
124a. Beckman Instruments, Inc., Technical data sheet DU-70-B (1957). F.s. d. Ca in serum.
125.* Beckmann, E., and P. Waentig, *Z. physik. Chem.*, **68**, 385 (1910). *CA*, **4**, 1936. New f.s., visual, 1% accuracy, many data; review.
126.* Beeghly, H. F., *Anal. Chem.*, **29**, 638 (1957). *CA*, **51**, 7219. Review on ferrous metallurgy, including f.p. anal.
127. Belcher, H., and T. M. Sugden, *Proc. Roy. Soc. (London)*, **A 202**, 17 (1950). *CA*, **45**, 944. Ionization equilibria with Li, Na, K, Rb, Cs in flame; equilibria with ions, hydroxides, oxides, halides.
128. Belke, J., and A. Dierkesmann, *Arch. exptl. Path. Pharmakol.*, **205**, 629 (1948). *CA*, **43**, 7070. F.p. d. Na, K, Ca in biol. fluids.
129. Benzon, V. M., and J. L. Kassner, Southeastern Regional Meeting, Amer. Chem. Soc., Birmingham, Ala., Oct. 1954. Abstr. in *Anal. Chem.*, **26**, 1855 (1954) F.s. d. Li in Li minerals.
130. Bernhardt, D., and R. Herrmann, *Ärztl. Wochschr.*, **10**, 61 (1955). *CA*, **49**, 8364. F.p. d. Ca in serum.
131. Bernstein, R. E., *S. African J. Med. Sci.*, **14**, 163 (1949). *CA*, **44** 5418. Int.-std. f.p. d. Na, K in serum.
132. Bernstein, R. E., *Nature*, **165**, 649 (1950). *CA*, **44**, 7917. Int.-std. f.p. d. K in serum.
133. Bernstein, R. E., *Biochem. et Biophys. Acta*, **9**, 576 (1952). *CA*, **47**, 2245. F.s. d. Na, K, Ca in biol. fluids, tissues.
134.* Bernstein, R. E., *S. African J. Med. Sci.*, **17**, 101 (1952). *CA*, **47**, 7014. F.s. d. Na, K in biol. fluids.
135. Bernstein, R. E., *J. Lab. Clin. Med.*, **40**, 707 (1952). *CA*, **47**, 3426. F.s. d. Na, K, Ca in gastric juice.
136. Bernstein, R. E., *S. African. J. Med. Sci.*, **18**, 99 (1953). *CA*, **48**, 5914. Preparative factors in f.p. d. K in serum, plasma.
137. Bernstein, R. E., *Am. J. Clin. Path.*, **23**, 933 (1953). *CA*, **47**, 12488. F.s. d. Na, K in serum, plasma, cerebrospinal fluid.
138. Bernstein, R. E., *Proc. Colloquium Spectroscopicum Internationale VI* (*Spectrochim. Acta*, **1957**), p. 50. Intf. in f.s. d. Ca, Mg in biol. fluids, tissues.
139.* Bernstein, R. E., *S. African J. Med. Sci.*, **20**, 57 (1955). *CA*, **50**, 5820. Atomization in f.p. anal. biol. fluids.
139a. Bernstein, R. E., *S. African J. Med. Sci.*, **23**, 103 (1958). *CA*, **53**, 6333. Li as int. std. in f.p. anal. biol. materials.
140. Berry, J. W., D. G. Chappell, and R. B. Barnes, *Ind. Eng. Chem., Anal. Ed.*, **18**, 19 (1946). *CA*, **40**, 2083. New int.-std. f.p. for Na, K.
140a. Beukelman, T. E., and S. S. Lord, Jr., *Appl. Spectroscopy*, **14**, 12 (1960). Std.-addition in f.s.

141. Biffen, F. M., *Anal. Chem.*, **22**, 1014 (1950). *CA*, **44**, 11051. Int.-std. f.p. d. Na, K in refractories.
142. Billings, D., *Glass Ind.*, **36**, 255 (1955). *CA*, **49**, 15196. F.p. d. Ca in glass.
143.* Bills, C. E., F. G. McDonald, W. Niedermeier, and M. C. Schwartz, *Anal. Chem.*, **21**, 1076 (1949). *CA*, **43**, 8943. Improved Perkin-Elmer f.p. for d. Na, K; intf.
144. Black, I. A., and E. Smith, *J. Sci. Food Agr.*, **1**, 201 (1950). *CA*, **45**, 795. Flame spectrography for Na, K, Sr, Mn in soils.
145. Bode, H., and H. Fabian, *Z. anal. Chem.*, **162**, 328 (1958). *CA*, **53**, 968. F.s. d. Cu by solvent extn.
146. Bode, H., and H. Fabian, *Z. anal. Chem.*, **163**, 187 (1958). *CA*, **53**, 1992. F.s. d. Cu.
146a. Bode, H., and H. Fabian, *Z. anal. Chem.*, **170**, 387 (1959). *CA*, **54**, 4260. F.s. d. Ga, In, Tl.
147. Boers, A. L., C. T. J. Alkemade, and J. A. Smit, *Physica*, **22**, 358 (1956). *CA*, **52**, 9752. Resonance fluorescence of Na in flame.
148. Boeschoten, F., J. M. W. Milatz, and C. Smit, *Physica*, **20**, 139 (1954). *Sci. Abstr.* **57A**, No. 7446. Photomultipliers for low light intensities.
149. Bohnstedt, R. M., R. Herrmann, R. Baumann, and H. Füller, *Ärztl. Forsch.*, **7**, 82 (1953). F.p. d. K, Ca in serum.
149a. Bolle-Jones, E. W., V. R. Mallikarjuneswara, and K. Ratnasingam, *J. Rubber Research Inst. Malaya, Commun.* **15**, 86 (1957). *CA*, **52**, 7401. F.p. d. K, Ca in rubber leaves.
149b. Bolt, W., H. v. Mallinckrodt, H. Valentin, and H. Venrath, *Z. ges. exptl. Med.*, **126**, 526 (1956). F.p. d. Na, K, Ca in serum; criticism of errors.
149c. Bond, R. D., and H. C. T. Stace, *Analyst*, **83**, 679 (1958). *CA*, **53**, 8916. Intf. filters for f.p.
149d. Bond, R. D., and J. T. Hutton, *Analyst*, **83**, 684 (1958). *CA*, **53**, 11106. H_2SO_4 to suppress Ca intf. in f.p. d. Na.
150. Bosch, H., *Tonind. Ztg. u. Keram. Rundschau*, **81**, 7 (1957). *CA*, **51**, 7936. F.p. and complexometric d. Ca, Mg in silicates.
151. Bovay, E., and A. Cossy, *Mitt. Gebiete Lebensm. u. Hyg.*, **46**, 540 (1955). *CA*, **50**, 8084. F.s. d. K, Ca, Mg, Na in grain, seeds, fertilizer, soil.
152. Boycks, E. C., *Instrumental Studies in Flame Spectrophotometry*, Ph.D. Thesis, Univ. Wisconsin, 1955. *Dissertation Abstr.*, **15**, 2395. Modifications of Beckman f.s. Spectra of various metals.
153. Boyle, A. J., T. Whitehead, E. J. Bird, T. M. Batchelor, L. T. Iseri, S. D. Jacobson, and G. B. Myers, *J. Lab. Clin. Med.*, **34**, 625 (1949). *CA*, **43**, 7071. Rotating-disk spectrographic d. Na, K, Ca, Mg, Fe in plasma, urine.
154. Brabson, J. A., and W. D. Wilhide, *Anal. Chem.*, **26**, 1060 (1954). *CA*, **48**, 10480. Sr int. std. for f.p.d. Ca in H_3PO_4.
155. Brabson, J. A., and W. D. Wilhide, *Anal. Chem.*, **26**, 1663 (1954). *CA*, **49**, 1472. Fe intf. in f.p. d. Ca in magnesite, H_3PO_4.
156. Braicovich, L., and M. F. Landi, p. 51 in *Proc. Colloquium Spectroscopicum Internationale VI*, Pergamon, London, 1957. *CA*, **54**, 152. F.s. d. Na in alumina.
157. Brealey, L., and R. E. Ross, *Analyst*, **76**, 334 (1951). *CA*, **45**, 7874. A new f.p. for Na, K, and a new photomultiplier f.s.
158. Brealey, L., *Analyst*, **76**, 340 (1951). *CA*, **45**, 7738. F.p. d. K in fertilizers.

159. Brealey, L., D. C. Garratt, and K. A. Proctor, *J. Pharm. and Pharmacol.*, **4**, 717 (1952). *CA*, **47**, 1333. F.p. in pharmaceutical anal.
160. Brech, F., Pittsburgh Conference on Anal. Chem. and Appl. Spectroscopy, Feb. 1956. Abstr. in *Spectrochim. Acta*, **8**, 125 (1956). Multi-channel f.s.
161. Brewster, D. A., and C. J. Clausen, Jr., *Iron Age*, **166**, No. 18, p. 88 (1950). *CA*, **45**, 493. Int.-std. f.s. d. Na in Al.
162. Bricker, C. E., W. A. Dippel and N. H. Furman, NYO-794 (1951). *Nuclear Sci. Abstr.*, **6**, 212 (1951). F.s. d. B; spectrum, intf.
163. Brite, D. W., *Anal. Chem.*, **27**, 1815 (1955). *CA*, **50**, 2363. F.s. d. organic P.
164.* Broderick, E. J., and P. G. Zack, *Anal. Chem.*, **23**, 1455 (1951). *CA*, **46**, 697. F.s. d. Li, Na, K in glass.
164a. Brody, J. K., *J. Opt. Soc. Am.*, **42**, 408 (1952). *CA*, **46**, 7925. F.s d. Li with grating and photomultiplier.
165. Brown, J. G., O. Lilleland and R. K. Jackson, *Proc. Am. Soc. Hort. Sci.*, **56**, 12 (1950). *CA*, **46**, 1391. F.s. d. K, Ca, Mg, Na in plants.
166. Brumbaugh, R. J., and W. E. Fanus, *Anal. Chem.*, **26**, 463 (1954). *CA*, **48**, 6906. F.p. d. Li. in spodumene.
167. Bryan, H. A., and J. A. Dean, *Anal. Chem.*, **29**, 1289 (1957). *CA*, **51**, 17573. F.s. d. Cr via solvent extn.
168. Buell, B. E., *Anal. Chem.*, **30**, 1514 (1958). *CA*, **52**, 19693. F.s. d. B in org. compounds.
169. Bulewicz, E. M., C. G. James, and T. M. Sugden, *Proc. Roy. Soc. (London)*, **A 235**, 89 (1956). *CA*, **50**, 12609. Equilibria with Na, Li in air–H_2 flame.
170. Bulewicz, E. M., *Nature*, **177**, 670 (1956). *CA*, **50**, 11108. Red flame bands of Ba.
171. Bulewicz, E. M., and T. M. Sugden, *Proc. Colloquium Spectroscopicum Internationale VI (Spectrochim. Acta*, 1957), p. 20. CuH, CuOH, H, O, OH concns. in H_2 flames.
172. Bunge, W., and R. Neuber, *Chem. Tech. (Berlin)*, **8**, 733 (1956). *CA*, **51**, 9399. Calcn. of results in routine f.p. anal.
173. Burriel-Martí, F., and J. Ramírez-Muñoz, *Mikrochemie ver. Mikrochim. Acta*, **1951**, 495. *CA*, **45**, 5054. Concn. of trace elements for spectrochem. anal.
174. Burriel-Martí, F., J. Ramírez-Muñoz, and M. C. Asunción-Omarrementería, *Mikrochim. Acta*, **1956**, 362. *CA*, **50**, 8372. F.s. d. Co, Cr, Ni, Mn.
175. Burriel-Martí, F., 15th Intern. Congr. Pure and Appl. Chem., Lisbon, 1956; *Experientia*, Suppl. No. 5, p. 71 (1956). *CA*, **51**, 7217. Current problems in trace anal.
176. Burriel-Martí, F., J. Ramírez-Muñoz, and M. C. Asunción-Omarrementería, *Anal. Chim. Acta*, **17**, 545 (1957). *CA*, **53**, 2934. Intf. among Cr, Co, Mn; anal. of ferrous alloys.
177. Burriel-Martí, F., J. Ramírez-Muñoz, and M. L. Rexach-M. de Lizarduy, *Anal. Chim. Acta*, **17**, 559 (1957). *CA*, **53**, 2934. Indirect f.p. d. sulfate.
178. Cadle, R. D., and P. L. Magill, *Ind. Eng. Chem.*, **43**, 1331 (1951). *CA*, **45**, 7830. Prepn. of solid and liquid aerosols.
179. Calker, J. van, p. 150 in H. Guyer (editor), *VIII. Colloquium Spectroscopicum Internationale* (Lucerne, 1959), Sauerländer, Aarau, 1960. Intf. alk. metals in f.p.; intf. in anal. of brass.
180. Calcote, H. F., *Combustion and Flame*, **1**, 385 (1957). *CA*, **52**, 3483. Mechanism of anomalous ionizn. in flames.

180a. Candler, C., *Spectrochim. Acta*, **8**, 262 (1956). *CA*, **51**, 5561. Linear plate characteristic for spectrographic trace anal.
181. Caraway, W. T., and H. Fanger, *Am. J. Clin. Pathol.*, **25**, 317 (1955). *CA*, **49**, 8367. Clinical ultramicro methods.
182. Caton, R. D., Jr., and R. W. Bremner, *Anal. Chem.*, **26**, 805 (1954). *CA*, **48**, 9257. Viscosity effects, intf. of org. compds. in f.p. d. Na, K, Ca; heated spray chamber.
182a. Tskhaĭ, N. S., and S. L. Mandel'shtam, *Optika i Spektroskopiya*, **7**, 141 (1959). *CA*, **54**, 8275. Concn. of Na, Sr atoms in air–C_2H_2 measured by anomalous dispersion; causes of intf.
183. Cencelj, J., *Proc. Colloquium Spectroscopicum Internationale VI* (*Spectrochim. Acta*, 1957), p. 62. *CA*, **54**, 152. F.p. d. Al by intf. Ca.
184. Chaney, A. L., and A. O. Beckman; same as 105.
185. Chaney, A. L., Application Data Sheet DU-9-B (1953) (Beckman Instruments, Inc., Fullerton, Calif.). F.p. d. Ca in biological materials.
186. Charton, M., and A. G. Gaydon, *Proc. Phys. Soc.* (*London*), **69A**, 520 (1956). *CA*, **51**, 78. Band spectra of Sr, Ba in flames.
187. Chen, P. S., Jr., and T. Y. Toribara, *Anal. Chem.*, **25**, 1642 (1953). *CA*, **48**, 2804. Ca in serum by Weichselbaum f.p.
188. Chen, P. S., Jr., and T. Y. Toribara, *Anal. Chem.*, **26**, 1967 (1954). *CA*, **49**, 4774. F.p. d. Ca in serum; oxalate pptn.
189. Chow, T. J., and T. G. Thompson, *Anal. Chem.*, **27**, 18, 910 (1955). *CA*, **49**, 5204, 11489. F.p. d. Ca, Sr in sea water, marine organisms.
190.* Christianson, G., R. Jenness, and S. T. Coulter, *Anal. Chem.*, **26**, 1923 (1954). *CA*, **49**, 4898. F.p. d. Na, K in milk.
191. Clark, R. O., and L. R. Pritchard, p. 1 in *Symposium on Flame Photometry*, Am. Soc. Testing Materials, Spec. Tech. Publ. No. 116 (1951). F.p. d. Na, K in cement (history in ASTM).
192. Close, P., W. E. Smith, and M. T. Watson, Jr., *Anal. Chem.*, **25**, 1022 (1953). *CA*, **47**, 11077. F.p. d. Mg in cement, limestone.
193. Close, P., and M. T. Watson, Jr., *J. Am. Ceram. Soc.*, **37**, 235 (1954). *CA*, **48**, 7861. F.p. d. K in glass.
194. Clouston, J. G., A. G. Gaydon, and I. I. Glass, *Proc. Roy. Soc.* (*London*), **A248**, 429 (1958). *Phys. Abs.*, **61**, No. 3952. Temp. of shock waves by Na line reversal.
195. Cobine, J. D., and D. A. Wilbur, *Electronics*, June 1951, p. 92. Electronic torch.
196. Coleman Instruments, Inc., Maywood, Ill., Bull. B-228, "Coleman Tools for Science", p. 26 (1954).
197. Collins, G. C., and H. Polkinhorne, *Analyst*, **77**, 430 (1952). *CA*, **48**, 8118 Anion intf. K, Na in f.p.
198. Conrad, A. L., and W. C. Johnson, *Anal. Chem.*, **22**, 1530 (1950). *CA*, **45**, 3153. F.p. d. Li, K, Ca, Sr, Ba additives in petroleum oils.
199. Conway, J. B., W. F. R. Smith, W. J. Liddell, and A. V. Grosse, *J. Am. Chem. Soc.*, **77**, 2026 (1955). Temp. of O_2–C_2N_2 flame.
199a. Cooley, M. L., *Cereal Chem.*, **30**, 39 (1953). *CA*, **47**, 3490. F.p. d. Ca in feeds; buffered with phosphate.
200. Cox, D. S., *Can. Paint & Varnish Mag.*, **29**, No. 10, p. 42 (1955). F.p. anal. of paint.
201. Cross, J. T., *J. Am. Water Works Assoc.*, **43**, 50 (Suppl., Mar. 1951). F.p. d. Li in water; Sr, Ca intf. suppressed by Al.

202. Curtis, G. W., H. E. Knauer, and L. E. Hunter, p. 67 in *Symposium on Flame Photometry*, Am. Soc. Testing Materials, Spec. Tech. Publ. No. 116 (1951). *CA*, **47**, 7363. Effect of org. solvents on f.p. of Ba.
203. Dallavalle, J. M., C. Orr, Jr., and H. G. Blocker, *Ind. Eng. Chem.*, **43**, 1377 (1951). *CA*, **45**, 8323. Bimodal particle-size distribution.
203a. David, D. J., *Analyst*, **83**, 655 (1958). *CA*, **53**, 8939. Absorption f.p. d. Zn, Mg, Cu, Fe in plants.
203b. Davis, S., *J. Biol. Chem.*, **216**, 643 (1955). *CA*, **50**, 1117. F.p. d. Mg in plasma; hydroxyquinoline pptn.
204. Davis, S., and J. H. Simpson, Jr., *J. Biol. Chem.*, **219**, 885 (1956). *CA*, **50**, 10846. F.p. d. bicarbonate in serum.
205. Deal, S. B., *Anal. Chem.*, **26**, 598 (1954). *CA*, **48**, 6908. F.s. d. Na, K in Zn–Cd sulfide phosphors.
206. Dean, J. A., and C. Thompson, *Anal. Chem.*, **27**, 42 (1955). F.s. d. B in O_2–C_2H_2 with MeOH.
207. Dean, J. A., *Anal. Chem.*, **27**, 1224 (1955). *CA*, **49**, 15623. F.s. d. Cu in alloys.
208. Dean, J. A., and J. H. Lady, *Anal. Chem.*, **27**, 1533 (1955). *CA*, **50**, 4706. F.s. d. Fe in alloys; solvent extn.
209. Dean, J. A., and J. C. Burger, Jr., *Anal. Chem.*, **27**, 1052 (1955). *CA*, **49**, 14563. F.s. d. Fe in siliceous materials.
210. Dean, J. A., and J. H. Lady, *Anal. Chem.*, **28**, 1887 (1956). *CA*, **51**, 3357. F.s. d. Cu in ferrous alloys; $CHCl_3$ extn.
211. Dean, J. A., and J. C. Burger, Jr., Pittsburgh Conference on Anal. Chem. and Appl. Spectroscopy, March 1957. Abstr. in *Spectrochim. Acta*, **9**, 172 (1957); *Appl. Spectroscopy*, **11**, 104 (1957). J. A. Dean, J. C. Burger, Jr., T. C. Rains, and H. E. Zittel, *Anal. Chem.*, **33**, 1722 (1961), *CA*, **56**, 4254. F.s. study of Ba.
212. Dean, J. A., and C. Cain, Jr., *Anal. Chem.*, **29**, 530 (1957). *CA*, **51**, 9411. F.s. d. Cu, Ni, Mn in Al alloys.
212a. Dean, J. A., and M. B. Carnes, Pittsburgh Conference on Anal. Chem. and Appl. Spectroscopy, Mar. 1959. Abstr. in *Spectrochim. Acta*, **1959**, 311. Org. solvents in f.p.
213. Debras, J., and I. A. Voinovitch, Colloquium Spectroscopicum Internationale VII, Liège, Sept. 1958; *Rev. universelle mines*, **15**, 408 (1959); *Anal. Abstr.*, **6**, No. 3367 (1959). EDTA and intf. Sr in O_2–C_2H_2.
214. Debras, J., and I. A. Voinovitch, *Bull. soc. franç. céram.*, **38**, 77 (1958). *CA*, **53**, 965. F.p. d. Li, Na, K in silicates.
215. Debras, J., and I. A. Voinovitch, *Compt. rend.*, **247**, 2328 (1958). *CA*, **53**, 19666. Intf. of P, S with Sr suppressed by EDTA.
216. Debras, J., and I. A. Voinovitch, *Chem. anal. (Warsaw)*, **3**, 303 (1958). *Anal. Abstr.*, **6**, No. 1998 (1959). Intf. Sr in O_2–C_2H_2; f.p. d. Sr in Si–Al minerals.
217. Debras, J., and I. A. Voinovitch, *Compt. rend.*, **248**, 77 (1959). *CA*, **53**, 12946. Intf. in f.s. d. Sr; releasing, protective agents.
218. Debras-Guédon, J., and I. A. Voinovitch, *Compt. rend.*, **248**, 3421 (1959). *CA*, **53**, 21401. F.s. d. Sr, Ca, Na, K, Li.
219. Debras-Guédon, J., and I. A. Voinovitch, p. 171 in H. Guyer (editor), *VIII. Colloquium Spectroscopicum Internationale* (Lucerne, 1959), Sauerländer, Aarau, 1960. F.s. d. Sr, Ca, K, Na, Li in silicates.

219a. Debras-Guédon, J., and I. A. Voinovitch, *Compt. rend.*, **249**, 242 (1959). *CA*, **54**, 2084. Direct f.s. d. Al in silicates.
220. Demoulin, R., and L. Léger, *Verres et réfractaires*, **7**, 331 (1953). Abstr. in *Rev. optique*, **33**, 306 (1954). F.s. d. Na in glass.
221. Denson, J. R., *J. Biol. Chem.*, **209**, 233 (1954). *CA*, **48**, 10826. F.p. d. Mg, Ca, K, Na in tissue, Ca in serum.
222. Diamond, J. J., and L. Bean, p. 28 in *Symposium on Flame Photometry*, Amer. Soc. Testing Materials, Spec. Tech. Publ. No. 116 (1951). *CA*, **47**, 7753, F.s. d. Na, K in cement.
223. Same as 222.
224. Diamond, J. J., and L. Bean, *Anal. Chem.*, **25**, 1825 (1953). *CA*, **48**, 3656. F.p. d. Na in cement; studies of Perkin-Elmer, Barclay, Beckman f.p.'s.
225. Diamond, J. J., *Anal. Chem.*, **27**, 913 (1955). *CA*, **49**, 11255. F.p. d. Sr in cement.
226. Diamond, J. J., *Anal. Chem.*, **28**, 328 (1956). *CA*, **50**, 7419. F.p. d. Mn in cement.
227. Diederichsen, J., and H. G. Wolfhard, *Proc. Roy. Soc. (London)*, **A236**, 89 (1956). *CA*, **50**, 15241. Spectrography of flames at high pressure.
228.* Dippel, W. A., *Analytical Flame Photometry*, Ph.D. Thesis, Princeton Univ., 1956. *Dissertation Abstr.*, **16**, 224 (1956); *CA*, **50**, 7649.
229. Dippel, W. A., C. E. Bricker, and N. H. Furman, *Anal. Chem.*, **26**, 553 (1954). *CA*, **48**, 6915. F.p. d. phosphate in rock by suppression of Ca.
230. Dippel, W. A., and C. E. Bricker, *Anal. Chem.*, **27**, 1484 (1955). *CA*, **49**, 15616. F.p. d. Mn by std. addition.
231. Dobner, W., *Zement-Kalk-Gips*, **10**, 382 (1957). F.p. d. Ca, Mg in slags.
232. Doiwa, A., *Welche Faktoren bestimmen die Genauigkeit der flammenphotometrischen Alkali- und Erdalkalianalysen?* Thesis, Univ. Frankfurt am Main, Feb. 1955.
233. Doiwa, A., *Das Problem der "dritten Partner" in der Flammenspektrometrie*. Dissertation, Univ. Frankfurt am Main, 1956.
234. Domange, L., and S. Longuevalle et Le Moan, *Rev. gén. sci. et Bull. philomath.*, **58**, 264. (1951). F.p. in biology and medicine.
234a. Dorche, J., M. Rollet, and C. Costet, *Ann. pharm. franç.*, **13**, 283, 288 (1955). *CA*, **49**, 16343. F.s. d. alk. metals in drugs.
235. Dreisbach, R. H., *Anal. Chem.*, **31**, 479 (1959). *CA*, **53**, 12759. Recording, scanning f.s. with photomultipliers.
235a. Dryer, R. L., *Clin. Chem.*, **2**, 112 (1956). *CA*, **50**, 13160. Semimicro f.p. d. Na, K in serum.
236. Dubbs, C. A., *Anal. Chem.*, **24**, 1654 (1952). *CA*, **47**, 11817. Hot spray chamber on Perkin-Elmer f.p.
237. Dulce, H. J., *Z. physiol. Chem.*, **302**, 102 (1955). *CA*, **50**, 16953. F.p. d. Ca, Mg in urine.
238. Dutina, D., *Anal. Chem.*, **30**, 2006 (1958). *Anal. Abstr.*, **6**, No. 2501. F.p. d. Li for estn. of B-10 burnup.
238a. Eckhard, S., and A. Püschel, *Z. anal. Chem.*, **172**, 334 (1960). *CA*, **54**, 15070. Study of Cr, Mn, Fe in O_2–H_2; height, vol., temp.
239. Edgcombe, L. J., and D. R. Hewett, *Analyst*, **79**, 755 (1954). *CA*, **49**, 3507. F.p. d. Ca in coal, coke ash.
240. Edgerton, J. H., H. G. Davis, L. C. Henley, and M. T. Kelley, *Anal. Chem.*, **28**, 557 (1956). *CA*, **50**, 9069. Filter for radioactive exhaust from f.s.

241. Eggertsen, F. T., G. Wyld, and L. Lykken, p. 52 in *Symposium on Flame Photometry*, Amer. Soc. Testing Materials, Spec. Tech. Publ. No. 116 (1951). *CA*, **47**, 7367. Acid, salt intf. in f.p. d. Na, K.
242. Ehrsam, H., *Zur Kenntnis der Flammenphotometrie von Alkalisalzen*, Dissertation, Zürich, 1951; Akademische Verlagsanstalt, Graz.
243. Eichhoff, H. J., and E. Mainka, *Mikrochim. Acta*, **1955**, 299. *CA*, **49**, 11486. Accuracy of self-standardization.
244. Anonymous, *Chem. Age (London)*, **62**, 857 (1950). F.p. d. Na, K in cement.
245. Elektronest, Ing. Anton Nest, Saarbrücken-Ottenhausen, Germany (also Electronest, or Sté. Électronest, S.A.R.L.). Instructions for Schuhknecht f.p., Model STA 58.
246. Elert, B. T., *Am. J. Med. Technol.*, **21**, 297 (1955). F.p. d. Ca in serum.
247. Ellestad, R. B., and E. Horstman, *Anal. Chem.*, **27**, 1229 (1955). *CA*, **49**, 15616. F.p. d. Li in silicate rocks.
248. Elliott, H. C., Jr., and H. L. Holley, *Am. J. Clin. Pathol.*, **21**, 831 (1951). *CA*, **45**, 10345. F.s. d. Na, K in serum.
249. Elliott, F. H., *Can. J. Technol.*, **29**, 111 (1951). *CA*, **45**, 6681. Prep. and intf. in f.p. anal. biol. materials.
250. Engstrom, R. W., *J. Opt. Soc. Am.*, **37**, 420 (1947). *CA*, **41**, 5795. Study of photomultipliers; low light levels.
251. Erdey, L., and G. Svehla, *Z. anal. Chem.*, **154**, 406 (1957). *CA*, **51**, 12739. F.p. titration of Ca.
252. Eshelman, H. C., J. A. Dean, O. Menis, and T. C. Rains, *Anal. Chem.*, **31**, 183 (1959). *CA*, **53**, 11101. F.s. d. Al by solvent extn.
253. Ettre, K., and J. Ádám, *Z. anal. Chem.*, **155**, 105 (1957). *CA*, **51**, 14472. F.p. d. and intf. Ba, Ca, Sr in cathodes.
254. Evans Electroselenium Ltd., Halstead, Essex; *J. Sci. Instr.*, **29**, 381 (1952); 30, 216 (1953). EEL f.p.
255. Exley, D., and D. Sproat, *J. Sci. Instr.*, **35**, 202 (1958). *Anal. Abstr.* **6**, No. 1583. Ultramicro integrating f.s.; data for Mg, etc.
255a. Fabrikova, E. A., *Zhur. Anal. Khim.*, **14**, 41 (1959). *CA*, **53**, 9893. F.p. d. Cs in minerals.
256. Fieldes, M., P. J. T. King, J. P. Richardson and L. D. Swindale, *Soil Sci.*, **72**, 219 (1951). *CA*, **47**, 10162. F.s. d. Ca, Mg, Na, K in soils.
257. Filcek, M., *Ärztl. Laboratorium*, **4**, 118 (1958). F.p. d. Ca in serum.
258. Filcek, M., *Z. Pflanzenernähr., Düng. u. Bodenk.*, **85**, 112 (1959), *CA*, **55**, 16277. Avoidance of P, S intf. in f.p. d. Ca.
259. Fink, A., *Mikrochim. Acta*, **1955**, 314. *CA*, **49**, 11486. Alc. solns. in f.p. d. Ca, Mg.
260. Fischer, J., *Glastech. Ber.*, **22**, 390 (1949). Anal. glass.
261. Fischer, J., and H. Zettler, *Chem.-Ing.-Tech.*, **24**, 146 (1952). *CA*, **46**, 4861. Study of Riehm–Lange f.p.
262. Fischer, J., and H. Zettler, *Angew. Chem.*, **65**, 569 (1953). Contributions to f.p. (talk).
263. Fischer, J., and A. Doiwa, *Proc. Colloquium Spectroscopicum Internationale VI* (*Spectrochim. Acta*, **1957**), p. 28. *CA*, **54**, 152. Anion, pH, dissocn., ionizn., compd. formation intf. in f.p. of Li.
264. Fischer, J., and A. Doiwa, *Mikrochim. Acta*, **1956**, 353. *CA*, **50**, 8371. Intf. in f.s.

264a. Fisher, A. M., and A. I. Finkel'shteĭn, *Zavodskaya Lab.*, **23**, 788 (1957). F.p. d. K in Na salts.
265. Foote, P. D., *J. Opt. Soc. Am.*, **42**, 886 (1952). *CA*, **47**, 2466. Emission spectroscopy in petroleum anal.
266. Ford, C. L., *Anal. Chem.*, **26**, 1578 (1954). *CA*, **49**, 2254. F.p. d. Na, K in cement, glass, clay, etc.
266a. Ford, C. L., *ASTM Bull.*, No. **233**, 57 (1958). *CA*, **53**, 5627. F.p. d. Mn, Na, K in cement.
267. Ford, O. W., *J. Assoc. Offic. Agr. Chemists*, **33**, 268 (1950); **34**, 660 (1951); **35**, 674 (1952); **36**, 649 (1953); **37**, 363 (1954); **38**, 445 (1955); **39**, 598 (1956); **40**, 722 (1957). *CA*, **44**, 8582; **45**, 10460; **47**, 2918; **48**, 9001; **49**, 10566; **50**; 13352; **51**, 13296. Reports on collaborative studies of K in fertilizers, including f.p. d.
268. Fornwalt, D. E., *Anal. Chim. Acta*, **17**, 597 (1957). *CA*, **53**, 2873. F.s. d. Ni, B in plating solns.
268a. Foster, W. H., Jr., and D. N. Hume, *Anal. Chem.*, **31**, 2028 (1959). *CA*, **54**, 6392. Study of atomizer-burners: intensity vs. flow rate, etc.
268b. Foster, W. H., Jr., and D. N. Hume, *Anal. Chem.*, **31**, 2033 (1959). *CA*, **54**, 6392. Ionizn. intf. alk. metals, Ca, Sr, Ba; Rb, Cs as int. std.
269. Fox, C. L., Jr., E. B. Freeman, and S. E. Lasker, p. 13 in *Symposium on Flame Photometry*, Amer. Soc. Testing Materials, Spec. Tech. Publ. No. **116** (1951). *CA*, **47**, 7264. Int.-std. f.p. for d. Na, K, Li, Ca in biol. fluids; intf.
270. Fox, C. L., Jr., *Proc. XII Intern. Congr. Pure Appl. Chem.*, **1951**, 39. Int.-std. f.p. for Na, K, Li in biol. fluids.
271. Fox. C. L., Jr., *Anal. Chem.*, **23**, 137 (1951). *CA*, **45**, 3661. Design of Janke int.-std. f.p.; d. Na, K in urine, plasma.
272. Frankenberg, B., V. Hospadaruk, and A. H. Neufeld, *Can. Med. Assoc. J.*, **65**, 388 (1951). *CA*, **46**, 2609. F.s. d. Na, K in blood, urine.
273.* French, J. R. U.S. Atomic Energy Comm., *ORNL-2001* (1955). *Nuclear Sci. Abstr.*, **10**, No. 7500. Bibliography on f.p. d. alk., alk.-earth metals.
273a. Freytag, H. E., *Z. anal. Chem.*, **136**, 161 (1952). *CA*, **46**, 9465. F.p. d. Rb in physiol. studies.
274. Fukushima, S., M. Shigemoto, I. Kato, and K. Otozai, *Mikrochim. Acta*, **1957**, 35. *CA*, **51**, 11157. Multiplicative intf. in f.s.; log-log translation rule.
275. Fukushima, S., K. Takahashi, S. Terasaka, and K. Otozai, *Mikrochim. Acta*, **1957**, 183. *CA*, **51**, 12734. Std.-addition method in f.s.
276. Fukushima, S., K. Yukawa, M. Shigemoto, and K. Otozai, *Mikrochim. Acta*, **1958**, 553. *CA*, **55**, 16261. Std.-addition and double-coef. method in f.s.
276a. Fukushima, S., *Mikrochim. Acta*, **1959**, 596. *Anal. Abstr.*, **7**, No. 37 (1960). Intf. of Al, B, P, S on Ca: study with dual atomizers.
277. Fuwa, K., *J. Chem. Soc. Japan, Pure Chem. Sect.*, **75**, 1257 (1954). *CA*, **49**, 13561. Spectroscopic (not f.p.) d. F by CaF bands.
278. Fuwa, K., *J. Chem. Soc. Japan, Pure Chem. Sect.*, **76**, 14 (1955). *CA*, **49**, 13828. Spectroscopic d. F by BaF bands.
278a. Fuwa, K., R. E. Thiers, and B. L. Vallee, *Anal. Chem.*, **31**, 1419 (1959). *CA*, **53**, 19480. O_2–C_2N_2 burner for f.s.
278b. Fuwa, K., R. E. Thiers, B. L. Vallee, and M. R. Baker, *Anal. Chem.*, **31**, 2039 (1959). *CA*, **54**, 6387. Intensity vs. sample flow rate in O_2–C_2N_2, O_2–H_2.

279. Gabsch, H. C. *Ärztl. Laboratorium*, **2**, 210 (1956). F.p. d. Na, K, Ca, Mg in biol. materials.
280. Gage, J. C., *J. Sci. Instr.*, **30**, 25 (1953). *CA*, **47**, 3621. Controlled-feed atomizer for bacteriology.
280a. Galloway, N. McN., *Analyst*, **83**, 373 (1958). *CA*, **52**, 18085. F.p. d. Ag in blister Cu.
281. Gammon, N., Jr., *Soil Sci.*, **71**, 211 (1951). *CA*, **46**, 675. F.p. d. K, Na in sandy soils.
282. García-Llauradó, J., *J. Clin. Pathol.*, **7**, 110 (1954). F.p. d. Ca in biol. materials.
283. Gaydon, A. G., *Nature*, **165**, 170 (1950). *Sci. Abstr.*, **53A**, No. 3043. Excitation mechanism in flames; abnormal excitation of Pb in oxyacetylene.
284. Gaydon, A. G., p. 1 in *Energy Transfer in Hot Gases*, Natl. Bur. Standards Circ. 523 (1954).
285. Gaydon, A. G., *Proc. Roy. Soc. (London)*, **A231**, 437 (1955). *CA*, **50**, 674. Band spectra of CaOH, CaOD, CaO.
286. Gaydon, A. G., *Mém. soc. roy. sci. Liège*, **18**, 507 (1957). *CA*, **51**, 16093. Oxide and hydroxide bands of Ca, Sr, Ba, Mg.
287. Geffcken, W., *Z. angew. Phys.*, **6**, 249 (1954). *Sci. Abstr.*, **57A**, No. 8211. High-efficiency interference band filter.
288. Gehrke, C. W., H. E. Affsprung, and E. L. Wood, *J. Agr. Food Chem.*, **3**, 48 (1955). *CA*, **49**, 3724. F.p. d. K; ion-exchange sepn. of interferents.
289.* Gehrke, C. W., and E. L. Wood, *Missouri, Univ. Agr. Expt. Sta., Research Bull.* No. 635 (1957). *CA*, **52**, 2319. F.s. d. K in fertilizers, using ion exchange.
290. Georgii, H. W., *Z. Aerosol-Forsch. u. Therap.*, **3**, 496 (1954). Evapn. rate of small drops.
291. Gerber, C. R., N. H. Ishler, and E. Borker, Pittsburgh Conference on Anal. Chem. and Appl. Spectroscopy, Mar. 1951. Abstr. in *Anal. Chem.*, **23**, 684 (1951). F.s. d. Fe, Mn, Cu, Co in mineralized salts.
292. Gergely, G., and P. F. Váradi, *Acta Phys. Acad. Sci. Hung.*, **5**, 51 (1955). *CA*, **50**, 7510. A.c. photomultiplier f.s. with O_2–H_2 atomizer-burner; d. Ca, Sr, Ba in cathodes, K in fluors.
293. Gettkandt, G., *Z. Pflanzenernähr., Düng. u. Bodenk.*, **74**, 135 (1956); **78**, 187 (1957). F.p. d. Ca in plants.
294. Gettkandt, G., *Landwirtsch. Forsch.*, **11**, 93 (1958). *CA*, **52**, 20821. F.p. d. Ca in soil extracts.
295. Giesen, K., and P. Kampa, *Forschungsbericht des Wirtschafts- und Verkehrsministeriums Nordrhein-Westfalen*, **59**, 29 (1954); *Tonind.–Ztg. u. Keram. Rundschau*, **77**, 383, (1953). F.p. d. Na, K in refractories.
296. Gilbert, P. T., Jr., Reprint R-26 (1948), Beckman Instruments, Inc., Fullerton, Calif. F.s. d. Na, Ca.
297. Gilbert, P. T., Jr., *Scalacs*, **4**, 263 (1949). Review on f.s.
298. Gilbert, P. T., Jr., R. C. Hawes, and A. O. Beckman, *Anal. Chem.*, **22**, 772 (1950). *CA*, **44**, 7592. Model 10300 Beckman f.s.
299. Gilbert, P. T., Jr., Pittsburgh Conference on Anal. Chem. and Appl. Spectroscopy, Mar. 1951; Symposium on Molecular Structure and Spectroscopy, Columbus, Ohio, June 1951. Abstracts in *Anal. Chem.*, **23**, 684, 1050 (1951). Atomizer-burner for f.s.

300. Gilbert, P. T., Jr., p. 77 in *Symposium on Flame Photometry*, Amer. Soc. Testing Materials, Spec. Tech. Publ. No. 116 (1951). *CA*, **47**, 7766. F.s. d. Pb in gasoline.
301. Gilbert, P. T., Jr., *Science*, **114**, 637 (1951). *CA*, **46**, 7373. Silicone water-repellents.
302. Gilbert, P. T., Jr., *Industrial Labs.*, **3**, No. 8, p. 41 (Aug. 1952). Review on f.s.; table of detection limits.
303. Gilbert, P. T., Jr., *Chemist-Analyst*, **41**, 52 (1952). *CA*, **46**, 11010. Silicone water-repellents.
304. Gilbert, P. T., Jr., Pittsburgh Conference on Anal. Chem. and Appl. Spectroscopy, Mar. 1958. O_2–C_2N_2 f.s.
305. Gilbert, P. T., Jr., *Spectrochim. Acta*, **12**, 397 (1958). *CA*, **53**, 5013. F.s. d. In.
306. Gilbert, P. T., Jr., *Anal. Chem.*, **31**, 110 (1959). *CA*, **53**, 5966. F.s. d. Cd.
307. Gilbert, P. T., Jr., Bulletin 753 (1959), Bulletin 753A (1961), Beckman Instruments, Inc., Fullerton, Calif. Flame spectra of the elements: wavelengths and intensities.
307a.* Gilbert, P. T., Jr., p. 73 in *Symposium on Spectroscopy*, Amer. Soc. Testing Materials, Spec. Tech. Publ. No. 269 (1960). New developments in f.p.; spectrograms, tables.
307b. Gilbert, P. T., Jr., *Analyzer* (Beckman Instruments, Inc.), **1**, No. 1, p. 3 (Jan. 1960). F.s. of rarer elements.
308. Gille, F., *Zement-Kalk-Gips*, **5**, 208 (1952). F.p. d. Na, K in cement, raw materials.
309. Gilliland, J. L., p. 33 in *Symposium on Flame Photometry*, Amer. Soc. Testing Materials, Spec. Tech. Publ. No. 116 (1951). *CA*, **47**, 7946. F.p. d. Na, K in silicates, water, metals, cement, soil, etc.
310. Same as 309.
311. Gleason, S. I., and G. Hold, *Beckman Bulletin* (Beckman Instruments, Inc.), No. 16, p. 4 (1955). F.s. anal. lubricating oil.
312. Glendening, B. L., D. B. Parrish, and W. G. Schrenk, *Anal. Chem.*, **27**, 1554 (1955). *CA*, **50**, 4278. Flame spectrographic d. Rb in plant and animal tissues; Li int. std.
313. Glick, D., R. H. Swigart, S. N. Nayyar, and H. R. Stecklein, *J. Histochem. Cytochem.*, **3**, 6 (1955). *CA*, **49**, 5619. F.p. d. K in animal tissues.
314. Goldstein, S. W., and D. P. Sanders, *Drug Standards*, **22**, 137 (1954). *CA*, **48**, 13161. F.p. d. Mg, Ca, K, Na in drugs.
314a. Gordon, G. D., *J. Appl. Phys.*, **30**, 1759 (1959). Mechanism of breakup of drops.
314b. Grove, E. L., C. W. Scott, and F. Jones, Pittsburgh Conference on Anal. Chem. and Appl. Spectroscopy, Mar. 1959. Abstract in *Spectrochim. Acta*, **1959**, 311. Acid and mutual intf. among alk. metals in f.s.
314c. Gundlach, H., *Z. anal. Chem.*, **171**, 9 (1959). *CA*, **54**, 3078. F.p. d. Sr in minerals; org. solvents.
315. Gurvich, I. G., and E. I. Khanaev, *Izvest. Akad. Nauk S.S.S.R., Ser. Geol.*, **1956**, No. 6, p. 101. *CA*, **51**, 6433. F.p. d. K in rocks.
315a. Häussler, A., and P. Hajdú, *Arch. Pharm.*, **292**, No. 5, p. 73 (1959). *CA*, **53**, 20223. F.p. d. Na, K, Ca in serum, urine.
316. Halperin, A., and S. Sambursky, *J. Opt. Soc. Am.*, **42**, 475 (1952). *CA*, **46**, 8569. Rb detn. by self-standarization and int. std. in arc; self-absorption.

317. Halstead, W. J., and B. Chaitken, *Public Roads*, **26**, No. 5, p. 99 (1950). *CA*, **45**, 796. F.p. d. Na, K in soil, rock, cement, etc.
318. Hansen, G., *Optik*, **1**, 227, 269 (1946). Principles of photoelec. optical instruments.
319. Hansen, G., *Optik*, **6**, 337 (1950). *Sci. Abstr.*, **54A**, No. 935. Imaging of volume radiators in spectrographs.
320. Hansen, G., *Mikrochim. Acta*, **1955**, 708. *CA*, **49**, 11487. Comparison of eye, photographic film and elec. photodetectors in spectrophotometry.
320a. Harris, E. J., *J. Sci. Instr.*, **36**, 369 (1959). *Anal. Abstr.*, **7**, No. 2570 (1960). Red-sensitive photomultiplier for f.p. d. Li, K, Rb.
321. Harrison, G. E., *Nature*, **182**, 792 (1958). *CA*, **53**, 4399. F.s. d. Sr in biol. materials.
322. Hourigan, H. F., and J. W. Robinson, *Anal. Chim. Acta*, **16**, 161 (1957). *CA*, **51**, 7233. F.p. d. Na in Al–Cu alloys.
323. van der Have, A. J., and H. Mulder, *Neth. Milk Dairy J.*, **11**, 128 (1957). *CA*, **52**, 3188. F.p. d. Na, K, Ca in milk, cheese.
324. Hegedüs, A. J., F. K. Fukker, and M. Dvorszky, *Magyar Kém. Folyóirat*, **59**, 334 (1954). F.p. d. Na in alumina.
325. Hegedüs, A. J., T. Millner, and E. Pungor, *Magyar Kém. Folyóirat*, **59**, 304 (1953). F.p. d. Ca, Sr, Ba in cathodes.
326. Hegedüs, A. J., and M. Dvorszky, *Mikrochim. Acta*, **1959**, 160. *Anal. Abstr.*, **6**, No. 3396. F.p. d. Na, K, Li in glass.
327. Hegedüs, A. J., J. Neugebauer, and M. Dvorszky, *Mikrochim. Acta*, **1959**, 282. *Anal. Abstr.*, **6**, No. 4378. F.p. d. Na, K, Ca in W and W oxide.
327a.* Hegedüs, A. J., *Mikrochim. Acta*, **1959**, 735. *Anal. Abstr.*, **7**, No. 857. Review on f.s.
328. Hegemann, F., and B. Pfab, *Glastech. Ber.*, **26**, 238 (1953). *CA*, **47**, 11073. F.p. d. Na with Zeiss f.p.
329. Hegemann, F., and B. Pfab, *Glastech. Ber.*, **27**, 189 (1954). *CA*, **49**, 8036. Intf. of Ca in f.p. d. Na.
330. Hegemann, F., V. Caimann and H. Zoellner, *Ber. deut. keram. Ges.*, **31**, 315 (1954). Detn. Na with Zeiss f.p.
331. Hegemann, F., and B. Pfab, *Glastech. Ber.*, **28**, 85 (1955). *CA*, **49**, 7822. F.p. d. Na, K in glass.
332. Hegemann, F., H. Kostyra and B. Pfab, *Glastech. Ber.*, **30**, 14 (1957). *CA*, **51**, 4873. F.s. d. Na, K, with Ba buffer.
333. Hegemann, F., W. Hert, and W. Schmidt, *Glastech. Ber.*, **31**, 81 (1958). *CA*, **52**, 10523. F.s. d. Al in glass.
333a. Hegemann, F., and L. Süss, *Glastech. Ber.*, **31**, 185 (1958). *CA*, **52**, 13209. F.p. d. Na, Ca in Al glasses.
334. Hegemann, F., and W. Hert, *Ber. deut. keram. Ges.*, **35**, 258 (1958). *CA*, **54**, 1816. F.s. d. K, Na, Ca, Al in kaolin.
335. Hegemann, F., W. Schmidt, and W. Hert, *Glastech. Ber.*, **32**, 15 (1959). *CA*, **53**, 8561. Intf. in f.s. d. Na, K in glass.
336. Heggen, G. E., and L. W. Strock, *Anal. Chem.*, **25**, 859 (1953). *CA*, **47**, 7936. A method of concn. and prepn. of all types of sample for arc spectrography.
337. Hemingway, R. G., *Analyst*, **81**, 164 (1956). *CA*, **50**, 7652. Phosphate intf. in f.p. d. Ca in plants.
337a. Henly, A. A., and R. A. Saunders, *Analyst*, **83**, 584 (1958). *CA*, **53**, 5965. EDTA d. Ca in biol. materials.

338. Henning, F., and C. Tingwaldt, *Z. Physik*, **48**, 805 (1928). *CA*, **23**, 1350. Temp. of O_2–C_2H_2 flame.
339. Herrmann, R., *Z. ges. exptl. Med.*, **118**, 187 (1952). *CA*, **47**, 1223. Filter f.p. for simultaneous detn. Na, K, Ca in serum.
340. Herrmann, R., and R. Baumann, *Z. ges. exptl. Med.*, **119**, 487 (1952). *CA*, **47**, 7020. Micro f.p. d. Na, K, Ca.
341. Herrmann, R., and H. Schellhorn, *Z. angew. Phys.*, **4**, 208 (1952). *Sci. Abstr.*, **55A**, No. 8537. Concentric atomizer for f.p.
342. Herrmann, R., *Z. ges. exptl. Med.*, **122**, 84 (1953). *CA*, **48**, 8858. Ultramicro f.p. d. Na, K, Ca in serum.
343. Herrmann, R., *Z. Aerosol-Forsch. u. Therap.*, **3**, 16 (1954). Atomization in f.p.
343a. Herrmann, R., *Optik*, **11**, 505 (1954). *Sci. Abstr.*, **58A**, No. 2558. Vibrating slit for oscillographic presentation of spectra.
344. Herrmann, R., *Optik*, **12**, 189 (1955). *Sci. Abstr.*, **58A**, No. 7487. Multichannel int.-std. f.p. for Na, K, Ca, Li.
345. Herrmann, R., *Z. ges. exptl. Med.*, **126**, 334 (1955). F.s. d. Cu in serum.
346. Herrmann, R., *Z. ges. exptl. Med.*, **126**, 371 (1955). F.s. d. Mg in serum.
347. Herrmann, R., and H. Schellhorn, *Z. angew. Phys.*, **7**, 572 (1955). *CA*, **50**, 5334. New atomizer-burner for f.s.; motor-driven syringe for sample pumping.
348. Herrmann, R., *Ärztl. Laboratorium*, **1956**, 229. F.p. in the clinical lab.
349. Herrmann, R., *Z. ges. exptl. Med.*, **129**, 55 (1957). *CA*, **54**, 12895. F.s. d. Na at 330 mμ in serum, etc.
350. Herrmann, R., and W. Rick, *Naturwissenschaften*, **46**, 492 (1959). *CA*, **54**, 4731. EDTA in f.s. d. Ca in serum.
350a. Herrmann, R., and W. Rick, Second Czechoslovakian Spectrographic Congress, Tatra-Lomnitz, Oct. 1959. F.p. anal. serum.
351. Hershenson, H., and D. F. Smith, *J. Am. Pharm. Assoc.*, **44**, 731 (1955). *CA*, **50**, 3712. F.p. d. Mg, Ca, K, Na in parenterals.
352. Hilger and Watts, Ltd., Camden Road, London, N.W.1. Catalog CH 318. Uvispek spectrophotometer, with flame source.
353. *Hilger Journal* (Hilger and Watts, Ltd.), **5**, No. 1, p. 12 (Aug. 1958). Absorption f.p.
354. Hilgers, A., *Z. physiol. Chem.*, **294**, 61 (1953). *CA*, **49**, 14860. F.p. d. Na, K, Ca in serum.
355. Hilgers, A., *Z. physiol. Chem.*, **304**, 193 (1956). *CA*, **51**, 536. F.p. d. Na, K, Ca in urine.
356. Hinnov, E., *J. Opt. Soc. Am.*, **47**, 151 (1957). *CA*, **51**, 4816. Optical cross-sections detd. from f.p. working curves.
357. Hinsvark, O. N., S. H. Wittwer, and H. M. Sell, *Anal. Chem.*, **25**, 320 (1953). *CA*, **47**, 5297. F.s. d. Ca, Sr, Ba in mixtures.
358. Hinze, J. O., *Appl. Sci. Research*, **A1**, 273 (1953). Critical speeds and sizes of liquid droplets.
359. Hitchcock, R. D., and W. L. Starr, *Appl. Spectroscopy*, **8**, 5 (1954). *CA*, **48**, 6902. Flame spectrography for sea water.
360. Holiday, E. R., and J. R. K. Preedy, *Biochem. J.*, **55**, 214 (1953). *CA*, **48**, 223. New f.p.; detn. Na, K in urine, serum.
361. Hollander, T., A. J. Borgers, and C. T. J. Alkemade, *Appl. Sci. Research*, **B5**, 409 (1956). *CA*, **50**, 10598. F.p. d. Sr, Ba, Li.

362. Honma, M., and C. L. Smith, *Anal. Chem.*, **26**, 458 (1954). *CA*, **48**, 6911. F.s. d. N by CN bands in org. compds.
363. Honma, M., *Anal. Chem.*, **27**, 1656 (1955). *CA*, **50**, 4718. F.s. d. Cl in sea water via CuCl bands.
364. Honold, R., *Über die Fehlergrösse und Fehlermöglichkeiten bei der Bestimmung von Na, K, und Ca im Harn bei Routinebestimmungen mit dem Flammenphotometer.* Dissertation, Freiburg i. Br., 1953.
365. Horstman, E. L., *Anal. Chem.*, **28**, 1417 (1956). *CA*, **50**, 16545. F.s. d. Li, Rb, Cs in rocks.
366. Hospadaruk, V., B. Frankenberg, and A. H. Neufeld, *Can. Med. Assoc. J.*, **65**, 264 (1951). *CA*, **46**, 2441. Optimization of f.s.
367. Hourigan, H. F., and J. W. Robinson, *Anal. Chim. Acta*, **13**, 179 (1955). *CA*, **50**, 2360. F.p. d. Na in Al.
368. House, H. P., and J. M. Rogers, Southeastern Regional Meeting, Am. Chem. Soc., Birmingham, Ala., Oct. 1954. Abstr. in *Anal. Chem.*, **26**, 1855 (1954). F.s. d. Na, K, Rb, Cs in mixtures.
368a. Howling, H. L., and P. E. Landolt, *Anal. Chem.*, **31**, 1818 (1959). *CA*, **54**, 3073. F.p. d. Li in minerals.
369. Hübener, H. J., *Z. physiol. Chem.*, **289**, 188 (1952). *CA*, **48**, 10092. F.p. d. Ca in serum, plasma.
370. Hübener, H. J., H. Maurer, and T. Walther, *Klin. Wochschr.*, **31**, 1095 (1953). *CA*, **48**, 4621. F.p. d. Ca in serum by difference.
370a. Huldt, L., *Arkiv Fysik*, **A33**, No. 5 (1946). *CA*, **41**, 913. Intf. among Sr, Ca, Al in air–C_2H_2.
371. Huldt. L., *Eine spektroskopische Untersuchung des elektrischen Lichtbogens und der Azetylen-Luftflamme*, Thesis, Uppsala, 1948; publ. by Almqvist & Wiksells. Ionization of Ba, Sr, Ca, Cs, K, Na in air–C_2H_2.
372. Huldt, L., and E. Knall, *Z. Naturforsch.*, **9a**, 663 (1954). *CA*, **49**, 52. Pressure broadening of Na line in air–C_2H_2.
373. Huldt, L., and E. Knall, *Naturwissenschaften*, **41**, 421 (1954). *CA*, **49**, 10054. Absorption by OH in air–C_2H_2.
374. Huldt, L., and A. Lagerqvist, *Arkiv Fysik*, **9**, 227 (1955). *Sci. Abstr.*, **58A**, No. 3567. Term diagrams of CaO, SrO, BaO, from O_2–H_2 flame.
375. Huldt, L., and E. Knall, *Arkiv Fysik*, **11**, 229 (1956). *CA*, **51**, 1729. Polymerization of Sr and Ca in flames.
376. Huldt, L., and A. Lagerqvist, *Arkiv Fysik*, **11**, 347 (1956). *CA*, **51**, 7850. Origin of band spectra of Ca, Sr.
377. Hultgren, R., *J. Am. Chem. Soc.*, **54**, 2320 (1932). *CA*, **26**, 3990. Spark in air–H_2 as spectrochem. source.
378. Hunter, F. R., and G. D. Berenda, *J. Biol. Chem.*, **192**, 701 (1951). *CA*, **46**, 2119. F.p. d. Na, K in blood, erythrocytes.
379. Ikeda, S., *Nippon Kagaku Zasshi*, **76**, 354 (1955). Micro f.s. d. Na in Al. *Ibid.*, 783. Micro f. s. d. Ca, Mg. *Ibid.*, 1122. F.s. d. Mg in Al alloys. *Ibid.*, 1258. (*CA*, **51**, 12739.) F.s. d. Ca, Mg in slag. *Ibid.*, **78**, 913 (1957) (*CA*, **51**, 5628.) F.s. d. Mn.
380. Inman, W. R., R. A. Rogers, and J. A. Fournier, *Anal. Chem.*, **23**, 482 (1951). *CA*, **45**, 5566. F.s. d. Na, K in Li.
381. Ishida, R., *Nippon Kagaku Zasshi*, **73**, 35 (1952). F.s. d. Mn. *Ibid.*, **76**, 56 (1955). (*CA*, **49**, 13012.) F.s. d. Na, K. *Ibid.*, 60. (*CA*, **49**, 13012.) F.s. d. La. *Ibid.*, **77**, 238 (1956). (*CA*, **51**, 933.) Line-reversal temp. detn.

of flame. *Ibid.*, 241 (*CA*, **51**, 933.) F.s. d. F by CaF bands. *Ibid.*, 242. (*CA*, **51**, 934.) Intensities of lines in air–C_2H_2, O_2–H_2.
382.* Ishida, R., *Bunko Kenkyu*, **4**, No. 2, p. 3 (1956). *CA*, **53**, 5953. Review of f.s.
383. Ishidate, M., Y. Mashiko, and Y. Kanroji, *J. Pharm. Soc. Japan*, **75**, 1492 (1955). *CA*, **50**, 5198. F.s. d. Na in mineral water.
384. Ivanov, D. N., *Zavodskaya Lab.*, **14**, 1136 (1949). *CA*, **43**, 4595. Spark in air–C_2H_2 as spectrochem. source.
385. Ivanov. D. N., *Zhur. Anal. Khim.*, **9**, 344 (1954). *CA*, **49**, 4445. F.p. d. Ca in soils, plants.
386. Ivanov, D. N., *J. Anal. Chem. U.S.S.R.*, **9**, 383 (1954). *CA*, **49**, 9433. F.p. d. Ca.
387. Ivanov, D. N., and B. Ya. Kaplan, *Zavodskaya Lab.*, **22**, 569 (1956). *CA*, **51**, 2460. F.p. d. Li.
388. Jackson, P. J., and A. C. Smith, *J. Appl. Chem. (London)*, **6**, 547 (1956). *CA*, **51**, 6978. F.p. d. K, Na in coal ash, etc.; Ca intf. removed by Al.
389. Jacquinot, P., *J. Opt. Soc. Am.*, **44**, 761 (1954). *CA*, **49**, 727. Luminosity of prism, grating and etalon spectrometers.
390. Jacquinot, P., *J. Opt. Soc. Am.*, **45**, 996 (1955). *Sci. Abstr.*, **59A**, No. 1035. Luminosity of spectrometers.
391. James, C. G., and T. M. Sugden, *Nature*, **171**, 428 (1953). *CA*, **47**, 6743. Working curves of alk. metals in air–H_2, theory, hydroxide formation.
392. James, C. G., and T. M. Sugden, *Nature*, **175**, 252 (1955). *CA*, **49**, 7989. NO–O continuum for estn. of O atoms in flames.
393. James, C. G., and T. M. Sugden, *Nature*, **175**, 333 (1955). *CA*, **49**, 10055. Equilibria of Ca, Sr, Ba in air–H_2; MOH bands.
394.* James, C. G., and T. M. Sugden, *Proc. Roy. Soc. (London)*, **A227**, 312 (1955). *CA*, **49**, 7375. Concn. of Li, Na in air–H_2.
395. James, C. G., and T. M. Sugden, *Proc. Roy. Soc. (London)*, **A248**, 238 (1958). *CA*, **54**, 14886. Alkali continuum in air–H_2.
396. Jarrell-Ash Co., Newtonville 60, Mass. Bulletin 82020.
397. Jefferson, J. H., *Flame spectrophotometric determination of magnesium*. Ph.D. Thesis, Univ. Wisconsin, 1956. *Dissertation Abstr.*, **16**, 231 (1956).
398. Jenaer Glaswerk, Schott u. Gen., Mainz. Bulletins on glass color filters and intf. filters.
399. Jentzsch, D., and G. Jacob, *Chem. Tech. (Berlin)*, **7**, 93 (1955). *CA*, **49**, 10795. F.p. d. Li, Na, K in salts, brines.
400. Johnson, A. F., and K. A. Jurbergs, Am. Chem. Soc. Symposium on New Methods for Analytical Characterization of Cellulose, Atlantic City, Sept. 1956. F.p. d. Na in pulp.
401. Johnston, B. R., C. W. Duncan, K. Lawton, and E. J. Benne, *J. Assoc. Offic. Agr. Chemists*, **35**, 813 (1952). *CA*, **47**, 5301. F.p. d. K in plants.
402. Jordan, J. H., Jr., *Petroleum Refiner*, **32**, No. 8, p. 139 (1953). *CA*, **50**, 8186. F.s. d. Pb in gasoline.
403. Jordan, J. H., Jr., *Petroleum Refiner*, **33**, No. 3, p. 158 (1954). *CA*, **50**, 8182. F.d. Cu in gasoline.
404. Établissements Jouan, 113 Blvd. St.-Germain, Paris 6ᵉ. *Chim. anal.*, **36**, 336 (1954). The Jouan f.p. (see Table 15).
405. Judd, W. C., and L. P. Pepkowitz, Pittsburgh Conference on Anal. Chem. and Appl. Spectroscopy, Mar. 1954. Abstr. in *Anal. Chem.*, **26**, 432 (1954). F.s. d. Na in $UO_2(NO_3)_2$ soln.

406. Kafka, J., and R. Herrmann, *Ärztl. Wochschr.*, **9**, 547 (1954). *CA*, **48**, 12858. Errors in f.p. d. Na, K, Ca in serum.
407. Kaiser, H., *Spectrochim. Acta*, **3**, 297 (1948). *CA*, **43**, 2539. Effect of background on spectrochem. working curves.
408. Kaiser, H., and G. Hansen, *Spectrochim. Acta*, **3**, 433 (1948). *Sci. Abstr.*, **52A**, No. 4624. Resolution and luminosity as functions of slit width.
409. Kaiser, H., *Spectrochim. Acta*, **4**, 351 (1951). *CA*, **46**, 2347. Calculator for spectrochem. anal.
410. Kaiser, H., and H. Specker, *Z. anal. Chem.*, **149**, 46 (1956). *CA*, **50**, 5445. Statistical errors of anal.
411. Kapuscinski, V., N. Moss, B. Zak, and A. J. Boyle, *Am. J. Clin. Path.*, **22**, 687 (1952). *CA*, **46**, 8701. F.s. d. Ca, Mg in serum; wet ashing.
412. Karchmer, J. H., and E. L. Gunn, *Anal. Chem.*, **24**, 1733 (1952). *CA*, **47**, 5669. F.p. d. Na, K in petroleum oil and ash.
413. Keirs, R. J., and S. J. Speck, *J. Dairy Sci.*, **33**, 413 (1950). *CA*, **44**, 8012. F.p. d. Na, K, Ca in milk.
414. Kelley, M. T., D. J. Fisher, and H. C. Jones, *Anal. Chem.*, **31**, 178 (1959). *CA*, **53**, 10858. High-sensitivity recording f.s.
414a. Ketelaar, J. A. A., C. Haas, F. N. Hooge, and R. Broekhuijsen, *Physica*, **21**, 695 (1955). *CA*, **50**, 8323. Far infrared flame emission; OH spectrum, temp.
415. Kick, H., *Z. Pflanzenernähr., Düng. u. Bodenk.*, **67**, 53 (1954). *CA*, **49**, 1472. F.p. d. Ca, Mg, Mn in plants, soils.
416. Kick, H., *Z. Pflanzenerhähr., Düng. u. Bodenk.*, **60**, 163 (1953). *CA*, **47**, 7325. F.s. d. Mn, Ca, K, Li, Mg, Na, Sr with photomultiplier.
416a. Kick, H., *Z. anal. Chem.*, **151**, 406 (1956). *CA*, **51**, 130. F.s. d. Mn in plants; intf. from Ga, K.
416b. Kick, H., and R. Bucher, *Landwirtsch. Forsch.*, **10**, 96 (1957). F.p. d. Mg in soil, plants.
416c. Kick, H., *Z. anal. Chem.*, **163**, 252 (1958). *CA*, **53**, 5016. F.p. d. Sr in soil and plants; intf. from Ca, Ba, Mg, Y.
417. King, W. H., Jr., and W. M. Priestley, Jr., p. 97 of *Symposium on Flame Photometry*, Am. Soc. Testing Materials, Spec. Tech. Publ. No. 116 (1951). *CA*, **47**, 7264. Recording f.s. with photomultiplier; data for Pb, Fe.
418. Kingsley, G. R., and R. R. Schaffert, *Science*, **116**, 359 (1952). *CA*, **47**, 2246. F.s. d. Na, K in serum, etc.; enhancement by org. solvents; protein effects.
419. Kingsley, G. R., and R. R. Schaffert, *Anal. Chem.*, **25**, 1738 (1953). *CA*, **48**, 2805. F.s. d. Na, K, Ca in serum, etc.
420. Kingsley, G. R., and R. R. Schaffert, *J. Biol. Chem.*, **206**, 807 (1954). *CA*, **48**, 5263. Micro f.s. d. Na, K, Ca in serum; org. solvents.
421. N.V. v/h P. J. Kipp en Zonen, Voorstraat 67–75, Delft. Kipp f.p. (Vlamfo 53).
422. Klyne, W., *Spectrochim. Acta*, **4**, 64 (1950). Clinical f.p.; detn. of Na, K in plasma, urine.
423. Knewstubb, P. F., and T. M. Sugden, *Trans. Faraday Soc.*, **54**, 372 (1958). *CA*, **52**, 17918. Ionizn. kinetics of alk. metals in flame.
424. Knewstubb, P. F., and T. M. Sugden, *Nature*, **181**, 474 (1958). *CA*, **52**, 10728. Mass-spectrometric observation of flame ions.
425. Knewstubb, P. F., and T. M. Sugden, p. 247 in *Seventh Symposium (International) on Combustion*, Butterworth, London, 1959. Same topic.

426. Knickmann, E., *Z. Pflanzenernähr., Düng. u. Bodenk.*, **54**, 117 (1951). *CA*, **46**, 8565. Filters for f.p. d. Na, K, Ca.
427. Knight, S. B., W. C. Mathis and J. R. Graham, *Anal. Chem.*, **23**, 1704 (1951). *CA*, **46**, 4417. F.p. d. Na, K, Li, Ca in rocks, glass; study of int. std.
428. Knight, S. B., and M. H. Peterson, *Anal. Chem.*, **24**, 1514 (1952). *CA*, **46**, 11029. Int.-std. f.s. d. Na in surfactants.
429. Knutson, K. E., *Analyst*, **82**, 241 (1957). *CA*, **51**, 11919. Rich O_2–C_2H_2 for f.p. d. Mg in plants.
430. Köhnlein, J., and K. E. Lücke, *Z. Pflanzenernähr., Düng. u. Bodenk.*, **57**, 114 (1952). *CA*, **46**, 11023. F.p. d. Ca in plants, soil; phosphate intf.
431. Konopicky, K., and P. Kampa, *Forschungsbericht des Wirtschafts- und Verkehrsministeriums Nordrhein-Westfalen*, **1955**, No. 149, p. 5. F.p.d. Ca.
431a. Konopicky, K., and W. Schmidt, *Z. anal. Chem.*, **173**, 358 (1960). *CA*, **54**, 17145. Enhancement of Li–Cs, Mg–Ba, Cr, Mn, V, Cu, B, Al, In, Tl, La, Fe, Co, Ni by BuOH.
431b. Konopicky, K., and W. Schmidt, *Z. anal. Chem.*, **174**, 262 (1960). Enhancement of Al by HF.
432. Konschak, M., *Aufklärungsblätter für Arbeitschutz*, **1952**, vol. 22, p. 9; vol. 23, p. 12. Safety in gas handling.
433. Kramer, H., *Anal. Chim. Acta*, **17**, 521 (1957). *CA*, **53**, 122. F.p. d. Ca in rocks.
434. Kropik, K., *Z. Pflanzenernähr., Düng. u. Bodenk.*, **70**, 138 (1955). *CA*, **49**, 16294. F.p. d. Ca in soil.
435. Kuemmel, D. F., and H. L. Karl, *Anal. Chem.*, **26**, 386 (1954). *CA*, **48**, 7482. F.s. d. Na, Li, Ca, Mg in cast Fe.
436. Kühns, K., and G. Müller, *Z. physiol. Chem.*, **294**, 86 (1953). *CA*, **49**, 14861. F.p. d. Na, K in serum, organs; errors.
437. Ladenburg, R., and F. Reiche, *Ann. Physik*, **42**, 181 (1913). Selective absorption.
438. Lady, J. H., *Applications of solvent extraction to flame spectrophotometry.* Ph.D. Thesis, Univ. Tennessee, 1955.
439. Lagerqvist, A., *Naturwissenschaften*, **40**, 268 (1953). *CA*, **48**, 443. Band spectra of CaO in violet and ultraviolet.
440. Lagerqvist, A., *Arkiv Fysik*, **8**, 83 (1954). *CA*, **48**, 11922. Anal. of CaO spectrum.
441. Lagerqvist, A., and L. Huldt, *Naturwissenschaften*, **42**, 365 (1955). *CA*, **50**, 11109. Alk. earth flame bands due to SrOH, CaOH, BaOH.
441a. Lagos, A. E., *Proc. Intern. Conf. Peaceful Uses Atomic Energy, Geneva, 1955*, vol. 8, p. 309; Columbia Univ. Press, New York, 1956. F.s. d. Ga, In, Tl in U.
442. Lane, W. R., *Ind. Eng. Chem.*, **43**, 1312 (1951). *CA*, **45**, 7827. Mechanism of atomization.
443. Dr. B. Lange, Hermannstr. 14–18, Berlin-Zehlendorf. Brochure 27/58. F.p. Model 6.
444. Langmuir, I., *Phys. Rev.*, **12**, 368 (1918). *CA*, **13**, 274. Evapn. of small spheres.
445. Lehmann, H., and W. Pralow. *Tonind.-Ztg. u. Keram. Rundschau*, **76**, 33 (1952). *CA*, **46**, 4424. F.p. d. Na, K, Ca in silicate materials (review).
446. Leppänen, V., and N. Hallman, *Scand. J. Clin. and Lab. Invest.*, **5**, 250 (1953). *CA*, **48**, 1078. Int.-std. f.p.

447. Lewis, B., and G. von Elbe, *J. Chem. Phys.*, 11, 75 (1943). *CA*, 37, 1909. Stability, structure of burner flames.
448. Leyton, L., *Analyst*, 76, 723 (1951). *CA*, 46, 1821. Filter f.p. with concentric atomizer; d. K, Ca.
449. Leyton, L., *Biochem. J.*, 50, *Proc.*, p. xl (1952). Same subject as 448.
450. Leyton, L., *Analyst*, 79, 497 (1954). *CA*, 48, 11973. Phosphate intf. in f.p. d. Ca.
451. Same as 282.
451a. Luck, C. P., and P. G. Wright, *Nature*, 183, 1595 (1959). *CA*, 53, 22340. F.p. d. Na, K in aqueous humor; dialysis test for intf.
452.* Malinowski, J., *Chem. anal. (Warsaw)*, 3, 549 (1958). *CA*, 53, 11093. Indirect methods in f.p.; S, P, Mo, W, V, As, Zr anion intf. with Ca, Sr. Malinowski, J., W. Rutkowski, and S. Szymczak, *Polish Acad. Sci., Inst. Nuclear Research, Rept.* No. 110/VIII, Part II (1959). *CA*, 54, 11814. Indirect methods in f.p.; Se, Te, Ti anion intf. with Ca, Sr. Malinowski, J., and D. Dancewicz, Part III of the same report. Indirect f.p. d. Be in bronze.
453.* Malinowski, J., and D. Dancewicz, Colloquium Spectroscopicum Internationale VII, Liège, Sept. 1958; *Rev. universelle mines*, 15, 405 (1959). *Anal Abstr.*, 6, No. 3367 (1959). Also at the Second Czechoslovakian Spectrographic Congress, Tatra-Lomnitz, Oct., 1959. Indirect. f.p. d. Be in Be bronze.
453a. Malinowski, J., D. Dancewicz, and S. Szymczak, *Polish Acad. Sci., Inst. Nuclear Research, Rept.* No. 113/VIII (1959). *CA*, 54, 8447. F.p. d. Ga, In, Tl.
454. Malinowski, J., p. 167 of *Proceedings, VIII. Colloquium Spectroscopicum Internationale* (Lucerne, 1959). Sauerländer, Aarau, 1960. EDTA to protect Ca, Sr from Al, P.
454a. Malmstadt, H. V., and W. E. Chambers, *Anal. Chem.*, 32, 225 (1960). *CA*, 54, 9386. Null-point absorption f.s.; Na, K.
455. Mandel'shtam, S. L., *Compt. rend. acad. sci. U.R.S.S.* 22, 403 (1939). *CA*, 33, 8120. Dissocn. Na, K, Rb halides in flame.
456. Manna, L., D. H. Strunk, and S. L. Adams, *Anal. Chem.*, 28, 1070 (1956). *CA*, 50, 13651. F.s. d. traces of Cu.
456a. Manna, L., D. H. Strunk, and S. L. Adams, *Anal. Chem.*, 29, 1885 (1957). *CA*, 52, 4392. F.s. d. traces of Mg.
456b. Marcinka, K., Second Czechoslovakian Spectrographic Congress, Tatra-Lomintz, Oct. 1959. Effect of alcohols on f.p. Mg.
457. Margoshes, M., and B. L. Vallee, Pittsburgh Conference on Anal. Chem. and Appl. Spectroscopy, Mar. 1955. Abstract in *Anal. Chem.*, 27, 320 (1955). Multichannel grating f.s. with dual slits for background correction, for Na, K, Ca, Sr.
458. Margoshes, M., and B. L. Vallee, *Anal. Chem.*, 28, 180 (1956). (*CA*, 50, 7650.) Multichannel f.s.; intf. *Ibid.*, 1066. (*CA*, 50, 13520.) Automatic background correction.
458a. Margoshes, M., and B. F. Scribner, *Spectrochim. Acta*, 15, 138 (1959). *Anal. Abstr.*, 7, No. 307 (1960). Plasma jet for spectrochem. anal.
458b. Marius, Ganzenmarkt 4–8, Utrecht, Holland. F.p. (see Table 15).
459. Marquardt, G. H., G. M. Cummings, M. L. Phillips, and C. I. Fisher, *Am. J. Clin. Pathol.*, 26, 1094 (1956). *CA*, 51, 541. F.p. d. Ca in plasma.

460. Marsh, G. E., *Appl. Spectroscopy*, **10**, 8 (1956). *CA*, **50**, 5953. F.s. d. Na, Mg, Ca, Fe in water.
461. Marsh, G. E., *Appl. Spectroscopy*, **12**, 113 (1958). *CA*, **53**, 4011. F.s. d. org. chloride via CuCl bands.
462. Martens, P.H., *Chim. et ind. (Paris)*, **7**, 135 (1956); *Chim. anal.*, **39**, 361 (1957). *CA*, **52**, 3238. F.p. d. K in fertilizers.
463. Mashiko, Y., and Y. Kanroji, *J. Pharm. Soc. Japan*, **76**, 441 (1956). (*CA*, **50**, 10311.) F.p. d. and sepn. of Li in mineral water. *Ibid.*, 689. (*CA*, **50**, 15007.) F.p. d. K in mineral water.
464. Mason, W. B., Pittsburgh Conference on Anal. Chem. and Appl. Spectroscopy, Mar. 1955. Abstract in *Anal. Chem.*, **27**, 320 (1955). F.s. d. Ca with Li int. std. and radioactive phosphor as reference.
465. Massey, H. F., *Anal. Chem.*, **29**, 365 (1957). *CA*, **51**, 8195. F.p. d. Cu in plants.
466. Mathis, W. T., *J. Assoc. Offic. Agr. Chemists*, **35**, 406 (1952). *CA*, **46**, 11474. F.p. d. K, Ca, Mg, Na in alfalfa (collaborative study).
467. Mathis, W. T., *J. Assoc. Offic. Agr. Chemists*, **36**, 411 (1953). *CA*, **47**, 12668. Collaborative study, including f.p.
468. Mathis, W. T., *J. Assoc. Offic. Agr. Chemists*, **38**, 387 (1955). (*CA*, **49**, 10446.) F.p. d. K in plants. *Ibid.*, **39**, 419 (1956). (*CA*, **50**, 9218.) F.p. d. K, Na in plants.
469.* Mavrodineanu, R., *Appl. Spectroscopy*, **10**, 51, 137 (1956). *CA*, **50**, 12727. Bibliography on anal. f.p. with index.
469a.* Mavrodineanu, R., *Appl. Spectroscopy*, **13**, 132, 149 (1959); **14**, 17 (1960). Continued bibliography with index.
470. Mavrodineanu, R., p. 15 in *Proceedings, VIII. Colloquium Spectroscopicum Internationale* (Lucerne, 1959), Sauerländer, Aarau, 1960. Review; hot flames, new sources, flame excitation, atomization, sensitive monochromators.
471. Mayer, F. X., *Spectrochim. Acta*, **4**, 539 (1952). Meeting report on spectroscopy in microchem. anal.; no f.p.
472. Mazzamaro, P., and G. Tatoian, *Anal. Chem.*, **26**, 1512 (1954). *CA*, **49**, 1476. F.d. p. Na in pure water, acid.
473. McCoy, W. J., and G. G. Christiansen, p. 44 in *Symposium on Flame Photometry*, Am. Soc. Testing Materials, Spec. Tech. Publ. No. 116 (1951). *CA*, **47**, 7753. F.p. d. Li in cement.
474. MacIntyre, I., *Rec. trav. chim.*, **74**, 498 (1955). *CA*, **49**, 16034. F.p. d. Ca in serum.
475. MacIntyre, I., and J. E. S. Bradley, *Biochem. J.*, **62**, 38P (1956). *Biol. Abstr.*, **31**, 7149 (1957). F.s. for Cu, Ca, Fe, Mg.
475a. MacIntyre, I., *Biochem. J.*, **67**, 164 (1957). *CA*, **51**, 18061. F.s. d. Ca in serum; errors in Kramer–Tisdall method.
476. McNally, J. R., Jr., G. K. Werner, and D. D. Smith, Opt. Soc. Am., Boston, Oct. 1952. Abstr. in *J. Opt. Soc. Am.*, **42**, 870 (1952). Spectroisotopic anal. of Li.
477.* Meggers, W. F., and B. F. Scribner, *Index to the Literature on Spectrochemical Analysis, 1920–1939*, 2nd ed., Am. Soc. Testing Materials, Philadelphia, 1941.
478. Mehlich, A., and R. J. Monroe, *J. Assoc. Offic. Agr. Chemists*, **35**, 588 (1952). *CA*, **47**, 2624. Report on f.p. d. K in soil.

479. Mehlich, A., and M. E. Harward, *J. Assoc. Offic. Agr. Chemists*, **36**, 227 (1953). *CA*, **47**, 12114. F.p. d. K in soil, plants.
480. Mehlich, A., *J. Assoc. Offic. Agr. Chemists*, **39**, 330 (1956). *CA*, **50**, 9218 F.p. d. K in soil.
481. Meine, W., *Erdöl u. Kohle*, **8**, 711 (1955). *CA*, **50**, 8186. F.p. d. Pb in gasoline.
482. Melaven, A. D., and A. J. Chadwell, Regional Conclave, Amer. Chem. Soc., New Orleans, Dec. 1953. Abstract in *Anal. Chem.*, **25**, 1934 (1953). Perrhenate intf. in f.p. d. K.
483. Meloche, V. W., p. 3 in *Symposium on Flame Photometry*, Am. Soc. Testing Materials, Spec. Tech. Publ. No. 116 (1951). *CA*, **47**, 6758. Review on f.p.
484. Meloche, V. W., and R. Shapiro, *Anal. Chem.*, **26**, 347 (1954). *CA*, **48**, 6053. Spectrographic and electrometric detn. Ca, Mg in lake water; no f.p.
485. Meloche, V. W., J. B. Ramsay, D. J. Mack, and T. V. Philip, *Anal. Chem.*, **26**, 1387 (1954). *CA*, **49**, 11496. F.s. d. In in Al bronze.
486.* Meloche, V. W., *Anal. Chem.*, **28**, 1844 (1956). *CA*, **51**, 3353. Review on f.p.
487. Meloche, V. W., and B. L. Beck, *Anal. Chem.*, **28**, 1890 (1956). *CA*, **51**, 3367. F.s. d. Ga in alloys.
488. Menis, O., H. P. House, and T. C. Rains, *Anal. Chem.*, **29**, 76 (1957). *CA*, **51**, 4873. Indirect f.p. d. halides.
489. Menis, O., T. C. Rains, and J. A. Dean, *Anal. Chim. Acta*, **19**, 179 (1958). *CA*, **54**, 1171. F.s. of La in alc. solns.
490. Menis, O., T. C. Rains, and J. A. Dean, *Anal. Chem.*, **31**, 187 (1959). *CA*, **53**, 12943. F.s. d. La by solvent extn.
491. Merker, W., and R. Herrmann, *Ärztl. Wochschr.*, **9**, 1196 (1954). *CA*, **49**, 4062. Errors in detn. Ca in serum.
492. Metrohm A.-G., Herisau, Switzerland. Model E 1006, E 1007 f.s.
493. Mitchell, R. L., and I. M. Robertson, *J. Soc. Chem. Ind.*, **55**, 269T (1936). *CA*, **30**, 8065. F.p. d. Al by intf. with Ca or Sr; Sr, Ca as buffer against Al in f.p. d. Ca, Sr, resp.
494. Mitchell, R. L., *Spectrochim. Acta*, **4**, 62 (1950). Review on f.p.'s; detn. Na, K, Ca in soil.
495. Mitscherlich, A., *Ann. Physik u. Chem.*, **116**, 499 (1862). Pt wire wick with Bunsen flame for spectroscopy.
496. Moberg, M. L., V. B. Waithman, W. H. Ellis, and H. D. DuBois, p. 92 in *Symposium on Flame Photometry*, Am. Soc. Testing Materials, Spec. Tech. Publ. No. 116 (1951). *CA*, **47**, 7767. F.s. d. Ca in lubricating oil.
497. Guérin de Montgareuil, P., *Contribution à l'étude des interactions chimiques dans les flammes*. Thesis, Univ. Paris, 1954.
498. Monvoisin, J., and R. Mavrodineanu, *Spectrochim. Acta*, **4**, 152 (1950). *CA*, **44**, 10487. Ultrasonic oscillator for f.p. atomizer. F.p. d. Na, K in refractories.
499. Monvoisin, J., and R. Mavrodineanu, *Spectrochim. Acta*, **4**, 396 (1951). *CA*, **46**, 2347. Ultrasonic jet and spark atomizer for flame spectrography; detn. alkalies, alk. earths in refractories; Mn, Cr, Ni, Co, Cu, Pb in steel; spectra of Sn, Pb, Al, Ag, etc.
500. Mosher, R. E., A. J. Boyle, E. J. Bird, S. D. Jacobson, T. M. Batchelor, L. T. Iseri, and G. B. Myers, *Am. J. Clin. Pathol.*, **19**, 461 (1949). *CA*, **43**, 5441. F.s. d. K, Na in plasma, urine.
501. Mosher, R. E., E. J. Bird, and A. J. Boyle, *Anal. Chem.*, **22**, 715 (1950). *CA*, **44**, 7183. F.s. d. Ca in minerals; Mg to buffer Al intf.

502. Mosher, R. E., M. Itano, A. J. Boyle, G. B. Myers, and L. T. Iseri, *Am. J. Clin. Pathol.*, **21**, 75 (1951). *CA*, **45**, 2533. F.s. d. Ca in plasma.
503.* Moss, M. L., *Anal. Chem.*, **29**, 670 (1957). Review on nonferrous metallurgy, mentioning f.p.
504. Müller, E. R., *Naturwissenschaften*, **40**, 442 (1953). *CA*, **48**, 7162. Binding of Ca in serum.
505. Mulligan, B. W., and A. F. Haught, *J. Research Natl. Bur. Standards*, **61**, 499 (1958). *CA*, **53**, 13866. Compensation of photomultiplier drift in f.p.
506. Mullin, H. R., and T. P. Sheehy, Pittsburgh Conference on Anal. Chem. and Appl. Spectroscopy, Mar. 1953. Abstract in *Anal. Chem.*, **25**, 529 (1953). F.s. d. Na, K, Li in Be solns.
506a. Murthy, G. K., and R. McL. Whitney, *J. Dairy Sci.*, **39**, 364 (1956). *CA*, **50**, 10939. F.p. d. Na, K, Ca in milk.
507. Muth, H. W., and W. Beckmann, *Ärztl. Wochschr.*, **7**, 961 (1952). Clinical f.p.
508. Muth, H. W., *Ärztl. Wochschr.*, **8**, 160 (1953). *CA*, **47**, 4946. F.p. d. Ca in serum.
509. Muth, H. W., *Ärztl. Laboratorium*, **2**, 243 (1956). Chem. and f.p. d. Ca, Mg.
510. Natelson, S., *Am. J. Clin. Pathol.*, **20**, 463 (1950). *CA*, **44**, 6907. Perkin-Elmer f.p. for detn. Na, K in serum, etc.
511. Neeb, K. H., and W. Gebauhr, *Z. anal. Chem.*, **162**, 167 (1958). *CA*, **53**, 126. F.p. d. K in Na, Na salts, enriching with tetraphenylboron.
512. Neeb, K. H., p. 176 in *Proceedings, VIII. Colloquium Spectroscopicum Internationale* (Lucerne, 1959). Sauerländer, Aarau, 1960. *Z. anal. Chem.*, **174**, 328 (1960). *CA*, **55**, 1278. F.p. d. Na in Al, sepg. Al with EtBr.
513. Netheler und Hinz G.m.b.H., Hamburg-Wellingsbüttel. *Handbuch Flammenphotometer Eppendorf*, 1957.
514. Gebrüder Netzsch Maschinenfabrik, Selb, Bavaria. Zoellner f.p., 1955.
515. Neuberger, A., E. Schöffmann and K. Herkenhoff, *Arch. Eisenhüttenw.*, **29**, 35, (1958). F.p. anal. slag.
516. Neuberger, A., E. Schöffmann, and K. Herkenhoff, *Arch. Eisenhüttenw.*, **29**, 547 (1958). *CA*, **53**, 976. F.p. anal. ores.
517. Nielsen, S. O., *Rev. Sci. Instr.*, **26**, 516 (1955). *CA*, **49**, 12133. Recording attachment for Beckman DU spectrophotometer.
517a. Nikonova, M. P., Eighth Mendeleev Congress for Pure and Applied Chemistry, Moscow, 1959. Integration in f.p.
518.* Noebels, H. J., Pittsburgh Conference on Anal. Chem. and Appl. Spectroscopy, Mar. 1954. Abstract in *Anal. Chem.*, **26**, 432 (1954). Publ. by Beckman Instruments, Inc., Reprint R-60. Review of applications of f.p.
519. North American Philips Co., Inc., Mount Vernon, N.Y., *Rev. Sci. Instr.*, **26**, 420 (1955); Bulletin 13–55a (1955). Norelco f.p.
520. Nukiyama, S., and Y. Tanasawa, *Trans. Soc. Mech. Engrs. (Japan)*, **5**, No. 18, p. 63 (1938). *CA*, **35**, 4258. Atomization of liquids.
521. Obolenskaya, L. I., *Pochvovedenie*, **1951**, 420. *CA*, **46**, 3195. Detn. K, Na in soil by Rusanov f.p.
522. Odeen, M. H., *Beckman Bulletin*, No. 3, p. 4 (Aug. 1950). Review of applications of f.s.
523. Oelschläger, W., *Z. anal. Chem.*, **149**, 190 (1956). *CA*, **50**, 8096. F.p. d. K, Na in hay, etc.
524. Oer, A. von, and H. J. Höfert, *Arch. exptl. Pathol. Pharmakol.*, **214**, 109 (1951). *CA*, **47**, 2240. F.p. d. Na, K, Ca in serum.

525.* Ohyagi, Y., *Japan Analyst*, **4**, 179 (1955). *CA*, **50**, 10594. Review of f.s.
526. Ohyagi, Y., *Sci. of Light (Tokyo)*, **7**, 14 (1958). F. s. anal. biol. materials; wet ashing; IO flame bands; Hitachi f.s.
527. Olney, J. M., and A. H. James, *Science*, **115**, 244 (1952). Air cleaner for Barclay f.p.
528. Opler, A., and J. H. Miller, *J. Opt. Soc. Am.*, **42**, 784 (1952). *Sci. Abstr.*, **56A**, No. 110. Recorder on Beckman f.s. to plot Na in ion-exchange eluate.
529. Optica S.p.a., Via Calatafimi 7–9, Milan; Optica United Kingdom Ltd., Gateshead-on-Tyne, 11; Optica, Inc., 830 Evarts St., N.E., Washington, D.C. Flame attachment for CF-4 spectrophotometer.
530. Optische Werke (formerly Carl Zeiss), Jena. Zeiss f.p. Model III.
531. Ørskov, S. L., *Acta Physiol. Scand.*, **27**, 321 (1953). *CA*, **47**, 5476. F.p. d. Na, K in blood, minced tissue, after electrolysis.
532. Osborn, G. H., and H. Johns, *Analyst*, **76**, 410 (1951). *CA*, **45**, 8393. F.s. d. Na, K in rocks; intfs.
532a. Ounsted, D., *J. Inst. Fuel*, **31**, 474 (1958). *CA*, **53**, 778. F.p. d. Na aerosol concn. in furnace gases.
533. Overman, R. R., and A. K. Davis, *J. Biol. Chem.*, **168**, 641 (1947). *CA*, **41**, 6595. F.p. d. Na, K in erythrocytes, etc.
534. Padley, P. J., and T. M. Sugden, p. 235 in *Seventh Symposium (International) on Combustion*, Butterworth, London, 1959. Chemiluminescence of Na in N_2–O_2–H_2 flames.
535. Page, F. M., *Discussions Faraday Soc.*, No. 19, 87 (1955). *CA*, **50**, 8330. Detn. electron affinity of OH by microwave measurements on flames.
536. Page, F. M., and T. M. Sugden, *Trans. Faraday Soc.*, **53**, 1092 (1957). *CA*, **52**, 7861. Mechanism of alk. ionizn. in H_2 flames.
537. Parks, T. D., H. O. Johnson, and L. Lykken, *Anal. Chem.*, **20**, 822 (1948). *CA*, **43**, 59. Intf. in d. Na, K with Perkin-Elmer f.p.
538. Penner, S. S., and R. W. Kavanagh, *J. Opt. Soc. Am.*, **43**, 385 (1953). *Sci. Abstr.*, **56A**, No. 4706. Curves of growth with Doppler and Lorentz broadening.
539. The Perkin-Elmer Corp., Norwalk, Conn.; Perkin-Elmer and Co. G.m.b.H., Überlingen/See (Bodensee), Germany, Bulletin 146. Model 146 f.s.
540. Perman, J., p. 177 in *Proceedings, VIII. Colloquium Spectroscopicum Internationale* (Lucerne, 1959). Sauerländer, Aarau, 1960. F.s. d. Li in nuclear materials.
540a. Perman, I., "*J. Stefan*" *Inst. Repts. (Ljubljana)*, **3**, 179 (1956). *CA*, **51**, 17465. Pneumatic atomizer for f.p.
541. Perry, J. W., *J. Opt. Soc. Am.*, **45**, 995 (1955). *Sci Abstr.*, **59A**, No. 1034. Luminosity of spectrometers.
542. Pfleiderer, T., P. Otto, and W. Hardegg, *Klin. Wochschr.*, **37**, 39 (1959). *CA*, **53**, 22365. F.p. d. K in serum, plasma.
543. Phifer, L. H., *Anal. Chem.*, **29**, 1528 (1957). *CA*, **52**, 1603. F.p. d. Ca in cellulose.
544. Pietzka, G., and H. Chun, *Angew. Chem.*, **71**, 276 (1959). *CA*, **53**, 14824. Intf. in f.p.; enhancement by org. solvents; releasing agents for Ca.
545. Pigor, E., *Zur flammenphotometrischen Bestimmung von Calcium, Natrium und Kalium im Blutserum. Methodik und Beeinflussung durch Viscosität und Oberflächenspannung.* Dissertation, Giessen, 1954.

546. Pinta, M., *J. recherches centre natl. recherche sci. Labs. Bellevue (Paris)*, No. 21, 260 (1952). *CA*, **47**, 8511. Spectra and f.s. d. Sc, Y, La–Lu.
547. Pinta, M., *Contribution à l'étude des spectres de flamme. Applications analytiques*. Thesis, Univ. Paris, 1952. Rare earths, etc.
548. Pinta, M., *Chim. anal.*, **36**, 126 (1954). *CA*, **48**, 8112. Design of f.p. Review of applications.
549. Pinta, M., and C. Bove, *Compt. rend.*, **243**, 179 (1956). *CA*, **50**, 16533. Intfs. in f.s. anal. plants.
550.* Pinta, M., and C. Bove, *Mikrochim. Acta*, **1956**, 1788. *CA*, **51**, 5624. F.s. d. Na, K, Mg, Ca in plants; study of intfs.
551. Pinta, M., and H. Aubert, *Compt. rend.*, **244**, 873 (1957). *CA*, **52**, 13164. F.p. d. Al in soil via Ca intf.
552. Piper, E., and H. Hagedorn, *Arch. Eisenhüttenw.*, **22**, 299 (1951). *CA*, **46**, 2223. F.p. d. K in fertilizer, K salts.
553.* Pisarev, V. D., *Zavodskaya Lab.*, **22**, 462 (1956). *CA*, **51**, 2450. Review on spectrochem. anal. of solns.
553a. Plagnol, H., and M. Bredillet, *Bull. soc. chim. biol.*, **38**, 563 (1956). *CA*, **50**, 15675. F.p. d. Na, K in serum.
554. Plyler, E. K., and C. J. Humphreys, *J. Research Natl. Bur. Standards*, **40**, 449 (1948). *CA*, **42**, 7634. Infrared emission spectra of flames.
555. Plyler, E. K., and C. J. Humphreys, *J. Research Natl. Bur. Standards*, **42**, 567 (1949). *CA*, **43**, 7822. C_2H_2 flame spectrum, 1–5 μ.
556. Plyler, E. K., *Mikrochim. Acta*, **1955**, 421. *CA*, **49**, 12127. Infrared flame spectra.
557. Pólit, I., and J. García-Llauradó, *Biochem. Z.*, **323**, 418 (1953). *CA*, **47**, 5174. Home-made f.p.
558. Police code regarding traffic with combustible liquids. Bonn, 1947.
559. Poluektov, N. S., *Zavodskaya Lab.*, **21**, 1045 (1955). *CA*, **50**, 9205. Review on f.p.
560. Poluektov, N. S., L. I. Kononenko, and M. P. Nikonova, *Zhur. Anal. Khim.*, **12**, 10 (1957). *CA*, **51**, 9410; **52**, 7941. F.s. d. Li in ores.
561. Poluektov, N. S., M. P. Nikonova, Ts. A. Leĭderman, and G. S. Lauer, *Zhur. Anal. Khim.*, **12**, 699 (1957). *CA*, **52**, 8834. F.s. d. Sr in ores.
561a. Poluektov, N. S., and M. P. Nikonova, *Zhur. Anal. Khim.*, **13**, 635 (1958). *CA*, **53**, 5834. Mechanism of intf. in f.p.; expts. with dual atomizers.
562. Poluektov, N. S., M. P. Nikonova, and R. A. Vitkun, *Zhur. Anal. Khim.*, **13**, 48 (1958). *CA*, **52**, 9856. F.s. d. Na, K in minerals.
562a. Poluektov, N. S., Eighth Mendeleev Congress for Pure and Applied Chemistry, Moscow, 1959. Compound formation as cause of intf. in f.p.
563. Porter, P., and G. Wyld, *Anal. Chem.*, **27**, 733 (1955). *CA*, **49**, 11485. Intfs. in f.s. d. Na, K.
564. Powell, F. J. N., *J. Clin. Pathol.*, **6**, 286 (1953). F.p. d. Ca in biol. fluids.
565. Pritchard, L. R., *Pit and Quarry*, **41**, No. 1, p. 83 (1948). *CA*, **43**, 373. F.p. d. Na, K in cement.
566. Pro, M. J., and A. P. Mathers, *J. Assoc. Offic. Agr. Chemists*, **39**, 225 (1956). (*CA*, **51**, 1530.) F.s. d. Sr as tracer in whiskey. *Ibid.*, 236. (*CA*, **51**, 1530.) Same for Li. Pro, M. J., R. A. Nelson, and A. P. Mathers, *Ibid.*, 506. (*CA*, **51**, 1530.) Same for Cs, Rb.
567. Process and Instruments Corp., 17 Stone Ave., Brooklyn 33, N.Y. Flame photometer Model 1B.

568. Proehl, E. C., and W. P. Nelson, *Am. J. Clin. Pathol.*, **20**, 806 (1950). *CA*, **44**, 10779. Clinical f.p. d. Na, K.
569. Pruvot, P., *Ingénieurs et Techniciens (Paris)*, **103**, 93 (1957). Flame spectrophotometers.
569a. Puffeles, M., and N. E. Nessim, *Analyst*, **82**, 467 (1957). *CA*, **51**, 16199. F.p. d. Na, K in soil, plants, waters, serum.
570. Puffeles, M., and N. E. Nessim, *Anal. Chim. Acta*, **20**, 38 (1959). *CA*, **53**, 21405. F.p. d. Ca in soil, plants, water, serum; intfs.
571. Pungor, E., and A. J. Hegedüs, *Magyar Kém. Lapja*, **9**, 178, 201 (1954). *Chem. Zentr.* **1955**, 10820. F.p.; oxyhydrogen atomizer-burner.
572. Pungor, E., and E. É. Zapp, *Acta Chim. Acad. Sci. Hung.*, **7**, 185 (1955). *CA*, **50**, 4632. *Magyar Kém. Folyóirat*, **61**, 117 (1955). *CA*, **50**, 671. F.s. d. Li, Na, K, Rb, Cs.
573. Pungor, E., and A. Hegedüs, *Magyar Kém. Folyóirat*, **61**, 308 (1955). *CA*, **52**, 8842. F.s. d. Ca, Sr, Ba in mixtures.
574. Pungor, E., Dissertation, *Magyar Tudományos Akad., Kém. Tudományok Osztályának, Közleményei*, 1956. Critical review on theoret. problems in f.p.
575. Pungor, E., A. J. Hegedüs, I. Konkoly-Thege, and E. É. Zapp, *Mikochim. Acta*, **1956**, 1247. *CA*, **50**, 8372. Mutual intfs. among K, Rb, Cs suppressed by cooling flame.
576. Pungor, E., and A. Hegedüs, *Z. anal. Chem.*, **151**, 375 (1956); *Magyar Kém. Folyóirat*, **61**, 308 (1955). *CA*, **52**, 8842. Micro f.p. d. Ca, Sr, Ba in mixtures.
577. Pungor, E., and E. É. Zapp, *Acta Chim. Acad. Sci. Hung.*, **10**, 1, 179 (1956); *Magyar Kém. Folyóirat*, **61**, 421 (1955). *CA*, **51**, 5625. F.p. d. Li, Na, K, Rb, Cs.
578.* Pungor, E., and I. Konkoly-Thege, *Magyar Kém. Folyóirat*, **62**, 17 (1956); *Acta Chim. Acad. Sci. Hung.*, **11**, 23 (1957). *CA*, **52**, 161. F.s. Mg; intfs.
579. Pungor, E., and I. Konkoly-Thege, *Magyar Kém. Folyóirat*, **62**, 228 (1956); *Acta Chim. Acad. Sci. Hung.*, **13**, 1 (1957). *CA*, **52**, 6924. Study of Cu flame bands.
580. Pungor, E., and I. Konkoly-Thege, *Magyar Kém. Folyóirat*, **62**, 231 (1956); *Acta Chim. Acad. Sci. Hung.*, **13**, 39 (1957). *CA*, **52**, 6925. F.p. study of B.
581. Pungor, E., and E. É. Zapp, *Mikrochim. Acta*, **1957**, 150. *CA*, **51**, 12738. F.p. d. Ba in samples contg. $BaSO_4$.
582. Pungor, E., and I. Konkoly-Thege, *Magyar Kém. Folyóirat*, **62**, 225 (1956); *Acta Chim. Acad. Sci. Hung.*, **13**, 235 (1958). *CA*, **52**, 7007. F.p. study of Ag.
582a. Pungor, E., Second Czechoslovakian Spectrographic Congress, Tatra-Lomintz, Oct. 1959. *Ann. Univ. Sci. Budapest Rolando Eötvös Nominatae, Sect. Chim.*, **2**, 431 (1960). *CA*, **56**, 3. New developments in f.p.
582b. Pungor, E., and I. Konkoly-Thege, *Mikrochim. Acta*, **1959**, 712. *Anal. Abstr.*, **7**, No. 897 (1960). F.p. d. traces Sr in presence of Ca, Ba.
582c. Pungor, E., and A. J. Hegedüs, *Mikrochim. Acta*, **1960**, 87. *Anal. Abstr.*, **7**, No. 3638. F.s. d. Ca, Sr, Ba in oxide cathodes.
582d. Püschel, A., and S. Eckhard, *Arch. Eisenhüttenw.*, **30**, 731 (1959). *CA*, **54**, 9611. Theory of org. solvents in f.s.; Cr, Mn, Fe in alcohols.
583. Radmacher, W., and W. Schmitz, *Brennstoff-Chem.*, **38**, 270, 308 (1957). *CA*, **52**, 1586. F.p. d. Na, K in fuel ash.
584. Ramsay, J. A., *J. Exptl. Biol.*, **27**, 407 (1950). *CA*, **45**, 3449. F.p. for detn. Na in biol. microsamples.

585. Ramsay, J. A., S. W. H. W. Falloon, and K. E. Machin, *J. Sci. Instr.*, **28**, 75 (1952); *Spectrochim. Acta*, **4**, 536 (1952). Integrating f.s., sample on Pt wire in flame; Tl, Na; detn. Na in serum.
586. Ramsay, J. A., R. H. J. Brown, and S. W. H. W. Falloon, *J. Exptl. Biol.*, **30**, 1 (1953). *CA*, **47**, 5842. Ultramicro f.p. for Na, K.
587. Rathje, A. O., *Anal. Chem.*, **27**, 1583 (1955). *CA*, **50**, 4709. F.s. d. Ag in CdS, ZnS phosphors.
588. Rausch, L., and E. H. Graul, *Ärztl. Wochschr.*, **4**, 524 (1949). (*CA*, **44**, 4065.) Anal. of errors in blood anal. Graul, E. H., and L. Rausch, *Ibid.*, 564, 591. (*CA*, **44**, 10030–1.) Statistical anal. of errors in d. K, Ca in blood.
589. Rauterberg, E., and E. Knippenberg, *Angew. Chem.*, **53**, 477 (1940). *CA*, **35**, 706. Modified Schuhknecht–Waibel f.p. for d. K.
590. Rauterberg, E., and E. Knippenberg, *Ernähr. Pflanze*, **37**, 73 (1941). *CA*, **36**, 4052. F.p. instrumentation for d. K.
591. Rauterberg, E., and E. Knippenberg, *Forschungsdienst*, **15**, 150 (1941). *Chem. Zentr.*, **41**, II, 1299. F.p. instrumentation for d. K.
592. Reich, H. F., and F. Grabbe, *Tonind. Ztg. u. Keram. Rundschau*, **79**, 127 (1955). *CA*, **49**, 10795. F.p. d. Na, K, Ca in refractories.
593. Rice, J. K., *Combustion*, **28**, 57 (1956). F. p. d. Na in boiler deposits.
594. Riehm, H., *Z. anal. Chem.*, **128**, 249 (1948). *CA*, **42**, 5795. Riehm–Lange f.p. for Na, K, Ca; filters, intf.
595. Riehm, H., *Z. Pflanzenernähr., Düng. u. Bodenk.*, **40**, 152 (1948). Lactate method.
596. Riethmüller, H. U., and A. Bretschneider, *Z. physiol. Chem.*, **293**, 49 (1953). *CA*, **49**, 9723. Errors in f.p. d. Ca in serum, urine.
596a.* Riethmüller, H. U., *Mikrochim. Acta*, **1953**, 178. *CA*, **48**, 782. Review on spectroscopic d. Ca in biol. materials.
597. Robinson, A. R., K. J. Newman, and E. J. Schoeb, *Anal. Chem.*, **22**, 1026 (1950). *CA*, **44**, 10772. Flame spectrograph, Co int. std., for Ag, Ba, Ca, Co, Cr, Cu, Fe, K, Li, Mg, Mn, Na, Ni, Sr in biol. material.
598. Robinson, A. M., and T. C. J. Ovenston, *Analyst*, **76**, 416 (1951). *CA*, **45**, 8356. Modified int.-std. Boon f.p., liquid filters.
599. Robinson, A. M., and T. C. J. Ovenston, *Analyst*, **79**, 47 (1954). *CA*, **48**, 3192. Int.-std. f.p. d. Li in Mg–Li alloys.
600. Rosendahl, F., *Eine spektrochemische Analysenmethode zur quantitativen Spurenbestimmung in Titandioxyd*. Dissertation, Dortmund/Bonn, 1955.
601. Rosendahl, A., *Tonind.-Ztg. u. Keram. Rundschau*, **20**, 223 (1956). Intfs. in f.p. d. alk. metals.
602. Rosendahl, F., *Spectrochim. Acta*, **10**, 201 (1957). Math. and errors of self-standardization; example from arc spectrography.
602a. Rothe, C. F., and L. A. Sapirstein, *Tech. Bull. Registry Med. Technologists*, **25**, 184 (1955); *Am. J. Clin. Pathol.*, **25**, 1076 (1955). *CA*, **49**, 14871. Self-standardization f.p. d. Ca in biol. materials.
603. Rothermel, D. L., p. 117 in *Symposium on Flame Photometry*, Am. Soc. Testing Materials, Spec. Tech. Publ. No. 116 (1951). Ionizn. intf. of Na, K in f.s.; anal. glass.
604. Rothermel, D. L., and M. E. Nordberg, *Am. Ceram. Soc. Bull.*, **31**, 324 (1952). *CA*, **46**, 10564. F.s. d. durability of glass, via Na, K, Li extn.
605. Roy, N., *Anal. Chem.*, **28**, 34 (1956). *CA*, **50**, 6003. F.p. d. Na, K, Ca, Mg. Mn in glass, etc.

606. Roynette, A., and J. Baron, *Bull. soc. franç. céram.*, No. 22, p. 5; No. 24, p. 3; No. 27, p. 5 (1954); *Glastech. Ber.*, **31**, 33 (1958). *CA*, **49**, 9433; **51**, 11915. F.s. d. Na, K in refractories, ash.
607. Rusanov, A. K., E. V. Gusyatskaya, and N. V. Il'yasova, *Zavodskaya Lab.*, **16**, 447 (1950). *CA*, **49**, 12189. F.p. d. Na, K.
608. Russell, B. J., J. P. Shelton, and A. Walsh, *Spectrochim. Acta*, **8**, 317 (1957). *CA*, **51**, 10966. Absorption f.s.
609. Ryssing, E., *Scand. J. Clin. & Lab. Invest.*, **5**, 321 (1953). *CA*, **48**, 5255. Effect of surface tension on f.p. anal. serum, urine.
610. Sage, S. J., Pittsburgh Conference on Anal. Chem. and Appl. Spectroscopy, Feb. 1956. Abstract in *Spectrochim. Acta*, **8**, 125 (1956). New Ca filter for f.p.
611. Sanduzzi, J., *Am. J. Med. Technol.*, **20**, 375 (1954). *CA*, **48**, 7035. Clinical f.p. d. Na, K.
611a. Schachtschabel, P., and U. Schwertmann, *Z. Pflanzenernähr., Düng. u. Bodenk.*, **82**, 38 (1958). *CA*, **54**, 11354. Effect of alc. on f.p. d. Mg. in soil.
612. Schäfer, K., and K. Staab, *Naturwissenschaften*, **39**, 375 (1952). *CA*, **47**, 7325. Intfs. of Na on K in f.p.
613. Same as 420.
614. Schall, E. D., and R. R. Hagelberg, *J. Assoc. Offic. Agr. Chemists*, **35**, 757 (1952). *CA*, **47**, 2918. F.p. d. K in fertilizers.
615. Schall, E. D., *J. Assoc. Offic. Agr. Chemists*, **36**, 902 (1953). *CA*, **48**, 9863. F.p. d. K in fertilizers.
616. Scharrer, K., and J. Jung, *Z. Pflanzenernähr., Düng. u. Bodenk.*, **67**, 240 (1954). *CA*, **49**, 2937. Intfs. in f.p. d. Ca in agricultural samples.
617. Scharrer, K., and K. Mengel, *Landwirtsch. Forsch.*, **9**, 204 (1956). *CA*, **51**, 7630. F.s. d. Mg in soil, plants.
617a. Scharrer, K., and K. Mengel, *Z. Tierphysiol. Tierernähr. u. Futtermittelk.*, **13**, 142 (1958). *CA*, **53**, 17747. Ion exchange removal of intfs. in f.s. d. Ca, K in plants.
617b. Scharrer, K., and K. Mengel, *Z. Tierphysiol. Tierernähr. u. Futtermittelk.*, **15**, 1 (1960). *CA*, **54**, 21290. F.s. d. K, Na, Ca, Mg in urine, milk, blood.
618. Schätzer, L., *Silikat Tech.*, **6**, 335 (1955). Abstr. in *Ber. deut. keram. Ges.*, **33**, 259 (1956). F.p. anal. feldspar, kaolin, clay, etc.
619. Schinkman, A., *Silikat Tech.*, **2**, 163 (1951). F.p. d. Na, K in ceramics.
620. Schläfer, R., p. 361 in *Proc. Colloquium Spectroscopicum Internationale VI*, Pergamon, London, 1957. Intf. filters.
621. Schlütz, G. O., *Schweiz. med. Wochschr.*, **83**, 383 (1953). *CA*, **48**, 4621. F.p. d. Ca in serum.
622. Schlütz, G. O., *Schweiz. med. Wochschr.*, **83**, 452 (1953). *CA*, **48**, 4030. F.p. d. Ca in .1 ml. whole blood.
622a. Schmauch, G. E., and E. J. Serfass, *Anal. Chem.*, **30**, 1160 (1958). *CA*, **52**, 14409. *Appl. Spectroscopy*, **12**, 98 (1958). *CA*, **53**, 3991. Perchloryl fluoride for f.s.
623. Schmidt, W., and K. Konopicky, *Ber. deut. keram. Ges.*, **35**, 317 (1958). *CA*, **53**, 4681. F.s. d. Al in Al silicates.
624. Schmitz, W., *Jahresber. Limnol. Flusssta. Freudenthal*, **1950**, 45. *CA*, **49**, 11212. F.p. d. Na, K, Ca in waters, urine.
625. Schneider, R., *Landwirtsch. Forsch.*, **6**, 200 (1954). *CA*, **48**, 10480. F.p. d. Ca in plants.

626.* Schönberg, W. D. von, *Arch. exptl. Pathol. u. Pharmakol.*, **214**, 358 (1952). F.p. in the clinic.
627. Schrenk, W. G., and F. M. Smith, *Anal. Chem.*, **22**, 1023 (1950). *CA*, **44**, 10488. Flame attachment for spectrograph.
628. Schrenk, W. G., Round-table discussion, Amer. Chem. Soc., Houston, Mar. 1950. Abstract in *Anal. Chem.*, **22**, 1202 (1950). Miscellaneous on f.p.
629. Schrenk, W. G., and B. L. Glendening, *Anal. Chem.*, **27**, 1031 (1955). *CA*, **49**, 11484. Intf. filters in f.p.
630. Schroeder, D., *Z. Pflanzenernähr., Düng. u. Bodenk.*, **73**, 86 (1956). *CA*, **51**, 2457. F.p. d. Ba in estg. exchange capacity of soil.
631. Schuhknecht, W., *Angew. Chem.*, **50**, 299 (1937). *CA*, **31**, 4921. F.p. d. K in leaves, etc.
632.* Schuhknecht, W., *Optik*, **10**, 245 (1953). *CA*, **48**, 7481. Review of f.p., flame spectrography.
633. Schuhknecht, W., and H. Schinkel, *Z. anal. Chem.*, **143**, 321 (1954). *CA*, **49**, 1475. Al for removing Ca intf. in f.p. d. K, Na, Li.
634. Schuhknecht, W., *Z. anal. Chem.*, **157**, 338 (1957). *CA*, **52**, 963. F.p. d. traces Ca in presence of Na.
635. Schuhknecht, W., and H. Schinkel, *Brennstoff-Chem.*, **38**, 275 (1957). *CA*, **52**, 1586. F.p. d. alkali in fuel ash.
636. Schuhknecht, W., and H. Schinkel, *Z. anal. Chem.*, **160**, 23 (1958). *CA*, **52**, 13529. F.p. d. Ca, Sr, Ba in mixtures.
637. Schuhknecht, W., and H. Schinkel, *Z. anal. Chem.*, **162**, 266 (1958), *CA*, **53**, 967. Mechanism of intf. of Al, Si, P in f.p. d. alk. earths.
638. Schütz, W., *Z. Physik*, **64**, 682 (1930). *CA*, **25**, 3240. Width and self-absorption of Cs blue lines.
639. Schwarz, G., and B. Krauss, *Kiel milchwirtsch. Forschungsber.*, **4**, 579 (1952). *CA*, **47**, 10144. F.p. d. K, Na, Ca in milk.
640. Scott, R. K., V. M. Marcy, and J. J. Hronas, p. 105 in *Symposium on Flame Photometry*, Am. Soc. Testing Materials, Spec. Tech. Publ. No. 116 (1951). *CA*, **47**, 7703. F.s. d. Na, K in water and deposits; intfs.
641.* Scribner, B. F., and W. F. Meggers, *Index to the Literature on Spectrochemical Analysis, 1940–1945*, Part 2. Am. Soc. Testing Materials, Philadelphia, 1947.
642.* Scribner, B. F., and W. F. Meggers, *Index to the Literature on Spectrochemical Analysis, 1946–1950*, Part 3. Am. Soc. Testing Materials, Philadelphia, 1954.
643.* Scribner, B. F., *Anal. Chem.*, **30**, 596 (1958). *CA*, **52**, 9846. Review on emission spectroscopy, including f.p.
644. Seay, W. A., O. J. Attoe, and E. Truog, *Soil Sci.*, **71**, 83 (1951). *CA*, **46**, 675. Removal of Ca intf. in f.p. d. Na in soils, plants.
645. Servigne, M., and P. Guérin de Montgareuil, *Chim. anal.*, **36**, 115 (1954). *CA*, **48**, 8113. Indirect f.p. d. Al.
646. Severinghaus, J. W., and J. W. Ferrebee, *J. Biol. Chem.*, **187**, 621 (1950). *CA*, **45**, 3451. F.s. d. Ca in serum, urine.
647.* Shaw, W. M., *Anal. Chem.*, **30**, 1682 (1958). *CA*, **53**, 2934. Indirect f.s. d. sulfate via Ba.
648. Shimp, N. F., *Methods and results of spectrochemical analysis of biological materials*, Thesis, Rutgers Univ. *Dissertation Abstr.*, **16**, 1058 (1956).

649. Shoji, K., and S. Nakata, *Proc. Am. Soc. Hort. Sci.*, **64**, 299 (1952). *CA*, **49**, 10120. F. s. d. K, Ca, Mg in plants.
650. Shorter, A. J., Soc. Anal. Chem., Glasgow, Mar. 1956. Abstract in *Anal. Chem.*, **28**, 1060 (1956). F.p. anal. silicates.
651. Shukers, C. F., *Am. J. Clin. Pathol.*, **22**, 606 (1952). *CA*, **46**, 8749. Review on f.p. d. Na, K in serum.
652. Davis, S., *J. Biol. Chem.*, **216**, 643 (1955). *CA*, **50**, 1117. F.p. d. Mg in plasma; oxine pptn.
653. Siebert, H., and S. Rapoport, *Biochem. Z.*, **326**, 413 (1955). *CA*, **50**, 5068. Enhancements in f.p. d. K.
654. Siebert, H., and S. Rapoport, *Deut. Gesundheitsw.*, **10**, 481 (1955). Enhancements in f.p. d. K in serum.
655. Siebert, H., *Über die Steigerung der Emission bei der Flammenphotometrie*. Dissertation, Humboldt Univ., Berlin, 1955.
656. Siebert, H., and S. Rapoport, *Z. anal. Chem.*, **150**, 81 (1956). *CA*, **50**, 9205. Enhancements of Na, K, Ca in f.p.
657. Sigrist und Weiss A.-G., Falkenstr. 23, Zürich 1/8. Flame photometer.
658. Sims, E. A. H., and L. Kaplow, *J. Lab. Clin. Med.*, **41**, 303 (1953). *CA*, **47**, 5175. Dust protection for f.p.
659. Smales, A. A., *Analyst*, **76**, 348 (1951). *CA*, **45**, 7466. F.p. and radiochem. d. Sr in sea water.
660. Smit, J. A., and A. J. H. Vendrik, *Physica*, **14**, 505 (1948). *CA*, **43**, 5303. Absolute conc. of Li, K in flame; ionizn., mol. formation with Li, K, Rb.
661. Smit, J., C. T. J. Alkemade, and J. C. M. Hattinga-Verschure, *Biochim. et Biophys. Acta*, **6**, 508 (1951). *CA*, **46**, 11302. Basic study of a.c. f.p.; ionizn. intf.; org. solvents; anal. of errors; d. Na, K in serum.
661a. Smith, G. W., and A. K. Palmby, *Anal. Chem.*, **31**, 1798 (1959). *CA*, **54**, 3928. F.s. d. Pb, Mn in gasoline.
662. Smith, H., and T. M. Sugden, *Proc. Roy. Soc. (London)*, **A211**, 31 (1952); *Ibid.*, 58. *CA*, **46**, 10803. Microwave detn. of ionizn. in air–H_2 flame contg. K, Na; stability of alkali hydroxides.
663. Smith, R. G., P. Craig, E. J. Bird, A. J. Boyle, L. T. Iseri, S. D. Jacobson, and G. B. Myers, *Am. J. Clin. Pathol.*, **20**, 263 (1950). *CA*, **44**, 5421. Spectrochem. values for Na, K, Fe, Ca, Mg in blood; f.p. d. Na, K.
664. Smithells, A., and F. Dent, *J. Chem. Soc.*, **65**, 603 (1894). C_2N_2 flame.
664a. Smoczkiewiczowa, A., *Nature*, **183**, 1260 (1959). *CA*, **53**, 20371. F.p. d. Na, K, Ca in biol. fluids.
665. Snelleman, W., and J. A. Smit, *Physica*, **21**, 946 (1955). *CA*, **50**, 10538. Photoelec. temp. measurement by line reversal.
666. Sobolev, N. N., E. M. Mezhericher, and G. M. Rodin, *Izvest. Akad. Nauk S.S.S.R., Ser. Fiz.*, **14**, 737 (1950). *CA*, **45**, 4134; **46**, 4359. Line profile and working curves of alk. and alk. earth metals in flame.
667. Sobolev, N. N., E. M. Mezhericher, and G. M. Rodin, *Zhur. Exptl. Teoret. Fiz.*, **21**, 350 (1951). *CA*, **46**, 4359. Same subject as 666.
668. Sobolev, N. N., p. 310 in *Proc. Colloquium Spectroscopicum Internationale VI*, Pergamon, London, 1957. *CA*, **54**, 1062. Line profiles for Na, Li, Ca, Tl in flame and arc; working curve of K.
669. Solomon, A. K., and D. C. Caton, *Anal. Chem.*, **27**, 1849 (1955). *CA*, **50**, 1378. F.p. for microdetn. Na, K.

670. van Someren, E., and W. H. Storey, discussion at Industrial Spectroscopy Group of Inst. of Physics and Photoelec. Spectrometry Group, London, Mar. 1949; *Spectrochim. Acta*, **4**, 66 (1950).
671. Sommer, H., *Monatsh. Veterinärmed.*, **11**, 154 (1956). F.p. d. K in pig serum, plasma.
672. Spector, J., *Anal. Chem.*, **27**, 1452 (1955). *CA*, **49**, 15604. Al for suppressing Ca intf. with alkalies in f.p.
673. Spencer, A. G., *Lancet*, **259**, 623 (1950). *CA*, **45**, 3017. Review; int.-std. f.p. d. Na, K in biol. fluids.
674. Spindler, F., and E. F. Wolf, *Landwirtsch. Forsch.*, **9**, 179 (1956). *CA*, **51**, 7936. Removal of intf. in f.p. d. Ca.
675. Staab, K., *Die Flammenphotometrie als analystisches Verfahren*. Dissertation, Technische Hochschule, Karlsruhe, 1953.
676. Stace, H. C. T., *J. Council Sci. Ind. Research (Australia)*, **21**, 305 (1948). *CA*, **43**, 8590. Flame spectrography for Mg in soil.
676a. Stace, H. C. T., and J. T. Hutton, *Soils and Fertilizers, Commonwealth Bur. Soil Sci.*, **22**, 157 (1959). F.s. in soil anal.
677. Standen, G. W., and C. B. Tennant, *Anal. Chem.*, **28**, 858 (1956). *CA*, **50**, 9212. F.p. d. Ca in slag.
678. Steens-Lievens, A., *J. pharm. Belg.*, **11**, 541 (1956). *CA*, **51**, 8187. Review; clinical f.p.
679. Strange, E. E., *Anal. Chem.*, **25**, 650 (1953). *CA*, **47**, 5842. F.p. d. Li in Mg alloys.
680. Strasheim, A., and J. P. Nell, *J. S. African Chem. Inst.*, **7**, 79 (1954). *CA*, **49**, 11758. F.p. d. Ca in plants, etc.
681. Strasheim, A., and R. J. Keddy, *Appl. Spectroscopy*, **12**, 29 (1958). *CA*, **52**, 13424. Math. treatment of spectrochem. precision.
682. Straub, W. A., S. H. Bauer, and W. D. Cooke, Pittsburgh Conference on Anal. Chem. and Appl. Spectroscopy, Mar. 1958; Abstr. in *Spectrochim. Acta*, **12**, 377 (1958). W. A. Straub, *Spark-in-flame spectroscopy*, Ph.D. Thesis, Cornell Univ., 1958; *CA*, **52**, 16116.
683. Straubel, H., *Naturwissenschaften*, **40**, 337 (1953). *CA*, **48**, 6291. Electrostatic atomization.
684. Straubel, H., *Mikrochim. Acta*, **1955**, 329. *CA*, **49**, 11329. Electrostatic atomizer for f.p.
685. Straubel, H., *Dechema Monograph.*, **32**, 153 (1959). *CA*, **54**, 16033. Evapn. rate and charge change on liq. droplets.
686. Streed, E. R., and U. W. Stoll, *ASTM Bull.*, No. 189, p. 58 (1953). *CA*, **47**, 6629, F.p. d. cement in soil–cement and pozzolan–cement mixtures.
687. Strickland, R. D., and C. M. Maloney, *Am. J. Clin. Pathol.*, **24**, 1100 (1954). *CA*, **49**, 417. F.s. d. sulfate in serum via Ba.
688. Strohmeyer, F. B., B. E. Leach, and R. G. Heath, Regional Conclave, Am. Chem. Soc., New Orleans, Dec. 1953. Abstr. in *Anal. Chem.*, **25**, 1935 (1953). F.s. d. Ca, Mg in biol. materials; enhancement by alc.
689. Stuart, W. A., M. Simpson, and W. H. Hardwick, *Analyst*, **82**, 200 (1957). *CA*, **51**, 10302. EEL f.p. d. Li, Na, K in mixtures.
690. Stumpf, K. E., *Angew Chem.*, **66**, 756 (1954). F.p. d. Sr in Ba compounds.
691. Stumpf, K. E., and T. Gonsior, p. 35 in *Proc. Colloquium Spectroscopicum Internationale VI*, Pergamon, London, 1957. *CA*, **54**, 152. Mutual intf. in f.p. d. Na, K.

692. Stumpf, K. E., and T. Gonsior, Colloquium Spectroscopicum Internationale VII, Liège, 1958; *Rev. universelle mines*, **15**, 404 (1959). F.p. anal. slag.
693. Sugawara, K., T. Koyama, and N. Kawasaki, *Bull. Chem. Soc. Japan*, **29**, 679, 683 (1956). *CA*, **51**, 2462. F.p. d. Na, K, Ca, Sr in waters.
694. Sugden, T. M., *Discussions Faraday Soc.*, No. 19, 68 (1955). *CA*, **50**, 8324. Ionizn. and hydroxide formation of alk. and alk. earth metals in flames.
695.* Sugden, T. M., and R. C. Wheeler, *Discussions Faraday Soc.*, No. 19, 76 (1955). *CA*, **50**, 8330. Ionizn., LiOH formation, stability of BaO, SrO in air–H_2 flame.
696. Same as 171.
697. Sugden, T. M., and P. F. Knewstubb, *Research Correspondence*, **9**, No. 8, p. S32 (1956). *CA*, **50**, 16299. Ionizn. of Pb in flames.
698. Sugden, T. M., *Trans. Faraday Soc.*, **52**, 1465 (1956). *CA*, **51**, 7114. Equilibria of H, OH; dissocn. of CuOH, KOH, LiOH, CuH in H_2 flames.
699. Sykes, P. W., *Analyst*, **81**, 283 (1956). *CA*, **50**, 10601. F.p. d. Li in ores.
700. Taylor, A. E., and H. H. Paige, *Anal. Chem.*, **27**, 282 (1955). *CA*, **49**, 8037. F.s. d. traces Sr; intfs.
700a. Teloh, H. A., *Am. J. Clin. Pathol.*, **30**, 129 (1958). *CA*, **52**, 17364. F.s. d. Mg in serum by std.-addition.
701. Thiele, W., *Engineer's Digest*, **19**, 526 (1958). Instrumentation and methods in f.p.
702. Todd, H. E., and H. M. Tramutt, *Anal. Chem.*, **26**, 1137 (1954). *CA*, **49**, 15275. F.s. d. rubber on cords, fabrics, via Na in the fabric.
703. Toribara, T. Y., P. A. Dewey, and H. Warner, *Anal. Chem.*, **29**, 540 (1957). *CA*, **51**, 9777. F.p. d. Ca in biol. materials.
704. Török, T., *Z. anal. Chem.*, **119**, 120 (1940). *CA*, **34**, 4687. F.p. titration of alk. earth with phosphate, using Zn + HCl in sample to generate spray.
704a. Same as 182a.
705. Regulations for accident prevention: Berufsgenossenschaft der Feinmechanik und Elektrotechnik VBG 15, Zentralstelle für Unfallverhütung, Reuterstr. 157–159, Bonn.
706. Unger, I., and L. Unger, *Glastech. Ber.*, **29**, 15 (1956). F.p. d. Na, K in glass.
707. Unicam Instruments Ltd., Arbury Works, Cambridge. *SP.900 Flame Spectrophotometer*, 1958.
708. Valencia, R., *Bull. soc. chim. biol.*, **38**, 1071 (1956). *CA*, **51**, 2921. F.s. d. Na, K, Ca, Mg in plasma, tissues.
709. Vallee, B. L., Pittsburgh Conference on Anal. Chem. and Appl. Spectroscopy, Mar. 1954. Abstr. in *Anal. Chem.*, **26**, 432 (1954). Multichannel grating f.s. for Na, K, Ca, Mg.
710. Vallee, B. L., *Nature*, **174**, 1050 (1954). *CA*, **49**, 5204. Multichannel f.s. for Na, K, Ca, Mg, Sr.
711. Vallee, B. L., and M. R. Baker, Pittsburgh Conference on Anal Chem. and Appl. Spectroscopy, Mar. 1955. Abstr. in *Anal. Chem.*, **27**, 320 (1955). O_2–C_2N_2 f.p.
712. Vallee, B. L., and M. Margoshes, *Anal. Chem.*, **28**, 175 (1956). *CA*, **50**, 7650 Multichannel f.s.
713. Vallee, B. L., and A. F. Bartholomay, *Anal. Chem.*, **28**, 1753 (1956). *CA*, **51**, 2455. O_2–C_2N_2 f.p.
714. Vallee, B. L., p. 49 in *Proc. Colloquium Spectroscopicum Internationale VI*, Pergamon, London, 1957. O_2–C_2N_2 f.p.

714a. Vallee, B. L., (paper given by R. E. Thiers), Syposium on Spectroscopy, Am. Soc. Testing Materials, San Francisco, Oct. 1959. O_2–C_2N_2 f.p.
715. Velken, S., *Tidsskr. Kjemi, Bergvesen Met.*, **17**, 20 (1957). Abstr. in *Stahl u. Eisen*, **77**, 673 (1957). Anal. of slags.
716. Voinovitch, I. A., and J. Debras, *Bull. soc. franç. céram.*, No. 32, p. 29 (1956). *CA*, **52**, 19703. Sample prepn. and intfs. in f.p. d. Na, K in ceramics.
717. Voinovitch, I., J. Debras-Guédon, and E. Vilnat, in *Proc. VIII. Colloquium Spectroscopicum Internationale* (Lucerne, 1959), Sauerländer, Aarau, 1960. (*Anal. Abstr.*, **7**, No. 5100). Arc-in-spray spectroscopy.
718.* Vukanović, D. D., *Bull. Inst. Nuclear Sci. "Boris Kidrich"* (*Belgrade*), **8**, 43 (1958). *CA*, **52**, 14424. F.p. d. Li in org. solvents.
719. Wagner, K. T., *Untersuchungen über die Verwendbarkeit der Ca-Linie bei 423 mμ für flammenspectrophotometrischen Serum-Calcium-Analysen.* Dissertation, Giessen, 1957.
720. Wallace, W. M., M. Holliday, M. Cushman, and J. R. Elkinton, *J. Lab. Clin. Med.*, **37**, 621 (1951). Janke f.p. for anal. biol. samples.
721. Walsh, E. G., *J. Sci. Instr.*, **29**, 23 (1952). *Sci. Abstr.*, **55A**, No. 3095. Color-wedge null-balancing int.-std. f.p. for Na, K.
722. Walsh, A., *Spectrochim. Acta*, **7**, 108 (1955). *CA*, **49**, 10792. Absorption f.p.: instrument, theory, data.
723. Warren, R. L., *J. Sci. Instr.*, **29**, 284 (1952). *CA*, **47**, 2547. Oxybutane atomizer-burner f.p. for Na, K in biol. samples.
724. Warren, R. L., p. 213 in *Proc., VIII. Colloquium Spectroscopicum Internationale* (Lucerne, 1959), Sauerländer, Aarau, 1960. High-luminosity prism f.s. with atomizer-burner, automatic scanning, for micro samples.
725. Wasilko, E. G., Pittsburgh Conference on Anal. Chem. and Appl. Spectroscopy, Feb. 1956. Abstr. in *Spectrochim. Acta*, **8**, 125 (1956). F.s. d. Pb, Cu in plating baths.
726. Watanabe, H., and K. K. Kendall, *Appl. Spectroscopy*, **9**, 132 (1955). Flame spectrograms of Al, Ba, Ca, Cs, Cr, Co, Cu, Fe, Pb, Li, Mg, Mn, Ni, K, Rb, Ag, Na, Sr.
727. Wehner, G., and W. Bunge, *Chem. Tech.* (*Berlin*), **5**, 251 (1953). *CA*, **48**, 6903. F.p. d. Na in K salts and Ca in Ba salts.
728. Weichselbaum, T. E., and P. L. Varney, *Proc. Soc. Exptl. Biol. Med.*, **71**, 570 (1949). *CA*, **43**, 8217. Atomizer-burner for f.p. d. Na, K.
729. Weichselbaum, T. E., P. L. Varney, and H. W. Margraf, Pittsburgh Conference on Anal. Chem. and Appl. Spectroscopy, Mar. 1951. Abstr. in *Anal. Chem.*, **23**, 684 (1951). Atomizer-burners for O_2–H_2, O_2–C_2H_2, O_2–gas; f.s. d. Ba, Ce, Cu, La, Li, Mn, Pb, Rb, Sr, Tl.
730. Werk, O., *Untersuchungen über den K- und Ca-Gehalt von Blättern feucht und trocken gezogener Pflanzen*. Dissertation, Institute of Technology, Hanover, 1953.
731. West, P. W., P. Folse, and D. Montgomery, *Anal. Chem.*, **22**, 667 (1950). *CA*, **44**, 8025. F.s. d. Na, K, Ca in water; intfs. removed by buffer.
732. Westerhoff, H., *Landwirtsch. Forsch.*, **7**, 128 (1955). *CA*, **49**, 8514. F.p. d. Ca in foods, soil.
733. Wever, F., W. Koch, and G. Wiethoff, *Arch. Eisenhüttenw.*, **24**, 383 (1953). *CA*, **48**, 1880. Detn. ferrite in steel via f.s. d. Fe, Cr, Mn in Na citrate solns.

734. Whisman, M., and B. H. Eccleston, *Anal. Chem.*, **27**, 1861 (1955). *CA*, **50**, 9210. Flame spectra of Ba, B, Ca, Cr, Co, Cu, Fe, Pb, Li, Mg, Mn, Ni, K, Rb, Ag, Na, Sr, Tl, Ti, V, petroleum oils.
735. White, C. E., M. H. Fletcher, and J. Parks, *Anal. Chem.*, **23**, 478 (1951). *CA*, **45**, 5564. Fluorimetric d. Li in rocks.
735a. White, J. U., *Anal. Chem.*, **24**, 394 (1952). *CA*, **46**, 4861. New int.-std. f.p. for Na, K, Ca.
736. Wiethan, K., *Anwendung der Flammenphotometrie im Eisenhüttenlaboratorium*. Dissertation, Technische Hochschule, Aachen, 1953.
737. Wiethoff, G., *Flammenspektralanalytische Untersuchung von Schwermetallen, besonders zur Bestimmung der Ferritzusammensetzung in Stählen*. Thesis, Max Planck-Institut für Eisenforschung, 1953. Same subject as 733.
738. Wilberg, E., *Z. anal. Chem.*, **131**, 405 (1950). *CA*, **45**, 1907. F.p. d. Li.
739. Wiley, J. T., and T. B. Smitherman, Am. Chem. Soc., Little Rock, Dec. 1952. Abstract in *Anal. Chem.*, **24**, 2016 (1952). Advances in f.p.
740. Will, E. G., and B. Schwarzkopf, *J. Am. Water Works Assoc.*, **47**, 253 (1955). *CA*, **49**, 7782. F.s. d. Ca in water.
741. Willebrands, A. F., Jr., *Rec. trav. chim.*, **69**, 799 (1950). *CA*, **44**, 8984. F.p. d. Na, K in serum, urine.
742. Willgallis, A., *Z. anal. Chem.*, **157**, 249 (1957). *CA*, **52**, 963, F.p. d. Na, K in rocks.
743. Williams, J. P., and P. B. Adams, *J. Am. Ceram. Soc.*, **37**, 306 (1954). *CA*, **49**, 4447. F.s. d. Li, Na, K, Rb, Cs in glass, ores.
744. Willits, C. O., and J. A. Connelly, *Anal. Chem.*, **24**, 1525 (1952). *CA*, **47**, 1433. Reflux atomizer for f.p.
745. Wilson, T. C., and N. J. Krotinger, *ASTM Bull.*, No. 189, p. 56 (1953). *CA*, **47**, 6629. F.p. d. Mg in cement.
746. Winer, A. D., and D. M. Kuhns, *Am. J. Clin. Pathol.*, **23**, 1259 (1953). *CA*, **48**, 1464. F.s. d. Ca in serum, urine, with photomultiplier and alc.
747. Winer, A. D., and K. F. Ernst, *US Armed Forces Med. J.*, **5**, 823 (1954). *CA*, **48**, 8850. Clinical f.p. d. Na, K, Ca.
748. Wirtschafter, J. D., *Science*, **125**, 603 (1957). *CA*, **51**, 11861. Protective chelation with EDTA to remove intf. in f.s. d. Ca.
749. Same as 735a.
750. Woldring, M. G., *Anal. Chim. Acta*, **8**, 150 (1953). F.p. d. Na, K in biol. fluids.
751. Woodcock, A. H., and A. T. Spencer, *J. Meteorol.*, **14**, 437 (1957). *CA*, **52**, 2462. Airborne f.p. for detn. Na in atm.
752. Wright, B. M., *Lancet*, **1950**, 883. Review of f.p.
753. Wurziger, J., *Deut. Lebensm.-Rundschau*, **51**, 124 (1955). *CA*, **49**, 11235. F.p. d. K, Na in wine.
754. Yofé, J., and R. Finkelstein, *Anal. Chim. Acta*, **19**, 166 (1958). *CA*, **54**, 1168. Releasing agents: La, Fe to remove P intf. with Ca.
755. Zagrodzki, S., and H. Zaorska, *Chem. Tech. (Berlin)*, **10**, 210 (1958). *CA*, **52**, 16659. Monitoring f.p. for Na, K.
756. Zaïdman, N., and D. Orechkin, *Novosti Neftyanoĭ Tekh.*, *Neftepererabotka*, **1955**, No. 5, p. 25. *CA*, **51**, 5625. F.p. d. Na, K, Li, Ca in petroleum wash water and catalysts.
757.* Zak, B., R. E. Mosher, and A. J. Boyle, *Am. J. Clin. Pathol.*, **23**, 60 (1953). *CA*, **47**, 3391. Review on f.p. d. Na, K, Ca, Mg in clinical lab.

758. Carl Zeiss, Oberkochen, Württemberg. Flame Photometer PF V; instruction manual G 50-620/1-d. Also flame attachment for the PMQ II spectrophotometer.
759. Carl Zeiss, index to the literature of flame photometry, 50-801-d.
759a. Carl Zeiss, *Flammenspektren*, Cat. No. A 50-812 (1961).
760. Zettler, H., *Die Methoden der Flammenspektrophotometrie und ihre Anwendungen*. Dissertation, Universität Frankfurt am Main, 1953.
761. Zoellner, H., *Glas-Email-Keramo-Tech.*, **2**, 290 (1951). *CA*, **46**, 5. Comparison of f.p.'s; detn. Na, K, Li, Rb in silicates.
762. Zoellner, H., *Glas-Email-Keramo-Tech.*, **2**, 349 (1951). *CA*, **46**, 3447. Accuracy of phys. and chem. measurements.
763. Zoellner, H., *Glas-Email-Keramo-Tech.*, **2**, 378 (1951). *CA*, **46**, 3446. F.p. d. Na, K in silicates.
764. Zoellner, H., *Glas-Email-Keramo-Tech.*, **5**, 164 (1954). New atomizer for f.p.
765. Zoellner, H., Deutsche keramische Gesellschaft, Düsseldorf, Oct. 1955. Review of experience with f.p., etc.
766. Zwetsch, A., *Sprechsaal*, **85**, 91 (1952). *CA*, **46**, 5481. F.p. d. Na, K in ceramics.

ATLAS OF
FLAME SPECTROGRAMS

ATLAS OF FLAME SPECTROGRAMS

The spectrograms represent 57 elements (listed in the index of emitting species, below), arranged in the order of the periodic system.

Equipment

Most of the spectrograms were obtained with a Beckman DU spectrophotometer with photomultiplier and spectral energy recording attachments and various strip-chart recorders. Below 700 mμ a 1P28 photomultiplier was used. At longer wavelengths the Beckman red-sensitive phototube was used where not otherwise specified; in a few cases an RCA C7160 or Farnsworth 16PMI photomultiplier was used instead. Spectrograms 19, 20, 24–33, and 35 were obtained with the Oak Ridge National Laboratory flame spectrophotometer,[414] using an RCA 6217 photomultiplier below 680 mμ and a Farnsworth 16PMI at longer wavelengths; the spectral slit width was 1.6 mμ. The oxycyanogen flame spectrograms were obtained with an indirect atomizer, a heated spray chamber and a small platinum-tipped burner operating on premixed oxygen and cyanogen.[304,307a] All the rest, including those from Oak Ridge, were obtained with a Beckman atomizer-burner (No. 4020), except Spectrogram 17, which was obtained with a modified 4020 burner operating on premixed oxyacetylene, and Spectrogram 5, obtained with a large-bore atomizer-burner (No. 4060). The sheath employed for Spectrograms 5, 9, 45, 75, 83, 85 and 92 is an attachment for the Beckman atomizer-burner that supplies a flowing atmosphere of oxygen (or air–oxygen mixture) surrounding the flame (Gilbert, Pittsburgh Conference, March 1962). Spectrogram 70 is not genuine—it was prepared by drawing the gallium lines in their proper intensities and profiles upon a recorded background.

Materials

The solutions were aqueous except as noted. The hexone solutions of the rare earths were 0.1 M in 2-thenoyltrifluoroacetone. In a few cases, acid present in the solution attacked the atomizer capillary, introducing lines of palladium, iron, nickel, manganese, chromium, and silver. In other cases impurities in the chemicals are revealed by their spectra. Note the lines of palladium in 36 and 49, the trace of gallium in 68, the trace of nickel in 55, the trace of tin in 88, and the impurities of beryllium in 10. The heavy sodium contamination in 68 was from the surface of the metallic aluminum sample. The mercury lines at 404.7, 435.8 and 546.1 mµ come from the fluorescent room lights. The silver and copper lines in the oxycyanogen spectra are due to the silver solder in the burner. The carbon line in 66, 74, and 87 occurs in the flame itself. A certain mixture of nine elements (Li, K, Mg, Sr, Cr, Fe, Ni, Cu, In) was used under various conditions in several of the spectrograms (Nos. 13, 17, 46, 47, 48, 50, 59).

Intensities

In all of the recordings the base line or wavelength scale represents the zero of intensity, except as noted (the stated zero suppression is the distance of the true zero line below the base line). The intensity (ordinate) scale measured from the zero line is adjusted arbitrarily in each spectrogram to give a convenient range. In a few cases recordings at different sensitivity adjustments (gain or amplification) or slit widths are included on the same spectrogram for comparison; the curves are marked as needed. The Oak Ridge spectrograms (Nos. 19, 20, 24–33, 35) are all at the same amplification (but with different zero suppressions) except for praseodymium, for which the amplification is twice as great. In most of the spectra the blank (usually water) is shown along with the spectrum of the element, to reveal the net line or band intensities, the continuum due to the element, and the background itself. When there is no measurable element continuum, as in 63 and 76, the blank is exactly superposed on the spectrum of the element except at the lines. In 56 the intensity scale of the blank is magnified with respect to that of the cobalt. In 7 and 8 the background is not entirely correct, since the recordings for the photomultiplier and the red-sensitive phototube were deliberately joined. In general, spectra obtained with the same kind of flame and solvent (oxyhydrogen and water, for example) can be compared with each other by using the blank intensity as a normalizing factor.

Wavelengths

These are indicated upon the lines and bands, except in the Oak Ridge spectrograms. For the latter, the bands can be identified by reference to Table 10, Beckman Bulletin 753-A,[307] or the tables in the 43rd edition of the *Handbook of Chemistry and Physics*, pp. 2804–33; most of the finer details listed in Table 13 cannot be seen on these rare-earth spectrograms. Unless otherwise labelled, the emission belongs to the element in question. Extraneous emissions and flame background bands are duly labelled as such, often in smaller print. The flame bands are frequently useful as landmarks, and many of them are identified on the spectrograms. The emitters of the flame bands are listed along with the elements in the index of emitting species. (Some of the molecules listed in this index are of very dubious identity; see the introduction to Table 13.) For diatomic molecules, the wavelengths of the band heads are indicated, and the vibrational transitions are often shown in parentheses. For headless or diffuse bands the peak is marked. For bands of sufficiently open structure, rotational lines are labelled as well. Most of the spectrograms of the atlas were prepared before Table 13, and consequently some discrepancies in wavelengths or identities between the spectrograms and the table will be noticed. In such instances, the data of the table should be accepted. Examples include the bands of cerium and InO and a few of the short-wavelength lines of cobalt.

Sources and Acknowledgments

Spectrograms 7, 8, 11, 14, 18, 38, 42, 49, 58, 62, 63, and 82 were prepared by Mr. Hideo Watanabe of Beckman Instruments, Inc., and are reproduced by permission of *Applied Spectroscopy*.[726] Spectrograms 36, 40, 65, 66, 73, and 77 were also prepared by Mr. Watanabe. Spectrograms 2 and 3 were prepared by Mr. Bruce E. Buell of the Union Oil Co., Brea, California, and are reproduced by permission of *Analytical Chemistry* (ref. 168 and a later paper*). Spectrograms 19, 20, 24–33, and 35 were made available by Dr. Oscar Menis, and have been published by T. C. Rains, H. P. House and O. Menis in *Analytica Chimica Acta* **22**, 315–327 (1960); they are reproduced by permission of that journal. The remaining spectrograms were prepared in the laboratories of Beckman Instruments by the translator, who also supplied the identifications, annotations, wavelengths, and wavelength scales on all of the spectrograms except those from Oak Ridge and No. 3. A few of the spectrograms were prepared originally for Mr. G. H. Wyatt of the British Transport Commission, to be published in *Comprehensive Analytical Chemistry*, Vol. II. Spectrograms 1, 10, 12, 13, 15, 17, 21, 23, 34, 36, 37, 39, 41, 46, 53, 54, 55, 56, 64, 67, 68, 71, 72, 73, 77, 81, 89, and 90 have been published by the American Society for Testing Materials.[307a]

* *Anal. Chem.* **34**, 635–640 (1962).

INDEX OF EMITTING SPECIES

Ag, 10, 37, 52, 57, 60, *63*, 81
Al, *68*, *69*
AlO, *68*
As, AsO, *87*
BO_2, *67*
Ba, BaO, *18*
Ba^+, BaOH, 13, *18*
Be, *9*
BeO, *10*
Bi, *89*, *90*, *91*
BiO, *89*, 90
C, *66*, *74*, 87
C_2, 2, 3, 66, 80, 84, *94*, *95*
CH, 2, 3, *80*, 84
CN, 37, 41, 44, 52, 57, 60, *94*, *95*
CO, *83*
CS, *92*
Ca, 1, 4, 13, *14*, 16, 17, 69
Ca^+, 13, 14, 16, *69*
CaO, *5*, *14*
CaOH, 1, 4, *5*, *13*, *14*, 17
Cd, *65*
CeO, *23*
Co, *53*, *54*, *55*, *56*, *57*
CoO, *56*
Cr, 10, *13*, *17*, 36, *38*, *39*, *47*, *48*, 59
CrO, *38*
Cs, *6*, *8*
Cu, 13, *17*, 37, *47*, 50, 52, 57, *62*, 81
CuH, CuO, CuOH, *62*
DyO, *30*
ErO, *32*
Eu, Eu^+, EuOH, *27*
Fe, 10, *13*, *17*, 36, *46*, *47*, *48*, *49*, *50*, *51*, *52*, 59, 68
FeO, 13, *49*
Ga, 68, *70*
GdO, *28*
Ge, *76*
HO, see OH
H_2O, 3, 5, 6, 22, 42
Hg, *66*
HoO, *31*

In, *13*, *17*, *47*, *48*, *71*, *72*
InO, *72*
K, *3*, *4*, 5, 6, 7, 13, 17, 22, 24, 25, *46*, *48*
LaO, *21*, *22*
Li, *1*, *5*, *13*, 17, 67
LuO, *35*
Mg, 4, *11*, *12*, *50*
Mg^+, 12
MgO, 4, *11*
MgOH, *4*, *11*, 13, 47, 48
Mn, 10, 13, 36, *42*, *44*, 46, 48, 57, 68
Mn^+, *43*
MnO, *42*, 44
MnOH, *42*
Mo, *41*
Mo continuum, *40*
NC, see CN
NH, 17, 52, 57, 60, *94*
NO, *74*, *85*, *91*, *92*, *94*
Na, 2, *4*, 5, 22, 68
NdO, *25*
Ni, 10, *13*, *17*, 36, *46*, *47*, *48*, 50, 55, *58*, *59*, *60*
NiO, *58*
OH, 3, 9, *12*, 17, 50, *52*, 53, 54, *57*, 60, *64*, 66, 68, 71, 77, 81, 86, 87, 88, 89, 90
PO, *86*
Pb, *82*, *83*, *84*
Pb^+, 83
Pd, 36, 49
PrO, *24*
Rb, *5*, *6*, *7*
Re, *45*
Rh, *61*
SC, see CS
Sb, SbO, *88*
ScO, *19*
Si, SiO, *75*
SmO, *26*
Sn, *77*, *78*, *79*, *80*, *81*, 88
SnO, *77*
Sr, 4, *13*, *15*, *16*, *17*

Sr⁺, 13, *15*, *16*, *17*, 48
Sr₂O₂, *15*
SrOH, 13, *15*, 17
TbO, *29*
TeO, *93*
Tl, *73*

TmO, *33*
V, V⁺, *37*
VO, *36*
YO, *20*
Yb, Yb⁺, YbO, YbOH, *34*
Zn continuum, 64

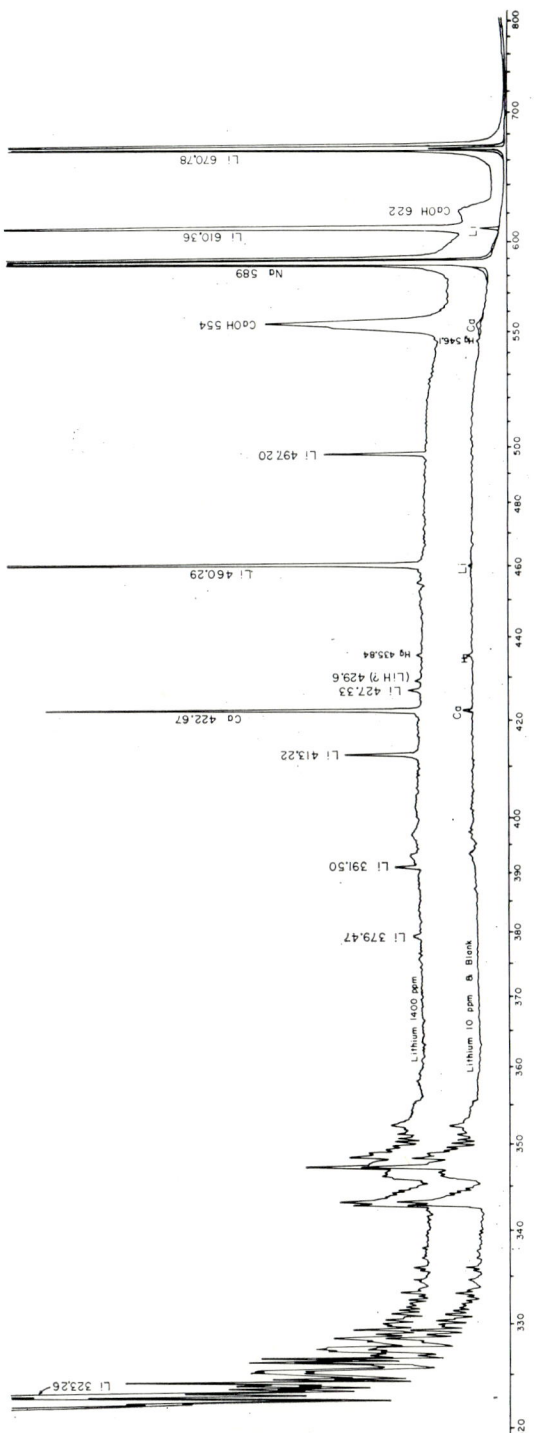

Spectrogram 1. Lithium, 10 and 1400 p.p.m.; oxyhydrogen; slit 0.02 mm.

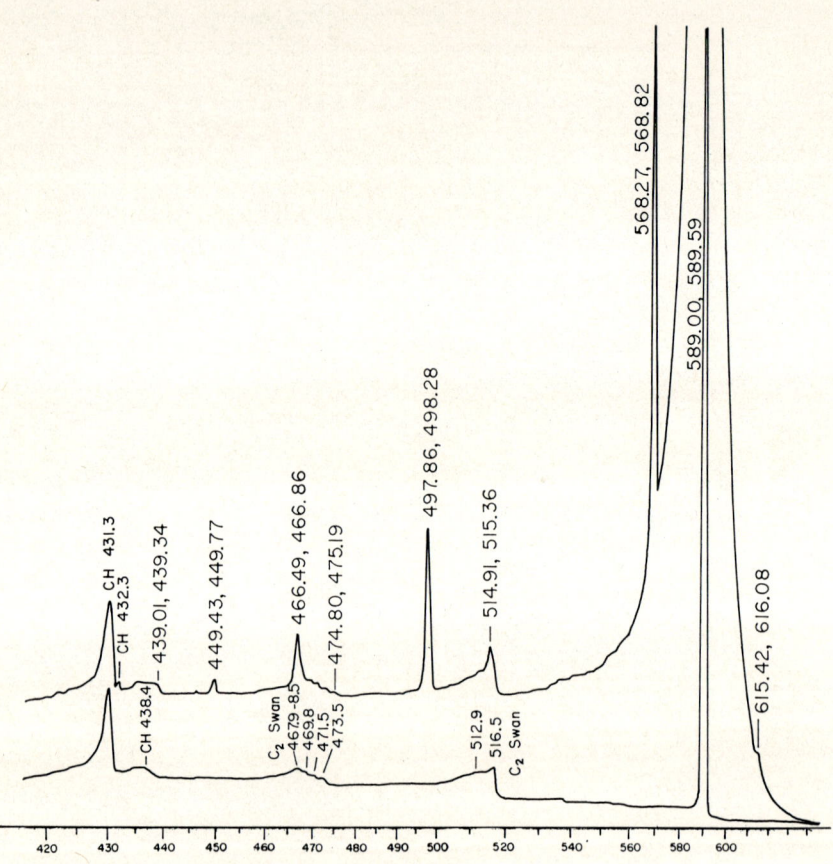

Spectrogram 2. Sodium, 800 mg./l. in naphtha–isopropanol; oxyhydrogen; slit 0.02 mm.

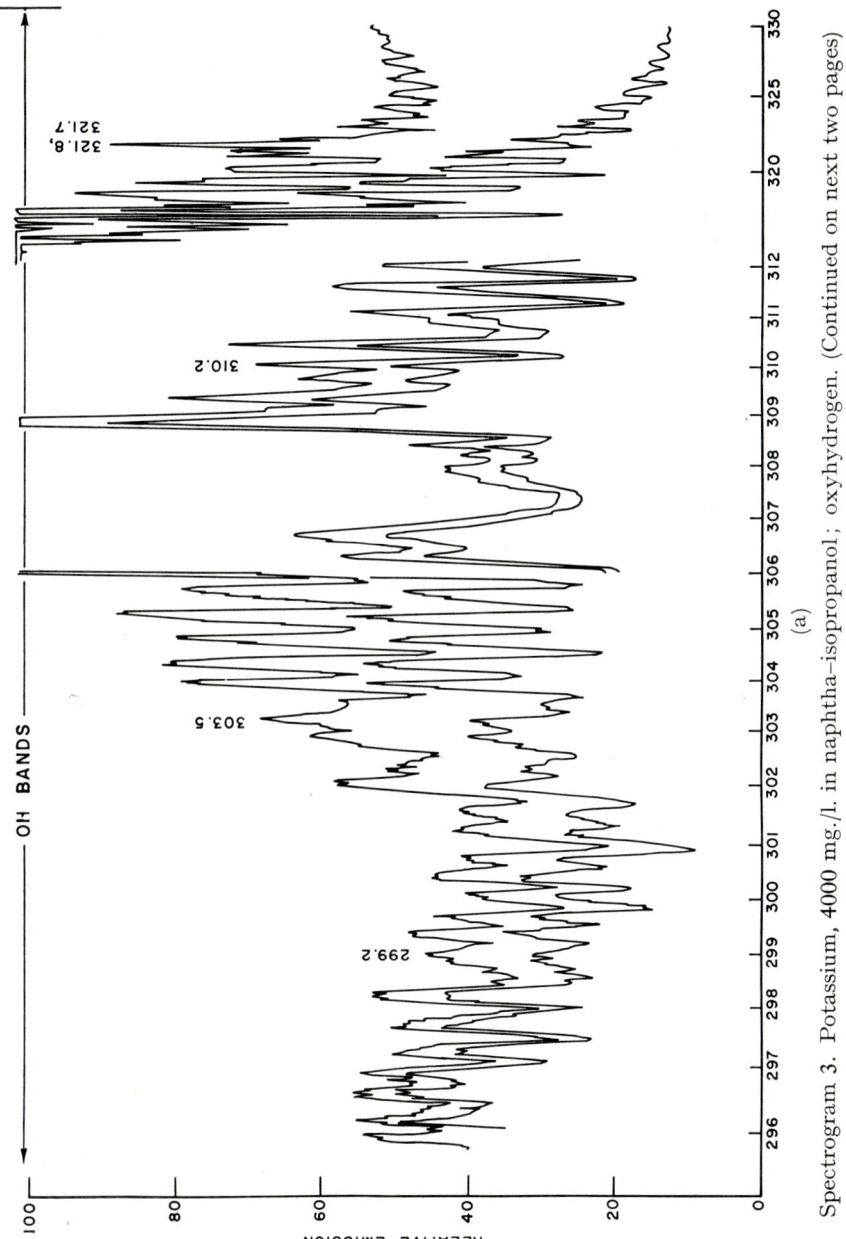

Spectrogram 3. Potassium, 4000 mg./l. in naphtha–isopropanol; oxyhydrogen. (Continued on next two pages)

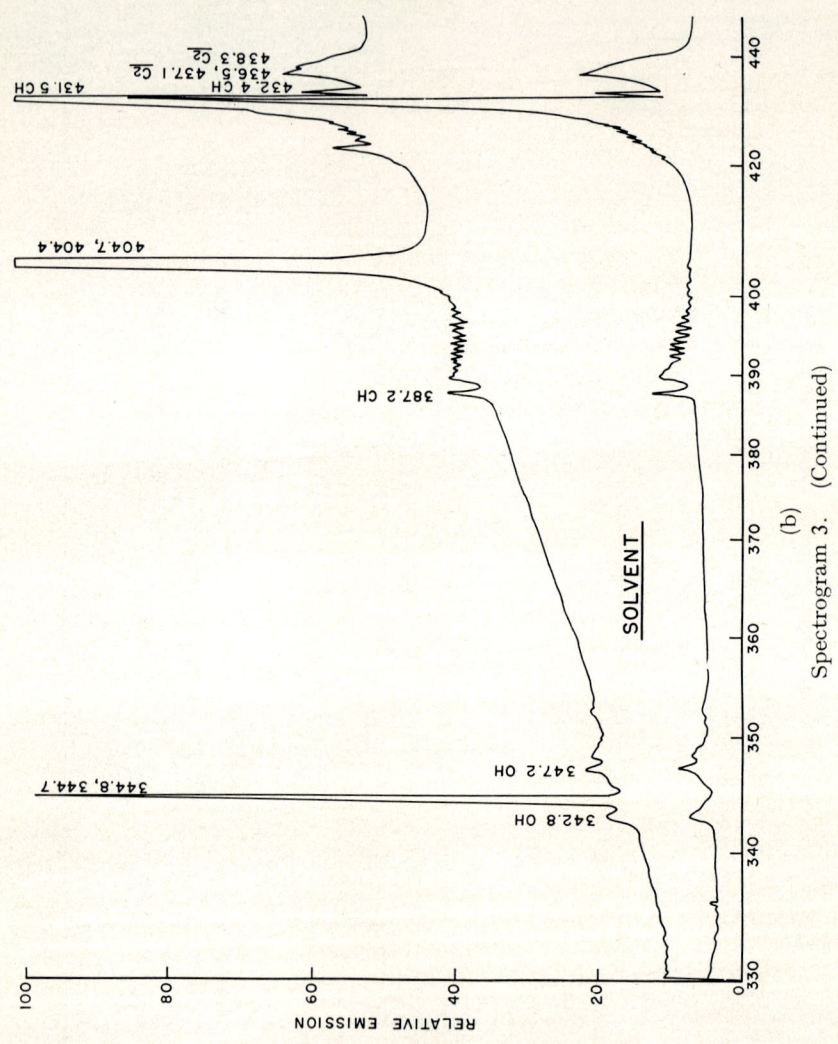

Spectrogram 3. (Continued)

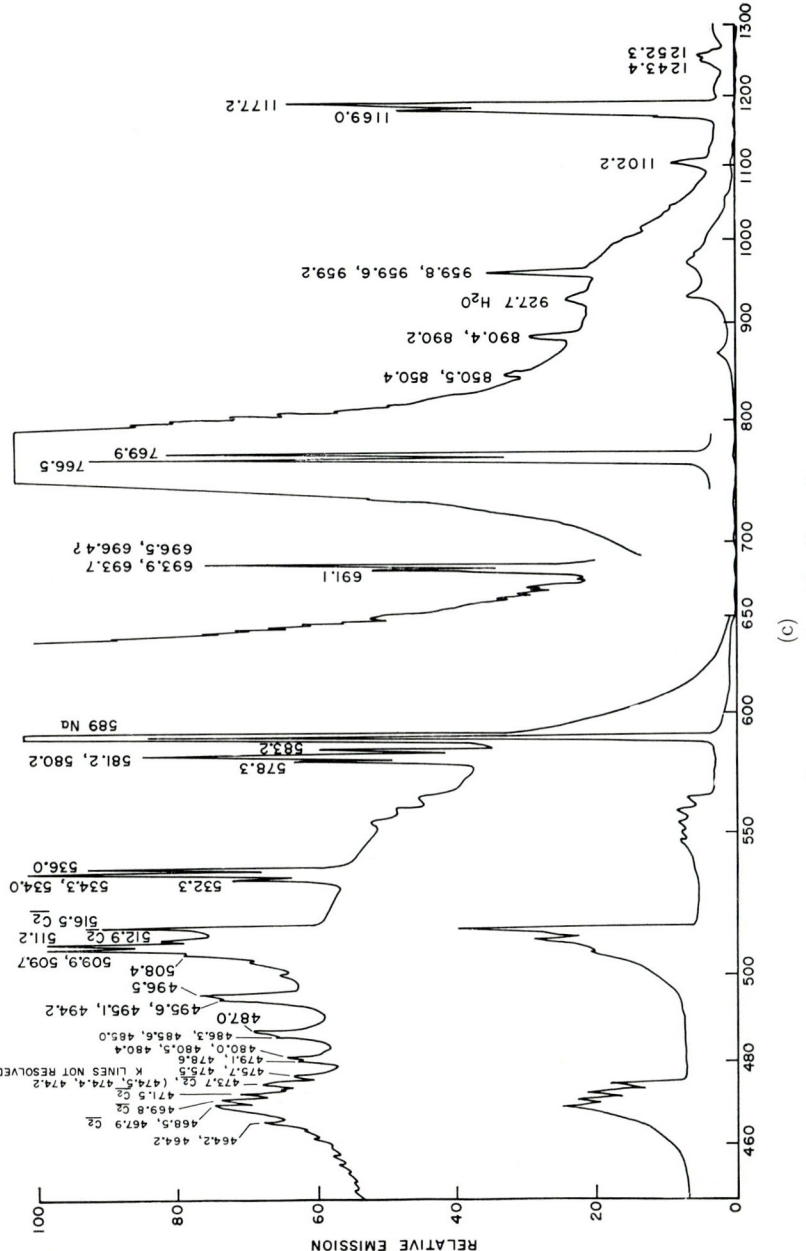

Spectrogram 3. (Continued)

483

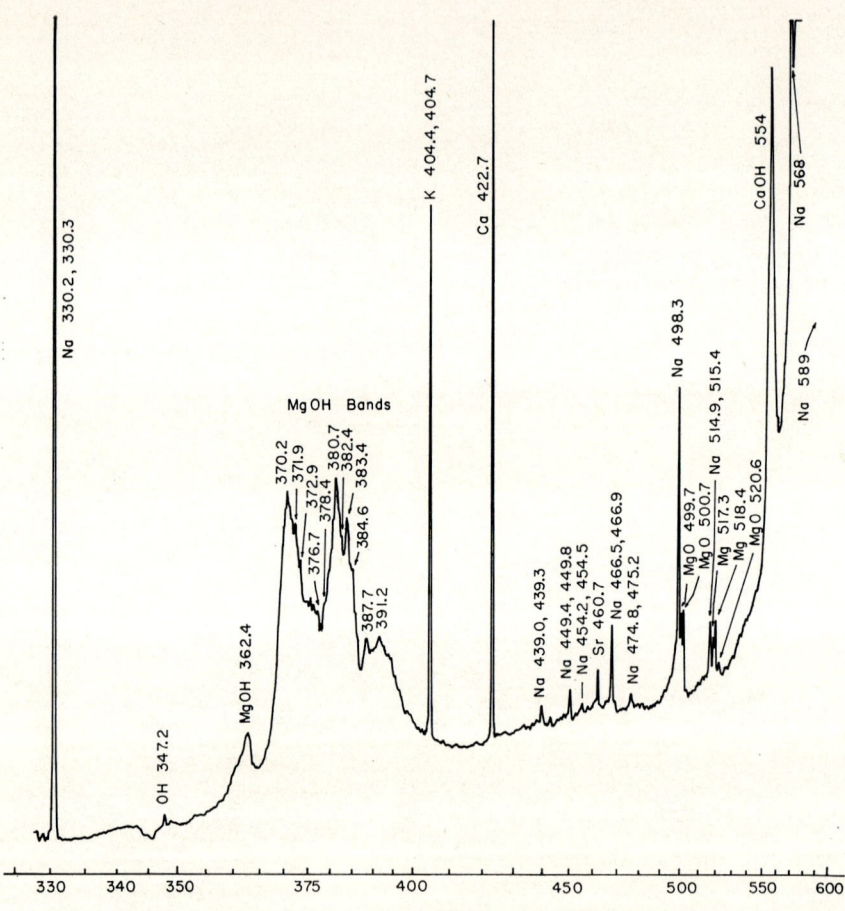

Spectrogram 4. Natural brine, diluted 1:10; oxyhydrogen; slit 0.02 mm.

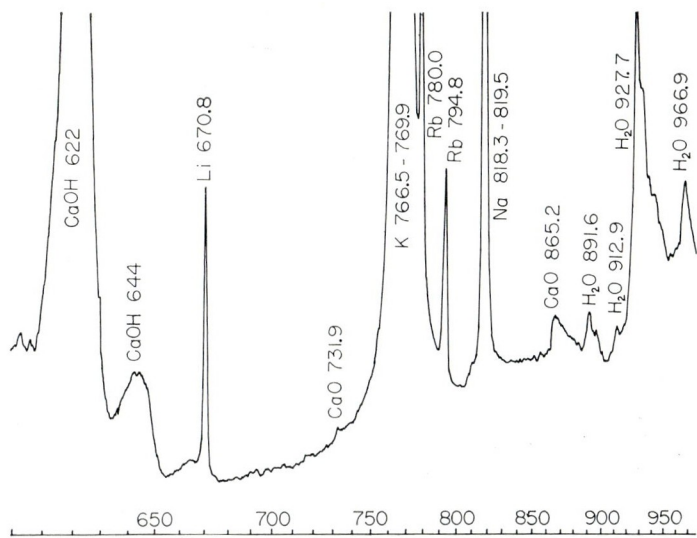

Spectrogram 5. Soil suspended in isopropanol–glycerine; sheathed oxyhydrogen; 16PMI photomultiplier; slit 0.02 mm.

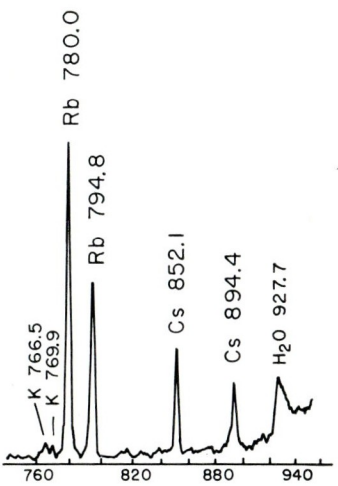

Spectrogram 6. Rubidium and cesium, 10 p.p.m. each; oxyhydrogen; RCA C7160 photomultiplier; slit 0.04 mm.

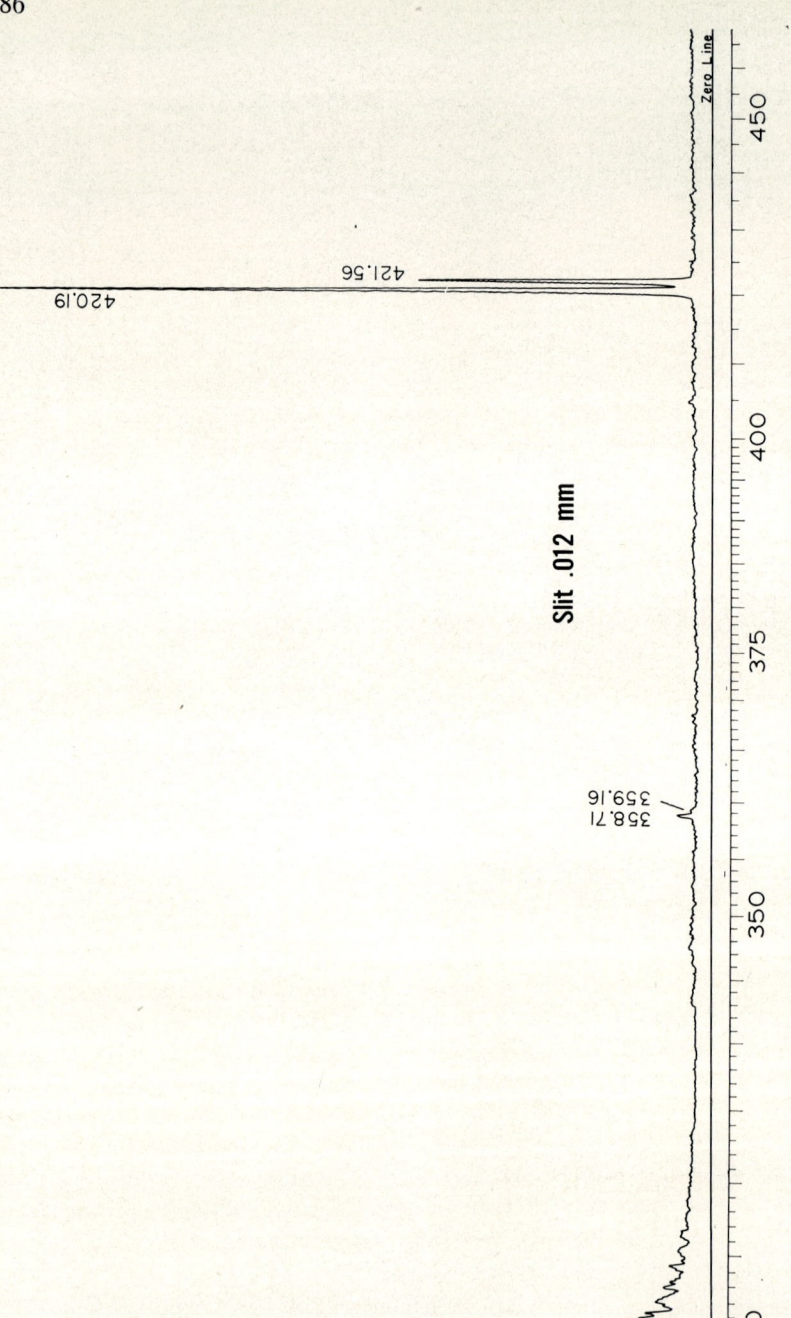

Spectrogram 7. Rubidium, 2000 p.p.m.; oxyhydrogen. (Continued on next page)

487

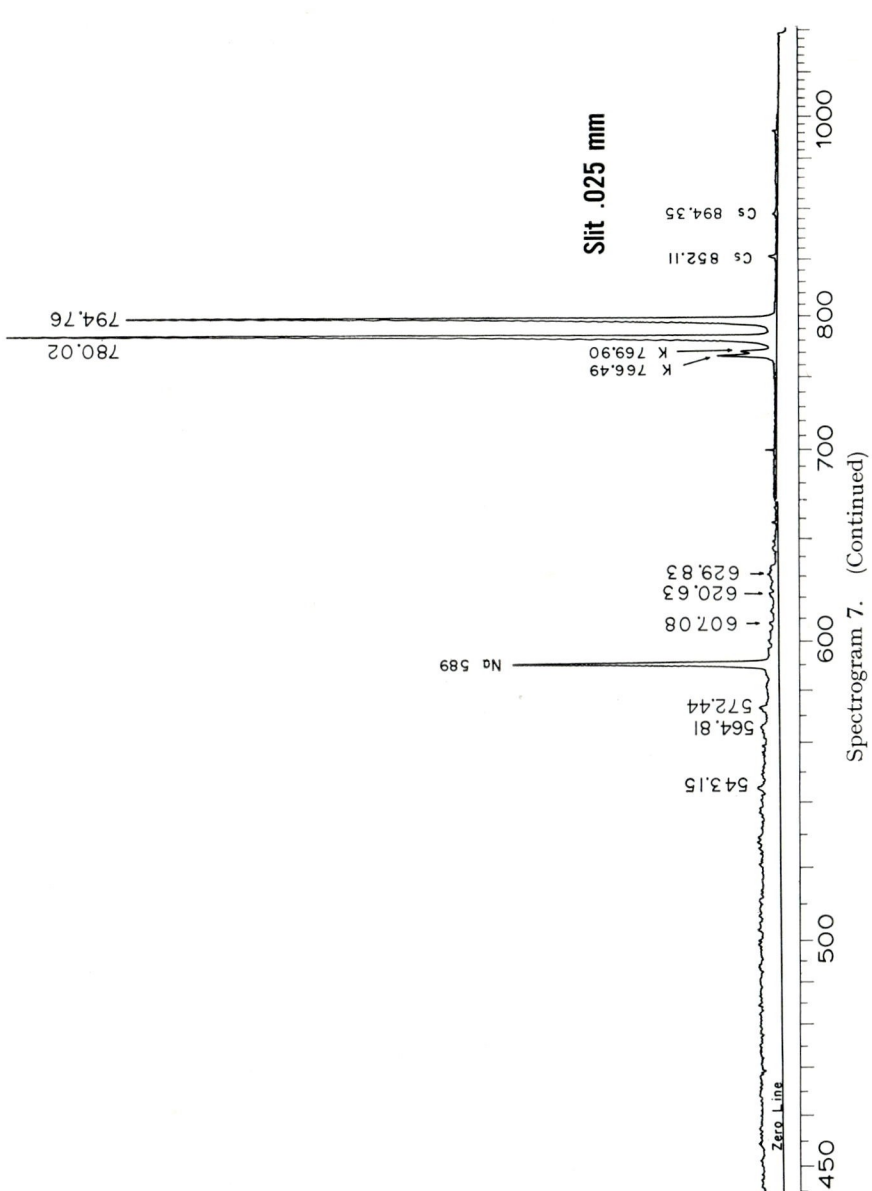

Spectrogram 7. (Continued)

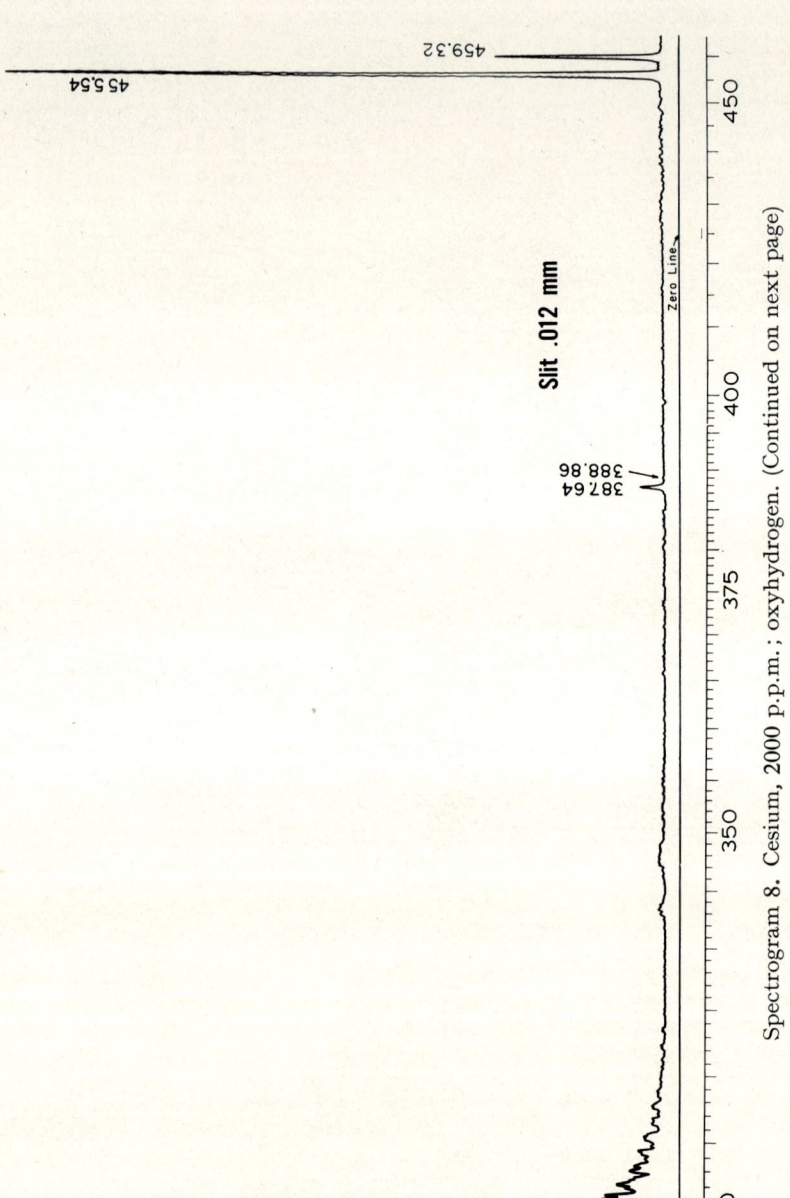

Spectrogram 8. Cesium, 2000 p.p.m.; oxyhydrogen. (Continued on next page)

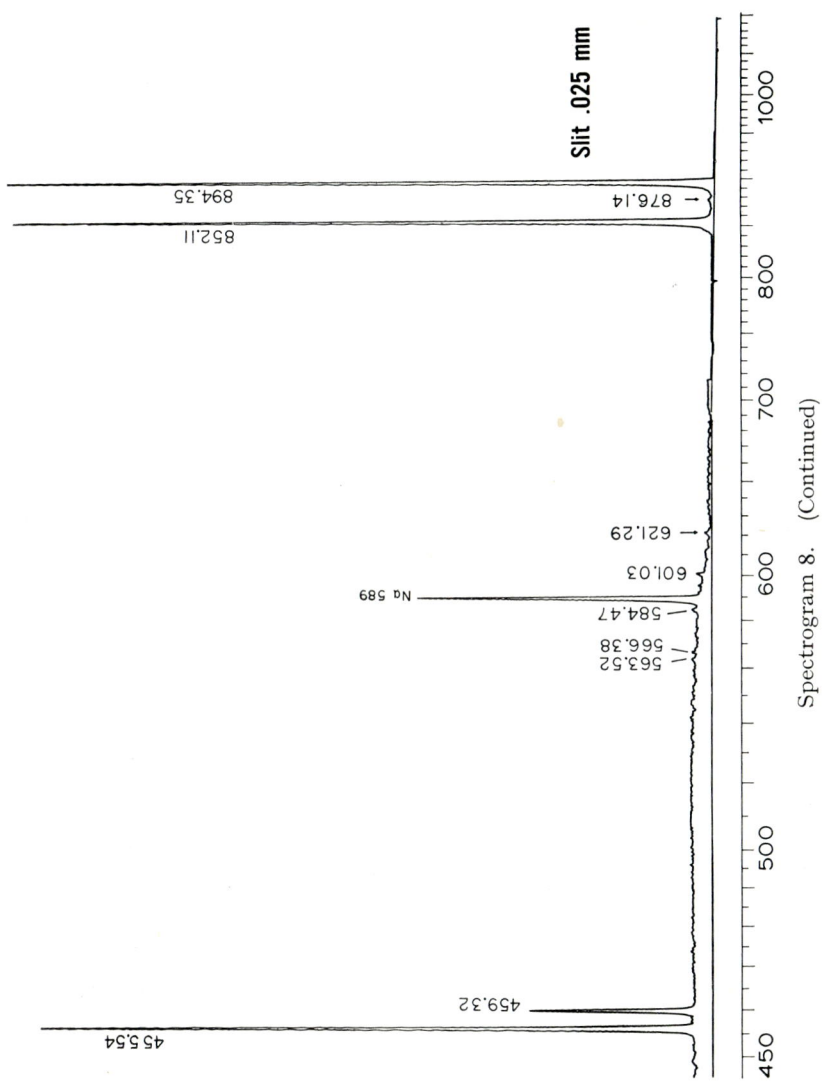

Spectrogram 8. (Continued)

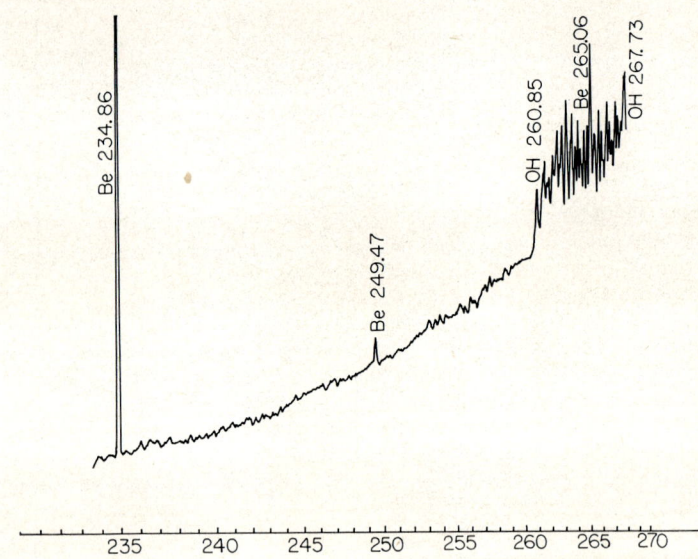

Spectrogram 9. Beryllium, 1000 p.p.m.; sheathed, fuel-rich (white) oxyacetylene; slit 0.02 mm.

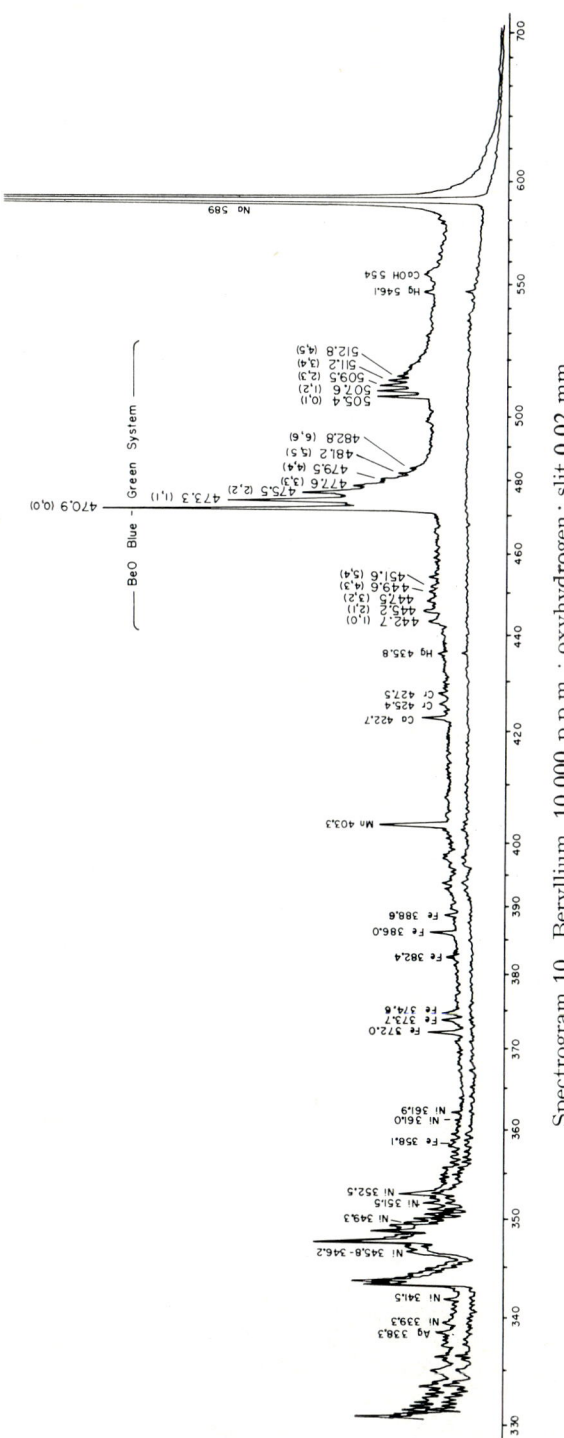

Spectrogram 10. Beryllium, 10 000 p.p.m.; oxyhydrogen; slit 0.02 mm.

492

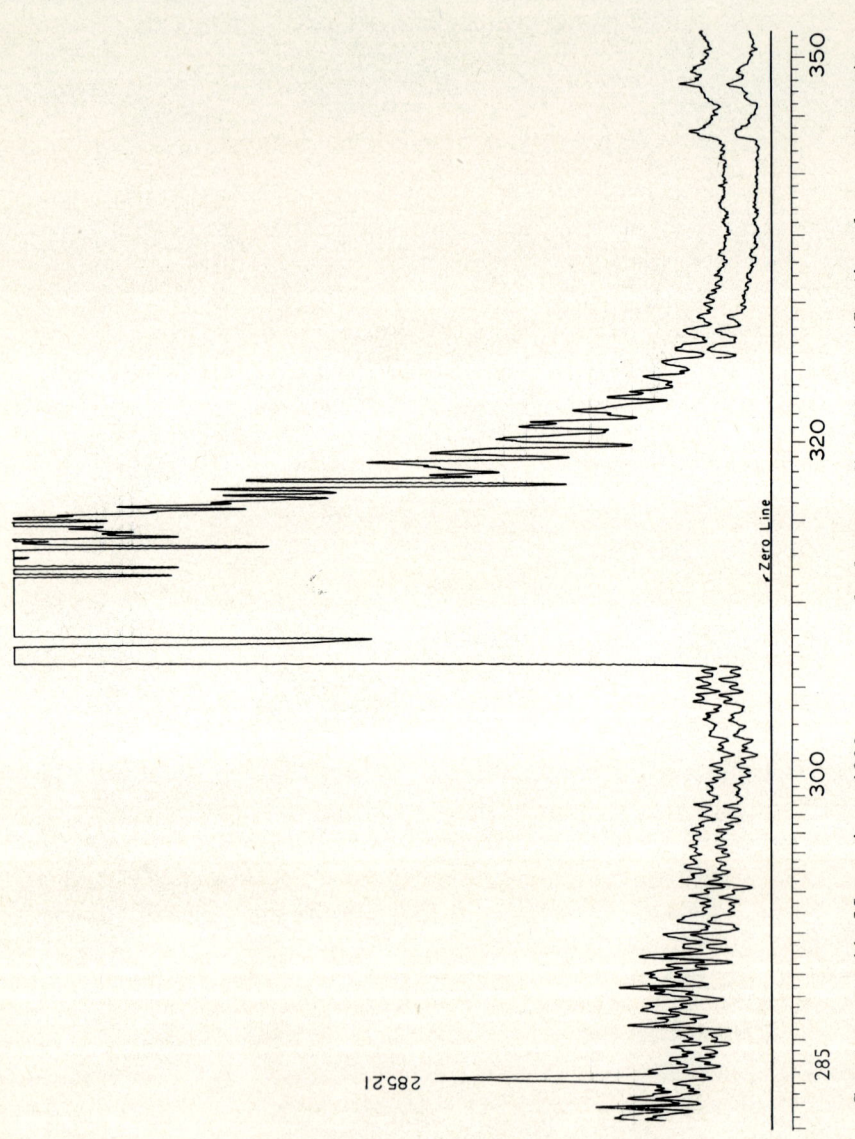

Spectrogram 11. Magnesium, 1000 p.p.m.; oxyhydrogen; slit 0.02 mm. (Continued on next page)

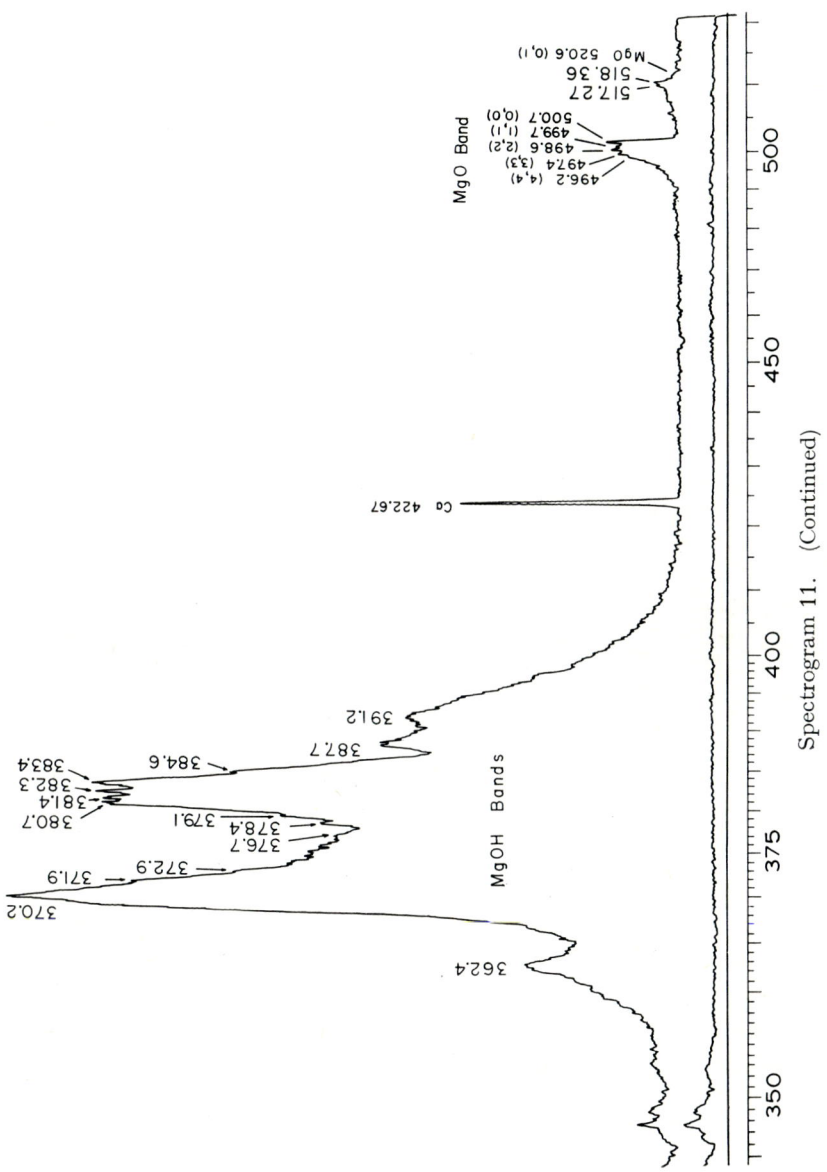

Spectrogram 11. (Continued)

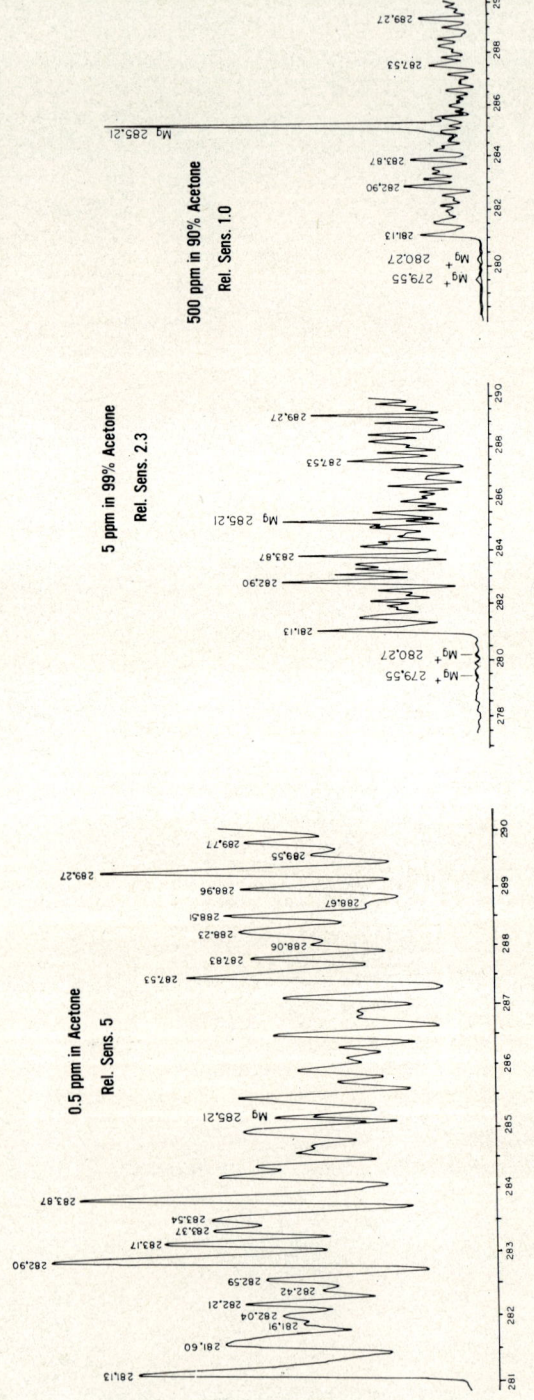

Spectrogram 12. Magnesium in acetone; oxyhydrogen; slit 0.02 mm.

495

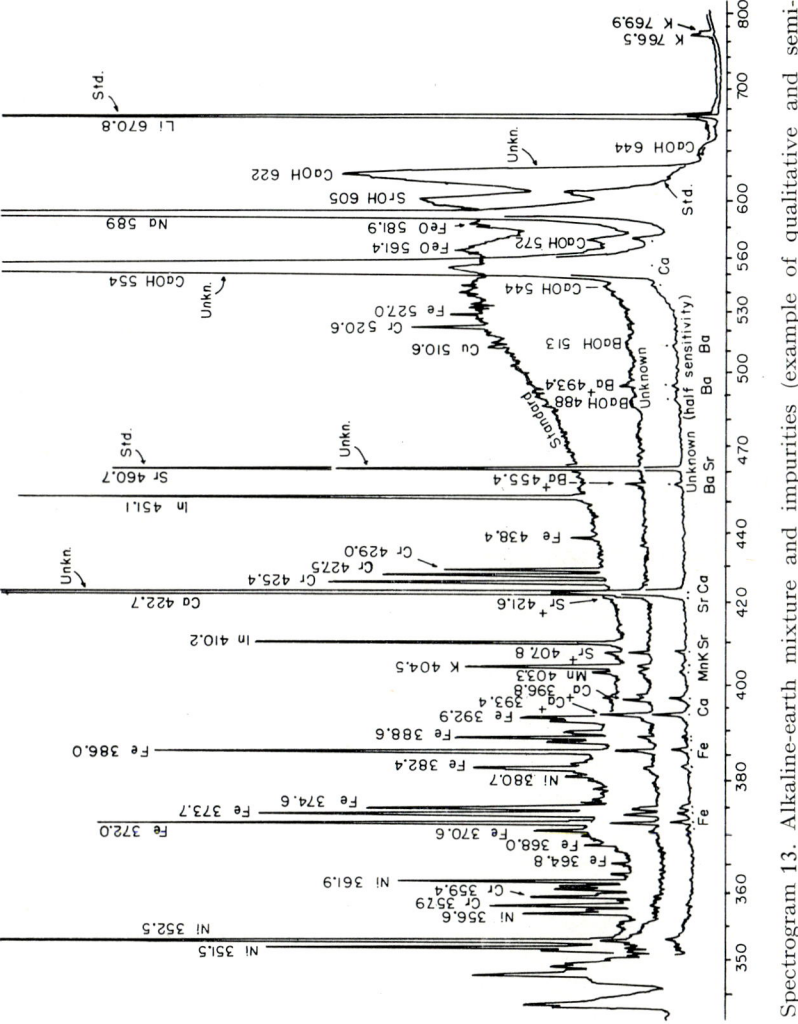

Spectrogram 13. Alkaline-earth mixture and impurities (example of qualitative and semi-quantitative analysis of pump trap residues from electron tube processing); standard same as in No. 17; oxyhydrogen; slit 0.04 mm.

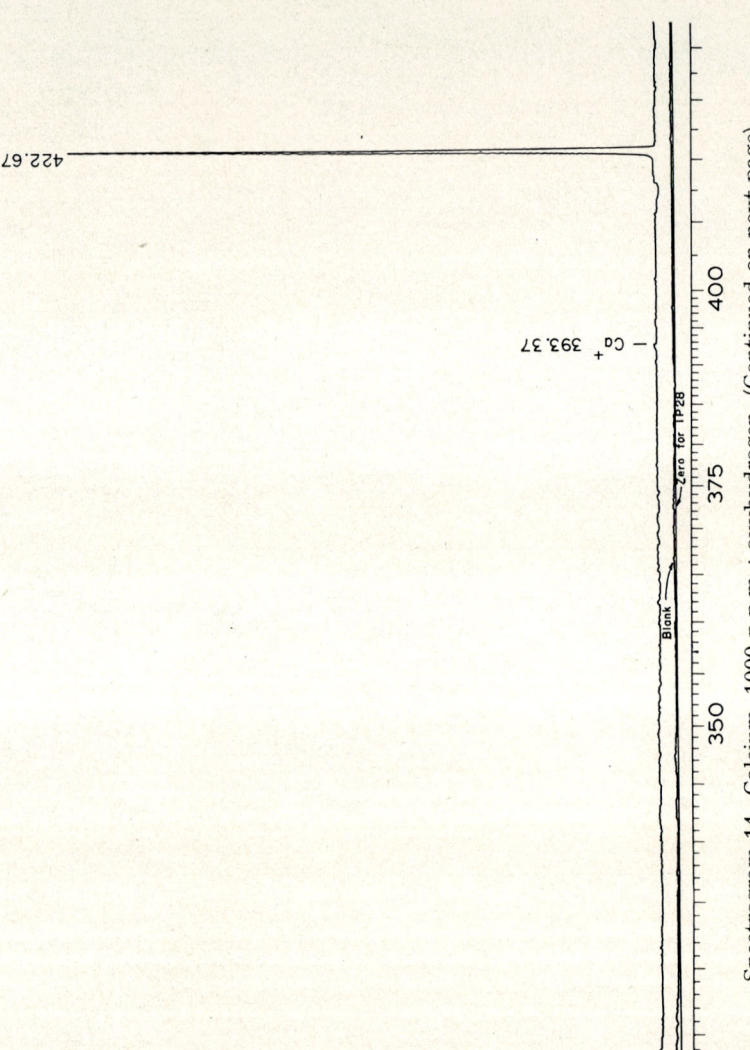

Spectrogram 14. Calcium, 1000 p.p.m.; oxyhydrogen. (Continued on next page)

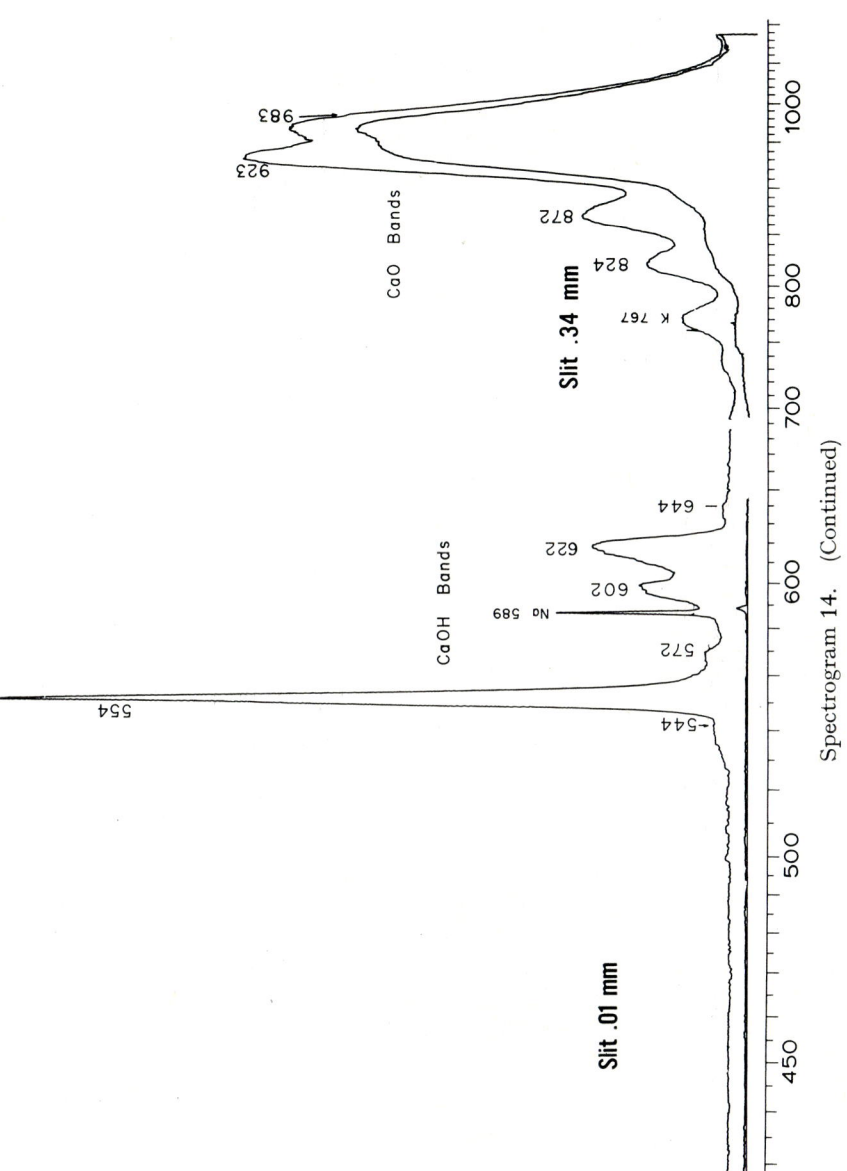

Spectrogram 14. (Continued)

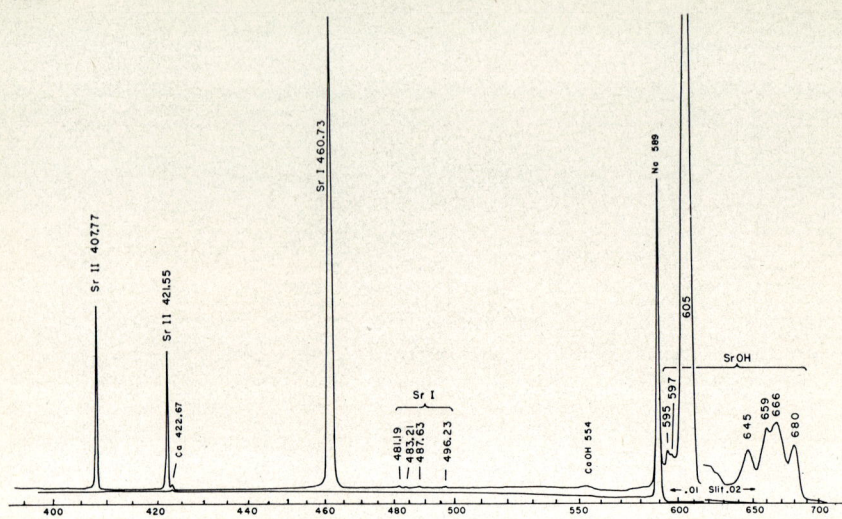

Spectrogram 15. Strontium, 100 p.p.m.; oxyhydrogen; true slit width 0.02 mm. below 615 mµ, 0.04 mm. above.

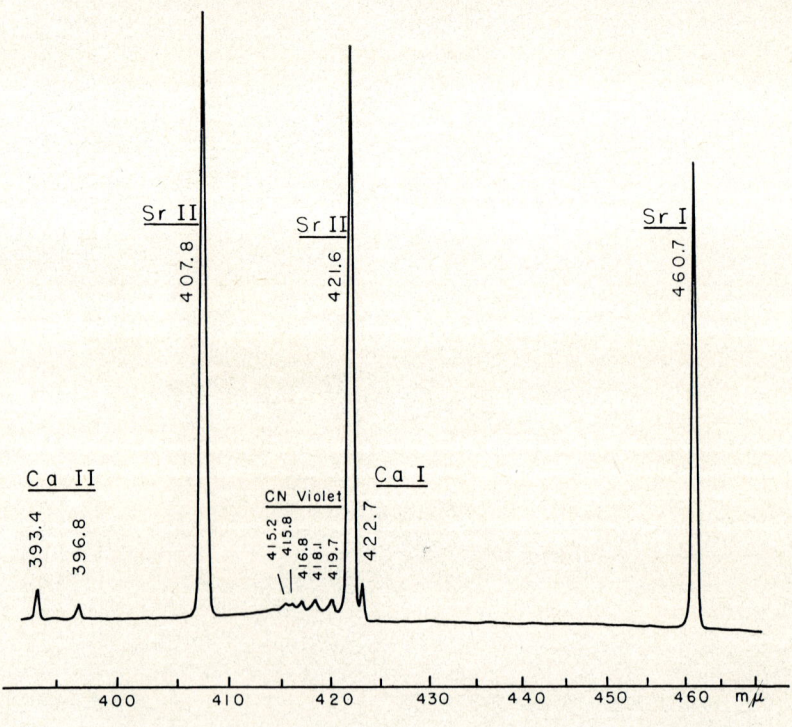

Spectrogram 16. Strontium, 100 p.p.m.; oxycyanogen; slit 0.02 mm.

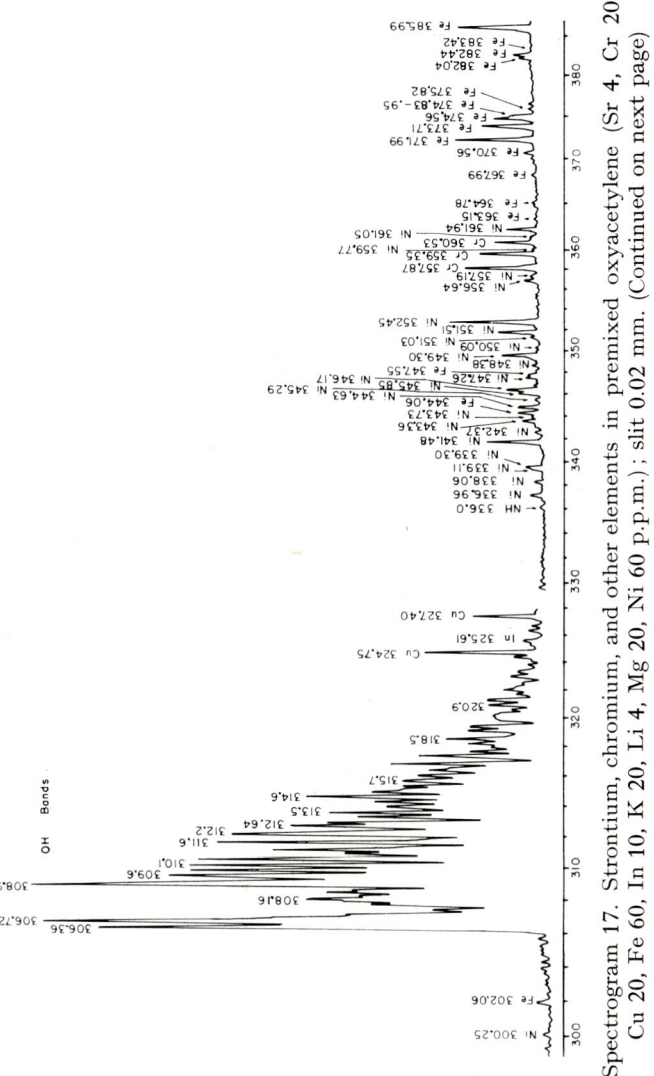

Spectrogram 17. Strontium, chromium, and other elements in premixed oxyacetylene (Sr 4, Cr 20, Cu 20, Fe 60, In 10, K 20, Li 4, Mg 20, Ni 60 p.p.m.); slit 0.02 mm. (Continued on next page)

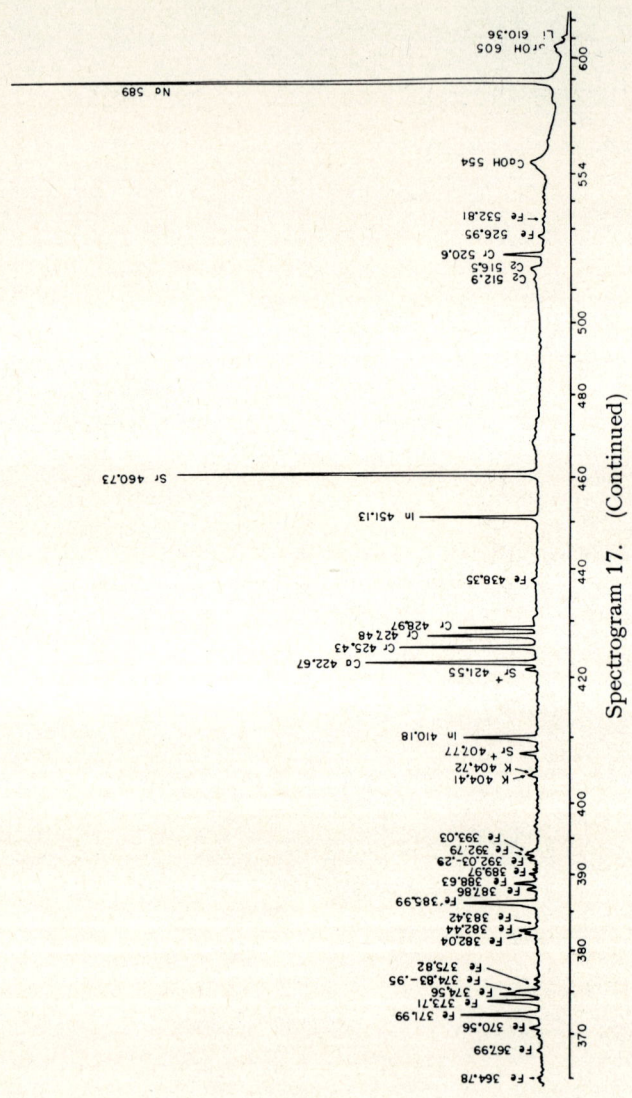

Spectrogram 17. (Continued)

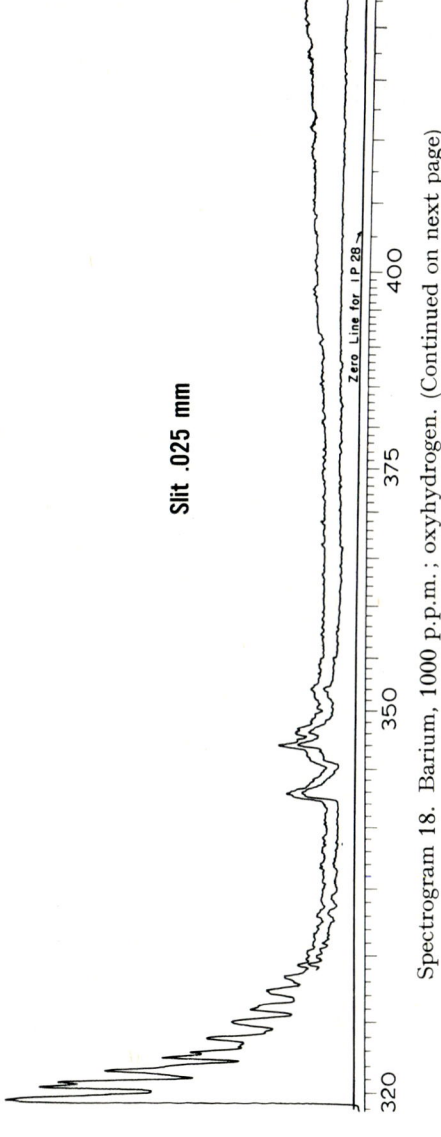

Spectrogram 18. Barium, 1000 p.p.m.; oxyhydrogen. (Continued on next page)

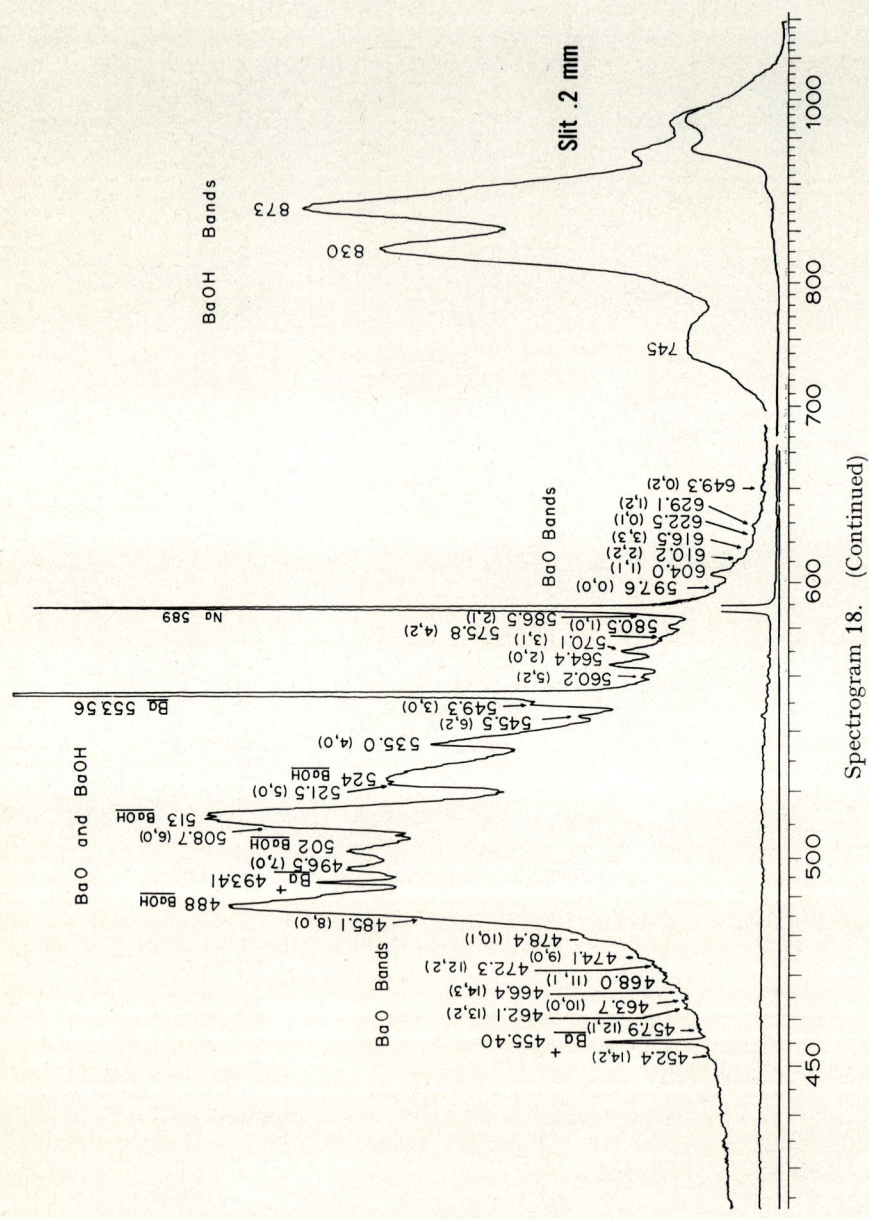

Spectrogram 18. (Continued)

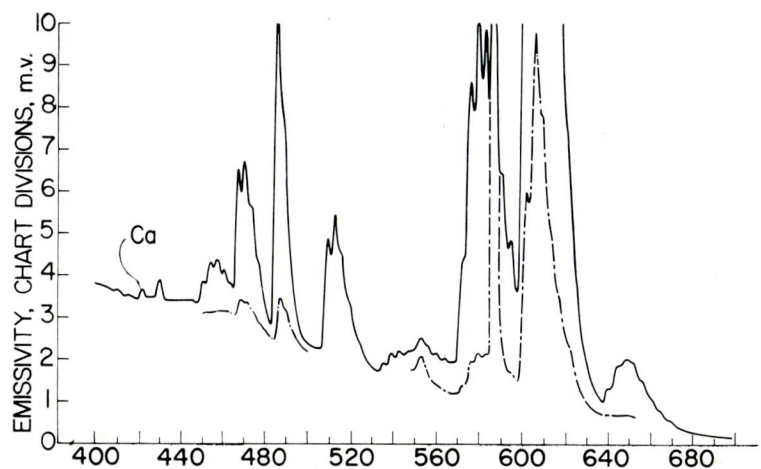

Spectrogram 19. Scandium, 5 and 50 mg./l. in hexone; oxyhydrogen; ORNL spectrophotometer, slit 0.25 mm.; zero suppressed 1.45 mv.

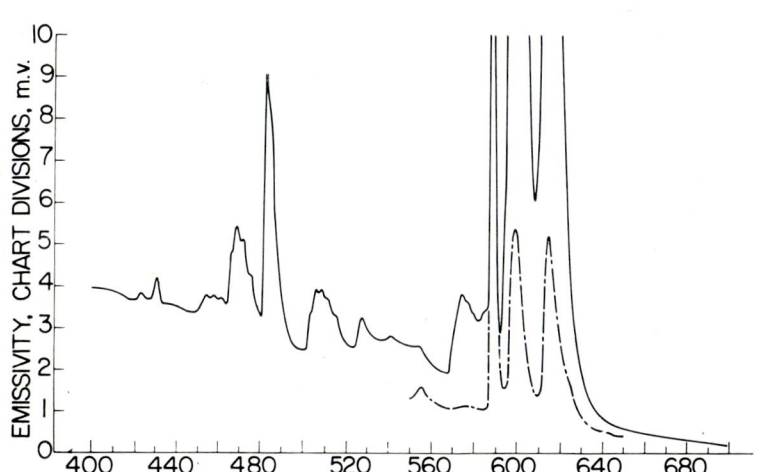

Spectrogram 20. Yttrium, 5 and 50 mg./l. in hexone; oxyhydrogen; ORNL spectrophotometer, slit 0.25 mm.; zero suppressed 1.25 mv.

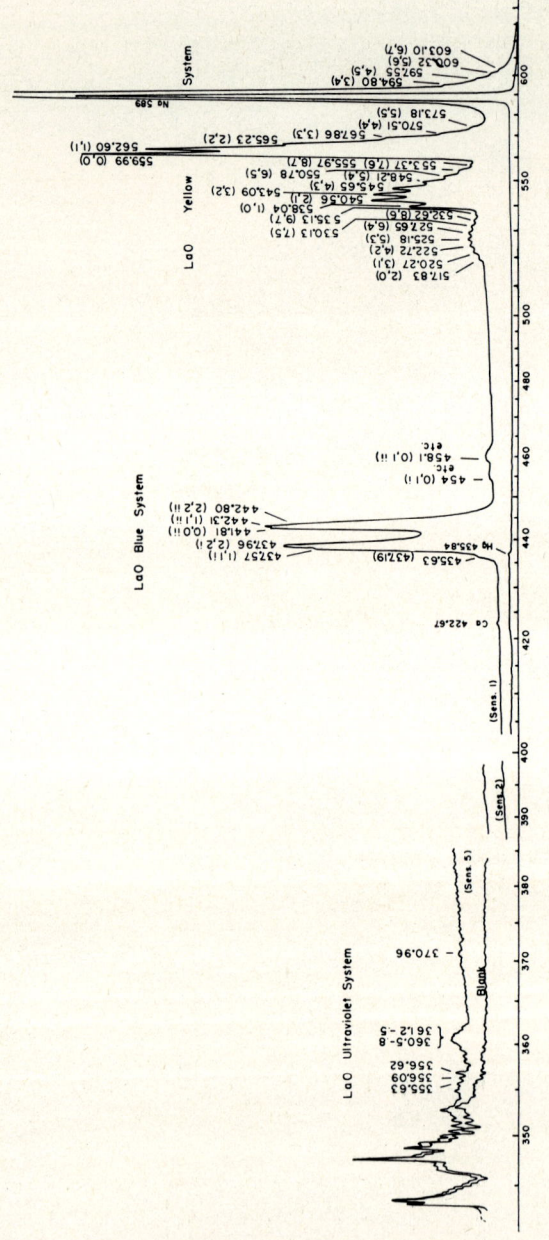

Spectrogram 21. Lanthanum, 1000 p.p.m.; oxyhydrogen; slit 0.02 mm.

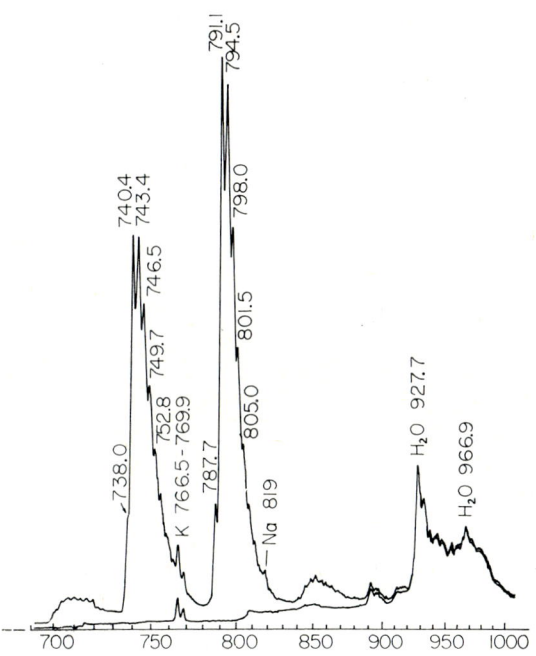

Spectrogram 22. Lanthanum, 1000 p.p.m.; sheathed oxyhydrogen; 16PMI photomultiplier; slit 0.02 mm.

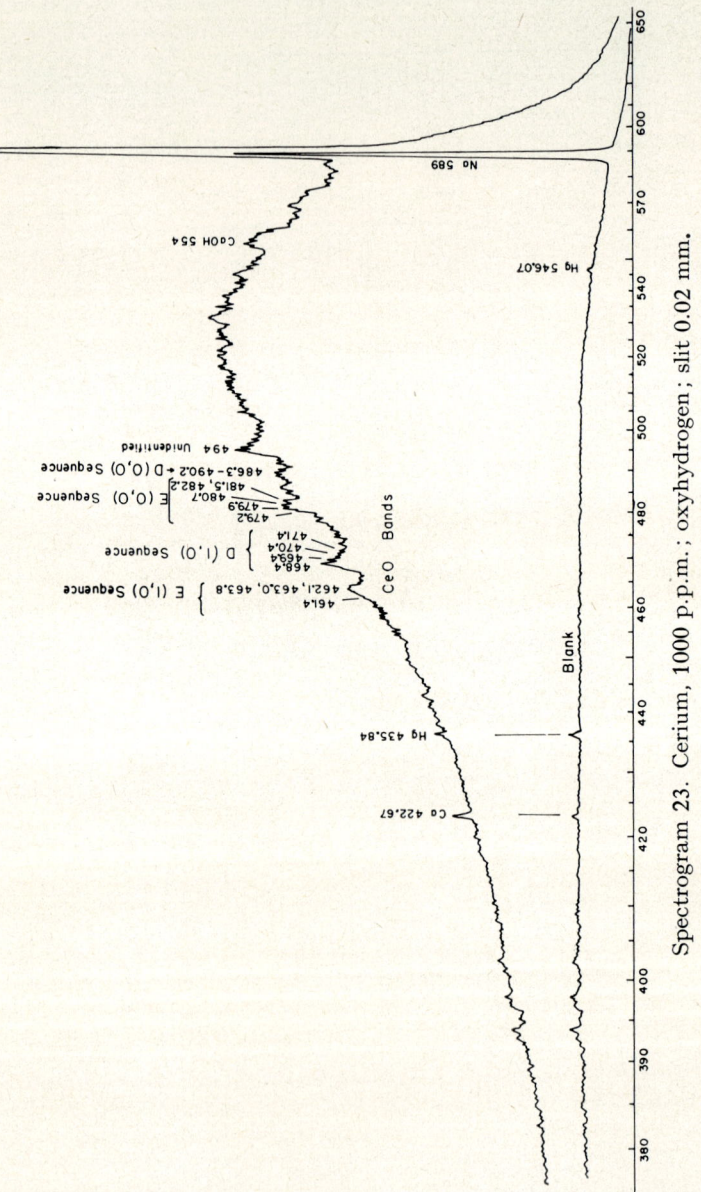

Spectrogram 23. Cerium, 1000 p.p.m.; oxyhydrogen; slit 0.02 mm.

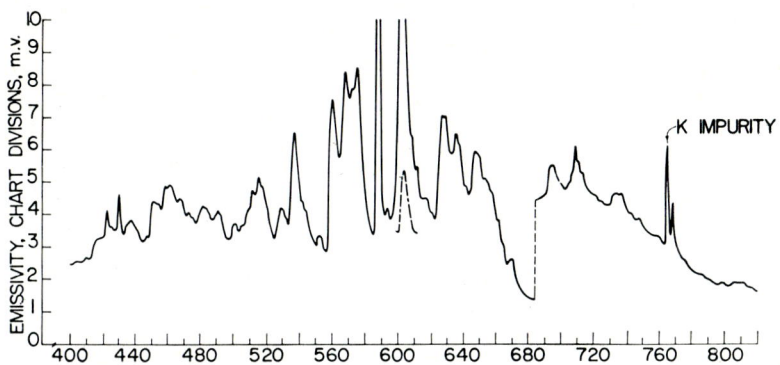

Spectrogram 24. Praseodymium, 100 and 500 mg./l. in hexone; oxyhydrogen; ORNL spectrophotometer with twice usual sensitivity; slit 0.25 mm.; RCA 6217 photomultiplier below 680 mµ, 16PMI above; zero suppressed 9.1 mv.

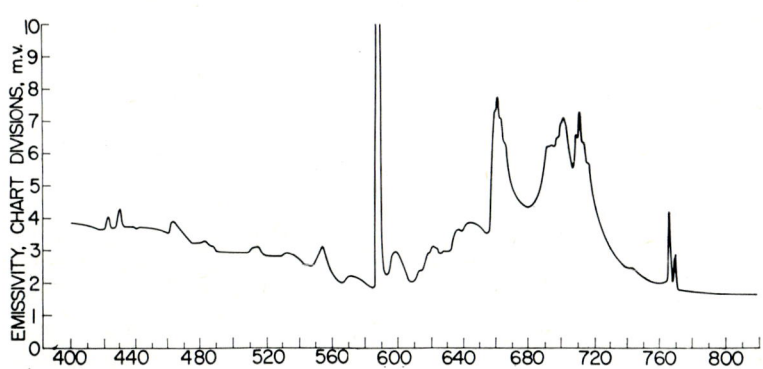

Spectrogram 25. Neodymium, 100 mg./l. in hexone; oxyhydrogen; ORNL spectrophotometer, slit 0.25 mm.; RCA 6217 photomultiplier below 680 mµ, 16PMI above; zero suppressed 1.25 mv.

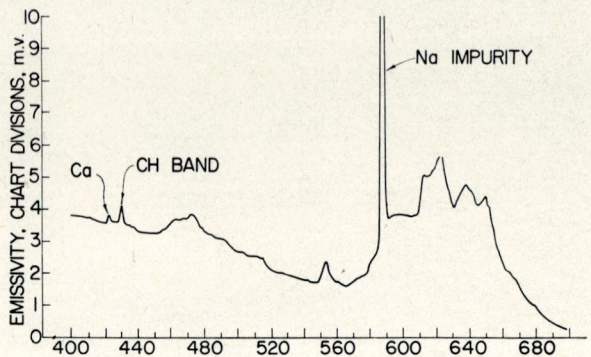

Spectrogram 26. Samarium, 50 mg./l. in hexone; oxyhydrogen; ORNL spectrophotometer, slit 0.25 mm.; zero suppressed 1.2 mv.

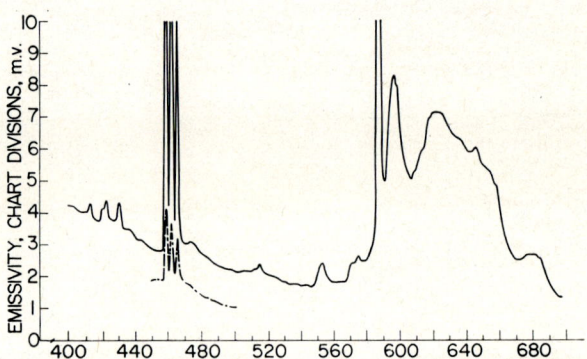

Spectrogram 27. Europium, 10 and 50 mg./l. in hexone; oxyhydrogen; ORNL spectrophotometer, slit 0.25 mm.; zero suppressed 1.2 mv.

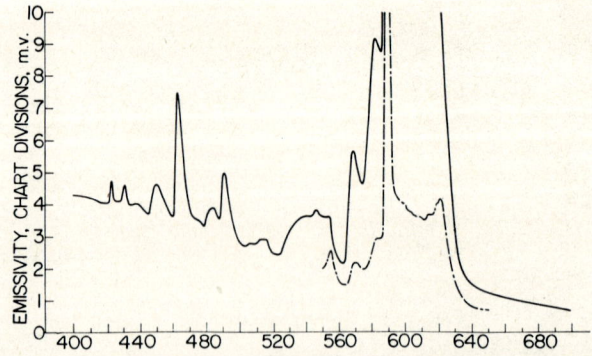

Spectrogram 28. Gadolinium, 10 and 50 mg./l. in hexone; oxyhydrogen; ORNL spectrophotometer, slit 0.25 mm.; zero suppressed 0.75 mv.

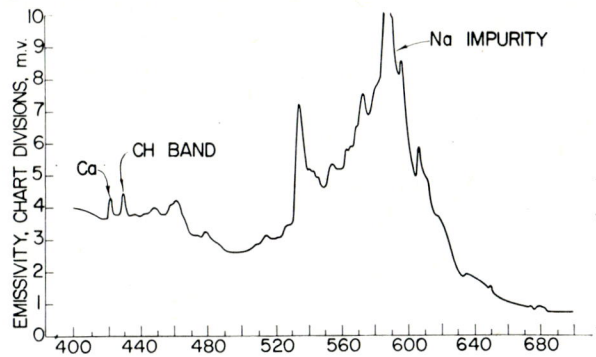

Spectrogram 29. Terbium, 50 mg./l. in hexone; oxyhydrogen; ORNL spectrophotometer, slit 0.25 mm.; zero suppressed 0.95 mv.

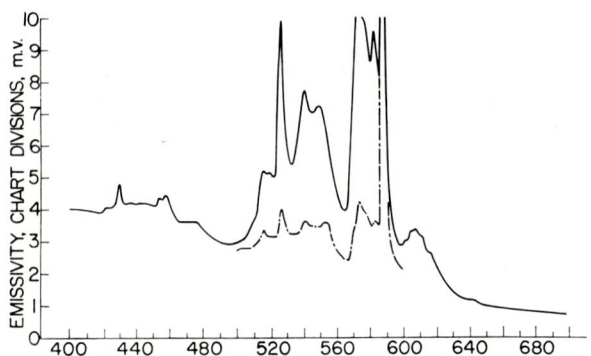

Spectrogram 30. Dysprosium, 10 and 50 mg./l. in hexone; oxyhydrogen; ORNL spectrophotometer, slit 0.25 mm.; zero suppressed 0.85 mv.

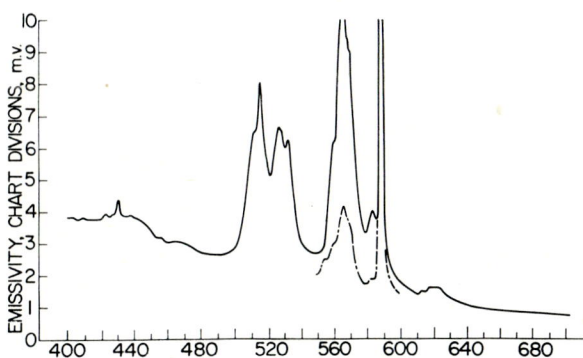

Spectrogram 31. Holmium, 10 and 50 mg./l. in hexone; oxyhydrogen; ORNL spectrophotometer, slit 0.25 mm.; zero suppressed 1.15 mv.

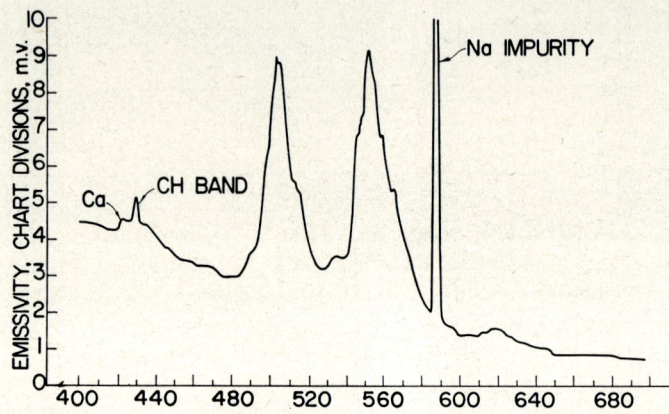

Spectrogram 32. Erbium, 50 mg./l. in hexone; oxyhydrogen; ORNL spectrophotometer, slit 0.25 mm.; zero suppressed 0.6 mv.

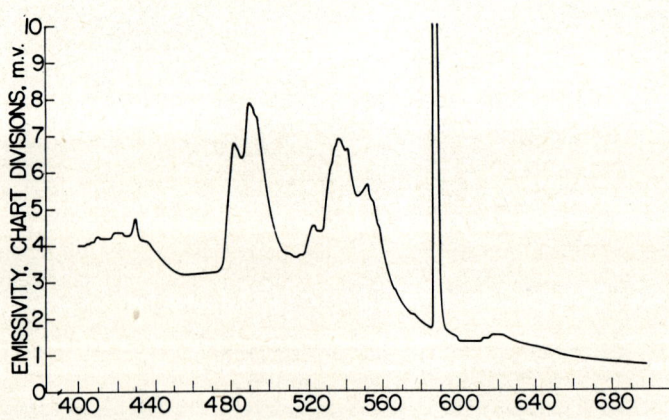

Spectrogram 33. Thulium, 50 mg./l. in hexone; oxyhydrogen; ORNL spectrophotometer, slit 0.25 mm.; zero suppressed 1.15 mv.

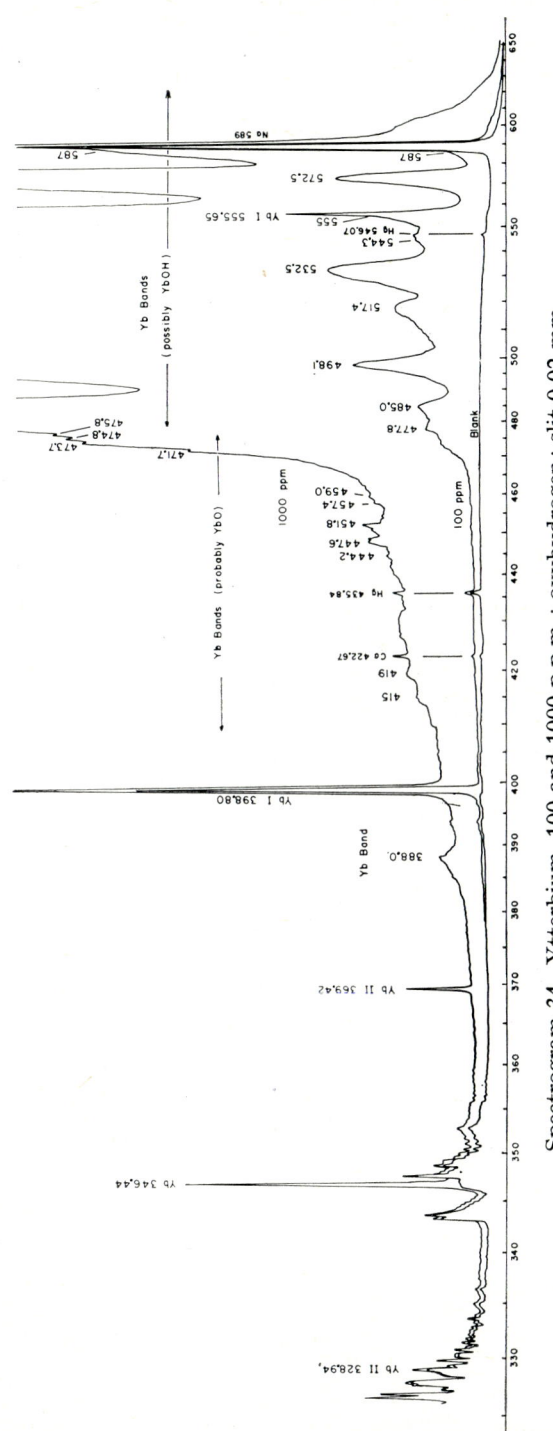

Spectrogram 34. Ytterbium, 100 and 1000 p.p.m.; oxyhydrogen; slit 0.02 mm.

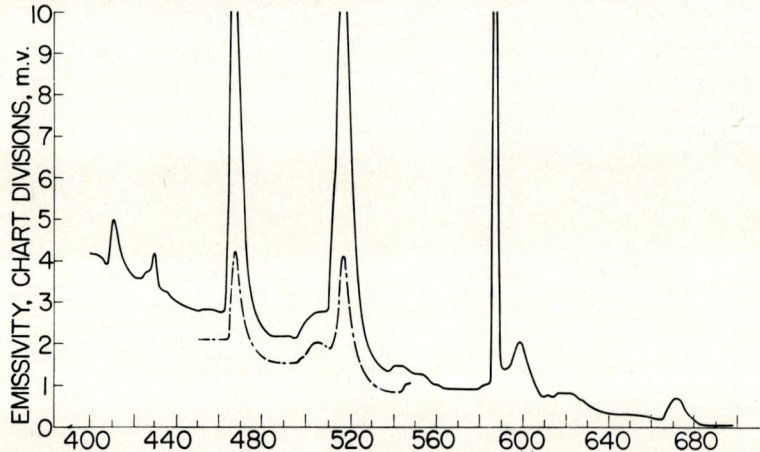

Spectrogram 35. Lutetium, 10 and 50 mg./l. in hexone; oxyhydrogen; ORNL spectrophotometer, slit 0.25 mm.; zero suppressed 1.25 mv.

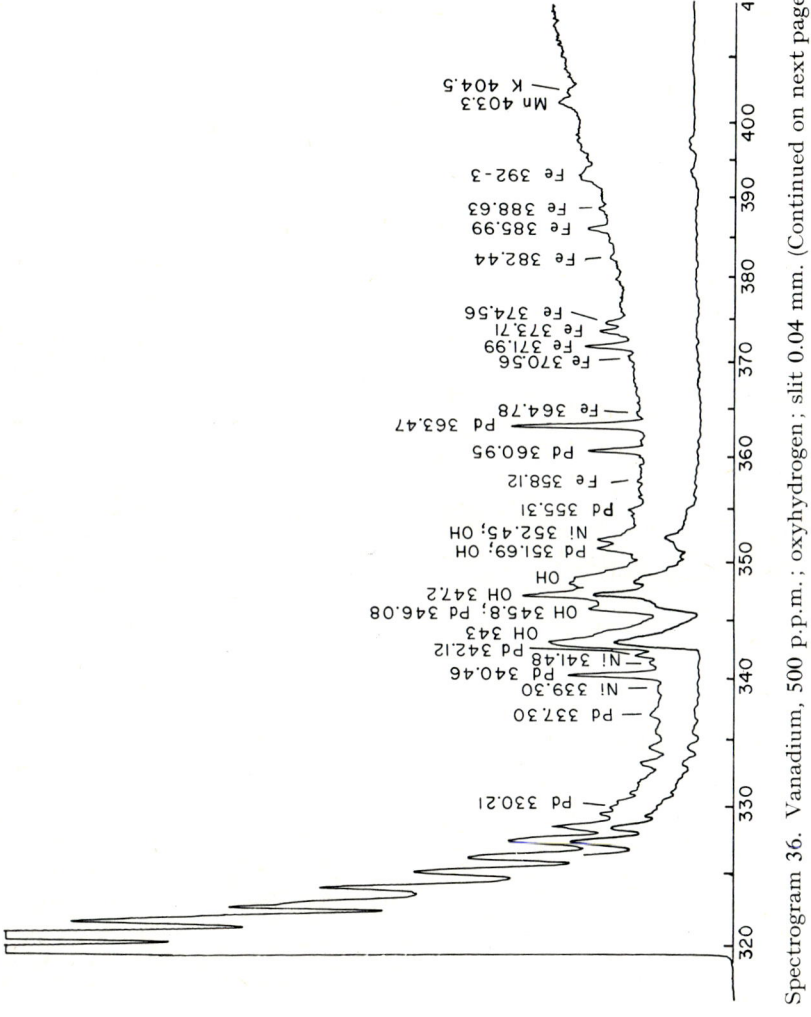

Spectrogram 36. Vanadium, 500 p.p.m.; oxyhydrogen; slit 0.04 mm. (Continued on next page)

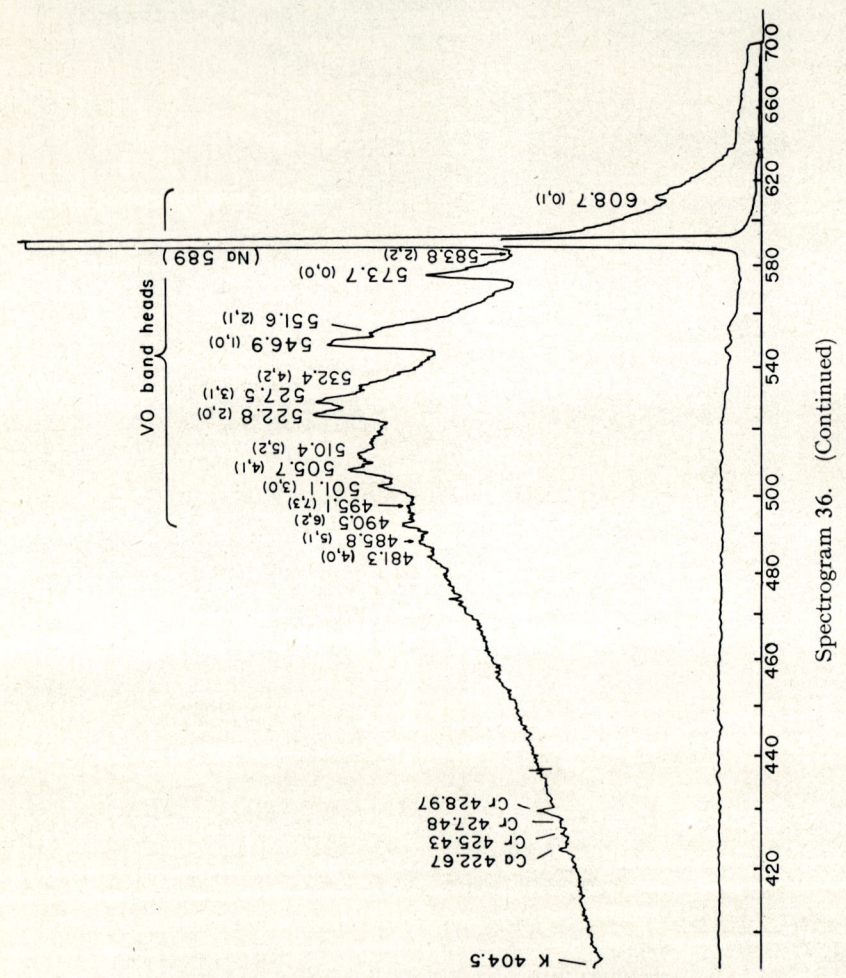

Spectrogram 36. (Continued)

515

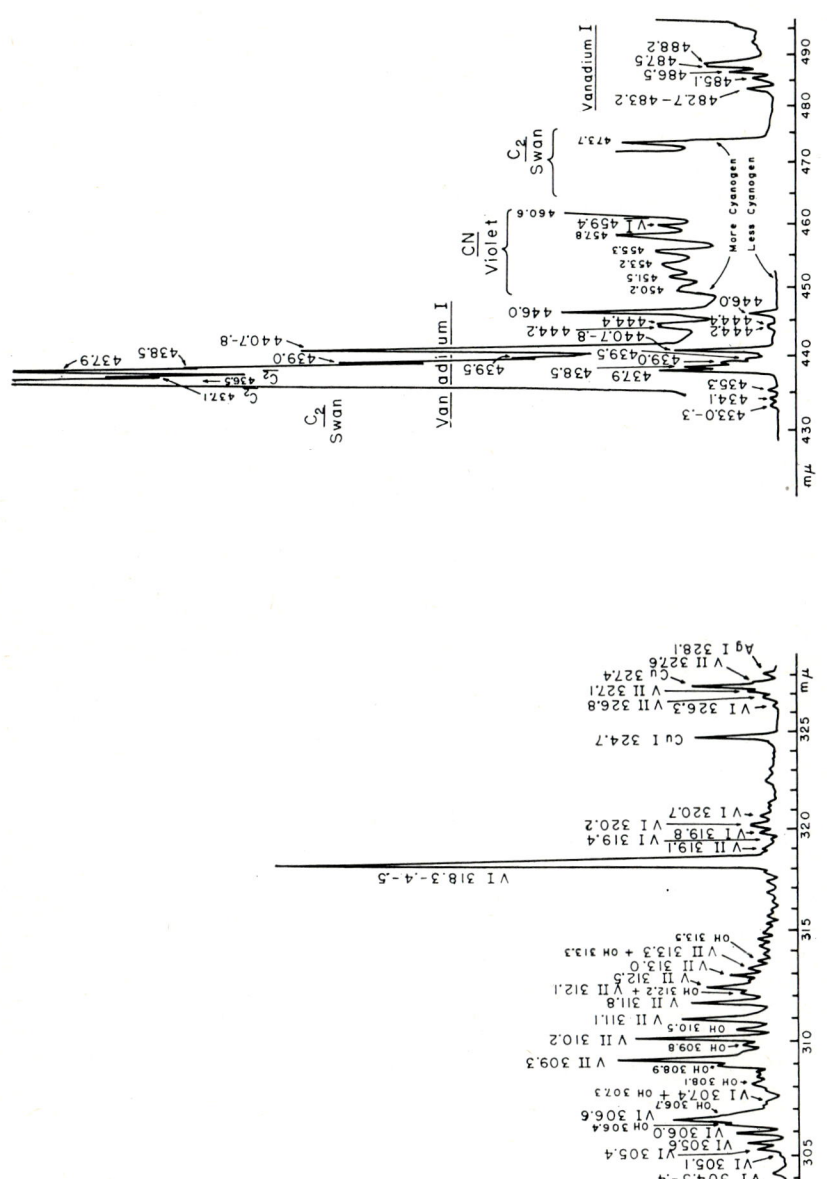

Spectrogram 37. Vanadium, 1000 p.p.m.; oxycyanogen; slit 0.02 mm.

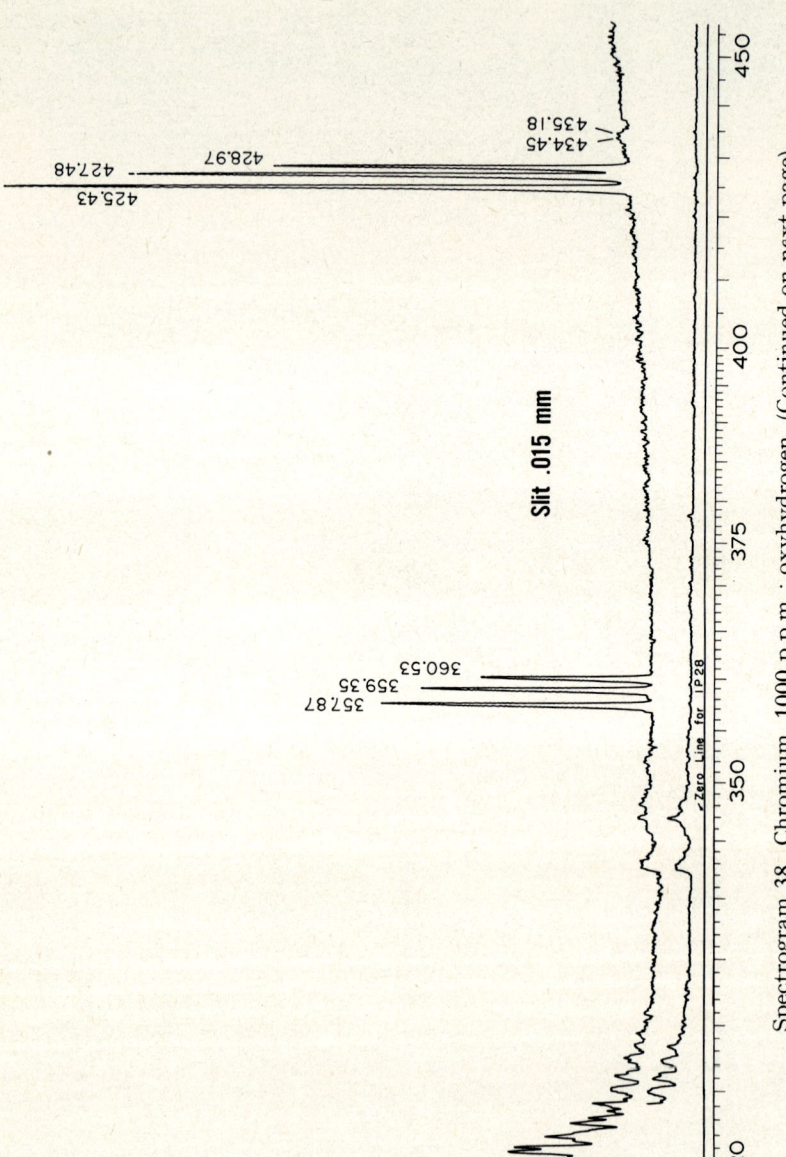

Spectrogram 38. Chromium, 1000 p.p.m.; oxyhydrogen. (Continued on next page)

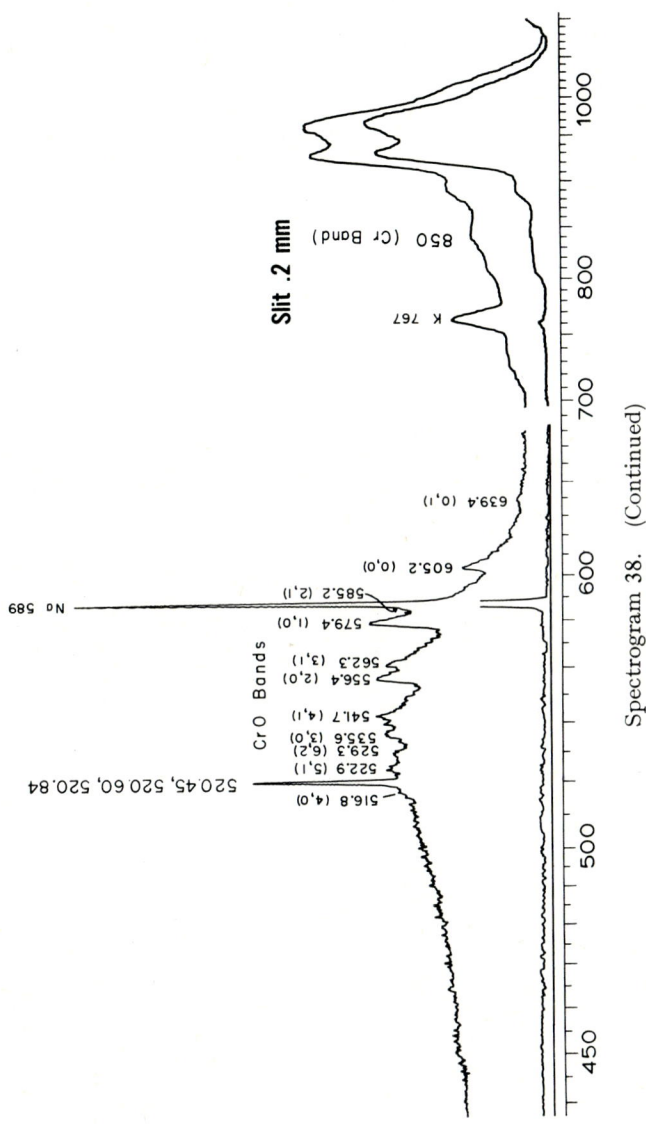

Spectrogram 38. (Continued)

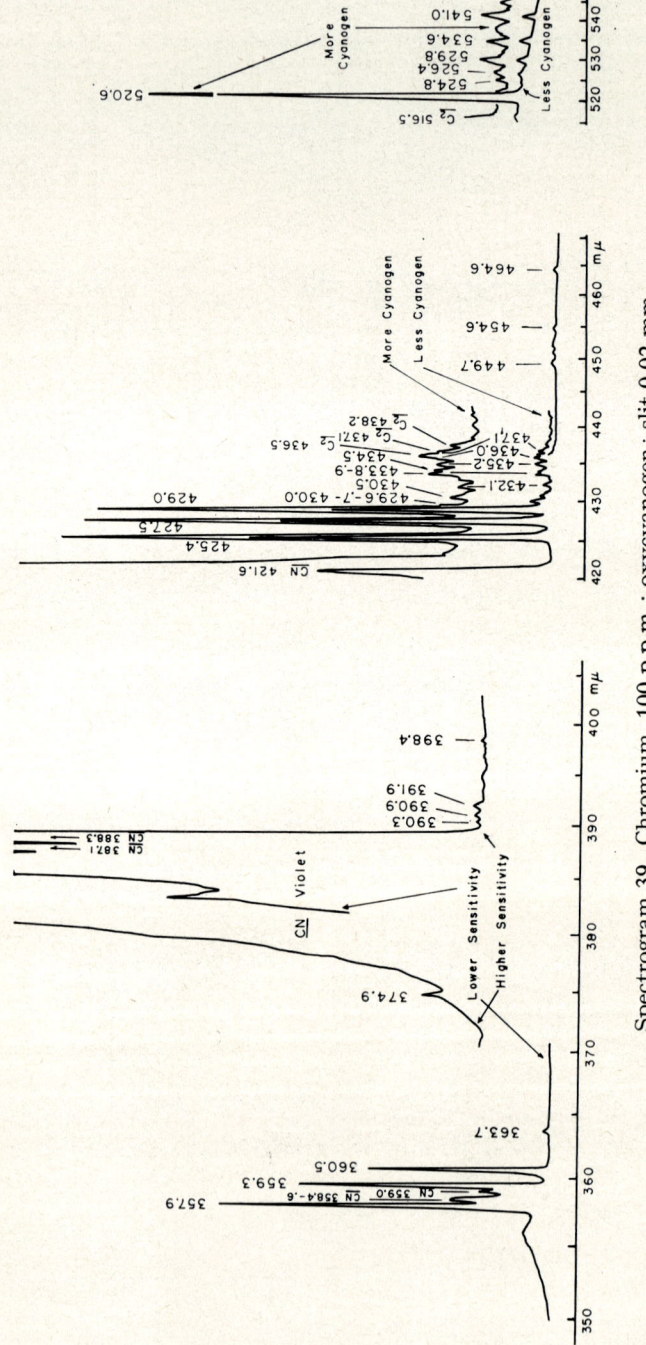

Spectrogram 39. Chromium, 100 p.p.m.; oxycyanogen; slit 0.02 mm.

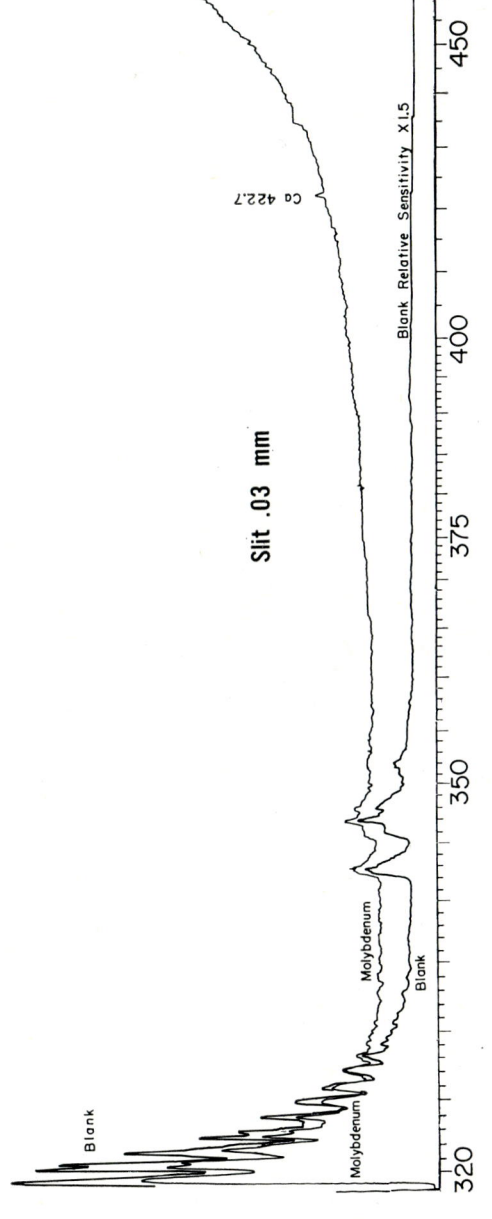

Spectrogram 40. Molybdenum, 5000 p.p.m.; oxyhydrogen. (Continued on next page)

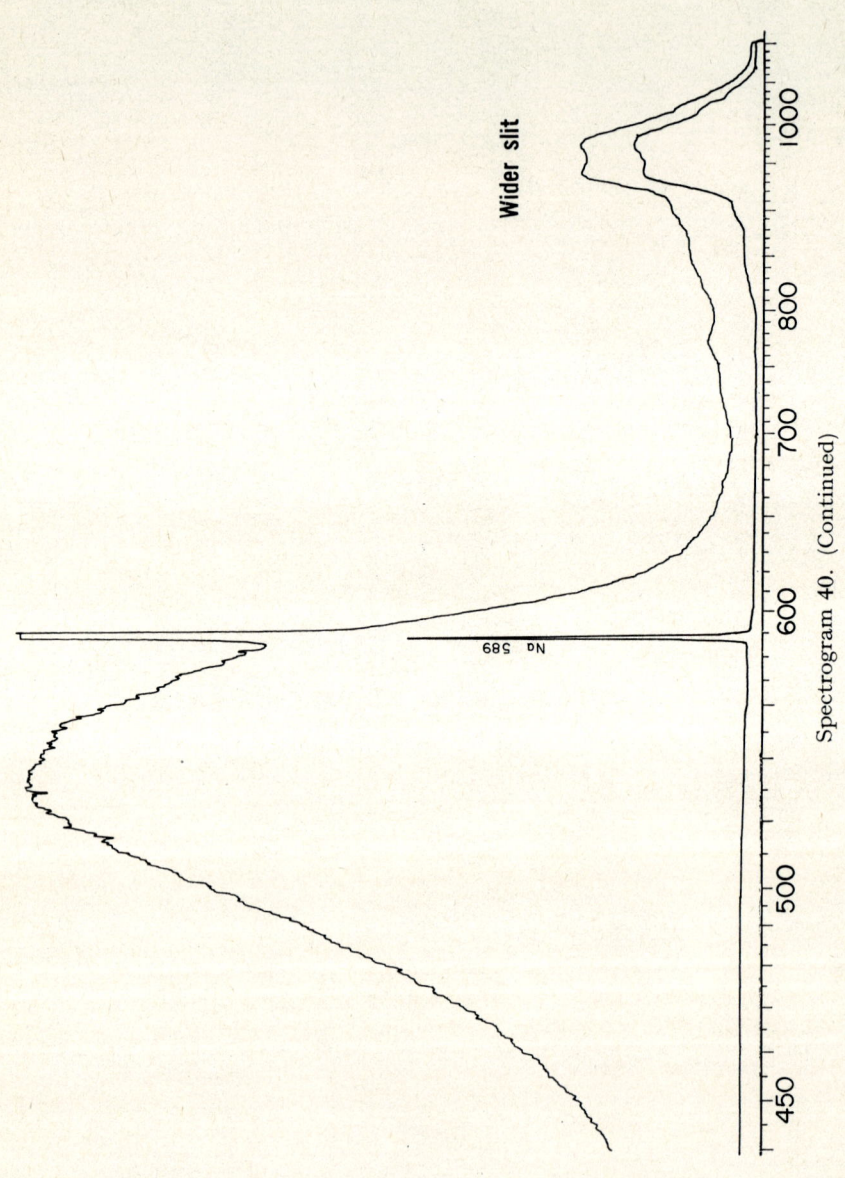

Spectrogram 40. (Continued)

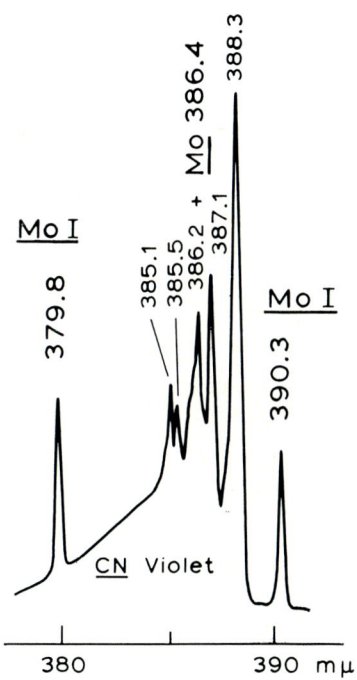

Spectrogram 41. Molybdenum, 1000 p.p.m.; oxycyanogen; slit 0.02 mm.

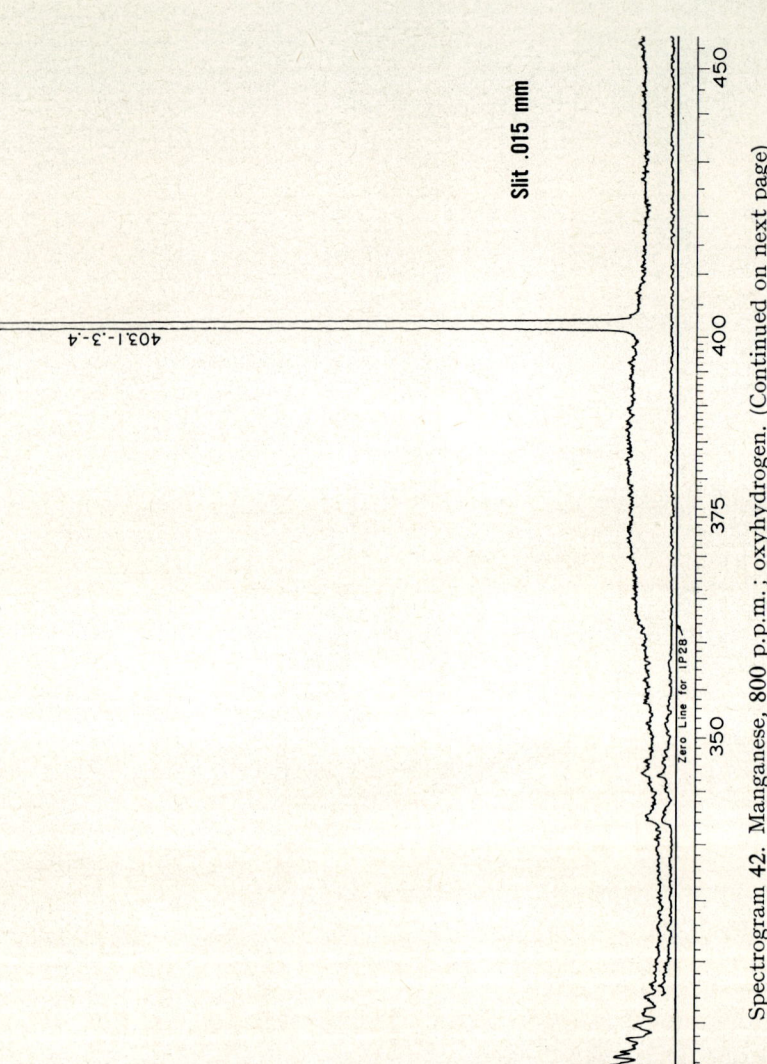

Spectrogram 42. Manganese, 800 p.p.m.; oxyhydrogen. (Continued on next page)

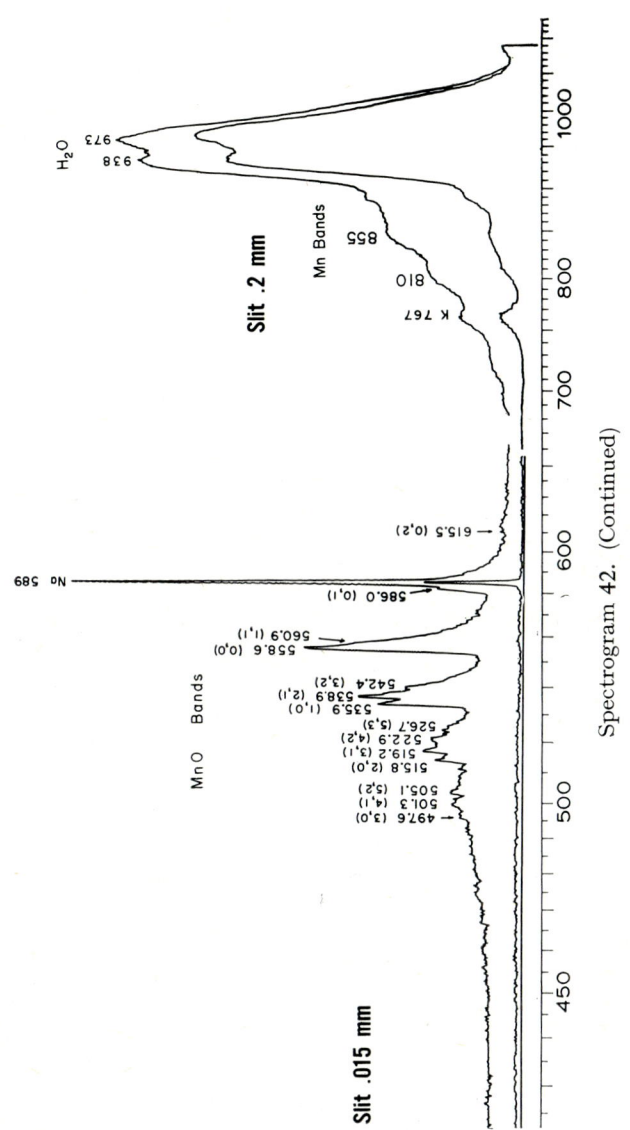

Spectrogram 42. (Continued)

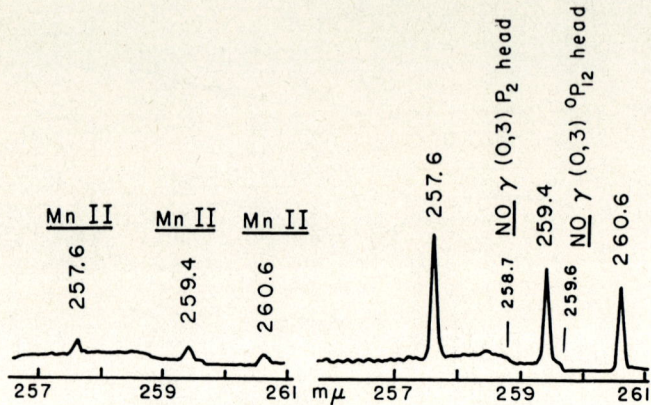

Spectrogram 43. Manganese, ion lines in oxycyanogen; 100 and 10 000 p.p.m.; slit 0.02 mm.

Spectrogram 44: See next page.

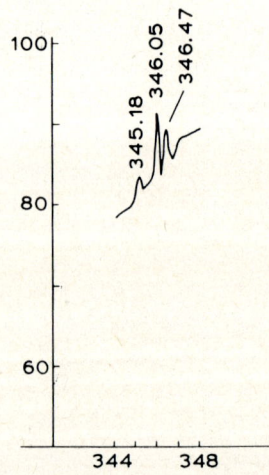

Spectrogram 45. Rhenium, 2000 p.p.m.; sheathed, fuel-rich (white) oxyacetylene; slit 0.02 mm.; zero suppressed 50 scale divisions.

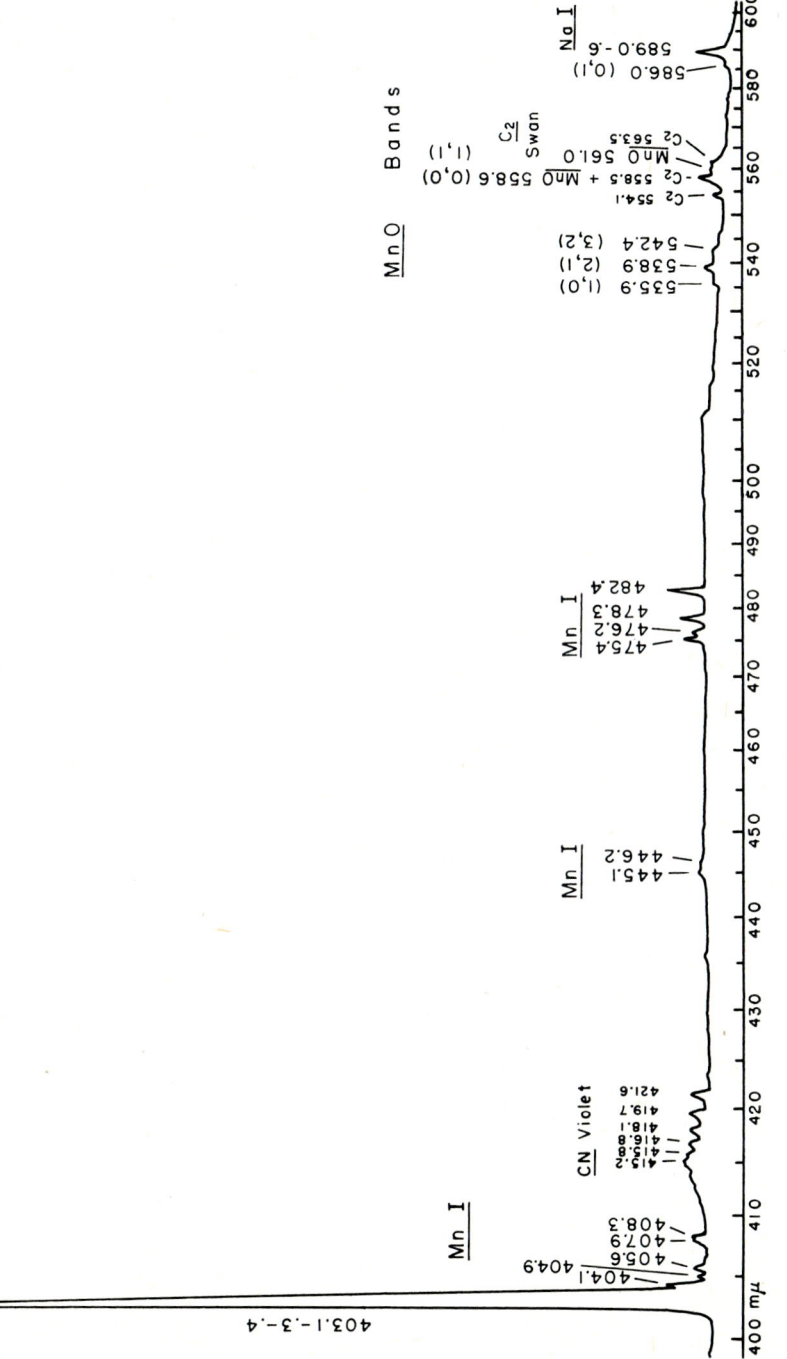

Spectrogram 44. Manganese, 10 000 p.p.m.; oxycyanogen; slit 0.02 mm.
Spectrogram 45: See preceding page.

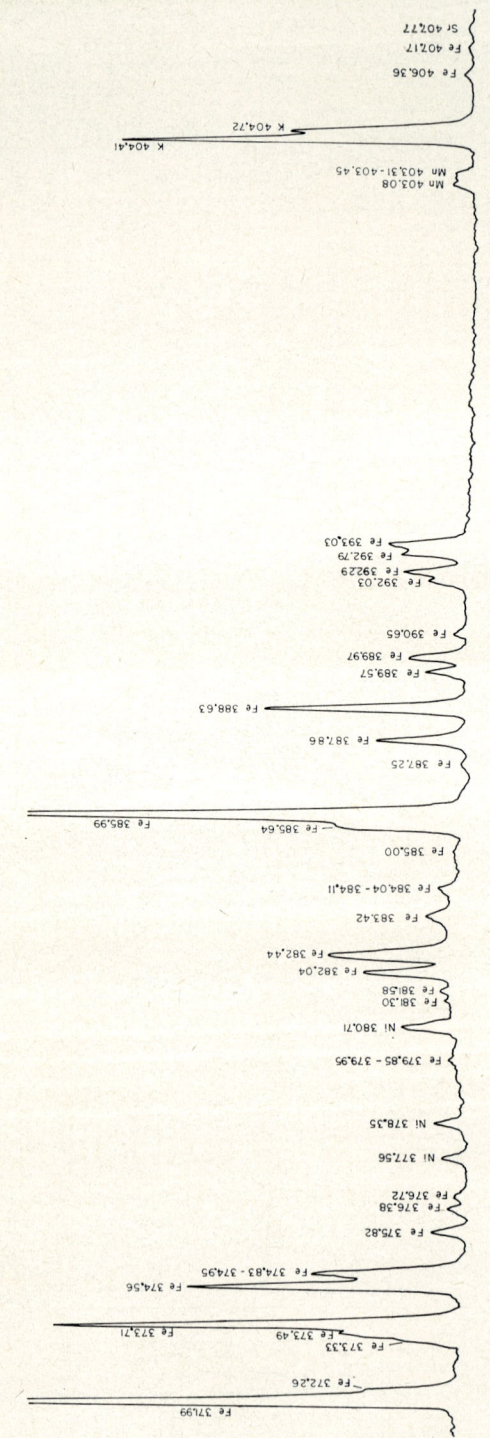

Spectrogram 46. Iron and other elements (Fe 600, Ni 600, K 200, Sr 40, Mg 200 p.p.m.); oxyhydrogen; slit 0.009 mm.

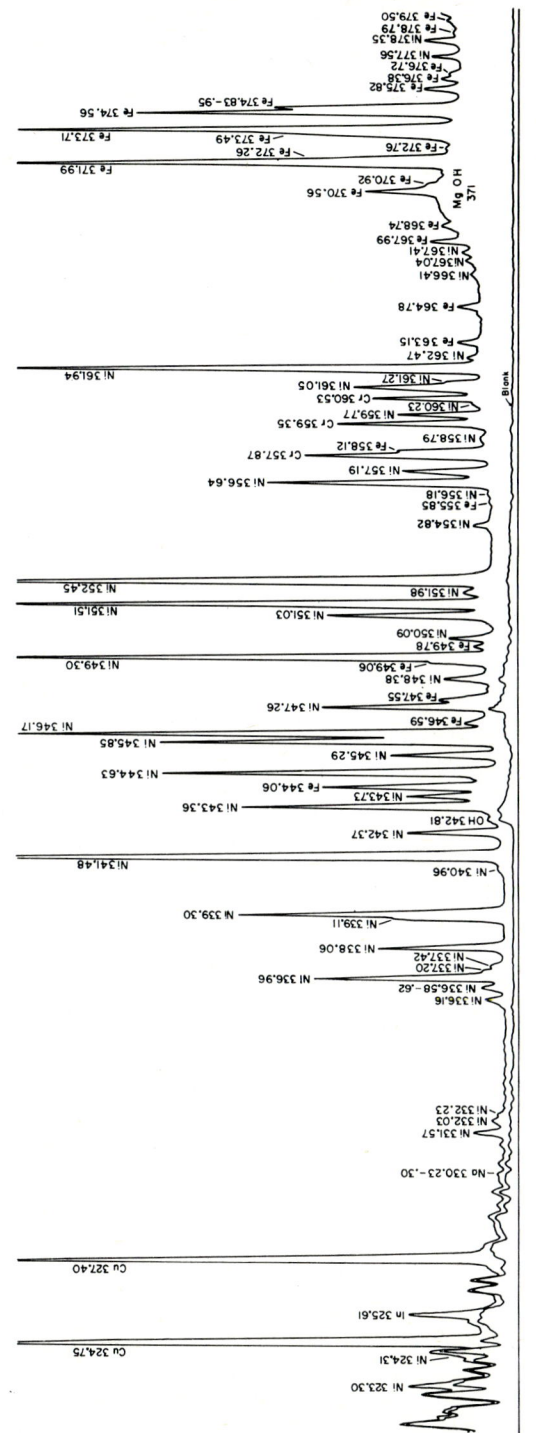

Spectrogram 47. Iron, nickel and other elements (Fe 600, Ni 600, Cr 200, Cu 200, In 100, Mg 200 p.p.m.); oxyhydrogen; slit 0.02 mm.

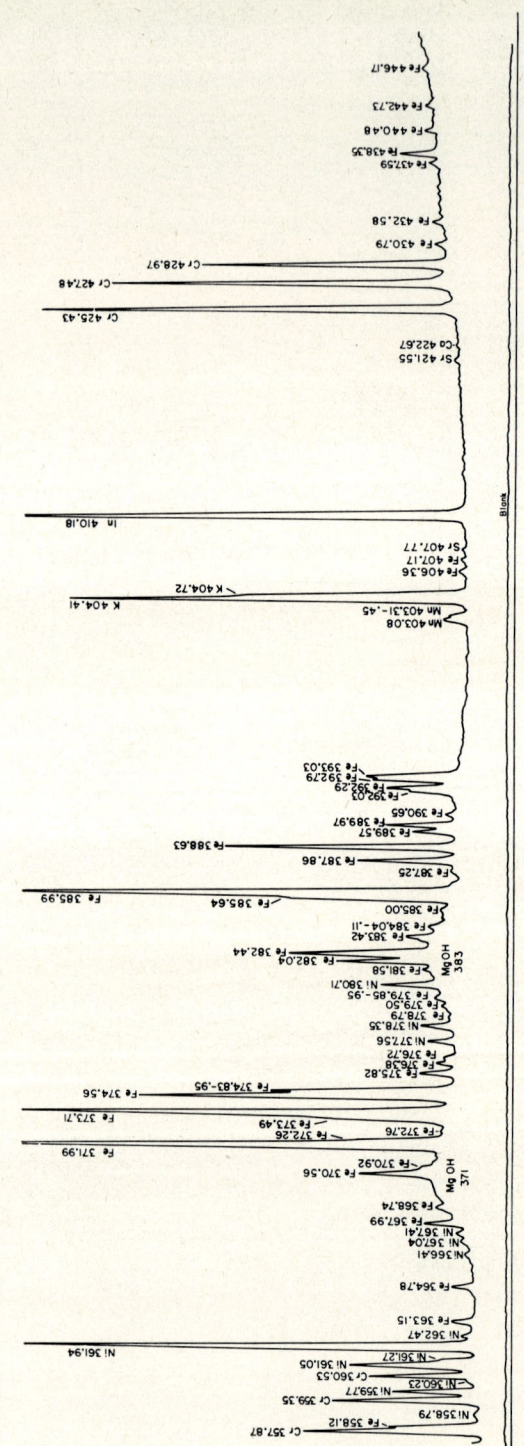

Spectrogram 48. Iron, nickel and other elements (Fe 600, Ni 600, Cr 200, In 100, Mg 200, Sr 40, K 200 p.p.m.); oxyhydrogen; slit 0.02 mm.

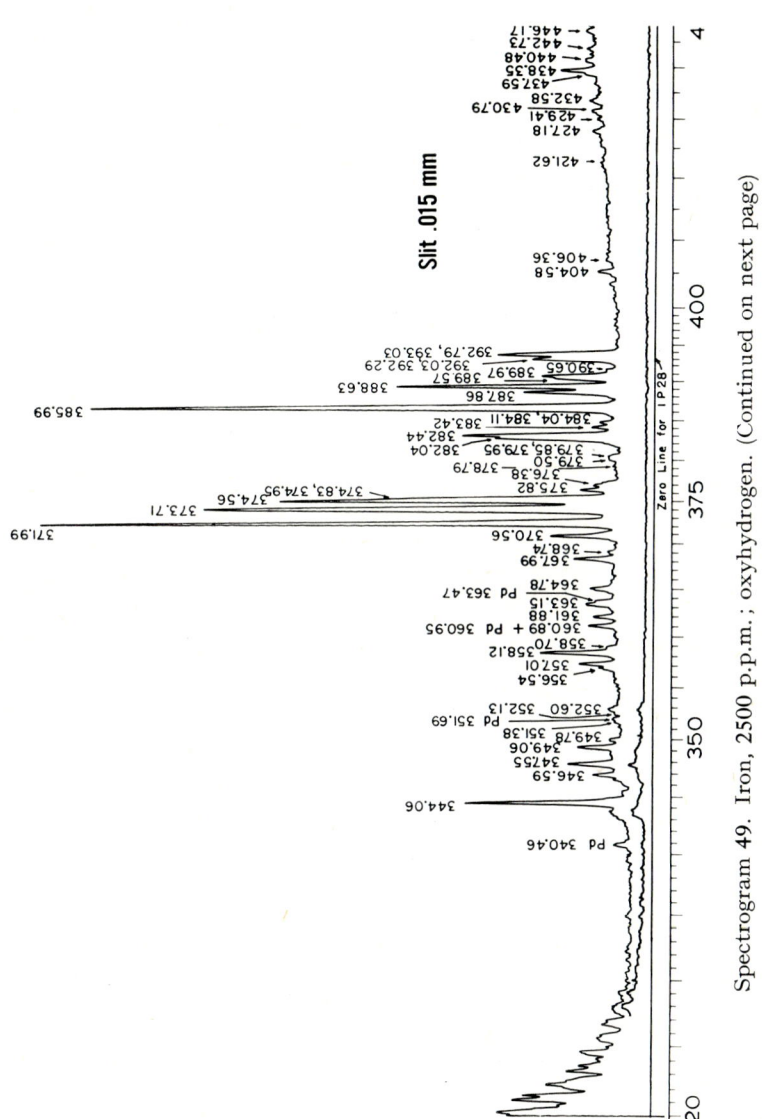

Spectrogram 49. Iron, 2500 p.p.m.; oxyhydrogen. (Continued on next page)

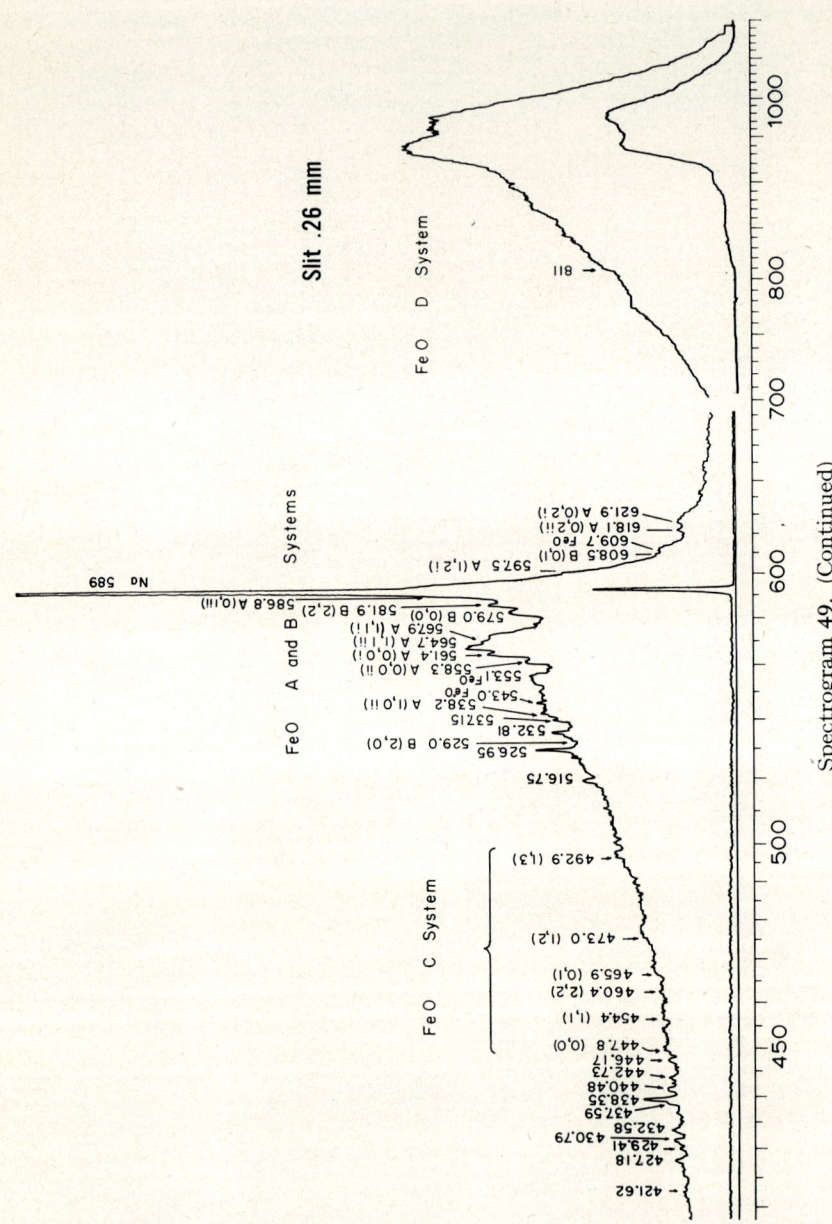

Spectrogram 49. (Continued)

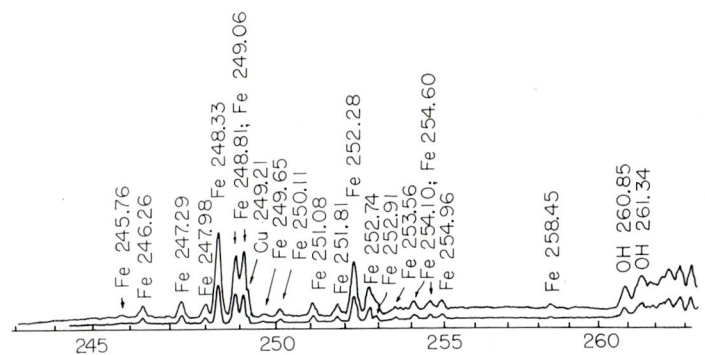

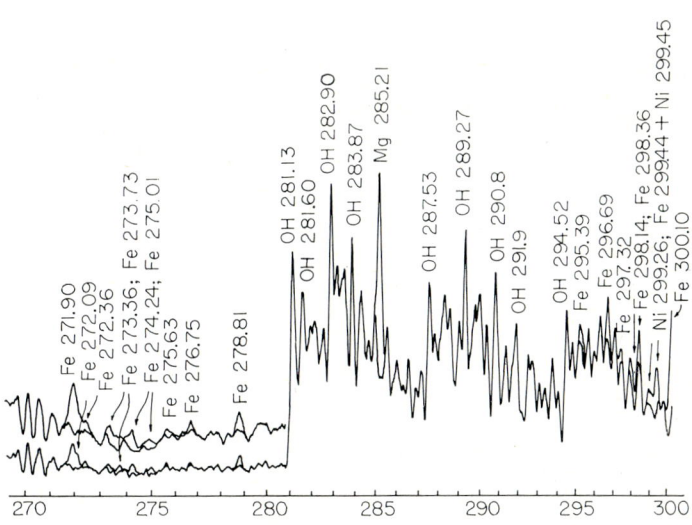

Spectrogram 50. Iron and other elements (Fe 150, Ni 150, Cu 50, Mg 50 mg./l.) in 75% isopropanol; air–hydrogen; slits 0.05 and 0.08 mm.; blank superposed on lower section.

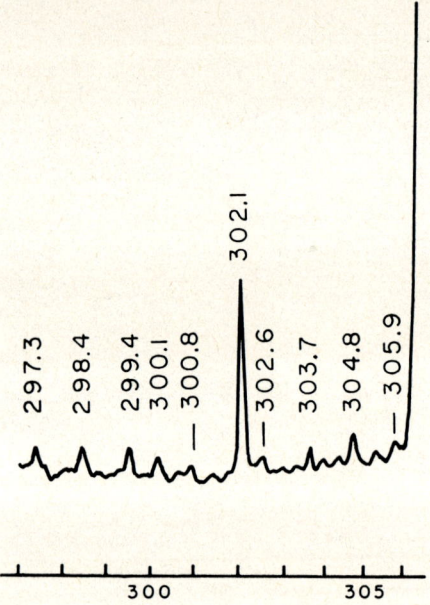

Spectrogram 51. Iron, 50 p.p.m.; oxycyanogen; slit 0.02 mm.

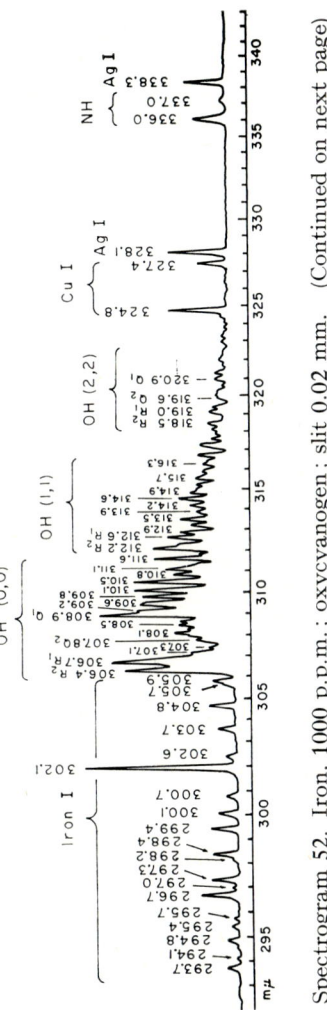

Spectrogram 52. Iron, 1000 p.p.m.; oxycyanogen; slit 0.02 mm. (Continued on next page)

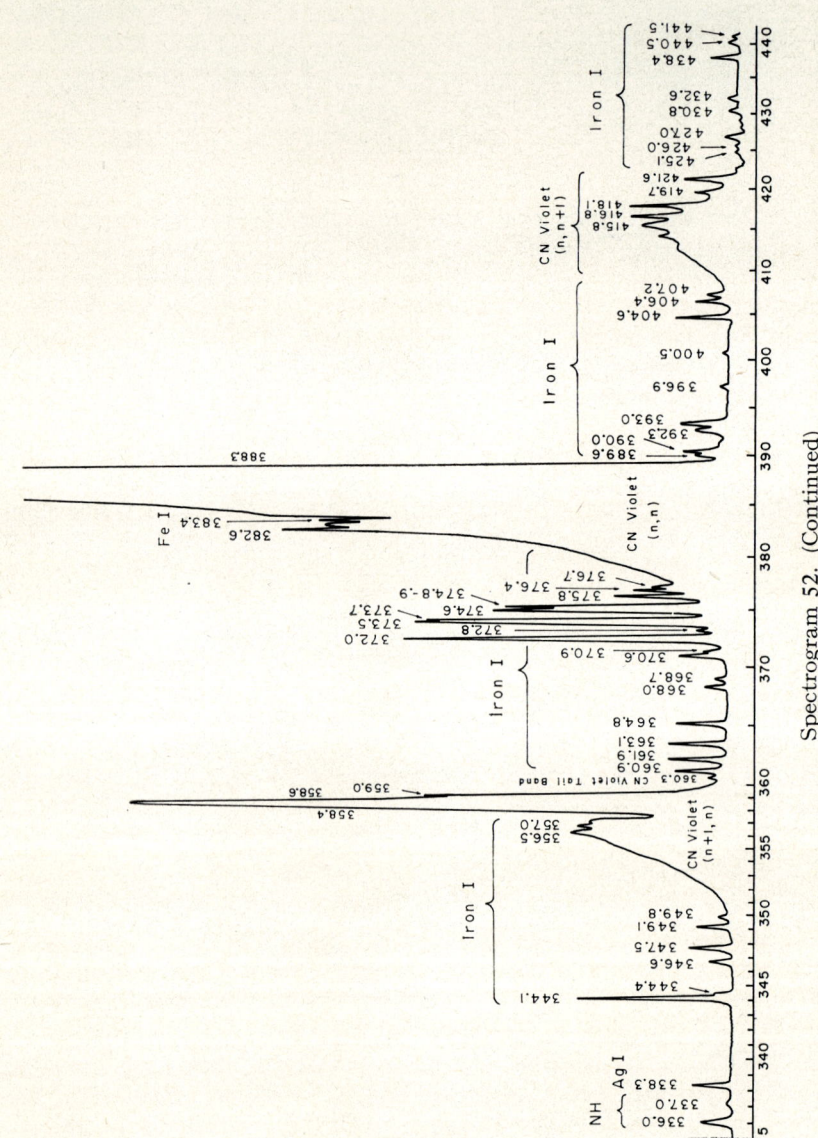

Spectrogram 52. (Continued)

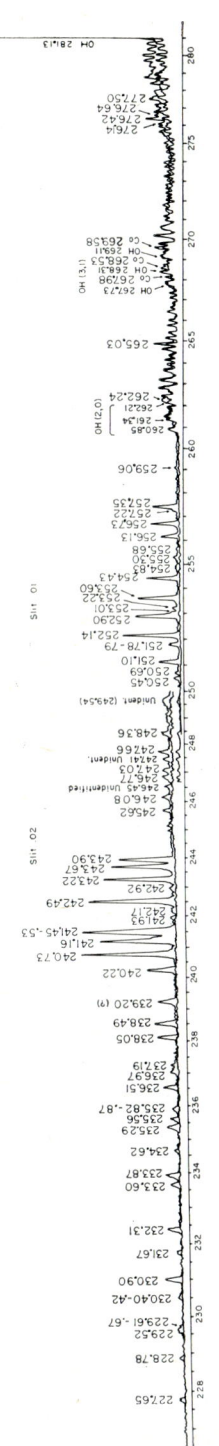

Spectrogram 53. Cobalt, 10 000 p.p.m.; oxyhydrogen; true slit 0.02 mm. above 247 mμ, 0.04 mm. below 250 mμ; blank superposed.

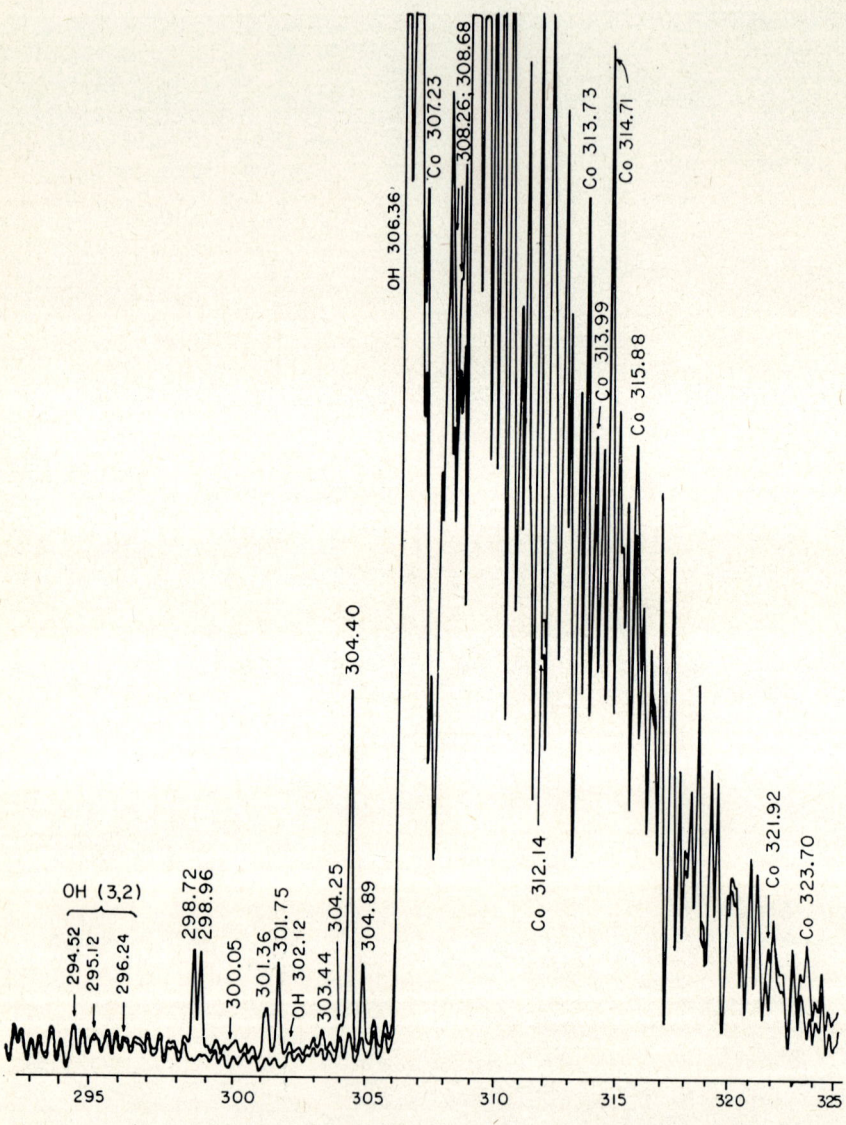

Spectrogram 54. Cobalt, 10 000 p.p.m.; oxyhydrogen; slit 0.02 mm.; blank superposed. The Co lines in the OH region are clearer if the spectrogram is examined from an angle so as to foreshorten the lines.

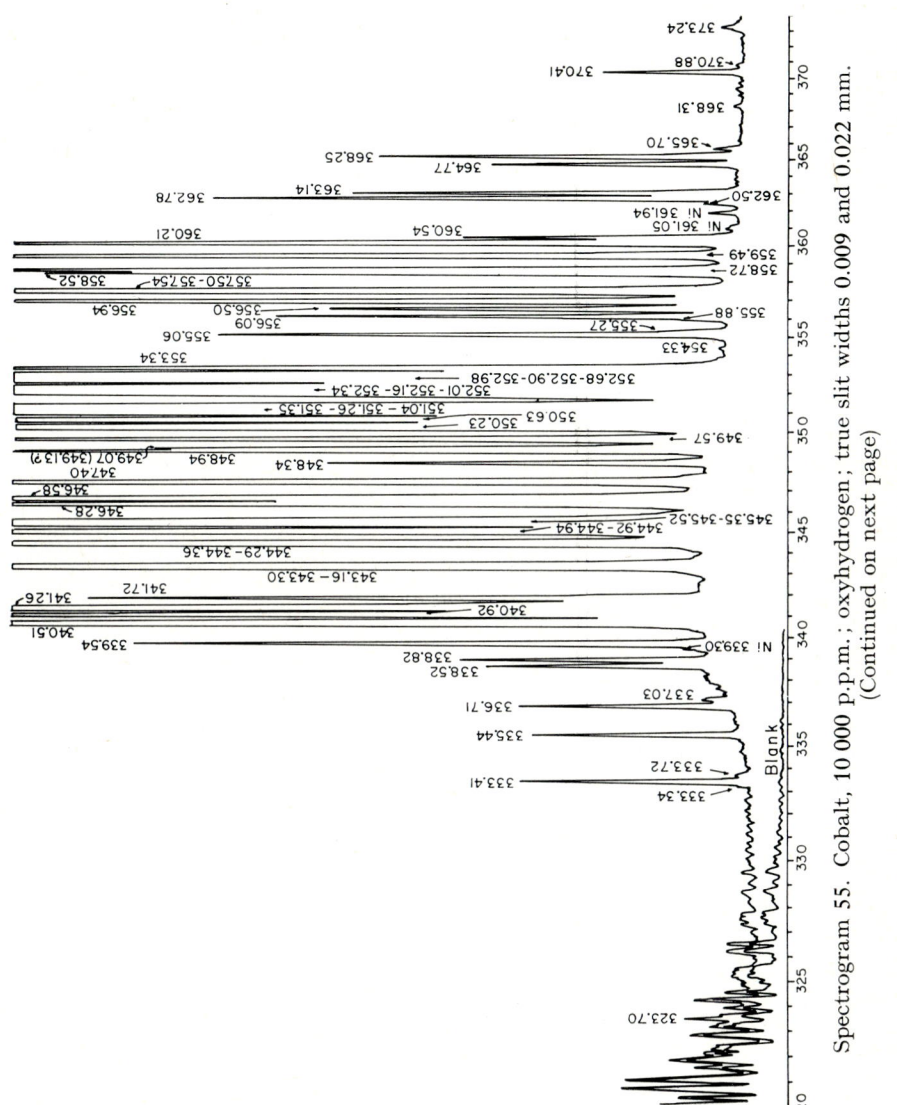

Spectrogram 55. Cobalt, 10 000 p.p.m.; oxyhydrogen; true slit widths 0.009 and 0.022 mm.
(Continued on next page)

538

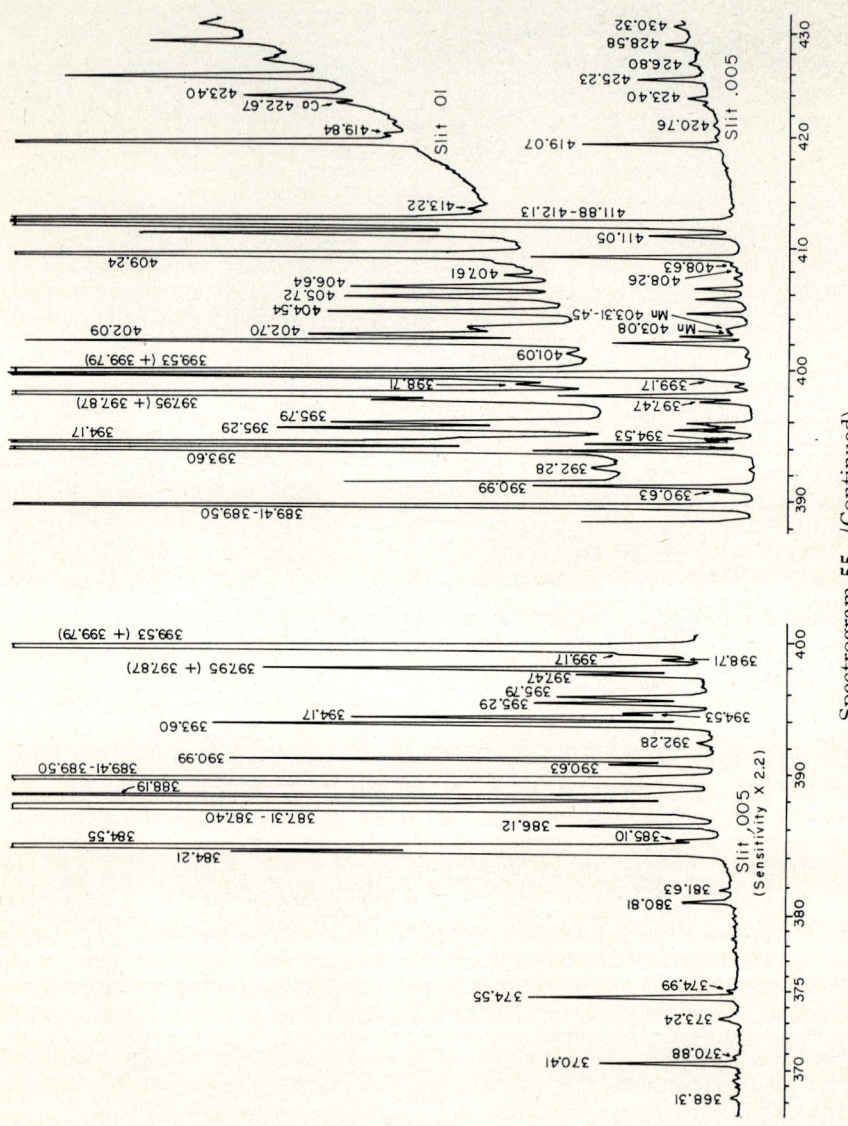

Spectrogram 55. (Continued)

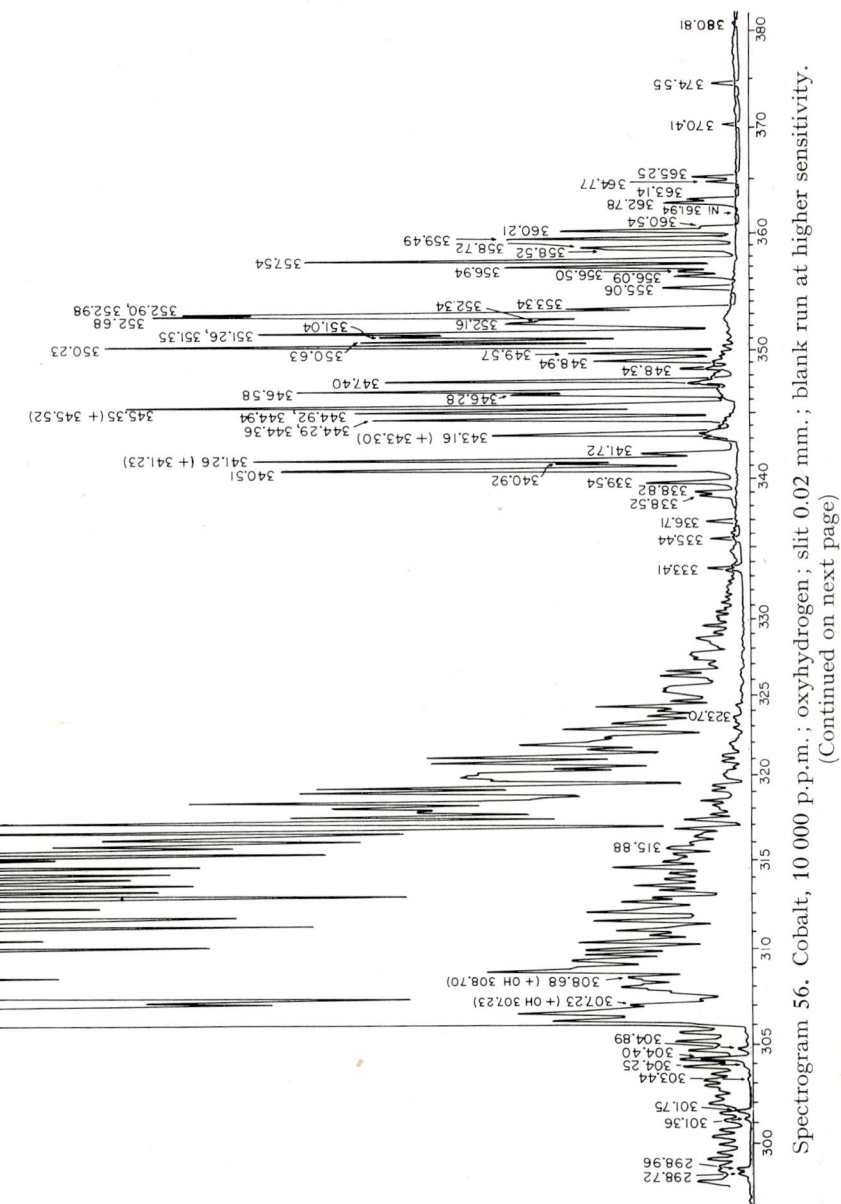

Spectrogram 56. Cobalt, 10 000 p.p.m.; oxyhydrogen; slit 0.02 mm.; blank run at higher sensitivity.
(Continued on next page)

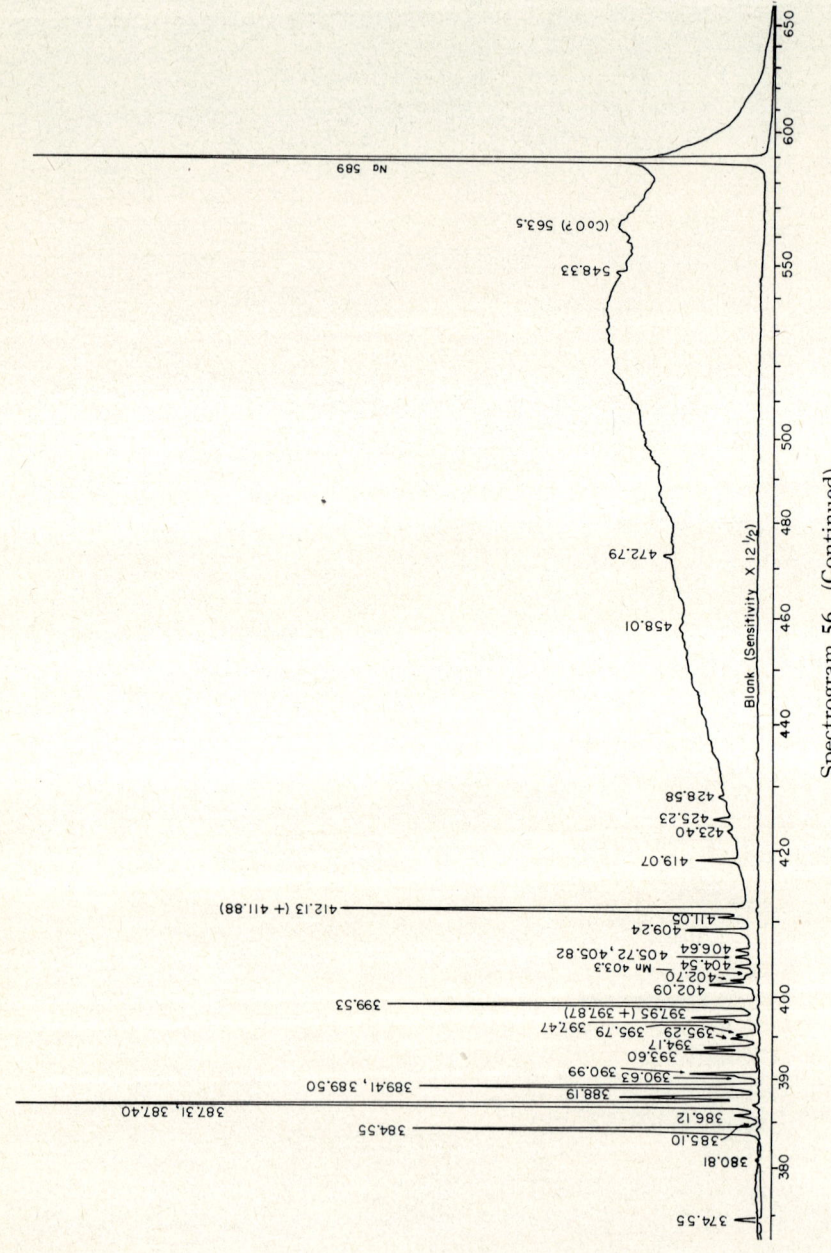

Spectrogram 56. (Continued)

541

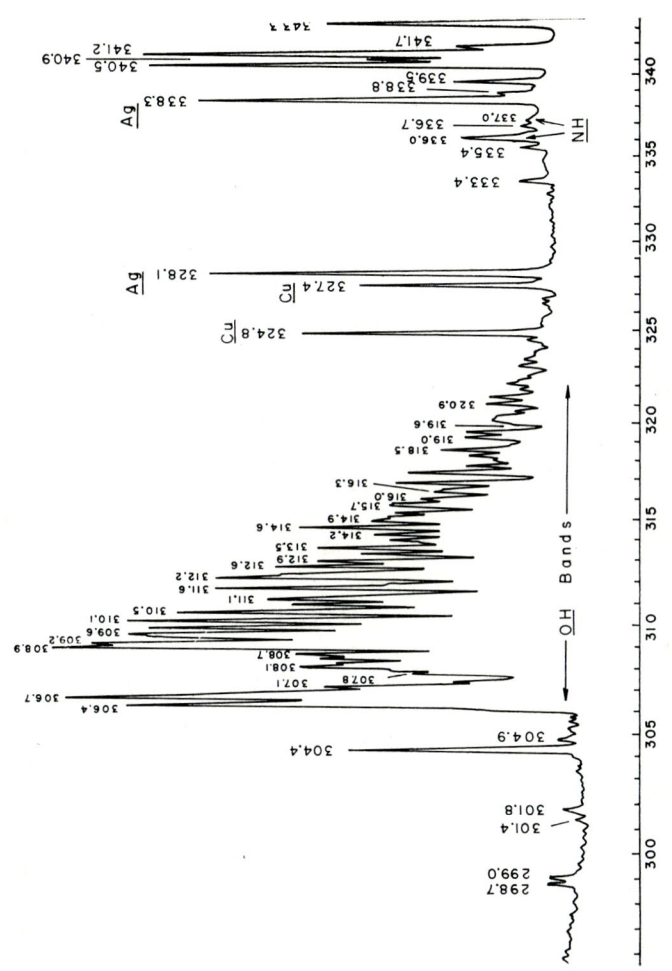

Spectrogram 57. Cobalt, 1100 p.p.m.; oxycyanogen; slit 0.02 mm. (Continued on next page)

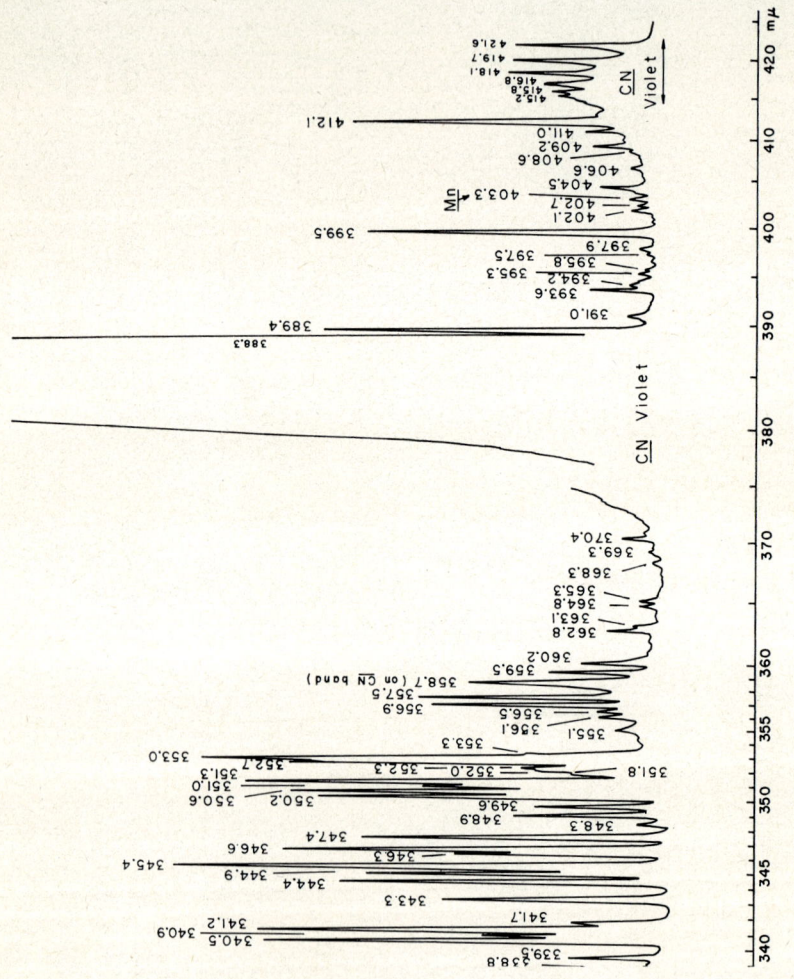

Spectrogram 57. (Continued)

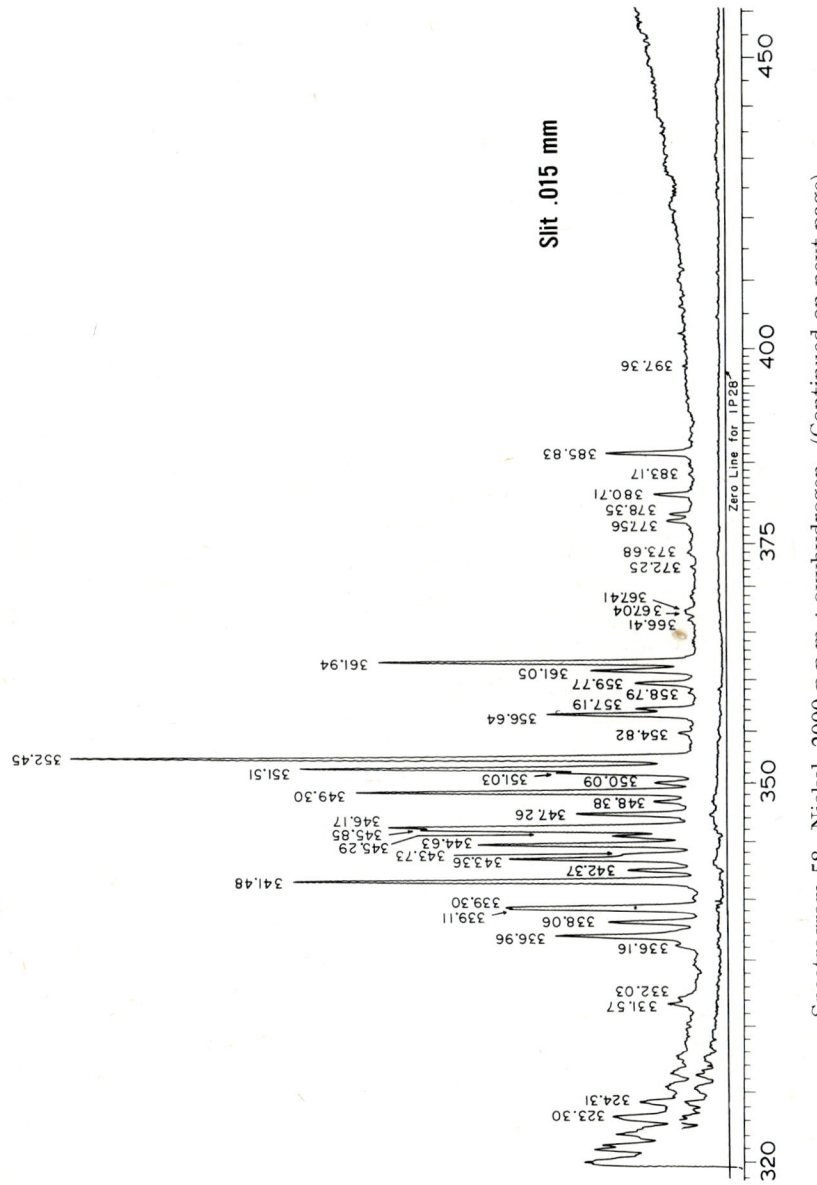

Spectrogram 58. Nickel, 2000 p.p.m.; oxyhydrogen. (Continued on next page)

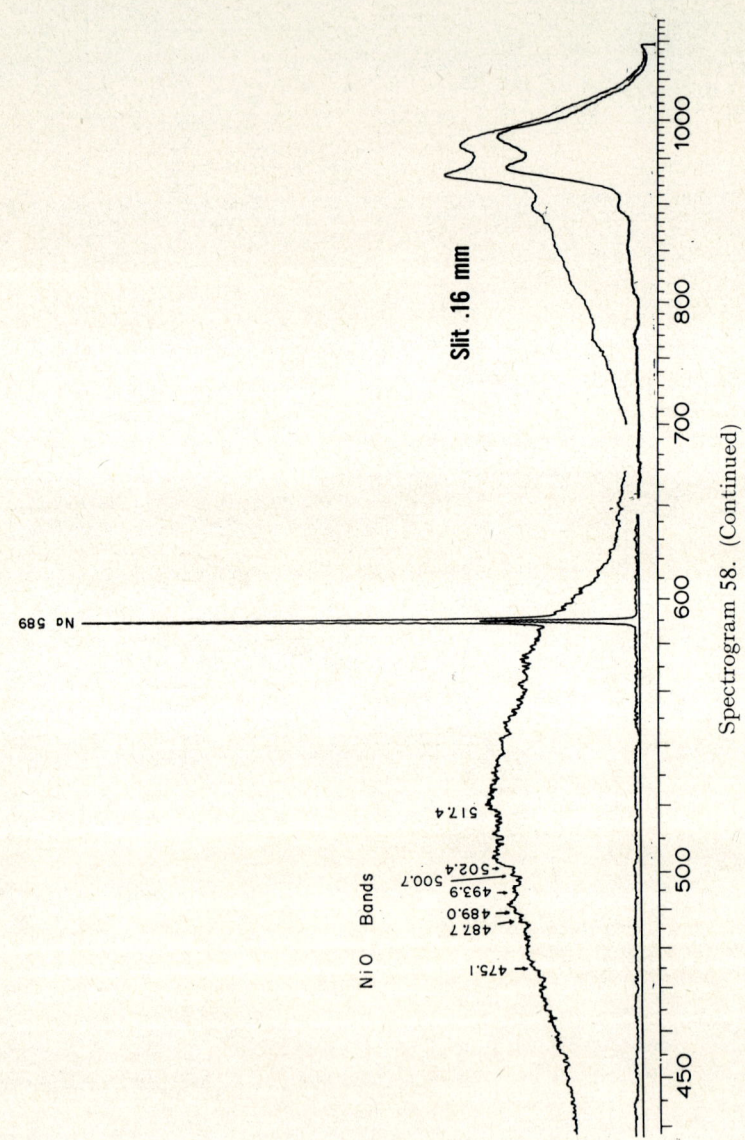

Spectrogram 58. (Continued)

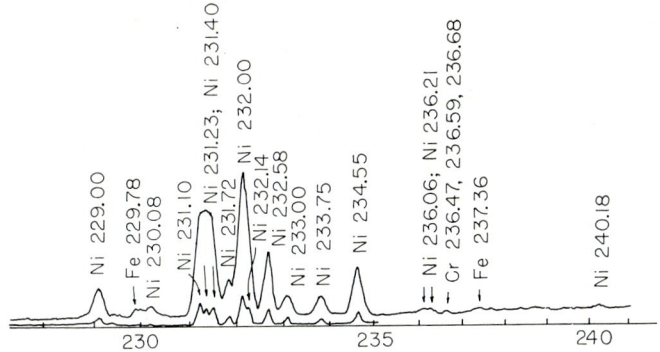

Spectrogram 59. Nickel and other elements (Ni 150, Fe 150, Cr 50 mg./l.) in 75% isopropanol; air–hydrogen; slits 0.05 and 0.14 mm.

Spectrogram 60: See next page.

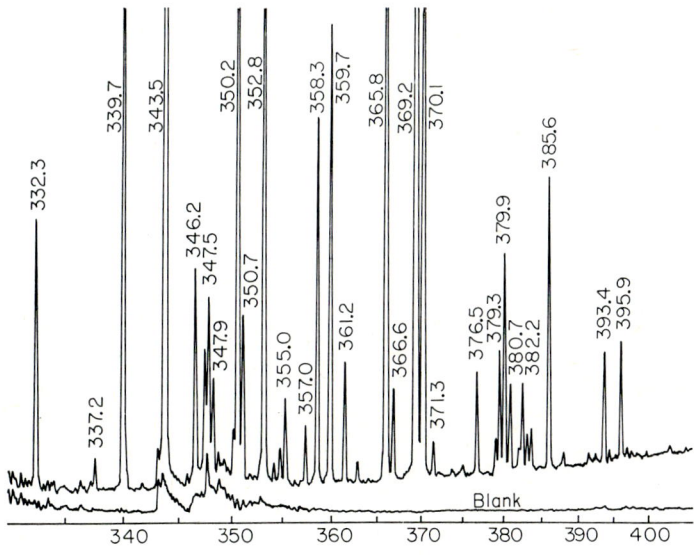

Spectrogram 61. Rhodium, 3000 p.p.m.; oxyhydrogen; slit 0.02 mm.

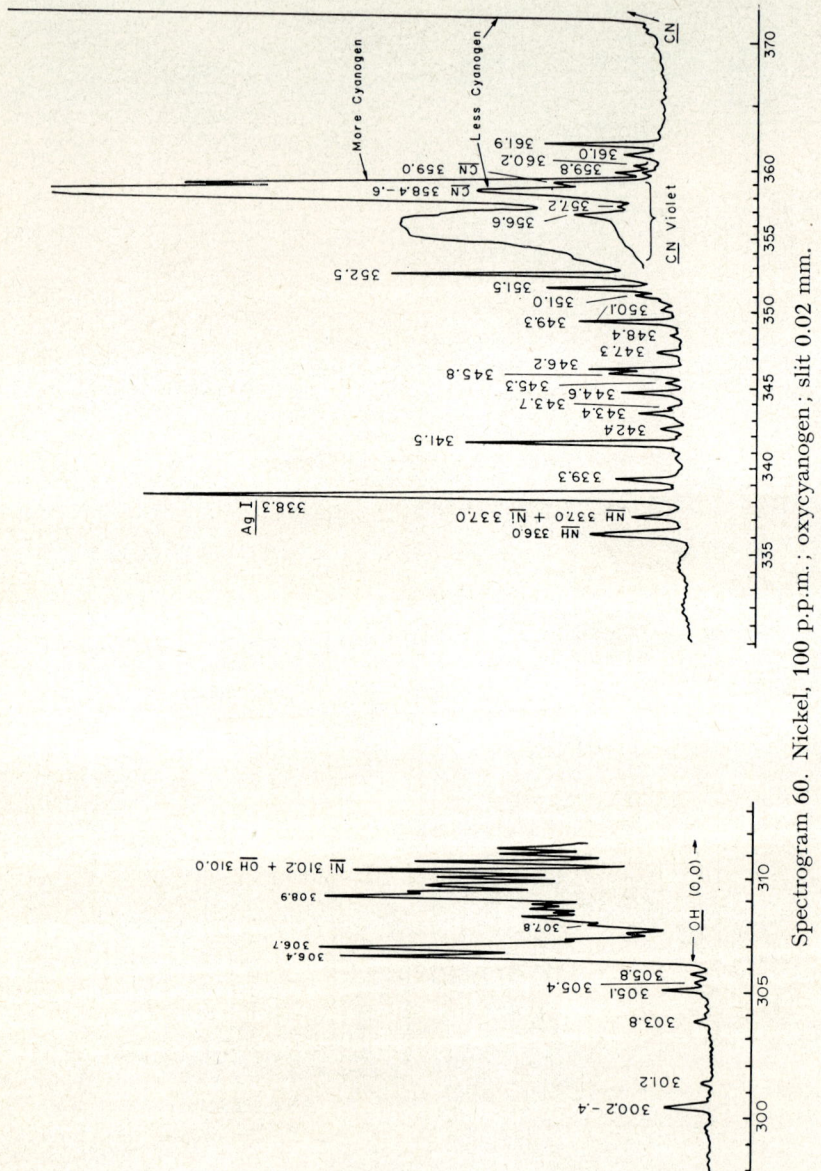

Spectrogram 60. Nickel, 100 p.p.m.; oxycyanogen; slit 0.02 mm. Spectrogram 61: See preceding page.

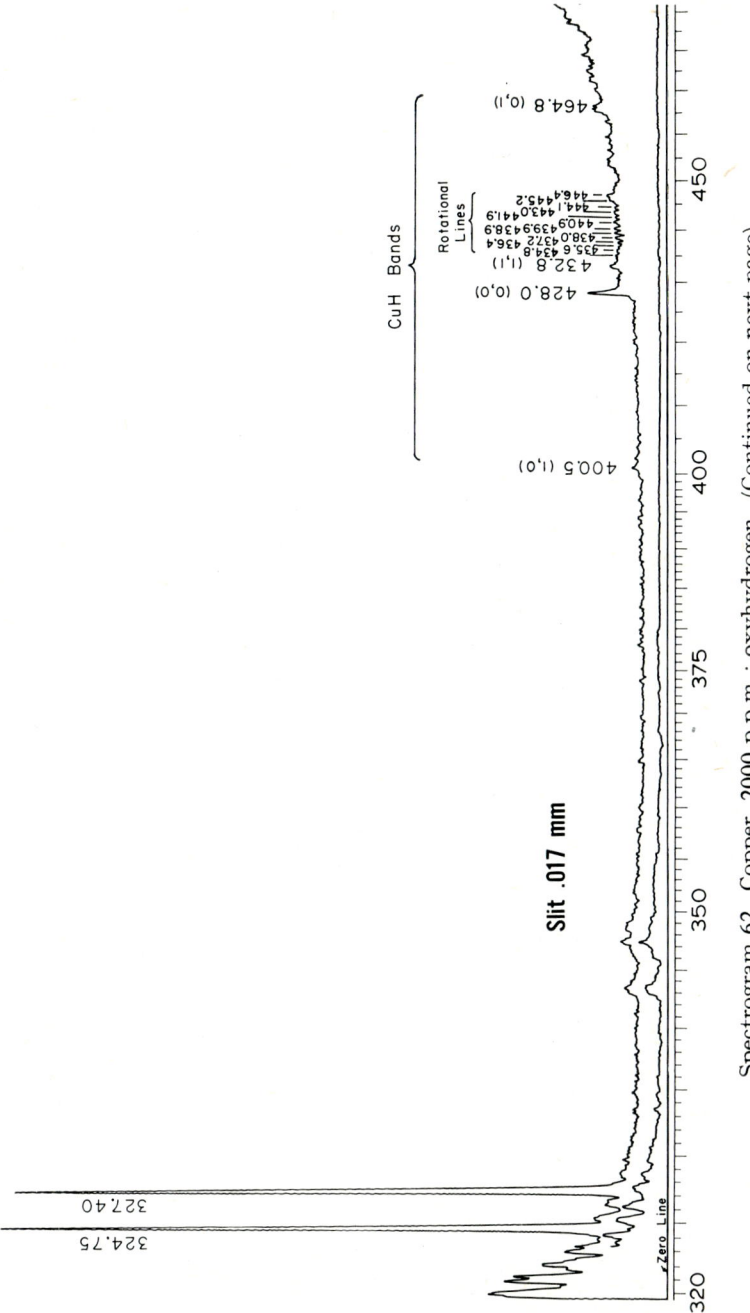

Spectrogram 62. Copper, 2000 p.p.m.; oxyhydrogen. (Continued on next page)

548

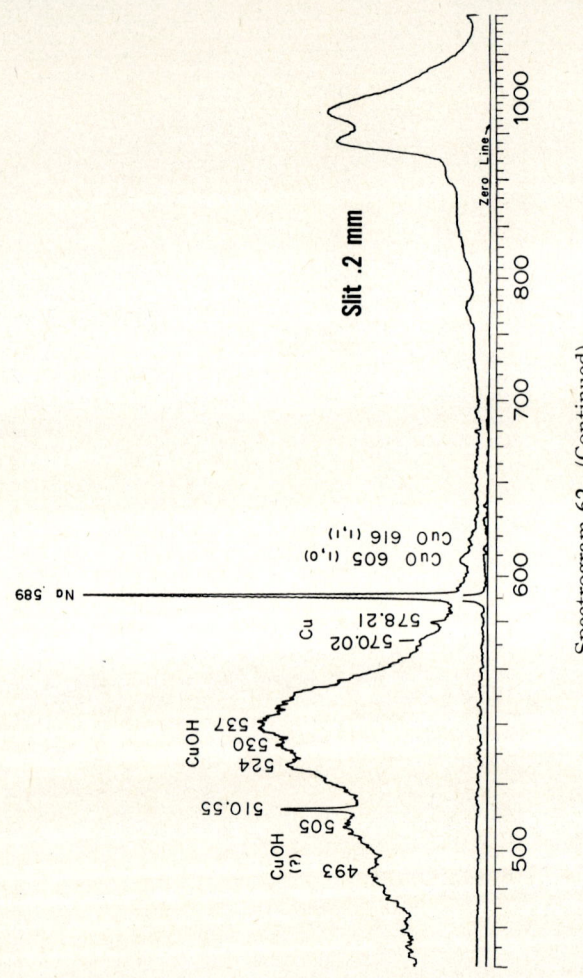

Spectrogram 62. (Continued)

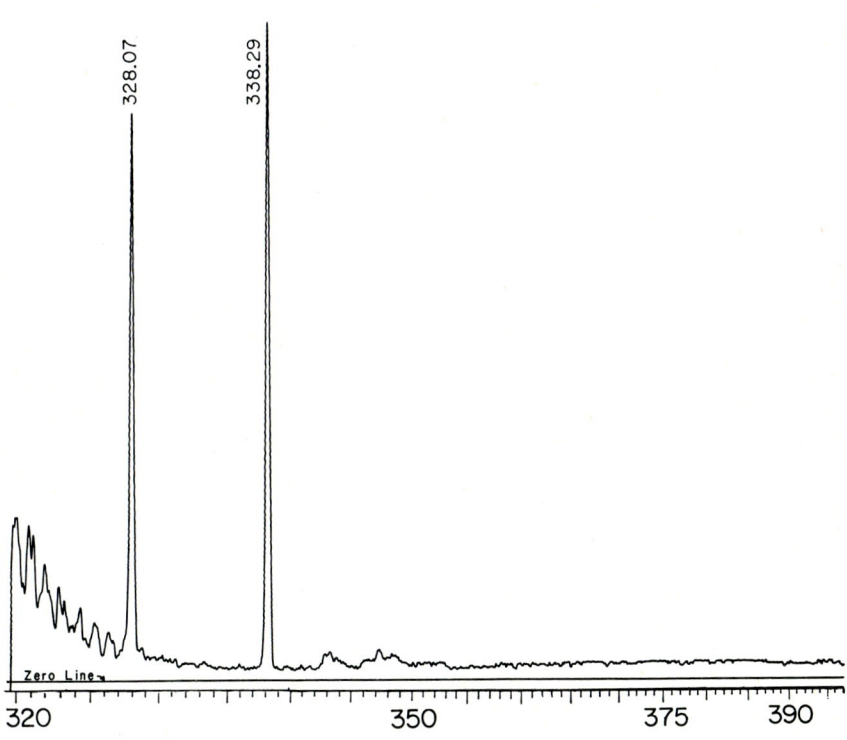

Spectrogram 63. Silver, 1000 p.p.m.; oxyhydrogen; slit 0.015 mm.

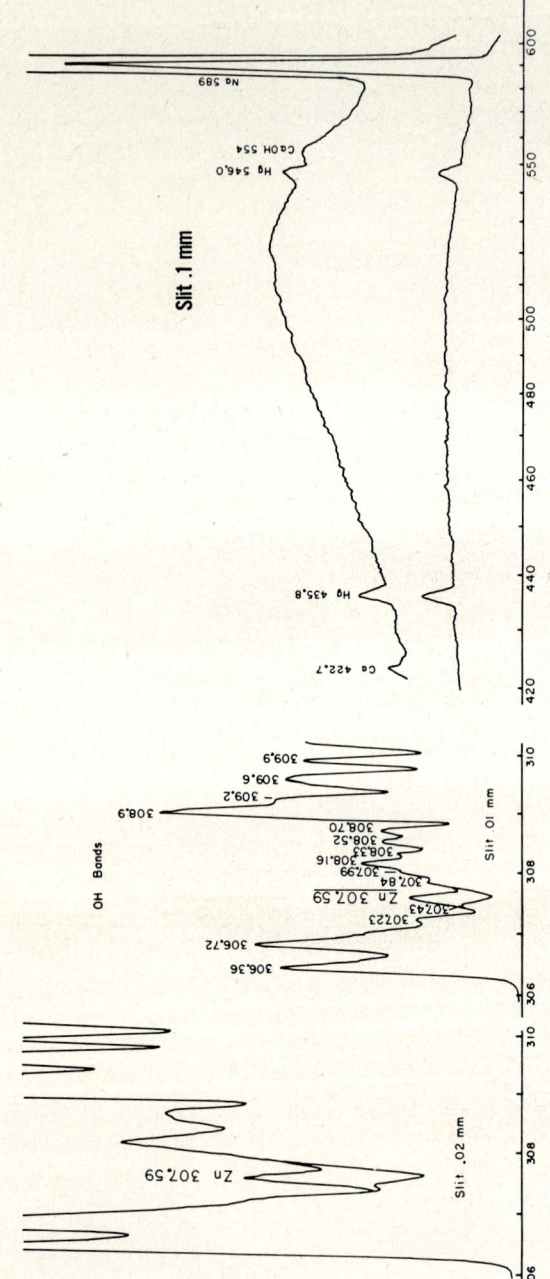

Spectrogram 64. Zinc, 5000 p.p.m.; air–hydrogen; true slit widths (left to right) 0.037, 0.022, 0.14 mm.

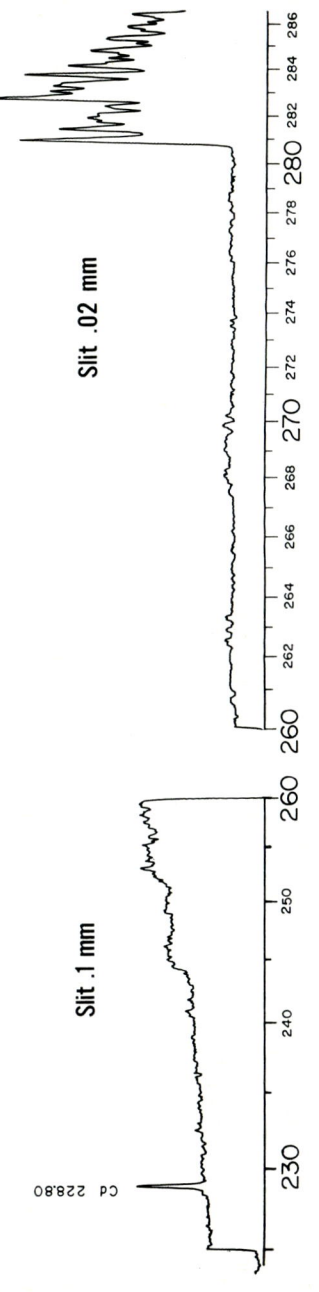

Spectrogram 65. Cadmium, 5000 p.p.m.; oxyhydrogen. (Continued on next page)

551

552

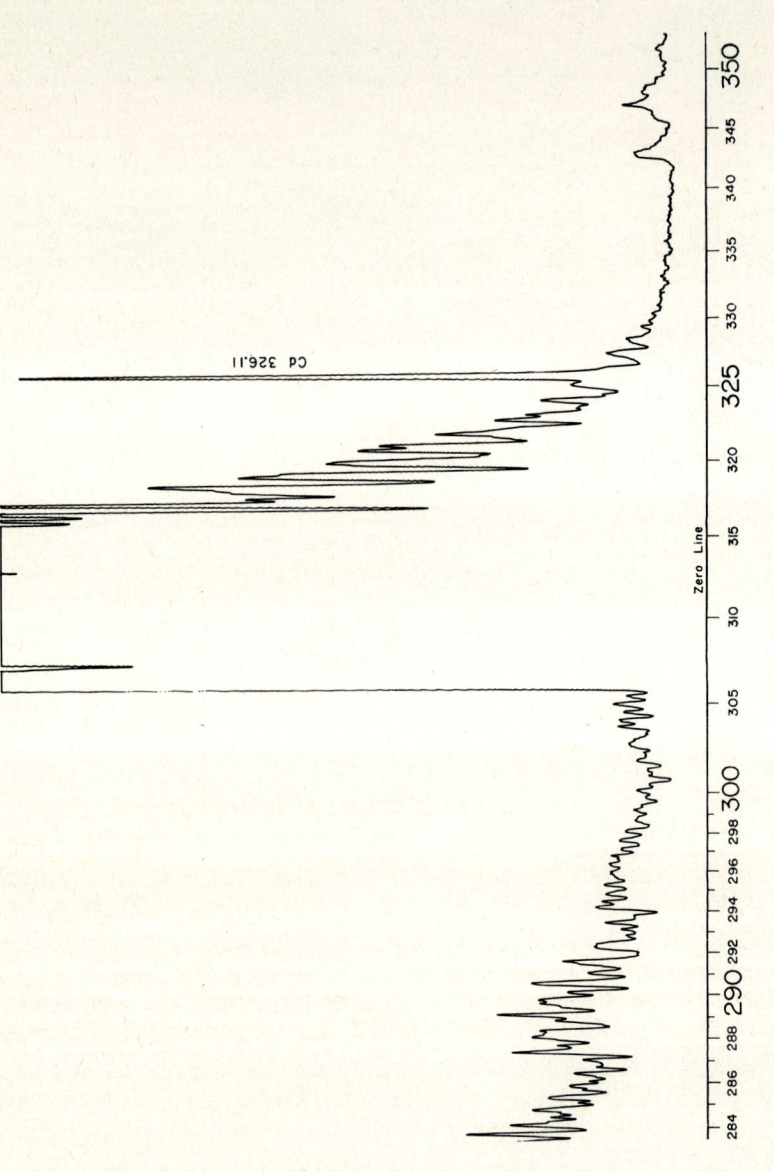

Spectrogram 65. (Continued)

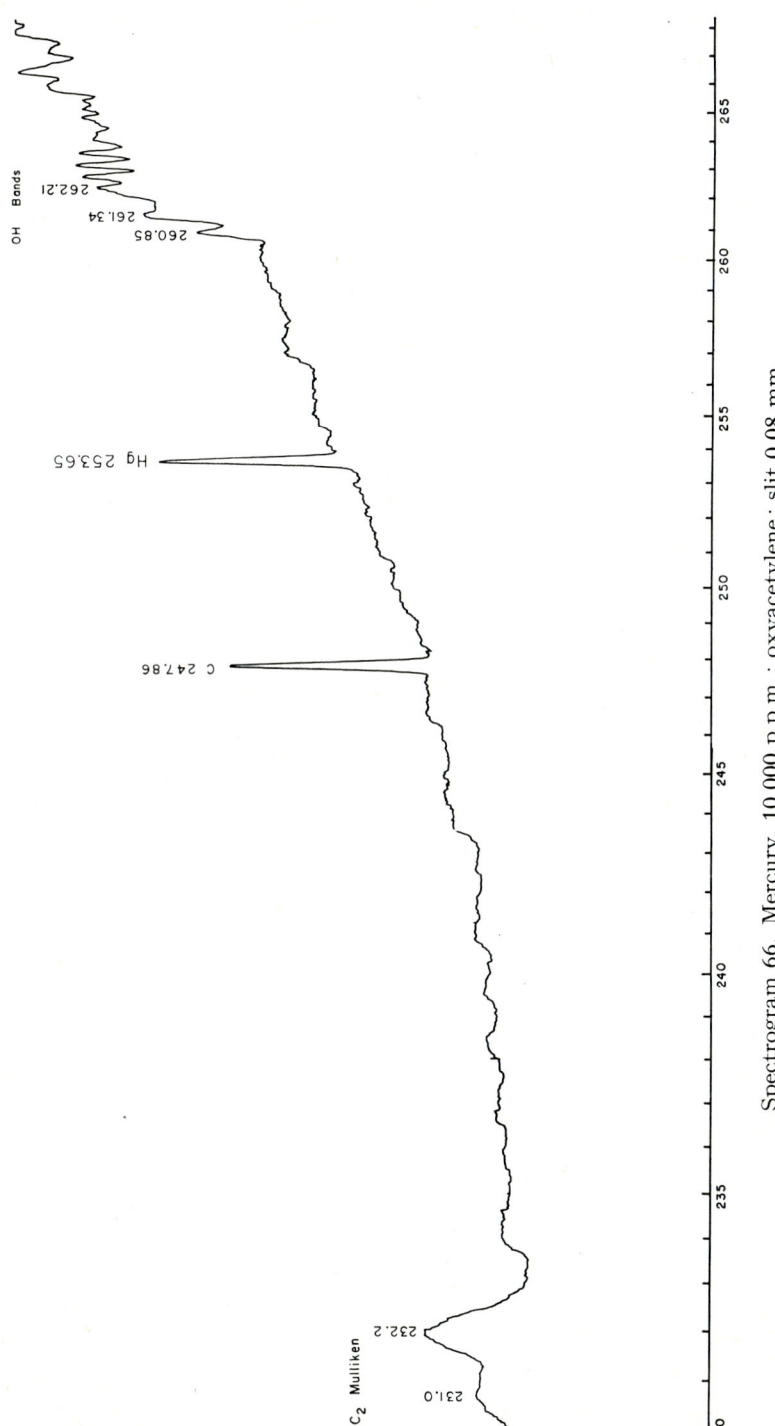

Spectrogram 66. Mercury, 10 000 p.p.m.; oxyacetylene; slit 0.08 mm.

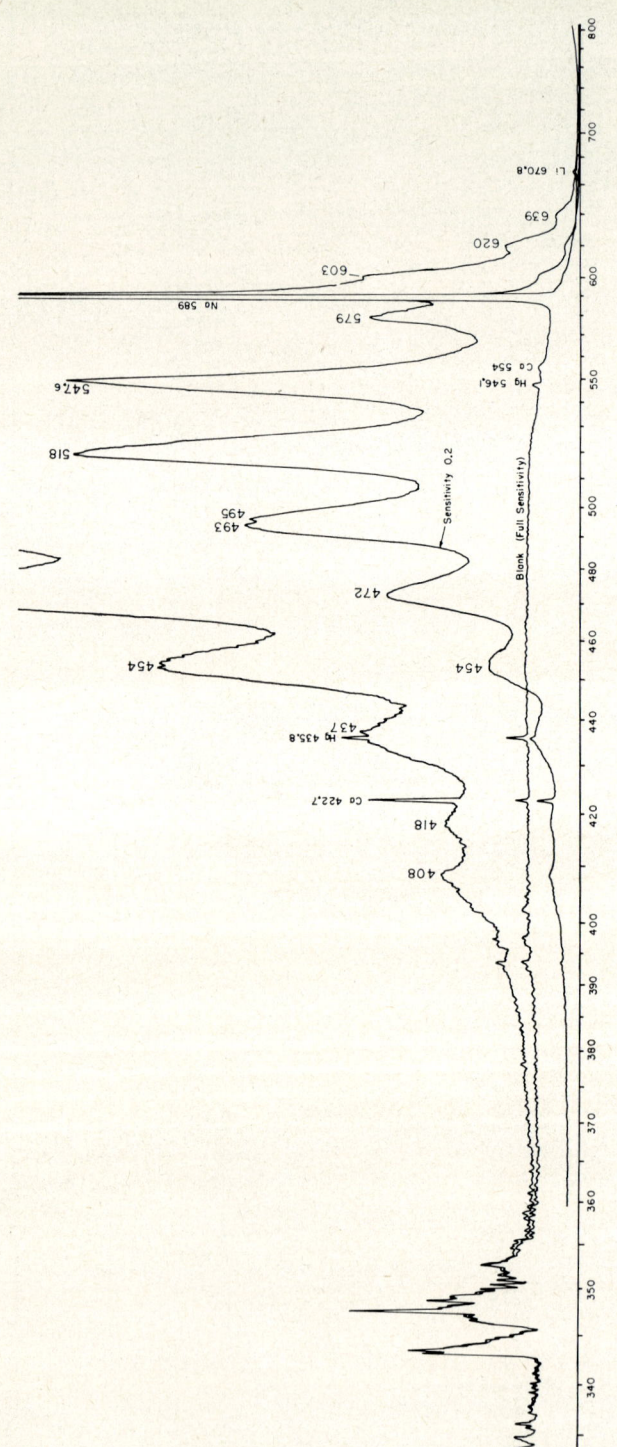

Spectrogram 67. Boron, 1000 p.p.m.· oxyhydrogen; slit 0.03 mm.

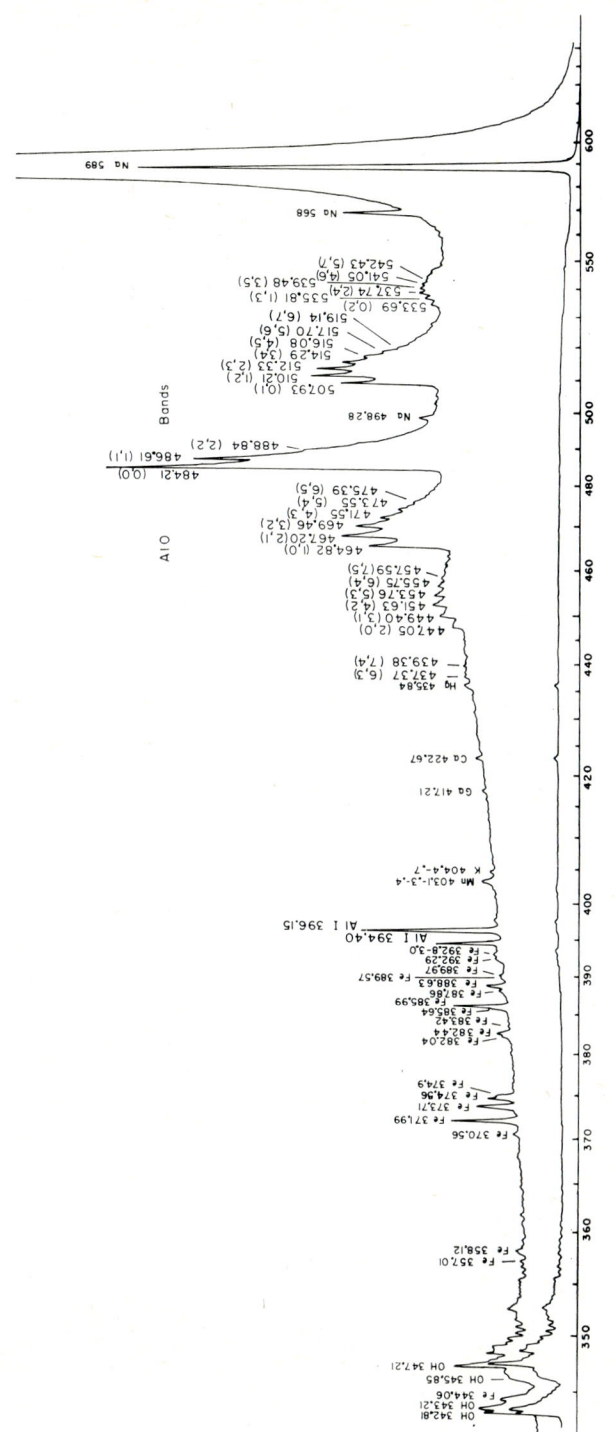

Spectrogram 68. Aluminum, 5000 p.p.m.; oxyhydrogen; slit 0.02 mm.

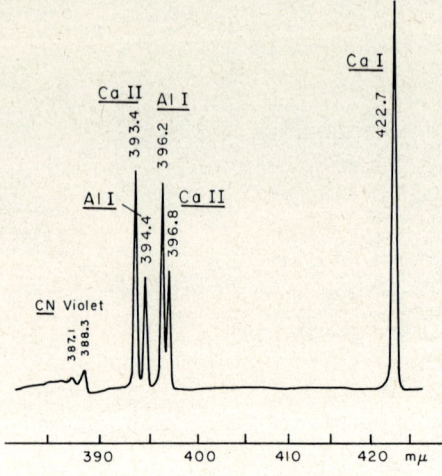

Spectrogram 69. Aluminum, 1000 p.p.m. (with calcium impurity); oxycyanogen; slit 0.02 mm.

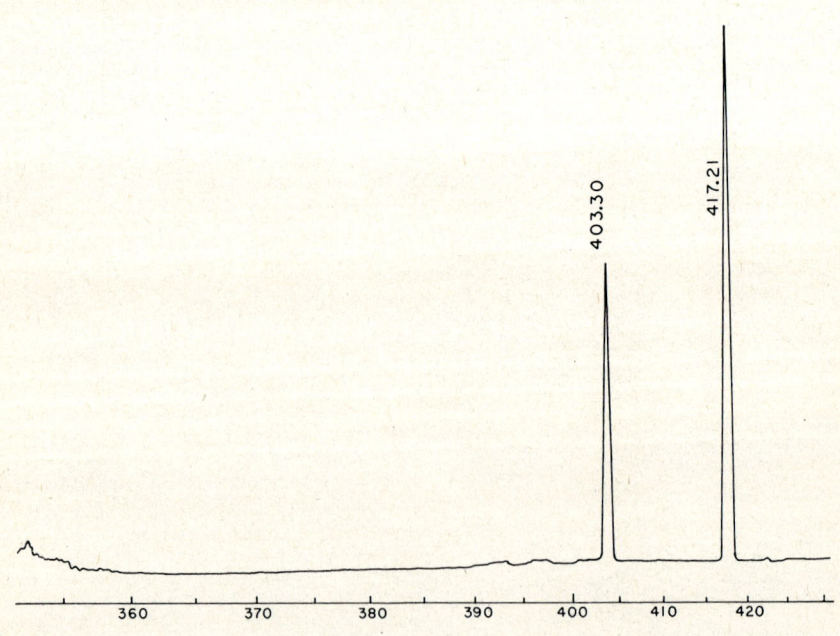

Spectrogram 70. Gallium, 100 p.p.m.; oxyhydrogen; slit 0.02 mm.

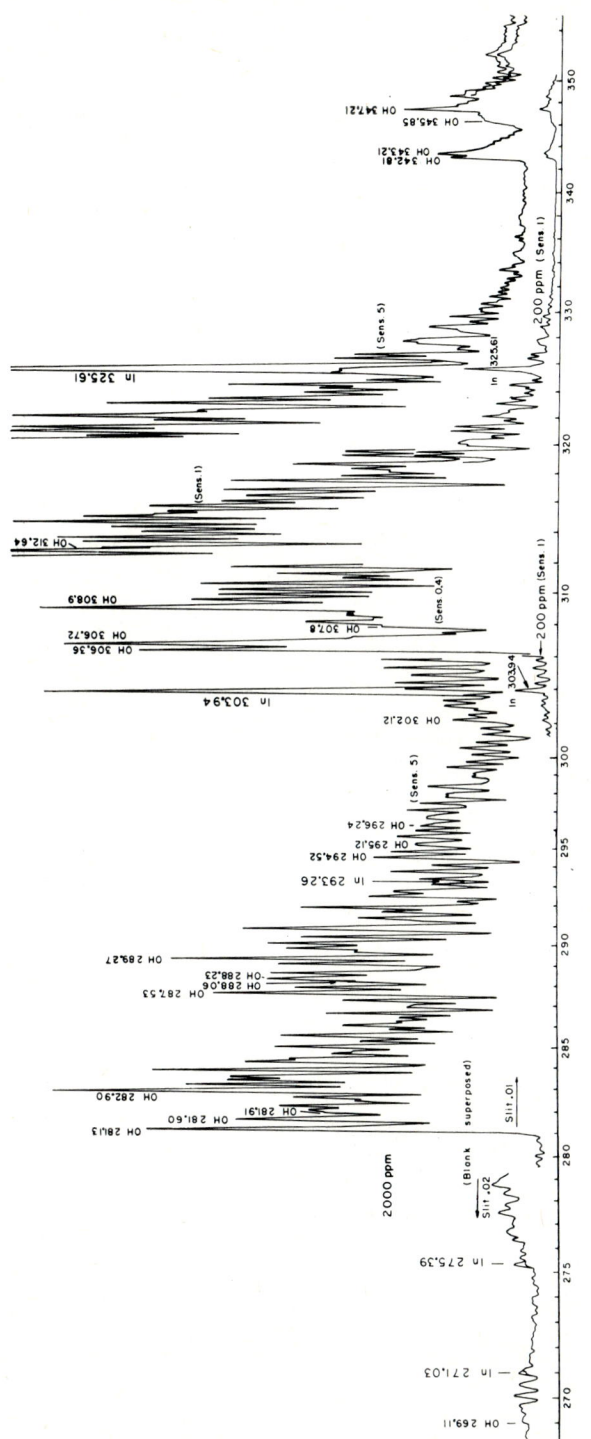

Spectrogram 71. Indium, 200 and 2000 p.p.m.; oxyhydrogen; true slit width 0.04 mm. below 280 mμ, 0.02 mm. above; different sections run at different instrument sensitivities, as shown.

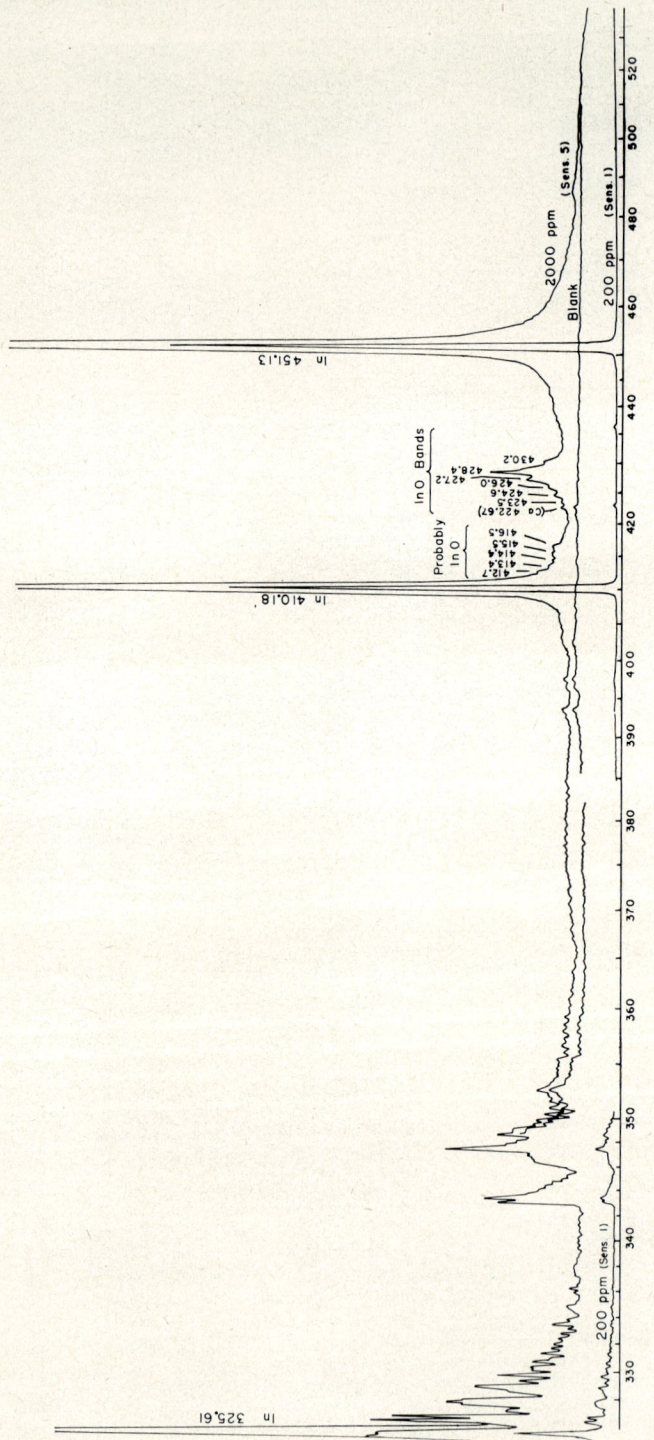

Spectrogram 72. Indium, 200 and 2000 p.p.m.; oxyhydrogen; slit 0.02 mm.; instrument sensitivities as shown.

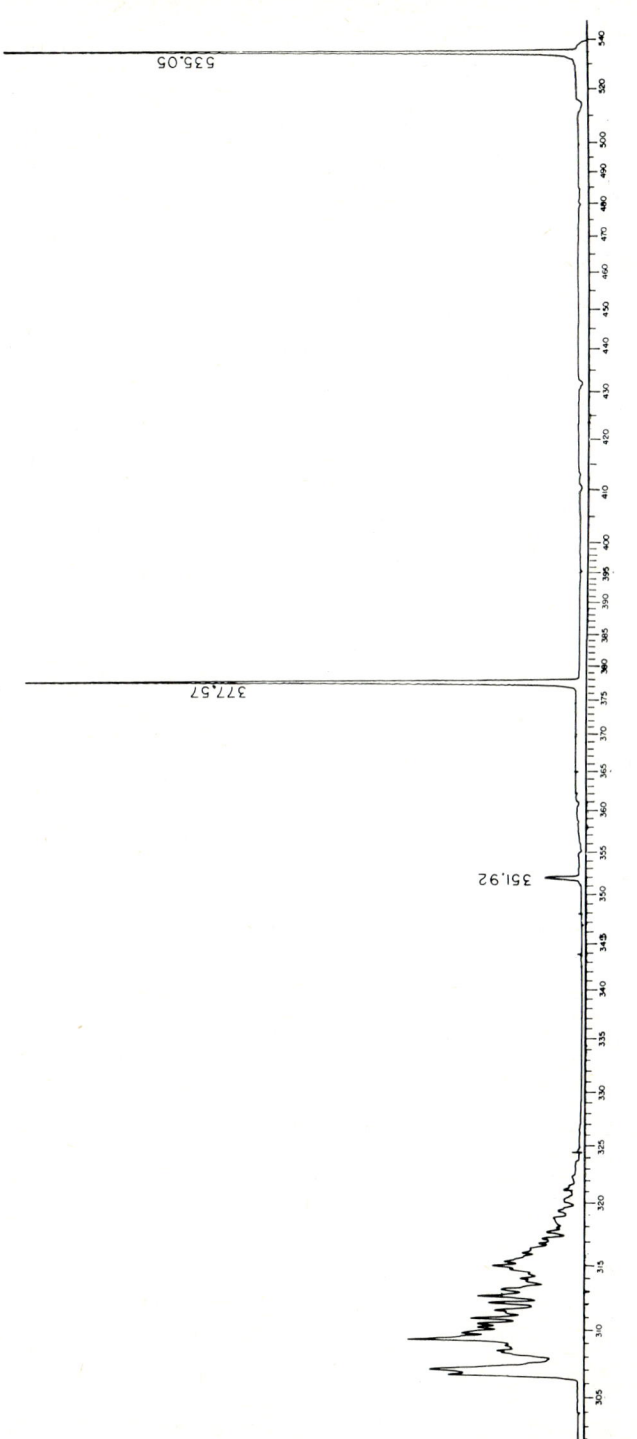

Spectrogram 73. Thallium, 5000 p.p.m.; oxyhydrogen; slit 0.015 mm.

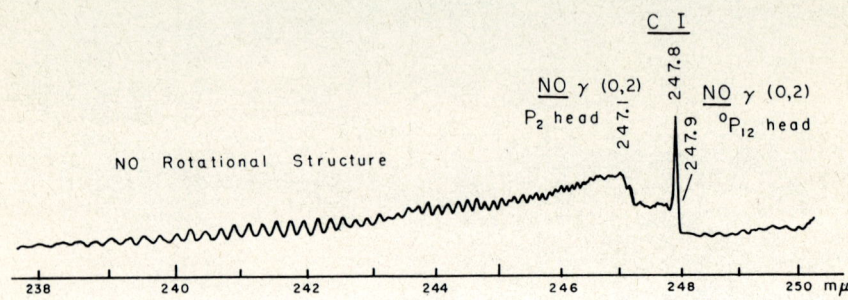

Spectrogram 74. Carbon line in the empty oxycyanogen flame; slit 0.02 mm.

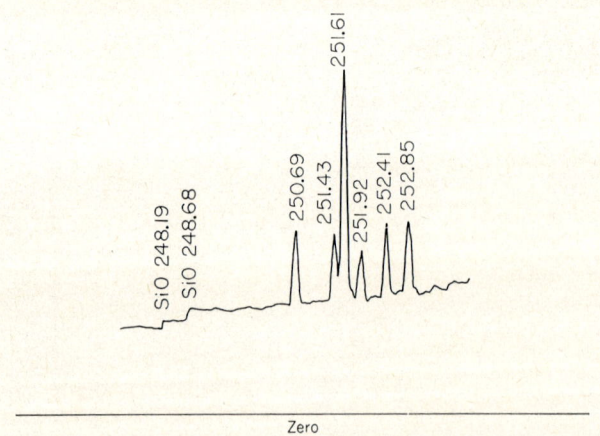

Spectrogram 75. Silicon, 1000 mg./l. in 80% methanol; sheathed, fuel-rich (white) oxyacetylene; slit 0.02 mm.

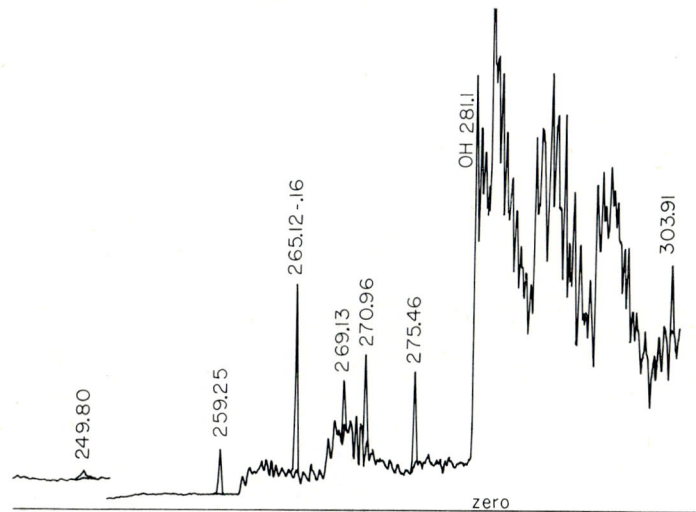

Spectrogram 76. Germanium, 1000 mg./l. in 50% isopropanol; air–hydrogen; slit 0.08 mm. (0.14 mm. at extreme left); blank superposed.

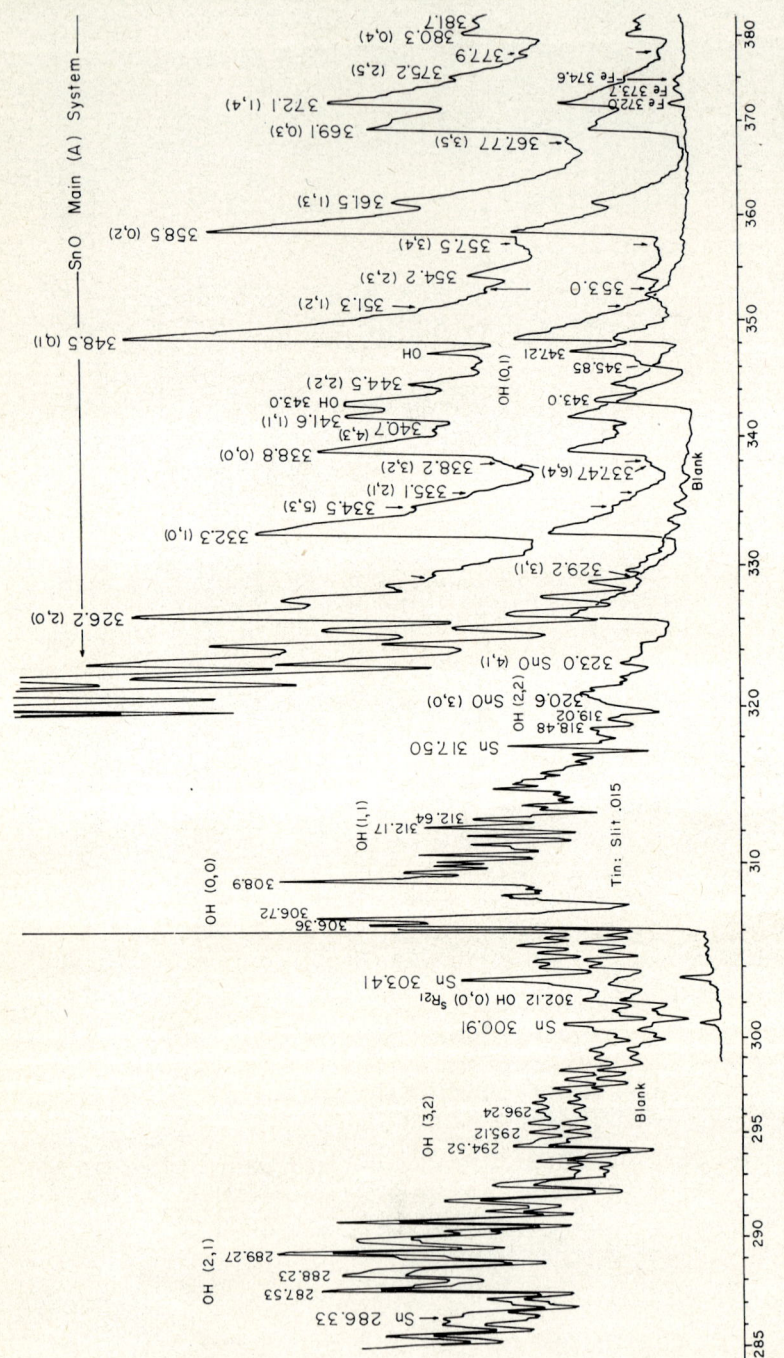

Spectrogram 77. Tin, 20 000 p.p.m.; oxyhydrogen; the two runs (at different slit widths) represent different samples and flame adjustments. (Continued on next page)

563

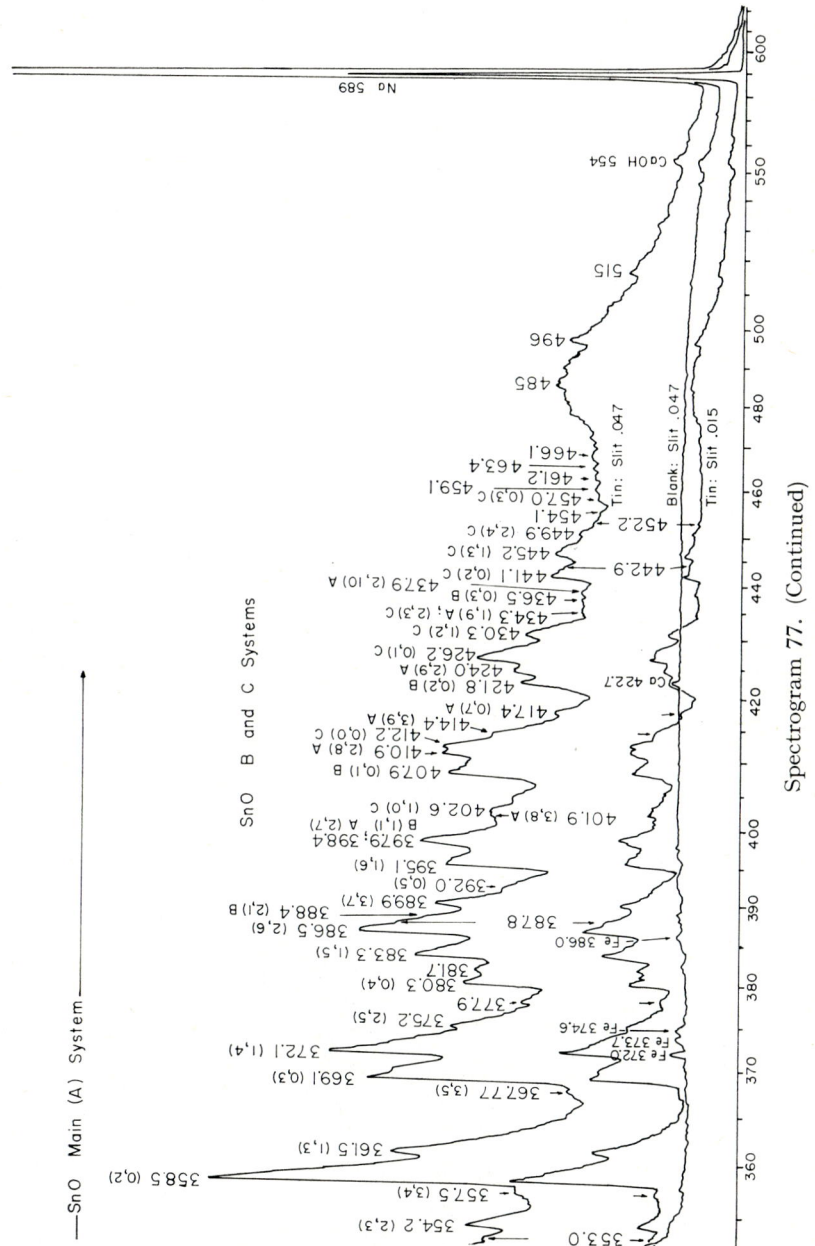

Spectrogram 77. (Continued)

19+C.A.F.P.

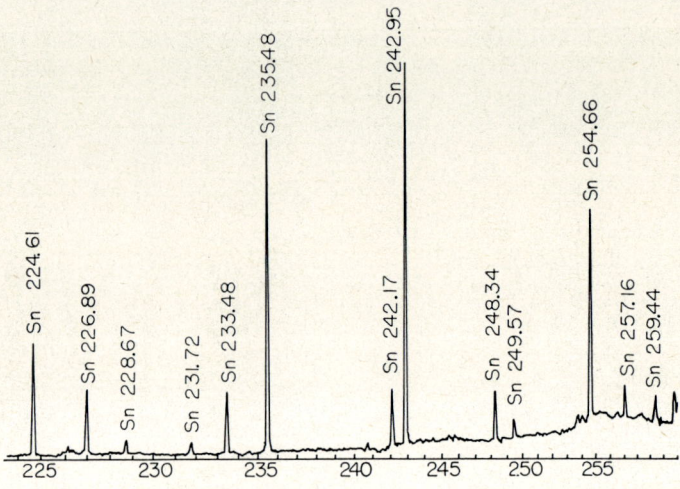

Spectrogram 78. Tin, 100 mg./l. in isopropanol; air–hydrogen; slit 0.04 mm.

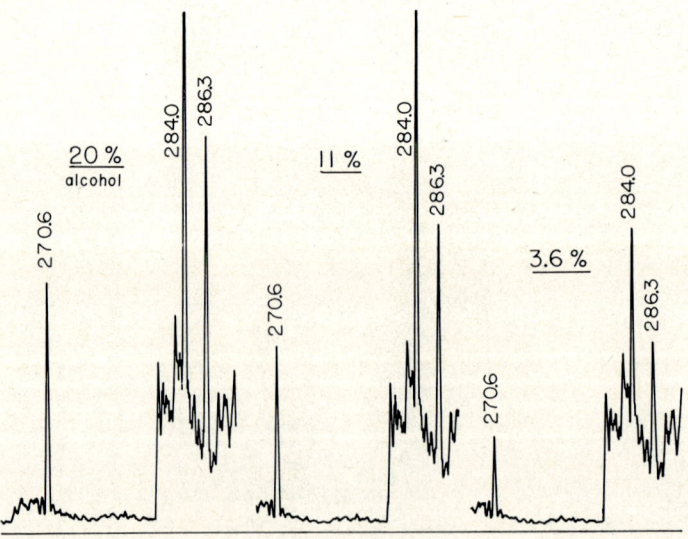

Spectrogram 79. Tin, 250 mg./l. in aqueous solutions containing various concentrations of isopropanol, as indicated; air–hydrogen; slit 0.08 mm.

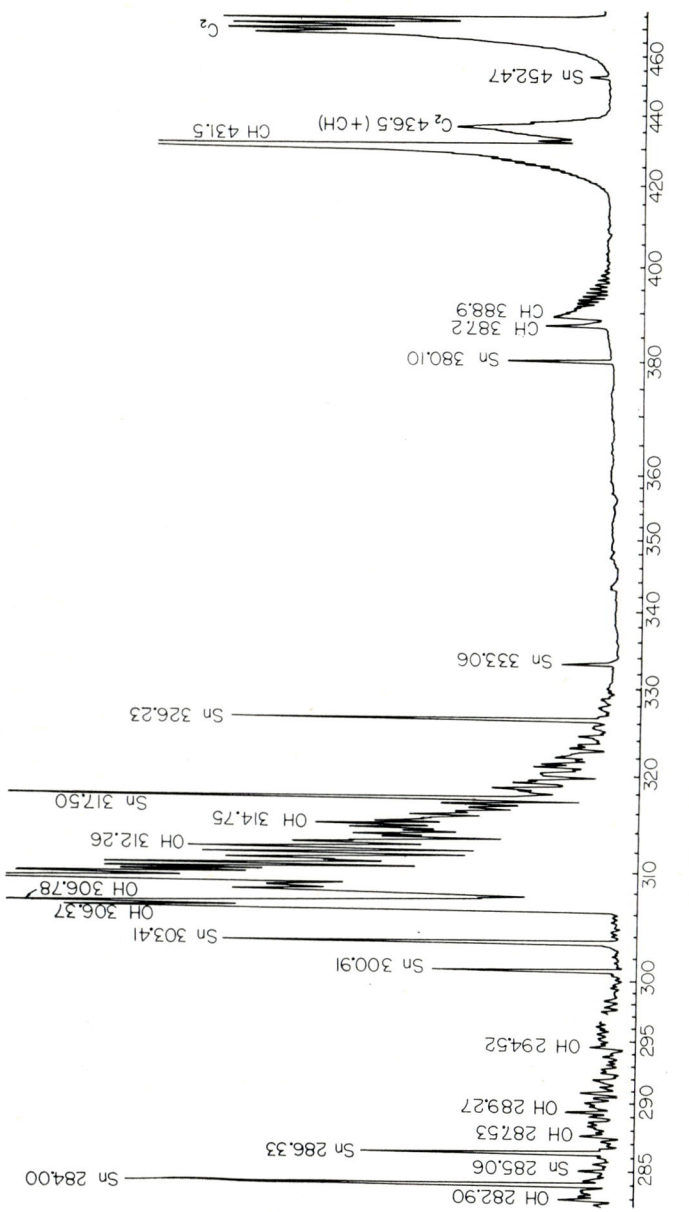

Spectrogram 80. Tin, 200 mg./l. in 92% isopropanol; air–hydrogen; slit 0.04 mm.

566

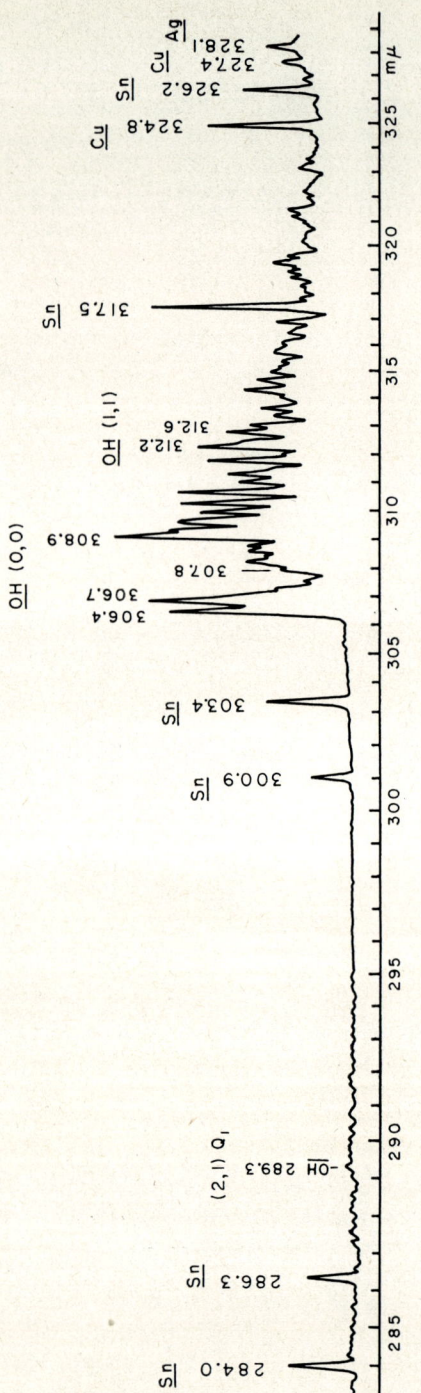

Spectrogram 81. Tin, 2000 p.p.m.; oxycyanogen; slit 0.02 mm.

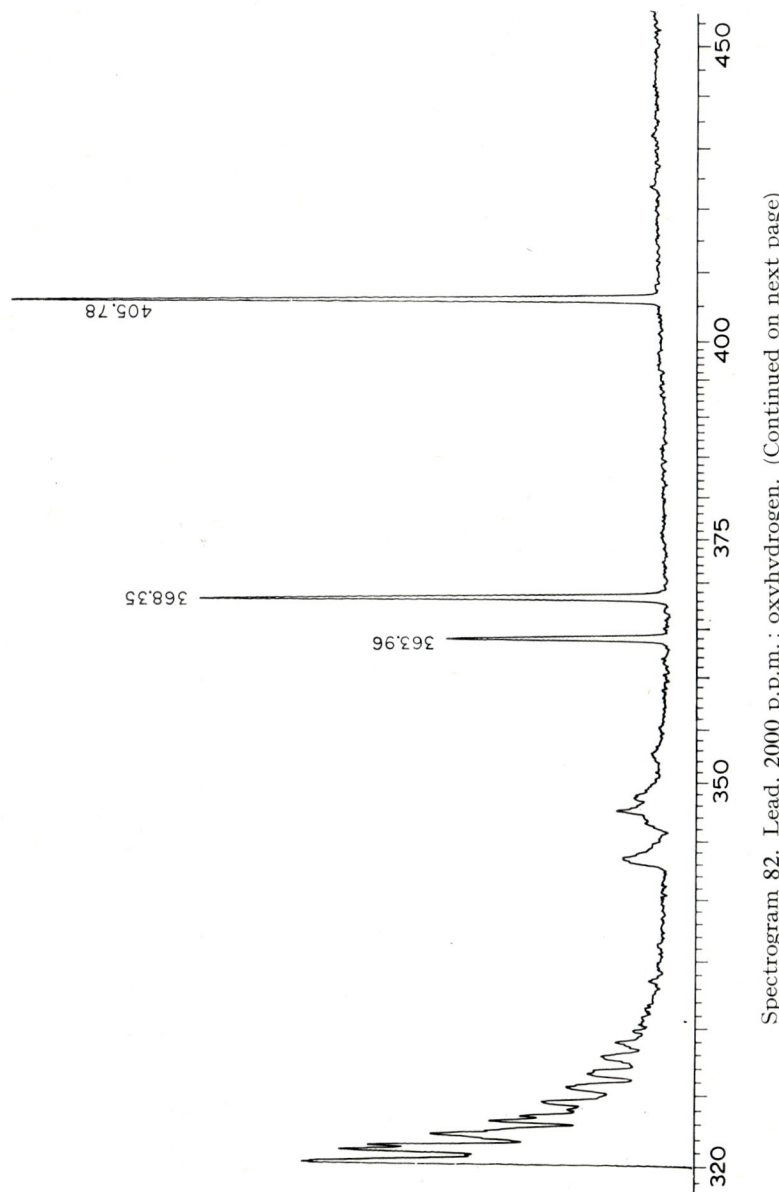

Spectrogram 82. Lead, 2000 p.p.m.; oxyhydrogen. (Continued on next page)

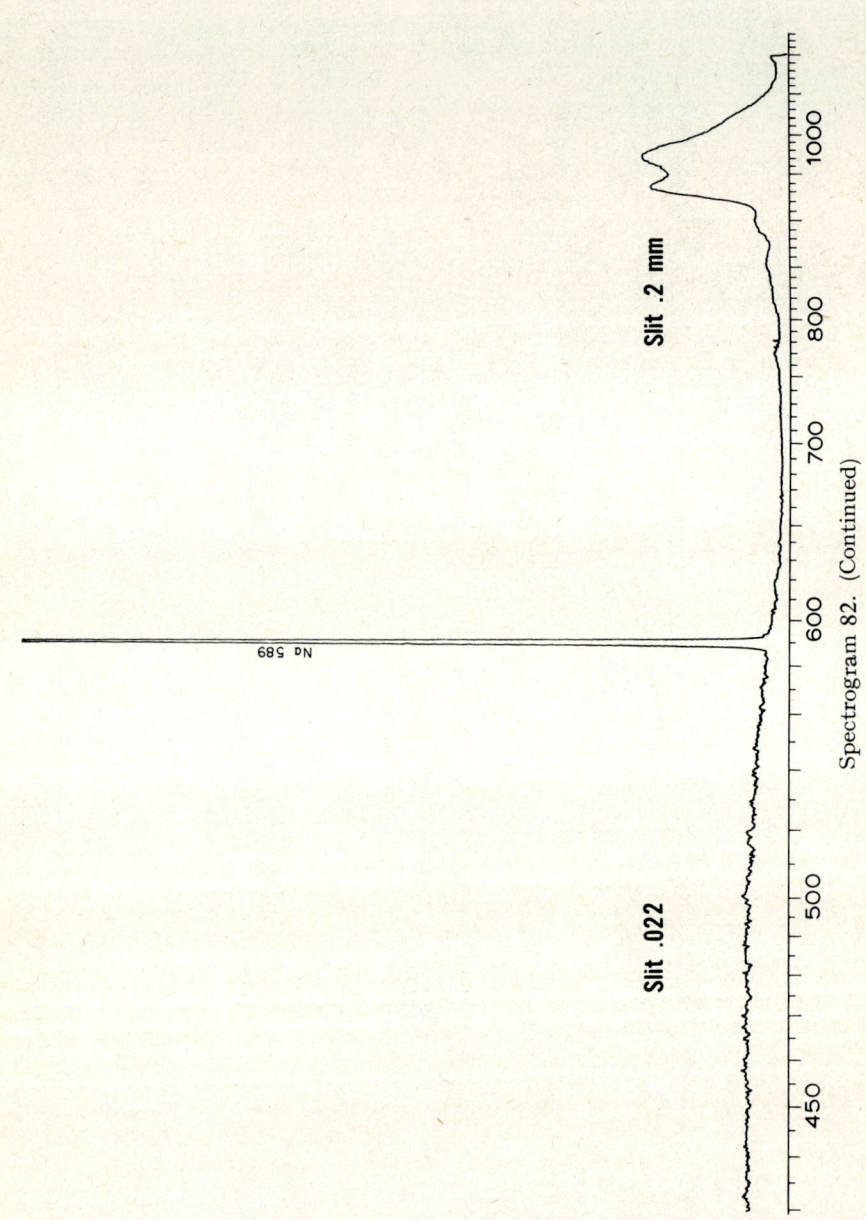

Spectrogram 82. (Continued)

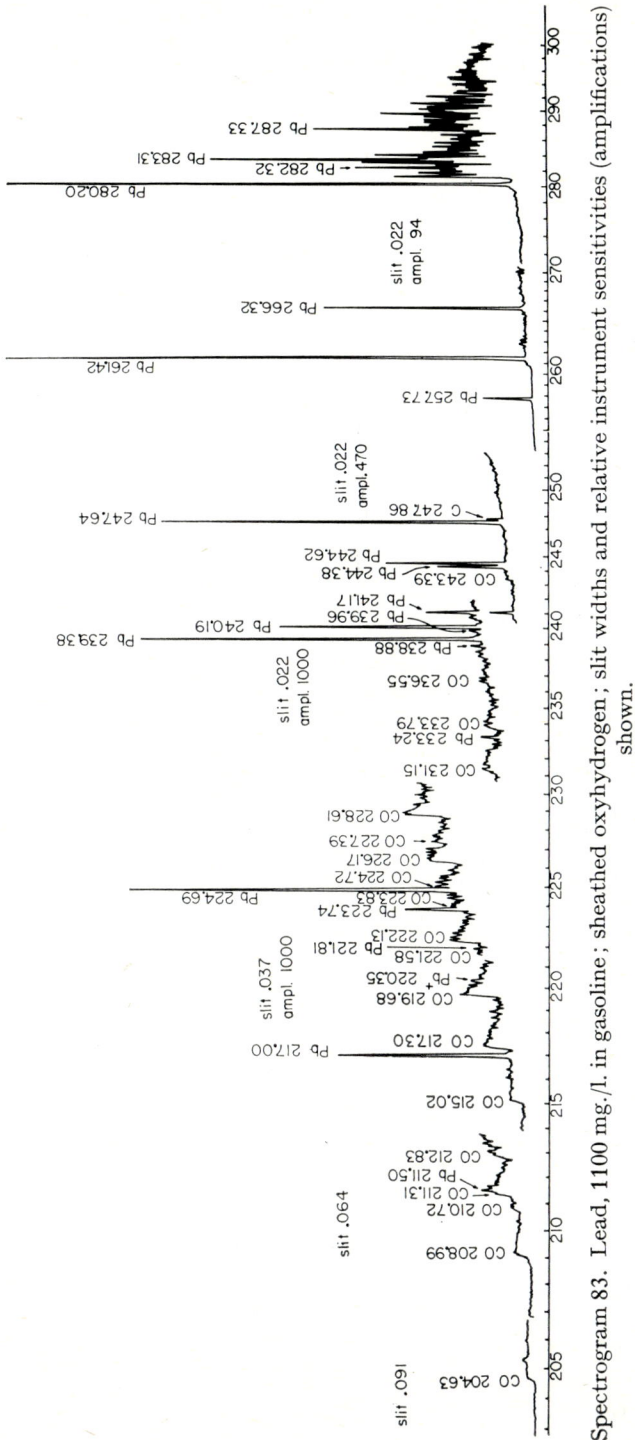

Spectrogram 83. Lead, 1100 mg./l. in gasoline; sheathed oxyhydrogen; slit widths and relative instrument sensitivities (amplifications) shown.

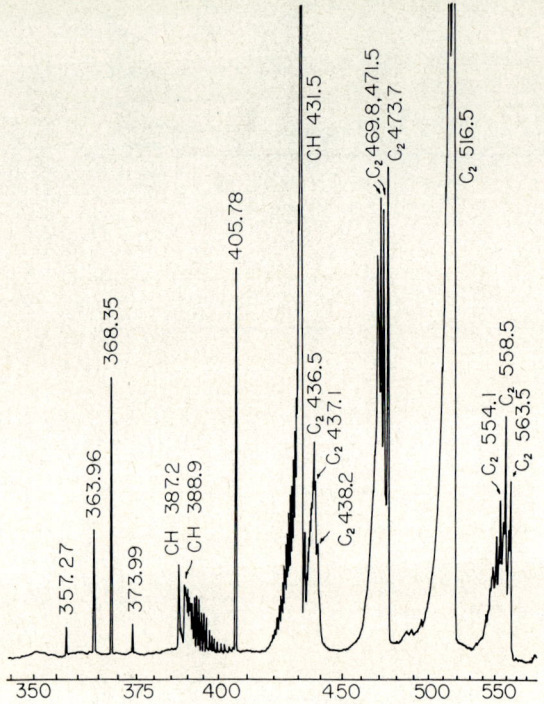

Spectrogram 84. Lead, 1100 mg./l. in gasoline; air–hydrogen; slit 0.02 mm.

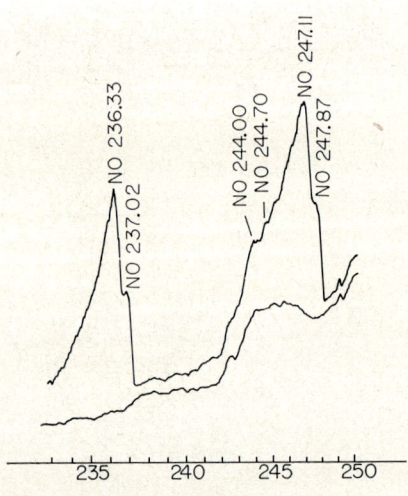

Spectrogram 85. Nitrogen, 10 000 p.p.m. (present as aqueous ammonia); sheathed air–oxygen–hydrogen–acetylene; slit 0.14 mm.

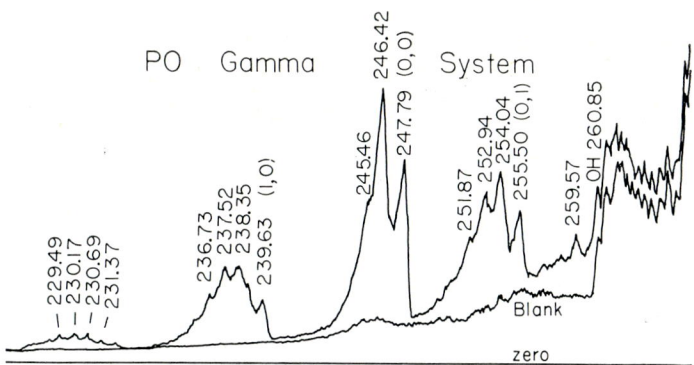

Spectrogram 86. Phosphorus, 800 mg./l. in 50% isopropanol (present as phosphoric acid); air–hydrogen; slit 0.14 mm.

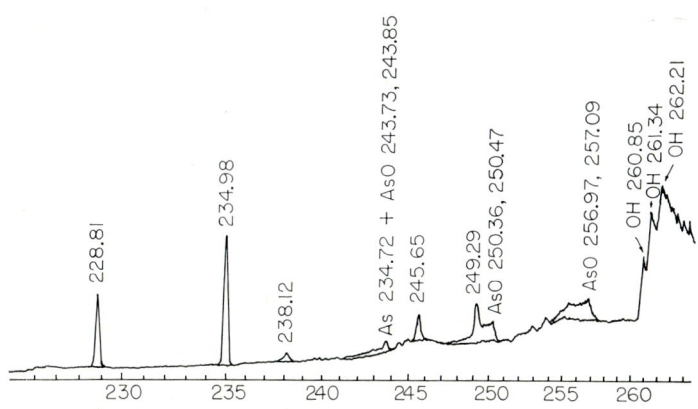

Spectrogram 87. Arsenic, 500 mg./l. in 50% isopropanol; air–hydrogen; slit 0.14 mm.; blank superposed.

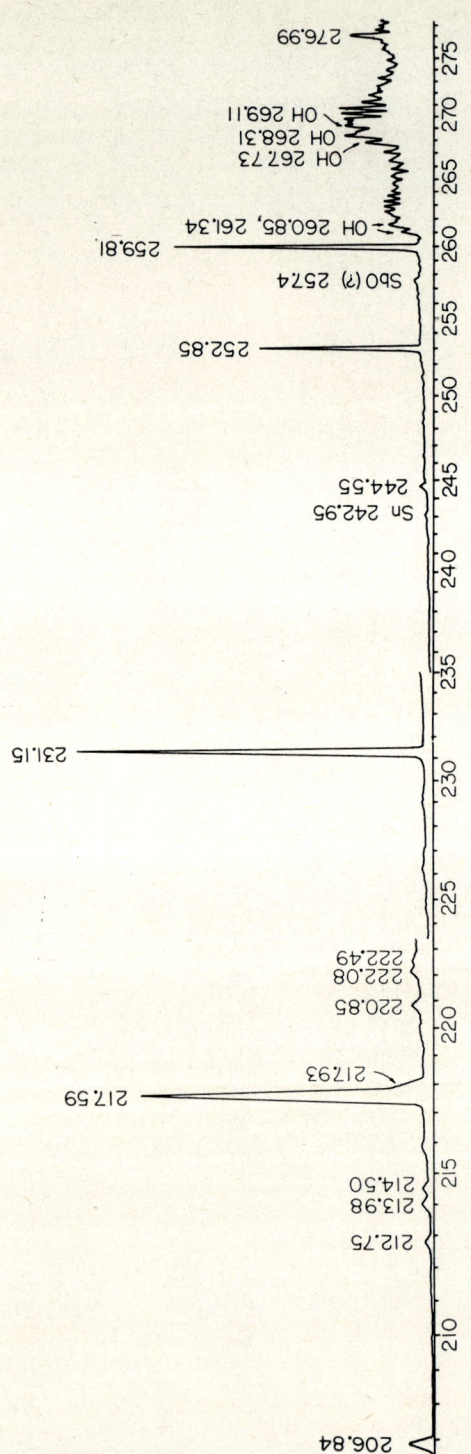

Spectrogram 88. Antimony, 600 mg./l. in 40% isopropanol; air–hydrogen; slit widths 0.26 mm. (left), 0.14 mm. (center), 0.08 mm. (right).

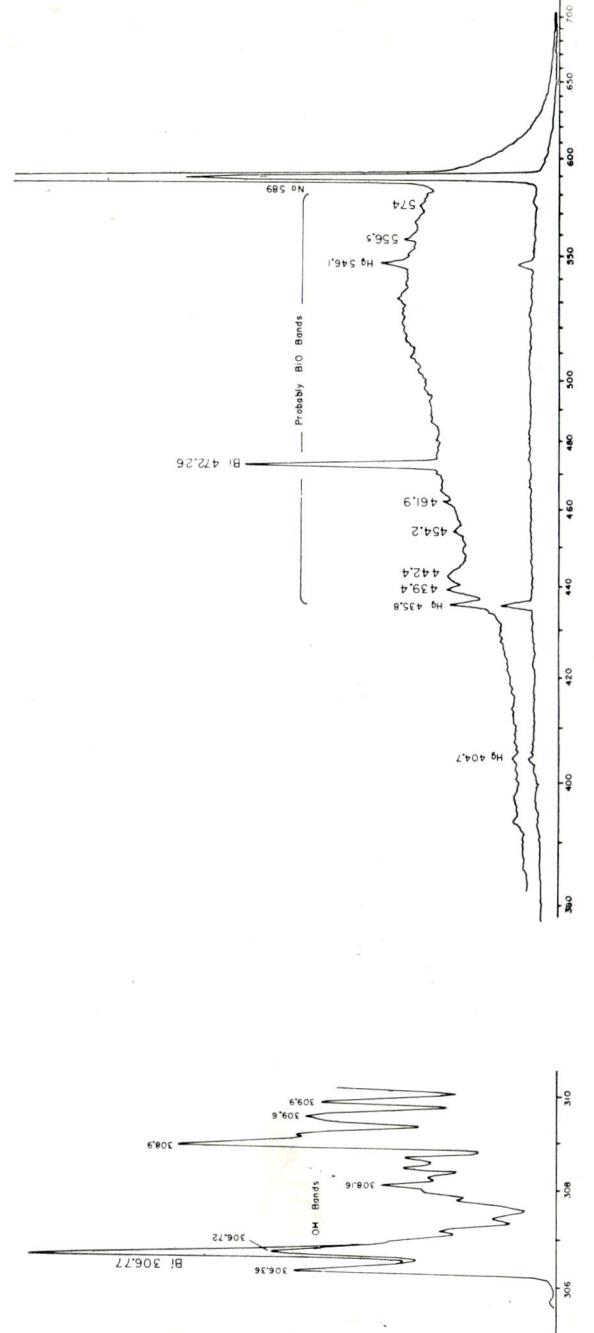

Spectrogram 89. Bismuth, 10 000 p.p.m.; air–hydrogen; slit widths 0.02 (left) and 0.08 mm.

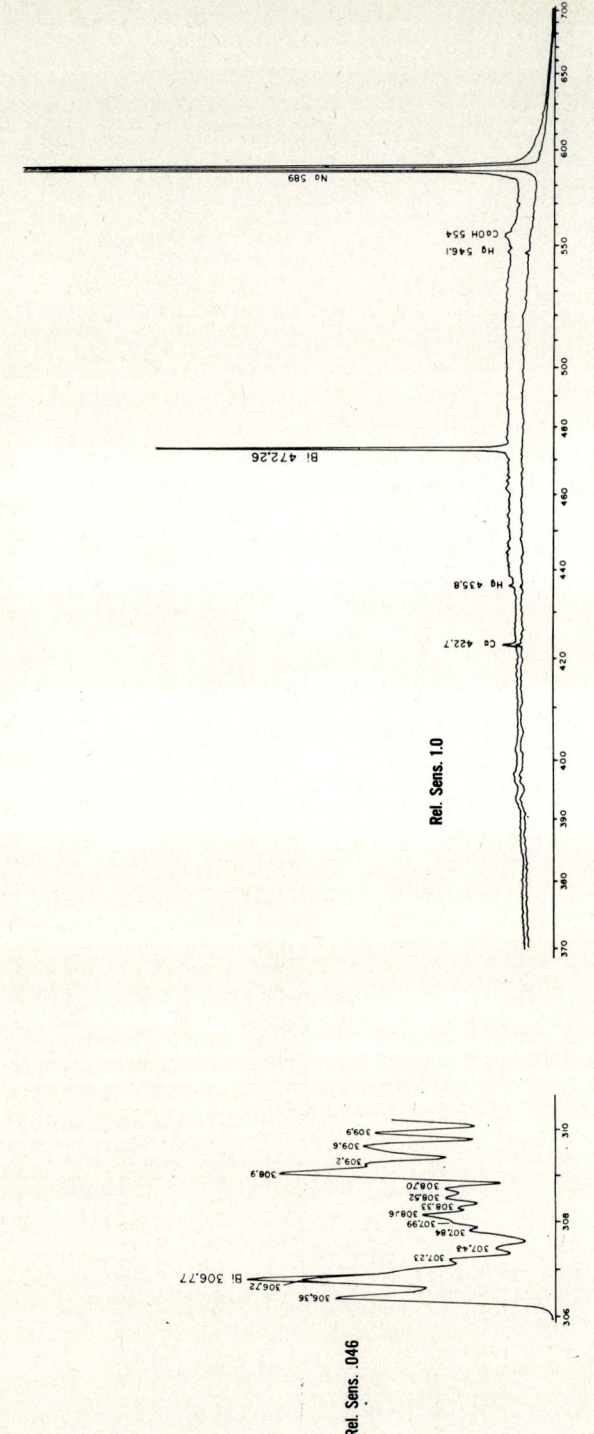

Spectrogram 90. Bismuth, 10 000 p.p.m.; oxyhydrogen; slit 0.02 mm.

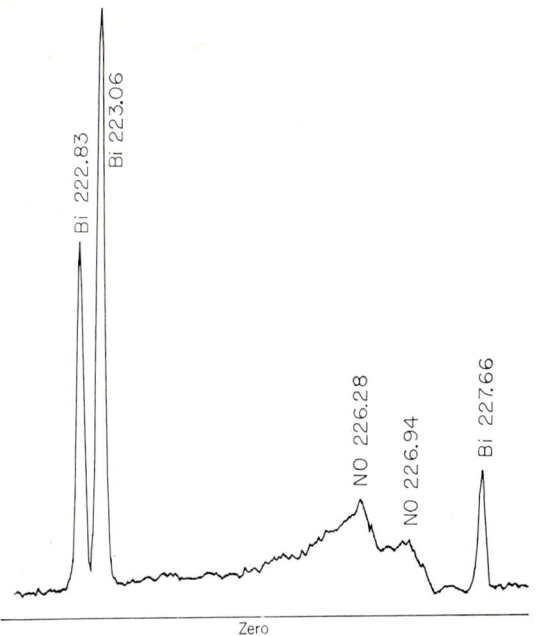

Spectrogram 91. Bismuth, 1000 mg./l. in 70% isopropanol; air–hydrogen; slit 0.04 mm.; NO bands due to the presence of nitric acid.

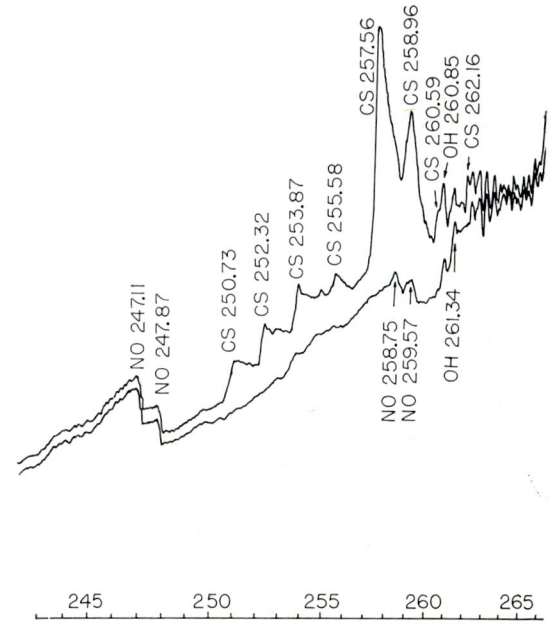

Spectrogram 92. Sulfur, 5000 mg./l. in 50% methanol (present as sulfuric acid); sheathed air–oxygen–acetylene; slit 0.05 mm.

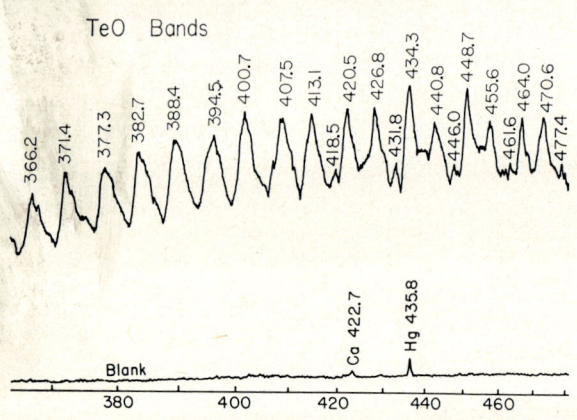

Spectrogram 93. Tellurium, 4000 p.p.m.; air–hydrogen; slit 0.08 mm.

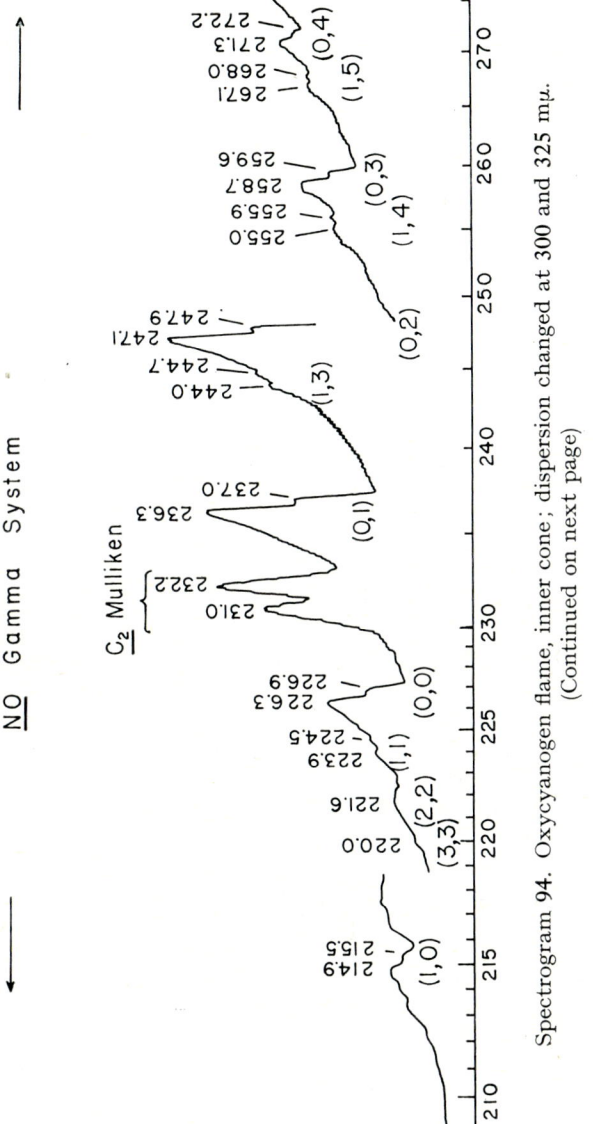

Spectrogram 94. Oxycyanogen flame, inner cone; dispersion changed at 300 and 325 mµ. (Continued on next page)

578

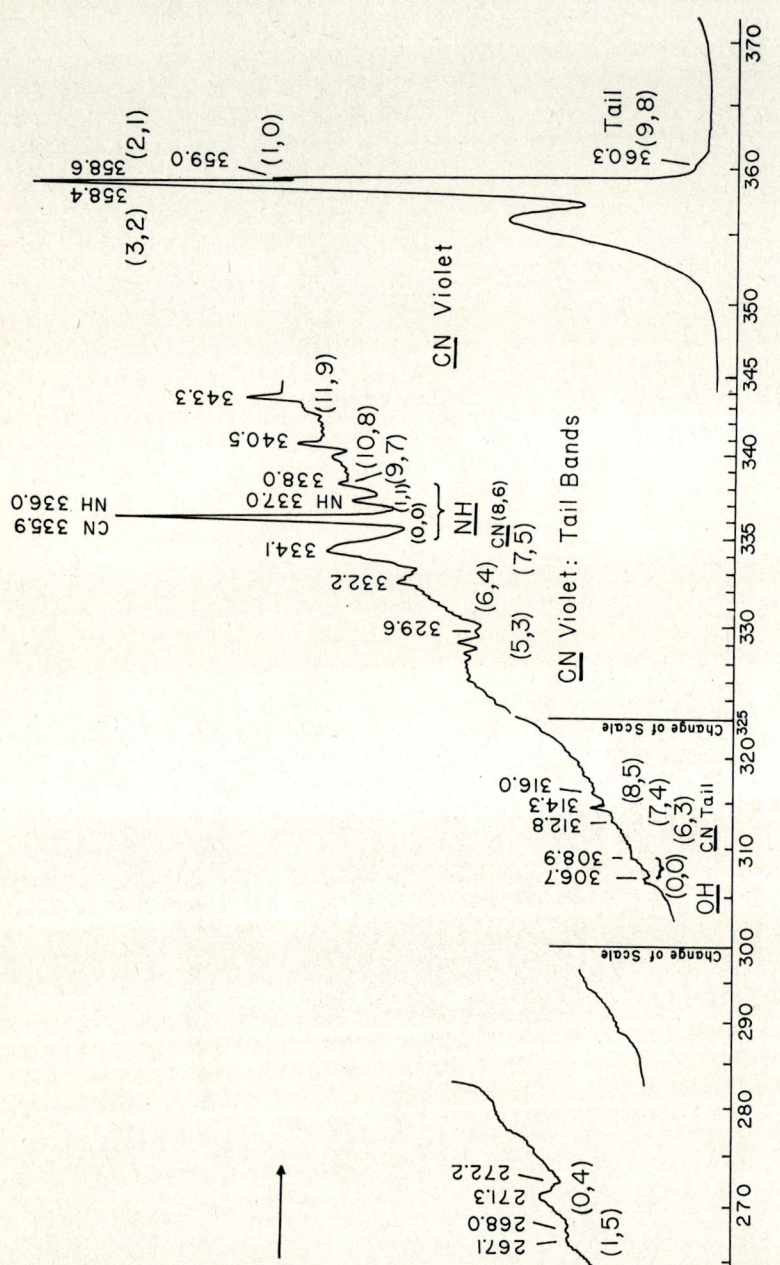

Spectrogram 94. (Continued)

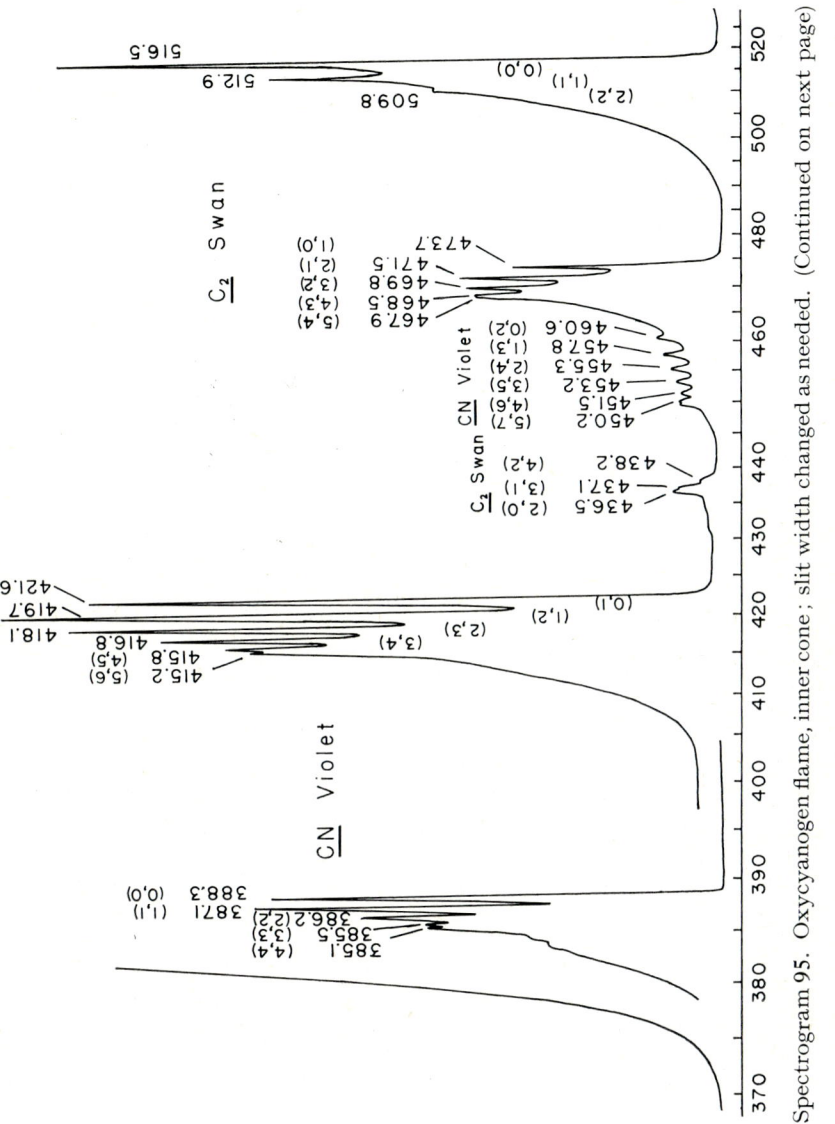

Spectrogram 95. Oxycyanogen flame, inner cone; slit width changed as needed. (Continued on next page)

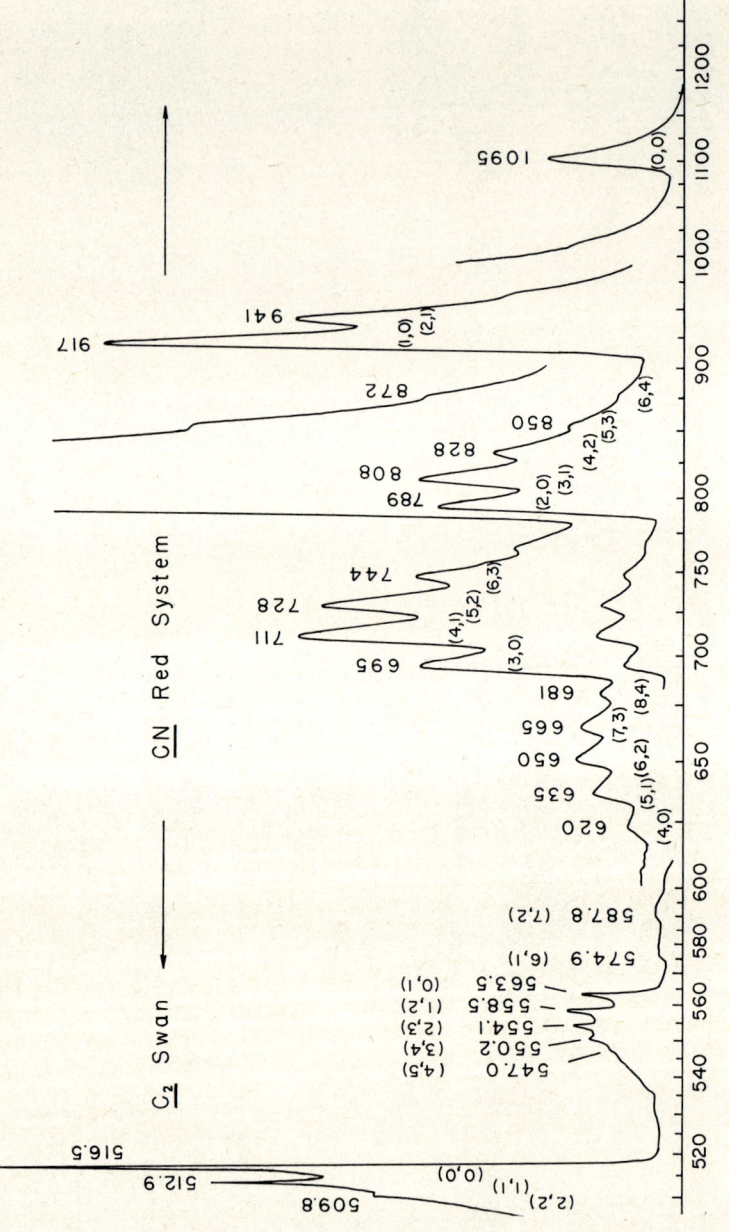

Spectrogram 95. (Continued)

INDEX

INDEX

This index includes information in the Appendix and Bibliography but not the Atlas. All names are indexed except senior authors' names occurring in their alphabetical places in the Bibliography. Abbreviations and symbols used in the principal tables and Bibliography are also indexed and defined. Since this is a book about flame photometry, all entries concerned with chemical analysis will be understood to pertain to flame photometry unless otherwise characterized. Entries referring to the Bibliography are indexed by reference number and are designated with an R.

a, line of neutral atom, 383
A, ethanol
A, see Einstein transition probability
AA, air–acetylene flame; acetylacetone
ab, arc line on a strong band (Table 10)
AB, air–butane flame
Abs., absorption flame photometry
Absorption, see Self-absorption, Self-reversal
    by hydroxyl, R373
    of radiation, 44ff, 49f
Absorption flame photometers, commercial, 428, 430, R353
    double-beam, R66
Absorption flame photometry, 11, 47f, R608, R722
    advantages of, 49, 66
    for alloy analysis, 368
    for bismuth, 354
    for cadmium, 354
    for copper, R203a
    differential, R454a
    interferences in, 48f, 66
    for iron, R72a, R203a
    for lead, 354
    for magnesium, 354, R72, R203a
    for manganese, R72a
    for mercury, 219, 354
    for nickel, 354
    for potassium, R454a
    reviews of, 48
    for sodium, R454a
    for zinc, 354, R203a
Absorptivity, 49
Ac, acetone

a.c., see Alternating current
Accident prevention, see Safety precautions
Accuracy, 2f, 4, 327ff, 331ff; see also Error, Precision
    requirements of, 278
Acetic acid, interference of, on calcium, 301
Acetone, in acetylene, 78
    effect of, on flame, 73, 80
    determination of, 247
    effect of, on flame temperature, 209
    evaporation of, 110
    as solvent, 208
Acetylacetone, for extraction, 367
Acetylene, 77ff, R60
    acetone content of, 73, 78, 80
    consumption rate of, 97
    copper tubing for, 85
    fire hazard of, 85f
    pressure of, 78
    purification of, 79
    purity of, 79
    solubility of, in acetone, 78
    stability of, 77f
    tanks for, 77ff
Acetylene flame, see Oxyacetylene
    background emissivity of, 53, 79
    burning velocity of, 77
    concentration of metals in, R182a
    electrons in, 35
    fuel flow, effect of, on, 94
    heat transfer from, to burner, 72
    impurities, effect of, on, 79, 80
    ionization in, R63, R371

584 INDEX

Acetylene flame—*contd.*
  soot from, 81f
  temperature of, 74, 77, 80, 94
  thickness of, 94
Acids, analysis of, for sodium, R472
  interference of, 65f, 300f, R241, R263, R314b
  organic, determination of, 248
Acoustic atomizer, 101
Actinides, 220
Activation, *see* Excitation
Ádám, J., R253
Adams, P. B., R743
Adams, S. L., R456, R456a
Addition method, *see* Self-standardization
Additive interference, 283f
Additives, determination of, in petroleum, R198
Adjustment method, 239, 240
Admixture method, 319f; *see* Self-standardization
Adsorption, of analyte, 110, 212, 250, 251, 343f
  studied by flame photometry, 356
Advanced Instruments, Inc., 427
Advantages of flame analysis, 2f, 4f, 10f
Aerosol, R55; *see* Atomization, Droplets (spray), Drop size distribution, Spray chamber
  analysis of, for sodium, 359, R532a, R751
  concentration of, 106
  electrostatic charging of, 114
  evaporation of, 24f, 27f, 107, 108, 305
  fluctuation of, 107f
  incandescent emission from, 42
  incomplete evaporation of, 21, 28, 123, 326
  mixing of, with gases, 115ff
  preparation of, R178
  recombination in, 24, 107
  temperature of, 25
  vaporization of, 27ff; *see* Vaporization interference
Affsprung, H. E., R288
AG, air–city gas flame
Agar-agar, for nonspecific interferences, 326
Aggregate error, 332

Aging, of emulsions, 196
  of photocells, 145
Agricultural analysis, 334ff, R22, R34, R39, R50, R121, R467, R616; *see* Alfalfa, Feeds, Fertilizer, Grain, Leaves, Nutrient solution, Plants, Seeds, Soil
AH, air–hydrogen flame
Air, *see* Compressed air
  cooling of, by atomization, 108
  cooling of, on expansion, 108
  entrainment of, by flame, 96f, 120f, 126, 199, 279
  flow of, in internal standardization, 166f
  purification of, 199, 279f, R527
  velocity of, in atomizer, 106
Airborne flame photometer, R751
Air compressors, 81ff, 200
  hazards of, 81f
  heating of, 108
Air flames, *see* the corresponding fuel, e.g., Acetylene
Al, alcohol
Albinati, J. F. Possidoni de, 249
Albumen, for simulation, 353
Alcohol, *see* Solvent, Transport interference
  effect of, 66, 208ff, 337f
Alcoholic beverages, *see* Whiskey, Wine
Alfalfa, analysis of, R466
Algae, 237
Alignment, *see* Entrance optics, Exit beam
Alkali metals, *see* Lithium, Sodium, etc.
  continuum from, 216, R395
  determination of, R232, R242, R273, R601
  determination of, in clay, R78
  determination of, in drugs, R234a
  dissociation interferences on, 30f
  flame spectra of, 216f
  hydroxides of, in flame, 32, 216f, R391, R694
  interferences on, 216, 294ff, R179, R314b, R364, R601
  interference of aluminum on, 303
  interference of phosphate on, 302
  ionization of, in flame, 33ff, R268b, R423, R536, R694
  line profiles of, R666, R667

Alkali metals—*contd.*
　oxides of, 32
　spectra of, origin of, 39
　working curves of, 217, R391, R666, R667
Alkaline-earth chlorides, spectra of, 42
Alkaline-earth hydroxides (MOH), formation in flame, R694
　identity of, 42; *see also* Calcium hydroxide, etc.
Alkaline-earth metals, *see* Calcium, etc.
　compound formation with, 32f, 42; *see also* Vaporization interference
　determination of, R232, R273
　　by flame-photometric titration, 252f, R704
　flame spectra of, 217ff
　interferences on, 300ff, R364
　ionization of, in flame, 218, R694
　ionization interference on, 300
　ions of, emission of, 39, 218
　line profiles of, R666, R667
　molecular spectra of, 39, 42
　spectra of, origin of, 39
　vaporization interference, 66, 300ff; *see also* Vaporization interference
　working curves of, 218, R666, R667
Alkaline-earth oxides, band spectra of, 42
　occurrence of, in flame, 218
Alkemade, C. T. J., 217, 228, 381, R81a, R147, R361, R661
Allan, J. E., 48
Alloys, analysis of, 367f; *see also* Metallurgical analysis
　for copper, R207
　for gallium, R487
　for iron, R208
Aloe Scientific Co., 214
Alternating current, *see* Chopped beam amplifiers, 149f, 185, 257
　flame photometer, R292, R661
　narrow-band systems for, 185f
　shot noise effect with, 185f
Alumina, analysis of, for sodium, R156
Aluminum, analysis of, for sodium, R161, R324, R367, R379, R512
　chemiluminescence of, 216
　continuum from, 252
　detection limits of, 261
　determination of, R252
　　in clay, R334
　　indirect, 246f, R183, R493, R551, R645
　　in glass, 361, R333
　　in silicates, R219a, R623
　　in soil, R551
　enhancement of, 220, 261, 361, R431a, R431b
　enhancement of, by complexing, 252
　flame spectrum of, 216, 220, 261, 385, R499, R726
　interference of, on alkali metals, 303
　　on alkaline-earth metals, 302ff, R637
　　on barium, 312
　　on boron, 312
　　buffering, 309; *see also* Releasing agents
　　on calcium, 246f, 302ff, 306f, 308f, 313, 360, R68, R183, R276a, R370a, R454, R493, R551, R672
　　on chromium, 314
　　on copper, 314
　　elimination of, 308f
　　on indium, 314
　　on lanthanum, 315
　　on lithium, 316, 368
　　on magnesium, 316
　　on manganese, 316
　　on potassium, 315
　　on sodium, 317
　　on strontium, 318, R370a, R454, R493
　　for suppressing spectral interference, 325, R201, R388, R493, R633, R672
　　on thallium, 318
　interference on, 312
　　from calcium, 221, 312
　ionization of, in flame, 298
　in oxycyanogen flame, 261
　separation of, 252, 325, R512
　vaporization of, in flame, 307
　working curve of, 222
Aluminum alloys, analysis of, for copper, 250, R212
　for indium, R485
　for iron, 367f
　for lithium, 368
　for magnesium, R379

Aluminum alloys, analysis of—*contd.*
  for manganese, 250, R212
  for nickel, 250, R212
  for sodium, 368, R322
Aluminum chloride, interference of, with calcium, 303, 309
Aluminum fluoride, in flame, 252
Aluminum nitrate, interference of, with calcium, 303, 308, 309
Aluminum oxide (AlO), flame spectrum of, 221, 261, 385, R499
Amidogen, spectrum of, 244
Amine, determination of, R97
Ammeter, alternating-current, R69
Ammonium salts, interference of, with potassium, 66
Amplification, control of, 157
  in photomultiplier, 145f, 182ff
Amplifiers, 149ff
  alternating-current, 149f
  bandwidth of, 183, 185f
  direct-current, 149ff
  linearity of, 154f, 159
  narrow-band, 185f
  for phototubes, 144f
  power supply for, 151
Analysis, *see* Flame analysis, Measurement, Standardization
  evaluation of, *see* Evaluation of results
  performance of, 201ff
  speed of, 276ff
Analyte, *see* Atoms, Molecules, Sample
  adsorption of, 110, 212, 250, 251, 343f
  complexing of, 251f; *see also* Protective chelation
  concentration of, in flame, 58, 106
  concentration of, in spray, 106
  definition of, 59
  enrichment of, 248ff
  loss of, *see* Adsorption
  precipitation of, 250
  separation of, 248ff
Analyte-to-background ratio, 53, 227, 286ff, 290
  in choice of fuel, 70
  and detection limit, 273, 274
  gas flow effect on, 96
  and height in flame, 132f
  slit width dependence of, 178f, 259
  temperature effect on, 75

Analytical curve, *see* Working curve
Analytical emission, 215ff; *see also* Band (spectral), Line (spectral)
  choice of, 226ff
Analytical line, 160; *see also* Line (spectral)
  choice of, 203f, 226ff
Analytical reading, *see* Measurement, Reading
Analytical technique, 199ff; *see also* Measurement, Standardization
Angle atomizer, 100, 103f, 105, 107
  cleaning, 111
Angular aperture, *see* Luminosity
Angular dispersion, 173
Anion interference, 66, R197, R263, R452; *see also* Acids, Vaporization interference
  on calcium, 300f
  removal of, R120
Anode of photomultiplier, 148
Anomalous dispersion, R182a
Antibiotic broth, analysis of, 359
Anticoagulants, 348
Antimony, chemiluminescence of, 216, 223, 261
  detection limits of, 261
  in microwave discharge, 76
  in oxycyanogen flame, 223, 261
  spectrum of, 216, 223, 261, 370, 401
Antimony oxide (SbO), spectrum of, 401
AOA, air–oxygen–acetylene flame
AOHA, air–oxygen–hydrogen–acetylene flame
AP, air–propane flame
Aperture, 181, 196; *see also* Luminosity
  ratio, 135
Applications, 333ff
  index of, R469, R469a, R477
Applied Physics Corp., 427
Aqueous humor, analysis of, R451a
Arc, ionization in, R371
  plasma jet, 76
  for sample atomization, 98
Arc spectroscopy, compared with flame analysis, 4ff
  history of, 8
  sample preparation for, R336
Arc-in-spray spectroscopy, R717
Ark, H. J. van, 94
Arkopal, 214

Arsenic, chemiluminescence of, 216, 223, 261
  continuum from, 244, 261
  detection limits of, 261
  determination of, indirect, R452
  flame spectrum of, 216, 223, 243, 261, 385
  interference of, with calcium and strontium, R452
  in microwave discharge, 76
  in oxycyanogen flame, 223, 261
Arsenic acid, interference of, with calcium, 300f, 313
  with strontium, 318
Arsenic oxide (AsO), spectrum of, 244, 372, 385
Ash, analysis of, 367, R412, R583, R606, R635; *see also* Coal
Ashing, 336, 337f, 341ff, 344
  of blood serum, 350
  dry, 342f
  wet, 342, 343, R411, R526
Aspiration, *see* Atomization
Astrophysics, R51
Asunción-Omarrementería, M. C., R174, R176
Atmosphere, analysis of, for sodium, R751
Atomic-absorption spectroscopy, *see* Absorption flame photometry
Atomic-hydrogen torch, 76
Atomic spectra, intensity of, 42ff
  origin of, fundamentals of, 36ff
  tables of, 370ff
Atomization, R343; *see also* Sample flow rate, Transport interference
  cooling on, 108
  drop size in, 21ff, 102f, 105f, R358; *see also* Droplets
  and flame flicker, 26
  loss of sample on, 21f, 25f
  pressure effect on, 106
  process, 20ff, 105f, R341a, R358, R442, R470, R520
  rate, *see* Sample flow rate
  recombination in, 24, 107
  review of, 26f
  separate, *see* Dual atomization
  shock waves in, 106
  spray shattering by, 105f
Atomizer, 98ff
  angle, 100, 103f, 105, 107, 111
  barrier for, 105, 107, 306, 308
  for biological analysis, R139
  capillary, *see* Capillary
  centrifugal, 101
  chemical, 101, R704
  classification of, 100
  cleaning of, 110f
  clogging of, 104, 110f, 202
  concentric, *see* Concentric atomizer
  controlled-flow, 100, 365, R280
  direct, *see* Atomizer-burner
  dual, *see* Dual atomization
  efficiency of, 25f, 58, 99, 105f, 115, 209
  electrolytic, 101
  electrostatic, 101f, 114, R683, R684
  glass, 99
  gravity-feed, 100, 109
  high-pressure, 95, 104, 106f, 114
  history of, 7f, 101
  hydraulic, 103, 105
  immersion, 100, 103, 104
  incrustation of, 111, 123, 239, 345
  indirect, *see* Indirect atomizer
  materials of construction for, 99f
  metallic, 100
  for microanalysis, 113ff, 253f
  molded, 99f
  nonpneumatic, 100, 101ff, 105
  nozzle of, *see* Nozzle, atomizer
  optimization of flow of, 102
  plastic, 100
  pneumatic, 20ff, 100, 103ff
  for powders, 98
  pressure of, 106f
  reflux, 104, 113, 114, R744
  requirements for, 98f
  rinsing of, 238f
  salting up of, 111, 123, 239, 345
  sample feeding to, 109ff
  spark, 98, R499
  spiral flow in, 104
  suction, 100, 109ff
  suction of, 21, 114
  switch for integration, 115, 155, 255
  testing of, 156
  ultrasonic, 101, R498, R499
  velocity of gas in, 95
  water-repellency effect on, 213
  Zoellner, R764
Atomizer-burner, 15, 100, 121ff, R268a, R292, R299
  advantages of, 15, 123

Atomizer-burner—*contd.*
  Beckman, *see* Beckman burner
  cleaning of, 110f
  commercial, 427ff
  disadvantages of, 123
  distance of, from slit, 181
  drop size distribution of, 23f, 124f
  flame, alignment of, 180f
    appearance of, 127
    for clinical analysis, 352
    cooling of, by solvent, 28, 73, 123f
    height of complete evaporation in, 28
    incomplete evaporation of spray in, 21, 114, 123, 326
    interferences in, 123
    nature of, 19f, 122
    noise of, 122, 128, 180
    turbulence in, 128
    wetting agents, effect of, on, 21, 65
  for flame attachments, 180
  for flammable samples, 363f
  Gilbert, *see* Beckman burner
  Herrmann–Schellhorn, 124f, R347
  high-pressure, 124f, R347
  history of, 9
  housing for, 122, 180
  incrustation of, 123
  for microanalysis, 114
  nonspecific interferences in, 63, 65, 327
  for oxycyanogen, R278a
  pilot flame of, 122
  Pungor–Hegedüs, R571
  sample flow rate effect in, 115; *see also* Sample flow rate
  sample temperature effect in, 108f
  sheathed, 260ff, 427, 431
  solvent, effect of, on emission from, 209
  for spraying solids, 5, 98
  Warren, R723, R724
  Weichselbaum, R728, R729
Atoms, concentration of, in flame, 58
  excited, *see* Excitation
  formation of, in flame, *see* Dissociation
Att., attachment
Attenuator, *see* Diaphragm
  of light, 154, 164
Attoe, O. J., R644
AutoAnalyzer, 430

Autocollimation, 173f
Automation, 277f
  for flame spectrography, 336
  for interference correction, 325
Azimuthal quantum number, 38f

b, peak of a narrow headless band, 383
Background, in spectrography, 196, 197
  subtraction of, 286ff, 290ff
    automatic, 187f, 255, 289, 292, 430, R457, R458
    in filter instruments, 290
    by slit variation, 289f
Background interferences, 52ff, 61, 229, 280, 282, 285ff; *see also* Flame background interference, Spectral interference
  classification of, 61f
  and cross-sensitivity, elimination of, 290ff
  elimination of, 285ff, 325
  numerical evaluation of, 290
  separation of interferent, 292
  suppression of, due to aluminum, 325
Background radiation, 52ff, 75; *see also* Infrared flame spectrum
  continuum in, 54
  emissivity of, 53
  flicker of, and detection limits, 184
  fuel–oxidant ratio, effect of, on, 126
  height dependence of, 132f
  impurity, effect of, on, 79
  as internal standard, 161
  monitoring and subtraction of, *see* Background
  solvent effect on, 288
  spectrum of, 373, R554, R555
Backing mirror, *see* Mirror
Baird-Atomic, Inc., 427, R82
Baker, M. R., R711
Balancing circuit, *see* Potentiometric photometric circuits
Baldwin Instruments Co. Ltd., 427, R87
Band (spectral), choice of, 226ff
  overlapping, measurement of, 241ff
  ratio of, to background, slit dependence of, 178, 275f
Band-pass filter, 141f, R287
Band-pass wedge filter, 143
Band spectra, *see* Molecular spectra
  intensity of, 374
  tables of, 370ff

Bandwidth, *see* Half-width
  of amplifier, 183, 185f
  spectral, of monochromator, 176, 178f, 275f; *see also* Band (spectral)
BaOH, *see* Barium hydroxide
Barclay flame photometer, R224, R527
Barium, as buffer, R332
  chemiluminescence of, 216
  continuum of, 218
  detection limits of, 261
  determination of, R211, R273, R357, R361, R573, R576, R636
    in biological material, R597
    in cathodes, 357, R253, R292, R325, R582c
    in petroleum oil, 366, R198
    in soil analysis, R630
    in sulfates, R581
  enhancement of, 208, R431a
  excitation potential of, 34
  flame spectrum of, 216, 218f, 261, 385, R186, R726, R734
  interferences of, on calcium, 313
    on copper, 314
    on lanthanum, 315
    on lithium, 316
    on potassium, 315
    on sodium, 317
    on strontium, 318, R416c
  interferences on, 312f
    by calcium, 219, 312
    by zirconium, 309, 313
  internal standardization of, 166
  ionic spectrum of, 218, 219, 261, 385
  ionization of, in flame, 218, R268b, R371
  ionization interference of, R268b
  ionization potential of, 34
  as normalizer, 298, 361
  organic solvents, effect of, on, 261, R202
  in oxycyanogen flame, 261
  for sulfate determination, 248, R647, R687
Barium chloride (BaCl), spectrum of, 372, 385
Barium compounds, analysis of, for calcium, R727
  for strontium, R690

Barium fluoride (BaF), for fluorine determination, 245, R278
  spectrum of, 385
Barium hydroxide (BaOH), enhancement of, by organic solvents, 208
  identity of, 42, R170, R186, R286, R393, R441
  ion of (BaOH$^+$), 35, 218
  isotope shift of, 42
  spectrum of, 42, 219, 261, 386, R393
Barium oxide, flame spectrum of, 219, 261, 385f, R286, R374
  stability of, in flame, R695
  term diagram of, R374
Barium sulfate, analysis of, for barium, R581
Barnes, R. B., 9, 163, 339, R140
Baron, J., R606
Barrier, atomizer, 105, 107, 306, 308
Barrier filter, *see* Filter (optical)
Barrier-layer cells, *see* Photocells
Bartholomay, A. F., R713
Batchelor, T. M., R153, R500
Batteries, for voltage supply, 151
Bauer, S. H., R682
Baumann, R., R149, R340
Beam, adjustment of, *see* Entrance beam, Entrance optics, Exit beam, Optics
  aperture of, 175
  chopping, *see* Chopped beam
  splitter, 164, 191f
Bean, analysis of, for sodium, 338f
Bean, L., R222, R224
Beck, B. L., R487
Beckman, A. O., R184, R298
Beckman B flame spectrophotometer, R115, R116
Beckman burner, 121ff, 127, 188, R99, R100, R109, R268a, R299
  for flame spectrography, 381
  sheathed, 427, 431
Beckman DU spectrophotometer, 175, 178f, 259ff, 427, R152, R517
Beckman flame attachment, 175, 259ff, 427, R99, R100, R109, R116, R118, R119, R152, R224, R528
Beckman flame photometer (model 4100), R111
Beckman flame spectrophotometer (model 10300), R114, R298

INDEX

Beckman Instruments, Inc., 222, 371, 372, 380, 381, 383, 427, R99–R124a, R185, R296, R307, R311, R518
   water-repellent, 211, 213
Beckmann, E., 8
Beckmann, W., R507
Belke, J., 339
Benne, E. J., R401
Berenda, G. D., R378
Berger, A. W., 372, 379
Berl, W. G., R7, R45
Berneking, A. D., R97
Berry, J. W., 9, 163, R90
Beryllium, detection limits of, 261
   determination of, indirect, 247, R452, R453
   in incandescent oxyacetylene, 261
   interference of, on alkaline-earth metals, 303
      on calcium, 313
      on lithium, 316
      on potassium, 315
      on sodium, 317
      on strontium, 318
   in microwave discharge, 76
   in oxycyanogen flame, 261
   spectrum of, 218, 261, 373, 386
Beryllium compounds, analysis of, R506
Beryllium oxide, spectrum of, 218, 261, 386
Beverages, analysis of, 337; *see also* Whiskey, Wine
   for copper and iron, 339, R103
Bibliographies of flame analysis, R273, R469, R469a, R477, R641, R642, R759
Bicarbonate, determination of, in blood, R204
Biochemical analysis, 339ff, R16, R20, R21, R22, R34, R36, R39, R49, R50; *see also* Agricultural analysis, Blood, Clinical analysis, Marine organisms, Tissues, Urine
   sample characteristics in, 340f
Biological fluids, analysis of, 344ff; *see also* Aqueous humor, Blood, Cerebrospinal fluid, Gastric juice, Milk, Urine
   for calcium, 349ff, R128, R133, R138, R269, R564, R664a

   instrumentation for, R139
   internal standardization in, R139a
   for lithium, R269, R270
   for magnesium, R138
   for potassium, 345ff, R128, R133, R134, R269, R270, R664a, R673, R750
   for sodium, 344f, R128, R133, R134, R269, R270, R664a, R673, R750
   preparation of, 341, 345, 354
Biological interferences, 346f
Biological materials, analysis of, R249, R526, R597, R648, R720
   for calcium, R185, R279, R282, R337a, R596a, R597, R602a, R688, R703
   for copper, 354, R597
   for lithium, R597
   for magnesium, 354, R279, R597, R688
   for potassium, R279, R597, R723
   for rubidium, R273a
   for sodium, R279, R584, R597, R723
   for strontium, 354, R321, R597
   ashing of, *see* Ashing
   characteristics of, 340f
   extraction of, 343
Bird, E. J., R153, R500, R501, R663
Bismuth, band spectrum of, 386
   chemiluminescence of, 216, 223, 262
   detection limits of, 262
   determination of, by absorption flame photometry, 354
   in microwave discharge, 76
   in organic solvents, 262
   in oxycyanogen flame, 223, 262
   spectrum of, 216, 223, 262, 370, 386
   working curve of, 223
Bismuth oxide, spectrum of, 386
Blackbody radiation, 46, 49, 50
Blank, composition of, 286
   definition of, 60, 288
   preparation of, 232f, 291f
Blank interference, 280ff; *see also* Background interference, Cross-sensitivity
   in choice of wavelength, 229
   classification of, 61f
   diagnosis of, 67, 189
   elimination of, 290ff

INDEX 591

Blank interference—*contd.*
  indirect, 285
  terminology of, 280
Blank reading, 239, 286, 288; *see also* Analyte-to-background ratio
  in atomizer-burner flame, 20
  determination of, 249, 257, 286ff
  errors of, 256f
  solvent effect on, 288
  subtraction of, *see* Background
Blast furnace slag, *see* Slag
Blaze, of grating, 173
Blocker, H. G., R203
Blocking filter, 139, 143, 290f
Blood (whole), analysis of, for calcium, R622
  coagulation of, 348
  composition of, 348, R663
  sampling of, 347ff
Blood cells, analysis of, for potassium, 346, R378, R533
  for sodium, R378, R533
  potassium exchange in, 346, 348f
Blood plasma, analysis of, 347f
  for calcium, R153, R369, R459, R502, R708
  for iron, R153
  for magnesium, R153, R203a, R652, R708
  for potassium, 346f, R136, R137, R153, R271, R500, R542, R671, R708
  for sodium, R137, R153, R271, R500, R708
  by spark spectrography, R153
Blood serum, analysis of, 339, R92, R136, R350a, R609
  for bicarbonate, R204
  for calcium, 348, 349ff, R86, R93, R94, R95, R124a, R130, R149, R149b, R187, R188, R221, R246, R257, R315a, R339, R342, R350, R354, R369, R370, R406, R411, R419, R420, R474, R475a, R491, R507, R508, R524, R545, R570, R588, R596, R617b, R621, R646, R719, R746, R747
  for copper, R345
  errors in, R588
  for magnesium, 354, R346, R411, R617b, R700a

  for potassium, 346f, R64, R93, R94, R95, R105, R123, R131, R132, R136, R137, R149, R149b, R235a, R248, R272, R315a, R339, R342, R354, R360, R378, R406, R418, R419, R420, R436, R510, R524, R531, R542, R545, R553a, R568, R569a, R588, R611, R617b, R651, R654, R661, R663, R671, R741, R747
  for sodium, R64, R93, R94, R95, R105, R123, R131, R137, R149b, R235a, R248, R272, R315a, R339, R342, R349, R354, R360, R378, R406, R418, R419, R420, R436, R510, R524, R531, R545, R553a, R568, R569a, R585, R611, R617b, R651, R661, R663, R741, R747
  for sulfate, R687
  ashing of, 350
  calcium binding in, 331, 341, 344, 350f, R504
  calcium content of, 331
  calcium separation from, 350f
  composition of, 348
  dilution of, 235f, 352
  hemolysis of, 346f
  phosphorus content of, 235
  potassium content of, 341, 345ff
  protein effects in, 235, 341, 351, R418
  protein separation from, 343f, 345, 351
  sampling of, 341, 346ff
  standardization of, 234f, 345, 353
  storage of, 348
  stratification of, 348, 350
  temperature effects in, R94
Blow-off of flame, 126
Body fluids, *see* Biological fluids, Blood, Urine
Bohr theory, 36ff
Boiler deposits, analysis of, 354ff, R593, R640
Boiler feed water, analysis of, 355f
Boiteux, H., 382, R37
Boltzmann equation, 43f, 49f
Boltzmann factor, 73
Bones, preparation of, 341
Boon flame photometer, R598

Borate, interference of, on calcium, 301, R276a
Borgers, A. J., R361
Borker, E., R291
Börnstein, R., R29
Boron, analysis of, for lithium, R238
　chemiluminescence of, 216
　detection limits of, 262
　determination of, R162, R206, R580
　　in organic compounds, R168
　　in petroleum fuels, 364
　　in plating solution, R268
　enhancement of, 262, R431a
　flame spectrum of, 216, 220, 262, R734; *see also* Boron oxide
　interferences of, 312f, 315ff
　interferences on, 312
　in irrigation water, 355
　oxycyanogen spectrum of, 10, 385
Boron oxide (BO_2), flame spectrum of, 244, 262, 385, R580
　isotope effect in spectrum of, 357, 374, 385
Bottled gas, *see* Tanks
Bottles, *see* Containers
Bourdon tube, 89
Bove, C., R549, R550
Boycks, E. C., 380
Boyle, A. J., R411, R500, R501, R502, R663, R757
Bracketing method, 233, 240f
Bradley, J. E. S., R475
Brass, analysis of, R179
Bredillet, M., R553a
Bremner, R. W., R182
Bretschneider, A., R596
Bricker, C. E., R229, R230
Bridge circuit, 163ff, 192; *see also* Potentiometric photometric circuits
Brightness, *see* Intensity
Brine, analysis of, R399
Brinkman Instruments, Inc., 429
Broadening of lines, 45f, R356, R372, R538, R638
Broekhuijsen, R., R414a
Bromide, indirect determination of, 248
　interferences of, 312, 315, 317
Bromine, spectrum of, 245
Bromine oxide (BrO), spectrum of, 245

Bronze, analysis of, for beryllium, R452, R453
Brown, R. H. J., R586
Bryan, H. A., 380
Bu, butanol (4% in water)
Bubble atomizer, 101
Bucher, R., R416b
Buell, B. E., 216, 370, 372, 380
Buffer, 309
Buffer flask, 107f
Buffering, 321, R731; *see also* Normalization, Parallel method, Releasing agents, Saturation
　by barium, R332
　of calcium, with phosphate, 309f, R199a
　against nonspecific interferences, 326
　against vaporization interference, 309, R493, R501
Buffer tank, 89f
Bugrim, E. D., 380
Building-block design, 169f
Bulewicz, E. M., 221, 380
Bunge, W., R727
Bunsen, R., 7
Bunsen burner, 118
Bunsen flame, 13ff
Burger, J. C., Jr., 380, R209, R211
Burner, 117ff; *see also* Atomizer-burner
　adjustment of, 131ff, 180f
　bunsen, 118
　cap, 16, 18, 117ff
　chimney, 120, 128f, 180
　cleaning of, 121, 128
　closed, 120, 121
　cooling of, 119
　design, 118ff
　distance of, from optics, 133, 181
　elongated, 119f
　enclosed, 120, 121, 129, 244
　fishtail, 118, 119f
　flashback in, 16, 81f, 116, 118f
　gas flow speed in, 118
　heat loss to, 72, 118
　height adjustment of, 117, 131ff, 134, 180
　housing for, 128f
　igniter for, 120
　Méker, *see* Méker burner
　for oxygen flames, 119
　port dimensions of, 16, 18, 118f
　requirements of, 117f

INDEX 593

Burner—*contd.*
  safety precautions with, 119, 128f
  for separated flames, 54, 121, 244
  shielded, 120, 129
  Teclu, 121
Burner-atomizer, *see* Atomizer-burner
Burning velocity, 15f, 77
Butane, as fuel, 80
Butane flame, burning velocity of, 77
  temperature of, 77
Butanol, for enhancement, 208, 361, R431a
Butter, analysis of, 337
BW_2, 214
Bz, benzene

c, continuum, 260
C, chloroform
C_2, *see* Carbon
CA, Chemical Abstracts
Cadmium, absorption flame photometry for, 48
  chemiluminescence of, 216
  detection limits of, 262
    in absorption, 48
  determination of, R306
  excitation potential of, 34
  flame spectrum of, 216, 219f, 262, 387f
  in incandescent oxyacetylene, 262
  interferences of, 312ff
  ionic spectrum of, 372, 388
  ionization potential of, 34
  in microwave discharge, 76
  in oxycyanogen flame, 262
  separation of, 250
Cadmium sulfide, analysis of, 357, R205, R587
Caimann, V., R330
Cain, C., Jr., R212
Calcium, for aluminum determination, 246f
  in blood serum, 331, 341, 350, R504
  as buffer against aluminum, 309, R493
  buffering of, with phosphate, 309, R199a
  Ca_2, occurrence of, 42
  for carbonate determination, 248
  chemiluminescence of, 216
  compound formation in flames with, 33, 66
  continuum, of, 218

  cross-sensitivities with, 282
  detection limits of, 262
  determination of, R218, R259, R273, R296, R340, R357, R379, R386, R416, R431, R448, R449, R464, R509, R573, R576, R634, R636, R674
  in alfalfa, R466
  in barium salts, R727
  in biological fluids, 349ff, R128, R133, R138, R269, R564, R664a, R757
  in biological materials, R185, R279, R282, R337a, R596a, R597, R602a, R688, R703
  in blood, R622
  in blood plasma, R153, R369, R459, R502, R708
  in blood serum, 348, 349ff, R86, R93, R94, R124a, R130, R149, R149b, R187, R188, R221, R246, R257, R315a, R339, R342, R350, R354, R369, R370, R406, R411, R419, R420, R474, R475a, R491, R507, R508, R524, R545, R570, R588, R596, R617b, R621, R646, R719, R746
  in catalysts, R756
  in cathodes, 357, R253, R292, R325, R582c
  in cellulose, R543
  in cement, 363
  in cheese, R323
  in clay, R334
  in coal, R239
  in feeds, R199a
  in ferrous metals, R59
  in fertilizer, R151
  by flame-photometric titration, 252, R251
  in foods, R732
  in gastric juice, R135
  in glass, R142, R333a, R427, R605
  in grain, R151
  instrument for, R269, R339, R344, R457, R475, R594, R709, R710, R735a
  in iron, R435
  in leaves, 336, R730
  in lubricating oil, 366
  in magnesite, R155
  in marine organisms, R189

594 INDEX

Calcium—*contd.*
    in milk, 339, R323, R413, R506a, R617b, R639
    in minerals, R501
    in nutrient solution, 359
    in petroleum, R198, R496
    in petroleum wash water, R756
    in pharmaceuticals, R314, R351
    in phosphoric acid, 358f, R154, R155
    in plants, R165, R293, R337, R385, R415, R430, R550, R570, R616, R617a, R625, R649, R680
    in refractories, R499, R592
    in rock, 360, R427, R432
    in sea water, R189
    in seeds, R151
    in silicates, R150, R219, R445
    in slag, 368, R230, R379, R677
    in soil, R151, R256, R294, R385, R415, R430, R434, R494, R570, R616, R732
    in tissues, R133, R138, R221, R708
    in tungsten, R327
    in urine, R153, R237, R315a, R355, R364, R596, R617b, R624, R646, R746, R747
    in water, R460, R484, R570, R624, R693, R731, R740
  enhancement of, R431a, R656, R688
  excitation potential of, 34
  filters for, 139f, 141f, R610
  flame spectrum of, 216, 218f, 262, 387, R726, R734
  interference of, 313ff, R370a
    automatic elimination of, 325
    on barium, 219, 312
    on lithium, 316, 363, R201
    on potassium, 315, R388, R644
    removal of, by aluminum, 363, R388, R633, R672
    on sodium, 284f, 317, R149d, R329, R388, R644
    spectral, 219, 284
    on strontium, 318, R416c
  interference on, 313f, 337, R138, R370a, R570, R616
    by acids, 301, 313
    by aluminum, 246f, 302ff, 306f, 308f, 313, 360, R68, R183, R276a, R370a, R454, R493, R501, R551, R633
    by anions, 300f, R83
    by borate, 313, R276a
    buffering of, 309, R199a; *see also* Releasing agents
    ionization, 300, R268b
    by iron, 313, R155
    in lines and bands, 309
    by molybdenum, tungsten, etc., 247, 313f, R452
    by perchloric acid, 66, 301, 304, 313
    by phosphate, 235, 245, 300ff, 304ff, 307ff, 313, 360, R70, R229, R258, R276a, R337, R430, R450, R452, R454, R754
    for phosphate determination, 245f
    by phosphate and sulfate, 301f, 306
    by potassium, 284, 313
    protective chelation for, 352, R350, R454, R748
    removal of, by ion exchange, R61
    by selenium, tellurium, titanium, 247, 313, R452
    by silicate, 309, 313, 368
    by sodium, 283f, 313, 349f, 363
    spectral, 219, 284
    by sulfate, 301f, 307, 309, 313, R276a, R452
    transport, R182
    vaporization, 300ff
  internal standardization of, 166
  ionic spectrum of, 218, 262, 387
  ionization of, in flame, 34, 218, R268b, R371
  ionization potential of, 34
  line profile of, R668
  loss of, by carbon dioxide, 212, 238
  in organic solvents, 262
  oxalate precipitation of, 250, 351, R188
  oxalate titration of, 351
  in oxycyanogen flame, 262
  for phosphate determination, 245f
  polymerization of, in flame, R375
  protein bonding of, 331, 341, 344, 350f, R504
  as releasing agent, 309
  separation of, 250, 325, 351, 361
  spectral interference of, with sodium, 138, 139, 283f

INDEX

Calcium—*contd*.
  standard, salts for, 237
  standardization of, 234f
  working curves of, 284, 301
Calcium aluminate, vaporization of, 306f
Calcium chloride (CaCl), flame spectrum of, 387
Calcium fluoride (CaF), flame spectrum of, 387
  for fluorine determination, 245, R277, R381
Calcium hydroxide (CaOH), identity of, 5, 42, R285, R376, R393, R441
  isotope shift of, 42, R285
  spectrum of, 42, 219, 262, 387, R285, R286, R393
Calcium oxide, spectrum of, 42, 219, 387, R285, R286, R374, R376, R439
  term diagram of, R374, R440
Calcium oxide (Ca$_2$O), occurrence of, 42
Calcium phosphate, in flame, 304; *see also* Calcium
Calculation of results, *see* Evaluation of results
Calibration, *see* Precalibrated scale, Standardization
  in spectrography, 197
Calibration curve, *see* Working curve
  in spectrography, 194, 198
CaOH, *see* Calcium hydroxide
Capillarity, for sample introduction, 97
Capillary, atomizer, annular, 104
  of atomizer-burner, 121f, 124
  bore of, 110, 114
  centering of, 122
  cleaning of, 110f
  clogging of, 104, 110f, 114, 202, 327
  design of, 110
  flow rate through, 22
  for high-pressure atomizer, 104
  for microanalysis, 114
  sieve for, 111
  as switch, 115, 155, 255
Carbon, band spectra of, 222
  determination of, 222, 247
  line emission of, 55, 222, 386
Carbon (C$_2$), for analysis, 247
  spectrum of, 54, 55, 222, 386
Carbonate, determination of, 248
  interferences of, 312, 313, 315

Carbon chloride (CCl), spectrum of, 245
Carbon dioxide, spectrum of, 54, 222
Carbon fluoride (CF), spectrum of, 245
Carbon monoxide, continuum of, 54
  flame spectrum of, 52, 54, 222, 247, 387
Carbon monoxide flame, spectrum of, 54
Carbon sulfide (CS), spectrum of, 224, 243, 244, 270, 387
Carnes, M. B., R212a
Carnes, W. J., 216
Carry-over of sample, 104f, 113, 233, 238f, 328
Cary spectrophotometer, 427
Catalysts, analysis of, R756
Cathode, *see* Photocathode
  analysis of, 357, R253, R292, R325, R582c
Cation interference, 65
Caton, D. C., R669
CCl, *see* Carbon chloride
Cellulose, analysis of, R543
Cement, analysis of, 362f, R98
  for calcium, 363
  for lithium, 363, R473
  for magnesium, 362f, R104, R192, R745
  for manganese, R226, R266a
  for potassium, R74, R191, R222, R244, R266, R266a, R308, R309, R317, R565
  for sodium, 363, R74, R191, R222, R224, R244, R266, R266a, R308, R309, R317, R565
  for strontium, R225
  determination of, in soil mixtures, R686
  sample preparation, 362
Centrifugal atomizer, 101
Centrifugal separation, of droplets, 107
  of suspended material, 110
Centrifuging, 343, 348
Ceramics, analysis of, 361f, R88
  for calcium, 362
  for potassium and sodium, R619, R716, R766
Cerebrospinal fluid, analysis of, R137
Cerium, detection limits of, 262
  flame spectrum of, 10, 220, 262
  interference of, on lanthanum, 315
  ionic spectrum of, 388
  in organic solvents, 262
  oxycyanogen spectrum of, 10, 388

20+C.A.F.P.

596 INDEX

Cerium oxide (CeO), spectrum of, 262, 388
Cerney, R. R., Jr., R96
Cesium, detection limits of, 262
  determination of, R273, R368, R572, R577
    in glass, R743
    in minerals, R255a
    in ores, R743
    in rocks, R365
    in whiskey, R566
  enhancement of, 262, R431a
  excitation potential of, 34
  flame spectrum of, 216f, 262, 383, 390, R726
  interferences of, 312, 314ff
  interferences on, 314
    spectral, 216
  as internal standard, R268b
  ionization of, in flame, R127, R268b, R371
  ionization interference of, 63, R268b, R575
  ionization potential of, 34
  line width of, R638
  as normalizer, 298
  in organic solvents, 262
  in oxycyanogen flame, 262
  self-absorption of, R638
  as saturator, 298
Cesium hydroxide, formation of, in flame, 32, 217
CF, *see* Carbon fluoride
CF-4 spectrophotometer, 430, R529
CH, *see* Methylidene
Chadwell, A. J., R482
Chaffee–Keyes flame photometer, 365, 429
Chaitken, B., R317
Chakrabarti, C. L., 222
Chalcogens, 224
Chambers, W. E., R454a
Champion, P., 7, 97
Chappell, D. G., 9, 163, R140
Characteristic curve, 194, 198, R180a
Chelation, *see* Protective chelation, EDTA
Chemical atomizers, 101, R704
Chemical equilibrium, *see* Equilibrium
Chemical industry, applications in, 358f

Chemical methods, combined with flame analysis, 247ff
  contrasted with flame analysis, 2ff
  as referee, 331
Chemiionization, 35, R180
Chemiluminescence, 49
  in flame analysis, 51, 215f, 381
  of flame radicals, 54
  mechanisms of, 51f, R283
  in outer cone, 54
  in reaction zone, 121, 216, 219, 220, 222ff, 389, 391f
  in rich flame, 126
  of sodium, R534
Chemische Werke Hüls, 214
Chimney, 120, 128f, 180
Chirnside, R. C., 380
Chloride, differential determination of, 247f
  interference of, 30f, 63, 66, 312ff
  ions in flame, 66
Chlorine, determination of, by CuCl bands, 245, R363, R461
  effect of, on chloride dissociation in flame, 30f
  spectrum of, 245; *see also* Copper chloride
Chlorine oxide (ClO), spectrum of, 245
Chopped beam, R661
  for alternating-current amplification, 150, 191
  for double-beam systems, 164
  for eliminating dark current and drift, 183f, 185f
Christiansen, G. G., R473
Chromium, chemiluminescence of, 216
  detection limits of, 263
  determination of, R167, R174
    in biological material, R597
    in ferrous alloys, R176
    in steel, R499, R733, R737
  enhancement of, 263, R431a, R582d
  flame spectrum of, 216, 224, 263, 389, R726, R734
  interference of, 314, R176
    on alkaline-earth metals, 303
    on other elements, 312ff
  in organic solvents, 263
  in oxycyanogen flame, 224, 263
  in oxyhydrogen flame, 263, R238a
  separation of, by oxine, 250

Chromium oxide, spectrum of, 224, 263, 389f
Chun, H., R544
Circuitry, see Electric circuitry
Citrus beverage, analysis of, R103
City gas, burning velocity of, 77
  composition of, 70, 77
  for flame photometry, 80f
  flame temperature of, 77
  regulation of, 92
Clark, G. L., 380
Clausen, C. J., Jr., R161
Clay, analysis of, 362, R618
  for alkali metals, R78, R266, R334
  for aluminum, R334
  for calcium, R334
Cleaning, of atomizer, 110f
  of burner, 121, 128
  of glassware, 211, 214f
Clinical analysis, 339ff, R16, R20, R21, R49, R81, R234, R348, R507, R611, R626, R678, R747, R757; see also Blood, Urine
  precision required in, 341
  sample characteristics in, 340f
  ultramicro, R181, R342
Clogging, 104, 110f, 114, 202, 327
CN, see Cyanogen
Coal, analysis of, for calcium, R239
  for sodium and potassium, R388
Cobalt, band spectrum of, 389
  continuum of, 225
  detection limits of, 263
  determination of, R174
    in biological material, R597
    in ferrous alloys, R176
    in mineralized salts, R291
    in steel, R499
  enhancement of, 263, R431a
  flame spectrum of, 225, 388f, R726, R734
  interference of, 312, 314f, 317, R176
    spectral, 225
  interferences on, 314
  as internal standard, 166
  in organic solvents, 263
  in oxycyanogen flame, 263
  separation of, by complexing agents, 250
Coefficient of variation, 329
Coke, analysis of, R239
Coleman Instruments, Inc., 427, R196

Collier, H. E., Jr., 380
Collimator, 174, 175, 188
Collision, in excitation, 49ff
Collision broadening, 45, R356, R372, R538, R666, R667, R668
Color filters, 137ff, 141, R398
Colorimetry, R28, R30
Color wedge, for null balancing, R721
Columbium, see Niobium
Combustible samples, 110, 123, 363f
Combustible solvents, see Organic solvents
Combustion, 13ff, R25, R31, R32, R35; see also Flame
  effect of organic substances on, 66
Combustion products, ventilation of, 199f
Combustion zone, see Reaction zone
Commercial flame photometers, 68, 333, 336, 427f
Comparison method, 310
Complexes, see Protective chelation
  for enhancing vaporization, 251f
  formed in flame, see Vaporization interference
  for separation, 250
Compounds, formed from analyte and flame gases, 31f
  formed from sample in flame, 28ff
  nonvolatile, see Vaporization interference
Compressed air, 81ff
  compressors for, 81ff
  consumption rate of, 96f
  cooling of, 108
  flow adjustment of, 93ff
  hazards of, 81f
  preheating of, 109
  purification of, 82f
  regulation of, 82, 89ff; see also Regulators (pressure)
  tanks for, 81
  temperature of, 82, 109
  velocity of, in atomizer, 95
Compressed gases, connections for, 84f
  installation of, 83f
  leak detection of, 84f
  regulation of, 89ff; see also Regulators (pressure)
  storage of, 83
  transport of, 83
Compressor residues, analysis of, 366

598 INDEX

Concentration correction factor, in internal standardization, 162, 338
Concentration limit, see Detection limit
Concentration of sample, adjustment of, 205ff, 210f
   for ionization interference, 299
   increasing of, 211f, 248ff
   range of, 206ff
Concentric atomizers, 100, 103, 104, 105, R341, R448
   cleaning of, 111
   drop size with, 23
   inverse (with annulus for sample), 104
   recombination of drop with, 24
Concomitant, see Interference, Interferent
   definition of, 59
   and detection limits, 275
   separation of, see Separation
Condensed-phase interference, see Vaporization interference
Condenser, 134ff, 181
Connelly, F. C., 380
Connelly, J. A., R744
Containers, 212ff
   for ashing, 342
   cleaning of, 214f
   contamination from, 212, 214, 215
   for sample, 109f
   for standards, 237
Contamination, see Air, Carry-over of sample
   of ambient air, 279f
   biological, 237
   from burned products, 199
   from cigarettes, 279
   in clinical analysis, 349
   from container, 212, 214f, 237
   from filtration, 110, 251, 343
   of flame, 121, 279
   from flame, 129, 279
   of gases, 279
   from reagents, 249, 282, 344
   from stirring rod, 343
Continuum, see Background interferences
   in background radiation, 54
   of carbon monoxide, 54
   causes of, 42, 51, 54, R395
   definition of, 42
   emission of, in flames, 42
   of nitric oxide, 54
   of potassium, R71, R395
   radiance of, 176
   slit width dependence of, 176, 178
Contrasty emulsion, 198
Control, see Process control
   of gases, see Regulators
Controlled-flow atomizer, 100, 365, R280
Controlled flow of sample, 21, 110, 114, 124, 228
   equipment for, 365, 429, 430, R347
   and interferences, 327
Control standard, 232
Cooke, W. D., 304, R682
Copper, analysis of, for silver, R280a
   chemiluminescence of, 216
   detection limits of, 263
   determination of, R146, R456, R579
      by absorption flame photometry, R203a
      in alloys, 368, R207
      in aluminum alloys, 250, R212
      for amino nitrogen, R97
      in biological material, 353f, R597
      in blood serum, R345
      in citrus beverage, 339, R103
      in ferrous alloys, R210
      in gasoline, 364, R403
      instrument for, R475
      in mineralized salts, R291
      in nutrient solutions, R107
      in plants, R203a, R465
      in plating bath, R725
      in sirup, 339
      by solvent extraction, R145
      in steel, R499
   enhancement of, 263, R431a
   excitation potential of, 34
   flame spectrum of, 216, 217, 262, 390, R579, R726, R734
   interferences of, 312, 314f, 317
   interferences on, 314
   internal standardization of, 166
   ionization potential of, 34
   in organic solvents, 263
   in oxycyanogen flame, 263
   protein bonding of, 344
   self-reversal of, 58
   separation of, 250
   working curve of, 58, 217

Copper alloys, analysis of, for sodium, R322
Copper chloride, for determination of chlorine, 245, R363, R461
  spectrum of, 372f, 390
Copper fluoride, spectrum of, 390
Copper hydride (CuH), equilibrium of, R171, R698
  spectrum of, 217, 370, 390
Copper hydroxide (CuOH), equilibrium of, R171, R698
  spectrum of, 217, 390, R579
Copper oxide, flame spectrum of, 217, 390, R579
Cork stoppers, 215
Corliss, C. H., 371, 373, 382
Corning filters, 137
Corpuscles, see Blood cells
Correction method, 321
  for ionization interference, 297
Corrosion, measured by flame analysis, 356
Cossy, A., R151
Costet, C., R234a
Coulter, S. T., R190
Counting, photon, 187
Craig, P., R663
Criminology, 368f
Critical damping, 153
Critical radius of droplet, 23, R358
Cross-sections, R356
Cross-sensitivity, 61, 138, 280ff
  and background, elimination of, 290ff
  elimination of, 283ff, 290ff
  and filters, 139, 141f, 280
  and spectral sensitivity, 147
  table of, 281f
  testing, 281ff
CS, see Carbon sulfide
CT, carbon tetrachloride
Cu, copper solution, 384
Cullum, D. C., 248
Cummings, G. M., R459
Curry, R. H., 215, 370, 372, 380
Curtis, G. W., 380
Curve of growth, R538; see also Working curve
Cushman, M., R720
Cyanogen (CN), for determination of nitrogen, 244, R362
  spectrum of, 222, 244, 267, 386f

Cyanogen (C_2N_2), see Oxycyanogen, Ozone–cyanogen
  availability of, 71, 77
  flame, R664
  toxicity of, 77
Cylinders, see Tanks

d, double; depression (Table 12)
d., determination (of)
D, dimethylformamide
Dade Reagents, Inc., 238, 345
Damping, see Time constant
  of aerosol fluctuations, 107f
  of meter, 153
  of pressure fluctuation, 89f
Dancewicz, D., R452, R453, R453a
Dangers, see Hazards
Dark current, in alternating-current systems, 185f
  causes of, 182
  drift of, 148f, 183
  elimination of, 150, 183ff
  noise of, 182ff
  of photocell, 144
  of photomultiplier, 148f, 182ff
  of phototube, 148
  of red-sensitive photomultiplier, 184
  voltage dependence of, 182ff
Davis, A. K., R533
Davis, H. G., R240
Davis, S., R652
Deactivation of atoms in flames, 45, 51
Deaeration of sample, 213
Dean, J. A., 216, 225, 280, 372, 380, 382, R56, R167, R252, R489, R490
Debras, J., R716
Debras-Guédon, J., R717
de Gramont, A., 381
Demodulators, 185
Densitometry, 197ff
Density, photographic, 196, 197f
  transformation of, 198
  of sample, effect on drop size, 23
  effect on flow rate, 22, 110
Dent, F., R664
Depression, see Interference, Specific interference, Vaporization interference
Deproteinization, 343f, 345, 351
Desiccant, for photodetector, 148
Desicote, 213

Detection limits, 255ff
  absolute (weight), 258
  and atomization, 274f
  calculated, 258f, 273
  and concomitants, 275
  and dark current noise, 182ff
  definition of, 256ff, 259, 273
  and flame adjustment, 274
  and flame flicker, 184, 273f
  in flame spectrography, 194
  and gas adjustment, 274
  and luminosity, 275
  measurement of, 257ff
  normalized, 258
  and photometric sensitivity, 275
  and recording, 276
  and spectral bandwidth, 275f
  tables of, 259ff, R110, R117, R302, R307
  and time constant, 257f
Detector, see Photodetector
Detergents, see Wetting agents
Determination limit, 256, 258
Deuteroxides, band spectra, 42, R285
Development, photographic, 197
Dewey, P. A., R703
Dialysis, to remove interference, R451a
Diaphragm, 143, 156f, 228
Diatomic molecules, see Molecular spectra
  dissociation energy of, R10
  excitation of, 40f
  spectra of, R19
Didymium filter, 139f
Dierkesmann, A., 339, R128
Diethyldithiocarbamate, 250
Differential absorption flame photometry, R454a
Differential methods, 247f
Diffuse series, 217
Diffusion, in blood analysis, 346, 348
Dilute solutions, losses from, 212
Dilution, for clinical analysis, 345
  for detecting interferences, 319
  effect of, on ionization interference, 321
  for eliminating interference, 320, 321, 326
  errors of, 165f, 210f, 236
  of sample, 165f, 205ff, 210f, 228, 236
  of standards, 236
  technique of, 211

Dippel, W. A., R162
Dir., direct (atomizer-burner)
Direct atomizer, see Atomizer-burner
Direct-reading photometry, 152, 158, 159
Dirt, see Contamination
Disadvantages of flame analysis, 3, 5f
Discharge lamps, in absorption flame photometry, 49
Discharges, electric, see Arc, Spark
  special, 76
Disk, for sample introduction, 97f
Dispersion (spectral), 172ff
  angular, 173
  linear, 173, 176
  of spectrograph, 196
Dissociation, degree of, 29f
  of sample in flame, 29ff, R455
  temperature effect on, 30, 74
  and working curve, 30, 56
Dissociation constant, 29f
Dissociation energy, R10, R694, R695, R698
  equation for, 29
  of salts, 30
Dissociation interference, 30, 61, 63, 294, R67, R263
Dithizone, book on, R24
DK spectrophotometer, 427
Dobner, W., R59
Doiwa, A., R263
Doppler broadening, 45, R538, R666, R667, R668
Double atomizer, see Dual atomization
Double band-pass filter, 141f
Double-beam systems, see Background, Internal standardization
  in absorption flame photometry, 48
  flicker elimination by, 164
Double-coefficient method, 298
Double monochromatic filter, 141f
Double monochromator, 171, 177, 430
Doublet splitting, 39
Doubling (of concentration), for detecting interference, 319f
  for eliminating interference, 323
Drainage, 111ff
Drift, in bracketing method, 241
  causes of, 148
  of dark current, 148f, 183f
  elimination of, 183f, 238, 239f, R505

INDEX 601

Drift—*contd.*
  in heated spray chamber, 115
  in photodetectors, 145, 146, 148
  in recording, 190
  from spray chamber evaporation, 108
Droplets (spray), *see* Atomization
  critical radius of, 23, R358
  critical speed of, R358
  electric charge of, R685
  evaporation of, 24f, 27f, R289, R444, R685
  mean diameter of, 23
  recombination of, 24, 107, 114
  separation of, in spray chamber, 24, 107
  shattering of, 105f
  size of, 102f, 105f, R358
Drop size distribution, 21ff, 105f, R358
  in atomizer-burner, 124f
  solvent effect on, 209
  in spray chambers, 24f, 107
  and vaporization interference, 306, 308
Drugs, *see* Pharmaceuticals
Dry ashing, 342f
Drying of sample, 342
Dual atomization, in studying vaporization interference, 67, 304f, 306, 307, R276a, R561a
DuBois, H. D., R496
Duncan, C. W., R401
DU spectrophotometer, 175
  line-to-background ratio of, 178f, 259
  recording, R417, R517, R528
  resolving power of, 178
  signal-to-noise ratio of, 259, 273
  slit width of, 178f, 273
Dust, flashes from, 153, 155f, 195, 199, 255, 279
  protection from, R658; *see also* Air
Dvorszky, M., R324, R326, R327
Dynamic pressure, in atomization, 22
Dynodes, 145f
  dark current from, 183
  voltage, 148, 182ff
Dysprosium, detection limits of, 263
  determination of, R546, R547
  flame spectrum of, 263, 390
  in hexone, 263
Dysprosium oxide (DyO), spectrum of, 221, 390f, R546, R547

EA, ether–alcohol–water, 260
Eccleston, B. H., 383, R734
Echelette grating, 173
Eckhard, S., R582d
Eder, J. M., 97
Edge fog, 196
EDTA, for aluminum enhancement, 220, 252
  as anticoagulant, 348
  for calcium determination, 352, R350, R454, R748
  for calcium titration, R337a
  for strontium determination, R213, R215, R454
  for vaporization interference, 310
EEL flame photometer, 428, R254, R689
Effective spectral bandwidth, 178f
Efficiency of atomizer, 25f, 58, 99, 105f, 115
Einstein transition probability, 43, 58
Elbe, G. von, R31, R447
Electric arc, *see* Arc
Electric circuitry, *see* Amplifiers, Meter, Power supply, Potentiometric photometric circuits
  for filter flame photometers, 149ff
  for internal standardization, 163f
  for photocell, 144f
  for phototube, 144f
Electrolysis, 251
Electrolytic atomizer, 101
Electron, electrons, *see* Photoelectrons, Quantum numbers, Secondary electrons
  angular momentum of, 38f
  excitation of, *see* Excitation
  excitation by impact of, 50
  in flames, *see* Ionization, Normalization, Saturation
  concentration of, 17
  effect of, on working curve, 35f, 57, 228
  from flame gases, 35, 299
  orbits of, 36ff
  quantum numbers of, 38f
  shells, 38f
  spin of, 38f
  transition energies of, 36ff
  valence, 36ff
Electronest, 428, R245

Electronics, R1a; see also Electric circuitry
 for alternating-current systems, 185f
Electronic torch, 76, R195
Electron volt, 37
Electroplating solution, analysis of, R268, R725
Electrostatic atomizer, 101f, 114, R683, R684
Électro-Synthèse, 428
Elektronest, see Electronest
Elements, number of, determinable by flame analysis, 3, 10
 spectra of, see Spectra
Elkinton, J. R., R720
Ellis, W. H., R496
Ells, V. R., 8
Em., emission flame photometry
Emission, 215ff; see also Background, Continuum, Height in flame, Intensity, Line (spectral), Radiation
 atomization characteristics, effect of, on, 106
 drift of, due to spray chamber, 108; see also Drift
 fluctuation of, and atomization, 26
 fluctuation of, and fuel flow, 94f
 fuel flow, effect of, on, 94
 ionization, effect of, on, 75; see also Ionization
 preheated air, effect of, on, 109
 sample flow rate, effect of, on, 22, 106, 109, 115, 156, 254, R85, R85a, R268a, R278a, R278b
 sample temperature, effect of, on, 108f
 temperature, effect of, on, see Temperature
Emission sources, see Arc, Spark
 special, 76f, 470
Emission spectrochemical methods, compared with flame analysis, 4ff
Emissivity, 49
 of background radiation, 53
Emitter, identity of, 371
Emulsion (photographic), contrasty, 198
 fog, 196
 graininess of, 198
 soft, 198
 spectral sensitivity of, 195, 196

Energy, see Vibrational energy, etc.
 conversion of, in excitation, 49
 transfer of, in excitation, 49ff
 kinetics of, 51
Energy level diagram, 37f, 40f, R15, R374
Energy levels, atomic, 37ff, 382
 molecular, 39ff
Enhancement, 253; see also Interference, Organic solvents
Enrichment, of analyte, 211f, 248ff, 325, 359
Entrainment, of air, 96f, 120f, 126, 199, 279
Entrance beam, 131ff
Entrance optics, 129ff; see also Filter (optical)
 distance of burner, 133
 in filter flame photometer, 134ff
 in flame attachment, 180f
 and flame flicker, 274
 height adjustment in, 131ff, 134
 for internal standardization, 163f
 mirror for, 133f
 for monochromator, 175
Epic, Inc., 429
Eppendorf flame photometer, 136, 429, R513
 burner of, 120f
 concentration ranges for, 205f
 cross-sensitivities in, 282
Equilibration time, see Time constant
Equilibrium, chemical, 17, 29, 51f, R127
 departures from chemical, 30, 32, 35, 51, 72
 departures from ionization, 35
 departures from thermal, 50
 ionization, 33, 35, R127; see also Ionization
 kinetics of attainment of, 51
 mechanism of attainment of, 49ff
 radiational, 17, 50
 thermal, 15, 16, 17, 43, 49, 64, R698
 thermodynamic, 43, 49
Equilibrium constant, 29f
Erbium, detection limits of, 264
 determination of, R546, R547
 in hexone, 264
Erbium oxide (ErO), spectrum of, 221, 264, 391, R546, R547
Ernst, K. F., R747

Error, 278ff, R661; *see also* Accuracy, Contamination, Interference, Precision
  aggregate, 332
  of blank, 256f
  causes of, 327f, 331
  of dilution, 165f, 210f, 236
  distribution of, 329
  estimation of, 204f
  instrumental, 278f, 327f
  in internal standardization, 165ff
  random, 256, 327ff
  of sampling, 204f, 327
  standard, 328f; *see also* Standard deviation
  statistical, 256f, R410
  subjective, 330
  systematic, 256, 327, 331f
Erythrocytes, *see* Blood cells
Eshelman, H. E., 380
Etalon spectrometer, luminosity of, R389
Ethylenediaminetetraacetic acid, *see* EDTA
Eucken, A., R17
Europium, detection limits of, 264
  determination of, R546, R547
  flame spectrum of, 220f, 264, 391, R546, R547
  in hexone, 264
  ionic spectrum of, 391
Europium hydroxide, spectrum of, 391
Evaluation of results, automatic, R409
Evans Electroselenium Ltd., 428, R254; *see also* EEL flame photometer
Evaporation, of analyte, *see* Vaporization
  in atomizer-burner flame, 28, 123, 326
  cooling of flame by, 28, 73, 123
  in flame, 27f
  incomplete, 21, 28, 123, 326
  of sample, 110
  in spray, 21, 24f, 27f, 107, 108, 305, R290, R685
  in spray chamber, 24f, 107, 108
  theory of, 24, R444
Evaporation interference, 62f, 325f; *see also* Transport interference, Vaporization interference
  elimination of, 326f
20*

Excitation, degree of, 43
  electronic, in atoms, 36ff
    in molecules, 39ff
  mechanism of, 49ff, R283, R470
Excitation energies, electronic, 36ff
  molecular, 39ff
  and population of excited states, 43
  and precision, 229
  table of, 34
Excitation interference, 61, 63f
  diagnosis of, 67
  resonance exchange, 64
Excitation sources, *see* Sources
Excited states, 37ff
  deactivation of, 45, 51
  population of, 43
  transition probability of, 43
Exit beam, adjustment of, 135f, 149
Explosion, R25, R31
Exposure, in spectrography, 194, 196, 197, 198
Extraction, for aluminum, 251, R252
  for chromium, R167
  for clinical samples, 343
  for copper, 250, R145, R210
  in flame photometry, 251, 325, 359, R438
  for iron, 251, 367f, R208
  for lanthanum, 251, R490
  for manganese, 250
  for nickel, 250
Eye, *see* Aqueous humor
  for photometry, R320

f, intersystem combination line
Fabian, H., R145, R146, R146a
Fabric, analysis of, R702
Falloon, S. W. H. W., R585, R586
Fanger, H., R181
Fanus, W. H., R166
Farbwerke Hoechst, 214
Farmilo, C. G., R91
Farnsworth (ITT Laboratories), 260
Fassel, V. A., 215, 370, 372, 380, 383
Fatigue, of photocells, 145, 148
  of photomultipliers, 146, 148
  of phototubes, 148
Feces, preparation of, 341
Feeds, *see* Alfalfa, Hay
  analysis of, for calcium, R199a
Feldspar, analysis of, R618
Ferrebee, J. W., R646

Ferrite, determination of, in steel, R733, R737
Ferrous metals, analysis of, 367f, R126, R736
  for calcium, R59
  for chromium, R176
  for cobalt, R176
  for copper, R210
  for magnesium, R59
  for manganese, R176
Fertilizer, analysis of, 335f
  for calcium, R151
  for magnesium, R151
  for potassium, 359, R151, R158, R267, R289, R462, R552, R614, R615
  for sodium, R151
Fettchemie Düsseldorf, 214
FH, fluorine–hydrogen flame
Field lens, *see* Condenser
Fil., filter
Film, *see* Photographic plate
Filter, for atomizer capillary, 111
Filter (optical), 136ff, R426, R594
  band-pass, 141f
  barrier, 139, 143, 290f
  beam divergence of, 142
  double band-pass, 141f
  double monochromatic, 141f
  exclusion, 143
  interference, 139ff; *see also* Interference filters
  interference wedge, 143
  liquid, 137, R598
  luminosity of, 143
  solid (color), 137ff, R398, R610
  stray light of, 280
  wavelength range of, 226
Filter flame photometers, background correction in, 290
  commercial, 336, 427ff
  condenser for, 134ff
  design of, 69
  electric circuitry for, 144f, 149ff
  filters for, 136ff
  limitations of, 170
  meters for, 152ff
  photodetectors for, 143ff
Filter flame photometry, 134ff
  history of, 8f
Filter paper, contamination from, 110, 251, 343
  for sample introduction, 98

Filtration, 251, 343
  of radioactive aerosols, 357, R240
Finkel'shteïn, A. I., R264a
Finkelstein, R., R754
Fisher, C. I., R459
Fisher, D. J., R414
Fishtail burner, 118, 119f
Fission products, analysis of, 356
Flame, adjustment of, 92ff
  alignment of, *see* Entrance optics, Exit beam
  of atomizer-burner, *see* Atomizer-burner
  background radiation of, *see* Background radiation
  blowoff of, 126
  books on, R11, R12, R25, R31, R32, R35
  bunsen, 13f
  burning velocity of, 15f, 77
  concentration of metals in, R182a
  contamination of, 121, 279
  control of, 86ff
  cooling of, by spray, 28, 73, 123
  distance of, from monochromator, 181
  electrons in, *see* Electron
  emission of, *see* Emission, Radiation
  emissivity of, 53
  entrainment of air by, 96f, 120f, 126, 199, 279
  equilibrium in, *see* Equilibrium
  flashback of, 16, 81f, 116, 118f, 123, 126
  flat, 120
  flicker of, *see* Flame flicker
  focusing of, *see* Entrance optics
  form of, 125ff
  height in, *see* Height in flame
  homogeneity of, 47, 58
  ignition of, 120, 201
  imaging of, 134f, 157, R319
  intermediate, 14
  ionization in, *see* Ionization
  laminar, 14ff, 127
  lighting of, 120, 201
  long, 119f
  luminous, 17, 126, 215, 243
  Méker, 16, 119
  noise of, 19f, 122, 128, 180
  premixed, 13ff, 120f
  pressurized, R227

INDEX 605

Flame—contd.
  radiation of, see Emission, Radiation
  richness of, 93f
  selection of, 372f
  separation of zones of, 54, 121, 244
  shape of, 125ff
  shape interference, 61, 63
  size of, 125ff
  in spark, 76, R377, R384, R682
  special, 76f
  spectra in, see Spectra
  spectrum of, see Background radiation
  stability of, 16, R447; see also Flame flicker
  stabilization of, 19
  structure of, 13ff, R447
  temperature of, see Temperature
  thickness of, 94
  transparency of, 53, 219
  turbulent, 14f, 18f, 20, 54; see also Atomizer-burner flame
  types of, 13ff
  unpremixed, 14f, 19f; see also Atomizer-burner flame
Flame accessories, see Flame attachment
Flame analysis, accuracy of, 2f, 4, 278, 327ff, 331f
  applications of, 333ff
  automatic, 277
  combined with chemical methods, 247ff
  compared with chemical methods, 2ff
  compared with spectrochemical methods, 4ff
  definition of, 2
  errors of, see Error
  history of, 6ff
  number of elements amenable to, 3, 10
  speed of, 276ff
Flame attachment, 179f
  alignment of, 180f
  Applied Physics, 427
  Beckman, see Beckman flame attachment
  Hilger, 428, R352
  Hitachi, 428
  Jarrell-Ash, 428
  Optica, 430, R529
  Perkin-Elmer, 430
  for spectrograph, R627
  Technicon, 430
  Zeiss, 180, 431, R758
Flame background, see Background
Flame background interference, 61, 285ff; see also Background radiation
Flame bands, see Background radiation
  spectra of, 370, 373
Flame flicker, 26, 53, 94f, 126, 273f, 328
  and detection limit, 184, 273f
  and entrance optics, 274
  and flame adjustment, 274
  and fuel–oxidant ratio, 274
  isotropy of, 164
Flame monitor, 168f, 358, 428, 429, 430, 431, R755
Flame photometer, see Atomizer, Burner, Electronics, Filter flame photometer, Gases, Optics, etc.
  accommodations for, 199ff
  airborne, R751
  choice of, 70
  commercial, 68, 333, 427ff, R761
  definition of, 2
  design of, 68ff, R548
  mounting of, 200
  multiple-use, 169f
  operation of, 200ff
  for sodium (spectronatromètre), 7
  standardization in design of, 64f, 71, 87f, 93
Flame-photometric titration, 252f, R251, R704
Flame photometry, apparatus for, 68ff
  definition of, 1, 2, 6, 8
  direct-reading, 152, 158, 159
  with filters, see Filter flame photometry
  history of, 8f
  multiple-beam, see Multiple-beam flame photometry
  null-reading, 153f
  technique of, 199ff
Flame separator, 54, 121, 244
Flame spectra, see Background radiation, Spectra
  book of, R12
  of the elements, 215ff
  recordings of, R307a, R726, R734, R759a
  tables of, 370ff, R307, R307a

Flame spectrograph, R597, R627
  commercial, 430
Flame spectrography, 192ff, R632
  advantages of, 155
  applications of, 336, R144, R359, R676
  automatic, 336
  with Beckman burner, 381
  definition of, 1, 68
  history of, 8, 9
  internal standardization for, 161
  with spark atomizer, R499
Flame spectrophotometer, *see* Flame attachment, Recording, Spectrophotometer
  Beckman, *see* Beckman flame attachment, etc.
  commercial, 427ff, R569
  Perkin-Elmer, 191; *see also* Perkin-Elmer Corp.
  photodetectors for, 181ff
  visual, R125
  Warren, R724
  Zeiss, 180; *see also* Zeiss flame spectrophotometer
Flame spectrophotometry, 170ff; *see also* Monochromators
  definition of, 1, 68
  history of, 9f
  internal standardization for, 190ff
  multiple-beam, 187f
  recording, 188ff
Flame tests, 1, 6, R495
Flammable samples, 110, 123, 363f
Flashback, 16, 81f, 116, 118f, 123, 126
Fletcher, M. H., R735
Flicker, *see* Flame, Flame flicker, Fluctuation
Flow (gas), calibration of, 88
  control of, 89ff
  damping of, 89f
  measurement of, 86f
  optimization of, 92ff
  pressure dependence of, 89f
  turbulence of, 18f
Flowmeter, 87f
  calibration of, 88
  construction of, 88
Flow rate, of gases, *see* Flow, Regulators
  constancy of, 116
  of sample, *see* Sample flow rate

Fluctuation, *see* Shot noise, Time constant
  damping of, 89f, 107f
  and detection limits, 256
  from droplets, 26
  of flame, *see* Flame, Flame flicker
  fuel flow, effect of, on, 94f
  of pressure, 89f, 328
  of spray density, 107f
Fluctuation bands, 220
Fluor, *see* Phosphor
Fluorescence in flame, 50, R147
Fluoride, interference of, 312, 315
Fluorine, *see* Barium fluoride, Strontium fluoride
  determination of, 245, R277, R278, R381
  flame spectrum of, *see* Calcium fluoride
Fluorine–hydrogen flame, spectra in, 385, 387, 390, 396, 404, R470
Fluosilicate, interference of, on barium, 313
Fluotitanate, interference of, on calcium, 313
  on strontium, 318
Flux transmitting power, *see* Luminosity
f/-number, 135, 181, 196; *see also* Luminosity
Focusing, *see* Entrance optics, Exit beam
Fog, in flame spectrography, 194, 196
Folse, P., R731
Foods, analysis of, 337ff
  for calcium, R732
  for magnesium, 354
  for sodium, 338
Forage, *see* Alfalfa, Hay
Formaldehyde, spectrum of, 247
Formánek, J., 380
Foryl, 214
Fournier, J. A., R380
f.p., flame photometer, flame-photometric, flame photometry
Frankenberg, B., R366
Fraunhofer, J. von, 6
Freeman, E. B., R269
Freeze-drying, 342
Frodl, V., 380, R8
Fructose, in blood analysis, 348, 353
Fruit, analysis of, 338

Fruit juice, analysis of, 337
f.s., flame spectrophotometer, flame-spectrophotometric, flame spectrophotometry
Fuel, see Acetylene, City gas, Gases, etc.
  burning velocity of, 77
  changing of, 117
  choice of, 69ff
  connections for, 84f
  consumption rate of, 96f
  density of, 77
  for flame photometry, 69ff, 83ff
  flame temperature of, 77; see also Temperature
  flow of, control of, 86ff, 116
    in internal standardization, 166f
    optimization of, 92ff
  hazards of, 71, 75, 80f, 83ff
  injection of, 115ff
  mixing of, 115ff
  nozzle for, 116f, 122
  pressure of, 116
  regulation of, 89ff; see also Regulators
  storage of, 83
Fuel ash, analysis of, see Ash, Coal
Fuel oil, analysis of, for sodium, R75
Fuel–oxidant ratio, effect on flame properties, 71f, 93ff, 126
  and flame flicker, 274
Fukker, F. K., R324
Füller, H., R149
Fumes, ventilation of, 199f
Fungi, 237
Furman, N. H., R162, R229
Furnace gases, analysis of, for sodium, R532a
Fuwa, K., 48, R85
f-value, 58
FW-118 photomultiplier, 184

G, gasoline
Gadolinium, atomic spectrum of, 393
  detection limits of, 264
  determination of, R546, R547
  in hexone, 264
Gadolinium oxide, spectrum of, 221, 264, 393, R546, R547
Gain, 145f, 157, 182ff; see also Sensitivity control
Gallenkamp and Co. Ltd., 428

Gallium, detection limits of, 264
  determination of, R146a, R453a
    in alloys, R487
    in uranium, R441a
  flame spectrum of, 220, 264, 393
  interference of, with manganese, 221, R416a
  interference on, 314
  in oxycyanogen flame, 10
  separation of, 250
  working curve of, 222
Galvanometers, 149, 152ff
  mounting of, 200
Gamma, in spectrography, 198
García-Llauradó, J., R557
Garratt, D. C., R159
Gases, see Fuel
  choice of, 69ff
  city, see City gas
  compressed, 83ff; see also Compressed gases
  consumption rate of, 96f
  for flame analysis, 69ff
  flow of, measurement of, 86ff
    optimization of, 92ff
  injection of, 115ff
  mixing of, 115ff
  nozzle for, 116f
  preheating of, 75
  pressure of, 116
  regulation of, 89ff; see also Regulators
  tanks for, 77ff, 81, 83; see also Tanks
  velocity of, in burner, 118
Gasoline, analysis of, for copper, 364, R403
  for lead, 364, R101, R300, R402, R481, R661a
  for manganese, 364, R661a
  base stock interference with, 365
  evaporation of, 110
  as fuel for flame photometer, 429
  lead emission from, 132
Gastric juice, analysis of, R135
Gatterer, A., 372, 380
Gaussian error distribution, 329
Gaydon, A. G., 380, 382, R41, R186, R194
Gebauer, J., 248
Gebauhr, W., R511
Gelatine, for nonspecific interferences, 326, 353

608 INDEX

Gelman Instrument Co., 428
Geoffroy, 6
Geological applications, 354, 359f
Gerlach, W., 8, 9
Germanium, chemiluminescence of, 264
    detection limits of, 264
    flame spectrum of, 222, 264, 393
    in microwave discharge, 76
    in organic solvents, 264
Gibson, J. H., 304
Gilbert, P. T., Jr., 10, 48, 51, 370, 380f
    atomizer-burner, see Beckman burner
Glass, I. I., R194
Glass, analysis of, 360f, R108, R260, R603
    for aluminum, 361, R333
    for calcium, R142, R333a, R427, R605
    for cesium, R743
    for lithium, R164, R326, R427, R604, R743
    for magnesium, R605
    for manganese, R605
    for potassium, 361, R102, R164, R193, R266, R326, R331, R335, R427, R604, R605, R706, R743
    for rubidium, R743
    for sodium, 361, R102, R164, R220, R266, R326, R331, R333a, R335, R427, R604, R605, R706, R743
    durability of, determination of, R604
    extraction of, 361
    preparation of, for analysis, 361
Glassware, cleaning of, 211, 214f
    contamination from, 212, 215
    for samples, 212f
    water-repellents for, 211, 212, 213
Glendening, B. L., R629
Glick, D., R36
Glycerine, for nonspecific interference, 326
Gold, band spectrum of, 370
    detection limits of, 264
    flame spectrum of, 217, 264, 385
    in hexone, 264
    in microwave discharge, 76
    working curve of, 217
Gonsior, T., R691, R692
Goto, H., 222
Gouy, A., 8, 30

Gr., grating monochromator
Grabbe, F., R592
Graham, J. R., R427
Grain, analysis of, 338, R151
Gramont, A. de, 381
Grating monochromators, 172f, 187f, 373, R164a, R457, R458, R709
    commercial, 427ff
    luminosity of, R389
    resolution of, 227
    wavelength range of, 227
Graul, E. H., R588
Gravity-feed atomizer, 100, 109
Grenat, H., 381
Grenier, M., 7, 97
Grille, burner, 16, 18, 117ff
Grosse, A. V., R199
Grossman, W. E. L., 304
Grotrian diagram, 37f
Ground state, 37, 40, 43, 45
Guérin de Montgareuil, P., R497, R645
Gunn, E. L., R412
Gusyatskaya, E. V., R607
Guyer, H., R71, R179, R219

h, with burner sheath, 260
H, hexone, 384
H, see Hydrogen, atomic
Haas, C., R414a
Hadjú, P., R315a
Hafnium, 222
Hagedorn, H., R552
Hagelberg, R. R., R614
Hagenbach, A., 381
Hagen–Poiseuille law, 22
Half-intensity width, of amplifier band pass, 185
    of filters, 140
    of monochromator, 178f
    of spectral lines, 45f
Half-width, 45f, 140
Halides, determination of, R488
    dissociation interference of, 30f, R455
    emission by, 31
    equilibria of, in flame, R127
Hallman, N., R446
Halogens, determination of, 245
    determination of, by halide emission, 31
    flame spectra of, 224, 243, 245
Hansen, G., R408

HAP, hexone–acetone–isopropanol–water, 260
Haranath, P. B., 381
Hardegg, W., R542
Hardwick, W. H., R689
Harrison, G. R., 98, 381
Hartley, W. N., 8, 243, 370, 381, 382
Harward, M. E., R479
Hasler, A., 8
Hattinga-Verschure, J. C. M., R64, R661
Haught, A. F., R505
Hawes, R. C., R297
Hay, analysis of, R523
Hazards, 199
  of air compressors, 81f
  in clinical analysis, 341
  of compressed gases, 83ff
  of fuels, 71, 75, 80f, 83ff
  in gasoline analysis, 364
  of toxic samples, 199
HCO, see Hydrocarbon flame bands
Head, of band, 41
  hydrostatic, of sample, 21f, 98, 109f, 327
Heated spray chamber, 114, 115, R182, R236
  of Jouan flame photometer, 429
Heath, R. G., R688
Hegedüs, A. J., R571, R573, R575, R576, R582c
Height of completed evaporation, 27f
Height in flame, see Entrance optics
  adjustment of, 117, 131ff, 134
  and alkali-metal emission, 216f
  and analyte–background ratio, 132f
  and chemiluminescence, 51
  effective, 60
  and hydroxyl concentration, 32, 217
  and interferences, 60, 320
  and internal standardization, 161
  and ionization, 35f, 217, 228
  and ionization interference, 299
  of lead emission, 132
  optimal, 126
  and temperature, 74
  and vaporization interference, 301f, 303, 305, 306f, 308
  and working curve, 57, 228
Height of suction, see Hydrostatic head
Held, van der, equation, 27

Hemolysis, 346f, 348
Henley, L. C., R240
Herkenhoff, K., R515, R516
Herman, L., 381
Herman, R., 381
Herman, R. C., 381
Herrmann, Roland, 9, 354, 372, 381, R92, R93, R94, R95, R130, R149, R406, R491
  atomizer-burner, 124
Herrmann, Rudolf, R50
Herschel, J. F., 6
Hert, W., R333, R334, R335
Hewett, D. R., R239
Hexone, 384
Heyes, J., 8, 9, 339
High-frequency torch, 76
High-pressure atomizer, 95, 104, 106f, 114
High-pressure atomizer-burner, 124f, R347
Hilger and Watts Ltd., 428, R352, R353
  absorption flame spectrophotometer, 428, R353
  Uvispek flame spectrophotometer, R352
Hinde, A., 325
History, 6ff, R26
Hitachi Ltd., 428, R526
Höfert, H. J., R524
Hofmann, K. A., 381
Hohlraum, 50
Hold, G., R311
Hollander, T., 228
Holley, H. L., R248
Holliday, M., R720
Holmium, detection limits of, 265
  determination of, R546, R547
  in hexone, 265
Holmium oxide (HoO), spectrum of, 221, 264, 393f, R546, R547
Hood, R. L., 9, R90
Hooge, F. N., R414a
Hornbeck, G. A., 381
Horstman, E., R247
Hose connections, for gases, 85
Hospadaruk, V., R272
House, H. P., 382, R488
Housing, for burner, 122, 128f, 180
Hp, heptane
Hronas, J. J., R640
Hughes, R. C., 76

Huldt, L., R441
Hume, D. N., R268a, R268b
Humphreys, C. J., R554, R555
Hunter, L. E., 380, R202
Hutton, J. T., R149d, R676a
Hydraulic atomizer, 103, 105
Hydraulic head, *see* Hydrostatic head
Hydrocarbon flame bands, 247
Hydrocarbon flames, spectrum of, 54
Hydrofluoric acid, for enhancement of aluminum, 220, 252, R431b
Hydrogen, atomic, concentration of, in flames, 17, 30, 32, 51, R171, R698
   excitation of sodium by, 51
   formation of hydrogen chloride by, 30
   Langmuir torch for, 76
   reaction of, with lithium hydroxide, 32
   recombination of, 18, 32, 51
Hydrogen chloride, formation of, in flame, 30
   interference by, 31
Hydrogen flame, *see* Fluorine–hydrogen flame, Oxyhydrogen flame, Perchloryl fluoride–hydrogen flame
   background spectrum of, 54
   detection limits in, 261ff
   equilibria in, R169, R171, R391, R393, R394, R662, R694, R695, R698
   spectrum of, *see* Hydroxyl, Water
Hydrostatic head of sample, effect of, on flow rate, 21f, 98, 109f, 327
Hydroxides, of alkali metals, 32, 216f, R391, R662
   of alkaline-earth metals, 32f, 42, R393, R441
   band spectra of, 42, R285, R286
   equilibria of, in flame, R127, R391, R662, R694
   formation of, in flame, 32, 216f, R694, R695
   identity of, 42, R285, R286, R441
   isotope shift of, 42
   recombination continuum of, 42
Hydroxyl, absorption by, R373
   effect of, on internal standardization, 161
   effect of, on working curves, 32
   electron affinity of, R535
   emissivity of, in acetylene flame, 53
   equilibrium of, in flame, R698
   height effect on, 32
   in interconal zone, 54
   occurrence of, in flames, 17, 32, 216f, R171, R373
   in oxycyanogen flame, 55
   recombination of, 18, 32
   spectrum of, 54, 370, 399, R414a
Hydroxyquinoline, 220, 250, 252

i, reaction zone, 260, 384
Identification of emitting species, 371, 374ff
Ignition of flame, 120, 201
Illuminating gas, *see* City gas
Illumination, 176, 177f, 178, 196
   of laboratory, 200
Il'yasova, N. V., 383, R41c, R607
Imaging of flame, 134f, 157, R319
Imidogen, flame spectrum of, 397
   nitrogen determination by, 244
Immersion atomizer, 100, 103, 104
Impact pressure, 22
Impurities, *see* Contamination
Incandescent flame, 126, 215, 243; *see also* Oxyacetylene
   radiation of, 17
Incrustation, of atomizer, 111, 123, 239, 345
   of burner, 128
Ind., indirect
Indirect atomizer, 107ff; *see also* Spray chamber
   definition of, 100
   drainage of, 111ff
   drop size with, 21ff
   efficiency of, 26f, 115
   for microanalysis, 114
   nonspecific interferences with, 63
   processes in, 20f
   sample temperature effect on, 108f
   suction of, 21
Indirect methods of analysis, 245ff
   for aluminum, 246f, R183, R493, R551, R645
   for arsenic, R452
   for beryllium, 247, R452, R453
   for bromide, 248
   for chloride, 247f
   differential, 247f

INDEX   611

Indirect methods of analysis—*contd.*
 for halides, R488
 interferences in, 246, 247
 for iodide, 248
 for molybdenum, 247, R452
 for organic compounds, 247
 for phosphate, 245f, R229, R452
 for selenium, 247, R452
 self-standardization in, 247
 by substitution, 248
 for sulfate, 248, R177, R452, R647, R687
 for tellurium, 247, R452
 for titanium, 247, R452
 for tungsten, 247, R452
 for vanadium, 247, R452
 by vaporization interference, 245ff
 for zirconium, R452
Indium, detection limits of, 265
 determination of, R146a, R305, R453a
  in aluminum alloys, R485
  in uranium, R441a
 enhancement of, 265, R431a
 flame spectrum of, 220, 265, 394
 interferences on, 314f
 in organic solvents, 265
 separation of, 250
 working curve of, 222
Indium hydride, spectrum of, 221
Indium oxide, spectrum of, 221, 394
Industrial applications, 354ff
Industrial process control, *see* Process control
Industrial wastes, analysis of, 359, R73
Infrared flame spectrum, 52, 53, 54, R414a, R554, R555, R556
Injector, 100, 103, 104
Inner cone, *see* Reaction zone
Inner quantum number, 38f, 43
Instability, *see* Fluctuations
Instrumental errors, 278f, 327f
Instrumentation, 68ff
 standardization of design of, *see* Standardization
Insulation, acoustic, 180
 electric, 182
Integration, 97, 115, 155f, 189f, 255, R255, R517a, R585
 of line profile, 189f
 in spectrography, 195
Intensity, *see* Emission

 estimation of, 372, 373ff
 normalization of, 373ff
 of radiation, 42ff, 73, 75, R381
 tables of, 370ff, R307, R307a
Intercombination line, 220, 224, 383
Interconal zone, 14, 16f
 background radiation of, 54
Interelement effects, *see* Specific interferences
Interference, 59ff, 278ff, R67, R71, R182a, R233, R264, R544, R561a; *see also* Anion interference, Background interference, Blank interference, Cation interference, Cross-sensitivity, Dissociation interference, Excitation interference, Flame background interference, Ionization interference, Radiation interference, Spectral interference, Transport interference, Vaporization interference
 in absorption flame photometry, 48f, 66
 classification of, 60f
 definition of, 59, 61
 diagnosis of, 66f, 190
 height in flame, effect of, on, 60
 identification of, 66f
 indirect, 285
 and internal standardization, 161, 165, 167
 multiple, 310
 multiplicative, 283f, 295ff, 302, 321ff, 326f, R274
 secondary, 60, 62
 testing of, 66f, 190
  by dialysis, R451a
Interference filters, 139ff, R149c, R287, R398, R620, R629
 beam divergence in, 142
 for reflection, 143
 wavelength range of, 226
Interference wedge, 143, 168
Interferent, definition of, 59
 monitoring of, 325
 precipitation of, 250
 separation of, 249ff, 310, 325, 360
Internal energy, in excitation, 49
Internal-standard element, 160f, 166; *see also* Lithium
 cesium as, R268b

Internal-standard element—*contd.*
  cobalt as, R597
  handling of, 165f
  as normalizer, 299, 327
  rubidium as, R268b
  silver as, 166, 368
  strontium as, 358f, R154
Internal-standard flame photometer, Advanced Instruments, 427
  Baird, 427
  Baldwin Instruments, 427
  Berry *et al.*, R140
  Boon, R598
  Électro-Synthèse, 428
  filter and prism, 191f
  Fox, R269, R270, R271
  Janke, R271
  Lange, 429
  Marius, 429
  National Instrument Laboratories, 429
  North American Philips, 429
  optical null-balancing, R721
  Perkin-Elmer, 191, 430
  Process and Instruments, 430
  Robinson, R598, R599
  Technicon, 430
  White, R735a
Internal standardization, 160ff, 320, 326f, R316, R427, R446
  advantages of, 165ff
  auxiliary (sequential) method of, 161ff, 190, R96
  by background, 161
  for biological analysis, 353, R139a, R597
  electric circuitry for, 163ff, 192
  errors of, 165ff, 330
  with flame spectrophotometer, 190ff, 338
  history of, 8, 9
  interferences in, 161, 165, 167
  for ionization interference, 299
  multiple-beam, 164f, 192, R344
  optics for, 163f
  precision of, 330
  in spectrography, 197, R597
  speed of, 277
  for vaporization interference, 310
Internal-standard line, 160f, 166
  isolation of, 191f
Interpolation, 240f

Interpretation, *see* Measurement
intf., interference
int. std., internal standard
Iodide, determination of, 248
Iodine, detection limits of, 265
  determination of, 245
  interference of, on phosphorus, 317
  in microwave discharge, 76
  spectrum of, 245
Iodine oxide (IO), spectrum of, 224f, 245, 265, 394, R526
Ion, *see* Ions
Ion exchange, control of, in industry, 355, 358
  separation by, 251, 337
    determination of sodium in, R528
    of interferents of calcium, R61
    of interferents of potassium, R288, R289
    of phosphate, 325
  substitution by, 252
Ionization, of analyte in flame, 33ff, R423, R662, R694, R695
  degree of, 33f, 229f, 323f
  equilibrium in, 35, R127
  in flame, basic studies of, 35, R127, R423, R424, R425, R694, R695
  flame electrons, effect of, on, 35f, 228
  of flame gases, 35, R424, R425
  and height in flame, 35f, 228
  kinetics of, 35, R423
  of lead, R697
  of potassium, R63, R127, R536, R662
  in reaction zone, 35, R180
  of sodium, R63, R127, R536, R662
  temperature, effect of, on, 34, 74f
  and working curves, 35f, 56f, 228
Ionization constant, 33, 323
Ionization energy, 33f, 75
Ionization interference, 33ff, 57, 63, 294ff, R67, R268b, R603, R661
  in absorption flame photometry, 48f
  concentration, effect of, on, 299, 321
  effect of, on working curve, 294f
  elimination of, 296ff, 320, 323f, 361
  and height in flame, 299
  internal standardization for, 299
  saturation of, 298
  second-order, 297
  substitution method for, 323ff
  temperature effect on, 75, 296, R575
  testing of, 296

Ionization potential, 33f, 75
ionizn., ionization
Ions, see Ionization
  molecular, 35
  negative, in flames, 35, 66
  recombination of, 35
  recombination continuum of, 42
Iridium, in flame, 225
  in microwave discharge, 76
Iris diaphragm, 143, 156f, 228
Iron, analysis of, R435; see also Ferrous metals, Steel
  chemiluminescence of, 216, 225
  detection limits of, 265
  determination of, R417
    by absorption flame photometry, R72a, R203a
    in alloys, 367f, R208
    in biological material, R597
    in blood plasma, R153
    in citrus beverages, 339, R103
    in ferrite, R733, R737
    instrument for, R475
    in lubricating oil, 366
    in mineralized salts, R291
    in nutrient solutions, R107
    in plants, R203a
    in silicates, R209
    in sirup, 339
    in water, R460
  enhancement of, 265, R431a, R582d
  flame spectrum of, 216, 225, 265, 391f, R726, R734
  interferences of, 312ff
    on alkaline-earth metals, 303
    on calcium, 313, R155
    spectral, 225
  interferences on, 314
  internal standardization for, 166
  in organic solvents, 265
  in oxycyanogen flame, 265
  in oxyhydrogen flame, 265, R238a
  as releasing agent, 309, R754
  separation of, 250, 367
Iron oxide, spectrum of, 225, 392f
Irrigation water, analysis of, 355
I.S., internal standard
Iseri, L. T., R153, R500, R502, R663
Ishida, R., 381
Ishler, N. H., R291
Isotopes, determination of, of lithium, R476

Isotope shifts, of boron bands, 357, 374, 385
  of hydroxide bands, 42, R285
Itano, M., R502
ITT Laboratories, 184
Ivanov, D. N., 325
I.W., interference wedge

j, inner quantum number, 38f
J, rotational quantum number, 40f
Jackson, R. K., R165
Jacob, G., R399
Jacobson, S. D., R153, R500, R663
James, A. H., R527
James, C. G., R169
Janke flame photometer, R271, R720
Jansen, W. H., 8, 9, 339
Jarrell-Ash Co., 428, R396
  absorption flame spectrophotometer, 428
Jenness, R., R190
Jet, see Nozzle
  cooling in, 108
Jet atomizer, 105
Jeuken, M. E. J., R68
Jobin-Yvon, 429
Johns, H., R532
Johns, J. W. C., 357
Johnson, H. O., R537
Johnson, L. H., R83
Johnson, W. C., R198
Jones, F., R314b
Jones, H. C., R414
Jouan, Établissements, 429, R404
Joule–Thomson effect, 108
Juice, analysis of, 337
Jung, J., R616
Junkes, J., 371, 380, R8, R9
Jurbergs, K. A., R400

k, detection limit with red-sensitive photomultiplier, 260
K, kerosene
Kampa, P., R295, R431
Kanroji, Y., R383, R463
Kaolin, see Clay
Kaplan, B. Ya., R387
Kaplow, L., R658
Karl, H. L., R435
Kaskan, W. E., 381
Kassner, J. L., R129
Kato, I., R274

Kavanagh, R. W., R538
Kawasaki, N., R693
Keddy, R. J., R681
Keenan, P. C., 381
Kelley, M. T., R240
Kel'tseva, Z. A., 325
Kendall, K. K., 383, R726
Keyes Scientific Co., 429
Khanaev, E. I., R315
Kinetics, of energy transfer, 51
  of ionization, 35, R423
  of recombination, 35
King, P. J. T., R256
Kinnear, J. E., 372, 381
Kipp en Zonen, P. J., 429, R420
Kirchhoff, G. R., 7
Kirchhoff's law, 49
Kleiman, H., 381
Klemperer, R. I., 8
Knall, E., R372, R373, R375
Knauer, H. E., 380, R202
Knewstubb, P. F., R697
Knickmann, E., R50
Knippenberg, E., R589, R590, R591
Kniseley, R. N., 215, 370, 380
Koch, H. J., R52
Koch, W., R733
Konen, H., 381, R26
Konkoly-Thege, I., R578, R579, R580, R582, R582b
Kononenko, L. I., R560
Konopicky, K., R623
Kostyra, H., R332
Koyama, T., R693
Kramer–Tisdall method, R475a
Krauss, B., R639
Krotinger, N. J., R745
Kuhns, D. M., R746

l, azimuthal quantum number, 38f
Laboratory, 199f
Lady, J. H., 380, R208, R210
Lagerqvist, A., R374, R376
Laminar flames, 14ff, 127
Laminar flow, 18f, 117, 119
Landi, M. F., R156
Landolt, P. E., R368a
Lang, K., R20
Lang, W., 354
Lange, B., 9, 429, R443
  flame photometers, 135, 429, R443, R594

Langmuir equation, 24, 27
Langmuir torch, 76
Lanthanides, see Rare earths
Lanthanum, detection limits of, 265
  determination of, 383, R381, R489, R490, R546, R547
  in uranium, 249
  enhancement of, 265, R431a
  interferences on, 315f
  in organic solvents, 265
  as releasing agent, 309, R754
Lanthanum oxide (LaO), spectrum of 221, 265, 394f, R546, R547
Lasker, S. E., R269
Lauer, G. S., R561
Lavèn, W. J., R69
Lawton, K., R401
Lax, E., R3
Leach, B. E., R688
Lead, chemiluminescence of, 216, 222, 266, R283
  detection limits of, 266
  determination of, R417
  by absorption flame photometry, 354
  in gasoline, 364, R101, R300, R402, R481, R661a
  in lubricating oil, 366
  in plating bath, R725
  in steel, R499
  emission and height in flame, 132
  flame spectrum of, 216, 222, 399, R499, R726, R734
  interferences of, 312, 314, 315, 317
  ionic spectrum, 399
  ionization in flame, 298, R697
  in organic solvents, 266
  in oxycyanogen flame, 266
  separation of, 250
Lead oxide, spectrum of, 222, 399
Leakage current, 182f
Leaks, in compressed-gas connections, 84, 85
Leaves, analysis of, 335f, R34, R730
  for calcium, 336, R730
  for potassium, 336, R631, R730
  rubber, analysis of, R149a
Léger, L., R220
Leĭderman, Ts. A., R561
Lens, 134ff, 181
Leonard, 8
Level of sample, effect of, on flow rate, 21f

INDEX

Liddell, W. J., R199
Light, *see* Emission, Photon, Radiance, Radiation, Stray light
  detection of, *see* Photodetectors
Light-gathering power, *see* Luminosity
Light trap, 134
Lilleland, O., R165
Limestone, analysis of, R192
Limiting sensitivity, *see* Detection limit
Line (spectral), advantages of, over band, 229
  –background ratio, *see* Analyte-to-background ratio
  broadening of, 45f, R356, R372, R538, R638
  choice of, 226ff
  net intensity of, determination of, 288ff
  net reading of, automatic, 187f
  origin of, 36ff
  profile of, 45f, 178f, R666, R667, R668
    recording of, 189
  response to, as function of slit width, 176, 178f
  reversal of, 73, R194, R381, R665
  slit function of, 178f
  wings of, 288
Linear dispersion, 173, 176
Linearity, of photocells, 144f
  of photodetector, 59, 228
  of phototubes, 145
  in spectrography, 198, R180a
  testing of, 154f
Line-reversal, 73, R194, R381, R665
Line spectra, *see* Atomic spectra, Line (spectral)
Lippincott, E. R., 381
Liquefied petroleum gas, 80
Liquor, *see* Whiskey, Wine
Lithium, analysis of, for potassium and sodium, 368, R380
  concentration of, in flame, 217, R394, R660
  cross-sensitivities with, 282
  detection limits of, 266
  determination of, R218, R263, R273, R361, R387, R416, R572, R577, R689, R718, R738
    in beryllium solutions, R506
    in biological fluids, R269, R270
    in biological material, R597
    in boron, R238
    in brine, R399
    in catalysts, R756
    in cement, 363, R473
    in glass, R164, R326, R427, R604, R743
    instrument for, R164a, R269, R270, R320a, R344
    in iron, R435
    in magnesium alloys, 368, R599, R679
    in minerals, R129, R166, R368a, R560, R699, R743
    in nuclear materials, R238, R540
    in petroleum oil, R198
    in petroleum wash water, R756
    in rocks, R247, R365, R427, R735
    in salts, R399
    in silicates, R214, R219, R247, R761
    in water, R201, R463
    in whiskey, R566
  enhancement of, 266, R431a
  excitation energy of, 34
  filters for, 141
  flame spectrum of, 216, 266, 395, R726, R734
  height in flame, effect of, on emission of, 32, 161
  hydroxide formation, *see* Lithium hydroxide
  interference of, 312, 313, 315, 317, 318
  interferences on, 316
    by acids, R263
    by aluminum, 368
    by anions, R263
    by calcium, 316, 363, R201, R633
    dissociation, R263
    ionization, R263, R268b
    by magnesium, 368
    by strontium, 316, 363, R201
  as internal standard, 161, 162, 163, 165f, 167f, 327, R139a, R312, R464
  internal standardization of, 166
  ionization of, in flame, R127, R268b, R660
  ionization potential of, 34
  isotopic analysis of, R476
  line profile of, R668
  as normalizer, 326f

Lithium—*contd.*
  in organic solvents, 266
  in oxycyanogen flame, 266
  as tracer, 355
Lithium alloys, analysis of, 368, R599
Lithium hydroxide, dissociation energy of, R698
  formation of, in flame, 32, 217, R169, R263, R394, R660, R695
Littrow prism, 174
Lizarduy, M. L. Rexach-M. de, R177
Load resistor, 145, 158
Lockyer, N., 8
Logarithmic slope, 322f
Logarithmic working curve, *see* Working curve
Log-log translation, 295f, 321, R274
Longuevalle et Le Moan, S., R234
Lord, S. S., Jr., R140a
Lorentz broadening, 45, R356, R372, R538, R666, R667, R668
LPG, 80
Lubricating oil, analysis of, 365f, R311, R496
  determination of additives in, R198
  spectrogram of, 366
Lücke, K. E., R430
Luminosity, and detection limits, 275
  of interference wedge, 143
  of monochromators, 173, 175f, R389, R390, R541
  of optics, 129f
  of spectrograph, 196, R408
Luminous flame, 17, 126, 215, 243
Lundegårdh, H., 8, 9, 97, 104, 107, 115, 161, 334ff, 339, 360, 382
Lutetium, detection limits of, 266
  determination of, R546, R547
  in hexone, 266
Lutetium oxide, spectrum of, 221, 266, 395, R546, R547
Lykken, L., R241, R537
Lyutyi, A. I., 380

m, multiple
M, methanol
M4Q monochromator, 180
Machin, K. E., R585
MacIntyre, I., R474, R475, R475a
Mack, D. J., R485
Magill, P. L., R178
Magnesite, analysis of, R155

Magnesium, absorption flame photometry for, 48
  chemiluminescence of, 216, 219
  detection limits of, 266
  in absorption, 48
  determination of, R255, R259, R273, R379, R397, R416, R456a, R456b, R509, R578
  by absorption flame photometry, R72, R203a
  in alfalfa, R466
  in aluminum alloys, R379
  in biological fluids, R138, R757
  in biological materials, 353f, R279, R597, R688
  in blood plasma, R153, R203b, R652, R708
  in blood serum, 354, R346, R411, R617b, R700a
  in cement, 362f, R104, R192, R745
  in ferrous metals, R59
  in fertilizer, R151
  in food, 354
  in glass, R605
  in grain, R151
  instrument for, R475, R709, R710
  in iron, R435
  in limestone, R192
  in milk, 339, R617b
  in nutrient solution, 359
  in pharmaceuticals, R314, R351
  in plants, R165, R203a, R415, R416b, R429, R550, R617, R649
  in refractories, R499
  in seeds, R151
  in silicates, R150
  in slag, 368, R230, R379
  in soil, R151, R256, R415, R416b, R611a, R617, R676
  in tissues, R138, R221, R708
  in urine, 354, R153, R237, R617b
  in water, R460, R484
  enhancement of, 266, R431a, R688
  excitation potential of, 34
  flame spectrum of, 216, 218f, 266, 396, R726, R734
  in incandescent oxyacetylene, 266, R429
  interferences of, 312ff
    on iron, 225
    on lithium, 316, 368
    on strontium, 318, R416c

Magnesium—*contd.*
  interferences on, 316, 321, 363, R138, R578
    releasing agents for, 321
  ionic spectrum of, 218f, 396
  ionization potential of, 34
  in organic solvents, 266
  oxine precipitation of, 250, R203b, R652
  in oxycyanogen, 266
  as releasing agent, 360, R501
  self-reversal of, 58
  separation of, 250, R203b, R652
  working curve of, 58
Magnesium alloys, analysis of, 368, R599, R679
Magnesium chloride (MgCl), spectrum of, 396
Magnesium fluoride (MgF), spectrum of, 396
Magnesium hydride (MgH), spectrum of, 370
Magnesium hydroxide (MgOH), spectrum of, 219, 396, R286
Magnesium oxide, spectrum of, 219, 396, R286
Magnetic shielding, 147f
Maher, J., 379
Mainka, E., R243
Malissa, H., 223, 383
Mallikarjuneswara, V. R., R149a
Mallinckrodt, H. v., R149b
Maloney, C. M., R687
Mandel'shtam, S. L., R182a
Manganese, chemiluminescence of, 51, 216
  detection limits of, 266
  determination of, R174, R230, R379, R381, R416
    by absorption flame photometry, R72a
    in aluminum alloys, 250, R212
    in biological materials, R597
    in cement, R226, R266a
    in ferrous alloys, R176
    in gasoline, 364, R661a
    in glass, R605
    in mineralized salts, R291
    in nutrient solution, R107
    in plants, R415, R416a
    in soil, R144, R415
    in steel, R499, R733, R737

  enhancement of, 266, R431a, R582d
  flame spectrum of, 216, 224, 266, 396, R726, R734
  interferences of, 312ff
  interferences on, 316f, R176
  ionic spectrum, 396
  in organic solvents, 266
  in oxycyanogen flame, 266
  in oxyhydrogen flame, 266, R238a
  separation of, 250
  spectral interference of, R416a
Manganese fluoride (MnF), spectrum of, 396
Manganese hydroxide (MnOH), spectrum of, 397
Manganese oxide, spectrum of, 224, 396
Manometers, 88f
Mantle, *see* Outer cone
Manufacturers, 68, 427f
Marcy, V. M., R640
Marggraf, A. S., 6
Margoshes, M., 371, 381, R712
Margraf, H. W., R729
Marine organisms, analysis of, R189
Marius, 429, R458b
Marshall, 8
Mashiko, Y., R383
Mass spectrometry, applied to flames, 35, R424, R425
Mathers, A. P., R566
Matheson Co., Inc., 77
Mathis, W. C., R427
Maurer, H., R370
Maurer photomultiplier, 147
Mavrodineanu, R., 76, 372, 382, R498, R499
McDonald, F. G., R143
Mean, 328
Measurement, analytical, *see* Indirect methods of analysis, Standardization
  adjustment method of, 239
  bracketing method of, 233, 240f
  combination (flame and chemical), 247ff
  errors of, *see* Error
  with overlapping emissions, 241ff
  quotient method of, 239f
  sequence of, 238ff
  speed of, 277
  statistical treatment of, 256f
  technique of, 238ff

618 INDEX

Medical analysis, 339ff; *see also* Clinical analysis
Meggers, W. F., 371, 373, 382, R641, R642
Meites, L., 381
Méker, G., 119
Méker burner, 16, 18, 118f
 cleaning of, 121
 elongated, 118, 119f
 flame of, 127
 heat transfer in, 72
 temperature distribution in, 73, 74
 turbulence in, 126
Melvill, 6
Mengel, K., R617, R617a, R617b
Menis, O., 372, 380, 382, R252
Merck, E., A. G., 238, 345
Mercury, absorption spectrum of, 219
 chemiluminescence of, 216, 220, 266
 detection limits of, 266
 determination of, by absorption flame photometry, 354
 flame spectrum of, 216, 219f, 266, 393
 in microwave discharge, 76
 in organic solvents, 266
Merritt, L. L., Jr., R56
Metallurgical analysis, 367f, R41a, R59, R503; *see also* Alloys, Aluminum, Brass, Copper alloys, Ferrous metals, Iron, Lithium, Magnesium, Steel, Tungsten
Meter, 152ff
 for alternating-current systems, 185f
 linearity of, 154f
 null, 157ff, 164
 overloading of, 158f
 for photocells, 145, 154
 reading of, 328f; *see also* Measurement
 sensitivity control of, 157
 time constant of, 150f, 153
Methane flame, burning velocity of, 77
 temperature of, 77
Methylidene, for determination of carbon, 247
 spectrum of, 54, 222, 370, 386
Metrohm A.G., 429, R492
Metzger, R., R95
Mezhericher, E. M., R666, R667
Microanalysis, 113ff, 253ff, R14, R24, R52, R93, R173, R175, R340, R379, R420, R471, R576, R584, R669; *see also* Ultramicroanalysis

atomizers for, 113ff
in biological applications, 340
integration for, 155, 168
scanning for, R724
spectrographic, R180a
Microatomizer, 113ff, 253f
Microphone effect, 147, 149, 150
Microwaves, in flame research, R535, R662
Microwave torch, 76
Milatz, J. M. W., R66, R148
Milk, analysis of, 337, 339
 for calcium, 339, R323, R413, R506a, R617b, R639
 for magnesium, 339, R617b
 for potassium, R190, R323, R413, R506a, R617b, R639
 for sodium, R190, R323, R413, R506a, R617b, R639
Miller, J. H., R528
Millikan, R. C., 381
Millimicron, 37
Millner, T., R325
Minerals, analysis of, 360; *see also* Clay, Feldspar, Magnesite, Ores, Silicates, Spodumene
 for calcium, R501
 for cesium, R255a, R743
 for lithium, R129, R166, R368a, R560, R743
 for potassium, R562, R743
 for rubidium, R743
 for sodium, R562, R743
 for strontium, R216, R314c, R561
Mineral water, analysis of, R383, R463
Mirror, for flame, 133f, 160, 181, 274
 semitransparent, for double-beam photometry, 164, 191f
Mirror analysis, 335
Mirror galvanometer, *see* Galvanometer
Mirror scale, 153
Mist, *see* Aerosol, Atomization, Droplets (spray), Drop size distribution
Mitchell, H. L., R97
Mitchell, R. L., 8, R34
Mitscherlich, A., 97, 382
Mixing chamber, for fuel gas, 117
Mode, 328
Modular design, 169f
Modulated light, *see* Alternating current, Chopped beam

Molecular bands, *see* Band (spectral), Molecular spectra
  working curves of, 58
Molecular dissociation, *see* Dissociation
Molecular spectra, *see* Background radiation
  atlas of, R9, R41
  data on, 382, R10, R19, R41
  identification of, 382, R41
  intensity of, 44
  isotope shifts of, 42
  origin of, 39ff
  tables of, 370ff
Molecules, dissociation energy of, R10
  excitation of, 39ff
  formation of, in flame, 28ff
Molybdenum, chemiluminescence of, 216, 266
  continuum of, 42, 266
  detection limits of, 266
  determination of, indirect, 247, R452
  flame spectrum of, 10, 215, 216, 224, 266, 397
  interferences of, 314f, 317f
    on alkaline-earth metals, 303
    on calcium, 313, R452
    on strontium, 318, R452
  in microwave discharge, 76
  in organic solvents, 266
  in oxycyanogen flame, 224, 266
  in reaction zone, 121, 224, 266
Monitoring, *see* Flame monitor, Process control
  of interferent, 325
Monochromatic emission, *see* Line (spectral)
Monochromatic wedge filter, 143
Monochromator, 170f, R470
  dispersion of, 172ff
  luminosity of, 175f
  requirements of, 176f
  resolving power of, 173f
  slit width of, *see* Slit width
Monroe, R. J., R478
Montgareuil, P. Guérin de, R497, R645
Montgomery, D., R731
Moore, C. E., 371, 382
Morgan, F., 382
Mosher, R. E., R757
Moss, N., R411
Motor oil, *see* Lubricating oil

Moving-coil galvanometer, *see* Galvanometer
Mulder, H., R323
Müller, G., R436
Mulliken bands, 386
Multicomponent analysis, with overlapping spectra, 241ff
Multiple-beam flame photometry, 159f, R160, R457, R458
  commercial equipment for, 430
  history of, 9
  with internal standardization, 164f, 192, R344
  optimization of gas flow for, 96
Multiple-beam flame spectrophotometry, 187f, R709, R710, R712
Multiplet tables, 382
Multiple-use flame photometer, 169f
Multiplicative interference, 283f, 295ff, 302, 326, R274
  correction of, 321ff, 326f
Multiplier phototube, *see* Photomultiplier
Mutual interference, 65, 294; *see also* Specific interference
Myers, G. B., R153, R500, R502, R663
Myers, R. G., 215, 370

n, nonaqueous solvent (Tables 10 and 14)
n, principal quantum number, 38
N, naphtha
Nakata, S., R649
Nanometer, 37
Naphtha, as solvent, 216
Narrow-band amplifier, 185f
Nash, J. B., 383
National Instrument Laboratories, Inc., 429
National Spectrographic Laboratories, 76
Nayyar, S. N., R313
Nebulizer, *see* Atomizer
Nell, J. P., R680
Nelson, W. P., R568
Neodymium, detection limits of, 267
  determination of, R546, R547
  in hexone, 267
  interference of, with lanthanum, 315
Neodymium oxide (NdO), spectrum of, 221, 267, 397f, R546, R547
Nessim, N. E., R569a, R570

Nest, A., *see* Electronest
Netheler und Hinz G.m.b.H., 120, 136, 238, 282, 429, R513
Netsch, Gebrüder, 429, R514
Neubauer, 336
Neuber, R., R172
Neufeld, A. H., R272, R366
Neugebauer, J. J., R327
Neutron dosimetry, R238
Newman, K. J., R597
NH, *see* Imidogen
NH$_2$, *see* Amidogen
Nickel, chemiluminescence of, 216
  detection limits of, 267
  determination of, R174
    by absorption flame photometry, 354
    in aluminum alloys, 250, R212
    in biological material, R597
    in plating solution, R268
    in steel, R499
  enhancement of, 267, R431a
  flame spectrum of, 216, 225, 267, 398f, R726, R734
  interferences of, 312ff, 317
  interferences on, 317
    spectral, 225
  in organic solvents, 267
  in oxycyanogen flame, 267
  separation of, 250
Nickel oxide, spectrum of, 225, 399
Niedermeier, W., R143
Nikonova, M. P., R560, R561, R561a, R562
N.I.L. flame photometer, 429
Niobium, continuum of, 267
  detection limits of, 267
  flame spectrum of, 10, 215, 223, 267
  in incandescent oxyacetylene, 267
  in microwave discharge, 76
Nitrate, interferences of, 66, 301, 312ff
Nitric acid, for wet ashing, 343
Nitric oxide, band spectrum of, 223, 243, 267, 373, 397
  continuous spectrum of, 54, 223, 244, R392
Nitrogen, detection limits of, 267
  determination of, by copper, R97
    by cyanogen bands, 244, R362
    in organic compounds, 244

emission of, in reaction zone, 121, 397
flame spectrum of, 244; *see also* Cyanogen, Imidogen, Nitric oxide
nm, nanometer, 37
Noise, *see* Flame flicker, Fluctuation, Shot noise, Signal-to-noise ratio
  of dark current, 182ff
  electronic, 149, 150
  of flame, 19f, 122, 128, 180
  frequency spectrum of, 258
  of gas discharge, 183
  of leakage current, 182f
  of photocurrent, 185f
  photometric, frequency spectrum of, 185
  photometric, limitation of, 149, 185
  of photomultipliers, 182ff
  of zero, 186
    in alternating-current systems, 185f
Nonaqueous solvents, 208ff; *see also* Organic solvents
  detection limits in, 260ff
Nonequilibrium, *see* Equilibrium, departures from
Nonionic detergent, 214
Non-Ion-Ox, 214
Nonlinearity, *see* Working curve
  of meters, 152ff
  of photocells, 144f
  of photodetectors, 59, 228
  of phototubes, 145
  in spectrography, 198, R180a
  testing of, 154ff
Nonmetals, determination of, 243ff
Nonpneumatic atomizer, 100, 101ff, 105, 114
Nonspecific interferences, in atomizer-burners, 327
  classification of, 61, 62f
  definition of, 62
  detection of, 319
  diagnosis of, 67
  elimination of, 322, 325ff
  in internal standardization, 167
Nonvolatile compounds, *see* Vaporization interference
Nordberg, M. E., R604
Norelco flame photometer, 429, R519

Normal error distribution, 329
Normalization, of detection limit, 258
  for eliminating interferences, 321
  by flame electrons, 299
  by internal standard, 299, 327
  for ionization interference, 298, 361
  by lithium, 167, 326f
  for nonspecific interference, 326
Normalizer, 298
North American Aviation, Inc., 380
North American Philips Co., Inc., 429, R519
Nozzle, atomizer, in atomizer-burner, 122, 124
  cleaning of, 111
  cooling in, 108
  design of, 104, 114
  fabrication of, 99f
  and flow measurement, 87f, 93
  incrustation of, 111
  and shock waves, 106
  standardization of, 87f, 93
  tangential injection in, 104
  velocity in, 106
Nozzle, gas, 116f
  in atomizer-burner, 122, 124
Nuclear materials, analysis of, 356f
  for lithium, R238, R540
Null meter, 157ff, 164
Null reading, for absorption flame photometry, R454a
  with color wedge, R721
  in potentiometric circuits, see Potentiometric photometric circuits
  time constant with, 153f
Nutrient solutions, analysis of, 359, R107
Nutrition, 337ff
Nuts, analysis of, 338

O, see Oxygen, atomic
OA, oxyacetylene
OAz, incandescent oxyacetylene, 384
OB, oxygen–butane
Observation height, see Height in flame
Oc, 2-octanone
OC, oxycyanogen flame
Odler, I., 248
OG, oxygen–city gas
OH, oxyhydrogen
OH, see Hydroxyl

Ohyagi, Y., 382
Oil, see Fuel oil, Lubricating oil, Petroleum
OP, oxygen–propane
Operation, 200ff
Opium, analysis of, 369, R91
Optica, 430, R529
  absorption flame spectrophotometers, 430
Optical axis, see Entrance optics
Optical cross-section, R356
Optical feedback, 182f
Optical null balance, R721
Optical transmittance, see Transmittance
Optics, 129ff, R318, R319; see also Entrance optics, Exit beam, Mirror, Monochromators, Spectral dispersion
  for filter flame photometers, 134ff
  for flame spectrophotometers, 187ff
  for internal standardization, 163ff
  luminosity of, see Luminosity
  transmittance of, 130
Optische Werke, 430, R530
Orechkin, D., R756
Ores, analysis of, R516, R560, R561, R699, R743
Organs, see Tissues
Organic acids, determination of, 248, 359
Organic compounds, analysis of, for boron, R168
  for chlorine, R461
  for nitrogen, 244, R362
  determination of, direct, 247
  indirect, 247
  interferences of, 65f, 312ff, R182
Organic solvents, background spectrum of, 55
  detection limits in, 260ff
  effects of, 312ff, R71, R582d
  on barium, 208, 313, R202
  on drop size, 125, 209
  on flame temperature, 73, 128, 209f
  on sodium, 317, R81a
  in flame photometry, 208ff, R212a, R544, R661
  theoretical studies of, 209f
Orr, C., Jr., R203
Oscillator strength, 58

Oscillographic spectroscopy, 278, R343a
Osmium, in flame, 225
  in microwave discharge, 76
  oxycyanogen flame spectrum of, 10, 399
Otozai, K., R274, R275, R276
Otto, P., R542
Outer cone, 14, 18f
  background radiation of, 54
  chemiluminescence of, 126
Ovenston, T. C. J., R598, R599
Overflow regulator, 90f
Overlap, see Spectral interference
Oxidant, see Compressed air, Oxygen
  ratio to fuel, see Fuel–oxidant ratio
Oxides, band spectra of, 42, R9, R285, R286
  equilibria of, in flame, R127
  formation of, in flame, 32
  and fuel–oxidant ratio, 93
Oxine, for aluminum, 220, 256
  for precipitations, 250
Oxyacetylene, background intensity of, 70, 79
  burning velocity of, 77
  carbon line in, 55
  detection limits in, 261ff
  flame spectrography in, 381
  flame spectrum of, 52
  gas consumption by, 97
  impurities, effect of, on, 79
  incandescent, for magnesium, R429
  incandescent, spectra in, 215, 243, 386, 401, 402
  reaction zone of, emission of, 216, 243
  temperature of, 72, 74, 77, R338
  as function of height, 74
Oxycyanogen flame, background spectrum of, 55
  burner for, R278a
  detection limits in, 261ff
  dissociation in, 33, 218
  elements determinable in, 10, 261ff
  for flame photometry, 76f, 381, R84, R85, R85a, R304, R470, R711, R713, R714, R714a
  flame spectra in, 218ff, 222ff, 261ff
  ionization in, 218
  pressure effect on, 77
  sample flow rate, effect of, on, R278b

  spectra in, 373, 385, 386, 388, 389, 392, 396, 397, 399, 400, 404, 405, 406, 407
  temperature of, 76, R199
Oxygen, atomic, concentration of, in hydrogen flame, R171
  reaction of, with nitric oxide, 54, R392
  recombination of, 32, 54
  band spectrum of, 54
  consumption rate of, 96f
  dissociation of, in flame, 72
  for flame photometry, 81
  flow optimization of, 93ff
  hazards of, 84
  nozzle, see Nozzle, atomizer
Oxyhydrogen flame, background intensity of, 53, 70
  burning velocity of, 77
  detection limits in, 261ff
  gas consumption by, 97
  sample flow rate, effect of, on, R278b
  size of, R238a
  temperature of, 77, R238a
Ozone–cyanogen flame, 77

p, resonance line, 383
P, isopropanol
Padley, P. J., 382
Paech, K., R22
Paige, H. H., R700
Paint, analysis of, R200
Palladium, continuum of, 226
  detection limits in, 267
  flame spectrum of, 225f, 267, 399f
  in hexone, 267
  interferences of, spectral, 225f
  in oxycyanogen flame, 10, 267
  separation of, 250
  working curves of, 226
Palmby, A. K., R661a
Paper, for sample introduction, 98
Paper pulp, analysis of, 359, R106, R400
Paraffin, for coating containers, 213
Parallax, 153
Parallel method, 309, 321
Parametric method, 233f, 283ff, 320, 345
  for ionization interferences, 296f
Parenterals, analysis of, R351
Parker, A. E., 382

Parks, J., R735
Parrish, D. B., R312
Particle size distribution, R203
Patterson, S., R55
Patwin flame photometer, 429
PC, photoconductive cell
Pearse, R. W. B., 382
Pease, R. N., R32
Pedology, 335
Pellet, H., 7, 97
Pennsalt Chemicals Corp., 382
Pepkowitz, L. P., R405
Perchloric acid, interferences of, 312ff
  on calcium, 66, 301, 304, 313
Perchloryl fluoride, for flame photometry, 382, R622a
  spectra in, 385, 387, 390, 396, 404
Period, see Time constant
  of meter, 153
Periodic acid, interference of, with calcium, 301, 313
Perkin-Elmer Corp., 430, R539
  absorption flame spectrophotometer, 430
  flame attachments, 430
  flame photometer, R143, R224, R236, R510, R537
    Model 52A, 191
    Model 146, 430, R539
Perrhenate, interference of, with potassium, 315, R482
Peterson, M. H., R428
Petri dish, as sample container, 110
Petroleum, see Fuel oil, Lubricating oil
  analysis of, 363ff, R265, R412, R734
  determination of additives in, R198
  wash water, analysis of, R756
Pfab, B., R328, R329, R331, R332
PH, perchloryl fluoride–hydrogen flame
Pharmaceuticals, analysis of, 359, R159
  for alkali metals, 359, R234a, R314, R351
  for alkaline-earth metals, 359, R314, R351
Phase-sensitive rectification, 185f, 187
Philip, T. V., R485
Phillips, M. L., R459
Phosphate, determination of, by flame-photometric titration, 253
  indirect, 245f, 248, R452
  in rock, R229

for flame-photometric titration, 252f, R704
interferences of, 66, 312ff
  on alkali metals, 302
  on alkaline-earth metals, 300ff, R637
  buffering of, 309, 353, R199a
  on calcium, 245f, 300ff, 304ff, 307, 308f, 313, 344, 360, R70, R258, R276a, R337, R430, R450, R452, R454, R754
  elimination of, 308ff
  protein effect on, 344
  releasing agents for, see Releasing agents
  on strontium 309, R215, R452, R454
  and sulfate, on calcium, 301f, 306
  separation of, 325
Phosphine, in acetylene, 79, 306
Phosphor, analysis of, for potassium and sodium, 357, R205, R292
  for silver, R587
  radioactive, as reference, R464
Phosphoric acid, analysis of, for calcium, 358f, R154
  for titrating calcium, 252
Phosphorus, in blood serum, 235
  continuum of, 66, 223, 244, 268, 302
  detection limits of, 268
  determination of, 244, R163; see also Phosphate
  indirect, 245f
  in rock, R229
  interferences of, see Phosphate
  interferences on, 317f
  in microwave discharge, 76
Phosphorus oxide (PO), spectrum of, 223, 243, 244, 268, 399
Photocathode, 145, 146
  illumination of, 149
Photocells, 144f
  amplifier for, 149ff
  fatigue of, 148
  galvanometer for, 154
  temperature effects of, 145
Photocurrent, see Photodetectors
  amplification of, see Amplifiers
  measurement of, see Measurement, Meters
  noise of, 185f
  saturation of, 145

Photocurrent—*contd.*
  and slit width, 176
  subtraction of zero from, in alternating-current systems, 186
  voltage dependence of, in photomultiplier, 182
Photodetectors, 143ff, R320; *see also* Photocells, Photomultipliers, Phototubes
  alignment of, 133, 149
  dark current of, *see* Dark current
  in double-beam systems, 164
  fatigue of, 145, 146
  imaging on, 135f, 149
  magnetic shielding of, 147f
  in monochromators, 181ff
  mounting of, 147
  nonlinearity of, 59, 144f, 155
  power received by, 176, 177f
  spectral sensitivity of, 138, 145
  thermal effects in, 145
Photoelectric cell, *see* Phototube
Photoelectric effect, R46
Photoelectrons, 145
Photoemissive tube, *see* Phototube
Photographic plate, 196f, R320
  characteristic curve of, 194, 198, R180a
Photography, 192ff
Photometry, R28, R318, R320; *see also* Alternating current, Electric circuitry, Measurement, Meter, Multiple-beam flame photometry, Photodetectors, Potentiometric photometric circuits, Sensitivity
  direct-reading, 152, 158, 159
  for internal standardization, 163f
  linearity testing in, 154f
  optical null-balancing, R721
  photographic, 194, 197f
Photomultipliers, 145ff, 182ff, R148, R250
  amplifiers for, 149ff
  cooling of, 184, 187
  counting by, 187
  dark current of, 148f, 182ff
  discharges in, 182f
  drift of, 148f, 183, R505
  FW-118, 184
  imaging of flame on, 149
  noise of, 182ff
  photocurrent gain function in, 182
  power supplies for, 151
  RCA 1P28, 182ff
  red-sensitive, 147, 148f, 184, R320a
    detection limits with, 260ff
  voltage of, 182ff
Photons, counting of, 187
  emission of, 37ff
Photoreceiver, *see* Photodetectors
Phototubes, 145, 181f, R58
  Beckman, 273
  electric circuitry for, 144f, 149ff
  fatigue of, 148
  gas-filled, 145
  spectral sensitivity of, 147
Photovoltaic cells, *see* Photocells
Physiological analysis, *see* Biochemical analysis, Biological materials, Clinical analysis
Pickering, W. F., 222
Pinta, M., 372, 382
Pipets, 211
Planck's constant, 37
Plant analysis, 335f, R549; *see also* Alfalfa, Hay, Leaves
  books on, R22, R34, R39
  for calcium, R165, R293, R337, R385, R415, R430, R550, R570, R616, R617a, R625, R649, R680
  for copper, R203a, R465
  for iron, R203a
  for magnesium, R165, R203a, R415, R416b, R429, R550, R617, R649
  for manganese, R415, R416a
  for potassium, R165, R401, R468, R479, R550, R569a, R617a, R649
  for rubidium, R312
  for sodium, R165, R468, R550, R569a, R644
  for strontium, R416c
  for zinc, R203a
Plantinga, O. S., 382
Plant nutrients, 335
Plant nutrient solutions, analysis of, 359, R107
Plant physiology, 335
Plasma, *see* Blood plasma
Plasma jet, 76, R458a
Plastic containers, 212f
Plasticizer, contamination from, 212f
Plate, *see* Photographic plate

Plate fog, 194
Plating solutions, analysis of, R268, R725
Platinum, chemiluminescence of, 216, 268
  detection limits of, 268
  flame spectrum of, 216, 225, 226, 268, 400
  in microwave discharge, 76
  in oxycyanogen flame, 268
Platinum wires, for sample introduction, 7, 97, R495, R585
Plugging, see Clogging
PM, photomultiplier
PMQII flame spectrophotometer, 431, R758
Pneumatic atomizer, 103ff, R540a; see also Atomizer
PO, see Phosphorus oxide
Poiseuille law, 22
Poisonous samples, venting of, 199
Polkinhorne, H., R197
Pollack, 8
Polonium, 224
Polychromator, 170, 187f
Polyethylene containers, 212
Polystyrene containers, 212
Portland cement, see Cement
Positive-displacement pumping, see Controlled flow
Possidoni de Albinati, J. F., 249
Pot., potentiometric
Potassium, biological sampling of, 346
  in blood, 346ff
  in blood cells, 346
  concentration of, in flame, R660
  continuum of, R71, R395
  cross-sensitivities with, 282
  detection limits of, 268
  determination of, R82, R218, R273, R288, R332, R340, R368, R381, R448, R449, R537, R563, R572, R577, R607, R689, R691
    by absorption flame photometry, R454a
  in alfalfa, R466
  in aqueous humor, R451a
  in ash, R583, R606, R635
  in beryllium solutions, R506
  in biological fluids, 345ff, R128, R133, R134, R269, R270, R664a, R673, R750, R757
  in biological materials, R279, R597, R723
  in blood cells, 346, R378, R533
  in blood plasma, 346f, R136, R137, R153, R271, R422, R500, R542, R671, R708
  in blood serum, 346f, R64, R93, R94, R95, R105, R123, R131, R132, R136, R137, R149, R149b, R235a, R248, R272, R315a, R339, R342, R354, R378, R406, R418, R419, R420, R436, R510, R524, R531, R542, R545, R553a, R568, R569a, R588, R611, R617b, R651, R654, R661, R663, R671, R741, R747
  in boiler scale, R640
  in brine, R399
  in catalysts, R756
  in cement, R74, R191, R222, R244, R266, R266a, R308, R309, R317, R565
  in ceramics, R619, R716, R766
  in cerebrospinal fluid, R137
  in cheese, R323
  in clay, R78, R266, R334
  in coal, R388
  in fertilizer, R151, R158, R267, R289, R462, R552, R614, R615
  in gastric juice, R135
  in glass, 361, R102, R164, R193, R266, R326, R331, R335, R427, R604, R605, R706, R743
  in grain, R151
  in hay, R523
  instruments for, R64, R111, R140, R143, R157, R269, R270, R271, R320a, R339, R344, R360, R458, R586, R589, R590, R591, R594, R669, R709, R710, R721, R728, R735a, R755
  in leaves, 336, R149a, R631, R730
  in lithium, 368, R380
  in milk, R190, R323, R413, R506a, R617b, R639
  in minerals, R562, R743
  in nutrient solution, 359
  in paper pulp, R106
  in petroleum oil, R198, R412
  in petroleum wash water, R756
  in pharmaceuticals, R314, R351

Potassium—*contd.*
  in phosphors, 357, R205, R292
  in plants, R165, R401, R468, R479, R550, R569a, R617a, R649
  in refractories, R141, R295, R498, R499, R592, R606
  in rocks, R315, R317, R427, R532, R742
  in salts, R399, R552
  in seeds, R151
  in silicates, R214, R219, R309, R445, R761, R763
  in sodium, R511
  in sodium salts, R264a, R511
  in soil, R144, R151, R256, R281, R309, R317, R478, R479, R480, R494, R521, R569a
  in sugar solutions, R96
  in tissues, R133, R221, R313, R436, R531, R708
  in tungsten, R327
  in urine, R124, R271, R272, R315a, R355, R364, R422, R500, R568, R611, R617b, R624, R741, R747
  in viscous samples, 162, R96
  in water, R309, R463, R569a, R624, R640, R693, R731
  in wine, R753
  dissociation interference on, 31, 294
  enhancement of, 268, R431a, R653, R654, R655, R656
  excitation potential of, 34
  filters for, 138ff, 141
  flame spectrum of, 216f, 268, 394, R726, R734
  hydroxide of, formation in flame, 32, 217, R660, R662
  interferences of, 312ff
    ionization, 294ff, 361, R268b, R575, R603
    on manganese, R416a
    on sodium, 297, 317, 361
  interferences on, 315, R143, R537, R563, R640
    by acids, 315, R241
    by ammonium salts, 66
    by anions, R197
    by calcium, 315, R633
    removal of, by ion exchange, R288, R289
    by rhenium, R482
    by salts, R241
    by sodium, 284, 294ff, 315, 361, R612, R691
    spectral, 216
    transport, R182
  as internal standard, 166
  internal-standard instrument for, 163
  internal standardization of, 166
  ionization of, 33f, 55, 56f, R63, R127, R268b, R371, R660, R662
  ionization potential of, 34
  in organic solvents, 268
  in oxycyanogen flame, 268
  in smoke, 199
  standardization of, 234f
  temperature for determination of, best, 75
  working curve of, 55, 56f, 295, 337, R668
Potassium chloride, dissociation energy of, 31
Potassium halides, dissociation of, in flame, R455
Potassium hydroxide, dissociation energy of, R698
  formation of, in flame, 32, 217, R660, R662
Potassium salts, analysis of, for potassium, R552
  for sodium, R77, R727
Potassium tetraphenylborate, R511
Potentiometric photometric circuits, 157ff
  for internal standardization, 163f
  for photocells, 145
Potentiometric recorders, *see* Recorders
Powdered samples, *see* Solid samples
Powder sprayer, 98
Power, at photodetector, 176, 177f
Powers, T. F., 379
Power supply, for amplifier, 151
  for photomultiplier, 145f, 151
  for phototubes, 151
Pozzolan, analysis of, for cement, R686
PP, photographic plate
Pr., prism monochromator
Pralow, W., R445
Praseodymium, detection limits of, 268
  determination of, R546, R547
  in hexone, 268
  interference of, with lanthanum, 316

Praseodymium oxide (PrO), spectrum of, 221, 268, 400, R546, R547
Precalibrated scale, 152, 233
Pre-cancellation method, 284
Precautions, see Safety precautions
Precipitation, for separations, 250
Precision, 327ff
  in clinical analysis, 341
  definition of, 329
  and excitation energy, 229
  in internal standardization, 165f, 330
  of potentiometric readout, 158
  of reading, 152, 256ff, 273f, 328ff
  requirements of, 278f
  in spectrochemistry, R681
  in spectrography, 198f
  and working curve curvature, 206, 228
Preedy, J. R. K., R360
Preheating, of compressed air, 109
  of fuel gases, 75
Premixed flames, 13ff
  partially, 120f
Preservatives, 237
Pressure, flames under high, R227
Pressure (of fuel and oxidant), adjustment of, 93ff
  control of, 89ff; see also Regulators
  effect of, on internal standardization, 166f
    on temperature of flame, 77
  fluctuation of, 89, 107f
Pressure broadening, see Lorentz broadening
Pressure gage, for flow measurement, 87f
  types of, 88f
Pressure regulators, see Regulators
Priestley, W. M., Jr., R417
Primary combustion zone, see Reaction zone
Principal quantum number, 38
Principal series, 216
Prism, dispersion of, 173
  Littrow, 174
  resolving power of, 174
Prism monochromator, 172ff, 191
  high-luminosity, R724
  luminosity of, 175f, R389
Prism spectrograph, 195
Pritchard, L. R., R191

Procedures, 200ff, 333ff; see also Measurement, Sample preparation, Standardization
  development of, 203ff
Process and Instruments Corp., 430, R567
Process control, 168f, 358, R755; see also Flame monitor
  commercial flame photometers for, 428, 429, 430, 431
Proctor, K. A., R159
Prokof'ev, V. K., 383, R57
Propane, 80f, R2
  burning velocity of, 77
  consumption rate of, 97
  flame temperature of, 77
  flow adjustment of, 95
  hazards of, 80f
  pressure of, 80f
  tanks for, 80
Propanol, as solvent, 209
Protective chelation, 252, 310, 352, R213, R215, R217, R350, R454, R748
Protein, bonding of, to calcium, 341, 344, 350f, R504
  bonding of, to metals, 344
  effect of, on blood serum analysis, 235, 341, 351, R418
    on phosphate interference, 344
  removal of, 343f, 345, 351
PT, phototube
Pugsley, L. I., R90
Pulp, analysis of, 359, R106, R400
Pulse counting, 187
Pump, sample, see Controlled flow
Pungor, E., R325
Purbrick, R. L., 382
Püschel, A., R238a
PV, photovoltaic cell
Pyrrolidine dithiocarbamate, 250

q, background emission
Quantometry, 187f, 277
Quantum of light, 37
Quantum numbers, atomic, 38f
  molecular, 40f
Quartz monochromator, 175, 177, 226f
Quenching, see Deactivation, Interference, Vaporization interference
Quotient method, 239f

Quotient readout, for internal standardization, 164, 165, 190
  for potentiometric photometric circuitry, *see* Potentiometric photometric circuits

r, head of band degraded to longer wavelengths
Radiance, 130, 175f; *see also* Intensity of continuum, 176
Radiant flux, *see* Illumination
Radiant power, *see* Power
Radiation, absorption of, 44ff
  atomic, fundamentals of, 36ff
  blackbody, 46, 49, 50
  collection of, *see* Entrance optics
  detector, *see* Photodetectors
  equilibrium of, in flame, 17, 50
  flux, *see* Power
  intensity of, 42ff, 73; *see also* Intensity
  molecular, fundamentals of, 39ff
  receiver, *see* Photodetectors
  transmission of, *see* Luminosity
Radiation interference, 292ff; *see also* Dissociation interference, Evaporation interference, Excitation interference, Ionization interference, Nonspecific interference, Specific interference, Transport interference, Vaporization interference
  classification of, 61, 62ff
  diagnosis of, 67, 282f
Radicals, *see* Hydroxyl, etc.
  radiation of, *see* Carbon (C_2), Hydroxyl, Methylidene, Molecular spectra, etc.
  in reaction zone, 17f
Radioactive materials, analysis of, 357
  filter for, 357, R240
  separation of, 249
Radioactive phosphor, as reference, R464
Radio Corp. of America photomultipliers, 146f, 149, 182ff, 273
  dark current of, 184
Radium, band spectrum of, 400
  detection limits of, 268
  flame spectrum of, 218, 219, 268, 372, 400
  ionic spectrum of, 218, 400

raie ultime, 44
Rains, T. C., 372, 380, 382, R211, R252, R488, R489, R490
Raĭskiĭ, S. M., 383, R57
Ramage, H., 382
Ramírez-Muñoz, J., R1, R173, R174, R176, R177
Ramsay, J. B., R485
Random error, 256, 327ff
Rao, P. T., 381
Rapoport, S., R653, R654, R656
Rare earths, determination of, R546, R547
  flame spectra of, 215, 220f, 379, 380, R546, R547
  in microwave discharge, 76
  spectra of, 380, R8
Rare elements, flame photometry for, R307b
Ratnasingam, K., R149a
RC, *see* Time constant
RCA, *see* Radio Corp. of America
Re, *see* Reynolds number
Reaction-free zone, *see* Interconal zone
Reaction zone, 13f, 15f; *see also* Equilibrium
  chemiluminescent line emission in, 121, 216, 219, 220, 222ff, 389, 391f, 396f, 399f
  excitation mechanism in, 50
  ionization in, 35
  in Méker burner, 119, 127
  separation of, 54, 121, 244
  spectrum of, 54, 222, 386, 387
  thickness of, 15
  in un-premixed flame, 19
Reading, *see* Measurement, Standardization
  rounding off, 329
  statistical evaluation of, 328f
Readout, *see* Electric circuitry, Measurement, Meter, Null reading, Photocurrent, Precision
  in alternating-current systems, 185f
  of net line, automatic, 187f
Rec., recording
Recombination, as cause of continuum, 42, 54
  of droplets, 24, 107, 114
  of ions and electrons, 35
  kinetics of, 35
  of radicals, 18, 72

Recorders, in alternating-current systems, 185
  as attachments, 156
  for process control, 169
Recording, 159, 168f, 188ff
  at fixed wavelength, 159, 168f, 189
  for flame spectrophotometry, 188ff
  history of, 10
  for integrating, 156, 189f
  for internal standardization, 190
  for microanalysis, 255
  for optimal adjustment, 189
  oscillographic, 278, R343a
  for process control, 168f
  of spectra, 189
  of spectral line, 178f, R724
  speed of, 278
Recording flame spectrophotometers, 427ff, R235, R414, R417, R517, R528, R724
Recovery, see Self-standardization
Rectifiers, 185f, 257
Reference standard, 232f, 234f, 239f
Reflux atomizer, 104, 113, 114, R744
Refractories, analysis of, 362; see also Silicates
  for calcium, R499, R592
  for magnesium, R499
  for potassium and sodium, R141, R295, R498, R499, R592, R606
Regulators (pressure), connections for, 84f
  design of, 91f
  explosion hazard of, 84
  oscillation of, 89
  overflow, 90f
Reiche, F., R437
Rejection, of interference filters, 141f
Relative aperture, see Luminosity
Releasing agents, 309f, 321, 353, 360, 368, R217, R493, R501, R544, R754
  mechanism of, 306
Reproducibility, see Precision
Research and Control Instruments, Inc., 430
Residual fuel oil, see Fuel oil
Resistor, load, 145, 158
Resolution, see Resolving power
  and slit width, R408
  in spectrum tables, 373

Resolving power, 173f, 176, 178
  in choice of wavelength, 226f
  of spectrograph, 195f
Resonance fluorescence, 50, R147
Resonance interference, 64
Resonance lines, in absorption flame photometry, 48
  definition of, 38, 383
  intensity of, 44
  self-absorption of, 45ff
  table of, 34
  working curves of, 57f
Resonance potential, see Excitation energy
Response, see Emission, Photocurrent, Readout
Response time, see Time constant
Results, analytical, see Measurement, Standardization
Reversal, see Line reversal, Self-reversal
Reviews of flame photometry, R1, R7, R12, R16, R26, R33, R36, R37, R49, R52, R65, R81, R125, R228, R273, R297, R302, R307a, R327a, R382, R470, R483, R486, R494, R518, R522, R525, R548, R553, R559, R574, R632, R643, R673, R675, R678, R701, R752, R757, R760, R765
Rexach-M. de Lizarduy, M. L., R177
Reynolds number, 18f, 22
Rhenium, band spectrum of, 370
  detection limits of, 268
  flame spectrum of, 215, 224, 268, 401
  in incandescent oxyacetylene, 268
  interference of, on potassium, R482
Rhodium, band spectrum of, 269, 401
  detection limits of, 269
  flame spectrum of, 225, 269, 401
  interferences of, spectral, 225
  in organic solvents, 269
Richardson, D., 9, R90
Richardson, J. P., R256
Richter, C., 8, 9, 339
Rick, W., R350, R350a
Riehm–Lange flame photometer, R261, R594
Rinsing, of atomizer, 238f
Risberg, P., 382
Robertson, I. M., R493

630 INDEX

Robinson, J. W., 10, 372, 382, R322, R367
Robot, for flame spectrography, 336
Rocks, analysis of, 360
  for calcium, 360, R427, R432
  for cesium, R365
  for lithium, R247, R365, R427, R735
  for magnesium, R192
  for phosphate, R229, R247
  for potassium, R315, R317, R427, R532, R742
  for rubidium, R365
  for sodium, R317, R427, R532, R742
Rodden, C. J., 382
Rodin, G. M., R666, R667
Rogers, J. M., R368
Rogers, R. A., R380
Rollet, M., R234a
Rosen, B., 372, 380, 382, R9
Ross, R. E., R157
Rossikhin, V. S., 380
Rotating disk, for sample introduction, 97f, R153
Rotational energy, 40f
Rotational lines, 370, 383
Rotational quantum number, 40f
Rounding off, 329
Rouse, R. D., R61
Rubber, determination of, in fabric, R702
  leaves, analysis of, R149a
Rubidium, cross-sensitivities with, 282
  detection limits of, 269
  determination of, R273, R316, R320a, R368, R572, R577
    in biological materials, R273a
    in glass, R743
    in ores, R743
    in plants, R312
    in rocks, R365
    in silicates, R761
    in tissues, R312
    in whiskey, R566
  enhancement of, 269, R431a
  excitation potential of, 34
  flame spectrum of, 216f, 269, 383, 400f, R726, R734
  interferences of, 313ff
    ionization, R575
    spectral, 139, 216
  interferences on, 318
    by sodium chloride, 30
  as internal standard, R268b
  ionization of, in flame, R127, R268b, R660
  ionization potential of, 34
  in organic solvents, 269
  in oxycyanogen flame, 269
Rubidium chloride, dissociation of, 30
Rubidium halides, dissociation of, R455
Rubidium hydroxide, formation of, in flame, 32, 217, R660
Rusanov, A. K., 383
  flame photometer, R521
Ruthardt, K., R44
Ruthenium, band spectrum of, 225
  detection limits of, 269
  flame spectrum of, 225, 269, 401
  in organic solvents, 269
  spectral interferences of, 225f
Rutkowski, W., R452

s, line of singly ionized atom
Safety precautions, 200, R42, R705; *see also* Hazards
  with burners, 119
  in clinical analysis, 341
  with compressed air, 81f
  in gasoline analysis, 364
Saha equation, 33, 35f, 75
Saidel, A. N., 383, R57
Salpeter, E. W., 372, 380, R9
Salts, analysis of, R291, R399
  interference of, 65f, R241
  vaporization of, in flame, 28ff
Samarium, detection limits of, 269
  determination of, R546, R547
  flame spectrum of, 402, R546, R547
  in hexone, 269
Samarium oxide (SmO), spectrum of, 221, 269, 403, R546, R547
Sambursky, S., R316
Sample, *see* Analyte
  carry-over of, 104f, 113, 254
  centrifugation of, 110
  changing of, 277
  concentration of, *see* Concentration of sample
  concentration of, in flame, 58
  container for, 109f, 212ff
  cooling of, in reflux atomizer, 113
  deaeration of, 213
  density of, 22, 23, 110

drainage of, 111ff
enrichment of, 211f, 248ff
evaporation of, 110
filtration of, 110
flammable, 110, 123, 363f
loss of, in containers, 212
microanalytical, 113ff, 155, 253f
pretreatment of, 248ff
small, atomizers for, 113ff, 253f
solid, 5, 97f
supplying of, to atomizer, 109ff
switching circuit by, 115, 155, 255
temperature of, 108f, 110, 113, 327
transport of, 109ff
Sample consumption, quantity, 115, 254
rate, see Sample flow rate
in spectral recording, 189
in spectrography, 193
Sample flow, laminarity of, 22
Sample flow rate, 21f, 105, 254
and atomization efficiency, 25f, 105f
density effect on, 22, 110
effect of, on emission, 22, 106, 109, 115, 156, 254, R85, R85a, R268a, R278a, R278b
effect of, on working curve, 56
pressure effect on, 106
testing of, 111
viscosity effect on, 22
Sample handling, 277
system for, 427, 429, 430
Sample introduction, 20ff, 97ff; see also Atomization, Atomizers, Dissociation, Evaporation, Vaporization
with controlled flow, see Controlled flow of sample
history of, 7f, 97f
hydrostatic head effect on, 21f, 98, 109f, 327
methods of, 97f, 109ff
transport in, 20ff
Sample preparation, 203, 248ff, 276; see also Ashing, Drying
automatic, 276
for clinical analysis, 341ff
for microanalysis, 253
for spectrography, R336
Sample pumping, see Controlled flow of sample
automatic, 276

Sample transport interference, see Transport interference
Sampling error, 204f, 327
in clinical analysis, 341, 346f
Sanders, D. P., R314
Sapirstein, L. A., R602a
Saturation, for eliminating interferences, 321
for ionization interference, 298
of photocurrent, 145, 146
in photomultipliers, 148
of vapor in flame, 225
for vaporization interference, 309
Saunders, R. A., R337a
Scale, see Boiler deposits
Scale, precalibrated, 152
reading precision of, 152
Scandium, detection limits of, 269
determination of, R546, R547
in hexone, 269
Scandium oxide (ScO), spectrum of, 221, 269, 402, R546, R547
Scanning, see Recording
automatic, R724
oscillographic, 278, R343a
speed of, 278
Scattering, see Random error
Schaffert, R. R., R418, R419, R420
Scheel, K., R13
Schellhorn, H., R341, R347
Schinkel, H., R633, R635, R636, R637
Schläfer, R., 139
Schmidt, W., 222, 372, 382, R333, R335, R431a, R431b
Schmitz, W., R583
Schoeb, E. J., R597
Schöffmann, E., 223, 383, R515, R516
Schön, 212
Schott und Gen., filters, 139, 141f, R398
Schrenk, W. G., R97, R312
Schroeder, L. W., 381
Schuffelen, A. C., 8
Schuhknecht, W., 8, 9
Schuhknecht (–Waibel) flame photometer, 131, 336, 428, R245, R589
Schwartz, M. C., R143
Schwartzschild exponent, 198
Schwarzkopf, B., R740
Schweitzer, E., 9
Schwertmann, U., R611a
Scott, C. W., R314b

Scribner, B. F., 371, 373, 382, R458a, R477
Sea water, analysis of, R89, R359
  for calcium, R189
  for chlorine, R363
  for strontium, 355, R189, R659
Secondary combustion zone, see Outer cone
Secondary electrons, 145, 146
Sector disk, 194, 197, 198
Seeds, analysis of, R151
Seidel transformation, 198
Selenium, determination of, indirect, 247, R452
  flame spectrum of, 224, 243, 370
  interferences of, 313, 318, R452
  in microwave discharge, 76
Selenium photocell, 9, 144f, 149, 154
Self-absorption, 44ff, 58, R67, R316, R638
  of bands, 58
  effect of, on working curve, 57f, 217
  flame thickness, effect of, on, 120
  and mirror intensification, 134
  temperature effect on, 75
  use of, in testing interferences, 67
Self-reversal, 47, 58
  effect of, on working curve, 58
Self-standardization, 319f, 321ff, 356f, 365, R62, R140a, R230, R243, R275, R276, R316, R602, R602a, R700a
  in arc spectrography, 323
  without blank, 291f
  double-coefficient method of, 298, 322f
  errors of, 323
  in indirect methods, 247
  for ionization interference, 297f
Sell, H. M., R357
Semiconductors, analysis of, 357
Sensitivity, see Detection limits
  adjustment of, 239
Sensitivity control, 156f
  in amplifier, 150
  automatic, 169
  of null meter, 159
  in Perkin-Elmer flame photometer, 191
  for photomultiplier, 145
  for phototube, 144

Separate atomization, see Dual atomization
Separation, of analyte, 248ff
  of droplets, 24, 107
  of interferents, 249ff, 310, 325, 360
  by ion exchange, see Ion exchange
Separator, Smithells, 54, 121, 244
Sequential internal standardization, 161ff
Serfass, E. J., R622a
Serum, see Blood serum
Sewage, analysis of, R73
SH, see Sulfur hydride
Shambon, A., 383
Shapiro, R., R484
Sheathed burner, 260ff, 427, 431
Sheehy, T. P., R506
Shelton, J. P., R608
Shielded burner, 120, 121, 129, 274
Shigemoto, M., R274, R276
Shimauchi, M., 383
Shock waves, 106
  temperature of, R194
Shot effect, 185
Shot noise, 149, 150, 185
  in alternating-current systems, 185f
  and detection limit, 273
  frequency range of, 185
  in photomultipliers, 182f, 184f
  suppression of, 183ff
Siemens flame photometer, 9
Signal, see Photocurrent, Readout, Response
Signal-to-noise ratio, 259, 273; see also Fluctuation, Noise, Shot noise
  in alternating-current systems, 186
Significance, statistical, 330
Sigrist und Weiss A.G., 430, R657
Silica, separation of, 368
Silicate, interference of, with alkaline earths, 303, R637
  with calcium, 303, 313, 368
Silicates, analysis of, R650
  for aluminum, R219a, R623
  for calcium, R150, R219, R445
  for iron, R209
  for lithium, R214, R219, R247, R761
  for magnesium, R150
  for potassium, R214, R219, R309, R445, R761, R763
  for rubidium, R761

Silicates—*contd.*
  for sodium, R214, R219, R309, R445, R761, R763
  for strontium, R216, R219
Silicon, chemiluminescence of, 216, 269
  detection limits of, 269
  flame spectrum of, 216, 222, 243, 269, 402
  in incandescent oxyacetylene, 269
  in microwave discharge, 76
  in organic solvents, 269
Silicone, protective coating of, 212, 213
Silicone water-repellents, 211, 213, R301, R303
  for blood samples, 348
  for deaeration, 213
Silicon oxide (SiO), flame spectrum of, 244, 269, 402
Silicon photocell, 144
Silker, R. E., R97
Silver, chemiluminescence of, 51, 216
  for chloride determination, 248
  detection limits of, 270
  determination of, R582
    in biological material, R597
    in copper, R280a
    in phosphors, R587
  excitation potential of, 34
  flame spectrum of, 216, 217, 270, 370, 385, R499, R726, R734
  interferences on, 312
  as internal standard, 166, 368
  ionization potential of, 34
  in organic solvents, 270
  in oxycyanogen flame, 270
  separation of, 250
  working curve of, 217
Simpson, J. H., Jr., R204
Simpson, M., R689
Simulation, 231f, 234f, 284, 297, 321, 337, 353; *see also* Parametric method
  for nonspecific interferences, 326
Siphon, drainage of, 111ff
Sirup, analysis of, 339
Sivaramamurty, V., 381
Slag, analysis of, R515, R677, R692, R715
  for calcium and magnesium, 368, R230, R379
Slit, curvature of, 173, 177
  imaging of, 172f

  of monochromator, 177
  size of, 175
  variation of, for background correction, 289f
  vibrating, R343a
Slit function, 178f
Slit width, choice of, 178
  effect of, on photocurrent, 176, 177f
  spectral, 176, 178f
  widening of, effective, 178
Slope, logarithmic, of working curve, 322
Slurry, spraying of, 5, 98
Smelting, 368
Smit, C., R148
Smit, J., R64
Smit, J. A., R147, R665
Smith, A. C., R388
Smith, C. L., R362
Smith, D. D., R476
Smith, D. F., R351
Smith, E., R144
Smith, F. M., R627
Smith, W. E., R192
Smith, W. F. R., R199
Smithells separator, 54, 121, 244
Smitherman, T. B., R739
Smoke, analysis of, for sodium, 359, R532a; *see also* Aerosol
Sodium, analysis of, for potassium, R511
  atomic structure of, 36ff
  chemiluminescence of, 51, 216, R534
  concentration of, in flame, R182a, R394
  contamination by, 214
  continuum of, 42, R395
  cross-sensitivities with, 282
  detection limits of, 190, 270
  determination of, R82, R218, R273, R296, R328, R329, R330, R332, R340, R368, R381, R416, R528, R537, R563, R572, R577, R585, R607, R689, R691
    by absorption flame photometry, R454a
    in acids, R472
    in alfalfa, R466
    in alumina, R156
    in aluminum, 368, R161, R322, R324, R367, R379, R512
    in aqueous humor, R451a
    in ash, R583, R606, R635

Sodium—*contd.*
  in atmosphere, R751
  in beans, R338f
  in beryllium solutions, R506
  in biological fluids, 344f, R128, R133, R134, R269, R270, R664a, R673, R750, R757
  in biological materials, R279, R584, R597, R723
  in blood cells, R378, R533
  in blood plasma, R137, R153, R271, R422, R500, R708
  in blood serum, R64, R93, R94, R95, R105, R123, R131, R137, R149b, R235a, R248, R272, R315a, R339, R342, R349, R354, R378, R406, R418, R419, R420, R436, R510, R524, R531, R545, R553a, R568, R569a, R585, R611, R617b, R651, R661, R663, R741, R747
  in boiler scale, R593, R640
  in brine, R399
  in calcium, 190
  in catalysts, R756
  in cement, 363, R74, R191, R222, R224, R244, R266, R266a, R308, R309, R317, R565
  in ceramics, R619, R716, R766
  in cerebrospinal fluid, R137
  in cheese, R323
  in clay, R78, R266, R334
  in coal, R388
  in copper, R322
  in fabric, R702
  in fertilizer, R151
  in foods, 338
  in fuel oil, R75
  in gastric juice, R135
  in glass, 361, R102, R164, R220, R266, R326, R331, R333a, R335, R427, R604, R605, R706, R743
  in grain, 338, R151
  in hay, R523
  instruments for, R64, R111, R140, R143, R157, R269, R270, R271, R339, R344, R360, R457, R584, R585, R586, R594, R669, R709, R710, R721, R728, R735a, R755
  in iron, R435
  in leaves, R149a
  in lithium, 368, R380
  micro-, 190, R340, R342, R420, R584, R586, R669
  in milk, R190, R323, R413, R506a, R617b, R639
  in minerals, R562, R743
  in nutrient solution, 359
  in paper pulp, R106, R400
  in petroleum oil, R198, R412
  in petroleum wash water, R756
  in pharmaceuticals, R314, R351
  in phosphors, 357, R205
  in plants, R165, R468, R550, R569a, R664
  in potassium, 190
  in potassium salts, R77, R727
  in refractories, R141, R295, R498, R499, R592, R606
  in rocks, R317, R427, R532, R742
  in salts, R399
  in seeds, R151
  in silicates, R214, R219, R309, R445, R761, R763
  in smoke, R532a
  in soil, R144, R151, R256, R281, R309, R317, R494, R521, R569a, R644
  in sugar solutions, R96
  in tissues, R133, R221, R436, R531, R708
  in tungsten, R327
  in uranium, 356f
  in uranium salts, R405
  in urine, R124, R271, R272, R315a, R355, R364, R422, R500, R568, R611, R617b, R624, R741, R747
  in viscous samples, R96
  in water, R309, R383, R460, R472, R569a, R624, R640, R693, R731
  in wetting agents, R428
  in wine, R753
  for determination of organic acids, 248
dissociation interference on, 30f
D line, *see* Sodium D lines
emission of, effect of acetylene flow on, 94
energy level diagram of, 37ff
enhancement of, by organic solvents, 270, R81a, R431a, R656
excitation of, degree of, 44

INDEX 635

Sodium—contd.
excitation potential of, 34
fluorescence of, R147
formation of atoms of, in flame, 29f
infrared line (819 mμ) of, 58, 138f, 270, 284, 345
interferences of, 312ff
  on calcium, 283f, 313, 349f, 363
  on potassium, 55, 57, 284, 294ff, 315, 361, R612, R691
interferences on, 30, 317, R143, R537, R563, R640
  by acids, R241
  by anions, R197
  by calcium, 284f, 317, R149d, R329, R633
  ionization, 294ff, 323f, 361, R268b, R603
  by potassium, 297, 317, 361, R691
  by salts, R241
  by sulfuric acid, 167f, 317
  transport, R182
as internal standard, 166
internal-standard instrument for, 163
internal standardization of, 166, 167f
ionization of, in flame, 34, R63, R127, R268b, R371, R662
ionization potential of, 34
line reversal of, R194
as normalizer, 298
in organic solvents, 270
in oxycyanogen flame, 270
reaction of, with hydroxyl, 42, R169
as saturator, 298
spectral interference of, on calcium, 139
spectral interference on, by calcium, 137
spectrum of, 37ff, 216f, 270, 397, R726, R734
standardization of, 234f
temperature for determination of, best, 75
ultraviolet line (330 mμ) of, 44, 58, 270, 345, R349
working curve of, 231
Sodium bromide, dissociation energy of, 30
Sodium chloride, dissociation of, in flame, 29f
dissociation energy of, 30, 31
interference of, with rubidium, 30

Sodium determinator (spectronatromètre), 7
Sodium D lines, exclusion of, by filter, 139
fluorescence of, in flames, 50
history of, 6
intensity of, 44, 270
isolation of, by filters, 137f, 140ff
origin of, 37ff
profile of, R668
self-reversal of, 47
width of, 47, R372, R668
Sodium halides, dissociation of, in flame, R455
Sodium hydroxide, formation of, in flame, 32, 217, R169, R394, R662
recombination continuum of, 42
Sodium iodide, dissociation energy of, 30
Sodium pyrrolidine dithiocarbamate, 250
Sodium salts, analysis of, for potassium, R264a, R511
Sodium tungstate, 344
Soft emulsion, 198
Soil, analysis of, 335f, R39, R50, R676a
  for aluminum, R551
  for calcium, R151, R256, R294, R385, R415, R430, R434, R494, R570, R616, R732
  for cement, R686
  for magnesium, R151, R256, R415, R416b, R611a, R617, R676
  for manganese, R144, R415
  for potassium, R144, R151, R256, R281, R309, R317, R478, R479, R480, R494, R521, R569a
  for sodium, R144, R151, R256, R281, R309, R317, R494, R521, R569a, R644
  for strontium, R144, R416c
Solid samples, introduction of, into flame, 5, 97f
Solubility, determination of, by flame photometry, 356
Solute vaporization interference, see Vaporization interference
Solvent, 20f, 24f, 27f, 108, 207ff; see also Evaporation interference, Nonaqueous solvents, Organic solvents

Solvent—*contd.*
  cooling of flame by, 28, 73
  effect of, on emission, 209f, 293
  evaporation of, *see* Evaporation
  extraction by, *see* Extraction
  purity of, 211
  selection of, 372f
Sooty flame, *see* Incandescent flame
Sound-proofing, 122, 180
Sources, *see* Arc, Spark
  special, 76f, R470
SP.900 flame spectrophotometer, 431, R707
Spark, for sample atomization, 98, R499
Spark-in-flame, 76, R377, R384, R682
Spark spectroscopy, 4ff
  applications of, R153
  history of, 8
Species, emitting, identification of, 371, 374ff
Specific interferences, *see* Dissociation interference, Ionization interference, Vaporization interference
  of alkali metals, 294ff
  automatic correction of, 325
  classification of, 61, 63ff, 293
  correction of, 320ff
  definition of, 63
  detection of, 319ff
  diagnosis of, 67, 282f, 319ff
  elimination of, 320ff
  history of, 293
  table of, 310ff
Specificity factor, 280
Speck, S. J., R413
Specker, H., R410
Spectra, atomic, 36ff, R5, R8, R15, R17, R26, R38, R41c, R47, R51
  of the elements, 215ff
  molecular, 39ff, R9, R10, R19, R41, R48
  tables of, 370ff
Spectral bandwidth, 176, 178f, 275f
Spectral dispersion, 172f
Spectral interference, 62, 216, 280ff, 285ff
  correction of, 241ff, 286ff
  and detection limits, 275
  and filters, 139
  and slit width, 289f

specificity factor for, 280
suppression of, by aluminum, 325, 363, R201, R388, R493, R633, R672
temperature effect on, 74
testing of, 283
Spectral line, *see* Line (spectral)
Spectral resolution, *see* Resolving power
Spectral scanning, *see* Recording, Scanning
Spectral sensitivity, effective, 139
  of emulsions, 195, 196
  of photocells, 145
  of photodetectors, 138, 146f
  of phototubes, 145
Spectral slit width, 176, 178f
  and detection limits, 275f
Spectrochemical emission methods, compared with flame, 4ff
  review of, R553
Spectrograms, atlas of, 215
  photographic, 193
  recorded, R307a, R726, R734, R759a; *see also* Recording
  for wavelength tables, 370ff
Spectrograph, 195f, R319
Spectrography, *see* Flame spectrography
  sample preparation for, R336
Spectronatromètre, 7
Spectrophotometer, *see* Monochromator
  Beckman, *see* DU spectrophotometer
  flame attachments for, *see* Flame attachments
  photodetectors for, 181ff
  Zeiss, 174, 180
Spectrophotometry, *see* Flame spectrophotometry
Spectrum, *see* Spectra
Speed, of flame analysis, 276ff
  spectroscopic, *see* Luminosity
Spencer, A. T., R751
Spex Industries, Inc., 76
Spin, 38f
Spodumene, analysis of, for lithium, R166
Spray, *see* Aerosol, Atomization, Droplets
Spray chamber, 20f, 107ff; *see also* Indirect atomizer

INDEX 637

Spray chamber—*contd.*
  barrier in, 105, 107, 306, 308
  cooling in, 25, 108
  damping of fluctuations in, 107f
  delay in, 99
  drainage of, 111ff
  drift in, 108, 115
  effect of, on drop size, 24f, 107
  evaporation in, 108
  heated, 114, 115, R182, R236
  loss of sample in, 25f, 99, 115
  pressure in, 112
  saturation of air in, 25, 108
Sprayer, *see* Atomizer
Sproat, D., R255
sr., steradian, 175
SrOH, *see* Strontium hydroxide
Staab, K., R612
Stabilization, *see* Damping, Fluctuation
Stace, H. C. T., R149c
Standard, *see* Internal standardization
  bracketing, 240
  reference, 232f, 234f, 239f
  stock, 235f, 237
Standard addition, *see* Self-standardization
Standard deviation (standard error), 257, 328ff
  of blank, 257
  determination of, 330
  of the mean, 329
Standardization, 236ff; *see also* Measurement
  of atomizers, 87f
  automatic, 169
  with cross-sensitivity, 283ff
  of manufactured equipment, 64f, 71, 93
  by referee chemical method, 331
  sequence of, 239f
  in spectrography, 197
  technique of, 238ff
  of working curve, 230ff
Standard solutions, chemicals for, 236ff
  choice of, 230ff
  for clinical analysis, 234f, 344f, 348, 350f, 353
  concentrations of, 232
  containers for, 237
  contamination of, 237
  dilute, 237
  dilution of, 236
  preparation of, 236ff
  preservation of, 237f
  simulation of sample by, 231f; *see also* Simulation
  suppliers of, 238, 345, 353, 427, 429f
  water for, 237
Stander, C. M., 223
Starr, W. L., R359
Statistical significance, 330
Statistical thermodynamics, *see* Boltzmann equation, Dissociation, Saha equation
Statistical weight, 43
Statistics, of analytical reading, 256f, 328ff
Stavinoha, W. B., 383
Stecklein, H. R., R313
Steel, analysis of, R499, R733, R737
Stellar spectra, R51
Step-sector disk, 194, 197, 198, 370
Sterin, Kh. E., R27
Stewart, 98
Stoll, U. W., R686
Storage, *see* Preservatives
  of clinical samples, 341
Storage battery, for photometry, 151
Storey, W. H., R670
Stray light, 147
  elimination of, 150
  in filter instruments, 280
  in monochromators, 177, 280
Strock, L. W., R336
Strontium, for aluminum determination, 246
  as buffer against aluminum, 309, R493
  chemiluminescence of, 216
  concentration of, in flame, R182a
  continuum of, 218
  detection limits of, 270
  determination of, R218, R273, R357, R361, R416, R573, R576, R582b, R636, R700
    in barium compounds, R690
    in biological materials, 354, R321, R597
    in cathodes, 357, R253, R292, R325, R582c
    in cement, R225
    instrument for, R457, R710
    in lubricating oil, 366
    in marine organisms, R189
    in minerals, R216, R314c, R56'

Strontium—contd.
   in petroleum oil, R198
   in plants, R416c
   in sea water, 355, R189, R659
   in silicates, R219
   in soil, R144, R416c
   in water, R693
   in whiskey, R566
  enhancement of, 270, R431a
  excitation potential of, 34
  flame spectrum of, 216, 218, 219, 270, 404, R186, R726, R734
  interferences of, 312ff
   on lithium, 363, R201
  interferences on, 318, R216, R217, R370a, R700
   by aluminum, 246, 318, R370a, R454, R493
   EDTA, effect of, on, R213, R215, R454
   ionization, R268b
   by molybdenum, tungsten, etc., 247, 318, R452
   by phosphate, 246, 309, 318, R215, R452, R454
   by selenium, tellurium, etc., 247, 318, R452
   spectral, 229, R416c
   by sulfate, 318, R215, R452
  as internal standard, 358f, R154
  ionic spectrum of, 218, 270, 404
  ionization of, in flame, 218, R268b, R371
  ionization potential of, 34
  in organic solvents, 270
  in oxycyanogen flame, 270
  for phosphate determination, 246
  polymerization of, in flame, R375
  as releasing agent, 309, 353, 368
  for sulfate determination, 248
Strontium chloride (SrCl), spectrum of, 372, 404
Strontium fluoride (SrF), for fluorine determination, 245
  spectrum of, 404
Strontium hydride (SrH), spectrum of, 404
Strontium hydroxide (SrOH), identity of, 42, R186, R286, R376, R393, R441
  isotope shift in, 42
  spectrum of, 42, 219, 270, 404, R393

Strontium oxide (SrO), spectrum of, 42, 372, 404, R286, R374, R376
  stability of, in flame, R695
  term diagram of, R374
Strontium oxide (Sr_2O_2), spectrum of, 404
Strunk, D. H., R456, R456a
Subsoil analysis, 335, 336
Substitution method, for indirect determination, 248, 252
  for interference elimination, 323ff
Successive dilutions, for removing interference, 320
Suction, of atomizer, 21f, 114
Suction atomizer, 100, 109ff
Suction height, see Hydrostatic head
Suction rate, see Sample flow rate
Sudo, E., 222
Sugar, analysis of, for potassium and sodium, 337f, R96
Sugar sirup, analysis of, 339
Sugden, T. M., 221, 380, 382, R127, R169, R171, R391, R392, R393, R394, R395, R423, R424, R425, R534, R536, R662
Suhrmann, R., R46
Sulfate, anion interference of, 66
  determination of, 248, R177, R452, R647, R687
   in water, 356
  dissociation interference of, 31
  interferences of, 312ff
   on calcium, 302, 307, 309, 313, R258, R276a, R452
   with phosphate on calcium, 301f, 306
   on strontium, 318, R215, R452
Sulfur, detection limits of, 270
  flame spectrum of, 224, 270; see also Carbon sulfide, Sulfur hydride, Sulfur oxide
  in microwave discharge, 76
Sulfur hydride (SH), spectrum of, 244
Sulfuric acid, interference of, see Sulfate
  on calcium, 300f
  on sodium, 167f
  for normalization, 326
Sulfur oxide (SO), spectrum of, 244
Supersonic gas flow, 106
Surface-active agents, see Wetting agents

INDEX 639

Surface tension, effect of, on atomization, 22f
  on drop size, 23
  on emission, 63, 65, R545, R609
  on emission with atomizer-burner, 21, 65
Suspended material, removal of, 110
Suspensions, spraying of, 5, 98
Süss, L., R333a
Svehla, G., R251
Swan bands, 54, 55, 386
Swigart, R. H., R313
Swindale, L. D., R256
Switch, atomizer, 115, 155, 255
Syringe pump, 124, R347
Systematic error, 256, 331f
Szymczak, D., R452, R453a

t, triple
Takahashi, K., R275
Talbot, W. H. F., 7
Tanasawa, Y., R520
Tanks, *see* Gases
  for acetylene, 77ff, 83
  for compressed gases, 83, 96
  connections for, 84f
  for propane, 80, 83
  valve of, 84f
Tantalum, flame spectrum of, 215, 223
  ionic spectrum of, 404
  in microwave discharge, 76
  oxycyanogen spectrum of, 10, 404
Tatoian, G., R472
Taylor, H. S., R32
Technetium, 224
Technicon Controls, Inc., 430, R80
Technique, 199ff; *see also* Measurement
Teclu burner, 121
Tellurium, chemiluminescence of, 224, 270
  detection limits of, 270
  determination of, indirect, 247, R452
  flame spectrum of, 224, 243, 270, 405
  interferences of, 314, 315, 318
    on calcium and strontium, 314, 318, R452
  in microwave discharge, 76
  in organic solvents, 270
Tellurium oxide (TeO), spectrum of, 224, 244f, 270, 405
Temperature, of aerosol, 108
  of compressed air, 109

  of flame, 71ff, R11, R414a
    acetone in acetylene, effect of, on, 73
    of atomizer-burner, 20, 73
    combustible solvents, effect of, on, 73, 209f
    effect of, on absorption, 49
    effect of, on dissociation, 30
    effect of, on intensity, 73ff, 229
    effect of, on interferences, 74
    effect of, on ionization, 34, 74, 75
    effect of, on self-absorption, 75
    effect of, on working curves, 74f
    and excitation interference, 63f
    flow rate, effect of, on, 22
    of fuels, various, 77
    fuel–oxidant ratio, effect of, on, 71f, 94
    heat losses, effect of, on, 72f
    and height in flame, 74
    homogeneity of, 47, 58, 73
    measurement of, 73, R381, R665
    nonequilibrium, effect of, on, 72
    optimization of, 73ff
    organic substances, effect of, on, 66
    preheating, effect of, on, 75
    premixed, 17
    solvent evaporation, effect of, on, 28, 73, 123
    testing of change in, 67
    un-premixed, 19
  of laboratory, 199
  of sample, 108f, 110, 200
  of shock waves, R194
Temperature coefficient, of line emission, 229
  of photocells, 145
Tennant, C. B., R677
Tenth-width, 140
Terasaka, S., R275
Terbium, detection limits of, 270
  determination of, R546, R547
  in hexone, 270
Terbium oxide, spectrum of, 221, 270, 404f, R546, R547
Term diagram, atomic, 37f
  molecular, 40f, R374
Term symbol, 39, 40
Tetraphenylboron, R511
Thallium, chemiluminescence of, 51
  detection limits of, 271

Thallium—*contd.*
  determination of, R146a, R453a, R585
    in uranium, R441a
  enhancement of, 271, R431a
  flame spectrum of, 220, 271, 405f, R734
  interference of, on sodium, 317
  interferences on, 318
    spectral, 221f
  line profile of, R668
  in organic solvents, 271
  in oxycyanogen flame, 271
  working curve of, 222
Thallium hydride (TlH), spectrum of, 221, 370
Thermal coefficient, of photocell, 145
Thermionic emission, 182f
Thege, I. K., R578, R579, R580, R582, R582b
Thierfelder, W., R21
Thiers, R. E., R85, R278a, R278b, R714a
Thomas, D. B., 248
Thompson, C., 380, R206
Thompson, T. G., R189
Thorium, flame spectrum of, 215
  interference of, with lanthanum, 316
  in microwave discharge, 76
Thrombocytes, analysis of, for potassium, 346
Thulium, detection limits of, 271
  determination of, R546, R547
  flame spectrum of, 220f, 271, 406
  in hexone, 271
Thulium oxide, spectrum of, 221, 271, 406, R546, R547
Time, for flame analysis, 276ff
Time constant, *see* Damping
  in alternating-current systems, 186
  and detection limit, 257f
  of meter, 150f, 153
  of null meter, 153f
  and speed of analysis, 277
Tin, chemiluminescence of, 216, 222f, 271
  detection limits of, 271
  determination of, 222
  flame spectrum of, 216, 222, 271, 403, R499
  in incandescent oxyacetylene, 271
  interference of, on copper, 314
  ionic lines of, 371

  in microwave discharge, 76
  in organic solvents, 271
  in oxycyanogen flame, 222, 271
  in reaction zone, emission of, 121, 271
Tingwaldt, C., R338
Tin hydride (SnH), spectrum of, 403
Tin oxide, spectrum of, 222f, 271, 403f, R499
Tissues, analysis of, 339
  analysis of, for calcium, R133, R138, R221, R708
    for magnesium, R138, R221, R708
    for potassium, R133, R221, R313, R436, R531, R708
    for rubidium, R312
    for sodium, R133, R221, R436, R531, R708
  preparation of, 341f
Titanium, chemiluminescence of, 216
  detection limits of, 271
  determination of, 222
    indirect, 247, R452
  flame spectrum of, 215, 216, 222, 271, R734
  interferences of, 315, 316, 317
    on calcium and strontium, 313, 318, R452
    spectral, 222
  in microwave discharge, 76
  in organic solvent, 271
  in oxycyanogen flame, 10, 405
Titanium dioxide, analysis of, R600
Titanium oxide (TiO), spectrum of, 222, 271, 405
Titration, *see* Flame-photometric titration
Torch, electronic, 76
  Langmuir atomic-hydrogen, 76
  plasma, 76
  welding, 74
Toribara, T. Y., R187, R188
Toxic samples, 199
Trace analysis, 337; *see also* Microanalysis
  concentration for, R173
  enrichment for, 250
  technique of, 249
Tracer, lithium as, in water, 355
  rare elements as, in whiskey, R566
Tracey, M. V., R22
Tramutt, H. M., R702
Transformed density, 198

Transition, electronic, 37ff
Transition probability, 43, 58
Translational energy, in excitation, 50f
  rate of conversion of, 51
Transmittance, in densitometry, 197f
  of filters, 140ff
  of monochromators, 173, 176
  of optics, 130
Transparency, of flame, 53
Transport interference, 21, 62f, 64, 65, 293, R182, R545
  with atomizer-burners, 65, 325f, 327
  and detection limits, 275
  effect of, on working curve, 56
  elimination of, 325ff
  testing of, 67, 210
Trap, drain, 111ff
Tremmel, C. G., 383
Trichloroacetic acid, for protein separation, 344, 345
  venting of, 199
Triple analysis, 335f
Truog, E., R644
Tskhaĭ, N. S., R182a
Tungsten, analysis of, R327
  determination of, indirect, 247, R452
  flame spectrum of, 215, 224
  interference of, with alkali metals, 315, 317
    with alkaline-earth metals, 303, 313, 318, R452
  in microwave discharge, 76
  oxycyanogen spectrum of, 10, 407
Tungsten oxide, analysis of, R327
Turbulent flames, 14f, 18f; *see also* Atomizer-burner flame
  background radiation of, 54
  research on, 20
Turbulent flow, 18f
  in Méker flame, 126

u, unclassified, 383
Ultramicroanalysis, 253ff
  clinical, R181, R342
  integration for, 155, R255, R586
  sample introduction in, 97, R255
Ultrasonic atomizer, 101, 102, R498, R499
Ultrasonic vibrations, in atomizer, 106
Ultraviolet, vacuum, 219
Ultraviolet spectrophotometry, 171
Unger, L., R706

Unicam Instruments Ltd., 431, R707
Union Oil Co., 216, 380
Un-premixed flames, 14f, 19f; *see also* Atomizer-burner flame
Unsteadiness, *see* Fluctuation
Uranium, analysis of, 356
  for gallium, etc., R441a
  for lanthanum, 249
  for sodium, 356f
  compound formation with, 33
  continuum of, 271
  detection limits of, 271
  determination of, 356
  flame spectrum of, 10, 33, 215, 220, 356
  interference of, on lanthanum, 316
  resonance potential of, 44
  separation of, 249, 356
Uranyl nitrate, analysis of, R405
Urea, for simulation, 353
Urine, analysis of, R609
  for calcium, R153, R237, R315a, R355, R364, R596, R617b, R624, R741, R747
  for iron, R153
  for magnesium, 354, R153, R237, R617b
  for potassium, R124, R153, R271, R272, R315a, R355, R360, R364, R500, R568, R611, R617b, R624, R741, R747
  for sodium, R124, R153, R271, R272, R315a, R355, R360, R364, R500, R568, R611, R617b, R624, R741, R747
  by spark spectrography, R153
  composition of, 341
  storage of, 341
Uvispek spectrophotometer, 428, R352

v, head of a band degraded to shorter wavelengths
v, vibrational quantum number, 40f
Vacuum phototube, *see* Phototube
Vacuum ultraviolet, for mercury line, 219
Valence electron, 36ff
Valenta, E., 97
Valentin, H., R149b
Vallee, B. L., 10, 48, R36, R84, R85, R85a, R278a, R278b, R457, R458

Valves, regulating, 91f; *see also* Regulators
  tank, 84f, 92
Vanadium, chemiluminescence of, 216, 272
  detection limits of, 272
  determination of, 223
    indirect, 247, R452
  enhancement of, 272, R431a
  flame spectrum of, 215, 216, 223, 272, 406, R734
  interference of, on calcium, 313, R452
    on indium and thallium, 315, 318
    spectral, 223
    on strontium, 318, R452
  interference on, 318
  ionic spectrum of, 223, 406
  in microwave discharge, 76
  in organic solvents, 272
  in oxycyanogen flame, 223, 272
Vanadium oxide (VO), spectrum of, 223, 272, 406f
van Ark, H. J., 94
van Calker, J., R179
van der Have, A. J., R323
van der Held's equation for drop evaporation, 27
van Someren, E., R670
Vaporization, *see* Evaporation
  enhancement of, by complexing, 251f
  of sample, 26, 27ff, 305
Vaporization interference, 64, 65, 66, R67, R452, R561a, R562a, R637; *see also* Aluminum, interference of, Dual atomization, Phosphate, interference of, Protective chelation, Releasing agents, Sulfate, interference of
  on alkali metals, 294
  on alkaline-earth metals, 300ff
  in atomizer-burner flame, 123, 301, 306, 308
  causes of, 304ff
  correction of, 310
  diagnosis of, 67
  double, 301f, 306
  drop size, effect of, on, 306, 308
  elimination of, 308ff, 320
  and height in flame, 301f, 303, 305, 306, 308
  for indirect determinations, 245ff
  knee in, 301ff, 304f, 307
  with lines and bands, 309
  plateau, in, 300ff, 304f, 307, 309
  ratios, stoichiometric, in, 304, 307
  saturation of, 309
  substitution method for, 323ff
  temperature effect on, 74, 301, 308
Vapor pressure, *see* Transport interference
Váradi, P. F., R292
Varney, P. L., R728, R729
Varnish, analysis of, 359
Vegetable products, 338f; *see also* Foods, Plants
Velocity of air, effect of, on drop size, 23, 106
  of air, in nozzle, 106
  of flame, *see* Burning velocity
  of gas, in burner, 118
Vendrik, A. J. H., R660
Venrath, H., R149b
Ventilation, 199f
Vibrational energy, 40f
  in collisional excitation, 50f
  rate of transfer of, 51
Vibrational quantum number, 40f
Vilnat, E., R717
Viscosity, effects of, R182, R545
  effect of, on atomization, 22
    on drop size, 23
    on sample flow rate, 22
  and internal standardization, 162
Visscher, M. B., R16
Visual flame photometer, R125
Vitkun, R. A., R562
Voinovitch, I. A., R213–R219a
Volatility, *see* Vaporization interference
Volumetric apparatus, for dilution, 211
von Elbe, G., R31, R447
von Oer, A., R524
Voorhuis, M. H., R70
Vreden-Kobetskaya, T. O., R27

w, peak of a wide or diffuse band
W, water
Wacker-Chemie, 213
Waentig, P., 8, R125
Waibel, F., 8, 9, 336, R589
Waithman, V. B., R496
Walsh, A., 48, R608
Walther, T., R370
Warm-up, 200f
Warner, H., R703

Warren atomizer-burner, R723, R724
Wastes, analysis of, 359, R73
Watanabe, H., 244, 372, 383
Watch glass, as sample container, 110
Water, analysis of, 354ff, R73
   for calcium, R460, R484, R570, R624, R693, R731, R740, R756
   for iron, R460
   for lithium, R201, R756
   for magnesium R460, R484
   for potassium, R309, R569a, R624, R640, R693, R731, R756
   for sodium, R309, R460, R472, R569a, R624, R640, R693, R731, R756
   for strontium, R693
   for sulfate, 356
  band spectrum of, 52, 54, 393
  boiler, analysis of, 355f
  deposits, analysis of, 354ff, R593, R640
  irrigation, analysis of, 355
  mineral, analysis of, R383, R463
  purification of, 237
  purity of, testing, 207f, 211
  sea, see Sea water
  as solvent, 207f
Water reading, 288
Water-repellent, R301, R303
  for blood samples, 348
  for deaeration, 213
  effect of, on atomizer, 213
  for glassware, 211, 212, 213
  paraffin as, 213
Waters Associates, 431
Watson, M. T., Jr., R192, R193
Watson, W. W., 383
Wavelength, best, 260ff
  choice of, 226ff
  determination of, 371f
  precision of, 373
  in quantum theory, 37, 39
  scale of monochromator, 172, 177
  scanning of, see Recording
  selection of, 370, 374ff
  setting of monochromator, 177
  tables, 260ff, 370ff, R307, R307a
Wedge filter, 143, 168
Weichselbaum, T. E., atomizer-burner, R728, R729
  flame spectrophotometer, R187
Weniger, S., 381
Werner, G. K., R476

Wet ashing, 342, 343, R411, R526
Wetting agents, analysis of, for sodium, R428
  for cleaning glassware, 214f
  effect of, on atomization, 63
    on emission in atomizer-burner flame, 21
  in plastic containers, 212f
  in serum analysis, 352f
Wheeler, J. A., 382
Wheeler, R. C., R695
Whiskey, analysis of, R566
Whisman, M., 383
Whitehead, T., R153
Whitney, R. McL., R506a
Wiberley, S. E., R18
Wiethoff, G., R733
Wilbur, D. A., R195
Wilhide, W. D., R154, R155
Wilkens-Anderson Co., 431
Willis, J. B., 219, 354
Wilson, C. L., 222
Wine, analysis of, R753
Wires, for sample introduction, 7, 97
Wittwer, S. H., R357
Wolf, E. F., R674
Wolf, K. L., R17
Wolfhard, H. G., R11, R227
Wollaston, W. H., 6
Wood, E. L., R288, R289
Working area, 199f
Working curve, 55ff, 230ff, R666, R667; see also Curve of growth, Measurement
  in absorption flame photometry, 48
  in choice of emission, 227
  concave, 206f
  dissociation, effect of, on, 30, 56
  equation for, 322
  establishment of, 230ff, 297
  families of, see Parametric method
  flame electrons, effect of, on, 35f
  flame thickness, effect of, on, 120
  height in flame, effect of, on, 57, 228
  hydroxide formation, effect of, on, 32, R391
  instrumental nonlinearity of, 59
  in internal standardization, 161, 162, 337
  interpolation on, 233, 240f
  ionization, effect of, on, 35, 55, 56f, 206f, 228, 294ff

Working curve—*contd.*
  knee of, 58, 120, 228, 231, R356
  logarithmic, 57f, 206, 227, 231, 240, 295f, 322, R356
  logarithmic slope of, 322
  precalibrated scale for, 152, 233
  and reading precision, 206, 227f
  sample transport, effect of, on, 56
  self-absorption, effect of, on, 45ff, 57f, 217, 228, R356
  self-reversal, effect of, on, 47, 58
  in spectrography, 194, R407
  straightening of, 227f, 324
  temperature effect on, 75
  vaporization interference, effect of, on, 301
  variations of, 230
Wright, P. G., R451a
Wyld, G., R241, R563

x, sharp peak in a headless band, 383

y, rotational line or group of lines, 383
Yoe, J. H., R52
Ytterbium, detection limits of, 272
  determination of, R546, R547
  flame spectrum of, 220f, 272, 407, R546, R547
  in hexone, 272
  ionic spectrum of, 407
Ytterbium hydroxide, spectrum of, 272, 407
Ytterbium oxide, spectrum of, 407
Yttrium, detection limits of, 272
  determination of, R546, R547
  in hexone, 272
  interference of, on strontium, R416c
Yttrium oxide, spectrum of, 221, 272, 407, R546, R547
Yukawa, K., R276

z, incandescent oxyacetylene, 260
Zack, P. G., R164
Zaĭdel', A. N., 383
Zak, B., R411
Zaorska, H., R755
Zapp, E. É., R572, R577, R581
Zeiss, Carl, 383, 431, R530, R758, R759, R759a
Zeiss flame photometers, 9, 336, 430, 431, R328, R330, R530, R758
  design of, 69

Zeiss flame spectrophotometer, M4Q, 180, 431
  PMQII, 431, R758
Zeiss monochromator, 174
Zemansky, M. W., R38
Zero, adjustment of, 150, 153
  displacement of, 185f
  drift, 150, 187
  effect of noise in alternating-current systems on, 185f
  establishment of, *see* Blank
  noise of, 186
  recovery of, 145
  suppression of, 241, 286
Zettler, H., R261, R263, R264
Zinc, absorption flame photometry for, 48, 354, R203a
  chemiluminescence of, 216, 272
  continuum of, 220, 272
  detection limits of, 272
    in absorption, 48
  determination of, by absorption flame photometry, 354, R203a
  excitation potential of, 34
  flame spectrum of, 216, 219f, 272, 407
  interferences of, 312, 314ff
  ionization potential of, 34
  in microwave discharge, 76
  in organic solvents, 272
  in oxycyanogen flame, 272
  self-reversal of, 58
  separation of, 250
  working curve of, 58
Zinc sulfide, analysis of, 357, R205, R587
Zirconium, detection limits of, 272
  determination of, indirect, R452
  flame spectrum of, 10, 222, 248, 272, 407
  interference of, on barium, 309, 313
    on calcium, 314, R452
    on lanthanum, 316
    on strontium, 318, R452
  ionic spectrum of, 407
  in microwave discharge, 76
  in oxycyanogen flame, 10, 407
  for phosphate determination, 248
Zirconium oxide (ZrO), spectrum of, 222, 272, 407
Zittel, H. E., R211
Zoellner, H., R330
Zones, of flame, *see* Flame
Zyderveld, G. R. D., 74

11/16/66

987